PETROLOGY
Igneous, Sedimentary, and Metamorphic

Second Edition

PETROLOGY
Igneous, Sedimentary, and Metamorphic

Second Edition

Harvey Blatt

The Hebrew University of Jerusalem
(formerly at the University of Oklahoma)

Robert J. Tracy

Virginia Polytechnic Institute
and State University

W. H. Freeman and Company
New York

About the cover: The inset image shows the dramatic igneous rock structures of the Palisades Interstate Park that cascade down to the Hudson River, north of New York City. The background image shows the contrast of the sedimentary rock cliffs falling into the Pacific Ocean at Torrey Pines State Park, Del Mar, California.

Photographs by Joshua Sheldon.

Library of Congress Cataloging-in-Publication Data

Blatt, Harvey.

Petrology: igneous, sedimentary, and metamorphic/ Harvey Blatt, Robert Tracy.

p. cm.

Includes index.

ISBN 0-7167-2438-3 (hardcover)

1. Petrology.   I. Tracy, Robert J.   II. Title.

QE431.2B57 1995

552—dc20   95-35545

CIP

CONTENTS

• •

PREFACE

∙∙∙∙∙∙∙∙∙∙∙∙∙∙∙∙∙∙∙∙∙∙∙∙∙∙∙

The preface to the first edition of this book was written almost 15 years ago. It pointed out that the subject of petrology had undergone a revolution during the previous decade (1970–1980), largely as a result of increasing awareness by geologists that plate tectonics has played a key role in the genesis and distribution of rocks. In the years since then, the ferment over plate tectonics has subsided; it has become standard theory, and petrology has firmly entered a consolidation phase. More than ever before, petrologists are interacting with earth scientists in other fields, notably geochemistry, geophysics, and structural geology, to document how petrologic processes and products fit into a tectonic framework.

The most striking aspect of the change in petrology since 1982 lies in the evolutionary, rather than revolutionary, quantification of techniques for the study of rocks—an evolution sparked by gradual improvements in techniques for indirect measurement of temperatures and pressures of formation, ages, and chemical and physical processes of petrogenesis. These techniques include elemental and isotopic chemical analyses, spectroscopy, and the ready availability of powerful computers to students and researchers. (Note that the explosion in personal computing postdates the writing of the first edition of this book!)

Petrology has long been one of the more quantitative of earth science fields, but the last 15 years have seen an acceleration of mathematical approaches, especially modeling of magmatic and metamorphic processes. Throughout this period, igneous, sedimentary, and metamorphic petrology have increasingly diverged in their fundamental approaches. Igneous petrologists now focus most strongly on geochemical studies of petrogenesis, on the physics of magmas, and on magmatism as a planetary process. Sedimentary petrologists examine the geochemistry of sediment formation and diagenesis, the details of the relationship of sedimentation patterns to plate tectonic regimes, and the use of sedimentary rocks in documenting long-term climatic patterns on the earth. In turn, metamorphic petrologists emphasize the refine-

ment and application of thermodynamic databases for calculation of phase equilibria, the development of more precise techniques for characterization of temperature-pressure-time paths during metamorphism, and the complex processes of fluid-rock interactions in the crust.

This broad diversity within the description of petrology and the rapidly increasing store of information about rocks and rock-forming processes have made writing this book a challenging task. Our goal has been as much to enable readers as to inform them. The book is aimed at giving undergraduate students an appreciation of the current fundamental intellectual approaches to the study of rocks, as well as the basic tools and information they need to make a start on reading and understanding the literature of petrology. We have not attempted the obviously impossible feat of creating an encyclopedic volume about rocks. Rather, we introduce readers to the way petrologists go about doing their work. Study exercises at the end of each chapter give students a way to test their understanding of the material, and the references and additional readings provide a pathway into a deeper pursuit of each topic. We have written the book as a text for a first course in petrology, at a level appropriate to college sophomores or juniors, and have assumed a basic understanding of elementary chemistry and mineralogy. We hope that all student readers will have an opportunity to work with thin sections and petrographic microscopes, and thus gain a greater appreciation for the inner workings of rocks.

We gratefully acknowledge the contributions of all the scientists whose work we have cited in this book and who have given us permission to reproduce their materials. In addition, we would especially like to thank colleagues who have inspired us as mentors, fellow teachers, or students and who in some cases have provided critical reviews or special materials; these include R. A. Badger, J. M. Christie, W. G. Ernst, R. L. Folk, R. A. Heimlich, P. C. Hess, D. A. Hewitt, J. F. Hogan, T. N. Irvine, D. M. Kerrick, John Longhi, the late H. O. A. Meyer, S. A. Morse, B. J. Munn, Peter Robinson, A. B. Thompson, J. B. Thompson, and E. W. Wolfe.

PETROLOGY
Igneous, Sedimentary, and Metamorphic

Second Edition

Introduction

....

WHY STUDY PETROLOGY?

L eaving aside the atmosphere and hydrosphere, rocks are the stuff of which the earth is made. Because of this, the study of rocks, petrology (from the Greek *petra*, "rock," and *logos*, "discourse or explanation"), occupies a central position among the earth sciences. The study of rocks is the source of virtually all of our ideas about the history of Earth. Knowledge about rocks, their origin and ages, and their distribution is potentially capable of contributing to the solution of a wide variety of problems that run the gamut of geological interests.

1. Most evolutionary biologists believe that living matter evolved from nonliving matter more than 3 billion years ago in a reducing atmosphere, because it is thought that primitive cells would have been defenseless against oxidation. This suggestion implies that there was little or no oxygen in the atmosphere of the early earth, a theory supported by the absence of free oxygen in the atmospheres of other planets in our solar system. However, early Precambrian (Archean) iron ores contain oxidized iron, in the form of magnetite, a fact that might be interpreted to mean that the early atmosphere contained substantial oxygen. Similarly, Precambrian weathering zones or soils have a red appearance caused by oxidized (hematitic) iron. Did these ores and soils contain magnetite or hematite from the beginning, or was more reduced iron gradually oxidized as free oxygen increased in the atmosphere in the later Precambrian and after? Can we use these rocks to infer how much oxygen was present in the early atmosphere?

2. The relative abundance of different sedimentary materials forming at present is drastically different from rock abundances found in the geologic record.

a. Dolostone is three times as abundant as limestone in Precambrian rocks. At present, formation of dolostone is rare, restricted to unusual environments such as the Persian Gulf or the Netherlands Antilles.

b. The Middle Precambrian stratigraphic column (about 2.5 billion years old) contains about 15 percent fine-grained silica in the form of chert. At present, chert formation is trivial outside the deep ocean basins.

c. Evaporites are extremely rare in the Precambrian as compared to more recent times. Why? Has the composition of seawater changed during the last 2.5 billion years? Has the proportion of rocks exposed to weathering, and thus the composition of material supplied to the oceans, changed since the Precambrian?

3. Igneous rocks have a wide variety of textures and compositions—in particular, a range in silica content from less than 45 to about 75 percent. Yet the great bulk of igneous rocks consists of either coarse-grained granitoid rocks (silica content of about 65 to 75 percent) or fine-grained basaltic rocks (silica content of 45 to 52 percent). What does this tell us about formation and evolution of magmas and about crust-forming processes?

4. Metamorphic rocks of identical chemical composition in different tectonic settings and at different crustal depths can consist of widely differing mineral assemblages. How, exactly, does this relate to metamorphic reactions and to different rates of temperature and pressure change during orogeny?

5. Some sandstones contain only quartz; others, 30 percent feldspar; and still others 90 percent volcanic rock fragments. Can these data be used to infer the types and proportions of rocks exposed on the earth's surface at different times and in different geographic locations? Can similar sedimentary rock types be identified as the precursors for metamorphic rocks in orogenic belts? Is the precise mineral composition of sandstones and other clastic rocks related to tectonic processes and crustal evolution—and, if so, how?

6. Geologic studies of the lunar surface and satellite studies of the Martian surface have given us a preliminary understanding of the early stages of planetary evolution in bodies that have been geologically inactive for billions of years. Can we use our developing understanding of other planetary bodies to interpret the much sparser information on the early earth preserved in rare, very old rocks that have survived later geologic activity and recycling?

THE MAJOR ROCK TYPES
• • • • • • • • • • • • • • •

Rocks are naturally occurring, mechanically coherent aggregates of minerals or mineraloids (coal, glass, opal), and most rocks consist of several different minerals. Rocks are traditionally divided into three groups: igneous, sedimentary, and metamorphic. In most outcrops and hand specimens, it is not difficult to apply these categories, and they serve the useful purpose of sorting rocks on the basis of observable characteristics that depend on the conditions of initial formation. The American Geological Institute's *Glossary of Geology* defines each group as follows:

Igneous rock: A rock that solidified from a molten or partially molten material, that is, from a magma.

Sedimentary rock: A rock resulting from the consolidation of loose sediment that has accumulated in layers. A clastic rock consisting of mechanically formed fragments of older rock transported from their source and deposited in water or from air or ice; or a chemical rock formed by precipitation from solution; or an organic rock consisting of the remains or secretions of plants and animals are examples.

Metamorphic rock: Any rock derived from preexisting rocks by mineralogical, chemical, and structural changes, essentially in the solid state, in response to marked changes in temperature, pressure, shearing stress, and chemical environment at depth in the earth's crust; that is, below the zones of weathering and cementation.

This fundamental classification scheme, as is true of all classification schemes, contains a flaw. Nature is a continuum; it is not segregated into discrete parts for our convenience. Hence, borderline or transitional rocks exist and end up in one or another of the categories because of historical precedence or the bias or whim of the classifier. For example, volcanic tuffs are rocks that originate in volcanoes. After explosive ejection as rock, mineral, or glass fragments into the atmosphere, they settle on land surfaces or in the water. They may even be transported some distance by water or wind. If they settle into layers (as is common), should they be classified as sedimentary or igneous? In most classification schemes, these rocks are classified as igneous. Another example is serpentinite, a rock mainly composed of minerals of the serpentine group. Many serpentinites are thought to originate as ultramafic magmatic rocks in small, shallow magma chambers near a mid-ocean ridge. During cooling, they undergo hydration from surrounding water-rich sediments that transforms most or all of the crystals into serpentine. Should this rock be classified as igneous or metamorphic? Most petrologists would classify it as metamorphic. A third example is migmatite, an outcrop-scale mixture of light and dark rocks thought to represent the onset of melting in the crust. Thus, migmatite is partly metamorphic and partly igneous.

In fact, the necessity of classifying rocks commonly forces petrologists into making critical observations and interpretations that aid in assessing the process of formation of rocks. In this book we discuss many of these observations and interpretations. Additional subdivisions within each of the major categories are introduced and discussed. Subdivision of rocks into the three general categories can be based to some extent on actual observation of processes of formation, particularly for sedimentary and volcanic rocks. For intrusive igneous rocks and metamorphic rocks, it is obviously impossible to observe their formation directly, and therefore indirect and inferential methods are required.

Table I-1 is designed to facilitate the subdivision of rocks into one or another of the major categories. The table lists some of the outcrop characteristics and structures, followed by textures and characteristic minerals. Note that any single characteristic may not be sufficient to categorize the rock but, rather, a number of such features may have to be used. Unfamiliar and technical terms that we use now or later are either defined in the text or found in the Glossary. Glossary terms are indicated by bold print in the text.

RELATIVE SURFACE AND CRUSTAL ABUNDANCES OF ROCK TYPES
• • • • • • • • • • • • • • •

The earliest geologic maps date from about 1800, and most parts of the earth have now been mapped either in detail or at a reconnaissance scale. Using such maps and related literature, we can determine the relative abundances of igneous, sedimentary, and metamorphic rocks on the continents. The method used to obtain the data is to generate random latitudes and longitudes by computer and then examine existing maps to determine the frequency of igneous, sedimentary, and metamorphic outcrops. Results indicate that Earth's surface is 66 percent sedimentary rocks and 34 percent igneous and metamorphic rocks. The bulk of the 34 percent is probably igneous, but large parts of some continents are mapped as "undifferentiated igneous and metamorphic rocks," so exact percentages are uncertain. As detailed geologic mapping proceeds worldwide, a better estimate may be possible.

TABLE I-1   General characteristics of igneous, sedimentary, and metamorphic rocks

IGNEOUS	SEDIMENTARY	METAMORPHIC
Outcrop characteristics and structures		
1. Volcanoes and related lava flows	1. Stratification and sorting	1. Distorted pebbles, fossils, or crystals
2. Cross-cutting relations to surrounding rocks, as in dykes, veins, stocks, and batholiths	2. Structures such as ripple marks, cross-bedding, or mud cracks	2. Parallelism of planar or elongate grains common over large areas
3. Thermal effects on adjacent rocks, such as recrystallization, color changes, reaction zones	3. Often widespread and inter-bedded with known sediments	3. Located adjacent to known igneous rocks, occasionally as a zoned aureole
4. Chilled (finer-grained) borders against adjacent rocks	4. The shape of the body may be characteristic of a sedimentary form, such as a delta, bar, river drainage system, and so on	4. Typically located in Precambrian or orogenic terranes
5. Lack of fossils and stratification (except for pyroclastic deposits)	5. The rocks may be unconsolidated or not	5. Rock cleavage related to regional structures
6. Generally structureless rocks composed of interlocking grains		6. Progressive change in mineralogy over a wide area
7. Typically located in Precambrian or orogenic terranes		7. Some are massive hard rocks composed of interlocking grains
8. Characteristic shapes and sizes, as in laccoliths, lopoliths, sills, stocks, batholiths, and lava flows		
Textures		
Porphyritic, glassy, vesicular, amygdaloidal, graphic, pyroclastic, or interlocking aggregate	Fragmental, fossiliferous, oolitic, pisolitic, stratified, interlocking aggregate	Brecciated, granulated, crystalloblastic, or hornfelsic
Characteristic minerals		
Amphibole	Abundant quartz, carbonates (especially calcite and dolomite), or clays	Amphibole
Feldspar abundant	Anhydrite	Andalusite
Leucite	Chert (microcrystalline quartz)	Cordierite
Micas	Gypsum	Epidote
Nepheline	Halite	Feldspar
Olivine		Garnet
Pyroxene		Glaucophane
Quartz		Graphite
Glass		Kyanite
		Sillimanite
		Staurolite
		Tremolite-actinolite
		Wollastonite
		Micas
		Quartz

The data for sedimentary rocks reveal the extent to which information is lost over time (Figure I-1). The data indicate that half of all outcropping sedimentary rocks are younger than 130 Ma (that is, Cretaceous or younger). Interpretations based on older rocks must be less comprehensive than those based on more recent rocks. Bear in mind that older sedimentary rocks are more likely to have been recycled into metamorphic rocks or even melted to form crustal magmas. It should also be remembered that the proportional rock estimates given here are *areal* estimates based on surface exposure, and are highly unlikely to correspond to *volumetric* estimates for the whole crust.

The crust is usually defined as the outer shell of the earth above the Mohorovicic discontinuity. The terms *crust* and *lithosphere* are not synonymous. The lithosphere contains both crust and a portion of upper mantle to an average depth of about 100 km, and generally constitutes the tectonic plates. The asthenosphere lies directly below the lithosphere. Geophysical data reveal that the mean thickness of the entire crust is about 17 km: Continental crust averages about 40 km and is dominated by granitoid rocks, whereas the oceanic crust averages less than 10 km and is basaltic. Mean sedimentary rock thickness on continental areas averages about 1.8 km, and these continental areas occupy about 35 percent of the earth's surface. In ocean basins, sediment cover is very thin, approximately 0.3 km. Combined data indicate that the average thickness of sedimentary rock on the earth's entire surface is about 0.8 km and that of igneous and metamorphic rocks about 16.2 km. Sedimentary rock thus forms about 4.8 percent of the entire crust and 0.013 percent of the whole earth.

As the whole earth has a radius of 6371 km, only the outer 0.27 percent of the radius is crust (0.74 percent of the volume of the earth). Nevertheless, the crust is the only part of the earth that is directly exposed for examination, and only a small part of that is actually seen. The subject matter of petrology is thus largely concerned with analysis of composition, structure, and origin of crustal rocks. However, we will examine how the problems of understanding the origin of some crustal rocks are inextricably linked to the nature of the material below the crust—the mantle. Petrologists eagerly exploit any opportunities to sample deeper levels below the surface, in the form of mines, drill holes, or unusual, deeply derived magmas. It is sobering to keep in mind, however, how little in petrology can be directly seen and how much is based on theory, experiment, and inference.

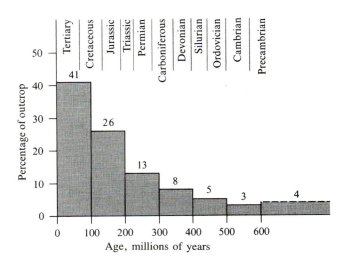

FIGURE I-1

The relationship between the age of sedimentary rocks and the amount of outcrop area. [From H. Blatt and R. L. Jones, 1975, *Geol. Soc. Am. Bull.*, 86, Fig. 1.]

REFERENCES AND ADDITIONAL READINGS

• • • • • • • • • • • • • •

American Association of Petroleum Geologists and U.S. Geological Survey. 1967. Basement map of North America between latitudes 24° and 60° N. U.S. Geol. Surv. Map.

Bates, R. L., and J. A. Jackson. 1980. *Glossary of Geology*, 2nd ed. Washington, DC: American Geological Institute.

Blatt, H. 1970. Determination of mean sediment thickness in the crust: A sedimentologic method. *Geol. Soc. Am. Bull.*, *81*, 255–262.

Blatt, H., and R. L. Jones. 1975. Proportions of exposed igneous, metamorphic, and sedimentary rocks. *Geol. Soc. Am. Bull.*, *86*, 1085–1088.

Dietrich, R. V., and B. J. Skinner. 1979. *Rocks and Rock Minerals*. New York: Wiley.

Periodicals containing significant petrologic literature include *American Journal of Science, American Mineralogist, Bulletin of the Geological Society of America, Canadian Mineralogist, Contributions to Mineralogy and Petrology, Journal of Geology, Journal of Metamorphic Geology, Journal of Petrology, Journal of Sedimentary Research,* and *Lithos.*

·······································

PART I

Igneous Rocks

·······································

Chapter 1

INTRODUCTION TO IGNEOUS ENVIRONMENTS

Before discussing the methods petrologists use to study and understand igneous rocks, it is useful first to establish some basic definitions. An *igneous rock* is defined as any crystalline or glassy rock that formed directly from a *magma*. **Magma** is a high-temperature molten substance that is chemically complex and contains the molecular building blocks for minerals (the nature and properties of magma are discussed in detail in Chapter 3). As it cools and crystallizes, a magma reorganizes itself into igneous minerals that individually have simpler chemistry than the parent magma. In any given igneous rock, these individual minerals (for example, quartz, feldspars, olivine) occur as either small or large crystals, both with variable perfection of crystal form. The minerals found in any igneous rock and the characteristics of these minerals depend almost entirely on three things: composition of the original magma, the rate at which it cooled; and the depth in the earth at which it solidified. The range of products is enormous, from glassy obsidians, which are volcanic rocks that cooled so rapidly that no crystals could form, to pegmatites, which can contain individual crystals a meter or more in size.

WHAT IS IGNEOUS PETROLOGY?

Igneous petrology is the field of geology in which the object of study is any igneous rock or any process involved in formation of igneous rocks. The purpose of studying igneous petrology is to gain an appreciation for how the final appearance and characteristics of igneous rocks are controlled by chemical and physical properties of magmas and their surroundings. Much of the fascination of petrology lies in its "detective story" aspect. Through geologic mapping and simple observations of igneous rocks collected with a hammer and viewed through a hand lens, a geologist can amass a wealth of clues to the origin of the rocks. By learning how to interpret these clues, we can create a sophisticated reconstruction of geologic history. Once some of the basic techniques of petrology, such as hand specimen identification and field examination of igneous rocks, have been mastered, the student moves on to **petrography.** In fact, petrography is commonly a major component of laboratory work in a petrology course. In this subfield of petrology, the techniques of optical mineralogy are used in the microscopic examination of ultrathin, transparent slices of rocks mounted on glass slides. Microscopic observation of rocks opens up a truly fascinating new world of detailed information.

The study of igneous rocks is an important component of a geology curriculum because these rocks not only are important throughout the crust of the earth but even dominate some crustal and upper mantle environments. For example, under the oceans, the thin crust is made up almost exclusively of **basalt** that originated in submarine volcanic and intrusive activity at mid-ocean ridges. Thick continental crust in many mountain belts is composed largely of huge intrusive bodies of **granite.** Igneous activity has continued since the earliest days of the earth; the oldest igneous rocks known have been dated at about 4 billion years. Planetary and meteoritic studies indicate that primitive basaltic rocks probably were the first rocks to form on Earth after it solidified from the solar nebula and remained the dominant rock type at the surface through the first several hundred million years of Earth history (before 4 billion years ago). Radiometric dating of the oldest granitic rocks suggests that these less primitive igneous rocks, as well as sedimentary and metamorphic rocks, appeared by about 4 billion years ago. In contrast to Earth's neighbors—planets Mars, Mercury, Venus, and the Moon—where surface and near-surface magmatic activity apparently ceased billions of years ago, the surface of Earth has remained active up to the present. Terrestrial magmatism

(A)

(B)

FIGURE 1-1
• • • • • • • • • • • • •

(A) A vertical dike (dark material) cutting a nearly horizontal segment of an older dike. The younger dike is discordant to the older one. Norfolk County, Massachusetts. [From N. E. Chute, 1966, *U.S. Geol. Surv. Bull., 1163-B*, Fig. 4.] (B) Concordant contact of the Beech Granite (massive rock at the upper left) with the Cranberry Gneiss (foliated rock at the lower right). Avery County, North Carolina. [From B. H. Bryant, 1970, *U.S. Geol. Surv. Prof. Paper 615*, Fig. 20.]

is the most obvious manifestation of the internal heat engine that drives virtually all geologic processes.

Igneous rocks crop out at the earth's surface (where they can be observed) as a result of two processes: extrusive magmatic activity related to volcanoes and erosional unroofing of igneous rocks that solidifed at various depths in the crust. A petrologist must make many observations in the field when faced with an outcrop of igneous rock. Its features can range in size from kilometer-scale relationships on a geologic map, through meter-scale characteristics such as layering, down to individual grains no more than a millimeter across and even to features within individual grains. All this information contributes to an understanding of the extent and shape of igneous bodies, the methods of emplacement, the relationships to adjacent country rock, the mineralogy of the igneous rocks, and the cooling history. Complete characterization of the mineralogy, geochemistry, and perhaps the ultimate origin of igneous rocks may require laboratory examination and analysis of samples collected in the field, but the best starting place in the consideration of igneous processes is an examination of their field aspects, including sizes and shapes of igneous masses, their relationship to the country rock, and their methods of emplacement.

IGNEOUS ENVIRONMENTS: INTRUSIVE VERSUS EXTRUSIVE
• • • • • • • • • • • • • •

One of the first things to determine about an igneous rock is whether it is intrusive or extrusive; that is, whether it formed below or upon the surface of the earth.

Intrusive igneous rocks result from the solidification of magma beneath the earth's surface at depths ranging from meters to tens of kilometers. Petrologists generally characterize intrusive bodies based on the depth of emplacement, the nature and geometry of contacts, and the size of the body. The term *pluton* is typically restricted to deeper intrusive bodies (greater than about 5 km), whereas *intrusion* is a more general term that may be used for both shallow and deep bodies. The term *hypabyssal* has commonly been used to describe very shallow intrusive bodies. The contact of an intrusive rock can be either concordant or discordant in relation to the country rock (Figure 1-1). If the country rock deforms in a brittle way during intrusion (typical of intrusion into relatively cool country rocks in the shallow crust), emplacement of the igneous rock may be accompanied by fracturing and faulting as the intruding magma

FIGURE 1-2

● ● ● ● ● ● ● ● ● ● ● ● ● ● ● ● ●

Xenoliths of hornblende diorite (dark fragments) in light-colored granodiorite, near Copper Lake, North Cascades National Park, Washington. [From M. H. Staatz, 1972, *U.S. Geol. Surv. Bull., 1359*, Fig. 15.B.]

makes room for itself. This process typically results in discordant contacts, because bedding or foliation is truncated at the margins of the igneous body.

Plutons are deeper intrusions that are typically emplaced into country rocks that are at higher ambient temperatures and thus deform in a ductile fashion. A magma may move into place by very slowly squeezing aside and plastically deforming the country rocks, much as wet clay is shaped by a potter's fingers. Bedding or foliation in the country rocks can be rotated during forceful emplacement until it is essentially parallel to the contact with the pluton. Some foliations in the country rock and the pluton can be the direct result of intrusive stresses. *Diapiric* intrusions characteristically have contact-parallel foliations. They are called diapiric because of their apparent similarity to salt diapirs, which are density-driven plastic intrusions of solid material, although salt diapirs (or *structural* diapirs) are commonly discordant. Concordant intrusive contacts are also found on a small scale in shallow sills, where the intrusive force of the magma has been able to force apart the country rock along bedding planes without producing appreciable fracturing.

Concordance or discordance of intrusive contacts is commonly a matter of scale of observation. Many intrusive rocks that appear to be concordant at individual isolated outcrops are actually broadly discordant when mapped on a regional scale. This situation is especially true for igneous rocks that have been exposed by erosion at the level within the crust at which they originated by melting. During deep crustal metamorphism, temperatures may exceed those required for partial melting of some metasedimentary rocks in a process called *anatexis*. When the anatectic granitic magma generated by this process remains at its site of origin or moves only locally, it produces complex patterns of concordance and discordance.

Careful examination of a contact reveals it to be either gradational or sharp. A gradational contact results from a strong chemical interaction between magma and country rock. This thermally driven interaction generally produces a zone of mixing between magma and the melted portion of the country rock. A sharp, precisely defined contact indicates a lack of chemical reaction between magma and country rock. This failure to interact is due either to the presence of a relatively unreactive country rock (such as quartzite) or to a rapid chilling of the magma against cool country rock. A large temperature contrast is commonly reflected by a significant decrease in grain size within the igneous rock near the contact. This region of reduced grain size is called a **chilled margin.**

Xenoliths and **xenocrysts** (inclusions of rock or mineral fragments from the surroundings) should always be searched for in the field and examined carefully. They typically occur near the margins of igneous bodies, but they can be present throughout. These alien bodies (their names are based on the Greek root *xeno-*, meaning "foreign") represent fragments of the rocks through which a magma has moved to its site of final emplacement and crystallization. In certain types of basalt and in related magmas like **kimberlite** that originate deep in the mantle, rare small fragments of mantle material (such as diamonds and fist-sized peridotite blocks) have been transported upward from near the source of magma formation or collected along the vertical ascent path. Granitic rocks derived within the crust more commonly contain large xenoliths of metamorphic or sedimentary rocks (Figure 1-2). Such xenoliths reflect the typical intrusive process that produces granites; in this process subsurface magma chambers expand and move upward by physically plucking country rocks from the walls and roof.

One of the most important things to determine in the field is the areal extent of an igneous body. This measurement can be accomplished by the traditional field mapping technique of "walking the contact"—a pleasant process on a beautiful spring day but perhaps not so much fun in a driving rainstorm in Labrador or on a sub-zero day in Antarctica or a July afternoon in Texas. Use of modern techniques such as remote sensing, satellite

(A)

FIGURE 1-3
● ● ● ● ● ● ● ● ● ● ● ● ● ● ●

(A) The eastern Pilbara region of Western Australia, seen in *Landsat* Image 8 114 801 282 500, showing petrologic and structural features. Color composites made from several spectral bands bring out even more features than those shown here in the black-and-white photo. The high degree of exposure shown by the photo reflects the extremely arid nature of the landscape. (B) Geologic features seen in the photo can be compared with the features labeled on the geologic map. [From R. P. Viljoen, M. J. Vilgoen, J. Grootenboer, and T. G. Longshaw, 1972, *Minerals Sci. Eng.*, 7, Fig. 9. The original photographic data were obtained by NASA and reproduced by the U.S. Geological Survey, EROS Data Center, Sioux Falls, South Dakota.]

(*Landsat*) imagery, and high-resolution aerial photography can ease the labor of mapping, but the most accurate location of contacts still requires field mapping. Topographic base maps and aerial photographs often prove useful, because many igneous rocks and their surrounding country rocks weather differently as a result of differences in chemistry and internal structure and, therefore, have contrasting elevation and drainage patterns.

The first *Earth Resources Technology Satellite, ERTS-1* (later renamed *Landsat-1*), was launched in 1972. *Landsat-2* was launched in 1975. These satellites and their successors have acquired hundreds of thousands of images that cover almost all of the earth. The images, taken at a number of different wavelengths within and outside the visible light spectrum, permit the production of either black-and-white or false-color images that highlight different properties of the material being photographed and are thus extremely valuable for mapping purposes. Satellite images reveal not only large-scale structural patterns but also differences in rock types not normally resolved with standard aerial photography. The usefulness of the technique for large-scale mapping is spectacularly revealed by Figure 1-3.

Extrusive rocks are formed by the flow of **lava** or the fall of **pyroclastic** debris on preexisting erosional surfaces, which can be either subaerial (exposed to air) or submarine. Because these surfaces tend to be irregular, the resulting contact surfaces commonly are also irregular. The contact can be parallel to the bedding or foliation in the country rock, in which case it is called *concordant*. More typically, the contact is at an angle to

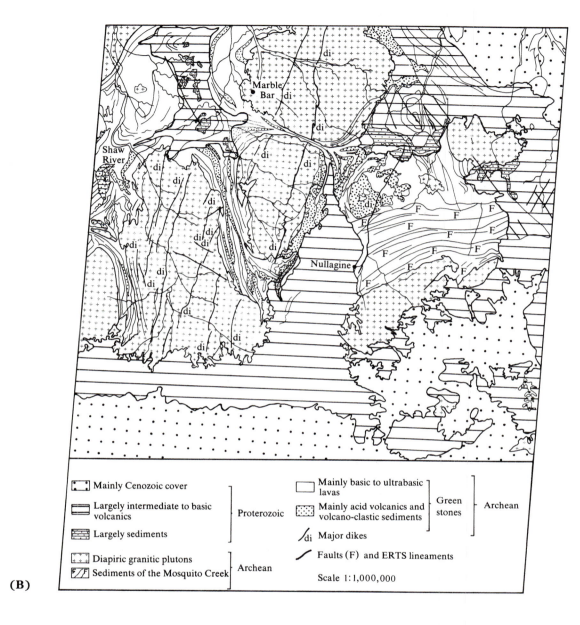

(B)

Legend:

Mainly Cenozoic cover

Largely intermediate to basic volcanics — Proterozoic

Largely sediments

Diapiric granitic plutons

Sediments of the Mosquito Creek — Archean

Mainly basic to ultrabasic lavas

Mainly acid volcanics and volcano-clastic sediments — Green stones — Archean

/di Major dikes

Faults (F) and ERTS lineaments

Scale 1:1,000,000

the bedding or foliation of the country rock and is called *discordant.* When an extrusive contact is located over an irregular surface of flat-lying sedimentary rocks, the contact may be concordant in some places and discordant in others.

In addition to observing geometric relationships, the petrologist should also carefully examine both the igneous and sedimentary rocks at the contact. The country rock below an extrusion might show the effects of weathering that predated the emplacement of the igneous rock, including soil formation, oxidation, or hydration; such effects demonstrate that the igneous rock was extruded on an erosional surface. If loose weathered material was present at the contact surface, some of it is often found within the lower part of the lava flow or ash layer. These foreign materials within any igneous rock

are termed *inclusions.* On occasion, lava flows extruded into a body of standing water such as a lake incorporate unconsolidated sedimentary material in their lower parts. Both inclusions and country rocks commonly show effects of the heating they have received. Beneath thin extrusive lava flows or ash falls, these effects consist of relatively minor baking, which produces some color change and perhaps a slight increase in hardness due to recrystallization.

The upper surfaces of lava flows or ash layers are commonly irregular as a result of the breakup of a solidified crust during flow or erosion following extrusion, thus producing a discordant contact with the overlying sedimentary or volcanic layer. In a succession of rock layers, the top of a buried lava flow or ash layer is sometimes indicated by the presence of **vesicles,** which are spherical or cylindrical cavities left by gas bubbles within solidifying lava. Vesicular structure commonly forms in the upper portions of lava flows (particularly highly fluid basaltic types) as low-density gas bubbles rise upward through the denser magma. The absence of vesicular structure does not necessarily disprove an extrusive origin, because erosion of the upper part of a lava flow or ash layer might have taken place before deposition of the overlying layer. Two persuasive lines of evidence of an extrusive origin of a layer are the development of a soil horizon at the top of the layer, indicating chemical weathering of igneous rock at the earth's surface, or the presence of eroded fragments of fine-grained igneous rock in the overlying sedimentary layer.

The formation of extrusive rocks is one of very few petrologic processes that are observable. Early humans may have been present to watch lava flows or the fall of pyroclastic debris several hundred thousand years ago in the East African Rift. As the records of Greeks and Romans show, Mediterranean peoples have long been affected by volcanoes such as Etna, Vesuvius, and Stromboli. Human populations in these and other volcanically active areas such as Iceland, the Japanese islands, Hawaii, Central America, and Indonesia have learned to live with volcanic hazards. The first recognition of a connection between modern volcanic activity and the rock record was made about 200 years ago. In the late eighteenth century, the Scotsman James Hutton was the first geologist to describe accurately the volcanic origin of ancient rocks in the vicinity of Edinburgh, and Charles Darwin made some remarkably sophisticated observations on volcanic phenomena in the Galápagos Islands in the 1830s. The origin of non-extrusive igneous rocks, however, was the subject of debate as late as the 1870s, because this process is not open to direct human observation.

SMALL-SCALE FEATURES
● ● ● ● ● ● ● ● ● ● ● ● ● ● ●

The microscopic features of igneous rocks are discussed in Chapter 2. However, many small-scale features of an igneous rock can be determined in hand specimens with either the unaided eye or a standard (10×) hand lens. It is very important for the geologist to be able to make these types of observations, for they can be critical in the field, where a petrographic microscope is not available.

The first major distinction to be made is whether or not individual mineral grains are visible. If primary mineral grains can be seen and identified without magnification (microscope or hand lens), the rock is classified as *phaneritic*; if not, it is *aphanitic*. These characteristics must be general in the rock; the presence of cavities filled with large secondary minerals in an otherwise aphanitic volcanic rock does not make it phaneritic. The grains in igneous rocks form an interlocking texture, in contrast to the granular texture found in many sedimentary rocks, such as sandstone.

Correlation of grain sizes with mode of occurrence (intrusive versus extrusive) reveals that most intrusive rocks are phaneritic and most extrusive rocks are aphanitic, although substantial overlap can occur, especially in intrusive chilled margins and very thick, extrusive lava flows. Furthermore, it has often been noted that both intrusive and extrusive rocks commonly show a reduction in grain size close to the country rock contact. Some rocks close to contacts (usually those of extrusive rocks) are even largely noncrystalline; that is, they consist mostly of glass. These observations suggest that grain size can be correlated with cooling rate, at least in a general way, with coarser grain sizes reflecting a slower rate of cooling. In some igneous rocks, a few crystals are conspicuously larger than the majority of matrix grains. These larger crystals, called *phenocrysts*, are generally considered to represent the earliest formed grains, which crystallized during a period of cooling slower than that of the bulk of the rock. Further examination of phenocrysts typically reveals that they consist of only one or two mineralogical types, whereas the finer groundmass contains a larger variety of minerals. This observation leads to the conclusion that all minerals in an igneous rock do not appear simultaneously and that the number of minerals simultaneously crystallizing continually increases as the magma cools, a conclusion borne out by studies of crystallization (Chapters 4 and 6).

If crystals are visible in a hand specimen, then the next observation concerns their orientation. Is it random or preferred (parallel, for tabular crystals)? This is, in fact, an observation best made in hand specimens. A random orientation usually indicates that crystallization occurred while the magma was at rest, because preferred orientations develop when crystals are aligned by the stresses generated by magma flow either during or after partial crystallization. Flow during crystallization is commonly indicated by the presence of elongated vesicles or by the parallel alignment of tabular or elongated phenocrysts or xenoliths.

Flow textures should be interpreted cautiously. Rocks formed from volcanic ash deposits commonly show a parallelism of grains or fragments as a result of their method of deposition and later compaction. In plutonic settings, some large mafic intrusions (the so-called **layered intrusions**) display a sinking or floating of certain crystals during solidification. This process produces horizontal layers in which the crystals show preferred orientation like that seen in some sedimentary deposits. Preferred orientation as a result of flow is most commonly found near the walls, tops, or bottoms of intrusive bodies, where chilling of early magma batches has encouraged early crystallization before all the magma has been finally emplaced in the chamber.

The color of an igneous rock is the result of the colors of the various minerals present. When a fresh specimen is obtained, the color is directly related to the primary igneous minerals. The color of a weathered fragment is a consequence of both the original igneous minerals and later weathering products (typically secondary oxidation and hydration minerals). Weathering tends to destroy the original igneous minerals partially or completely (ferromagnesian minerals are most susceptible) and to form secondary minerals that are thermodynamically stable under atmospheric conditions (see Chapter 12). A good example is the rock **dunite,** which is composed almost entirely of olivine. Fresh dunite has a deep olive or straw green color, the color of olivine. The term dunite derives from the type locality, Mt. Dun in New Zealand, which is probably named for the tan to reddish brown color (dun) of the weathered surfaces of olivine-rich rock. These weathering rinds consist mostly of clay minerals and iron oxides. Typically, weathering creates dull-looking rocks that have a dull or hollow sound when struck with a hammer. Unaltered igneous rocks have a somewhat vitreous (glassy) appearance and produce a ringing sound when struck.

Assuming that a fresh specimen can be obtained, a crude subdivision of igneous rocks can be made on the basis of color. Rocks that are rich in silica (the chemical component SiO_2) typically contain abundant quartz and light-colored feldspar. Rocks rich in magnesium and iron tend to contain high concentrations of the mafic minerals olivine, pyroxene, amphibole, and biotite, which are typically strongly colored. On the basis of the amount of colored versus white (or colorless) minerals, the geologist can set up a *color index* that relates in a rough way to the rock bulk composition. A rock that contains less than 30% colored ferromagnesian minerals is defined as light-colored and is called *leucocratic* (from the Greek *leuko-* meaning "light"). With 30 to 60% ferromagnesian minerals, the rock is *mesocratic;* and with more than 60%, *melanocratic* (from the Greek *mela-* meaning "dark"). More commonly, light-colored rocks are called *felsic* and dark ones *mafic.* Felsic and mafic are actually mineralogic terms that refer to quartz and feldspars and to ferromagnesian minerals, respectively.

The identification of the minerals present in an igneous rock can be carried out very successfully in a phaneritic rock by using a hand lens to observe color, cleavage, and grain shape. The mineralogic composition of aphanitic rocks is more difficult to establish, but can be approximated on the basis of the color, the identification of any phenocrysts that might be present,

the specific gravity, and the petrologist's experience with these rocks.

TYPES OF INTRUSIONS

Sills

Sills are concordant, tabular bodies that are emplaced essentially parallel to the foliation or bedding in the country rock (Figure 1-4); they commonly occur in relatively unfolded country rock at shallow crustal levels. A high degree of fluidity (that is, low viscosity) is required to produce this sheetlike form. (Viscosity is discussed in detail in Chapter 3; for now, simply note that *low viscosity* means "highly fluid" and *high viscosity* means "very stiff or pasty.") A majority of sills are basaltic, because basaltic magmas are considerably more fluid (less viscous) than granitic ones and therefore can more easily intrude in this fashion. Sills are either simple, multiple (more than one injection of magma), or differentiated. In a differentiated sill, the denser, earlier formed crystals (typically olivine) have settled into a zone near the chilled base; this process produces a variable composition within the sill, from base to top. Differentiation was first observed and described by Charles Darwin in the Galápagos Islands in the 1830s.

Although a sill is typically thin, sill thickness ranges from a few meters to several hundred meters. The Triassic Palisades Sill in New Jersey and New York, located along the west side of the Hudson River near New York City, is a classic differentiated sill that is about 300 m thick and crops out over an area 80 km long and 2 km wide. The Peneplain Sill of the Jurassic Ferrar diabase swarm in Antarctica is up to 400 m thick and crops out over at least 20,000 km^2. Thicker sills are much more likely to be differentiated than thin ones, because the thicker sills retain heat more efficiently and thus remain fluid long enough for crystal settling to occur. For similar thermal reasons, the textures of thick sills are more commonly coarser than those of thin sills, because the crystals have had more time to grow.

Laccoliths

Laccoliths are concordant, mushroom-shaped intrusions that range in diameter from 1 to 8 km and have a maximum thickness of about 1000 m (Figure 1-5). They occur in relatively undeformed sedimentary rocks at shallow depths. Laccoliths are created when magma that is rising upward in a cross-cutting dike through essentially horizontal layers in the earth's crust reaches a more resistant layer. The magma then spreads out laterally beneath this layer, thereby forming a dome in the overlying strata. If

FIGURE 1-4
• • • • • • • • • • • • • • •

Columnar-jointed (light-colored) Cape Aiak gabbro sill intruded between flat-lying lavas and sediments of the Unalaska Formation. The cliff is about 400 m high. Unalaska Island, Aleutian Islands, Alaska. [U.S. Geological Survey Photo by G. L. Snyder, 1954.]

the magma encounters low resistance to horizontal spreading, a laccolith may grade into a sill. Most laccoliths are created by relatively silica-rich magmas. Silicic magmas have a much higher viscosity than mafic magmas and thus have much greater resistance to the uniform lateral spread required for sill formation. Furthermore, chilling at the leading thin edges increases viscosity and encourages thickening and doming near the initial vertical magma conduit.

Decisive proof of the intrusive origin of laccoliths was first provided in the 1870s by G. K. Gilbert of the fledgling U.S. Geological Survey in his classic studies of the Henry Mountains of Utah. This area consists of flat-lying sedimentary rocks containing scattered mushroom-shaped bodies of igneous rock that have compositions about halfway between granites and basalts. These intrusions are floored by more or less horizontal sedimentary layers and have roofs consisting of domed sedimentary rocks. Gilbert concluded that magma forced its way between the sedimentary layers and pushed up the upper layers.

His proof that the igneous rocks were intrusive rather than extrusive included several lines of evidence:

1. Some of the igneous rocks cut across sedimentary bedding as tabular sheets (dikes), thus indicating that they postdated the sedimentary layers.
2. The sedimentary rocks near both the floor *and the roof* of the intrusion were baked by the heat of the magma.
3. No volcanic features such as gas cavities or fragmentation of the upper surfaces were found in the igneous rocks.

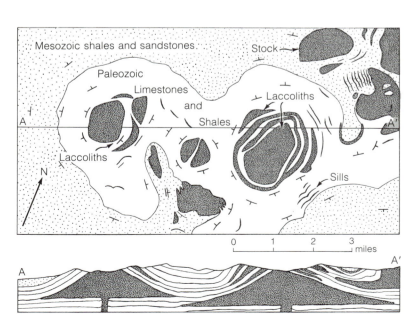

FIGURE 1-5
• • • • • • • • • • • • • • • •

Map and cross section of laccoliths and associated features in the Judith Mountains, Montana. [From F. Press and R. Siever, 1978, *Earth* (San Francisco: W. H. Freeman), Fig. 15-17. After W. H. Weed and L. V. Pirsson, U.S. Geological Survey.]

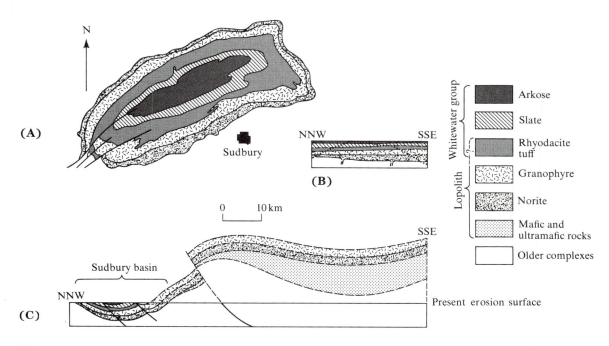

FIGURE 1-6

● ● ● ● ● ● ● ● ● ● ● ● ● ● ● ● ● ●

Map and cross sections of the Sudbury lopolith, Ontario. **(A)** Map of the Sudbury lopolith. **(B)** Cross section prior to deformation. **(C)** Present cross section showing preerosion extent of the lopolith. [From W. B. Hamilton, 1960, *Can. Mineral.*, *6*, Fig. 1.]

4. No eroded fragments of the upper part of the igneous rocks were found in the overlying sedimentary layers as would be expected in a buried volcanic terrane.

5. The dip angles of the upper contacts commonly exceed the maximum angle of repose (roughly 30°) at which sediments can be deposited on a sloping surface.

Although most geologists accepted this overwhelming array of evidence that Gilbert assembled, it took several decades for universal acceptance.

Many of the laccoliths in the Henry Mountains, Utah, deviate from the ideal mushroom shape and are tongue-shaped forms. The common situation in each peak of the Henry Mountains involved injection of a large central mass of magma several kilometers in diameter and about a thousand meters high. Forceful intrusion resulted in the creation of a wide zone of intensely shattered country rock and bulging of the roof and walls. Fracturing and weakening of adjacent rocks outside the shattered zone resulted in the injection of peripheral laccoliths and a wide variety of other intrusive bodies around the central cross-cutting mass. Typical map features associated with laccoliths are shown in Figure 1-5.

Lopoliths

A lopolith, as originally defined by F. F. Grout for the Duluth Complex in Minnesota, consists of a large, lenticular, centrally sunken but generally concordant basin- or funnel-shaped intrusive mass. Most lopoliths are found in undeformed or gently folded regions. The thickness is generally one-tenth to one-twentieth of the width. The diameter ranges from tens to hundreds of kilometers, with thicknesses up to thousands of meters. Lopoliths are almost always composed of well-layered mafic to ultramafic rock types, a characteristic giving rise to the commonly used term *layered intrusion*, although a few have a very thin cap of granitic material at their tops. When compositions of all the layers are averaged, total bulk compositions of lopoliths are close to basalt.

Commonly cited examples of lopoliths include the Duluth Complex (along the north shore of Lake Superior in Minnesota), the Sudbury intrusion (Ontario), the Bushveld Complex (South Africa), and the Skaergaard Complex (East Greenland). These examples range in age from Archean to Tertiary, and at least one (the Sudbury) may be the result of melting and crustal fracturing due to major meteorite impact. If so, the Sudbury intrusion was formed as a result of melting caused by the heat produced by shock waves and by rapid decompression, which was in turn caused by sudden loss of near-surface material as a crater was excavated. Pressure calculations or estimations for various layered intrusions have demonstrated that they typically crystallize at depths less than about 10 km. Figure 1-6

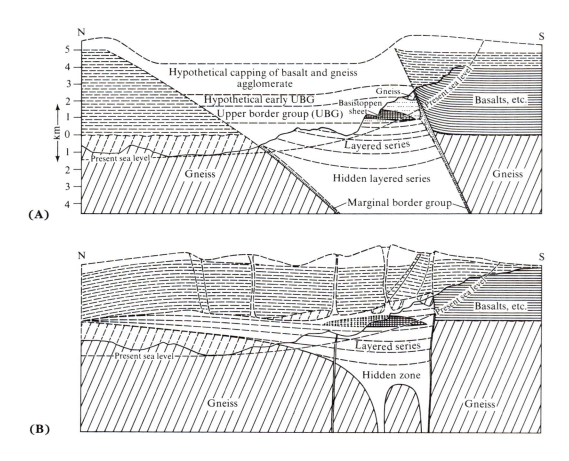

FIGURE 1-7
• • • • • • • • • • • • • • • •

The Skaergaard intrusion in its original orientation. **(A)** The Wager and Brown
interpretation. The solid irregular line indicates the present erosion surface. The
intrusive body is considered to be funnel-shaped, with a large, unexposed, hidden
zone at the base. [From L. R. Wager and G. M. Brown, 1967, *Layered Igneous
Rocks* (San Francisco: W. H. Freeman, Fig. 8.] **(B)** The McBirney interpretation.
The present level of erosion is indicated by a solid line. Much of the hidden zone
has been eliminated, and the extent of the middle and lower zones has been considerably increased. [From A. R. McBirney, 1975, *Nature, 253,* Fig. 7.]

shows the Sudbury lopolith, and Figure 1-7 shows the
Skaergaard lopolithic intrusion of East Greenland. The
Sudbury intrusion is seen both in plan view and cross
section in Figure 1-6. The cross section indicates that the
body is probably basin-shaped. Because the nature of the
lower contacts of lopoliths is not well known, it is possible that many are actually funnel-shaped. The funnel-
shaped character of the Skaergaard Complex is easy to
see in Figure 1-7.

Dikes and Veins

Dikes are thin, tabular, discordant intrusive bodies that
cut across the foliation or bedding of the country rock.

They range in thickness from less than a meter to
several hundred meters, and a few have been traced
along strike for tens of kilometers. Typically emplaced
into already existing fracture systems, they may occur
singly or in swarms (Figures 1-3A and 1-8). In some
areas, dikes occur as radiating swarms centered on an
intrusion or on the flanks of a volcano, where they
represent the feeders for eruptions distant from the
summit crater. Dikes very commonly are more resistant
to erosion than surrounding country rock and therefore
form residual ridges (Figure 1-9). In rare cases, vertical-
or outward-dipping **ring dikes** or inward-dipping **cone
sheets** are distributed in oval or circular patterns
around an intrusion. This arrangement appears to be

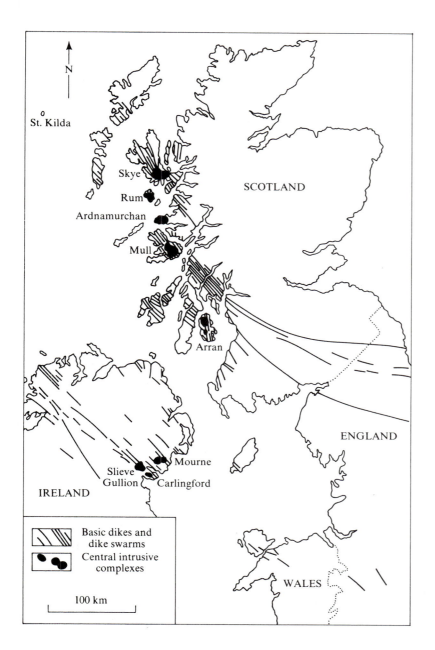

FIGURE 1-8
● ● ● ● ● ● ● ● ● ● ● ● ● ● ● ●

Regional dike swarm in eastern Scotland. [After
J. E. Richie, 1961, *Scotland: The Tertiary Volcanic
Districts*, 3rd ed., rev. by A. G. MacGregor and
F. W. Anderson (Edinburgh: Department of
Scientific and Industrial Research, Geological
Survey and Museum).]

related to fracturing that is associated with a doming
igneous body and the release of pressure that occurs
during extrusion and subsequent subsidence of the
overlying country rock due to partial emptying of
the magma chamber. A classic example of both phenomena is found in the Tertiary ring dike complex of
the Ardnamurchan peninsula on the western coast of
Scotland.

    Veins are not strictly igneous phenomena but in many
cases are created by hydrothermal activity related to
nearby igneous bodies. They are small, discordant, dikelike fracture fillings in country rock and are commonly
associated with replacement or alteration of the host
rock. Most veins appear to have formed by means of

mineral precipitation from hydrothermal fluids. The most
common sort are quartz veins, in which massive white
quartz forms the bulk of the vein, sulfide minerals are a
minor component, and native metals such as gold or silver are present in trace amounts. Quartz veins can commonly be mapped into granitic intrusions and probably
formed from fluids given off by the granitic magma as it
crystallized.

Batholiths and Stocks

Batholiths are large plutons that have steeply outward-
dipping walls and generally lack a floor or sharp lower

FIGURE 1-9
● ● ● ● ● ● ● ● ● ● ● ● ● ● ●

Dike and volcanic neck exposed by erosion at Shiprock, New Mexico. The exposed dike is about 8 km long. [From J. S. Shelton, 1966, *Geology Illustrated* (San Francisco: W. H. Freeman), Fig. 19.]

contact (Figure 1-10). They are commonly composed of silica-rich igneous rocks (granites and similar rocks) and range in outcrop area from 100 to several thousand square kilometers. An arbitrary lower limit on area separates them from **stocks,** which are similar to batholiths in shape but by definition have a maximum surface outcrop of 100 km². Stocks and batholiths range from completely concordant to completely discordant. Many batholiths conform to regional structure generally and thus are broadly concordant but may be highly discordant locally. Large plutons containing coarse-grained quartz-feldspar rocks are often described as granite during field mapping, although most commonly the actual rock types are not strictly granite but related rock types including granodiorite and quartz diorite (described in Chapter 3).

Composite plutons are a special and rather common class of batholithic or stocklike intrusive bodies that represent multiple pulses of intrusion. Diverse igneous rock types occur in sharp or gradational contact with one another in composite plutons. Gradational contacts commonly contain well-developed deformational fabrics such as foliations and lineations. Intrusive rock types grading from diorite to granite occur together in such plutons. Simultaneous intrusion of granites (or syenites) and basalts are more uncommon. Good examples of such so-called bimodal magmatism and of composite plutons are found along the coast of New England from northern Massachusetts to Maine, in the Sierra Nevada of California, and in the Virginia Dale Complex of the Boulder Creek Batholith in northern Colorado. Generally, composite plutons represent relatively short intervals of intrusion, and display features or processes characteristic of

coexisting (or *coeval*) magmas, including magma mixing or hybridization and gradational contacts.

Plutons of batholithic dimensions can be found in several different geologic environments and probably formed in different ways. Many granitic rocks within Precambrian (especially Archean) shield areas were created very early in Earth's history and probably crystallized from residual liquids remaining after partial crystallization of basaltic magmas. In contrast, most petrologists believe that the younger granites are the products of partial melting of preexisting sedimentary or metamorphic rocks. If an anatectic source region is exposed by uplift and erosion, many of the granitic plutons are seen in a generally concordant association with surrounding high-grade metamorphic rocks. Other granite magmas probably moved considerable distances from their source areas, and they show quite different relationships with the country rocks.

There are certain structural characteristics that are helpful in field classification of pluton emplacement depth. A. F. Buddington (1959) proposed a broad classification scheme based on the different characteristics of plutons emplaced at various depths. *Catazonal plutons* are completely surrounded by high-grade metamorphic rocks containing mineral assemblages indicative of the high temperatures and pressures required for anatexis. Contacts tend to be gradational because of significant chemical interaction between magma and country rocks, and broad zones of migmatites are present (Figure 1-11). **Migmatites** (from the Greek root *migma-*, meaning "mixture") are layered, mixed rocks that consist of light-colored granitic layers alternating with clear-

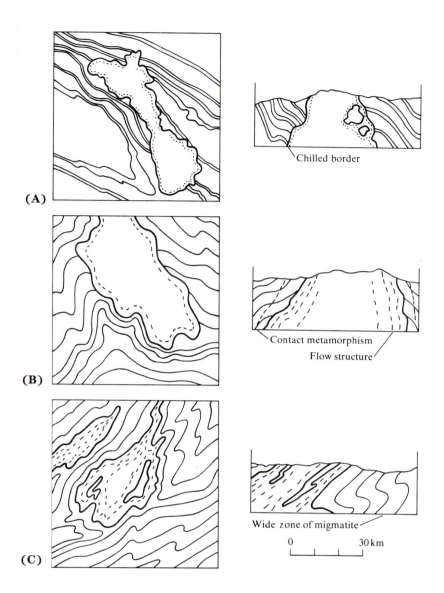

FIGURE 1-10
● ● ● ● ● ● ● ● ● ● ● ● ● ● ● ● ●

Structural patterns of batholiths and stocks, viewed from above (left) and in cross section (right). **(A)** Shallow (epizonal, less than 5 km deep), with strongly discordant contacts and chilled border. Usually very limited zone of contact metamorphism, if any; baked contacts may occur. Average size, about 10–100 km². **(B)** Middle-depth (mesozonal, 5–15 km), with contacts that are partially discordant, partially concordant. A wider zone of contact metamorphism is present, and flow structure is common within the pluton. Average size, 100–500 km². **(C)** Deep (catazonal, greater than 15 km), with generally concordant contacts, but contacts can be discordant locally if pluton has moved from the melting region. Flow structure or gneissic banding parallels the contact. A wide zone of migmatites and contact metamorphism is typically present. Average size, 50–1000 km².

ly foliated, dark-colored metamorphic rocks (schists or gneisses). Although migmatites have been a source of much petrologic controversy, there is now a consensus among petrologists that they represent local partial melting and melt segregation into layers that are interspersed with the residual unmelted material. In catazonal environments, there is a general conformity of foliations within both country rocks and igneous rocks that indicates essentially simultaneous metamorphism, plutonism, and deformation. Some migmatites show extreme ductile deformation and mylonitization (see Figure 1-11B).

Mesozonal plutons are surrounded by metamorphic rocks in which both mineral assemblages and textures indicate only low- to medium-grade metamorphism and therefore temperatures well below those required for melting. Contacts are fairly sharp and can be either concordant or discordant. Migmatites are typically absent or only minor. Moderate deformation is commonly present in the country rock, and flow structure can be present within the granite.

Epizonal plutons are largely discordant with the country rock and with the dominant regional structure, and most do not possess internal flow structure. Contacts with the country rock are sharp, and metamorphism of the country rocks tends to be minimal. Angular xenolithic fragments are common along pluton margins and most likely represent brittle fracturing of the country

(A)

FIGURE 1-11
• • • • • • • • • • • • • • • • •

Migmatites. **(A)** Migmatite showing typical wavy foliation. Fine-grained dark layers are rich in biotite, plagioclase, and garnet. Coarse light layers are mainly quartz, microcline, and plagioclase, and are of roughly granitic composition. Summit County, Colorado. [From M. H. Bergendahl, 1971, *U.S. Geol. Surv. Prof. Paper 652*, Fig. 6.] **(B)** Migmatitic gneisses from Glen Beg in the Western Highlands, Scotland. Deformed coarsely crystalline granitoid layers are parallel to depleted schist layers marked by notable biotite concentrations. Later coarse granitic fractions have crosscut the entire assemblage.

rock during emplacement. Chilled margins are commonly present along pluton borders, as indicated by reduced grain size. Many epizonal plutons are associated with volcanic rocks and collapse structures that are thought to be the results of the same magmatic event.

Buddington's broad classification scheme was developed at a time prior to the availability of precise quantitative methods of temperature and pressure estimation in both igneous and metamorphic rocks. Research geologists now use these techniques, which depend on detailed knowledge of rock and mineral chemistry. Nevertheless, Buddington's broad scheme is still useful as a field tool for general classification.

Batholiths are important igneous rocks from a tectonic viewpoint. They are the igneous intrusions that form the cores of the major mountain belts of the world. These intrusions can extend for hundreds of kilometers along strike in the major structural trend, and some are tens or even hundreds of kilometers wide. Analysis of most major mountain chains has revealed that these igneous cores do not represent a single large intrusion but consist of many hundreds of related intrusions that differ in size, composition, and time of emplacement. Together they represent one of the major

products of the mountain-building cycle in which interactions of Earth's plates recycle older materials. Understanding the igneous processes of batholith formation is a key part of plate tectonics.

METHODS OF EMPLACEMENT OF INTRUSIVE ROCKS
• • • • • • • • • • • • • • • •

Some methods of emplacement of intrusive igneous rocks were touched on earlier. Here it is useful to review some of the field evidence for emplacement mechanisms. Forceful injection and magmatic stoping are the two mechanisms for which the most evidence is available. Formation of melts by anatexis and retention of the magmas near their source is not really considered emplacement, because the magma has moved little, if at all.

Evidence of *forceful injection* is abundant around many intrusive bodies. In this process, magma under pressure is forced out of the magma chamber and into cracks and fractures in surrounding country rock. This effect is most easily seen where small dikes of intruding

(B)

magma have clearly forced apart the adjacent dike walls. The geologist should look for the presence of one or more structures that cross the intrusive dike at angles other than about 90°. Dilation or forceful opening by intrusive stresses will displace these features perpendicular to the dike walls. It is even better to find two country rock structures crossing the intrusion at different angles. When these features are found, it is obvious that the two opposing segments of dike wall have not been subjected to faulting (relative sideways displacement), but only to spreading or dilation. Matching of the particular contours of the opposite sides of the dike aids the argument. Forceful injection in the brittle regime can also be documented by finding fracturing, doming, and faulting of overlying rocks (as occurs in the Henry Mountains of Utah).

On occasion, a magma is substantially crystallized when emplaced. In this situation, a shear zone develops at the country rock contact, with displacement and intermixing of country rock and solidified igneous material near the edges of the intrusive body. Internal shear zones separating large regions of the interior of batholith- and stock-sized granitic bodies also occur. On other occasions, magmatic pressures may be sufficient to raise the roof of the magma chamber, thereby forming an inward-dipping conical fracture with its apex at the top of the intrusion. Magma (typically basaltic) injected along the fracture then crystallizes to form what is called a *cone sheet.*

Magmatic stoping is a second important mechanism of intrusion. In mining terminology, "overhand stoping" decribes a process whereby miners tunnel under an ore body and mine it from below by drilling and blasting upward, causing chunks of the roof rock to fall. Magmatic stoping is initiated during intrusion when overlying country rock fractures. Fragments sink into the magma chamber, thus allowing magma to move upward into the vacated space. The foreign fragments (xenoliths) may react with the melt, depending on their compositions and the magma temperature. Near-total assimilation of xenoliths produces diffuse streaks, bands, or clots either of minerals that are not ordinarily present in the igneous rock or of igneous minerals (typically dark-colored mafic ones in granites) that are locally present in anomalously high concentrations.

Stoping that involves the incorporation of many small blocks of country rock is often called *piecemeal stoping.* But when a large mass of the magma chamber roof collapses and sinks into the magma below, the process is called *cauldron subsidence.* It occurred at the classic locality, the Mt. Ascutney pluton in eastern Vermont. Magma that moves upward along the associated fracture crystallizes to form a more or less vertical *ring dike.*

The most extreme case of magmatic stoping is that of *roof foundering.* Here the magma chamber is large and close to the surface, and has a thin roof. Collapse of the roof can result in inclusion of huge blocks of country rock that have been tilted at different angles and perhaps have partially settled into the magma below. The collapse of overlying rocks can also allow easy passage of magma to the surface and thus can cause extensive volcanism to accompany the foundering. This mechanism has been proposed to explain the unroofing of the Idaho Batholith.

TYPES OF EXTRUSIONS
• • • • • • • • • • • • • • •

Extrusive rocks tend to to be either layered or conical, depending on the nature and amount of the erupted material and the mechanism of eruption. Silica-poor magmas that have low gas content and relatively low viscosity (a characteristic typical of basalts) commonly produce lava flows. The term *lava* is used both for magma that has been erupted onto the earth's surface and for the rock that has solidifed from it. Many lavas include minor volumes of pyroclastic material—loose fragmentary material such as cinders or small volcanic bombs—that has been ejected from the volcanic vent. Silica-rich magmas with high gas content and high viscosity (rhyolite, for example) virtually always form pyroclastic layers (ash falls) on the surface. Conical extrusions (discussed in detail later) form typical volcanoes such as cinder cones (Figure 1-12).

Most lava flows are basaltic, although they can also be andesitic. Subaerial lavas are classified as pahoehoe (pronounced pa-ho-e-ho-e); aa (pronounced ah-ah)—both Hawaiian words; or block lava. **Pahoehoe lavas** have glassy, smooth, billowy, and occasionally ropy surfaces (Figure 1-13). The shape of the upper surface commonly resembles irregular folds in cloth and results from distortion of a thin, chilled, but still plastic surface zone by differential movement of an underlying, more fluid material within the interior of the flow. Many pahoehoe flows also contain large areas that are relatively smooth.

Aa lavas have rough, fragmented surfaces (Figure 1-14). The upper portions of aa flows are vesicular and resemble furnace or slag clinkers, whereas the inner portions are relatively dense. The clinker fragments are very rough and irregular in shape and may have projecting spinelike points on the surface. Individual fragments are

FIGURE 1-12
• • • • • • • • • • • • • • •

Cerro Negro volcano, near Managua, Nicaragua, in 1968. This volcano is a classic cinder cone built up on older lava flows. [Mark Hurd Aerial Surveys.]

typically less than 15 cm in diameter but can be as large as a few meters. Space between the larger fragments is commonly partially filled by finely ground fragments produced by abrasion of the larger fragments as they collide during flow movement. Below the immediate surface, fragments become welded together.

Block lavas are composed of fragments that are smooth relative to the irregular aa clinkers. Surfaces of block lava flows are very irregular, sometimes with relief of several meters (Figure 1-15). Ridges perpendicular to the direction of flow often develop as a result of lava surges from the vent. Although a central massive layer is typically present, it is much less extensive than in aa lavas. Flow movement is slow and irregular, with a definite tendency for some layers to shear over others; thereby producing considerable volumes of crushed fragmental material. Very high viscosity (relative to pahoehoe lavas) and occasional rapid flow cause breakage of the cooled lava into angular fragments that become rounded or fragmented. The rigidity of the whole flow mass causes some sliding over the surface beneath it.

FIGURE 1-13
● ● ● ● ● ● ● ● ● ● ● ● ● ● ● ●

Classic pahoehoe lava in a small
cascade that has overflowed a lava
lake at Halemaumau, Kilauea, Hawaii.
Foreground width is about 20 m.
[From J. Green and N. M. Short, 1971,
*Volcanic Landforms and Surface
Features* (New York: Springer-Verlag),
Plate 110A. Photo courtesy of
Geophysical Laboratory, Carnegie
Institution, Washington, DC.]

A complete gradation exists between pahoehoe, aa, and block lavas. Commonly, flows issue from a volcanic vent as pahoehoe and change to aa or block lavas during downslope movement. The transition is principally a function of increasing viscosity and the amount of agitation of the flowing lava. As lava moves downslope, it cools, partially solidifies, and, most important, loses gas. All of these factors, but particularly the last, tend to increase the lava's viscosity and convert pahoehoe to aa or block lava. If a magma sufficiently outgasses before or during eruption, aa or block lavas are produced directly from the vent. Block lavas are typically more siliceous (a composition contributing to higher viscosity) and are associated with greater volumes of pyroclastics.

A single eruption can produce several contemporaneous outpourings of lava. These can vary locally in type (pahoehoe or aa), but all are considered to be a single *flow unit*. Flow units vary from a few centimeters to

FIGURE 1-14
● ● ● ● ● ● ● ● ● ● ● ● ● ● ● ●

Rough, rubbly aa lava surface at Mt.
Vesuvius, Italy. [From J. Green and
N. M. Short, 1971, *Volcanic Landforms
and Surface Features* (New York:
Springer-Verlag), Plate 140. Photo
courtesy of Geophysical Laboratory,
Carnegie Institution, Washington, DC.]

FIGURE 1-15
● ● ● ● ● ● ● ● ● ● ● ● ● ● ●

Block lava of the Yatsugatake
volcano, Japan. Most fragment
diameters range from 0.5 to 1 m.
[From J. Green and N. M. Short,
1971, *Volcanic Landforms and
Surface Features* (New York:
Springer-Verlag), Plate 139B.
Photo courtesy of Geophysical
Laboratory, Carnegie Institution,
Washington, DC.]

about 200 m in thickness, although most are considerably less than 100 m. For example, the thicknesses of
individual flow units in the Columbia River plateau basalts in Washington and Idaho vary from 14 to 35 m
(Figure 1-16). The total thickness of the multiple flow
units of the Columbia River basalts sometimes reaches
almost 2 km, with a total estimated volume of 160,000 km^3
(see Chapter 10). The areal extent of the Deccan Traps
(basaltic flows) in India is 160,000 km^2 (Figure 1-17),
approximately 20 percent of the surface area of that
country.

Pillow lavas consist of masses of ellipsoidal or
pillow-shaped bodies (Figure 1-18). The base of each
pillow takes the shape of the one below as pillows pile
up. The internal structure of each pillow consists of
a fine-grained crystalline core with a glass crust, all
with radial jointing. This arrangement contrasts with the
concentric structures caused by a radial arrangement of
vesicles that is sometimes found in pahoehoe lavas.
Pillows generally range in size from 10 cm to 6 m. The
composition of pillow lava is typically basaltic or andesitic, but these lavas commonly are enriched in sodium as a result of seawater alteration. Observations of
active underwater lava flows at Hawaii by divers indicate
that pillow structure results from the quenching of hot
basaltic lava by seawater. Pillow lavas have also been

FIGURE 1-16
• • • • • • • • • • • • • • • •

Columnar basalt exposed in the upper 200 m of the east wall of Grand Coulee, Washington. Flow units came different distances from different vents and are separated by the obvious structural breaks. Some erosion and soil layers are present. Average thickness of each flow unit is about 15 m. [From J. S. Shelton, 1966, *Geology Illustrated* (San Francisco: W. H. Freeman), Fig. 316.]

observed in mid-ocean ridge central rifts by deep submersible vehicles. Because pillow lavas are the most common type of lava on the seafloor and because the oceans cover about 70 percent of Earth's surface, this type of lava is undoubtedly the most abundant of all the flow types on Earth. The presence of preserved pillow lava in the rock record is also an important clue to the existence of a marine environment. Rarely, pillow lavas can form in nonmarine environments. In Iceland, eruptions under glacier ice produce pillow basalts. Eruption of lava from fissures near playa lakes in continental rifts can result in pillow lavas when the lava flows into the lakes. Good modern examples exist in the East African Rift, and ancient examples occur in the Triassic and Jurassic flows of the Mesozoic rift basins of eastern North America.

Lava either emerges from a single eruptive center or simultaneously erupts along a long, narrow fissure.

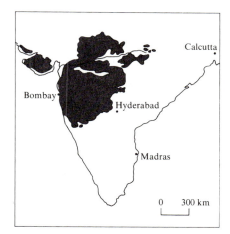

(A)

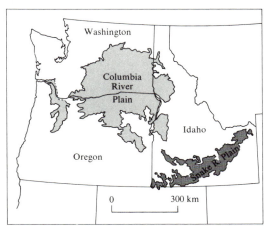

(B)

FIGURE 1-17
• • • • • • • • • • • • • • • •

Maps showing areal extent of major plateau basalt provinces. **(A)** Deccan basalts of India. **(B)** Columbia River and Snake River Plain basalts of the northwestern United States. [From G. A. MacDonald, 1972, *Volcanoes* (Englewood Cliffs, NJ: Prentice Hall), Fig. 11-2, after H. S. Washington, 1922, *Geol. Soc. Am. Bull.*, *33*, 765–803, and A. C. Waters, 1955, *Geol. Soc. Am. Special Paper* *62*, 703–722.]

FIGURE 1-18
● ● ● ● ● ● ● ● ● ● ● ● ●

Pillow lava at Nemour, Hokkaido, Japan. Foreground width is about 8 m. [From
J. Green and N. M. Short, 1971, *Volcanic Landforms and Surface Features* (New
York: Springer-Verlag), Plate 155B. Photo courtesy of K. Yagi, Sapporo, Japan.]

Whereas eruptive centers tend to produce at least small volcanic cinder cones, fissure eruptions typically produce huge volumes of relatively fluid basaltic magma that spreads out to form lava plateaus. Eruption of less fluid lavas along a fissure produces a linear chain of spaced volcanic cones. Central vent eruptions give rise to a variety of volcanoes, depending on the degree of fluidity of the melt and the amount of pyroclastic material thrown from the vent. Both of these factors reflect the overall magma composition and especially its gas content.

There are four principal types of volcanoes: shield volcanoes, composite volcanoes, cinder cones, and lava domes. At one extreme are the **shield volcanoes,** which are broad flat cones (named for their resemblence in cross section to ancient Germanic rounded shields) composed mainly to exclusively of lava flows. The sides typically have slope angles of 2° to 10°, and the base makes a smooth transition to the surroundings. Shield volcanoes range from small (with summit elevations of only 1000 m) to large (for example, some of the Hawaiian shield volcanoes, whose summits approach 10,000 m above the seafloor—equivalent to the height of Mt.

Everest). Many of the shield volcanoes are not circular in map view but are somewhat elongate, with the elongations controlled by the trend of the rift zone beneath. The crest of a shield volcano is generally capped by either a cinder or a spatter cone. The summits of the Hawaiian shield volcanoes commonly contain a large depressed central area—a **caldera**—formed when the roof of the magma chamber collapses following magma eruption or subterranean withdrawal of magma from the subsummit magma chamber. Later eruptions often fill the caldera with lava, thereby creating a lava lake. Alternatively, the caldera becomes quiescent, and later eruptions occur on the flanks of the shield along rifts.

Volcanic eruptions can be violent, throwing rock fragments and molten magma from the vent. Forceful ejection of liquids that fall and solidify immediately adjacent to the vent produces *spatter cones* (Figure 1-19); these are commonly found associated with the eruption of very fluid basalts. **Ejecta** are the solids thrown from the vent; accumulations of ejecta are called pyroclastic rocks or *tephra*. Explosive volcanic eruptions that eject many discrete larger blocks and bombs

FIGURE 1-19

• • • • • • • • • • • • • • • •

Small spatter cone forming on road during the 1955 eruption of Kilauea volcano, Hawaii. Cone height is about 7 m. [U.S. Geological Survey photo by G. A. MacDonald.]

are classified as **strombolian** (after the volcano Stromboli in the Mediterranean Sea). Strombolian ejection velocities of individual particles have been estimated to be close to the speed of sound. Tephra cones or **cinder cones** are more or less circular cones resulting from the accumulation of cinders around the central vent. More and larger cinders accumulate adjacent to the vent, thus creating the cone morphology. The sides slope outward at about 30° (close to the angle of repose for loose irregular materials), and the crater is generally bowl-shaped, having been formed by moderately explosive eruptions. Excellent examples of recent cinder cones are found near Flagstaff, Arizona, in the Rio Grande Rift in New Mexico and Arizona, and at Craters of the Moon National Monument in Idaho (see Chapter 10).

Many volcanic structures consist of mixtures of lava and tephra. Larger volcanic cones produced by both flows and tephra in alternating layers are called **composite volcanoes** or **stratiform volcanoes.** The vent is generally smaller than that of a shield volcano and its sides are considerably steeper. Well-known examples are Mt. Rainier and Mt. St. Helens in Washington, Mt. Shasta in California, Fujiyama in Japan, and Mt. Vesuvius in Italy.

Lava domes are bulbous, steep-sided masses of high-viscosity lava that commonly form in the craters of larger volcanoes of the composite type. The most widely known young lava dome is the one that formed in the blown-out crater at Mt. St. Helens, Washington, following the major eruption there. This dome took several years to form and as of this writing is several hundred meters high (Figure 1-20).

Pyroclastic materials are classified according to fragment size: bombs and blocks (greater than 32 mm in diameter), lapilli (4–32 mm), and ash (less than 4 mm). **Blocks** are solid when ejected and therefore form angular, blocky fragments, whereas **bombs** are fluid when ejected and have their shapes modified by air friction during flight. Although it is tempting to focus on the occasional special forms such as ribbon or cow-dung bombs, the great majority of bombs are simply irregular vesicular lumps known as cinder, scoria, or pumice. **Scoria,** essentially synonymous with cinder, is dark, relatively dense mafic rock with abundant vesicles and occurs most commonly as crusts on cooled flows. **Pumice** is a light-colored vesicular glass of siliceous composition in which the volume of closed gas- or air-filled cells is commonly so large that its low density allows the rock to float on water. **Lapilli** are formed either from rock fragments or by consolidation of very fluid drops of lava that solidified in the air. They are typically drop-shaped, but they can be spherical or rod-shaped. **Ash** is usually composed of glass (vitric ash), but it can also consist of rock or mineral fragments, in which case it is called lithic or crystal ash.

Consolidated (typically thermally welded) layers of ash are called **tuff.** Vitric tuff commonly consists of triangular fragments with concave sides, which are formed by the rapid production and breakup of glass- or melt-walled bubbles in the melt. Because ash is deposited mainly by vertical fall, it forms a continuous blanket over the countryside, unlike flows, which tend to follow the topographic lows of the area. Ash and tuff commonly show some size sorting, both laterally and vertically;

FIGURE 1-20
● ● ● ● ● ● ● ● ● ● ● ● ● ● ● ● ●

The lava dome in the blown-out crater at Mt. St. Helens, Washington, May 1982, viewed from the north. [Photo by Lyn Topinka, U.S. Geological Survey, David A. Johnston Cascades Volcano Observatory, Vancouver, Washington. Drawing from F. Press and R. Siever, 1994, *Understanding Earth* (New York: W. H. Freeman), Fig. 5.11.]

Ring of pyroclastic ejecta

Volcanic dome within crater

their distribution is dependent on the size of the fragments as well as on prevailing wind velocities and directions. Particularly violent eruptions such as those at Krakatau (Java) in 1883, Mt. St. Helens (Washington) in 1980, and Mt. Pinatubo (Philippines) in 1991 throw ash into the stratosphere, where it is carried around the globe, producing especially brilliant sunsets throughout the world. Sequential photos of the final, cataclysmic eruption of Mt. St. Helens on May 18, 1980, are shown in Figure 1-21. Figure 1-22 illustrates the downwind areal distribution of the ash fall; ash was deposited up to 10 cm deep on much of Washington State, northern Idaho, and eastern Montana. Ash falls can be spread over thousands of square kilometers; they are typically deposited within a few days and thus provide an excellent "time plane" for correlation of ancient rock sequences.

The "glowing avalanche," or **nuée ardente,** is intermediate in character between a lava flow and an ash fall. A nuée ardente caused the total destruction of the city of St. Pierre on the island of Martinique in 1902 (Figure 1-23), killing about 30,000 inhabitants. A glowing cloud of incandescent rock fragments and gases at temperatures between 700° and 1000°C swept down from Mt. Pelée at about 150 km/hr, destroying everything in its path. A small nuée ardente photographed on Mt. Unzen in Japan in 1991 is shown in Figure 1-24. Several small nuée ardentes were observed in the May 1980 eruption of Mt. St. Helens, Washington; these descended the south flank and flowed into the forest in less than 3 minutes. Release of pressure in a magma chamber that contains a volatile-rich magma causes rapid vesiculation (boiling) of the melt. If an extreme amount of vesiculation occurs rapidly enough, the molten material will be blasted from the vent to form an ash fall. In less extreme cases, a cloud of ash is produced, with the simultaneous formation of a glowing (basal) avalanche. The glowing avalanche consists of gas, frothy particles, liquids, and solids. The particles are presumed to be continually releasing gas, which maintains the turbulence and prevents particle-to-particle abrasion. Gas emission allows the mass to behave as a fluidized material that proceeds downslope for long distances with little loss of heat from the interior. An individual nuée ardente consists either of material shot out of the vent at a low angle in an explosion or, as in the case of Mt. Soufrière

on the island of St. Vincent, by material falling from the edges of a thick gas cloud. When the avalanche or ash flow comes to rest, the particles are either liquid or solid, or a mixture. A deposit formed from mainly solid particles will typically be unconsolidated and is called an **ash flow tuff.** If enough heat has been retained to weld the still plastic fragments together, the deposit is called a **welded ash flow tuff.** The term *ignimbrite* is used for any rock formed by an ash flow, regardless of its degree of welding.

The fragments within ash falls and ash flow tuffs consist mainly of glass **shards** that, when erupted, either have irregular shapes or are somewhat triangular with concave sides. If the ash fall or ash flow tuff is unconsolidated, these shapes will be retained. In thick deposits, welding occurs as very hot fragments are compacted and adjacent fragments develop a continuous glassy junction. Shapes of the still ductile fragments are commonly distorted by this compaction (Figure 1-25). Irregular masses

(A)

(B)

(C)

(D)

FIGURE 1-21
● ● ● ● ● ● ● ● ● ● ● ● ● ● ● ●

Final major eruption of Mt. St. Helens on May 1980. **(A)** 8:27:00 A.M., view looking southwest. **(B)** 8:32:37 A.M. The north slope of the volcano collapses as an earthquake triggers a massive landslide and debris flow. **(C)** 8:32:41 A.M. High-pressure gas and steam explode horizontally out of the north face of the breach with hurricane force. **(D)** 8:32:51 A.M. The gas-steam jet extends outward, leveling forests in its path. It was followed by a surge of pyroclastic flows and debris. [From F. Press and R. Siever, 1986, *Earth*, 4th ed. (New York: W. H. Freeman), Fig. 16-36. Photos copyrighted by G. L. Rosenquist, 1980.]

of pumice are commonly compacted in the same way, producing flattened elongate fragments containing similarly flattened vesicles. The ultimate in welding produces **obsidian,** a massive glassy rock with occasional fine-scale "ghost" laminations. In obsidian formation, compaction and welding have proceeded to such a degree that it is not uncommon for individual shards to be almost completely obliterated, because the interstitial pore spaces were eliminated. Members in a sequence of layered ash flow tuffs can commonly be distinguished and mapped on the basis of degree of welding. The base of each unit is typically chilled and nonwelded. The degree of welding increases above the base, as do secondary reactions from the presence of fluids. The amount of welding decreases toward the top of the simple cooling unit; and the uppermost layers, commonly unwelded and easily and quickly removed by erosion, are generally absent within multiple cooling units.

SUMMARY
● ● ● ● ● ● ● ● ● ● ● ● ● ●

Igneous rocks are the rock products of solidification of magma, or molten silicate material. Igneous rocks are important throughout the crust and even dominate some environments, such as the seafloor. Deciphering the origin of igneous rocks requires observations on all scales, from geologic mapping to outcrop observations of structures and hand specimen identification and, finally, to laboratory studies. A rock body in an outcrop or series of outcrops can be identified as igneous on the basis of both large-scale and small-scale features. Its extent, shape, and contact relationships with country rock can be used to determine whether the igneous rock

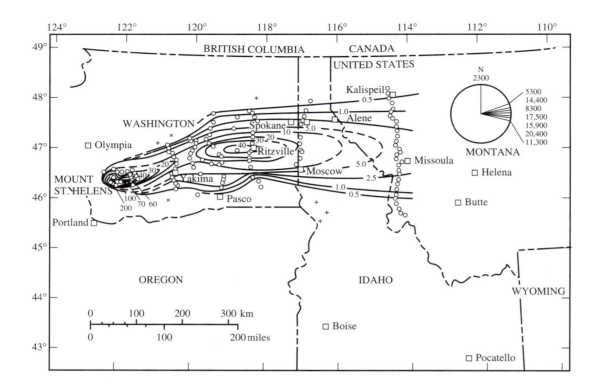

FIGURE 1-22

• • • • • • • • • • • • • • • • •

Map showing thickness (in millimeters) of air-fall ejecta from the Mt. St. Helens eruption of May 18, 1980. Contours represent uncompacted thickness of ash (+ indicates light dusting of ash; ×, no ash observed). The prevailing wind was generally from the west during the morning of May 18. Similar thickness maps for other Mt. St. Helens eruptions show very different ash plume directions, which depend on the wind direction. [From A. M. Sandra-Wojcicki, S. Shipley, R. B. Waitt, Jr., D. Dzurisin, and S. H. Wood, 1981, *The 1980 Eruptions of Mount St. Helens, Washington, U.S. Geol. Surv. Prof. Paper 1250*, P. W. Lipman and D. R. Mullineaux, eds., Fig. 336.]

is intrusive or extrusive. Small-scale features such as grain shapes, sizes, and manner of aggregation can be used to verify the method of origin.

Intrusive rocks solidify at depths in the earth ranging from only a few meters to tens of kilometers. Intrusive bodies either have thin tabular shapes like dikes and sills or form stocks and batholiths, which are large to very large irregular masses tens to thousands of square kilometers in outcrop area. Extrusive igneous rocks solidify on the surface as lava flows or sheets of pyroclastic material. Many of the features of extrusive rocks, including lava flow forms, shapes and sizes of volcanoes, and even the distinction between formation of lava flows and

volcanic ash falls, depend on the properties, and thus the chemistry, of the magmas. Succeeding chapters provide more detailed descriptions of small-scale features, particularly microscopic ones, and show how their origin correlates with field occurrences.

STUDY EXERCISES

• • • • • • • • • • • • • • •

1.   What is igneous petrology, and why is it an important part of an undergraduate geology curriculum?

(A)

(B)

FIGURE 1-23
● ● ● ● ● ● ● ● ● ● ● ● ● ● ●

St. Pierre, Martinique, shortly before **(A)** and after **(B)** the 1902 nuée ardente eruption.
[From T. A. Jagger, 1945, *Volcanoes Declare War* (Honolulu: Paradise of the Pacific, Ltd.),
Plates 26a, 26c.]

2. As a field geologist, what criteria would you use to determine whether an outcrop of igneous rock represented an old lava flow or an igneous intrusion?
3. Briefly describe the different forms of volcanic vents, and explain what genetic factors determine the external forms.
4. What petrologic or structural factors might be important in determining whether a batch of magma in the crust was finally emplaced as a dike, as a sill, or as a bulbous, stock-sized pluton?
5. What chemical or physical factors are important in causing a magma to erupt explosively and create a pyroclastic deposit?

REFERENCES AND ADDITIONAL READINGS
● ● ● ● ● ● ● ● ● ● ● ● ● ●

Barker, D. S. 1983. *Igneous Rocks.* Englewood Cliffs, NJ: Prentice Hall.

Buddington, A. F. 1959. Granite emplacement with special reference to North America. *Geol. Soc. Am. Bull.*, *70*, 671–747.

Hunt, C., P. Averitt, and R. L. Miller. 1953. *Geology and Geography of the Henry Mountains Region, Utah, U.S. Geol. Surv. Prof. Paper 228.*

FIGURE 1-24
● ● ● ● ● ● ● ● ● ● ● ● ● ● ●

A pyroclastic flow plunges down the slopes of Mt. Unzen, Japan, in June 1991. Note the fireman and fire engine in the foreground, trying to outrun the hot ash cloud descending on them. Three scientists who were studying this volcano were killed when they were engulfed by a similar flow. [From F. Press and R. Siever, 1994, *Understanding Earth* (New York: W. H. Freeman), Fig. 5.8. Photo by AP/Wide World Photos.]

Kilburn, C. R. J., and G. Luongo, eds. 1993. *Active Lavas.* London: U.C.L. Press.

Lipman, P. W., and D. R. Mullineaux, eds. 1981. *The 1980 Eruptions of Mount St. Helens, Washington. U.S. Geol. Surv. Prof. Paper 1250.*

McBirney, A. R. 1975. Differentiation of the Skaergaard Intrusion. *Nature, 253,* 691–694.

McBirney, A. R. 1993. *Igneous Petrology,* 2nd ed. Boston: Jones and Bartlett.

Newell, G., and N. Rast, eds. 1970. *Mechanism of Igneous Intrusion.* Liverpool: Gallery Press.

Shirley, D. N. 1987. Differentiation and compaction in the Palisades Sill, New Jersey. *J. Petrol., 28,* 835–866.

Wager, L. R., and G. M. Brown. 1967. *Layered Igneous Rocks.* San Francisco: W. H. Freeman.

Walker, K. L. 1969. *The Palisades Sill, N.J.: A Reinvestigation. Geol. Soc. Am. Spec. Paper 111.*

Wood, C. A., and J. Kienle, eds. 1990. *Volcanoes of North America: The United States and Canada.* Cambridge: Cambridge University Press.

FIGURE 1-25
● ● ● ● ● ● ● ● ● ● ● ● ● ● ●

Compaction of hot, plastic shard fragments deforms individual shards and welds the mass together. This photomicrograph in plane-polarized light shows the deformed shards and fragmental crystalline material (mainly feldspars, quartz, and amphiboles). [From J. C. Olson, 1968, *U.S. Geol. Surv. Bull., 1251-C,* Fig. 9.]

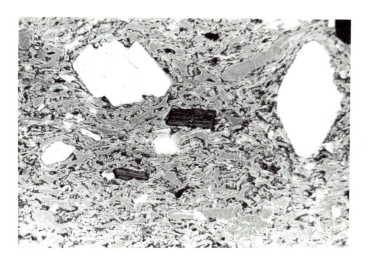

Chapter **2**

IGNEOUS MINERALS AND TEXTURES

The proper description of an igneous rock requires information on its mineralogy, its texture, and the large-scale structures indicative of environment. The nature of large-scale structures was presented in Chapter 1. This chapter reviews the common igneous minerals, textures, and small-scale structures and then provides brief descriptions of the more common igneous rocks. Textures are commonly best recognized through petrography, the microscopic examination of rocks in thin section. A detailed discussion of petrographic techniques and criteria is beyond the scope of this book, but some of the more fundamental aspects of mineral textural relationships as seen in thin section are introduced. The characterization of igneous rocks in outcrop, hand specimen, or thin section provides key information for development of ideas about how melts crystallize and a basis for interpreting igneous phase diagrams, which are introduced in Chapter 4.

MINERALS OF IGNEOUS ROCKS

Igneous rocks contain a significant variety of silicate and nonsilicate minerals, which fall into two broad categories: high-temperature *primary* minerals, which crystallize directly from a magma; and so-called *secondary* minerals, which form at temperatures below those at which any melt can exist (called the *subsolidus* regime). Secondary minerals typically reflect partial or total replacement of high-temperature primary minerals by low-temperature ones. This replacement commonly involves hydration or oxidation of the primary minerals by late fluids, which are either magmatic or due to near-surface weathering. In the great majority of igneous rocks, primary minerals dominate and secondary miner-

als occur only in minor amounts. The most important igneous minerals are reviewed here, with an emphasis on the properties that most strongly affect their presence in igneous rocks. It is most important that any student of petrology have an appreciation of which physical and chemical properties of minerals affect their behavior in rocks, because minerals are the foundation of the most important igneous rock classifications. Further information on these minerals can be found in Deer, Howie, and Zussman (1993).

Silica Minerals

The silica minerals have the composition SiO_2. They are framework silicates in which each of the silicon atoms is coordinated by four oxygen atoms and each SiO_4^- group (tetrahedron) is linked to four others. Silica minerals included in this group are α- and β-quartz, tridymite, and cristobalite. The latter two are relatively uncommon high-temperature polymorphs and are typically restricted to volcanic rocks. Trigonal α-quartz and hexagonal β-quartz are the polymorphs of quartz related by a phase transformation at roughly 600°C (β-quartz on the high-temperature side). Because virtually all magmas crystallize at temperatures greater than 600°C, β-quartz is the form that actually crystallizes, but all β-quartz spontaneously inverts to α-quartz when cooled below the transformation temperature.

Feldspars

The feldspar group constitutes several individual chemical species and is unquestionably the most abundant mineral group both in the earth's crust and in igneous rocks. Structurally, the feldspars are framework aluminosilicates based on a three-dimensional continuous framework of tetrahedrally coordinated silicon and

FIGURE 2-1

• • • • • • • • • • • • • •

The monoclinic structure of high sanidine and orthoclase, projected on the (201) plane. The rotational axes (twofold) and mirror planes appropriate to this structure are shown. T_1 and T_2 indicate the two geometrically distinct tetrahedral sites, K indicates the position of the potassium cation, and b indicates the length and orientation of the crystallographic b axis. Note that most other feldspars (microcline, plagioclase series) have reduced symmetry as a result of the ordering of aluminum and silicon between the T_1 and T_2 tetrahedral sites. [After J. J. Papike and M. Cameron, 1976, *Rev. Geophys. Space Phys.*, *14*, Fig. 43.]

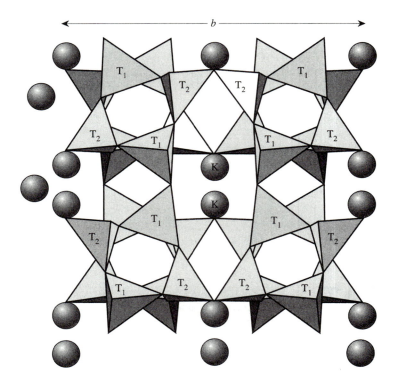

aluminum. All four corner oxygens in each SiO_4 or AlO_4 tetrahedron are linked to adjacent tetrahedra. Such a framework consisting only of SiO_4 tetrahedra would be electrostatically neutral (as in the silica minerals such as quartz); introduction of some AlO_4 tetrahedra requires addition of alkali or alkaline earth ions to balance the charge (Figure 2-1).

Beyond silicon and aluminum, the dominant cations in feldspars are sodium, potassium, and calcium, with greatly subordinate amounts of barium, strontium, rubidium, and other elements. The more common feldspar end members are

$KAlSi_3O_8$	Orthoclase (also microcline, sanidine)
$NaAlSi_3O_8$	Albite
$CaAl_2Si_2O_8$	Anorthite

The multiple names for potassic feldspar refer to polymorphs that have slightly different crystal structures because of different conditions of origin. Orthoclase is the typical potassium feldspar that crystallizes in plutonic rocks, although it commonly inverts to microcline as the rock cools. Sanidine is considerably rarer than the other two polymorphs and is a very high temperature mineral essentially restricted to volcanic rocks. The related volcanic mineral *anorthoclase* is effectively a sodium-rich sanidine. The potassium feldspar polymorphs may not be distinguishable in hand specimen, but they are easily distinguished microscopically.

Melting and crystallization temperatures of all natural feldspars are determined by the properties of the end members; these effects are examined in Chapters 4 to 6.

Feldspars, like many other minerals, rarely occur in a chemically pure end member form or in compositionally homogeneous crystals. There are various *solid solutions* in which the end members mix in either limited or complete amounts. The extent of mixing is controlled largely by the match of the structural and chemical properties between the end member lattices. For example, the most important feldspar solid solution is the *plagioclase series*, which involves complete mixing of albite and anorthite end members at magmatic temperatures. Compositions within the plagioclase series are typically indicated by petrologists in terms of molecular percentage of the anorthite component, for example, An_{50}, which represents 50% albite and 50% anorthite. There are a number of commonly used names for plagioclase compositional subranges within the series: albite (An_{0-10}), oligoclase (An_{10-30}), andesine (An_{30-50}), labradorite (An_{50-70}), bytownite (An_{70-90}), and anorthite (An_{90-100}). Similarly, there is complete mixing at high temperatures (and a moderate extent of mixing at lower temperatures) between orthoclase and albite in the *alkali feldspar series*. These two extensive solid solutions are the results of relatively good matches between lattices in which either the sizes or the formal charges of the two end member cations are close: Ca^{2+} (ionic radius, 1.0 Å) and Na^+ (ionic radius, 1.0 Å) in plagioclase; Na^+ (ionic radius, 1.0 Å); and K^+ (ionic radius, 1.3 Å) in alkali feldspar. The third possible pairing, between orthoclase and anorthite, shows much less solid solution because neither size nor charge of calcium and potassium cations matches. The approximate limits of the solid solutions in the albite-anorthite-orthoclase ternary system at igneous temperatures are illustrated in Figure 2-2. Feldspars and other minerals commonly show *chemical zoning*, the variation of chemical composition within single crystals. In many minerals, especially plagioclase, this zoning is concentric and is readily apparent in thin sections. Zoning in plagioclase is addressed in detail in Chapter 4.

Pyroxenes

The pyroxene group is a broadly diverse mineral family whose various members constitute the most abundant and widespread of the ferromagnesian minerals in igneous rocks. Pyroxenes occur as major rock-forming minerals across the whole spectrum of igneous rocks from peridotites and gabbros to certain granites. Although most pyroxenes contain significant ferrous iron and magnesium, there are certain unusual members of the family that are free of these elements, such as jadeite ($NaAlSi_2O_6$) and aegerine ($NaFe^{3+}Si_2O_6$). Structurally, pyroxenes are all single-chain silicates based on the tetrahedral linkage. They contain essentially endless one-dimensional chains made up of SiO_4 tetrahedra in which two apical oxygens are shared with

TABLE 2-1　　Typical pyroxene end members
• •

Name	Composition
Calcic Clinopyroxenes	
Diopside	$CaMgSi_2O_6$
Hedenbergite	$CaFeSi_2O_6$
Tschermak's molecule	$CaAl_2SiO_6$
Low-Calcium Clinopyroxenes	
Pigeonite	$(Mg,Fe,Ca)Si_2O_6$
Low-Calcium Orthopyroxenes	
Enstatite	$Mg_2Si_2O_6$
Ferrosilite	$Fe_2Si_2O_6$
Sodic Pyroxenes	
Jadeite	$NaAlSi_2O_6$
Aegerine	$NaFe^{3+}Si_2O_6$
Ureyite	$NaCrSi_2O_6$

adjacent tetrahedra (Figure 2-3). A relatively minor proportion of the tetrahedra can be filled with aluminum rather than silicon. The individual chains are bound together by interstitial cations in roughly octahedral coordination. The three most important of these cations are Ca^{2+}, Mg^{2+}, and Fe^{2+}, but others, such as Al^{3+}, Ti^{4+}, Mn^{2+}, Cr^{3+}, and Na^+, may also be present.

The broadest chemical subdivision of the pyroxenes separates them into the calcic and low-calcium varieties (Table 2-1). This subdivision largely correlates with a structural subdivision into monoclinic (*clinopyroxene*)

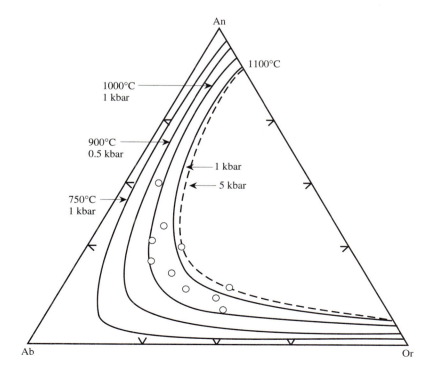

An

1100°C

1000°C
1 kbar

900°C
0.5 kbar

1 kbar

5 kbar

750°C
1 kbar

Ab　　　　　　　　　　　　　　　　　　　　Or

FIGURE 2-2
• • • • • • • • • • • • • • •

Isotherms at 750°C, 1 kbar; 900°C, 0.5 kbar; 1000°C, 1 kbar; and 1100°C, 1 and 5 kbar, for ternary feldspars. Ab, albite; An, anorthite; Or, orthoclase. The area left of the isotherm at each temperature represents possible compositions of feldspar solid solutions. Compositions in the area right of the isotherm can only stably exist as coexisting feldspar pairs with individual phase compositions on the isotherm. [Modified after M. L. Fuhrman and D. H. Lindsley, 1988, *Am. Mineral.*, Fig. 3, and other data.]

and orthorhombic (*orthopyroxene*) varieties. However, a variety of clinopyroxene called *pigeonite*, which is important in some igneous rocks, particularly basaltic volcanics, is a low-calcium variety. Clino- and orthopyroxenes are widely distributed in both igneous and high-grade metamorphic rocks. Solid solutions between end members are common within the pyroxenes and tend to be complete when end members are either both clinopyroxene or both orthopyroxene, but are less complete for mixed orthopyroxene and clinopyroxene end members.

The most common calcic pyroxene solid solution series is the complete series between *diopside* ($CaMgSi_2O_6$) and *hedenbergite* ($CaFeSi_2O_6$). (Many ferromagnesian mineral series show this complete mixing of magnesium and ferrous iron in their crystal structures.) Natural igneous calcic pyroxenes, *augites*, are rarely restricted to this pure compositional system and have minor amounts of aluminum, titanium, iron(III), and sodium, as well as other elements, in solution. In addition, augites are commonly somewhat deficient in calcium, with this cation site partially filled with iron or magnesium as a solid solution toward the *pigeonite* series (Figure 2-4). The solid solution between *enstatite* ($Mg_2Si_2O_6$) and *ferrosilite* ($Fe_2Si_2O_6$) constitutes the orthopyroxene series. Orthopyroxenes coexist with clinopyroxenes in many pyroxene-bearing igneous rocks, from peridotites to granites. The only sodic pyroxene that occurs in igneous rocks is *aegerine*, which can occur in relatively pure end member form or commonly in solid solution with augite as *aegerine-augite*. An essential requirement for the occurrence of pyroxenes in igneous rocks is that the magmas have a low water content. In water-rich magmas, hydrous minerals such as amphiboles or biotite typically crystallize instead of pyroxenes.

FIGURE 2-3

● ● ● ● ● ● ● ● ● ● ● ● ● ● ● ● ●

Crystal structure of the C2/*c* pyroxenes, projected down [001] onto the (001) plane; most natural monoclinic pyroxenes crystallize in this space group. Shaded areas outline so-called I-beam units consisting of two oppositely pointed tetrahedral chains and an octahedral strip. [After M. Cameron and J. J. Papike, 1980, *Rev. Mineral.*, 7, Fig. 6.]

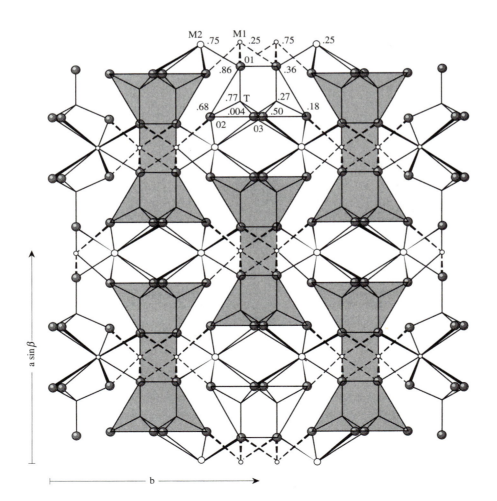

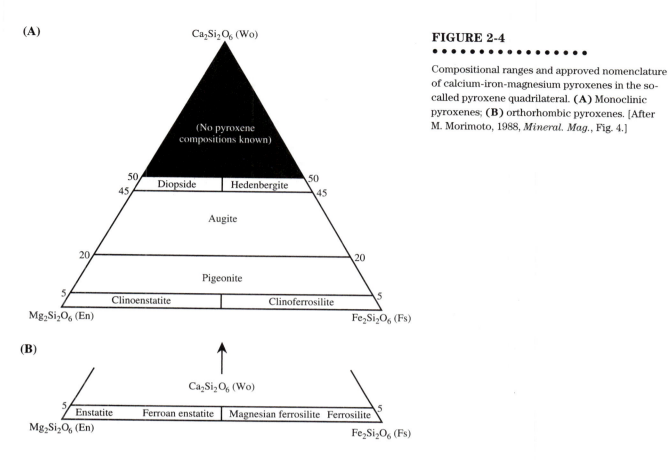

(A)

(B)

FIGURE 2-4

• • • • • • • • • • • • • • • •

Compositional ranges and approved nomenclature of calcium-iron-magnesium pyroxenes in the so-called pyroxene quadrilateral. **(A)** Monoclinic pyroxenes; **(B)** orthorhombic pyroxenes. [After M. Morimoto, 1988, *Mineral. Mag.*, Fig. 4.]

Olivine

Olivine is the characteristic ferromagnesian mineral of low-silica igneous rocks. Its structure is that of isolated silica tetrahedra linked together by iron and magnesium cations. Naturally occurring olivines are solid solutions of the magnesium end member forsterite (Mg_2SiO_4) and the iron end member fayalite (Fe_2SiO_4). Compositions within this solid solution series are indicated as molecular percentages of the forsterite component, for example, Fo_{50}. Magnesium-rich olivines are far more common than iron-rich ones and are essential constituents of ultramafic rocks, olivine basalts, and olivine gabbros. Fayalitic olivines are generally restricted to highly fractionated granitic rocks. This last observation illustrates an interesting principle: Whereas intermediate and magnesian olivines form only in low-silica rocks and are incompatible with quartz, fayalite can and does coexist with free quartz.

Feldspathoids

Feldspathoid minerals, like olivine, are characteristic of low-silica magmas. They are framework silicates like the feldspars but are even more deficient in silicon in the aluminosilicate framework. Interstitial cations are alkali metals or, less commonly, alkaline earth metals. Although no feldspathoids can be regarded as common, the most common of them is nepheline ($NaAlSiO_4$), which can have some potassium replacing sodium. It occurs across a spectrum of magma types from unusual alkaline ultramafic rocks to the silica-poor feldspathic plutonic rocks called syenites and their volcanic equivalents phonolites. Other typical feldspathoids are leucite ($KAlSi_2O_6$) and sodalite ($Na_3Al_3Si_3O_{12} \cdot NaCl$).

Amphiboles

The amphibole group consists of a very large number of possible end member compositions, all based on a double-chain silicate structure. The basic stoichiometry of amphiboles is represented by the formula $AX_2Y_5Si_8O_{22}(OH)_2$ in which the A site is either vacant or occupied by sodium or potassium; the X site is occupied by calcium in the calcic amphiboles, by sodium in the sodic amphiboles, or by iron and magnesium in the low-calcium amphiboles. The Y site is occupied by Fe^{2+}, Fe^{3+}, Mg^{2+}, Ti^{4+}, or Al^{3+}; and aluminum commonly substitutes for silicon in the tetrahedral site. Many different solid solution series are possible within the compositionally complex amphibole family. Table 2-2 lists the more common amphibole end member compositions.

TABLE 2-2 Typical amphibole end members

Name	Composition
Calcic Amphiboles	
Tremolite	$Ca_2Mg_5Si_8O_{22}(OH)_2$
Ferroactinolite	$Ca_2Fe_5Si_8O_{22}(OH)_2$
Tschermakite	$Ca_2(Fe,Mg)_3Al_2Si_6Al_2O_{22}(OH)_2$
Edenite	$NaCa_2Mg_5Si_7AlO_{22}(OH)_2$
Low-Calcium Clinoamphiboles	
Cummingtonite	$(Fe,Mg)_7Si_8O_{22}(OH)_2$
Grunerite	$Fe_7Si_8O_{22}(OH)_2$
Low-Calcium Orthoamphiboles	
Anthophyllite	$(Fe,Mg)_7Si_8O_{22}(OH)_2$
Gedrite	$(Fe,Mg,Al)_7(Si,Al)_8O_{22}(OH)_2$
Sodic Amphibole	
Riebeckite	$Na_2Fe^{2+}_3Fe^{3+}_2Si_8O_{22}(OH)_2$

The most common amphiboles in igneous rocks are calcic amphiboles. The fundamental calcic amphibole solid solution is the *tremolite* ($Ca_2Mg_5Si_8O_{22}(OH)_2$)–*ferroactinolite* ($Ca_2Fe_5Si_8O_{22}(OH)_2$) series, in which the name actinolite applies to most natural igneous compositions, which are in the middle of the series. Sodium in the A site and aluminum in both Y and tetrahedral sites generate the most common calcic amphibole, *hornblende*. Hornblende very commonly contains other minor elements. Calcic amphiboles can crystallize directly from melts but typically do so only in the intermediate and granitic compositions that are more likely to contain sufficient H_2O to form amphibole. They also occur in igneous rocks as postsolidification alterations of primary calcic pyroxenes. The only common low-calcium amphibole is cummingtonite, $(Fe,Mg)_7Si_8O_{22}(OH)_2$, but it occurs in igneous rocks only as a secondary alteration. The sodic amphiboles are a minor but significant group of amphiboles, of which the most common is riebeckite (Table 2-2), which tends to occur only in the most alkaline (and oxidized) igneous rocks such as alkali granites and syenites, and their volcanic equivalents.

Micas and Other Sheet Silicates

Micas are sheet silicates based on a two-dimensional unlimited linkage of silica tetrahedra. Two-sheet units are linked together by either aluminum-rich layers (muscovite group) or iron-magnesium layers (biotite group). These two-sheet units are in turn linked by large cations, typically potassium. *Muscovite* has the composition $KAl_3Si_3O_{10}(OH)_2$ in which minor amounts of sodium substitute for potassium. *Biotite* is a solid solution between the iron end member annite, $KFe_3AlSi_3O_{10}(OH)_2$, and the magnesium end member phlogopite, $KMg_3AlSi_3O_{10}(OH)_2$. The micas are richer in water than amphiboles (roughly 5 wt% versus 2 wt%) and thus typically form from more hydrous as well as more potassic magmas.

Two alkali-free sheet silicates commonly found as secondary alteration minerals in igneous rocks are chlorite and serpentine. Both occur as hydration-alteration products of anhydrous or less hydrous minerals; chlorite is commonly associated with pyroxenes or amphiboles and serpentine with olivine or orthopyroxene.

Other Silicates

Although the preceding compilation contains all the major rock-forming igneous minerals, there are a number of important silicates that occur as *accessory minerals*. These include *garnet*, $(Fe,Mg,Ca,Mn)_3Al_2Si_3O_{12}$, which, along with muscovite, is characteristic of unusually aluminous magmas, and *tourmaline*, a complex hydrous aluminoborosilicate containing sodium, magnesium, and iron. Tourmaline is the common repository for boron in rocks. This group also includes *zircon* ($ZrSiO_4$), which is most important for uranium-lead dating because of its common incorporation of uranium isotopes, and *sphene* or *titanite* ($CaTiSiO_5$), the only common titanium-rich silicate. Finally, *epidote*, $Ca_2Al_2Fe^{3+}Si_3O_{12}(OH)$, which is much more characteristic of metamorphic rocks, occurs rarely as a primary mineral in some high-pressure granites and as such it is an important indicator of depth of emplacement.

Oxides, Sulfides, and Phosphates

Numerous oxide minerals occur in igneous rocks, and some serve as key indicators of petrogenetic processes. Among the so-called iron-titanium oxide group are *magnetite* (Fe_3O_4) and the solid solution between *ilmenite* ($FeTiO_3$) and *hematite* (Fe_2O_3). The actual compositions of coexisting magnetite and ilmenite-hematite in igneous (and metamorphic) rocks constitute a very important monitor of oxygen content of the magma, an important clue to magmatic processes. Another important accessory oxide mineral is *spinel*, a complex solid solution containing Fe^{2+}, Fe^{3+}, Mg^{2+}, Al^{3+}, Cr^{3+}, and other cations. Magnetite is one important member of this group. Spinels characteristic of mafic to ultramafic igneous rocks are generally rich in aluminum and chromium, however, and include the solid solution spinel ($MgAl_2O_4$)–hercynite ($FeAl_2O_4$) and the chromites [$(Fe,Mg)Cr_2O_4$].

Many sulfides have been observed in igneous rocks, but the most common of these are the relatively simple iron-copper-nickel sulfides. Among them are *pyrrhotite*, *pyrite*, *chalcopyrite*, and *pentlandite*. The presence and composition of sulfide minerals yield important clues to magmatic sulfur content. If sufficiently concentrated, they form ore deposits.

Two phosphates occur commonly in igneous rocks, and they are important petrologically. *Apatite* is a calcium phosphate [$Ca_5(PO_4)_3(OH,F,Cl)$] and is virtually ubiquitous. In some rare cases, it occurs in concentrations high enough to make it a major rather than a minor mineral. The second phosphate is *monazite*, which nominally is $CePO_4$ but commonly contains substantial concentrations of rare earth and radiogenic actinoid elements (for example, uranium and thorium) replacing the cerium. Monazite is nearly ubiquitous as the small crystals surrounded by haloes of radiation damage (*pleochroic haloes*) in micas or amphiboles. Because of its high uranium and thorium content, monazite, like zircon, is an important mineral for dating igneous rocks.

IGNEOUS ROCK TEXTURES AND STRUCTURES
• • • • • • • • • • • • • • •

Igneous *textures* involve degree of crystallinity, grain size and shape, and the geometric arrangement of individual mineral grains. This last characteristic is also commonly referred to as *fabric*. Textures are inherently small-scale aspects of the rock most easily recognized in a hand specimen or a thin section. Igneous structures are typically large-scale features (for example, layering,

jointing, lineation, and preferred orientation) best recognized in the field, where their orientation can be determined by use of a geologic compass during geologic mapping. There is some overlap between texture and structure, especially on the hand specimen scale. Textural observations provide information on the order of mineral crystallization, crystallization rates, overall conditions of magma cooling and solidification, and magma viscosity. Structures primarily provide clues to magma movement or flow and to movement of crystals within magma by gravitational settling or flotation.

Degree of Crystallinity, Grain Size, and Grain Shape

The most obvious textural features are degree of crystallinity and grain size. Igneous rocks that consist entirely of crystals are called **holocrystalline** (from the prefix *holo-* meaning "entirely"). Conversely, there are rocks that consist of either glass alone (**holohyaline;** for example, obsidian) or a mixture of glass and crystals (**hypocrystalline;** for example, many young lavas or pyroclastic rocks). The restriction of partially or wholly glassy rocks to volcanic environments reflects the requirement for very rapid cooling to produce natural glasses, although magma chemistry also plays a role; viscous rhyolitic magmas are much more likely than low-viscosity basalts to form glass. Grain size is closely allied to degree of crystallinity and also is very much a function of cooling rate. The grain size in igneous rocks is enormously variable, ranging from crystals that can be meters in length in pegmatites to individual crystals so small that they cannot be seen with the naked eye, or even through a hand lens, in some fine-grained rocks. A rock is called *aphanitic* if the constituent crystals are so small that they cannot be seen without magnification (10× hand lens or microscope). All coarser grained rocks are referred to as *phaneritic*. As suggested by Williams, Turner, and Gilbert (1982), a rough subdivision based on grain size can be defined as follows when grain size is roughly uniform: *fine-grained* if the average size of grains is 1 mm or less; *medium-grained* if grains are between 1 and 5 mm; *coarse-grained* if they are between 5 mm and 3 cm; and *very coarse grained* if they exceed 3 cm. Very small crystals embedded in glass in volcanic rocks are called *microlites* or *crystallites*. The physical and chemical controls of grain size are discussed in more detail in Chapter 6.

Fabrics

Fabric is the term applied by petrologists to the geometric interrelationships of the grains in a rock. The most important controlling factor in development of fabrics is

(A)

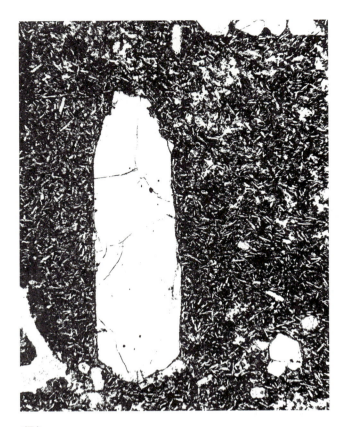

(B)

the order of crystallization of the minerals. A fundamental fabric characteristic is *grain shape*, which is determined by the crystallography and typical growth habit of each mineral. For example, unconstrained growth of pyroxenes typically leads to roughly equigranular, blocky crystals, whereas amphiboles more commonly have elongated or prismatic forms and micas have tabular shapes. When crystals in an igneous rock have such ideal shapes and are completely bounded by well-developed crystal faces, they are referred to as **euhedral.** Euhedral crystals are relatively uncommon, however, except as the earliest phenocryst minerals that form when the magma is still largely liquid. Most igneous crystals grow in a crystal-liquid mush, and their growth habit is constrained by adjacent crystals. When they grow in forms similar to their ideal forms but are only partly bounded by crystal faces, they are referred to as **subhedral.** Highly irregular grains that give no indication of ideal crystal form are **anhedral.** Most crystals in the coarser igneous rocks are subhedral to anhedral, but most phenocrysts in volcanic rocks and some accessory minerals

like zircon, apatite, and titanite are very commonly euhedral, probably indicating a tendency to form early in the crystallization history of a magma. Early-formed phenocrysts in volcanic rocks can also show *embayment*, a condition in which otherwise euhedral crystals are pitted or etched by partial dissolution in the magma. Figure 2-5 illustrates the various crystal habits.

Observation of fabrics in many igneous rocks has led to the formulation of general (although certainly not rigorous) rules regarding the typical order of crystallization of minerals. These were first propounded in the late nineteenth century by the famous petrologist Rosenbusch, but they have their most recognizable form in **Bowen's Reaction Series,** which is familiar to most students. The Reaction Series (Figure 2-6) is a simplified view of magmatic crystallization in which there are separate parallel trends of early to late crystallization among the ferromagnesian minerals (olivine-pyroxene-amphibole-biotite) and the feldspars (calcic plagioclase-sodic plagioclase-potassic feldspar), with both trends merging and culminating in crystallization of quartz. These general

(C)

FIGURE 2-5

• • • • • • • • • • • • • • • •

Photomicrographs illustrating euhedral, subhedral, and anhedral grain shapes in rock thin sections. Long dimension of field of view is about 2.5 mm for all three photos. **(A)** Euhedral plagioclase phenocrysts in trachyte, Tahiti. Plane-polarized light. **(B)** Subhedral olivine phenocryst in basalt, Hawaii. Plane-polarized light. **(C)** Anhedral quartz in granite, adjacent to subhedral plagioclase. Cross-polarized light.

rules are based on the thermal and chemical controls of silicate melt crystallization, as discussed in Chapters 4 through 6. Textural and fabric analyses have exposed many interesting exceptions to the generalized rules and have prompted much of the research in igneous petrology over the last 50 years.

A number of terms are used to describe fabric elements in igneous rocks. If grain size in a rock is more or less uniform, the rock is referred to as *granular*. *Subhedral granular* or **granitic texture** is a common granular texture in which a few minerals are euhedral, some are subhedral, and the rest anhedral, reflecting the increasing difficulty of crystallizing well-formed crystals as the magma became less liquid. The texture of a rock with notably nonuniform grain size is called **porphyritic;** the few larger grains are *phenocrysts* and the rest of the rock is the *groundmass* (Figure 2-7). In the relatively unusual **glomeroporphyritic** fabric, some phenocrysts, most commonly tabular plagioclase crystals, cluster together or form radiating bunches. Where one mineral constitutes all the phenocrysts, for

example, pyroxene or plagioclase, the rock is referred to as pyroxene-phyric or plagioclase-phyric. This terminology has especially been applied to basalts, in which the distinction between augite and calcic plagioclase as the first mineral to crystallize can be a critical indication of chemical character.

Special geometric textural relationships between minerals have been recognized and named. **Graphic texture** is relatively common in granites and pegmatites and involves an intergrowth of quartz and alkali feldspar in which thin blebs of quartz lie in crystallographically controlled orientations within large alkali feldspar crystals (Figure 2-8). This texture was named for its resemblance to ancient cuneiform or runic writing. When graphic texture is visible only under the microscope, it is referred to as *micrographic*. Graphic texture involving minerals other than quartz and alkali feldspar occurs but is rare. Another common, but less regular, quartz-feldspar intergrowth is *myrmekitic* texture, which is the intergrowth of very fine wormlike blebs of quartz and sodic plagioclase (Figure 2-9). All the above textures commonly occur when two minerals (or any materials) crystallize simultaneously; materials scientists and metallurgists call these textures *eutectoid*.

Exsolution texture represents the chemical decomposition of a homogeneous solid solution mineral into two more nearly end member minerals during cooling. It occurs most commonly in alkali feldspars that are solid solutions of albite and potassic feldspar; this texture is called **perthite** (Figure 2-10). Oriented lamellae of the exsolved phase (typically albite) lie within the host grain (orthoclase or microcline). **Microperthite** texture refers to very fine lamellae, and **mesoperthite** texture involves roughly equal proportions of coarse lamellae and host.

Ophitic texture is common in fine- to medium-grained mafic rocks, and refers to the enclosure of plagioclase

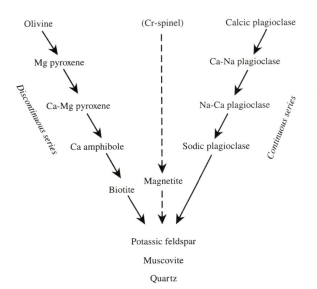

FIGURE 2-6
● ● ● ● ● ● ● ● ● ● ● ● ● ● ● ●

Bowen's Reaction Series, deduced by N. L. Bowen from the
evidence for crystallization order based on fabrics and chemical
fractionation relationships. The minerals on the left side are
related by a series of discontinuous reactions. Those on the right
form a continuous series that is due to solid solution relationships
within the plagioclase series. The reader is cautioned that the
Reaction Series is a useful conceptual tool but is not rigorous;
exceptions can occur.

laths by larger, subhedral augite grains (Figure 2-11).
If the augite grains are not large enough to enclose the
laths completely, the texture is **subophitic.** It has been
established that this texture actually represents simultaneous crystallization of plagioclase and pyroxene and
develops because of differences in the crystallization
properties of the two minerals. When a late-crystallizing
mineral completely encloses numerous smaller grains
of other minerals, the texture is **poikilitic** and indicates
a crystallization sequence in which the enclosed minerals crystallized earlier. The most common enclosing
minerals in poikilitic textures are the micas, which are
typically among the last minerals to crystallize from most
magmas.

Trachytic texture is commonly found in the volcanic
rocks called trachyte. This texture has a strong parallel
alignment of plagioclase laths that reflects compaction
or flow of the magma during crystallization (Figure 2-12).

Although they are more common in metamorphic
rocks, *coronas* also develop in magmatic rocks. In this
reaction texture, a crystal of an early-formed mineral is
surrounded by variably fine grained reaction product or
products, which typically represent reactions between
the crystal and a chemically evolving melt that is no
longer in equilibrium with the early-formed mineral
(Figure 2-13). Most commonly, this texture involves a
central grain of olivine that is surrounded—and separated from the matrix—by a rim or overgrowth of orthopyroxene. Feldspathoid grains can similarly be enclosed
by feldspars. In both cases, the texture reflects the
attainment of silica saturation by the magma after the
crystallization of the silica-undersaturated mineral and is
therefore an important clue to magmatic evolution.

Other textural features include *vesicles*, which are
open cavities in volcanic and some very shallow plutonic

FIGURE 2-7
● ● ● ● ● ● ● ● ● ● ● ● ● ● ● ●

Photomicrograph of a dacite from Rio
Grande County, Colorado, illustrating
porphyritic texture. Large phenocrysts of
plagioclase (Plag; An_{35-40}), augite (Aug),
and biotite (Biot). The matrix consists of
much smaller phenocrysts of the same
minerals within a cryptocrystalline groundmass. [From P. W. Lipman, 1975, *U.S. Geol.
Surv. Prof. Paper 852*, Fig. 43-A.]

FIGURE 2-8
Graphic texture in granite and granitic pegmatites shown by intergrowth of quartz (dark) and microcline (light). The sample is approximately 10 cm long.

0   5 cm

FIGURE 2-9
● ● ● ● ● ● ● ● ● ● ● ● ● ● ● ●

Myrmekite (in center of photo), consisting of a vermicular (wormlike) intergrowth of plagioclase and quartz. This texture is apparently a subsolidus reaction texture formed during cooling after solidification. Sample from Main Zone cumulate rock of the Bushveld Intrusion, South Africa. [From L. R. Wager and G. M. Brown, 1967, *Layered Igneous Rocks* (San Francisco: W. H. Freeman), Fig. 213.]

FIGURE 2-10
● ● ● ● ● ● ● ● ● ● ● ● ● ● ● ●

Perthitic feldspar intergrowth in granite, with potassic feldspar host (dark) and plagioclase lamellae (light blebs). Crossed nicols; width of field of view is about 0.8 mm. Perthite typically consists of a microcline host with albite or oligoclase (An_{0-30}) lamellae. A variety of specific names are used to describe the proportions of host and lamellae: mesoperthite (shown here) has roughly equal proportions of host and lamellae; microperthite has small lamellae; cryptoperthite has lamellae that are at the lower limits of optical resolution; antiperthite consists of the plagioclase host with potassic feldspar lamellae.

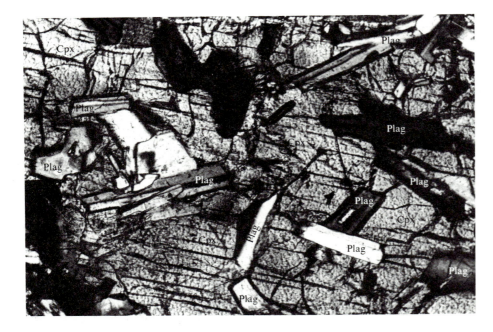

FIGURE 2-11
● ● ● ● ● ● ● ● ● ● ● ● ● ● ●

Ophitic texture, showing plagioclase laths (Plag) enclosed by augite (Cpx). Diabasic texture is the opposite of this, with interlocking large plagioclase laths surrounding separate pyroxene crystals. Crossed polarizers; width of field of view is about 0.8 mm.

FIGURE 2-12
● ● ● ● ● ● ● ● ● ● ● ● ● ● ●

Trachytic texture, showing a euhedral phenocryst of augite set in a matrix of strongly aligned potassic feldspar laths. Note the flow alignment of feldspar laths around the phenocryst and the typical 90° cleavage in the augite. Sample is a trachyte from the island of Tahiti. Plane-polarized light; width of field of view is about 1.7 mm. Trachytes commonly have phenocrysts of augite, aegerine-augite, or hornblende.

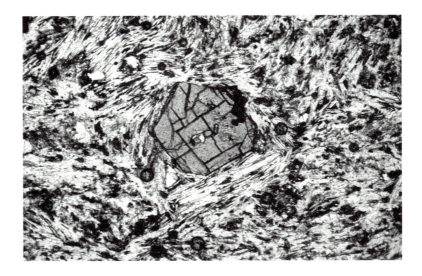

rocks. Vesicles reflect the presence of gas bubbles trapped in the magma as it completed its solidification. These cavities can be filled with minerals of secondary deuteric (hot postmagmatic fluids) or of groundwater-deposited origin, in which case they are called **amygdules.** *Miarolitic* texture occurs in plutonic rocks, particularly granites, and consists of cavities into which euhedral crystals project. They are believed to represent larger fluid bubbles into which unconstrained and rapid crystal growth occurred. *Orbicular* texture is a rare feature of some granitoids in which the rock is filled with medium to large ovoid structures that may be up to many centimeters in diameter. The ovoids contain alternating concentric shells with concentrations of felsic and mafic minerals, commonly nucleated around a small xenolith in the core.

Structures in igneous rocks include lineations, foliations, compositional layering, planar alignment of tabular minerals, mineral aggregates, slabby xenoliths, and flattened vesicles. Several types of lineations are possible, including parallel alignment of elongate prismatic minerals, flow-fold axes, tubelike vesicles, and elongate xenoliths. Layering and foliation are typically the result of inhomogeneous crystallization processes, such as crystal settling and accumulation on a magma chamber floor, but they can also be the result of magmatic flow or other syn- or postcrystallization deformation. Lineations are most commonly the result of magma flow.

FIGURE 2-13
• • • • • • • • • • • • • • • •

Photomicrograph of typical corona texture. Sample is a peridotite xenolith from Tahiti that is enclosed by alkali basalt. Orthopyroxene (large crystal at right) is surrounded by a fine-grained zone of olivine, spinel, and glass, succeeded outward by a zone of Ti-rich augite. Enclosing basalt can be seen at left. Corona textures are typically the result of reaction and commonly have a zonal structure. Plane-polarized light; width of field of view is about 3.5 mm.

COMMON PLUTONIC ROCKS
• • • • • • • • • • • • • • •

In this section and in the following one on volcanic rocks, mineralogic and textural descriptions of the most common rocks are introduced. Precise definitions of the rock names used are presented in Chapter 3, and readers are referred to the classification section of Chapter 3 (or to the Glossary) if they have questions regarding the names.

Granite and Alkali Granite

See Figures 2-14 and 2-20A for representative hand specimen and thin section appearance.

Granites have diverse appearances, from light-colored to dark and from fine-grained to rather coarse. The alkali feldspar is orthoclase or microcline, both of which are commonly perthitic. Plagioclase is typically oligoclase (An_{10} to An_{30}). Both feldspars are anhedral to subhedral. In porphyritic varieties, the most common phenocryst type is alkali feldspar. Quartz is anhedral and molds itself around and between the shapes of adjacent grains; very small, round fluid-filled bubbles (fluid inclusions) are common in quartz crystals, either in random orientation or along planes. Muscovite and dark green or brown subhedral to euhedral flakes of biotite are common in some varieties; biotite is very common in other varieties and shows pleochroic haloes around tiny radioactive zircon or monazite inclusions. Green to brown subhedral hornblende prisms are common, except in peraluminous varieties. Sodium-rich amphiboles such as riebeckite or arfvedsonite are common in alkali granites. Pyroxenes are unusual, except for aegerine or aegerine-augite in some alkali granites, and augite or orthopyroxene (or iron-rich olivine) in rare, particularly dry granites in high-grade metamorphic terranes. Accessory minerals include apatite (tiny prisms and needles), magnetite, ilmenite-hematite, zircon, titanite, tourmaline, fluorite (particularly in alkalic granite), and rarely, the aluminum-rich minerals cordierite and garnet. Hydrothermal alteration is common: Biotite alters to chlorite and sphene; potassic feldspar to sericite and kaolinite; plagioclase to epidote, zoisite, sericite, and kaolinite. The general texture is subhedral granular (granitic); other textures include quartz-feldspar graphic intergrowths, phenocrysts, myrmekite, and orbicular and miarolitic.

Syenite and Alkali Syenite

See Figures 2-15 and 2-20B.

The principal alkali feldspar is subhedral to tabular and is sanidine, orthoclase, microcline, or anorthoclase. Perthitic textures in the feldspar are common. Plagioclase is generally subhedral, with compositions between An_{20} and An_{40}. Other minerals include rare muscovite, brown or green biotite, subhedral prismatic green hornblende, augite, sodium-rich amphiboles and pyroxenes, and minor quartz or olivine. Quartz can be interstitial, micrographic, or myrmekitic. Accessory minerals include apatite, zircon, magnetite, titanite, sulfides, fluorite, zeolites, and calcite. Common alterations are feldspar to sericite and kaolinite; biotite to chlorite and titanite; pyroxene and amphibole to chlorite, calcite, or iron oxides. The typical texture is subhedral granular

[*Text continues on page 52*]

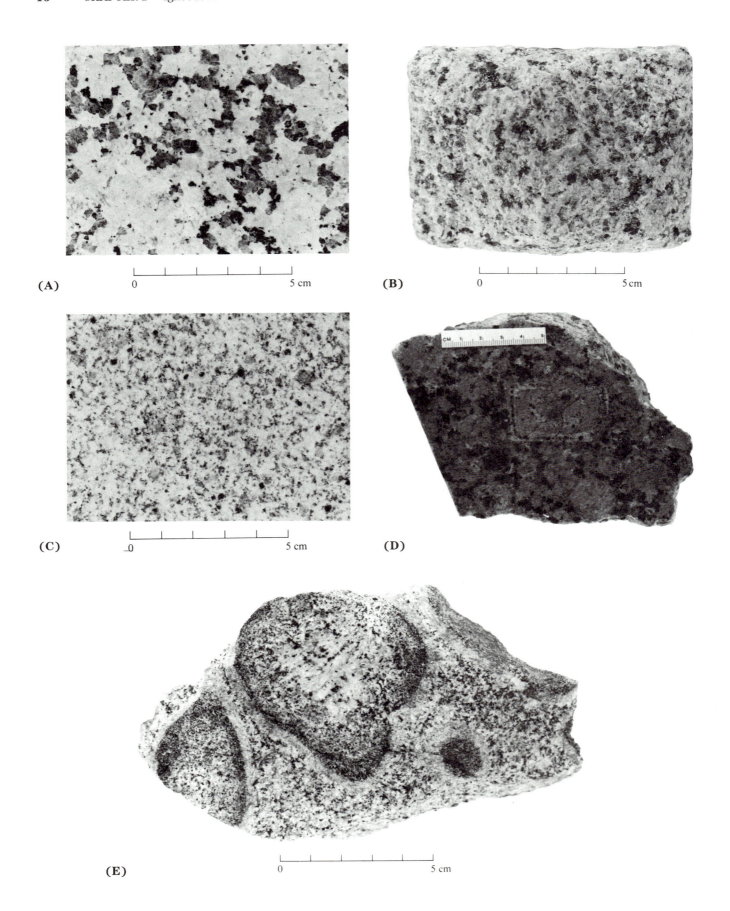

(A)   ⊢─┴─┴─┴─┴─┤
       0   5 cm

(B)   ⊢─┴─┴─┴─┴─┤
       0   5 cm

(C)   ⊢─┴─┴─┴─┴─┤
       0   5 cm

(D)

(E)   ⊢─┴─┴─┴─┴─┤
       0   5 cm

FIGURE 2-14 *(opposite page)*

A variety of granites with different appearances. (**A**) Polished surface of a medium-grained granite from West Chelmsford, Massachusetts. Dark gray areas are mostly quartz, and black specks are biotite. Light-colored areas consist of pink potassic feldspar and pale gray sodic plagioclase. Relative amounts of quartz, biotite, and feldspars can easily be estimated in hand specimen. (**B**) Biotite granite. Black areas are biotite, easily distinguished by reflections from basal cleavage surfaces. Lighter areas are mainly potassic feldspar, and gray areas are quartz. Feldspar can easily be distinguished from quartz by the presence of two cleavages, whereas quartz shows only an irregular fracture. (**C**) Polished surface of a fine-grained, light-colored granite. White areas are mainly potassic feldspar and sodic plagioclase, gray areas are quartz, and black specks are minor ferromagnesian minerals. (**D**) Polished surface of a dark, coarse-grained granite with large, medium-gray grains of potassic feldspar, dark gray quartz, and black ferromagnesian minerals. Note the large, rectangular potassic feldspar grain in the center that shows strong chemical zoning and incorporation of ferromagnesian mineral inclusions in several bands. (**E**) Granite showing orbicular texture. The rounded masses consist mainly of biotite and plagioclase, whereas the granite matrix contains quartz, potassic feldspar, and biotite. The origins of orbicular texture are widely debated, but the texture is commonly considered to result from incomplete assimilation of xenoliths.

(A)   0   5 cm

(B)   0   5 cm

(C)   0   5 cm

FIGURE 2-15

A variety of syenites. (**A**) Medium-grained hornblende syenite. Dark grains are hornblende and the lighter grains are mainly potassic feldspar with minor sodic plagioclase. The feldspars are distinguished typically by slight color differences and the presence of striations on plagioclase. The absence of quartz distinguishes this rock from similar appearing granites. (**B**) Fine-grained, medium-gray syenite that consists largely of potassic feldspar (light-colored) with minor biotite and hornblende (black). The larger dark xenoliths (left and right edges) are aphanitic basalt fragments. (**C**) Polished surface of a strongly porphyritic syenite. Large phenocrysts of orthoclase show a definite flow structure. The fine-grained dark matrix is composed mainly of orthoclase and biotite.

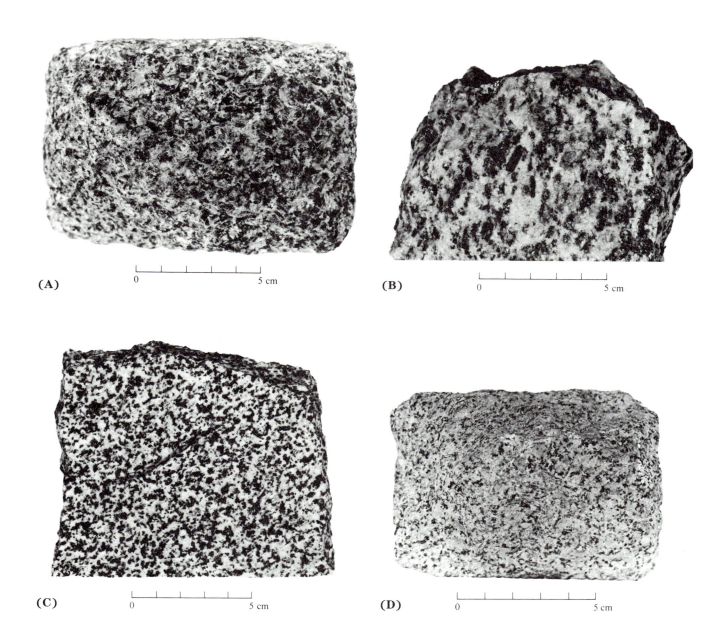

FIGURE 2-16
● ● ● ● ● ● ● ● ● ● ● ● ● ● ● ● ●

Intermediate to mafic plutonic rocks. **(A)** Quartz-bearing monzonite porphyry. The large, light-colored phenocrysts are composed of both potassic feldspar and sodic plagioclase. Plagioclase is typically distinguished by slightly paler color and striations on some cleavage surfaces. The matrix is a mixture of fine-grained feldspars with minor hornblende. **(B)** Coarse-grained quartz diorite. The dark minerals, hornblende and biotite, show a parallel orientation. Light-colored constituents are quartz and plagioclase. Sample from Clearwater County, Idaho. [From A. Hietanen, 1963, *U.S. Geol. Surv. Prof. Paper 433-D*, Fig. 5A.] **(C)** Hornblende biotite quartz diorite from the Bucks Lake Pluton, Plumas County, California. Dark minerals are hornblende and biotite; lighter minerals are quartz and plagioclase. Note the classic "salt-and-pepper" appearance in this medium-grained rock. [From A. Hietanen, 1973, *U.S. Geol. Surv. Prof. Paper 731*, Fig. 27.] **(D)** Medium-grained augite-orthopyroxene gabbro. Light areas are intermediate-calcic plagioclase. Dark areas are augite and orthopyroxene. Pyroxenes are generally distinguished on the basis of dark to black color and right-angle cleavages. Distinction between augite and orthopyroxene is usually best done on slightly weathered surfaces: augite weathers green to black, whereas orthopyroxene weathers brown.

FIGURE 2-17
● ● ● ● ● ● ● ● ● ● ● ● ● ● ● ● ● ●

(A) Mantle peridotite. Consists mostly of pale green olivine (lighter color) with darker grains of bright green chromium-bearing augite and brown-black orthopyroxene. Most peridotites contain minor amounts of garnet or spinel. Sample from Tahiti occurs as xenolith in gray alkali basalt. (B) Hornblende-bearing pyroxenite. The darker masses consist of hornblende in a matrix of green clinopyroxene. Such fine-grained ultramafic rocks are relatively rare. [From A. Hietanen, 1963, *U.S. Geol. Surv. Prof. Paper 344-D*, Fig. 4A.] (C) Medium-grained pyroxenite. This rock consists almost entirely of augite with minor orthopyroxene. Such rocks are typically dark gray-green to black. (D) Kimberlite. This is an altered porphyritic mica peridotite containing olivine, augite, phlogopite, and a variety of exotic minor minerals. The large darker areas are xenoliths of country rock acquired during ascent from the upper mantle. These can be of any composition from ultramafic mantle rock to lower crustal gneisses.

(B)

(A)

(C)

(D)

(A)

0   5 cm

(B)

0   5 cm

(C)

0   5 cm

(D)

0   5 cm

FIGURE 2-18
● ● ● ● ● ● ● ● ● ● ● ● ● ● ● ● ●

A variety of silica-rich volcanic rocks. **(A)** Rhyolite. This fine-grained, light-colored rock has a granitic composition. The original volcanic character of the rock is verified by the presence of a small number of vesicles and flow structure or flow layering. **(B)** Obsidian. This dark-colored volcanic glass has a roughly granitic composition. The glassy nature of obsidian is reflected by the homogeneous vitreous luster and conchoidal fracture. The front surface of this specimen is somewhat weathered, emphasizing the layering that has resulted from strong compaction of original glassy shards of welded volcanic ash. **(C)** Vitrophyre. This name is given to rocks having crystals of quartz and alkali feldspar set in a distinctly glassy matrix. The color of the glass is typically quite dark, like obsidian, and the composition of the rock is granitic. **(D)** Scoria. This is an extrusive, highly vesicular, glassy rock having the appearance of a clinker; it represents solidifed frothy lava. Scoriaceous texture is usually found in ejected fragmental volcanic material or within crusts of lava flows.

(A)

(B)

(C)

(D)

0   5 cm

FIGURE 2-19
● ● ● ● ● ● ● ● ● ● ● ● ● ● ● ●

(A) Latite, collected from a dike in Salt Lake County, Utah. The euhedral to subhedral phenocrysts are composed of equal proportions of potassic feldspar and plagioclase. The mineralogy of the very fine grained groundmass cannot be determined in hand specimen. [From W. J. Moore, 1973, *U.S. Geol. Surv. Prof. Paper 629-B*, Fig. 27-A.] **(B)** Andesite porphyry. Anhedral phenocrysts consist of intermediate plagioclase and minor ferromagnesian phases (largely hornblende and biotite). The groundmass mineralogy cannot be determined in hand specimen. The matrix of andesite is typically dark green or gray, rather than the black that is characteristic of basalt. **(C)** Hornblende andesite. The large, darker phenocrysts can be identified as hornblende rather than augite from the cleavage angles. Many similar andesites have a gray-green matrix color and abundant black ferromagnesian mineral phenocrysts. **(D)** Vesicular basalt. Groundmass is black and very fine grained; mineralogy cannot be determined from hand specimen. Vesicles are partially to completely filled with secondary minerals such as epidote, calcite, and zeolites. Some basalts contain larger phenocrysts such as augite, olivine, or plagioclase that can be identified in hand specimen.

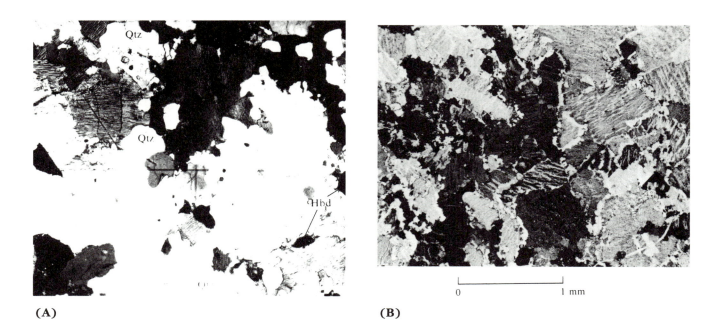

(A)   **(B)**

FIGURE 2-20

Photomicrographs of silica-rich to intermediate plutonic rocks. **(A)** Hornblende granite from Dover, New Jersey. The larger areas consist of rather obvious perthite (see Figure 2-10) with orthoclase hosts and sodic plagioclase lamellae. Clear quartz (Qtz) grains are intergrown with and included within the perthite. A few dark grains of hornblende (Hbd) are indicated in the lower right. Width of field of view is about 5 mm; photo was taken with crossed polarizers. [From P. K. Sims, 1958, *U.S. Geol. Surv. Prof. Paper 287*, Plate 16-B.] **(B)** Syenite from Gilpin County, Colorado. This rock consists almost entirely of large grains of perthite that have almost continuous rims of albitic plagioclase. Width of field of view is about 3 mm; crossed polarizers. [From W. A. Braddock, 1969, *U.S. Geol. Surv. Prof. Paper 616*, Fig. 15-D.] **(C)** Monzonite from Archuleta County, Colorado. This coarse-grained rock contains euhedral laths of plagioclase (Plag; about An$_{40}$) and darker augite (Aug) and orthopyroxene (Hyp). All these minerals are surrounded by large orthoclase (Orth), quartz, and biotite grains as large as 5 mm. [From P. W. Lipman, 1975, *U.S. Geol. Surv. Prof. Paper 852*, Fig. 49-A.]

[*Continued from page 45*]

(granitic); finer grained varieties can be porphyritic or trachytic.

Nepheline Syenite

Potassic feldspar can be either orthoclase, sanidine, microcline or anorthoclase and coexists with albitic plagioclase. It is commonly perthitic but can also occur as nonperthitic orthoclase or microcline grains with albite. Both feldspars are typically subhedral but can have a tabular habit. Nepheline is the most common feldspathoid. It varies from anhedral to stubby euhedral prisms and can be poikilitic. Sodalite, when it occurs, is present as euhedral dodecahedra but can also be anhedral and interstitial, or skeletal. Sodium-rich, subhedral, prismatic, or interstitial amphibole and pyroxene are common, along with lesser amounts of green to brown biotite, analcite, iron-rich olivine, and calcite. Accessory miner-

als include zircon, apatite, titanite, magnetite, sulfides, and fluorite. Common alteration reactions include biotite going to sericitic muscovite; nepheline altering to cancrinite, sodalite, analcite, and calcite; sodalite altering to analcite, calcite, and cancrinite; potassic feldspar altering to nepheline, sericite, kaolinite, calcite, and sodalite. The typical texture is subhedral granular, but porphyritic texture is also common. Cataclastic (sheared, crushed, or granulated) textures occur locally.

Monzonite

See Figures 2-16A and 2-20C.

Plagioclase is typically subhedral to euhedral, and compositions range from oligoclase to andesine. Potassic feldspar is most commonly orthoclase (rarely, microcline); chemical zoning, and poikilitic and microperthitic

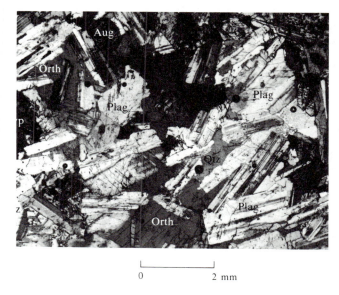

```
 0            2 mm
```

(C)

granular. Minerals can be phenocrystic, poikilitic, or myrmekitic.

Gabbro

See Figures 2-16D and 2-21B and C.

Plagioclase varies in composition from An_{50} to An_{100}, although it is most commonly around An_{65}. It is typically unzoned, anhedral, and equidimensional, although it is occasionally subhedral-tabular. Orthopyroxene of intermediate iron-magnesium ratio is common; it is typically prismatic with minor {100} clinopyroxene exsolution and shows prominent pale green to pale pink pleochroism in thin section. The most common pyroxene is augite, which is anhedral to subhedral and commonly shows twinning and exsolution. Subhedral magnesian olivine also may be present. Hornblende, when present, is green or brown and anhedral. It commonly forms larger, poikilitic grains or rims pyroxenes. Less abundant minerals that may be present include biotite, quartz, potassium feldspar, nepheline, and analcite. Accessory minerals include magnetite, ilmenite, sulfides, apatite, zircon, titanite, rutile, green or brown spinels, and garnet. Common alterations are plagioclase to sericite, epidote, zoisite, calcite, and albite; olivine to serpentine, talc, and amphibole; orthopyroxene to serpentine, amphibole, talc, and chlorite. Textures are variable and include subhedral granular, subophitic, and poikilitic. Some gabbros exhibit well-developed compositional layering.

Ultramafic Rocks

See Figures 2-17 and 2-21D and E.

Olivine is typically abundant and magnesium-rich. It may be euhedral to anhedral; irregular fractures are common, and crystals may show deformation twins as well. Both ortho- and clinopyroxenes are commonly large and poikilitic. Both pyroxenes are typically also magnesium-rich and can have a high aluminum content. In some cases, augite is bright green as a result of a high chromium content. Amphibole (green or brown hornblende) is not common and is typically due to secondary hydration of pyroxenes; sodic amphiboles are also known. Pale brown phlogopite and magnesium-rich garnet may also be present. Accessory minerals include green aluminous spinels, brown chromian spinels, apatite, sulfides, calcic plagioclase, magnetite, ilmenite, and chromite. Minerals produced by alteration include serpentine, brucite, chlorite, talc, magnesite, various amphiboles, and magnetite. The typical texture is anhedral-granular, but cataclastic texture is characteristic of many ultramafic rocks because of their tectonic mode of emplacement. Peridotites and pyroxenites commonly show layering on a variety of scales.

textures are common. Other minerals include light brown biotite, green hornblende, and pale green augite. Minor quartz, orthopyroxene, or olivine can be present. Accessory minerals commonly include zircon, titanite, apatite, magnetite, and ilmenite-hematite. Alteration products include epidote, chlorite, titanite, sericite, kaolinite, calcite, and serpentine. The texture is most commonly subhedral granular. Monzonite porphyries generally contain chemically zoned phenocrysts of plagioclase or, less commonly, orthoclase or perthitic potassic feldspar.

Diorite

See Figures 2-16B and C and 2-21A.

Plagioclase is subhedral and chemically zoned, with an average composition of An_{40-45}. Biotite is commonly present in brown, euhedral, somewhat poikilitic plates. Green or brown hornblende can be either subhedral or prismatic. Orthopyroxene or augite, when present, is typically subhedral, prismatic, and poikilitic. Minor amounts of orthoclase (or microcline), quartz, or olivine can be present. Accessory minerals include apatite, zircon, titanite, magnetite, ilmenite, sulfides, and occasional xenocrysts of metamorphic minerals. Common alteration reactions include pyroxene going to biotite or amphibole; hornblende going to biotite; biotite going to chlorite and sphene; plagioclase going to epidote, zoisite, kaolinite, and sericite. Texture is typically subhedral

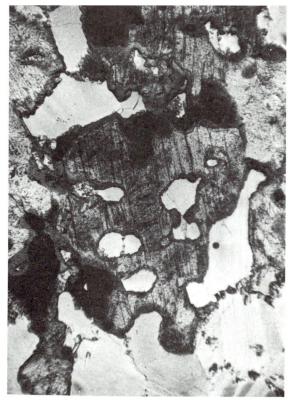

(A)

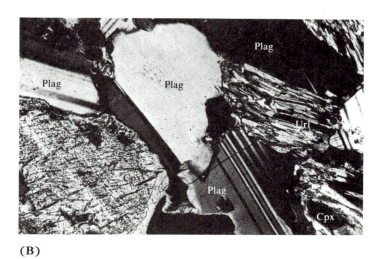

(B)

(C)

(D)

(E)

FIGURE 2-21
● ● ● ● ● ● ● ● ● ● ● ● ● ● ● ●

Photomicrographs of intermediate to mafic and ultramafic plutonic rocks. **(A)** Diorite, Mt. Hope, Maryland. The large, irregularly shaped grains are augite, being altered to hornblende (as indicated by dark rims). The lighter areas are plagioclase. Partially crossed polarizers; long dimension of field of view is about 2.3 mm. **(B)** Gabbro. The large twinned grains are calcic plagioclase (Plag). The large grain at lower right is augite (Cpx) with a narrow rim of hornblende. The other fibrous-looking grains are hornblende alteration of clinopyroxene (Url; uralitization). Crossed polarizers; width of field of view is about 1.2 mm. **(C)** Gabbro. Light grains are laths of calcic plagioclase. Larger gray crystals are pale brown augite, and black grains are magnetite. Some alteration to chlorite is present (cloudy areas). Plane-polarized light; long dimension of field of view is about 3 cm. [From W. B. Hamilton, 1956, *Geol. Soc. Am. Bull.*, *67*, Plate 2.] **(D)** Peridotite. The rock consists mostly of olivine (grains showing no cleavage) with some augite (central grain with good cleavage). Crossed polarizers; long dimension of field of view is about 2.3 mm. **(E)** Dunite. All grains are magnesium-rich olivine. Note the well-annealed texture, with grains forming 120° triple junctions (as in center). Crossed polarizers; long dimension of field of view is about 3.0 mm.

COMMON VOLCANIC ROCKS
● ● ● ● ● ● ● ● ● ● ● ● ● ● ●

Rhyolite, Dacite, Obsidian, Vitrophyre, Pumice, Scoria

See Figures 2-18, 2-7, and 2-22A.

Glass is more abundant in these volcanic rocks than in any other. Rhyolitic glasses have low refractive indices (about 1.49) and are reddish to dark green. They commonly contain highly angular or sharp broken glassy fragments (*shards*). Alkali feldspars are sanidine, orthoclase, or, rarely, anorthoclase and generally occur as phenocrystic and lath-shaped or tabular crystals. The most common plagioclase is oligoclase or andesine. The most common silica mineral is quartz, but cristobalite and tridymite also occur. Any of these can be intergrown with feldspar, in phenocrysts or as late fillings in cavities. Quartz is usually euhedral and bipyramidal, and slightly embayed or resorbed. Euhedral biotite phenocrysts are common. Other minerals include augite, intermediate to iron-rich orthopyroxene, iron-rich olivine, and hornblende. Sodium-rich pyroxenes and amphiboles are common in alkalic varieties. Accessory minerals are apatite, zircon, magnetite, ilmenite, topaz, and fluorite. Feldspars alter to sericite and clays, and ferromagnesian minerals go to chlorite, epidote, sericite, and calcite, but alterations can also include devitrification and silicification. Texture shows all variations from holocrystalline to holohyaline, and the rocks can be vesicular, crypto- or microcrystalline (containing unusually small crystals surrounded by glass), micropoikilitic, or porphyritic. In glassy rocks, cracks and microlites are common. Compositional layering may be present.

Trachyte

See Figure 2-22B.

The potassic feldspar is typically sanidine and, less commonly, orthoclase or anorthoclase. It can occur as Carlsbad-twinned tabular phenocrysts and can also be perthitic and chemically zoned. Plagioclase is generally oligoclase but can be more sodic in alkalic trachyte. Like

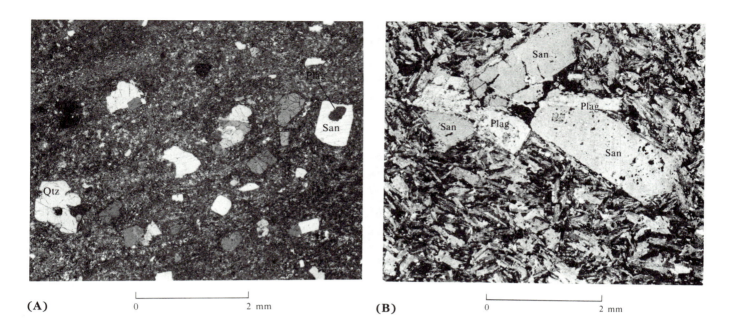

(A)

0   2 mm

(B)

0   2 mm

FIGURE 2-22

• • • • • • • • • • • • • • • • •

Photomicrographs of silica-rich and intermediate volcanic rocks. **(A)** Rhyolite, Rio Grande County, Colorado. Phenocrysts consist of embayed quartz (Qtz), sodic sanidine (San; $Or_{40}Ab_{60}$), and rare plagioclase, biotite, magnetite, and ilmenite. Quartz is distinguished in thin section by lack of good cleavage and uniaxial positive optical character. Feldspars show at least one direction of good cleavage and are commonly twinned. The groundmass is microcrystalline and devitrified. [From P. W. Lipman, 1975, *U.S. Geol. Surv. Prof. Paper 852*, Fig. 54-B.] **(B)** Sanidine trachyte, Gilpin County, Colorado. The larger crystals are mainly sanidine (San) or plagioclase (Plag). Groundmass consists primarily of lath-shaped sanidine, with minor quartz, in a flow structure. Crossed polarizers; width of field of view is about 3 mm. [From W. A. Braddock, 1969, *U.S. Geol. Surv. Prof. Paper 616*, Fig. 15-F.] **(C)** Latite dike rock, Salt Lake County, Utah. Subhedral plagioclase (Plag), orthoclase (Orth), and biotite (black) phenocrysts in a very fine-grained groundmass. Plane-polarized light; width of field of view is about 3 mm. [From W. J. Moore, 1973, *U.S. Geol. Surv. Prof. Paper 629-B*, Fig. 28-A.]

alkali feldspar, it shows a wide size variation, from microlites to phenocrysts. Rarely, glass is also present. Brown or reddish brown biotite is common and occurs as euhedral crystals. Other common minerals in trachytes are hornblende, quartz, tridymite, and iron-rich olivine; alkalic trachytes can also contain sodium-rich amphiboles or pyroxenes. Accessory minerals include apatite, zircon, magnetite, ilmenite, and titanite. Textures vary from holocrystalline to hypocrystalline (partially glassy). Phenocrysts of plagioclase, pyroxene, amphibole, or biotite are common, and elongate or tabular phenocrysts (generally plagioclase) may exhibit parallel alignment (trachytic texture).

Phonolite

In phonolite, a silica-undersaturated rock, alkali feldspar is sanidine, anorthoclase, or orthoclase (often microper-

thitic), occurring as phenocrysts, laths, or microlites. Nepheline is usually present as euhedral groundmass grains or rarely as phenocrysts. Sodic plagioclase is rare. Other minerals in phonolite include sodalite, hauyne, leucite, analcite, biotite, sodium-rich pyroxenes and amphiboles, and iron-rich olivine. Accessory minerals are titanite (common), apatite, corundum, zircon, magnetite, and ilmenite. Most phonolites are porphyritic-holocrystalline. The groundmass is trachytic or granular, rarely glassy or vesicular.

Latite

See Figures 2-19A and 2-22C.

Plagioclase phenocrysts are typically euhedral and chemically zoned, and have an average composition of andesine. Smaller laths of plagioclase in the groundmass are commonly slightly more sodic. Sanidine, orthoclase,

(C)

or anorthoclase are usually present as groundmass laths and microlites. Biotite phenocrysts and glass are both common. Minor amphibole is commonly present as either hornblende or basaltic hornblende (an unusual amphibole rich in Ti^{4+} and Fe^{3+}). Augite is the most common pyroxene, but aegerine-augite and orthopyroxene also occur. Quartz or tridymite occur as minor or accessory minerals. Other accessory minerals are magnetite, ilmenite, apatite, and zircon. The texture is typically porphyritic, with phenocrysts set in a holocrystalline to hypocrystalline matrix.

Andesite

See Figures 2-19B and C and 2-23A and B.

Plagioclase varies in composition from anorthite to oligoclase but averages about An_{40}. Highly to spectacularly zoned plagioclase phenocrysts are very typical of andesites. Plagioclase in the groundmass as laths or microlites is usually more sodic than the phenocrysts. Pyroxenes are augite, subcalcic augite, pigeonite (low-calcium clinopyroxene), and orthopyroxene, and any of these can occur both as prominently zoned phenocrysts and in the groundmass. Hornblende or oxyhornblende may be present as prismatic phenocrysts. Less common phases are glass, magnesium-rich olivine, and biotite. Accessory minerals include apatite, zircon, magnetite, ilmenite, hornblende, biotite, and garnet. The texture is typically porphyritic and the groundmass holocrystalline. Andesites may exhibit vesicular, amygdaloidal, trachytic, and either ophitic or subophitic textures.

Basalt

See Figures 2-19D and 2-23C and D.

Basalts are dominated by pyroxenes and calcic plagioclase. The principal pyroxene is augite or subcalcic augite, in phenocrysts or in the groundmass. Euhedral to subhedral pigeonite or orthopyroxene more commonly occurs in groundmass. In some varieties, olivine is an important phase. It forms euhedral to anhedral crystals that are commonly strongly zoned; it occurs both in phenocrysts and the groundmass. The plagioclase composition is moderately calcic, varying from bytownite to labradorite and averaging about An_{55}. Plagioclase in groundmass and phenocrysts is euhedral to subhedral, often as laths, and is commonly zoned. Minor phases in basalt include interstitial glass, alkali feldspar, silica minerals, nepheline, or analcite. Accessory minerals are most commonly magnetite, ilmenite, and apatite. Basalt textures are holocrystalline to hypocrystalline, porphyritic, vesicular, amygdaloidal, and subophitic to ophitic (especially in coarse-grained varieties).

SUMMARY
• • • • • • • • • • • • • • •

Only a few (about 50 or so) of the thousands of known minerals are common enough in igneous rocks to be important for students to know well. These fall into the categories of primary minerals, which crystallize directly from magma, and secondary minerals, which reflect

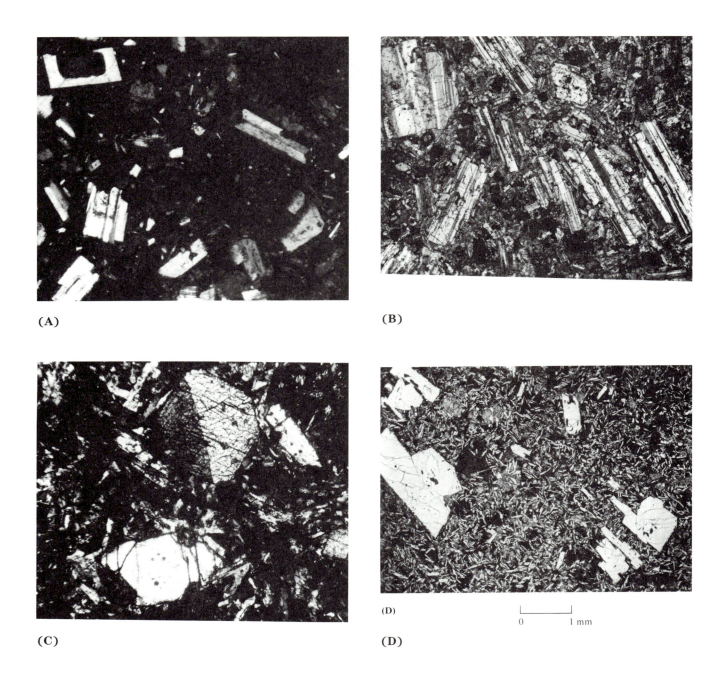

(A)

(B)

(C)

(D)

FIGURE 2-23
● ● ● ● ● ● ● ● ● ● ● ● ● ● ● ● ●

Photomicrographs of mafic volcanic rocks. **(A)** Typical andesite. Phenocrysts consist of zoned and unzoned plagioclase and dark hornblende with oxidized rims. Plagioclase in andesites commonly shows oscillatory zoning. Crossed polarizers; width of field of view is about 2.3 mm. **(B)** Andesite containing abundant large phenocrysts of strongly zoned plagioclase, small phenocrysts of augite and orthopyroxene, and sereptine pseudomorphs after olivine. Fine-grained groundmass consists of plagioclase, pyroxenes, and Fe-Ti oxides. [From P. W. Lipman, 1975, *U.S. Geol. Surv. Prof. Paper 852*, Fig. 38-B.] **(C)** Basalt. Phenocrysts of euhedral pyroxene (twinned grain, upper center) and subhedral olivine (lower center). Matrix consists mainly of plagioclase, laths, pyroxene, and opaque oxides. Crossed polarizers; width of field of view is about 2.3 mm. **(D)** Porphyritic basalt (plagioclase-phyric), Power County, Idaho. Phenocrysts consist of plagioclase. Groundmass contains microlites of plagioclase enclosed in pyroxene, with minor opaque oxides. Plane-polarized light; width of field of view is about 5.5 mm.

postsolidification alteration of primary minerals. The most important minerals include quartz, plagioclase and potassic feldspars, feldspathoids, pyroxenes, amphiboles, micas and other sheet silicates, olivine, and a variety of volumetrically minor silicates, oxides, and other minerals such as garnet, tourmaline, epidote, zircon, magnetite, ilmenite, spinel, sulfides, and apatite. Igneous textures reflect the character of individual crystals and their geometric relationships to one another, known as fabrics. Important aspects of texture include grain size and shape, relative uniformity of grain size, preferred orientations such as linear or planar alignments, overgrowths or inclusion relationships, and intracrystalline fabrics such as exsolution. This chapter includes general mineralogic and textural descriptions of the most important plutonic and volcanic rocks, along with both hand specimen and thin section photographs of these common rock types.

STUDY EXERCISES

1. What is the distinction between texture and fabric? Which texture elements are directly related to cooling rates and which are independent of cooling?
2. Lineations are parallel linear textural elements that are important because they commonly reveal direction of magma flow. List the number of different textures and structures of igneous rocks that might produce lineations.
3. What is the most ubiquitous mineral of igneous rocks? Why is it so abundant?
4. Based on mineral chemistry, why are sodic and potassic feldspars typical of lower-T magmas, such as granites, and calcic feldspars typical of higher-T magmas, such as basalt?

REFERENCES AND ADDITIONAL READINGS

Bard, J. P. 1986. *Microtextures of Igneous and Metamorphic Rocks*. Boston: Reidel.
Cameron, M., and J. J. Papike. 1980. Crystal chemistry of silicate pyroxenes. *Rev. Mineral.*, *7*, 5–92.
Cameron, M., and J. J. Papike. 1981. Structural and chemical variations in pyroxenes. *Am. Mineral.*, *66*, 1–50.
Deer, W. A., R. H. Howie, and J. Zussman. 1993. *Introduction to the Rock-Forming Minerals*, 2nd ed. New York: Wiley.
Fuhrman, M. L., and D. H. Lindsley. 1988. Ternary feldspar modeling and thermometry. *Am. Mineral.*, *73*, 201–215.
Klein, Cornelis, and C. S. Hurlbut, Jr. 1993. *Manual of Mineralogy*, 21st ed. New York: Wiley.
Leake, B. E. 1978. Nomenclature of amphiboles. *Am. Mineral.*, *63*, 1023–1052.
MacKenzie, W. S., and C. Guilford. 1980. *Atlas of Rock-Forming Minerals in Thin Section*. New York: Wiley.
MacKenzie, W. S., C. H. Donaldson, and C. Guilford. 1982. *Atlas of Igneous Rocks and Their Textures*. New York: Wiley.
Morimoto, M. 1988. Nomenclature of pyroxenes. *Mineral. Mag.*, *52*, 535–550.
Papike, J. J., and M. Cameron. 1976. Crystal chemistry of silicate minerals of geophysical interest. *Rev. Geophys. Space Phys.*, *14*, 37–80.
Ribbe, P. H., ed. 1983. *Feldspar Mineralogy. Rev. Mineral.*, *2* (2nd ed.).
Veblen, D. R., and P. H. Ribbe, eds. 1982. *Amphiboles: Petrology and Experimental Phase Relations. Rev. Mineral.*, *9B*.
Williams, H., F. J. Turner, and C. M. Gilbert. 1982. *Petrography: An Introduction to the Study of Rocks in Thin Sections*, 2nd ed. San Francisco: W. H. Freeman.

Chapter 3

CHEMISTRY AND CLASSIFICATION OF IGNEOUS ROCKS

Magma is any partially or completely molten natural material. The only magmas for which there is *direct* evidence of a molten nature are those that occur as modern lava flows. For ancient volcanic rocks and all plutonic rocks exposed at the surface by erosion, a magmatic origin can only be inferred or assumed, although there typically is persuasive structural and textural evidence for such an origin. One of the best indirect lines of evidence comes from the laboratory, where petrologists melt rock samples at geologically reasonable temperatures, then cool and crystallize these artificial magmas to produce synthetic igneous rocks. Most magmas are silicate liquids, containing about 45 to 75% SiO_2 (silica) by weight; but some rare igneous rocks can be produced by crystallization of very silica-poor magmas. For example, carbonatite is an intrusive or extrusive rock that is dominated by sodium, calcium, or magnesium carbonate minerals with subordinate silicates; nelsonite is a virtually SiO_2-free magnetite-ilmenite-apatite rock that is known to occur in lava flows or as segregations in plutons.

PHYSICAL PROPERTIES OF MAGMA

The properties of silicate magmas are largely dictated by their chemistry. For example, *density* is controlled by the relative concentrations of chemical components with different atomic weights. Magmas rich in heavy elements such as calcium, titanium, and iron have substantially higher densities than magmas rich in light elements such as silicon, aluminum, and sodium.

Many silicate minerals have individual melting points well above 1000°C, thus requiring magmas to be at high temperature to remain molten. Of all the fundamental properties of magmas, temperature is the only one that is not directly controlled by magma chemistry. Because there is no single temperature at which magmas go from being completely molten to being totally solid, magma bodies commonly consist of mixtures of silicate liquid and crystals, and perhaps of gas as well. For most magma compositions, there is a *crystallization interval* of up to several hundred degrees Celsius between the appearance of the first high-temperature crystals and the final crystallization of the last few percent of liquid. Both the crystallization interval itself and its bounding temperatures vary with pressure as well as with magma composition. Further complicating the picture, many magmas contain a substantial concentration of volatile species such as carbon dioxide and water. At the high temperatures at which magma is essentially totally molten, the volatiles are completely dissolved in the melt. As the amount of liquid is reduced during crystallization, magma becomes supersaturated with volatiles, which exsolve as a separate gas phase (at low pressure) or fluid phase (at high pressure) in a physical process like boiling. The typical total *magmatic system*, therefore, consists of coexisting liquid, solid, and gas phases during crystallization. The exact mix of the three exerts a critical control over magmatic properties and behavior.

Viscosity is one of the most important properties of any fluid, including magma. **Viscosity** is typically defined as resistance to flow during application of differential or shear stress. All liquids flow under differential stress (part of their fundamental definition), and those with higher viscosity flow more slowly. One can demonstrate viscosity by placing puddles of several liquids on a horizontal wooden board. As long as the board remains horizontal, no movement or deformation of the puddles occurs. If the board is tilted, however, the liquids all start to flow downhill. In effect, tilting the board applies an equal gravitational shear stress on each liquid puddle. Suppose that one puddle contains water, the second one contains heavy oil, and the third contains asphalt. Clearly the three will flow downhill at very different rates because of their different viscosities, with water flowing most rapidly. In fact, the asphalt may not flow at all under gravity alone but might require a physical push to deform. The viscosity of pure water at room temperature is low and is used as the viscosity standard, defined as 1 *poise* (the unit of viscosity).

In a similar fashion, various magmas have very different viscosities. The silica content of any magma is the most important factor controlling viscosity, because of the tendency of silica in magmas to polymerize or link together as SiO_4 tetrahedra joined at the corners into chains, sheets, or even frameworks like those in quartz or feldspar crystals. Silica and alumina (both capable of forming tetrahedra) have been called *network formers* because they promote tetrahedral polymerization. Cations that typically form octahedra or larger coordination polyhedra, for example, Fe^{2+}, Fe^{3+}, Mg^{2+}, Ca^{2+}, K^+, and Na^+, retard polymerization and reduce viscosity. These cations have been referred to as *network modifiers*. Water has a strong tendency to disrupt tetrahedral polymerization by weakening Si–O bonds, so a high water content serves to reduce the viscosity of magma. As demonstrated by Shaw (1965), 4 wt% H_2O dissolved in dry granitic magma reduces the viscosity from about 10^8 to 10^5 poise. The reader is referred to Hess (1989) for a comprehensive discussion of melt structure and viscosity.

Various laboratory studies have shown that control of polymerization and viscosity by silica content is typical of silicate magmas. Dry silica-rich granitic magmas (70–75 wt% SiO_2) at 800°–1000°C have very high viscosities (10^8–10^{10} poise), whereas andesite magma (52–55 wt% SiO_2) at 1200°C has moderate viscosity (10^4–10^5 poise) and silica-poor olivine basalt magmas (45–48 wt% SiO_2) at 1200°–1400°C have relatively low viscosity (100–300 poise). Temperature also controls viscosity: All other things being equal, a higher temperature magma will have a lower viscosity than a lower temperature magma of similar chemistry.

Many magmas spend most of their existence at temperatures in the crystallization interval between the **liquidus** (the temperature at which the first crystals appear during cooling) and the **solidus** (the temperature at which the magma becomes totally solidified), and thus consist of melt, suspended crystals, and perhaps gas bubbles as well. Such polyphase mixtures have complicated viscosity behavior but, in general, have higher viscosity than pure liquids of similar chemistry. Their viscosities also vary with applied stress, so once flow is initiated, continued flow may require less force.

As shown in Chapter 1, physical behavior of magma is strongly influenced by chemistry, viscosity, gas content, and the proportion of suspended crystals. Extrusive low-viscosity mafic magmas (basalt, andesite) tend to form lava flows, whereas extrusive high-viscosity intermediate and felsic magmas produce extensive sheets of pyroclastic material such as ash, tuff, or vitrophyre. Degassing of basaltic lava flows during downslope movement causes a shift from pahoehoe to aa lava types. Intrusive mafic magmas have such low viscosities that they can exploit thin cracks as conduits, a characteristic resulting in a wide range of scales and geometries of intrusions. High-viscosity granitic magmas rarely occur as thin dikes or sills but much more commonly form large, rounded or lobate masses such as stocks and batholiths. A substantial proportion of suspended crystals in any magma makes that magma much more viscous, possibly inhibiting mobility.

CHEMICAL CONSTITUENTS OF IGNEOUS ROCKS

Igneous rocks, like all other materials, consist of chemical elements. The two most important of these are the two most abundant elements in the earth's crust, oxygen and silicon (Table 3-1), but many others also occur in at least trace amounts in magmas and igneous rocks. The chemical constituents of igneous rocks generally fall into three categories of elements: the major elements, the minor elements, and the trace elements. Major elements are typically defined as those occurring in concentrations greater than 2 weight percent (wt%), and minor elements as those between 0.1 and 2 wt%. Trace elements are present at concentrations below 0.1 wt% and are typically reported as parts per million (ppm) or parts per billion (ppb). In addition to oxygen and silicon, the most abundant elements include aluminum, titanium, iron, manganese, magnesium, calcium, sodium, potassium, and phosphorus. The normal manner of reporting abundances of the major and minor elements is in the form of their simple oxides, that is, SiO_2, TiO_2, Al_2O_3, FeO, MnO, MgO, CaO, Na_2O, K_2O, and P_2O_5. Other major elements are sulfur, fluorine, and chlorine in their elemental form. All other elements are reported as trace elements in parts per million or parts per billion. The usual convention is to report compositional data on a weight basis.

Although rock and mineral analyses are commonly presented as weight percents of oxides, *molar* amounts are used for many petrologic purposes. The conversion from weight percent to mole percent is straightforward and utilizes the molecular weights of the oxides, which can be calculated from a periodic table or found in many mineralogy texts such as Deer, Howie, and Zussman (1993). To convert, simply divide the weight percent of

TABLE 3-1 Chemical composition (wt%) of Earth, mantle, and crust

Element	Earth[a]	Oxide	Mantle[a]	Oceanic crust[b]	Continental crust[c]
Fe	31	SiO_2	45.2	49.4	60.3
O	30	TiO_2	0.71	1.4	1.0
Si	18	Al_2O_3	3.54	15.4	15.6
Mg	16	FeO^d	8.48	10.1	7.2
Ni	1.7	MnO	0.14	0.3	0.1
Ca	1.8	MgO	37.48	7.6	3.9
Al	1.4	CaO	3.08	12.5	5.8
Na	0.9	Na_2O	0.57	2.6	3.2
		K_2O	0.13	0.3	2.5
		P_2O_5	—	0.2	0.2

[a] Ringwood (1975).
[b] Ronov and Yaroshevskiy (1976).
[c] Taylor (1964).
[d] Total iron oxide ($FeO + Fe_2O_3$).

each oxide by its molecular weight. Add all these values, then normalize to 100%. Table 3-2 shows results of a sample calculation for a simple rock analysis.

Two other important chemical constituents are water and carbon dioxide. In an igneous rock, these compounds occur in an analysis only if there are hydrous or carbonate minerals that contain them. Major amounts of these two constituents are dissolved in many magmas but are mostly to completely lost in the crystallization process. It is very important to keep in mind that the absence of these chemical components from an igneous rock chemical analysis does not imply their absence from the original magma. Both water and carbon dioxide can play highly significant roles in the origin, transport, and crystallization of magmas through control of phase equilibrium processes and magma physical behavior, as discussed later.

Isotopes of several elements form another important category of igneous rock chemical constituents. The isotopes of a single element are atoms that have different numbers of neutrons but the same number of protons in the nucleus. Thus an element's isotopes have the same atomic number but different atomic weights. There are two broad categories of isotopes, the light stable isotopes and the radiogenic isotopes. The *light stable isotopes* are various (but not all) isotopes of hydrogen, carbon, nitrogen, oxygen, and sulfur. None of these isotopes is radioactive or created by radioactive decay of other elements. Their amounts therefore stay the same throughout time, a critical advantage in using them to interpret igneous

processes such as fractionation. The *radiogenic isotopes* are isotopes of certain elements that are either depleted in amount over time through radioactive decay (parent elements) or are created over time through decay of other elements (daughter products). These isotopes include the well-known isotopes of potassium, argon, rubidium, strontium, uranium, thorium, and lead, as well as other less well known isotopes. The principal use of the radiogenic isotopes is in determining the age of petrologic events such as magma crystallization, but radiogenic isotopes have also been used for monitoring igneous processes.

Igneous petrologists have long recognized the desirability of measuring the chemical compositions of igneous rocks. From the inception of analyzing igneous rocks in the nineteenth century until well into the twentieth, the principal purpose of measuring major and some trace elements was for use in classification. Although classification is still important, particularly in field igneous petrology, the focus of petrologic research has shifted to understanding igneous processes. Data on chemistry of magmas and magmatic rocks are now widely used to decipher many of the details of processes involved in melting, chemical evolution, and crystallization of magmas. In particular, data on trace elements and especially the rare earth elements (from cerium to lutetium in row 6 of the periodic table), as well as both radiogenic and stable isotopes, have enabled very sophisticated tracing and modeling of magma source areas and magma mixing.

TABLE 3-2 Sample conversion of weight percent oxides to mole percent oxides

Oxide	Weight percent	Molecular weight	Wt%/MW	Mole %
SiO_2	72.04	60.09	1.1989	77.95
TiO_2	0.30	79.90	0.0038	0.25
Al_2O_3	14.42	101.935	0.1415	9.20
FeO	2.90	71.85	0.0404	2.63
MgO	0.71	40.31	0.0176	1.14
CaO	1.82	56.08	0.0324	2.11
Na_2O	3.69	61.85	0.0597	3.88
K_2O	4.12	94.20	0.0437	2.84
Total	100.00		1.5380	100.00

METHODS OF CHEMICAL ANALYSIS

The methods for determining the chemical compositions of rocks and minerals are diverse, ranging from traditional "wet chemical" quantitative analysis to sophisticated modern spectroscopic and mass spectrometer techniques. The range of elements or concentrations appropriate for analysis by each of these techniques is large, and complete analysis of a rock may require application of two or more methods (Table 3-3). A review of the most widely used methods follows.

Wet Chemical Analysis

Wet chemical analysis is a method of quantitative analysis that uses reagent titrations of solutions produced by total dissolution of rock or mineral samples in acids. Techniques of dissolution depend on rock type, but generally a very finely ground or powdered sample is dissolved in an acid solution of some combination of nitric, phosphoric, and hydrochloric acids. Long a standard analytical technique of chemistry, this method is slow and meticulous, and is therefore largely obsolete, although some geologists claim that it remains potentially the most accurate method of quantitative rock analysis. Used almost exclusively for determining concentrations of major and minor elements, this technique does not have the resolution required for analysis of most trace elements. The very large majority of rock analyses (and many mineral analyses as well) reported in the petrologic literature prior to about 1975 were done with this technique.

Atomic Absorption Spectrophotometry

Similar to wet chemical analysis, this technique also requires the total dissolution of the sample in an acid solution. The solution is then vaporized in a gas arc and the optical emissions of the excited atoms are quantitatively measured and compared with standard solutions of known concentration of that particular element. This technique is generally more accurate than wet chemical analysis for major elements at low concentrations, but can be less accurate at high concentrations. It is a very useful technique for analyzing natural or experimental fluids or solutions, or solid materials that are easily dissolved. It is, however, less practical for materials such as rocks that are difficult to dissolve and is typically used only in the absence of more appropriate laboratory methods.

X-Ray Emission or X-Ray Fluorescence (XRF) Spectroscopy

All materials emit X-rays (that is, fluoresce) from their individual constituent atoms when these atoms are excited by an intense focused energy source, in this case, a high-energy X-ray source. Atoms of a particular element emit an X-ray spectrum in which certain characteristic frequencies are intense. Quantitative analysis involves measurement of these intensities for the unknown rock and comparison to a standard material of known concentration. Certain corrections based on the physics of X-ray emission and absorption are applied. This method requires less meticulous sample preparation than the previous ones and is generally considerably quicker.

Except for fluorine, major elements and many trace elements can be measured by XRF. This technique became one of the standard analytical techniques in the 1980s.

Electron Microprobe (EMP) and Proton-Induced X-Ray Emission (PIXE)

Both techniques are fundamentally similar to XRF, with two exceptions: the source of excitation energy and the resolvable analytical area. The EMP technique uses high-energy electrons to excite the sample and can excite an area on the sample surface as small as one-thousandth of a millimeter (1 micrometer or 1 μm). The PIXE technique uses high-energy protons for the energy source and excites a slightly larger area. Both techniques require only simple sample preparation: a flat, smoothly polished surface on a slab of rock or a thin section. Like XRF, both techniques use a comparison of X-ray intensities from materials of known and unknown chemical concentration to arrive at a quantitative analysis. The resolution of very small areas allows single mineral grains or even parts of grains to be analyzed. In fact, by far the widest use of these techniques is for analysis of single mineral grains in situ in a polished rock thin section. Electron or proton beam sizes can also be expanded to allow analysis of larger polycrystalline areas. Rock analysis by these techniques has typically involved powdering and melting the sample and analyzing the resulting glass (special care is required to produce homogeneous glass). Analysis of all major elements and selected trace elements is readily accomplished with both EMP and PIXE.

Inductively Coupled Plasma (ICP) Spectroscopy

The newest technique for whole-rock analysis of major and selected trace elements uses a powdered, dissolved rock sample, which is vaporized into a plasma (a very high temperature ionized gas). The plasma is examined and compared with standards by using emission spectroscopy. Although some sample preparation is required, this technique is accurate and relatively fast. It has become the standard method employed by commercial laboratories that perform rock and mineral analyses. A further advantage of this technique is that the plasma can be directed into a mass spectrometer and analyzed for isotopic composition, as described below, in the section on MS methods.

Instrumental Neutron Activation Analysis (INAA)

INAA involves irradiation of a powdered rock sample by using a high-flux neutron source, typically in a synchrotron or a nuclear reactor, for analysis of most trace elements, especially the rare earth elements. Short-lived nuclides are generated from each element during irradiation. Concentrations of trace elements of interest can be measured by monitoring the emission of electromagnetic radiation from α, β, and γ decay of the short-lived nuclides in the sample.

Mass Spectrometry (MS) Methods

Isotopic analyses for both stable and radiogenic isotopes are done by using a mass spectrometer, which is capable of discriminating atomic or molecular particles of different masses. Material is introduced into the mass spectrometer either as a gas or as a liquid evaporated onto a filament. In most cases, the latter method is used, and a stream of particles enters the mass spectrometer from the heated filament. Individual atoms of the same element that have different atomic weights are resolved by mass in the instrument, and relative abundances of the different isotopes of an element can be calculated. When an ICP unit is the gas source, the technique is referred to as ICP-MS.

TABLE 3-3 Analytical methods for analyzing igneous rocks

Method	Elements analyzed
Wet chemical	Major, minor
Atomic absorption spectrophotometry	Major, minor, selected trace
X-ray fluorescence spectroscopy	Major, minor, selected trace
Electron microprobe	Major, minor, selected trace
Inductive coupled plasma spectroscopy	Major, minor
Instrumental neutron activation analysis	Trace
Mass spectrometry	Isotopes

TABLE 3-4 Average chemical compositions (wt%) of common igneous rock types

Component	Granite[a]	Granodiorite	Diorite	Syenite	Anorthosite	Gabbro	Basalt
SiO_2	72.04	66.80	58.58	57.49	51.05	51.06	50.06
TiO_2	0.30	0.54	0.96	0.82	0.63	1.17	1.87
Al_2O_3	14.42	15.99	16.98	17.23	26.57	15.91	15.94
Fe_2O_3	1.22	1.52	2.55	3.05	0.99	3.10	3.90
FeO	1.68	2.87	5.13	3.22	2.07	7.76	7.50
MnO	0.05	0.08	0.12	0.13	0.05	0.12	0.20
MgO	0.71	1.80	3.73	1.84	2.14	7.68	6.98
CaO	1.82	3.92	6.66	3.54	12.76	9.88	9.70
Na_2O	3.69	3.77	3.60	5.48	3.18	2.48	2.94
K_2O	4.12	2.79	1.81	5.03	0.62	0.96	1.08
P_2O_5	0.12	0.18	0.29	0.29	0.69	0.24	0.34
No. of analyses	2485	885	872	517	104	1451	3796

[a]Rock names are those applied by the original investigator, not by the compiler.

Source: LeMaitre (1976).

THE CHEMICAL COMPOSITIONS OF IGNEOUS ROCKS

During 150 years of performing chemical analyses of igneous rocks, petrologists have demonstrated several fundamental patterns. For example, the mafic rocks such as basalt are richer in calcium, iron, and magnesium and poorer in sodium, potassium, and silicon relative to felsic rocks. The term *mafic* refers to rocks rich in dark ferromagnesian minerals (olivine, pyroxene, amphibole), and *felsic* includes rocks rich in light-colored minerals (quartz, feldspars, feldspathoids). These chemical trends reflect the mineral contents of the various rock types and are inevitable results of processes of magma formation and evolution. Table 3-4 reports data on average chemical compositions of a number of rock types. LeMaitre (1976) calculated these compositions from a compilation of over 20,000 chemical analyses. It is interesting to note the numbers of individual analyses within each category. The granite and basalt categories show the largest number of analyzed rocks, a situation reflecting the fact that these two igneous rock types are the most abundant in the exposed crust of the earth. For some of the other types, however, the numbers can be misleading because rare rock types attract disproportionate interest from petrologists and thus are more often analyzed.

The first thing in Table 3-4 that should strike the reader as important is the limited range in concentrations of some oxides, particularly silica. Igneous rocks rarely contain less than 45% by weight (wt%) silica or more than 75 wt%. Igneous rocks are overwhelmingly made up of silicate minerals, which, except for quartz, have a range of silica concentrations of 35 to 70 wt%. Most rocks are volumetrically dominated by feldspars, which contain between 55 and 68 wt% silica, so this limited range of silica values is expected. After all, rock compositions are nothing more than weighted averages of the compositions of their individual constituent minerals. This simple and fundamental principle is the basis for the correlation between chemistry and mineral content, and for one of the most important petrologic calculations, the CIPW norm (see Appendix 1).

MEASURING AND ESTIMATING MINERALOGY

The mineralogic composition of an igneous rock is an important characteristic because it is used for classification and for interpretations of the origin and evolution of magmas.

Weight and Volume Modes

The most direct measure of the mineral content of a rock is called the **mode**, which is typically expressed as a volume or weight percent of each mineral constituent.

Modes can be measured in various ways, but the most common is also the most traditional: use of a mechanical microscope stage to *point-count* mineral grains in a thin section of rock (Williams, Turner, and Gilbert 1982). This process involves moving the thin section systematically along a grid and identifying and tallying the mineral grains at each grid intersection. Anywhere from 500 to 5000 counts, depending on the grain size and the degree of precision required, are normalized to 100%. The resulting percentages are, of course, areal percents but are considered to be equivalent to volume percents in texturally homogeneous rocks. Semiquantitative modes are measured by visual estimation of mineral proportions in either thin section or sawed, polished slabs. For some petrologic purposes, *weight modes* are required, for example, for comparison of actual mineral content to phase diagrams plotted with weight percent scales. A weight mode is calculated from volume modes by multiplying the volume percent of each mineral by its average specific gravity, then normalizing the sum of these new values to 100%. (Because specific gravity equals weight divided by volume, weight percent must equal normalized volume percent times specific gravity.) It is most important to notice whether a mode is being reported as a volume or weight mode; if this is unspecified, it is typically assumed to be a volume mode. Mineral modes are the basis of the widely accepted International Union of Geological Sciences (IUGS) plutonic rock classification system.

The CIPW Norm

Proportion of minerals in an igneous rock is a fundamental property used for comparison and classification. In some cases, however, it is impractical or impossible to measure modal mineralogy by using traditional thin section techniques, for example, for the fine-grained or glassy volcanic rocks. For some plutonic rocks, vagaries of the magmatic crystallization process, such as variable gas contents or depths of emplacement, have produced minerals that cannot be directly compared chemically, for example, pyroxenes versus amphiboles or amphiboles versus biotite. For this reason, four early twentieth-century petrologists—Whitman Cross, J. P. Iddings, L. V. Pirsson, and H. S. Washington (CIPW)—devised a scheme for using the analytical results pertaining to an igneous rock to calculate an *idealized* or hypothetical mineralogy on the basis of a number of standard rules. These rules allow the analyst to calculate the *normative minerals* by allocating chemical constituents to the high-temperature, first-to-crystallize minerals before lower-temperature minerals, thus simulating the actual order of crystallization of an idealized magma (akin to Bowen's Reaction Series). Most important, the norma-

tive minerals are anhydrous and thus allow hydrous magmatic rocks to be directly compared with less hydrous ones.

Therefore, there is commonly no direct obvious correlation between the modal and normative mineralogies for a particular rock, but algebraic correspondences between natural and idealized minerals are usually apparent. For example, one unit of magnesium biotite is chemically equivalent to one unit of potassium feldspar plus three units of the pyroxene enstatite, minus three units of quartz:

$$\underset{\text{biotite}}{KMg_3AlSi_3O_{10}(OH)_2} =$$

$$\underset{\text{K-feldspar}}{KAlSi_3O_8} + 3\ \underset{\text{enstatite}}{MgSiO_3} - 3\ \underset{\text{quartz}}{SiO_2} + H_2O$$

Similarly, one unit of actinolite is equivalent to two units of diopside plus three units of enstatite and one unit of quartz:

$$\underset{\text{actinolite}}{Ca_2Mg_5Si_8O_{22}(OH)_2} =$$

$$2\ \underset{\text{diopside}}{CaMgSi_2O_6} + 3\ \underset{\text{enstatite}}{MgSiO_3} + \underset{\text{quartz}}{SiO_2} + H_2O$$

Direct comparison of modal and normative mineralogies is not simple, because normative mineralogy is calculated on an atomic rather than on a volumetric or weight basis. Conversion of normative mineralogy to a form compatible with either volume or weight modes can be done by using molecular volumes or molecular weights of the appropriate minerals. An excellent discussion of the philosophy and mechanics of the CIPW norm can be found in Philpotts (1990). A list of rules for calculating a CIPW norm is given in Appendix 1, with a calculated example. Students are strongly encouraged to practice several of these calculations to see how they work.

MINERALOGIC CLASSIFICATION
• • • • • • • • • • • • • • • •

For at least 200 years, petrologists have been attempting to identify, characterize, or classify igneous rocks. Igneous rocks have been named on the basis of mineral content, chemistry, locality, and texture or with no apparent basis at all. Because of this proliferation of naming, a huge historical nomenclature exists for igneous rocks. Much of this earlier approach has been summarized by Johannsen (1931, 1937, 1938) in his four-volume set of books entitled *A Descriptive Petrography of the Igneous Rocks*. Johannsen ably summarized the problem:

Many and peculiar are the classifications that have been proposed for igneous rocks. Their variability depends in part upon the purpose for which each was intended, and in part upon the difficulties arising from the characters of the rocks themselves. The trouble is not with the classifications but with Nature which did not make things right. (1931, p. 51)

The obvious approach to the classification of igneous rocks is one built around mineralogy and texture. These characteristics supply a great deal of information about the origin and history of the rock, are easily described in the field, and are only refined rather than refuted by later thin section study. The mineralogical data can be elaborated, if deemed useful, by chemical or isotopic analyses. For microcrystalline rocks such as rhyolites and basalts, chemical analyses are typically required for classification.

The accuracy of mineralogic identification from field and hand specimens is quite variable. It depends on the size of the crystals, their degree of alteration, and the quality of observation by the geologist. This last factor is often as important as the others, because differences in ability among geologists are as great as differences in crystal size between granite and rhyolite. In any event, it is obvious that descriptions such as "myrmekitic albite-riebeckite granite" cannot be based solely on hand specimen examination. Thin sections are required for the most accurate classification. In the field, descriptions such as "basalt (?)" for a dark-colored microcrystalline rock, or "lithic rhyolite tuff (?)" for a light-colored, low-density, layered rock containing fragments of uncertain character are often used. Students' characterizations of rocks will improve in direct proportion to their knowledge of mineral characteristics and mineral associations and to their experience.

Most igneous rocks contain only a few minerals in large abundance and a wide variety of minor ones. Because of ease of identification and their significance in petrogenesis, the more common minerals are usually chosen as the basis of classification. The most abundant crustal igneous rocks contain significant concentrations of feldspars, along with a silica mineral such as quartz, or (to a considerably lesser extent) a feldspathoid mineral indicative of silica deficiency. Igneous rocks rich in these light-colored minerals are commonly referred to as *felsic* or *leucocratic*. Conversely, many igneous rocks contain abundant and mostly dark-colored ferromagnesian minerals such as pyroxenes, amphiboles, olivine, and biotite in addition to the light-colored minerals. Such rocks are commonly classed as *intermediate* or *mesocratic*. When dark-colored minerals predominate, the rock is called *mafic* or *melanocratic*. If dark minerals are virtually the only constituents, the rock is *ultramafic*. The igneous rock types anorthosite and dunite are exceptions: Anorthosite consists mostly of plagioclase and is thus light-colored or leucocratic, and dunite consists mostly of olivine and thus is light to medium green. Anorthosite is not regarded as a felsic rock, however, but most commonly is associated with mafic rocks such as gabbro. Most igneous classifications are based on the relative amounts of light and dark minerals, and on grain size, which reflects cooling rate and manner of emplacement. Some classification schemes have depended on a so-called color index or CI (see later), which is essentially a darkness scale from 0% (white) to 100% (black) roughly based on the percentage of dark minerals in the rock.

The IUGS Classification System

To meet the need for a single rational classification system of igneous rocks for world use, Albert Streckeisen published a generally accepted preliminary classification scheme for plutonic rocks in 1967. The International Union of Geological Sciences then established a commission of geologists from around the world, headed by Streckeisen, to elaborate and formalize his proposal for plutonic rocks, later adding a classification system for volcanic rocks. There is now an internationally accepted, comprehensive system for igneous rock classification that permits nomenclature application to any desired degree of precision. Most important, given accurate mineralogic determinations by individual investigators, it allows a reader of geologic literature to be certain that a hornblende-biotite granodiorite in Russia is the same rock type as a hornblende-biotite granodiorite described from Texas. The IUGS recommendations have been published in book form along with a glossary of terms (LeMaitre 1989), and the reader is referred to this book for details of classification and nomenclature.

To classify a rock correctly on the basis of its mineral composition, the percentages of five minerals (or mineral groups) must be determined: quartz, plagioclase (with anorthite content greater than 5%), alkali feldspar (including albite), ferromagnesian minerals, and feldspathoids. In hand specimen work, quartz is identified by its translucency, vitreous luster, and lack of obvious cleavage; plagioclase by its cleavages and polysynthetic twinning striations on cleavage surfaces; potassic feldspar by its cleavages, lack of twinning striations, and common pink to tan color; ferromagnesian minerals by their brown, green, or black color; and feldspathoids by individual characteristics possessed by each. The only difficult feldspathoid to recognize is nepheline, which can easily be mistaken for quartz in hand specimen work. In thin sections, discrimination of quartz and nepheline is reasonably straightforward.

As noted earlier, the mineralogy of a rock reflects its chemical composition (see Table 3-4). Rocks that con-

FIGURE 3-1

● ● ● ● ● ● ● ● ● ● ● ● ● ● ● ●

The technique of finding a point by using a triangular coordinate system. Lines parallel to the base indicate percentages of quartz (Q). Lines parallel to the left side indicate percentages of plagioclase (P), and those parallel to the right side indicate alkali feldspar (A). For every point, percentages of Q, A, and P must total 100. Points that fall on the edges indicate that only two constituents are present. The point plotted represents a rock that is 30% Q, 40% A, and 30% P, and falls within the field of "granite" in Figure 3-2.

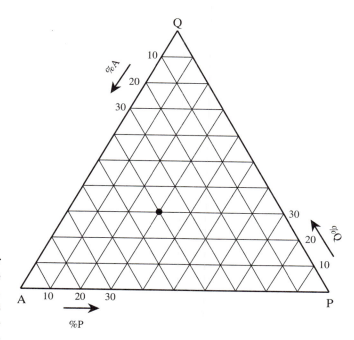

tain free quartz are typically relatively rich in silica (for example, granites); those with calcic plagioclase as the dominant feldspar are usually rich in CaO (for example, diorite and gabbro); those dominated by mafic minerals contain considerable MgO and FeO (for example, peridotite). An igneous rock classification based on mineralogy therefore is a direct reflection of magma chemistry. The IUGS classification distinguishes the common rocks first on the basis of grain size. Phaneritic rocks are classified as plutonic, and aphanitic rocks are classified as volcanic. Within each of these broad categories, the rocks are named on the basis of mineral percentages. The classification categories defined are based on common usage and abundance of natural mineralogic groupings. Therefore most rocks fit well within their categories and relatively few ambiguities occur.

FELSIC AND MAFIC ROCKS.   Triangular diagrams are very commonly used by petrologists. The simple geometrical procedure for finding or plotting an individual point within any triangular diagram is given in Figure 3-1. The IUGS classification shows the common plutonic rocks in two triangular diagrams that permit a rock to be classified in terms of four constituents: quartz (Q), alkali feldspar (A), plagioclase (P), and feldspathoid (F) (Figure 3-2A). These triangles share a common edge (A–P), and the incompatibility of quartz and feldspathoid (foid) minerals means that any rock will plot either above (Q-bearing) or below (F-bearing) this line. Remember that albite, by convention, is plotted at the A corner (as an alkali feldspar); the P corner is reserved for plagioclase with greater than about 5% anorthite content. Within the IUGS rules for use of the diagram of Figure 3-2A, plutonic rocks with anywhere between 10 and 100% Q + A + P (*leucocratic* or light-colored minerals) can be classified (the rest of the rock is assumed to consist of dark-colored mafic minerals); rocks with less than 10% Q + A + P are classified as ultramafic and a different classification diagram is used (see later).

The classification technique involves determination of the volumetric percentages of each of the A, P, and Q or F constituents, along with the amount and type of mafic constituents. Assume that a rock has 50% mafic minerals, 15% Q, 20% A, and 15% P. Because the mafic minerals are not included in the triangle, Q, A, and P are recalculated to equal 100% (a process called normalization of a subset of constituents), thus giving 30% Q, 40% A, and 30% P. When plotted in Figure 3-2A, this point falls within the defined field of granite (refer to Figure 3-1 for instructions on how to plot points within triangular diagrams). If a plutonic rock contains only a small proportion of unspecified mafic minerals, this should be indicated by the prefix *leuco-* (as in leucogranite); with large proportions of mafic minerals, the name should be prefixed by *mela-* (as in melagranite). If the identities of the mafic mineral or minerals are known, they should be used in the name, for example, hornblende-biotite granodiorite; the less or least abundant of the mafic minerals is listed first. The full use of the QAPF classification diagram requires fairly precise knowledge of mineral modes, information typically acquired through laboratory examination. When precise mineral percentages cannot be determined, for example, during field mapping or routine hand specimen examination, generalized group names are commonly used (Figure 3-2B).

Each of the plagioclase-rich areas of the QAPF diagram contains two or more rock types whose names depend on the anorthite content of plagioclase or the abundance or identity of mafic minerals. Anorthosite is a plutonic rock that contains more than 90% plagioclase and less than 10% mafic minerals. The distinction between gabbro

(A)

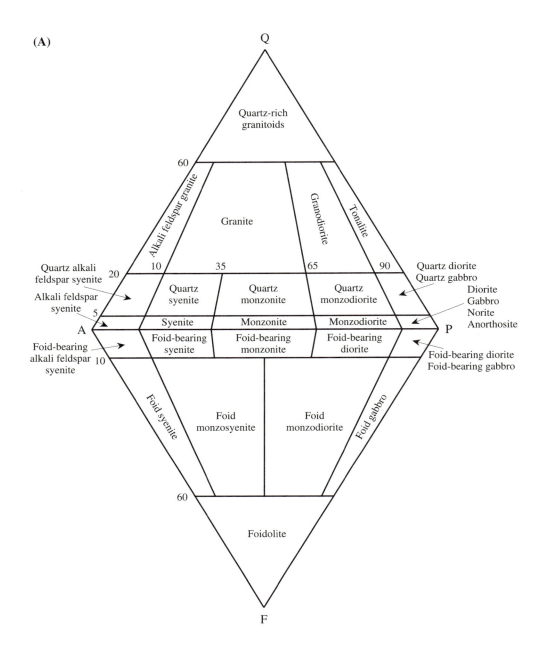

FIGURE 3-2
●●●●●●●●●●●●●●●●●

(A) IUGS classification of plutonic (phaneritic) rocks with mineralogic compositions that fall within the QAPF diagram. Q, quartz; A, alkali feldspar; P, plagioclase; F, feldspathoid (foid). The rock must be less than 90% mafic minerals. Nomenclature has been simplified somewhat in this figure; full details can be found in LeMaitre (1989). See text for the criteria for subdivision of the fields in the P-rich corner. [From R. LeMaitre, 1989, *A Classification of Igneous Rocks and Glossary of Terms* (Oxford: Blackwell), Fig. B.4.] **(B)** Generalized group names (for field use) when mineral percentages cannot be determined with precision. When a feldspathoid mineral is present in fields along the A–P line, its name should be used as a qualifier, for example, "nepheline syenitoid." [From A. L. Streckeisen, 1976, *Earth Sci. Rev.*, *12*, Fig. 1a.]

(B)

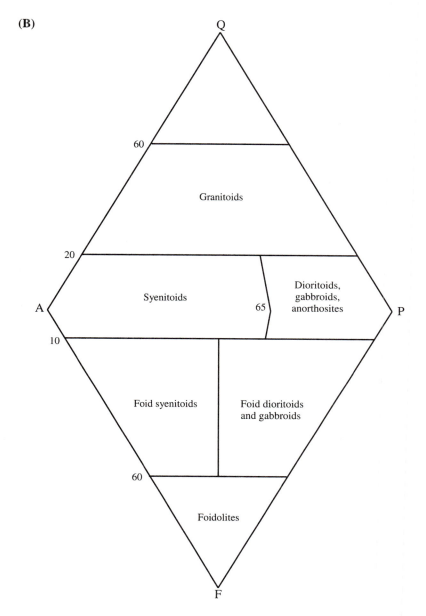

and diorite (and their volcanic equivalents basalt and andesite) can be based on either criterion. The plagioclase in gabbro has a composition more calcic than An_{50}, whereas plagioclase in diorite is less calcic than An_{50}. If a rock is to be classified by means of hand specimens only, distinction between gabbro and diorite obviously cannot be based on plagioclase composition, and the proportion of mafic minerals can be used instead. Gabbro typically contains more than 35% by volume of the mafic minerals olivine, augite, or orthopyroxene. Diorites contain less than 35% mafic minerals by volume and generally contain hornblende as well as, or instead of, pyroxenes. A further subdivision of gabbroic rocks is based on which mafic minerals are present in addition to calcic plagioclase. Gabbro contains clinopyroxene, norite contains orthopy-

roxene, and gabbronorite has subequal proportions of orthopyroxene and clinopyroxene. The classification of plagioclase-rich rocks is summarized in Figure 3-3.

ULTRAMAFIC ROCKS. Ultramafic rocks are virtually always phaneritic and have $Q + A + P + F$ content of less than 10%, that is, mafic minerals make up more than 90% of the rock. The major mafic minerals in ultramafic rocks are magnesian olivine, augite, orthopyroxene, and hornblende. A wide variety of minor minerals can occur, the most common of which are Al, Cr-spinel, magnetite, ilmenite, garnet, phlogopite, and plagioclase. Typically the rarer hornblende-bearing and more common hornblende-free ultramafic rocks have little overlap, and most hornblende-bearing varieties are simply

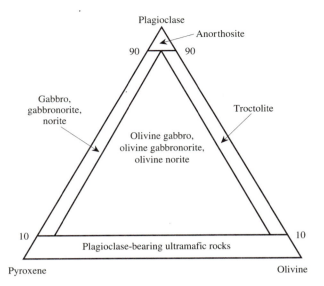

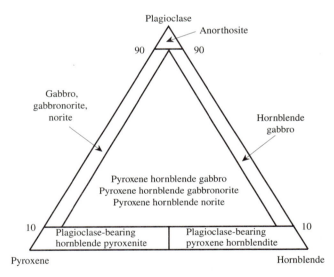

FIGURE 3-3

• • • • • • • • • • • • • • •

Classification and nomenclature for the gabbroic rocks, expressed in terms of the proportions of plagioclase, pyroxene, and olivine (top) and plagioclase, pyroxene, and hornblende (bottom). Multiple names in pyroxene-rich fields relate to the clinopyroxene (Cpx) to orthopyroxene (Opx) ratio: Cpx > 50%, gabbro; Opx > 50%, norite; Cpx ≈ Opx, gabbronorite.

chromium-bearing calcic pyroxene. Commonly contains a minor aluminous mineral, either Cr, Al-spinel or garnet.

Harzburgite: Specific term for an olivine-rich, olivine-orthopyroxene rock. Most commonly contains Cr, Al-spinel as a minor mineral.

Dunite: A peridotite containing 90–100% olivine, with most of the remainder composed of pyroxene.

Websterite: Named for its type locality at Webster, North Carolina, this is a rock composed mostly of sub-equal proportions of orthopyroxene and clinopyroxene, with the small remainder either olivine or hornblende.

Kimberlite: (Not in Figure 3-4.) Rare porphyritic ultramafic rock with excess potassium, and thus containing phlogopite or potassic amphibole phenocrysts — effectively a mica peridotite. Contains olivine (commonly altered to serpentine or carbonate), phlogopite (possibly altered to chlorite), pyroxenes, and chromite. Characteristic accessory minerals are monticellite, magnesium-rich garnet, and titanium-rich minerals. Some kimberlites contain diamonds.

Lamproites: (Not in Figure 3-4.) Similar to kimberlites and lamprophyres (see later), lamproites are ultramafic and have total alkalis in excess of alumina, making them peralkaline (see following alumina content of granites classification scheme). Lamproites can occur as lava flows. They commonly contain rare minerals, including diamond.

called *hornblendites*, with minor mineral names used as qualifiers, for example, *garnet hornblendite*. By far the most common ultramafic rocks are those dominated by olivine and pyroxenes, and their classification scheme uses a triangular diagram with olivine, calcic clinopyroxene, and orthopyroxene at the corners (Figure 3-4). The more commonly encountered varieties or names are

Peridotite: A general or field term for a rock containing 40–100% olivine, with the remainder mostly pyroxene.

Lherzolite: Very important rock type, which has been postulated to constitute most of the earth's mantle. Lherzolite, named for its occurrence in an ultramafic body at Lherz in the French Pyrenees, is an olivine-rich rock with substantial orthopyroxene and minor

VOLCANIC ROCKS.   Volcanic rocks are named on the basis of a diagram similar to that used for plutonic rocks (Figure 3-5). The distinction between basalt and andesite is made mainly on the basis of color index and silica content (a rock with less than 52 wt% SiO_2 is a basalt, and a rock with more than 52 wt% SiO_2 is an andesite, as shown in Figure 3-5C) or, less accurately, on the basis of plagioclase compositions (a rock with plagioclase more sodic than An_{50} is an andesite). Plagioclase composition in many volcanic rocks is difficult to use as a criterion because of the very common presence of strong compositional zoning within crystals. Basalt, which falls near the P corner of the triangle, is subdivided into *tholeiite*, *olivine tholeiite*, *high-Al basalt*, and *alkali basalt*. Distinguishing mineralogic and chemical characteristics of basalts are listed in Table 3-5. Note that many basalts

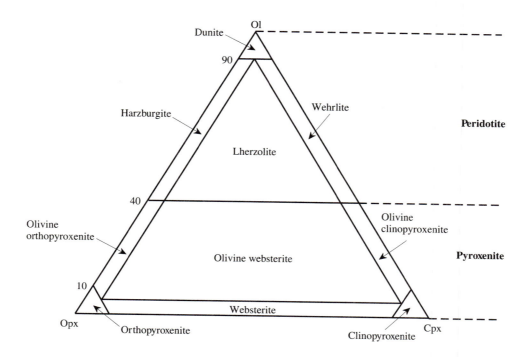

FIGURE 3-4

• • • • • • • • • • • • • • • •

IUGS classification scheme for ultramafic rocks. Ol, olivine; Opx, orthopyroxene; Cpx, clinopyroxene. The general name "pyroxenite" is used when olivine content is 0–40%, whereas "peridotite" refers to olivine content of 40–100%. Minor minerals are listed as follows: If less than 5%, use *-bearing* (for example, "garnet-bearing lherzolite"). If greater than 5%, use the mineral name as a qualifier (e.g., "garnet lherzolite"). [After R. LeMaitre, 1989, *A Classification of Igneous Rocks and Glossary of Terms* (Oxford: Blackwell), Fig. B.8.]

cannot be placed in a particular category by microscopic analyses but require chemical analyses or calculation of normative minerals. The common feldspathoid-bearing volcanic rocks can be rich in alkali feldspar (phonolite), plagioclase (tephrite), and nepheline or leucite (nephelinite or leucitite).

Other Aspects of Classification

Unfortunately, the IUGS classification does not include a textural input to classification, other than the obvious distinction between phaneritic and aphanitic rocks. However, there are a few igneous rocks that are named on the basis of textural criteria, with mineral content being a secondary consideration, including the following:

Pegmatite: A very coarse rock (most grains are greater than 1 cm and may approach or exceed 1 m) with interlocking grains. The composition is typically granitic, and pegmatites most commonly contain large alkali feld-

spars (albite or sodic plagioclase plus microcline) and quartz crystals. Unusual characteristics of pegmatite magmas (particularly the very high water content) permit transport of rare or unusual elements and promote crystallization of minerals that contain them. Unusual mafic pegmatites also exist, commonly as lenses or enclaves within gabbro or diabase bodies, and consist of exceptionally coarse intergrown pyroxene and plagioclase grains.

Obsidian: A black, dark brown, or dark green glassy volcanic rock, essentially nonvesicular and typically of rhyolitic composition. Obsidians are characterized by conchoidal fracture and may show incipient to advanced devitrification with production of very fine grained secondary minerals.

Tuff: Compacted volcanic ash and dust that may contain up to 50% sedimentary material.

Breccia: Similar to tuff, but containing large angular fragments (>2 mm) in a finer matrix.

Aplite: A fine-grained rock of granitic composition, characterized by a homogeneous, sugary-grained appearance and a lack of mafic minerals. Typically

FIGURE 3-5

• • • • • • • • • • • • • • • • •

(A) IUGS classification of volcanic (aphanitic) rocks with mineralogic compositions that fall within the QAPF diagram. The rock must be less than 90% mafic minerals. Nomenclature has been simplified somewhat in this figure; full details can be found in LeMaitre (1989). Appropriate modifying terms are based on mafic minerals or distinctive texture. [From R. LeMaitre, 1989, *A Classification of Igneous Rocks and Glossary of Terms* (Oxford: Blackwell), Fig. B.4.]
(B) Generalized group names (for field use) when mineral percentages cannot be determined with precision. When a feldspathoid mineral is present in fields along the A–P line, it should be used as a qualifier, for example, "nepheline syenitoid." [From A. L. Streckeisen, 1976, *Earth Sci. Rev.*, *12*, Fig. 1a.]
(C) Distinction between basalt and andesite is based on color index (volume percent of mafic minerals) and silica content.

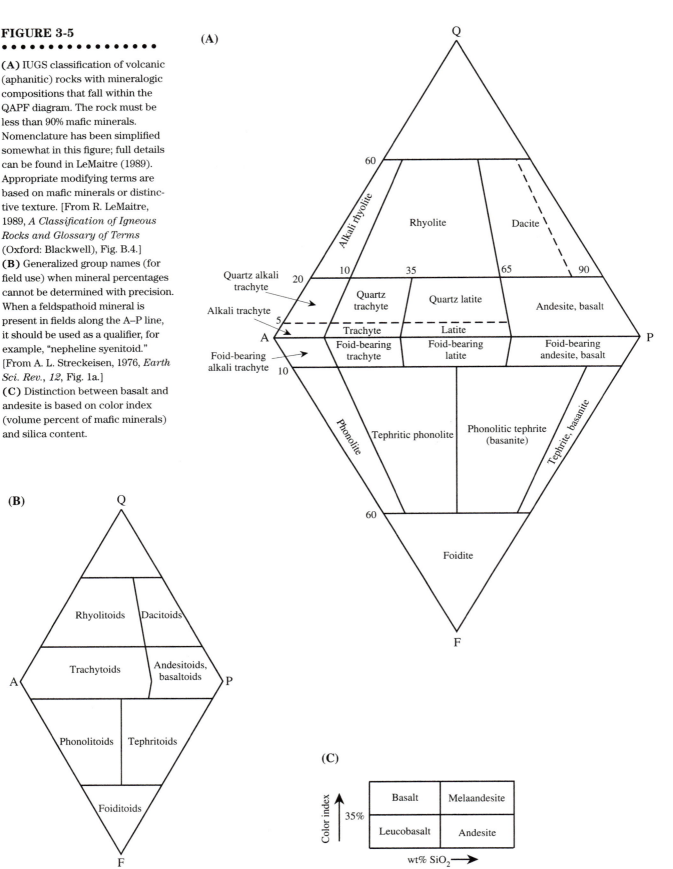

associated with pegmatites or the margins of coarse-grained granites.

Porphyry: Fifty percent or more of the rock is made up of coarse phenocrysts (commonly plagioclase) that are notably larger than the matrix of the rock.

Some well-recognized igneous rocks are defined and named on the basis of highly unusual bulk chemistry or the presence of characteristic postmagmatic alteration. Compositional modification is an integral part of the origin or emplacement of these rocks rather than the result of later weathering. Igneous rocks in these two categories include

Carbonatite: Very unusual and rare magmatic rocks known to be both intrusive and volcanic (typically found in continental rift-type environments such as the East African Rift). They contain a high proportion of calcium and sodium carbonate minerals that have crystallized from a melt along with silicate and oxide minerals, commonly exotic ones.

Spilite: An altered submarine basalt, commonly vesicular and exhibiting pillow structure. All plagioclase has typically been converted to albite and is usually accompanied by secondary chlorite, calcite, epidote, chalcedony, or prehnite. Spilites are thought to have been subjected to submarine hydrothermal seawater alteration with accompanying sodium increase.

Serpentinite: A rock consisting almost entirely of serpentine, derived from alteration of preexisting pyroxene and olivine in a rock rich in these minerals, typically an ultramafic rock. Although they are considered to be metamorphic rocks, very young serpentinites have been dredged from the Mid-Atlantic Ridge where serpentinization by hydrothermal processes must have been virtually synchronous with crystallization from mid-ocean ridge magma.

Lamprophyres: A family of igneous rocks that normally occur in dikes and are characterized by common porphyritic textures and moderately to highly potassic bulk compositions. Lamprophyres are generally mafic but rarely ultramafic, and have K_2O contents much higher than typical for such mafic magmas. This composition commonly results in large biotite (an essential constituent) or potassic amphibole phenocrysts.

TABLE 3-5   Types of basalts, based on mineralogic and chemical characteristics

Varietal names	Pyroxenes	Other minerals[a]	Remarks
Tholeiite	Augite and orthopyroxene or pigeonite	Interstitial siliceous glass, tridymite, or quartz; possibly minor olivine with pyroxene rims	Oceanic varieties have >2.5 wt% TiO_2; continental varieties about 1.0 wt% TiO_2
Olivine tholeiite	Augite and orthopyroxene or pigeonite	Abundant olivine, possibly with pyroxene rims	Oceanic varieties have >2.5 wt% TiO_2; continental varieties about 1.0 wt% TiO_2
High-Al basalt	Augite, rarely pigeonite or orthopyroxene	Olivine common, with or without pyroxene rims	Al_2O_3 > 17 wt%; TiO_2 about 1.0 wt%; intermediate character between tholeiite and alkali basalt
Alkali basalt	Augite (commonly Ti-enriched), never low-Ca pyroxene	Olivine abundant; feldspathoids, alkali feldspar, phlogopite, or kaersutite may be present	Total alkali content higher and silica content lower than above types

[a] Plagioclase and iron-titanium oxides are always present.

CHEMICAL CLASSIFICATION
• • • • • • • • • • • • • • •

A great variety of chemical classifications of igneous rocks have been proposed, some based on complete chemical analysis of a rock and others on only a portion of rock chemistry. Many of these classifications are used for quite sophisticated petrogenetic schemes and are well beyond the scope of this book. Several broad chemical classifications are widely used, however, and a student of petrology should be familiar with their general applicability.

Silica Saturation

Because igneous rocks crystallize from magma, it is convenient to think of the magma as a liquid that is *saturated* with a particular mineral. Chemical rules specify that liquids cannot ordinarily contain more than a certain amount of a particular dissolved component. If a cooling magma is saturated with a particular mineral, then that mineral will precipitate and be part of the resultant rock. This concept is commonly used for silica in a wide variety of igneous rocks from basalts to granitoids. Thus a rock is described as *silica-oversaturated* if it contains quartz or an equivalent silica mineral. It is *silica-undersaturated* if it contains silica-deficient minerals such as feldspathoids or magnesian olivine that cannot exist in the presence of quartz. Some magmas form silica-deficient minerals early but tend to become saturated with silica late in their crystallization. Under such conditions, earlier grains of olivine will combine with silica to form orthopyroxene:

$$Mg_2SiO_4 + SiO_2 = 2\ MgSiO_3$$

Feldspathoids combine with silica to form feldspars, for example

$$KAlSi_2O_6\ (leucite) + SiO_2 = KAlSi_3O_8\ (orthoclase)$$

An igneous rock that contains neither free quartz nor a silica-deficient mineral is described as *silica-saturated*.

Alumina Content of Granites

Granites are defined in the IUGS classification system as rocks that contain between 20 and 60% quartz and in which alkali feldspar constitutes more than 35% of the total feldspar. However, petrologists have discovered that magmatic rocks that fit this mineralogic criterion can have quite variable contents of Al_2O_3, as reflected

by accessory mineral content. For example, some granites high in Al_2O_3 contain minerals such as garnet or muscovite, whereas granites low in Al_2O_3 contain minerals such as riebeckite or aegerine-augite. The alumina content of granites can be a very direct monitor of the character or type of crustal rocks that melted to form the granite magma, so a classification of granites based on Al_2O_3 content has been devised. This classification in effect recognizes the influence of relative amounts of alumina, alkalis, and calcium on the CIPW normative mineralogy, with particular reference to the ratio of alumina to alkalis plus calcium in feldspars.

A hydrous granite magma will not crystallize muscovite, for example, unless the amount of Al_2O_3 (on a *molar* basis) exceeds the amount of $Na_2O + K_2O + CaO$. (A molar basis means that proportions of chemical constituents are calculated on the basis of molecular percents rather than weight percents.) Such granites have more alumina than needed to make feldspars and are referred to as *peraluminous*. If molar $Al_2O_3 < Na_2O + K_2O$, then the excess of alkalis (or deficiency of alumina) will probably result in the presence of a sodium-rich mineral such as aegerine-augite or a sodic amphibole. These granites are called *peralkaline*. If there is an excess of neither alumina nor alkalis—that is, $Na_2O + K_2O < Al_2O_3 < Na_2O + K_2O + CaO$—then neither muscovite nor a sodic ferromagnesian mineral will result and the granite is referred to as *metaluminous* (the Greek prefix *met-* meaning "middle" or "medium"). It should be remembered that there are typical mineralogic consequences to these chemical distinctions. The chemical subdivision of granites into peraluminous, metaluminous, and peralkaline is exceptionally important in granite petrology and is a major basis for assessing tectonic environments of granite magma derivation.

CHEMICAL TRENDS
• • • • • • • • • • • • • • •

Igneous rock suites that crop out over limited geographic areas commonly show substantial ranges in bulk chemistry. To decipher possible genetic relationships between the different magmas and to facilitate comparisons of igneous rocks from different areas, igneous petrologists have developed a number of chemical and mineralogic indices and graphical schemes to portray igneous rocks. Some of these rely on measured modal mineralogy, some use estimated mineralogy (such as CIPW norms), and others use various ratios of chemical components. A common type of diagram is called the *variation diagram*, in which a large amount of chemical data can be plotted to assess the possible effects on magma chemistry of removal of crystals during magma evolution.

FIGURE 3-6

● ● ● ● ● ● ● ● ● ● ● ● ● ● ● ●

Data for a hypothetical suite of related basalts and andesites plotted on a Harker variation diagram. The linear trends shown are approximately correct for fractional removal of olivine from crystallizing parental magma.

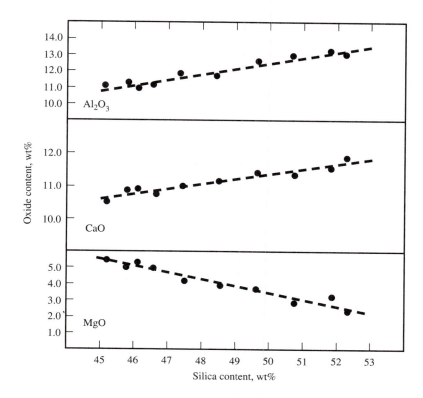

Some of the most widely encountered of these indices are summarized here. In Chapters 8 through 10, these diagrams are used in the discussion of igneous suites.

Harker Diagrams

One of the most widely used of all variation diagrams, the typical Harker diagram plots the weight percent of all other oxides as a function of the weight percent of one oxide. Weight percent SiO_2 most commonly forms the abscissa, because it is a very useful indicator of magma evolution in many cogenetic suites where various magmas are all descended from a single parent: The more primitive members of the series commonly have lower silica content than the more advanced ones. This distinction occurs because the earliest minerals to crystallize (the ones highest in Bowen's Reaction Series) are typically low in silica relative to the later ones. Removal of these early crystals (fractionation) thus enriches the remaining magma in silica. Oxide pairs are commonly correlated in a linear fashion in fractionation processes, as illustrated in the Harker diagrams shown in Figure 3-6.

Harker-like variation diagrams can use the content of any oxide on the abscissa as long as this parameter shows sufficient variation throughout the igneous suite. One drawback to silica content variation diagrams is that simultaneous removal of pyroxene, plagioclase, and

olivine in early stages of differentiation of basaltic suites may not have much of an effect on the SiO_2 content of magma, because these minerals have a *net* silica content about the same as the magma. For this reason, MgO content is commonly used on the abscissa in basalt-andesite suites because it is strongly controlled by the fractional removal of olivine and pyroxene together. Correlations or trends in Harker diagrams indicate various igneous processes operating in a series of related magmas, for example, fractionation and mixing. Overall, linear correlation trends indicate the chemical evolution of cogenetic magmas and can be used to predict the exact processes quantitatively.

AFM (or FMA) Diagrams

The AFM diagram is another type of variation diagram and is useful for discriminating fractionation processes. The various oxides—$Na_2O + K_2O$ (A), $FeO + Fe_2O_3$ (F), and MgO (M)—are plotted on the corners of a triangular diagram. The shape of the trend of a cogenetic suite in the AFM diagram indicates whether olivine or clinopyroxene removal is controlling magmatic evolution (Figure 3-7). This diagram allows simultaneous observation of two parameters that mark fractionation: Fe:Mg ratio and total alkali content. AFM diagrams are especially useful in assessing magmatic evolution trends

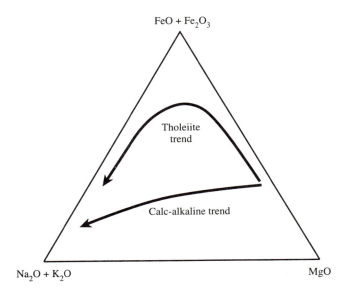

FIGURE 3-7
● ● ● ● ● ● ● ● ● ● ● ● ● ● ● ●

FM diagram showing the calc-alkaline and tholeiite trends. The tholeiite trend is due to fractional removal of olivine during crystallization, whereas the calc-alkaline trend is caused by removal of clinopyroxene + calcic plagioclase.

in convergent plate environments and are discussed in detail in Chapter 9.

Differentiation Index

The differentiation index is a measure of differentiation or total magmatic evolution. It is calculated by summing the normative minerals at the bottom of Bowen's Reaction Series: orthoclase, albite, and quartz. Because it is based on normative mineralogy expressed as percentages, its possible range is from 0 to 100. Basalts generally have a low differentiation index (<25), whereas granites commonly have indices exceeding 75.

Alkali-Lime Index

In most cogenetic suites of igneous rocks, total content of alkali oxides increases and CaO content decreases with increasing SiO_2 content. In a Harker variation diagram, where both $Na_2O + K_2O$ content and CaO content are plotted against SiO_2 content, the two linear trends have a crossover point where wt% ($Na_2O + K_2O$) = wt% CaO. The value of wt% SiO_2 at this point is called the *alkali-lime index*. In general, igneous rock suites derived from an alkali-rich parent (alkaline series) have low values of alkali-lime index (<50), whereas low-alkali suites (calcic series) have high indices (>60).

Larsen Index

The Larsen index was developed for use in Harker diagrams in which SiO_2 content alone does not show much

change in basaltic suites because pyroxene and plagioclase are removed along with olivine. Defined as $\frac{1}{3} SiO_2 + K_2O - (FeO + MgO + CaO)$, the Larsen index replaces SiO_2 content on the abscissa of a Harker diagram and produces an exaggerated shift in early fractionation trends in many basaltic suites.

Spider Diagrams

The patterns of trace element concentrations can be very distinctive for igneous rocks. These patterns can be used to distinguish the source areas of different magmas or to resolve magma fractionation or magma mixing. To provide a baseline for comparisons, values for individual rocks must be normalized to some reference chemical reservoir. The two most commonly used reservoirs are the primordial mantle (Wood et al. 1979) and the primordial earth modeled by chondrite meteorites (Sun 1980). In spider diagrams, normalized element abundance is plotted vertically and a selection of elements is plotted horizontally. The order of these elements is arbitrary but is designed to produce a smoothly increasing pattern for normalized element abundances in mid-ocean ridge basalt (MORB; see Chapter 8). Figure 3-8 illustrates a typical spider diagram.

Assimilation and Fractional Crystallization (AFC)

It is now widely accepted by igneous petrologists that magmas commonly evolve chemically in magma chambers by simultaneous operation of processes that selectively remove components from the melt by fractional crystallization and add others by selective extraction from magma chamber walls or xenoliths (see Chapter 6 for a detailed discussion). DePaolo (1981) has developed a set of equations and plots to assess the implications of assimilation and fractional crystallization (AFC) processes for both isotopic and trace element compositions of magmas.

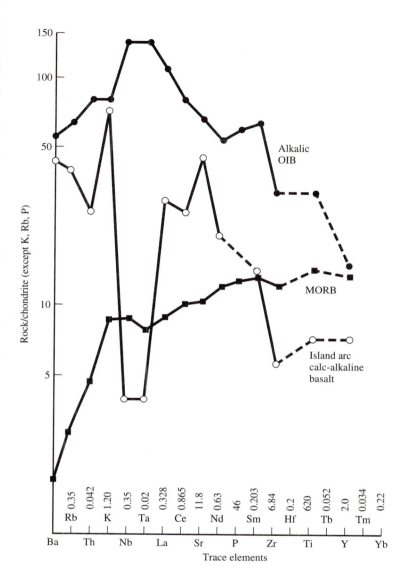

FIGURE 3-8

● ● ● ● ● ● ● ● ● ● ● ● ● ●

A typical spider diagram pattern for mid-ocean ridge (MORB), ocean island (OIB), and island arc basalts. The ordinate scale is element abundance normalized to abundance in chonditic meteorites (see text). Normalization constants along the abscissa relate to estimates of abundances of each element in the primordial earth. [From M. Wilson, 1989, *Igneous Petrogenesis: A Global Tectonic Approach* (New York: Chapman & Hall), Fig. 2.4(a).]

SUMMARY

● ● ● ● ● ● ● ● ● ● ● ● ●

Igneous rocks are composed of minerals (except for the rare glassy varieties such as obsidian), and the identity and relative proportions of these minerals directly reflect the bulk composition of the original magma. There are a variety of techniques that petrologists use to characterize all aspects of igneous rock chemistry, from whole rock major, minor, and trace element concentrations, to chemistry of individual minerals, and finally to ratios of stable and radiogenic isotopes. All of this chemical information is critical for creating and testing hypotheses of the origin and solidification of magmas. However, many observations of igneous rocks do not require sophisticated laboratory techniques but still yield important information for characterizing magmatic rocks. For example, estimation or measurement of the mineral proportions in an igneous rock enables a petrologist to classify the rock and to use phase diagrams to construct scenarios for crystallization of the rock. Assessment of the tens of thousands of igneous rock chemical analyses obtained over the last century or more reveals that rock chemistry is less diverse than the wide range of igneous rock appearances (colors, grain sizes, mineralogy, etc.) would suggest. For example, virtually all igneous rocks have silica concentrations between 45 and 75 wt%. The dominant role of feldspars in most igneous rocks controls this relative chemical uniformity.

Objective and rigorous classification is an important part of igneous petrology because it standardizes use of names and ensures that two geologists mapping what each calls a granite are indeed both working with the same rock type. Commissions working under the authorization of the International Union of Geological Sciences have developed standardized classifications for most igneous rocks. Grain size and mineral content are the foundations of classification and of naming schemes both for plutonic and volcanic rocks that are dominated by quartz, feldspars, and feldspathoids and for ultramafic rocks. Mineral content is generally measured by using thin sections and the petrographic microscope. Bulk compositional information can be used to create a "synthetic mineralogy" for all igneous rocks by performing a CIPW norm calculation, which partitions rock chemical constituents into general hypothetical minerals. This technique is especially valuable when a volcanic rock is so fine-grained that optical measurements are not feasible. Although they play little role in the IUGS classifications, textural or special compositional characteristics can also be used for naming rocks. A wide variety of special diagrams or indices have been developed by petrologists to utilize chemical and mineralogical data on igneous rocks to trace genetic processes, particularly fractionation.

STUDY EXERCISES

1. Viscosity is one of the most important physical properties of magmas. What physical and chemical factors are most important in controlling viscosity?

2. How do trace elements differ from major elements in magmatic rocks? How do stable isotopes differ from radioactive or radiogenic isotopes? Are the same analytical techniques used to measure the content of each of these in a rock?

3. Volume modes are easily calculated from measurements made in thin sections. How is this done? For some petrologic purposes, weight modes are required. They are messy to measure directly because the rock must be broken up into individual grains and each distinct mineral fraction weighed. A weight mode can be calculated from the more easily measured volume mode. How is this done?

4. What is the purpose for calculating a normative mineralogy such as the CIPW norm?

5. Using the rules presented in Appendix 1, calculate by hand a CIPW norm for any one of the average igneous rock compositions given in Table 3-4. If a computer program is available, use it to calculate CIPW norms for all the compositions given in Table 3-4.

REFERENCES AND ADDITIONAL READINGS
• • • • • • • • • • • • • •

Bowen, N. L. 1928. The *Evolution of the Igneous Rocks.* Princeton, NJ: Princeton University Press.

DePaolo, D. J. 1981. Trace element and isotopic effects of combined wallrock assimilation and fractional crystallization. *Earth Planet. Sci. Lett., 53,* 189–202.

Hess, P. C. 1989. *Origins of Igneous Rocks.* Cambridge, MA: Harvard University Press.

Johannsen, A. 1931, 1937, 1938. *A Descriptive Petrography of the Igneous Rocks.* Chicago: University of Chicago Press, 4 vols.

LeMaitre, R. W. 1976. Some problems of the projection of chemical data into mineralogical classifications. *Contrib. Mineral. Petrol., 56,* 181–189.

LeMaitre, R. W., ed. 1989. *A Classification of Igneous Rocks and Glossary of Terms.* Oxford: Blackwell.

Philpotts, A. L. 1990. *Igneous and Metamorphic Petrology.* Englewood Cliffs, NJ: Prentice Hall.

Ringwood, A. E. 1975. *Composition and Petrology of the Earth's Mantle.* New York: McGraw-Hill.

Ronov, A. B., and A. A. Yaroshevsky. 1969. Chemical composition of the earth's crust. *Am. Geophys. Union Monogr. 13,* 37–57.

Shaw, H. R. 1965. Comments on viscosity, crystal settling and convection in granitic magmas. *Am. J. Sci., 263,* 120–152.

Streckeisen, A. L. 1976. To each plutonic rock its proper name. *Earth Sci. Rev., 12,* 1–34.

Streckeisen, A. L. 1978. Classification and nomenclature of volcanic rocks, lamprophyres, carbonatites and melilitic rocks. *Neues Jahrb. Mineral. Abh., 134,* 1–14.

Sun, S. S. 1980. Lead isotopic study of young volcanic rocks from mid-ocean ridges, ocean islands and island arcs. *Phil. Trans. R. Soc. London, A297,* 409–445.

Taylor, S. R., and S. M. McLennan. 1985. *The Continental Crust: Its Composition and Evolution.* Oxford: Blackwell.

Williams, H., F. J. Turner, and C. M. Gilbert. 1982. *Petrography: An Introduction to the Study of Rocks in Thin Section,* 2nd ed. San Francisco: W. H. Freeman.

Wilson, M. 1989. *Igneous Petrogenesis: A Global Tectonic Approach.* London: Unwin-Hyman.

Wood, D. A., J. Tarney, A. D. Saunders, H. Bougault, J. L. Joron, M. Treuil, and J. R. Cann. 1979. Geochemistry of basalts drilled in the North Atlantic by IPOD Leg 49: Implications for mantle heterogeneity. *Earth Planet. Sci. Lett., 42,* 77–97.

Chapter 4

CRYSTALLIZATION OF MAGMAS

The magmatic origin of basalt and rhyolite was first recognized in the late 1700s in Italy and France through observation of active volcanoes and correlation of their known products with similar looking ancient rocks. As obvious as this interpretation now seems, one alternative eighteenth-century theory of the origin of basalt held that it was precipitated from a primeval ocean, much as evaporites and some limestones are now known to form. Observation of volcanoes in Italy and France greatly impressed James Hutton (a Scotsman) with the importance of underground heat. He had observed a baked coal seam in Scotland that was adjacent to basaltic dikes, and he had also noted an example of forceful injection of basalt into overlying sediments. In the nineteenth-century spirit of empiricism, and to obtain further information on the nature and origin of basalt, James Hall (an Englishman) heated basalt samples in crucibles in his laboratory and found that they melted between 800° and 1200°C. Rapid cooling of the melt produced a glass, whereas slower cooling produced a fine-grained crystalline aggregate similar to basalt. Subsequent studies of igneous rocks revealed that most are composed of silicate minerals whose individual melting points are well over 1000°, some as high as 1800°C. Investigations into the formation of basaltic melts have shown that complete melting is typically achieved at 1100°–1200°C, whereas granitic melts form at a lower temperature range of 650° – 800°C.

Much vital information about igneous rocks comes from field, hand specimen, and thin section investigations, as already discussed. Petrologists determine the identity and proportions of the minerals in igneous rocks and note special textures and structures that give clues to the processes that formed them. An understanding of an igneous rock is formed mainly from observations in the field, especially those regarding areal variation in mineralogy or texture; from laboratory studies of thin sections, which establish mineral content; and from analyses of chemical composition. However, the processes of formation of all plutonic rocks and of many volcanic rocks cannot be observed directly. A petrologist is thus forced to rely on laboratory experiments and the application of crystallization and melting theory to infer the actual processes and the physical conditions of magma formation and crystallization. The *phase diagram* is the fundamental tool used by petrologists, materials scientists, and metallurgists (in fact, any scientist who deals with molten materials) to illustrate crystallization and melting. This type of diagram is nothing more than a graphical portrayal of the stability ranges of minerals and melts as functions of bulk composition, temperature, and pressure. In this chapter and the next two, we present the fundamentals of construction and interpretation of igneous phase diagrams, the principal technique for understanding the origin, crystallization, and evolution of magmas.

LABORATORY EXPERIMENTS ON MAGMAS

The experiments performed in 1792 by James Hall are one of the earliest examples of experimental igneous petrology. Hall's studies were necessarily restricted to the melting of existing rocks because he was unable to synthesize rock compositions from chemical reagents. However, modern chemical knowledge and equipment allow such syntheses to be carried out and the results to be analyzed with great precision. In experimental laboratories it is possible to reproduce the entire range of compositions, temperatures, and pressures that might exist in nature during the formation and crystallization of magmas. The phase diagrams that petrologists use are constructed and calibrated on the basis of the results of these experiments.

In principle, the experimental method in petrology seems simple and analogous to preparing a meal (perhaps inspiring the phrase "cook and look" for reconnaissance experiments)—mixing the ingredients, heating them up, and enjoying the result. In fact, the procedures are more complicated and the experimental plan very carefully thought out. The desired initial starting compositions are made from either pure reagents or analyzed minerals, carefully weighed, mixed, and homogenized by grinding. For high-temperature experiments (called *runs*), the starting materials are commonly fused (melted), cooled to form a glass, and then ground to a uniform fine grain size to ensure compositional homogeneity. Experiments at low temperatures commonly use reactive chemical gels to increase ordinarily slow reaction rates. Crystalline seeds can be added to provide favorable nucleation sites for minerals that are slow to nucleate. Because processes studied in the laboratory must necessarily be made to proceed at rates higher than those in nature, the experimentalist must skillfully exploit rate-enhancing shortcuts and technical tricks that do not compromise the applicability of the results to natural rocks. Heating equipment used in experimental petrology is quite varied, ranging from one-atmosphere furnaces wound with platinum wire (much like megatoasters) to internally heated hydraulic presses capable of reproducing the pressure and temperature conditions of the earth's mantle. Other equipment includes hydrothermal reactors, which can maintain samples at moderate temperatures and pressures in the presence of selected fluid species (generally either water or carbon dioxide).

After the sample has been subjected to the desired pressure-temperature conditions for a sufficient time (typically hours to weeks), the pressure and temperature are suddenly reduced to quench or "freeze in" the mineral-melt assemblage that equilibrated during the run. Supercooled melts commonly solidify without crystallization to form glasses. Examination of the glass and embedded crystals with a petrographic microscope and X-ray diffraction tells the experimental petrologist what minerals coexisted with melt at the run conditions and the relative proportions of crystals and melt. More detailed examination involves single-crystal X-ray or electron microprobe techniques. All the resultant data on the mineral-melt assemblages for each selected bulk composition at various pressures and temperatures are the input data for construction of phase diagrams. Thousands of such diagrams have been produced for geologic and industrial purposes; many have been collected and published in a multivolume collection by the American Ceramic Society (*Phase Diagrams for Ceramists*). Because phase diagrams are in essence the technical language of igneous petrogenesis, a number of them are examined here as we develop the straightforward principles used in their interpretation by the geologist.

EQUILIBRIUM AND THE PHASE RULE
• • • • • • • • • • • • • • •

One of the most important late nineteenth and early twentieth century advances in petrology was the appreciation that rocks and magmas typically behave rather rigorously according to the laws of physics and chemistry, with the proviso that enough time is available for the processes that "enforce" these laws. In particular, the laws of thermodynamics have been found to have great applicability to petrologic processes. The word *thermodynamics* tends to intimidate many students; and, in fact, there is a substantial body of complicated theory that has been developed by scientists and engineers over the last 150 years or so. However, thermodynamics is simply the codification of a number of fairly obvious principles, such as the flow of heat from a hot body to a cool body and conservation of energy and mass. At least the more basic concepts of thermodynamics cannot be avoided in explaining how magmas form and crystallize. The key to appreciating thermodynamic concepts is to focus on applications, and we firmly believe these concepts, as applied to rocks and igneous processes, are perfectly accessible and understandable to the typical student.

Equilibrium

The first important concept is **equilibrium,** which simply means that a geologic system (rock or magma) is in a state where there is no driving force for change; that is, the temperature, pressure, and proportions of minerals and melt (if present) remain fixed. If the physical conditions of the system are changed, for example, by altering either the temperature or pressure (or both), the system is no longer in equilibrium and must typically shift the type or proportions of the minerals (or amount and composition of melt) by reaction to achieve a new state of equilibrium. As with all physical and chemical processes, the spontaneous attainment of equilibrium in petrology is a process of energy minimization. If the system is not in its energy-minimum state, something generally happens (crystallization or melting in igneous systems) until the system achieves a new energy minimum. For example, if a magma cools through the temperature where the first crystals appear and then remains at constant temperature, the proportions of melt and crystals will remain unchanged once equilibrium is established. But if the magma is then allowed to cool a bit more, additional crystallization will occur and the proportion of minerals relative to melt will increase. In most petrologic cases, perturbation of equilibrium is caused by changes in temperature or pressure and only rarely by bulk compositional changes.

It is convenient to consider the concepts of open and closed systems and of mass balance in the context of equilibrium. A **closed system** is one in which thermal and mechanical energy can be exchanged with the surroundings, but mass cannot be exchanged. Therefore, an igneous rock or magma that is regarded as a closed system can undergo variation in temperature and pressure, but the content of chemical constituents remains fixed. Conversely, an **open system** can exchange both energy and mass with the surroundings. The process of mass transfer is commonly referred to as metasomatism, although this is more common in metamorphic than igneous systems. Petrologists generally regard water, carbon dioxide, and other volatile constituents such as fluorine and chlorine as mobile constituents and do not typically include them in a discussion of open versus closed system behavior. **Mass balance** is the concept of preservation of a constant bulk composition during operation of chemical or physical processes in rocks. Thus the shift of chemical constituents from one physical state to another (as in crystallization of melts) or from one mineral to another must be balanced around maintenance of constant overall composition (excluding volatiles). Mass balance is the chemical principle that lies behind the algebraic technique of reaction balancing, and its application to interpretation of crystallization sequences by using phase diagrams is discussed in this and the following two chapters.

Phases

The equilibrium relationships of minerals and melts can be described most easily by using phase diagrams. A *phase* is usually defined as a physically distinct part of a system (for example, a melt or a particular species of mineral) that is mechanically separable from other phases in the system, at least theoretically. Chemically, phases can have either a fixed composition (for example, quartz, SiO_2), or a variable composition (for example, melts or solid solution phases such as plagioclase). More than one phase can have the same composition (for example, polymorphs such as quartz, tridymite, and cristobalite, or sillimanite and andalusite) as long as each phase has distinctive physical properties (for example, different crystal structures, densities, or viscosities). A magmatic system can contain a melt phase (or, rarely, more than one), a gas or supercritical fluid phase, and

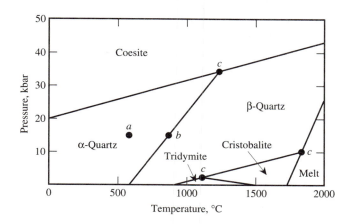

FIGURE 4-1
● ● ● ● ● ● ● ● ● ● ● ● ● ● ● ● ●

P-T diagram for the one-component system SiO_2, showing the divariant, single-phase stability fields for α- and β-quartz, tridymite, cristobalite, coesite, and molten SiO_2. Labeled points are discussed in the text. [After D. T. Griffin, 1992, *Silicate Crystal Chemistry* (New York: Oxford University Press), Fig. 1-1.]

one or more crystalline phases such as olivine, augite, or plagioclase. A phase diagram is a graphical means of showing the pressure-temperature-composition stability regions of different phases or groups of phases (Figure 4-1). Because only two variables can be shown in the two dimensions on a piece of paper, a petrologist must choose which two of these three variables to portray.

Components

With the one-component system shown in Figure 4-1, the choice is easy: There are zero compositional variables (the system is always fixed as pure SiO_2), and pressure and temperature are the only variables. **Components** are defined as the minimum number of chemical constituents that are required to describe the compositions of all phases in the system. Note that a component is *not* a mineral, although components can have the same compositions as mineral phases. Components are principally used for bookkeeping in describing the composition of the system. The one fundamental rule for defining components is, *There can never be more components than phases in a system.* If a rock consists of three minerals, then only three components are required to be defined. These commonly can be the actual compositions of the minerals, but other choices may be more appropriate. If a fluid or a silicate melt is present, it is also considered to be a phase.

When components are defined in the most traditional way as the simple major-element oxides that make up a chemical analysis, most igneous rocks end up with 10 or 12 components. This number of components probably exceeds the number of solid minerals (plus melt) and is incorrect if it does so. The correct alternative is to recombine simple oxides into larger mineral building blocks. Examples include two molar units of MgO plus one molar unit of SiO_2 to form one molar unit of Mg_2SiO_4 (forsterite) or one unit of CaO plus one unit of MgO plus two units of SiO_2 to form one unit of $CaMgSi_2O_6$ (diopside). This subdivision of a system into mineral components makes no implications about the actual minerals present and is thus independent of the actual physical conditions that dictate the actual minerals. The most commonly used units for relative amounts of components are *moles*, the standard chemical unit, and *mole percents* (abbreviated mol%) or *mole fractions* (abbreviated X) for ratios of components. For example, in the two-component system forsterite-diopside, the compositional midpoint is designated as 50 mol% diopside (or forsterite). The application of these units will be apparent in the following discussion of phase diagrams.

The Phase Rule

In the SiO_2 system, there are seven known phases (six solids and one melt, all with the composition SiO_2) whose stability fields can be shown within the pressure-temperature (P-T) limits of a P-T diagram (Figure 4-1). The six solids (α-quartz, β-quartz, tridymite, cristobalite, coesite, and stishovite) are all polymorphs of SiO_2, and each, along with the melt, occupies a unique stability area on the diagram that reflects the P-T range over which it is thermodynamically stable (the ultrahigh-pressure phase stishovite has not been shown in Figure 4-1, because it only occurs above about 80 kbar). The stability areas are called *divariant areas*, because both pressure and temperature can be independently changed a small amount and the same phase remains stable. Divariant areas are separated by lines that are called phase boundaries or *univariant lines*. Two phases can coexist only along one of these special lines that indicate a specially restricted range of P-T conditions. These lines are univariant because only one variable can be independently changed while maintaining the system at two phases. Three phases can coexist only at points where three lines and areas come together (labeled c in Figure 4-1). Because no change in pressure or temperature is possible while three phases coexist, these are called *invariant points*.

Point a on the diagram, at a pressure of 15 kilobars (abbreviated as kbar; 1 kbar $\approx$ 1000 atmospheres) and

a temperature of 500°C, falls within the area labeled α-quartz. If equilibrium is maintained (that is, if pressure and temperature remain constant), then the component SiO_2 should exist only as the α-quartz form. It is possible that chemical components can crystallize as a phase not described by the equilibrium phase diagram at the pressure and temperature of formation. In this case, the system is in a state of disequilibrium, and the phase is *unstable* if it is in the process of changing into a more stable phase. A *metastable* phase has higher energy than the stable equilibrium phase but persists for an extended time without changing, as a result of very slow reaction rates. (In fact, igneous and metamorphic rocks exposed at the earth's surface consist mostly of metastable minerals that have P-T stability fields appropriate to their formation deep in the earth.)

Returning to Figure 4-1, consider what happens when stable α-quartz at the P-T point a is heated while pressure remains constant at 15 kbar. The α-quartz remains stable until the α-quartz–β-quartz phase boundary at b is encountered. Here α-quartz begins the polymorphic phase transformation to β-quartz. If the temperature were held constant at b, both phases would coexist in a "steady state," with β-quartz transforming to α-quartz at exactly the same rate that α-quartz transforms to β-quartz. This point is made to emphasize that equilibrium is actually a dynamic process. The same atoms do not remain in one phase or the other for unlimited periods; the key to the concept of equilibrium is that the *proportions* of phases do not shift. Instead, the rates of transformation in both directions are equal. At 15 kbar and temperatures above b, all α-quartz would spontaneously and irreversibly transform to β-quartz, whereas below b, all β-quartz would similarly become α-quartz.

To reiterate: If there is only one phase in this one-component system, then both pressure and temperature can be varied independently without altering the phase assemblage. This behavior is called *divariance* because the values of two variables can be modified. If two phases exist in equilibrium with each other, then only one *environmental variable*, as pressure and temperature are called, can be varied independently, a constraint resulting in *univariance*. If three phases coexist, then the system has no variability in pressure and temperature and is *invariant*. There is no limit to the *total* number of phases that can exist within a particular chemical system over all pressures and temperatures, and therefore no limit to the number of divariant regions, univariant lines, or invariant points. But as discussed, there are very definite limits to how many phases can *coexist in a stable fashion* at a single pressure and temperature. At equilibrium, the P-T point labeled a in Figure 4-1 has one stable phase; point b has two stable phases; and the points c have three coexisting phases.

Derivation of the Phase Rule

The general relationships among phases, components, and the environmental variables pressure and temperature were described in the 1870s by J. W. Gibbs as the *phase rule*. The phase rule provides a means of classifying systems. It also describes the maximum number of phases that can coexist in a system of any complexity, thus providing a means to test whether the minerals in a rock are coexisting in a stable equilibrium. Derivation of the phase rule can be made in terms of the variance of a system or the number of degrees of freedom (F); or, in other words, how many variables can be independently varied without changing the phase assemblage. The variable F is equal to the number of unknown relationships minus the number of known (dependent) relationships. The unknown relationships include the number of phases and their compositions. Assume that there are p phases, and each phase contains some of each of the components c that describe the system. Just as the composition of each phase can be given in terms of atomic or mole fractions or weight percentages of each of the components, the number of unknowns in the composition of each phase is $c - 1$; that is, if all the concentrations but one are known, the last is known by difference. Because p phases are present, the total number of compositional unknowns is $p(c - 1)$. In addition to compositional variables, the system is subject to a particular (but unknown) pressure and temperature that are the same for all phases in the system. The number of unknown variables in the system is therefore $p(c - 1) + 2$.

With regard to the known or dependent variables or restrictions, thermodynamics has established that the tendency of each component to react (known as the chemical potential) is the same in all coexisting phases. If it were not, there would be a lower energy configuration for the system, which would thus not be in equilibrium, and reaction would occur. If the chemical potential for a component were to be determined for one phase, it would therefore be known for all phases. Hence for each component there are $p - 1$ dependent relationships, and the total for the system is $c(p - 1)$.

Subtracting the known or dependent relationships from the unknown ones gives the phase rule, a statement of the variance of a system:

$$F = p(c - 1) + 2 - c(p - 1)$$
$$F = c - p + 2$$

If the system is restricted by fixing an environmental variable such as pressure (known as an *isobaric* condition) or temperature (known as an *isothermal* condition), the number of unknown relationships is reduced by 1, and the phase rule becomes $F = c - p + 1$. This discussion assumes the system to be closed, that is, no material

enters or leaves. If the system is open to movement of mobile components (commonly fluid components such as water or carbon dioxide), the number of dependent variables is increased by the number of mobile components m. Thus

$$F = p(c - 1) + 2 - [c(p - 1) + m] = c - m - p + 2$$

Let us apply the phase rule to the one-component system already discussed (see Figure 4-1). If it is stated that three phases coexist within the system (as at points c), the phase rule indicates that the assemblage is invariant:

$$F = c - p + 2$$
$$F = 1 - 3 + 2$$
$$F = 0$$

Without reference to the diagram, it is obvious that a three-phase assemblage has no degrees of freedom and therefore can exist only at one specific temperature and pressure; this is, of course, true for all three-phase assemblages. Because there cannot be negative degrees of freedom, stable coexistence of more than three phases in a one-component system is impossible.

A two-phase assemblage (as at point b) can be shown to be univariant:

$$F = c - p + 2$$
$$F = 1 - 2 + 2$$
$$F = 1$$

The assemblage is maintained with change in one variable (either pressure or temperature) if a dependent change is made in temperature or pressure, respectively, so as to stay on the univariant line (as at b).

Divariant equilibrium is present in the one-phase regions of the diagram (such as point a), because

$$F = c - p + 2$$
$$F = 1 - 1 + 2$$
$$F = 2$$

Both pressure and temperature can be changed independently and the same one-phase assemblage is maintained.

TWO-COMPONENT SYSTEMS
• • • • • • • • • • • • • • • •

Geometric-Topologic Properties

The basic two-component or *binary* system with congruently melting phases is of significant interest to petrologists. **Congruent melting** means that the pure solid phase melts completely at the melting temperature to produce a liquid of the same composition as the solid. The simplest binary systems of this type are bounded at the ends by two pure solids and their behavior can be portrayed in a two-dimensional diagram that plots either temperature as a function of composition (isobaric diagram) or pressure as a function of composition (isothermal diagram), with the other environmental variable held constant. The exact arrangement of divariant areas, univariant lines, and invariant points is referred to as the *topology* of the diagram.

Igneous petrologists typically use the isobaric temperature-composition diagram (commonly referred to as an *isobaric T-X diagram*, where X refers to mole fraction, the compositional variable) because many magmas crystallize under constant pressure. Figure 4-2 shows the 1-bar isobaric phase relations in the binary system $CaMgSi_2O_6$–$CaAl_2Si_2O_8$ (diopside-anorthite), which is a useful analog for a simple basalt. Temperature increases upward on the ordinate, and composition is shown horizontally. Any point within the diagram represents a particular ratio (X) of diopside *component* to anorthite *component* at a particular temperature (T). The ratios of components are given as mole percentages in this diagram, but a weight percent scale can also be used. In Figure 4-2, X_{an} is the mole percentage of anorthite component, and therefore $X_{an} = n\ CaAl_2Si_2O_8/(n\ CaAl_2Si_2O_8 + n\ CaMgSi_2O_6)$, where n is the number of moles. On this scale, the mineral diopside has $X_{an} = 0$, and the mineral anorthite has $X_{an} = 1.0$.

There are three phases on this diagram: diopside (pure $CaMgSi_2O_6$) on the left side, anorthite (pure $CaAl_2Si_2O_8$) on the right side, and melt (variable ratio of the two components) in the middle. It is exceptionally important to keep in mind that $CaMgSi_2O_6$ and $CaAl_2Si_2O_8$ can exist *either* as mineral phases or as "molecular" chemical components mixed together in a melt. It must be specified whether minerals (phases) or components are meant when either the chemical compositions or the names diopside and anorthite are used. The crystal structures of the minerals control their compositions and determine that the *mineral* diopside is $CaMgSi_2O_6$ and the *mineral* anorthite is $CaAl_2Si_2O_8$. There is no such compositional restriction of the relatively structureless melt. Pressure is not shown on this isobaric diagram but could be represented by an extra coordinate at right angles to T and X; graphically, this could be shown in a more complicated three-dimensional perspective drawing.

The phase assemblages in the different areas of the diagram are labeled in Figure 4-2 for different combinations of T and X. The fields for diopside and anorthite solids are represented by the vertical lines at either end of the diagram, and the liquid field is bounded at lower temperatures by the two curved lines. The bulk composition shown at a (80% $CaAl_2Si_2O_8$ and 20% $CaMgSi_2O_6$) consists entirely of homogeneous liquid at 1600°C. At 1370°C, this same composition (b) is composed of liquid (at e) plus crystals of anorthite (at d); and at 1200°C (c), it contains no liquid but is a mixture of crystals of diopside (at f) and anorthite (at g). The determination of the phase assemblage on a binary phase diagram is simply a question of finding the proper T-X point on the diagram and reading off the phases that occur at either end of a horizontal line going to the nearest phase boundaries.

If this process is carried one step further, percentages and compositions of each of the phases present at points a, b, and c can be determined quantitatively. Point a at 1600°C is located in the one-phase field of liquid. Because the entire system consists of liquid, it is obvious that the liquid composition must equal the bulk composition, that is, 80% anorthite and 20% diopside. Point b at 1370°C represents a two-phase assemblage: liquid plus

FIGURE 4-2

• • • • • • • • • • • • • • • •

The simplified two-component system $CaMgSi_2O_6$–$CaAl_2Si_2O_8$ (diopside-anorthite) at 1 bar pressure. The melting point of diopside is at 1392°C, that of anorthite is at 1553°C, and the eutectic point is at 1274°C. Liquidus and solidus curves and eutectic point are indicated. Labeled points are described in the text.

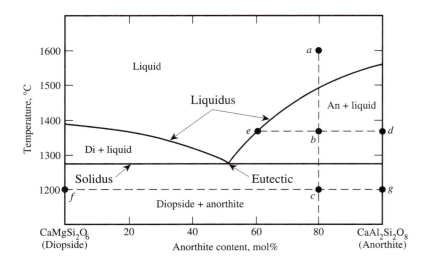

anorthite. The horizontal line (commonly called a *tie line*) through *b* connects the phase boundaries of liquid at *e* and anorthite at *d*. Remember that liquid can have any composition within the liquid field, whereas diopside and anorthite are constrained to be pure phases at the ends of the horizontal coordinate. The percentages of liquid and anorthite crystals can be determined from this horizontal line, keeping in mind two of the most fundamental properties of equilibrium phase diagrams. First, all phases in equilibrium must be at the same temperature and pressure and thus must lie on a horizontal line at a specific temperature. Second, *mass balance* in a closed system demands that the compositions of all the phases added together must equal the original unique bulk composition (Baldy's law: Some of it plus the rest of it equals all of it). Qualitatively, if point *b* were slightly to the right, the system would contain more anorthite crystals at 1370°C; if it were slightly to the left, there would be a greater proportion of liquid.

Exact percentages of liquid and crystals can be obtained by using the so-called *lever rule*, a geometric property of any tie line with fixed end points that represents a mass balance around an intermediate compositional point. The intermediate point represents the bulk composition (analogous to a fulcrum in physics) and either falls closer to one end than the other or is in the exact middle. The length of the tie line segment from the intermediate point to one end of the line is proportional to the amount of the phase on the *other* end. This principle is illustrated in Figure 4-2: The tie line segment from *b* to *d* divided by the total tie line length (*de*) represents the proportion of liquid, and the segment from *b* to *e* divided by length *de* represents the proportion of anorthite. The lengths of the tie line segments are perfect analogs of the compositional units used on the horizontal axis. In this case, the bulk composition is given by *b* (80 mol% $CaAl_2Si_2O_8$, 20 mol% $CaMgSi_2O_6$), the liquid composition by *e* (62 mol% $CaAl_2Si_2O_8$, 38 mol% $CaMgSi_2O_6$), and the anorthite composition by *d* (100 mol% $CaAl_2Si_2O_8$). The weight percentages of liquid *e* and anorthite *d* are given by where *b* falls on the line between them:

mol% anorthite (at *d*) = (length *be*/length *de*) × 100
mol% liquid (at *e*) = (length *bd*/length *de*) × 100

If we use actual measurements from Figure 4-2 (given as arbitrary length units, *l.u.*), then the result is

mol% anorthite = (1.6 *l.u.*/3.6 *l.u.*) × 100 = 44.4 mol%
mol% liquid = (2.0 *l.u.*/3.6 *l.u.*) × 100 = 55.6 mol%

(The reader is encouraged to double-check this by measuring the figure with a millimeter scale and converting arbitrary length units to millimeters.)

The lower limit of the divariant area representing the field of liquid alone is indicated in Figure 4-2 by the two curved lines that are known as the *liquidus lines*. They represent the lowest temperatures at which liquid of any composition can exist without beginning to crystallize. The single horizontal line that forms the upper boundary of the region in which only solids exist is called the *solidus*. Note that the solidus temperature is equal to the temperature at which the lowest temperature liquid can exist in the system. The single point where the two liquidus lines meet the solidus line is called the **eutectic point** and represents both the temperature and composition of the lowest temperature melt. To either side of the eutectic, the areas between liquidus and solidus lines are regions where liquid and one solid coexist in equilibrium if a *T-X* point falls within them.

Equilibrium Crystallization

The temperature at which crystallization begins in various compositions is indicated by the liquidus lines. The end points of these lines are the congruent melting points (or, alternatively, freezing points) of the pure phases that bound the system. The liquidus lines themselves represent the freezing point depression caused by adding a second component to a pure compound. A well-known example is the effect of adding salt to water. A little salt added to water causes the freezing point to be depressed slightly below 0°C; more salt lowers the freezing point even more. This freezing point lowering occurs on both sides of the system, of course, and the two liquidus curves are generated. The point where they cross, the eutectic point, represents the lowest temperature at which liquid (at the eutectic composition) can exist. This point is fixed in *T-X* space on an isobaric diagram and is therefore an invariant point. Why is it invariant? Recall that the phase rule for an isobaric binary system is

$$F = c - p + 1$$

In this situation, there are two components (a binary system) and three phases (anorthite, diopside, and melt). Therefore

$$F = 2 - 3 + 1$$
$$F = 0$$

The decrease in freezing temperatures due to admixture of a second component can be best understood in terms of the dynamics of crystal formation and dissociation on an atomic scale.

Equilibrium crystallization is the crystallization process that will be analyzed geometrically for various

FIGURE 4-3

• • • • • • • • • • • • • • • • •

Equilibrium crystallization within the simplified binary system $CaMgSi_2O_6$–$CaAl_2Si_2O_8$ (diopside-anorthite) at 1 bar pressure. Labeled points are described in the text.

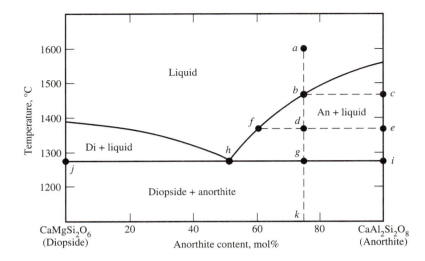

binary systems in this chapter. This must not be confused with the basic thermodynamic concept of equilibrium. Instead, it refers to maintenance of a mass balance in a closed system; that is, the crystals that have formed remain in contact with melt and continually equilibrate with it. Crystallization paths on T-X diagrams are therefore constrained by the bulk composition, especially the end points where liquid is all used up and crystallization ceases. In the first example, the binary system of two congruently melting minerals, any liquid within the system must end up at the eutectic before it finishes crystallizing, but we show later in more complicated systems that this is not always the case. In Chapter 6 we examine another more complicated process called **fractional crystallization** in which crystals are physically separated from the liquid, thus causing the system to become *open*, that is, to have a changing bulk composition.

The easiest approach to demonstrating an equilibrium crystallization path in this simplest binary system is to draw a vertical line down from a selected bulk composition (Figure 4-3); this line of constant composition and variable temperature (*ak*) is called an *isopleth*. The isopleth goes through the various phase fields of the diagram, and the sequence of assemblages that must occur during the crystallization of the magma can be read progressively downward. At any temperature in a two-phase field, the percentages of the phases present can be determined by using tie line calculations. The progressive assemblages encountered along the *ak* isopleth are liquid, anorthite + liquid, diopside + anorthite + liquid (at the eutectic temperature), and finally diopside + anorthite below the solidus.

The liquid a is cooled to the liquidus at b, where the first crystals of anorthite appear at c. With further cooling, the liquid composition moves down along the liquidus curve toward the eutectic, all the time crystallizing

anorthite. The movement of the liquid composition toward the eutectic can be appreciated by understanding that the removal of $CaAl_2Si_2O_8$ component from the liquid as anorthite crystals must drive the composition of remaining liquid toward enrichment in $CaMgSi_2O_6$. Note carefully that this effect is due solely to decrease in the amount of liquid and removal of $CaAl_2Si_2O_8$, *not* to any addition of $CaMgSi_2O_6$. The proportions of liquid and anorthite could be continuously monitored (for example, when the isopleth is at d) by examining the relative lengths of tie line segments from the isopleth to anorthite (*de*, amount of liquid) and from the isopleth to the liquidus (*df*, amount of anorthite). As cooling reaches the eutectic temperature at g, the anorthite crystals at i are in equilibrium with liquid at h, the eutectic composition. Diopside now begins to precipitate from the liquid as anorthite continues to crystallize. The proportions of anorthite and diopside crystallizing at the eutectic are given by the relative lengths of *jh* (anorthite) and *ih* (diopside), that is, the same as the eutectic liquid composition. These compositions must be the same because the liquid does not and cannot change its composition as crystals are precipitated at the eutectic. Crystallization continues without any further temperature change (because both pressure and temperature are fixed at an invariant point) until the liquid is completely consumed at 1280°C. The solid assemblage diopside + anorthite then cools from g to k, and ultimately to room temperature. The ratio of diopside to anorthite in the final rock is given by Di/An = gi/gj.

Some pedagogic concepts that substantially help in understanding the evolution of liquids and solids in phase diagrams have been developed by S. A. Morse (1980). These concepts are liquid composition (LC), total solid composition (TSC), and instantaneous solid composition (ISC). Figure 4-4 shows how a crystallization

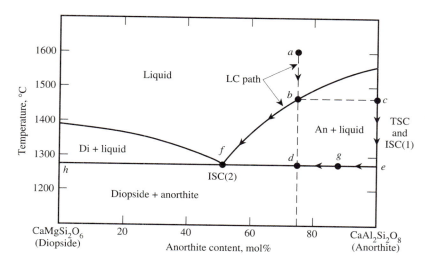

FIGURE 4-4

• • • • • • • • • • • • • • • •

Equilibrium crystallization sequence within the simplified binary system CaMgSi$_2$O$_6$–CaAl$_2$Si$_2$O$_8$ (diopside-anorthite) portrayed with the LC-TSC-ISC terminology (Morse 1980). The labeled points and paths are described in the text.

path like that already examined can be interpreted by using these new concepts. A magma at a cools until crystallization of anorthite begins at b, when the liquidus is reached. The LC follows the isopleth until point b and then proceeds along the liquidus with further cooling. When the liquid is at b, the ISC(1) is at c because only anorthite is crystallizing. For the entire cooling interval through the anorthite + liquid field, tie lines connect the LC with both ISC(1) and TSC, which are coincident between c and e. When the liquid (LC) reaches the eutectic, the ISC and TSC diverge, as ISC(2) is now at the eutectic composition and TSC is still at e. As eutectic crystallization proceeds, LC does not change but TSC moves horizontally across the solidus as diopside joins anorthite in the solid assemblage. The limit on TSC during equilibrium crystallization is point d, because the total solid assemblage obviously cannot become more diopside-rich than the isopleth. When TSC reaches d, the liquid is exhausted and crystallization is complete.

The TSC can be used to determine proportions of liquid, diopside, and anorthite during eutectic crystallization. For example, consider that TSC has reached point g. There are two pairs of tie lines that can be used. The position of the isopleth (d) relative to the TSC (g) and the eutectic liquid composition (f) provides information on proportions of liquid (tie line gd) and total solids (df)—in this case, roughly 50% of each. The relative proportions of diopside and anorthite within the total solids can be calculated from the position of the TSC (g) on the tie line from diopside to anorthite (he). The resultant tie line segments yield approximately 84% anorthite and 16% diopside in the 50% of the system that is solid. The total system composition is therefore 50% liquid, 42% anorthite, and 8% diopside.

TWO-COMPONENT SYSTEMS WITH AN INCONGRUENTLY MELTING PHASE
• • • • • • • • • • • • • •

Geometric-Topologic Properties

Incongruent melting occurs when a solid phase does not simply melt but breaks down to form both a liquid and another solid phase, neither having the same composition as the original phase. The compositions of the melt and new solid must, of course, add up to that of the melting phase to maintain mass balance. There are several binary systems of considerable petrologic interest in which an intermediate compound melts incongruently. The bounding solid phases are congruent in these systems, as was also the case in the basic binary system diopside-anorthite. In this section, we discuss the crystallization behavior of melts in binary systems that contain an incongruent solid.

The system forsterite-quartz (Mg_2SiO_4–SiO_2) is shown in Figure 4-5. The congruent bounding solid phases are forsterite and one of the SiO_2 polymorphs, cristobalite, tridymite, or quartz, depending on temperature. Enstatite ($MgSiO_3$) is the intermediate compound that melts incongruently. The composition and temperature limits of this phase are shown as the vertical solid line bc. Going from left to right across the diagram, the possible solid phase assemblages are forsterite alone, forsterite + enstatite, enstatite alone, enstatite + tridymite (or quartz), and tridymite (or quartz) alone. Notice that there is no stable divariant field for forsterite + tridymite (or quartz). These phases, when mixed, will react to form enstatite plus whichever of them is in excess.

FIGURE 4-5

• • • • • • • • • • • • • • •

Schematic, simplified *T-X* diagram for the binary system Mg_2SiO_4–SiO_2 (forsterite-quartz, cristobalite, tridymite). Note the incongruent melting behavior of enstatite and thus the generation of a peritectic, or reaction, point at *p*. Labeled points are described in the text.

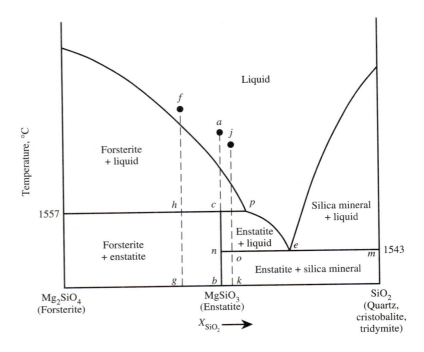

The topologic significance of divariant regions, univariant lines, and invariant points is the same as in previous figures. However, now there are two invariant points involving liquid: the eutectic (*e*) (1543°C) between enstatite and tridymite, and another point (*p*) at 1557°C. This new point (*p*) is called a **peritectic point.** It is quite different from the eutectic, because liquid can leave as cooling proceeds, moving down the liquidus to the eutectic point, which is always the lowest temperature liquid composition in the system. This peritectic liquid composition at (*p*) is the special composition produced by the incongruent melting of enstatite to liquid plus forsterite crystals. Alternatively, if crystallization is considered, it is possible to describe the process at the peritectic point as the reaction of liquid (*p*) with forsterite crystals to form enstatite crystals. The presence of a peritectic point therefore has considerable significance for the crystallization sequence.

Equilibrium Crystallization

Before crystallization paths in this type of system are discussed, it is important to stress one critical principle that applies to equilibrium crystallization in all the more complicated binary and ternary systems. In any of these systems with peritectic points, the final liquid composition will be at an invariant point but will *not necessarily* be at a eutectic. The final liquid must be the one that can coexist with the final crystalline assemblage, according to the rules for equilibrium crystallization. Therefore, the first thing to be done in interpreting crystallization in a

liquidus phase diagram is to examine what the isopleth indicates about the final assemblage.

For example, in Figure 4-5 there are three isopleths indicated: *ab, fg,* and *jk.* The ultimate solid phase assemblage for each can be predicted by examining each isopleth below its appropriate solidus. For isopleth *ab,* the solidus is at the temperature of the peritectic point, and the ultimate solid assemblage is enstatite alone. For *fg,* the solidus is the same, and the solid assemblage is forsterite + enstatite. The last liquid for each of these isopleths must be at the peritectic point, *p.* If any liquid were to become more silica-rich than *p,* it would have to move down the enstatite liquidus (curve *pe*) and ultimately crystallize some tridymite at the eutectic, and there can be no tridymite along either of these first two isopleths. The third isopleth, *jk,* has a solidus at the eutectic temperature, and the solid phase assemblage predicted is enstatite + tridymite. This liquid will pass through the peritectic point as it cools, but it must leave the peritectic and move down to the eutectic to crystallize tridymite before the liquid is exhausted. Interpretation of crystallization paths is straightforward if this basic principle of using the isopleth to predict the final solid assemblage, and thus the final liquid composition, is kept in mind.

Now consider the actual equilibrium crystallization sequence for each of the isopleths in Figure 4-5. For *ab,* the liquid will cool until it intersects the liquidus curve for forsterite and crystals begin to form. As cooling proceeds and forsterite crystals precipitate, the liquid composition shifts continuously down the liquidus until it finally reaches the peritectic point *p.* Peritectic points are commonly referred to as "reaction points" because that is exactly what now occurs. The liquid at p has become so relatively rich in silica that it cannot coexist with forsterite, and it therefore reacts with all the existing forsterite crystals, dissolving them and forming new crystals of enstatite. Because the isopleth has exactly the right proportions of Mg_2SiO_4 and SiO_2 to make a rock composed of 100% enstatite, the liquid must remain at p until it is used up at exactly the same instant that the last bit of forsterite is converted into enstatite.

The early crystallization sequence for isopleth fg is much the same as for ab, except that crystallization begins at a slightly higher temperature. The same reaction begins when the liquid reaches the peritectic point. However, this isopleth indicates that there must be some forsterite in the final "rock." Therefore the reaction does not proceed to completion as before, but the liquid is used up well before all the forsterite is consumed, and the final solid assemblage is approximately 25% forsterite and 75% enstatite.

For the third isopleth, jk, crystallization of forsterite begins at a lower temperature and continues until the peritectic is reached. Again, reaction begins and forsterite is converted to enstatite. But this isopleth shows that the final assemblage has no forsterite; instead it is enstatite + tridymite. The peritectic reaction must therefore go to completion, at which point the remaining liquid coexists only with enstatite. The liquid composition leaves the peritectic point and shifts down the enstatite liquidus toward the eutectic. At the eutectic, tridymite joins enstatite in crystallizing until the liquid is consumed and the final solid assemblage is about 87% enstatite and 13% tridymite. This estimate is derived from the position of the jk isopleth relative to the compositions of enstatite and silica. The proportion of enstatite is thus represented by the length of line om divided by the length of nm; the proportion of tridymite is represented by on divided by nm.

The ISC, TSC, and LC concepts of Morse also can be used to describe these sequences. The use of these concepts for simple equilibrium crystallization may seem redundant, but the value will become apparent when more complex systems and melting behavior are encountered. For all three isopleths, ISC and TSC move down along the forsterite composition line while the three liquids are crystallizing only forsterite. Again, for all three isopleths, the ISC jumps to the enstatite composition when the LC reaches the peritectic because that is the instantaneous composition of crystalline material being produced. Adding this new ISC to the TSC at d pulls the TSC to the right along the de line until it reaches either the isopleth or the enstatite composition. For isopleth fg, the liquid is used up when the TSC reaches h; and for isopleth ab, liquid is totally consumed when TSC

reaches c. For the third isopleth, jk, there is still liquid left when the TSC reaches c. As the liquid moves down the enstatite liquidus, the TSC and ISC move downtemperature along the line cb, because the only solid coexisting with liquid is enstatite. When the liquid reaches the eutectic, the ISC now jumps to the eutectic composition because both enstatite and tridymite are crystallizing. The addition of this new ISC to the TSC at n shifts the TSC to the right along the nm line until it intersects the isopleth at o, and the final liquid is consumed.

TWO-COMPONENT SYSTEMS WITH COMPLETE SOLID SOLUTION

The systems discussed so far contained no solid solution between the various mineral phases, but most natural minerals show at least partial and even complete solid solution behavior. The plagioclase system $NaAlSi_3O_8-CaAl_2Si_2O_8$ shows essentially complete mixing of the end members, at least at high temperatures. A bulk composition chosen between the end members will crystallize as a single phase. As noted earlier in the book, such intermediate compositions are indicated as molecular percentages of the end members, as in $Ab_{10}An_{90}$, or more simply An_{90}. The crystallization behavior of liquids within this chemical system is shown in Figure 4-6. There are two regions containing only one phase—a liquid field at high temperature and a plagioclase field at low temperature—separated by a two-phase crystalliquid "solid solution" loop. A melt of composition a is a single-phase liquid of composition An_{65}; a crystal at point b is a single-phase plagioclase of composition An_{65}. Point c along this isopleth consists of about 40% liquid of composition d and 60% plagioclase of composition e. The upper boundary of the two-phase region is the liquidus and the lower one is the solidus; they represent the lower and upper temperature limits, respectively, for a single phase of any composition.

Equilibrium crystallization within this system is straightforward. For the isopleth shown, liquid a will cool until it intersects the liquidus at f. The plagioclase composition in equilibrium with this liquid is g, as indicated by the intersection of a tie line from the liquidus to the solidus. Removal of this more calcic plagioclase from liquid at f will drive the liquid toward enrichment in $NaAlSi_3O_8$ as it cools along the liquidus. As liquid compositions progressively move down the liquidus, the coexisting plagioclase must become more sodic as well, as its composition moves along the solidus. Proportions of liquid and crystals can be determined by the relative tie line segment lengths. When the plagioclase composition

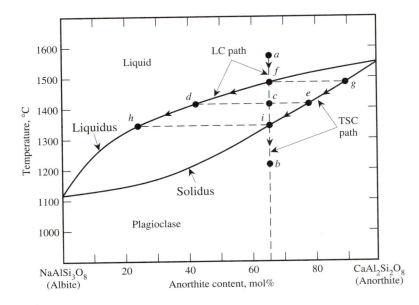

FIGURE 4-6

● ● ● ● ● ● ● ● ● ● ● ● ● ● ● ●

The binary system NaAlSi$_3$O$_8$–CaAl$_2$Si$_2$O$_8$ (albite-anorthite) at 1 bar pressure. Liquidus and solidus curves are labeled. Albite melts at 1118°C and anorthite at 1553°C. Labeled points and LC and TSC paths are explained in the text. [Based on experiments reported by N. L. Bowen, 1913, *Am. J. Sci.*, *4*, 35, Fig. 1.]

reaches the isopleth at *i*, the last bit of liquid is consumed at composition *h*. Using the alternative way of describing the crystallization path, we say that the LC follows the isopleth from *a* to *f*, then proceeds along the liquidus to *d* and *h*. The TSC starts at *g*, then moves down along the solidus to *e* and finally *i*. The interpretation of an ISC in this type of diagram is tricky, but it could be considered to be infinitesimally further along the solidus than the TSC, and thus "pulls" the TSC along toward the isopleth.

Ideal equilibrium crystallization in this type of diagram assumes a constant shift in the composition of plagioclase from the first calcic seeds to the more sodic final composition and implies compositionally homogeneous plagioclase crystals during the whole process. Such re-equilibration of the interior plagioclase compositions in larger crystals would require diffusion of the plagioclase components through the crystal structures in the solid state. This diffusion process is very slow in plagioclase, and thus ideal or perfect equilibrium crystallization behavior may actually be unachievable except under conditions of unrealistically slow cooling rates, although some plutonic rocks have nearly homogeneous plagioclase. However, many natural plagioclases, and particularly those in volcanic or hypabyssal rocks such as andesites, show chemical zoning that is due to the inability of the diffusion process to homogenize the crystals (Figure 4-7). Only the outermost portion of the crystal

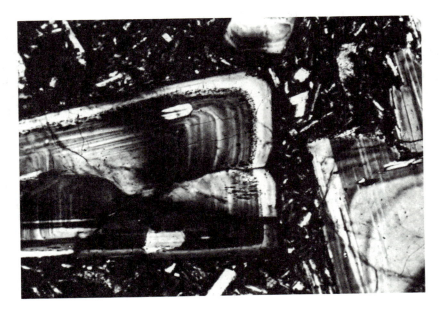

FIGURE 4-7

● ● ● ● ● ● ● ● ● ● ● ● ● ● ● ●

Oscillatory zoning of plagioclase phenocrysts in andesite. Compositional zoning as revealed by differences in extinction angles shows changes in external grain shape during growth. Polysynthetic twin bands parallel the longer grain edges. Sample from Washoe, Nevada; width of field of view is about 0.3 mm.

is truly in equilibrium with the liquid at any point during crystallization, and a more calcic core remains. This zoning in plagioclase from a calcic core to sodic rim is referred to as normal zoning. Reversed zoning can also occur; in this case, rims are more calcic than cores. Oscillatory zoning occurs when plagioclase crystals show bands of alternating Ca:Na ratio (see Figure 4-7). Interpretations of reversed and oscillatory zoning generally involve models of injections of new batches of more calcic magma into a magma chamber.

TERNARY SYSTEMS

Simple Ternary Systems

Ternary systems present a special problem in graphical representation because an extra dimension is needed to represent the extra compositional variable. Compositions of the three components in ternary systems are typically represented at the corners of an equilateral triangle (Figure 4-8). Each corner represents 100% of that com-

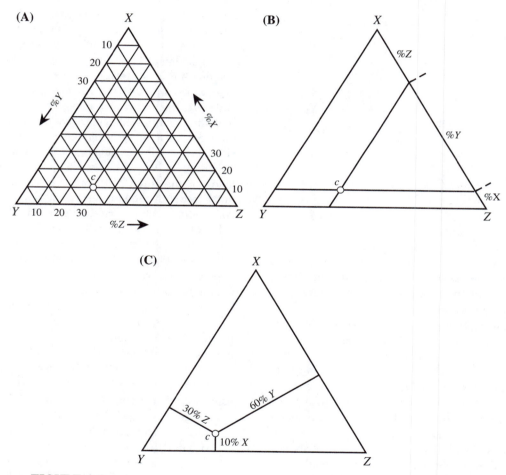

FIGURE 4-8

Methods for plotting the position of a point c in a triangular diagram. **(A)** Plotted on triangular graph paper. Percentages of components X, Y, and Z are read directly from the labeled coordinate axes. **(B)** Determination of c by the two-line method. Two lines are drawn through c parallel to any two of the sides of the triangle (here XY and YZ). The intersection of these two lines with the third side (XZ) divides that side into three parts whose lengths are proportional to the relative amounts of components X, Y, and Z. **(C)** Determination of point c by construction of perpendicular lines through c to each of the sides of the triangle. The relative lengths of these three perpendicular lines yield the percentages of components X, Y, and Z.

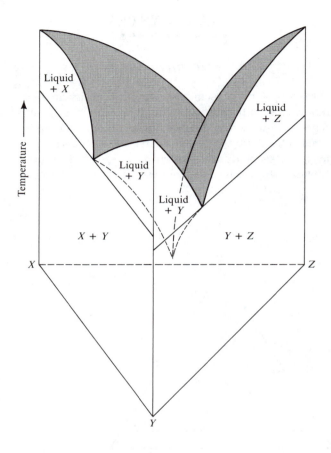

FIGURE 4-9
● ● ● ● ● ● ● ● ● ● ● ● ● ● ● ● ●

Temperature-composition model of the general ternary system
XYZ. Each of the three vertical sides is a binary *T-X* system. The
shaded top shows the ternary liquidus surfaces. Dashed lines are
the three cotectics that lead downward to the ternary eutectic.

ponent; points along the edges consist of mixtures of the
two components at the ends of the line; points within the
triangle indicate mixtures of all three components.
Ternary systems can be viewed as three individual binary
systems joined together, for example, the systems *XY*,
XZ, and *YZ* in Figure 4-9.

If both dimensions on paper are used for compositional representation in a triangle, a temperature coordinate can be erected perpendicular to the composition
triangle. A perspective view of this is shown in Figure
4-9, with the liquidus for each mineral now expressed as
a *surface* rather than as a curved line because of the
extra compositional dimension. The liquidus and solidus
lines for *XY* and *YZ* are shown on their respective sides
of the triangle. The liquidus lines of the binaries are connected in the third dimension to form the liquidus *surfaces* that are shown as the shaded upper portion of
Figure 4-9.

Accurate representation of phase relations in such
perspective drawings would be very difficult, so ternary
liquidus systems are shown with the liquidus surfaces
contoured with lines of equal temperature (isotherms) in
the same manner that a topographic map is contoured
for elevation. The representation is as if the observer is
looking down at a three-dimensional object. Figure 4-9
shows this for the simplest ternary system, which is com-

posed of three simple binary eutectic systems, in this
case *XY*, *XZ*, and *YZ*.

When the liquidus surfaces are contoured for temperature, one obtains the standard ternary representation that is usually used in the petrologic literature.
Such a system is shown in Figure 4-10, the system
$CaMgSi_2O_6 - Mg_2SiO_4 - CaAl_2Si_2O_8$ (diopside-forsterite-
anorthite) at 1 bar, which represents an idealized olivine
basalt. The simplified ternary phase diagram in Figure
4-10 shows the three liquidus surfaces of diopside, forsterite, and anorthite separated by heavy lines where the
liquidus surfaces intersect. These lines are called *boundary curves* or **cotectic lines.** Their high-temperature
ends are the individual binary eutectics in the three
bounding binary systems. (Using the topographic meta-
phor for the contoured liquidus surfaces, we can consider these lines as analogs for stream valleys.) The
boundary curves separate the primary crystallization
fields of the three solid phases, which are labeled to
indicate which solid phase crystallizes first for any liquid
composition falling within the field. When cooled, a
liquid of composition *a* will intersect the liquidus surface
for forsterite at 1700°C and begin to crystallize forsterite.
Liquid *b* will begin crystallization of anorthite at 1400°C.
A liquid that coincidentally lies exactly on one of the
boundary curves will simultaneously begin crystallizing
both phases — in this case, diopside and forsterite. A
liquid at *e*, the intersection of all three primary phase
fields, will simultaneously begin to crystallize all three
phases. This minimum-temperature point is called the
ternary eutectic. Note that the isotherm intersections on
the boundary curves point upward (thermally) in the
same way that topographic contours "V" upstream on a
map. Arrows are typically added to the boundary curves
to indicate the direction of decreasing temperature.

Consider the cooling and equilibrium crystallization
of a ternary melt such as *a*. It remains completely liquid
until it has cooled to 1700°C; the melt intersects the
forsterite liquidus surface and crystals of forsterite begin
to form. The liquid is depleted in forsterite component
and thus moves directly away from the forsterite corner
toward the diopside-forsterite boundary curve along line
ah. This *liquid descent line* is drawn by extending a
line from the forsterite corner through the original melt
composition and on to the boundary curve. When the
evolving melt composition reaches the boundary curve
of the diopside field at *h*, diopside joins forsterite in crys-

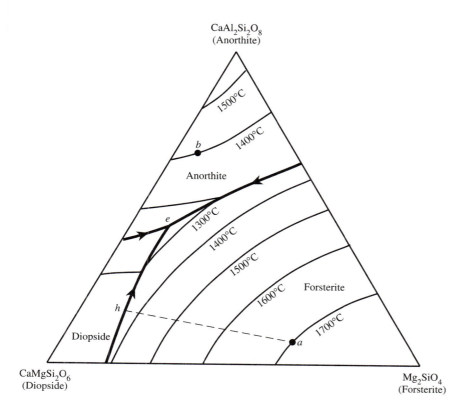

FIGURE 4-10

Simplified ternary diagram for the system $CaMgSi_2O_6$–Mg_2SiO_4–$CaAl_2Si_2O_8$ (diopside-forsterite-anorthite) at 1 bar pressure, eliminating the pseudoternary field of spinel. Labeled points are described in the text. The eutectic point at e is at 1270°C. [Adapted from E. F. Osborn and D. B. Tait, 1952, *Am. J. Sci.*, Bowen Volume, Fig. 4.]

tallizing. The melt now moves "down" (toward a lower temperature) the boundary curve and directly away from a point on the diopside-forsterite binary that represents the proportions of these two phases that are coprecipitating. The degree of curvature of a boundary curve reflects the changing proportions of **coprecipitating** phases; at any instant, the tangent line to the curve at the melt composition projects back to the instantaneous solid composition. The liquid further evolves by cooling and moving toward the ternary eutectic, which it eventually reaches. At this point, all three solid phases coprecipitate, and their proportions are exactly those of the eutectic composition. The system remains at eutectic temperature until all the liquid is consumed, and the proportions of solid phases in the crystallized material are those of the initial magma composition a — in this case, roughly 60% forsterite, 35% diopside, and 5% anorthite.

In the alternative view of crystallization, the LC is at a until the melt cools to 1700°C. When crystallization of forsterite begins, both the ISC and the TSC are at the forsterite corner of the ternary (Figure 4-11) and remain there until the liquid intersects the boundary curve at h. When diopside begins crystallizing, the ISC jumps to the point on the diopside-forsterite edge where the tangent to the boundary curve at h intersects. Addition of this ISC to the TSC draws the TSC from the forsterite corner toward diopside. As liquid (LC) moves down the boundary curve, the ISC will shift slightly and progressively

toward the diopside corner as a result of the slight curvature in the curve. While LC moves toward the eutectic, TSC moves further along the diopside-forsterite edge. The limit on its movement is the line from the eutectic through a to its intersection with the diopside-forsterite edge, because the tie line that connects the TSC with the LC must equal the bulk composition and therefore must always pass through point a (see the two tie lines in Figure 4-11). Once the liquid reaches the eutectic, the ISC now jumps to the eutectic composition and begins to "pull" the TSC along the tie line toward a. Of course, when the TSC reaches a, there is no remaining eutectic liquid and the system is entirely crystallized.

Complex Ternary Systems

Ternary systems are commonly more complex than the relatively simple one just illustrated. Some of these complexities arise from the presence of compounds of intermediate composition (either binary or ternary), incongruent melting, and the effects of solid solution. A reader interested in the details of advanced ternary systems is referred to Morse (1980). A useful introduction to the characteristics of more complicated diagrams can be obtained from consideration of the ternary system $CaMgSi_2O_6$–$NaAlSi_3O_8$–$CaAl_2Si_2O_8$ (diopside-albite-anorthite) (Figures 4-12A and B). This system is a

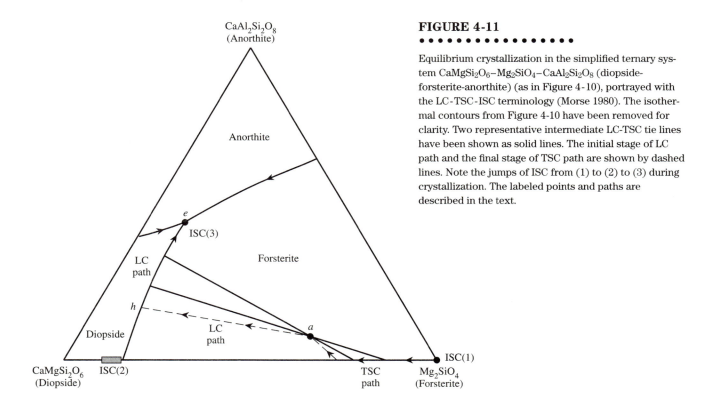

FIGURE 4-11
● ● ● ● ● ● ● ● ● ● ● ● ● ● ● ● ● ●

Equilibrium crystallization in the simplified ternary system $CaMgSi_2O_6$–Mg_2SiO_4–$CaAl_2Si_2O_8$ (diopside-forsterite-anorthite) (as in Figure 4-10), portrayed with the LC-TSC-ISC terminology (Morse 1980). The isothermal contours from Figure 4-10 have been removed for clarity. Two representative intermediate LC-TSC tie lines have been shown as solid lines. The initial stage of LC path and the final stage of TSC path are shown by dashed lines. Note the jumps of ISC from (1) to (2) to (3) during crystallization. The labeled points and paths are described in the text.

good approximation of simple basaltic magmas, from which the minerals diopside and plagioclase crystallize in the P-T range characteristic of the earth's crust.

In the three bounding binary systems (Di-An, Di-Ab, and An-Ab), there is complete solid solution between albite and anorthite end members, but none between diopside and plagioclase of any composition. Therefore, any melt bulk composition within the ternary should cool and crystallize to two solid phases, diopside plus plagioclase of appropriate Ca:Na ratio. The final proportions of diopside and plagioclase are calculated (as in binary diagrams) by the relative lengths of the tie lines from the original liquid composition to composition points of the two phases crystallized. For example, liquid e in Figure 4-12A will crystallize to a final solid assemblage of about 80% plagioclase (An_{50}) and 20% diopside.

Note that there is only one boundary curve in the ternary diagram, which has this geometry because there can be only one two-phase boundary when there are only two solid phases possible within the system. In other words, with a maximum of three coexisting phases possible (diopside + plagioclase + melt), the lowest phase rule variance is one. The phase rule for an isobaric system is defined as

$$F = c - p + 1$$

There are three components ($CaMgSi_2O_6$, $NaAlSi_3O_8$, and $CaAl_2Si_2O_8$) and three phases (diopside, plagioclase, melt), so

$$F = 3 - 3 + 1 = 1$$

There cannot be any invariant points such as eutectics or peritectics in the interior of this type of system. Keep in mind, however, that the lowest temperature point on the boundary curve (labeled as 1085°C) could be considered invariant because a liquid there is fixed in both composition and temperature. The trick to the phase rule analysis of this point is that the 1085°C point is actually the *binary* eutectic on the albite-diopside join, and therefore $F = 2 - 3 + 1 = 0$. Because this system is isobaric, any liquid composition on the boundary curve can only coexist with a single plagioclase composition (plus diopside); thus liquid a coexists with plagioclase b and diopside (Figure 4-12B). The actual compositions of plagioclase coexisting with liquids on the plagioclase liquidus or on the boundary curve (as shown by the dashed lines in Figure 4-12) can only be determined by experiment; they cannot be predicted geometrically from the diagram. If liquid a is the liquid in stable coexistence with plagioclase b, then initial melts such as e or f must have their final liquid composition at a, as discussed below.

(A)

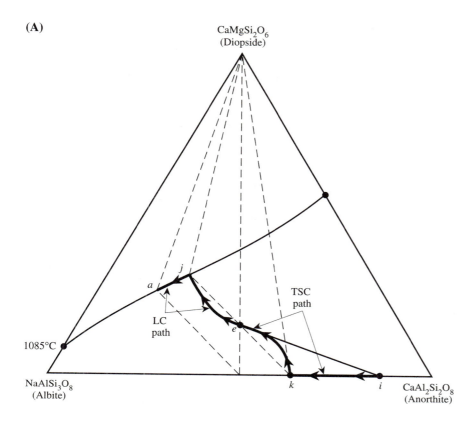

(B)

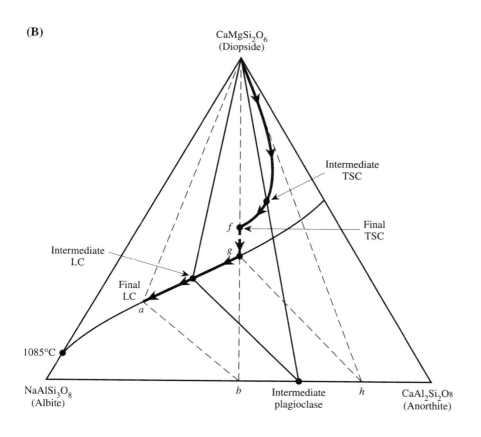

FIGURE 4-12

● ● ● ● ● ● ● ● ● ● ● ● ● ● ● ● ●

(A) Equilibrium crystallization in the complex ternary system $CaMgSi_2O_6$–$CaAl_2Si_2O_8$–$NaAlSi_3O_8$ (diopside-anorthite-albite), portrayed with the LC-TSC method (Morse 1980) for a melt of composition e initially saturated with plagioclase. LC path is from e to j to a; TSC path is from i to k to e. Solid line shows tie line from melt to initial plagioclase (i), and the two dashed triangles show the initial (right) and final (left) three-phase triangles. **(B)** Equilibrium crystallization of a melt composition (f) saturated with diopside. LC path is from f to g to a; TSC path is from diopside to f. Dashed triangles are the initial (right) and final (left) three-phase triangles. The solid triangle shows one intermediate three-phase triangle, along with the LC, TSC, and plagioclase composition for this stage in crystallization.

The crystallization of a liquid at *e* in Figure 4-12A is somewhat complicated because of plagioclase solid solution effects during primary crystallization on the plagioclase liquidus surface. Cooling of liquid *e* results in intersection with the plagioclase liquidus and the beginning of crystallization of a very calcic plagioclase at *i* (determinable only by experiment). As this plagioclase is removed, the liquid moves directly away from it down the liquidus surface (labeled LC path). The liquid descent path must be curved, because, as the liquid cools and becomes more sodic, the plagioclase composition is also constantly shifting to the left toward *k* (labeled TSC path). The convex-downward curved LC path shown in Figure 4-12A is thus composed of a very large number (approaching infinity) of infinitely short, straight increments that progressively become more nearly vertical as the boundary curve (cotectic) is approached. The LC path must intersect the boundary curve at *j* because liquid of this composition and plagioclase at *k* constitute a tie line passing through the initial liquid composition *e*. When the liquid reaches *j*, diopside (in addition to plagioclase) begins to crystallize and the LC path now follows the cotectic from *j* toward *a*. As diopside joins plagioclase in the solids, the TSC path leaves the plagioclase join at *k* and shifts toward *e*. The TSC path is curved as shown because it constantly trends toward the ISC, which is approximately at the midpoint of the plagioclase-diopside tie line. The end of the TSC path must be at *e*, because this is the bulk composition of the initial magma; when TSC arrives at *e*, the LC is at *a* and crystallization is complete.

Now consider the equilibrium crystallization of liquid *f* (Figure 4-12B). Cooling to the liquidus surface, this melt begins to crystallize diopside. As diopside component is removed, the liquid composition shifts directly away from the diopside corner in a straight line and down-temperature to the cotectic at *g*. When LC arrives at *g*, plagioclase joins diopside in crystallizing. Experimentally determined tie lines from cotectic liquids to plagioclase indicate that plagioclase at *h* coexists with liquid at *g*. The triangle Di*hg* is the initial three-phase triangle for cotectic crystallization. The ISC now being removed from the liquid lies about midway along a line from plagioclase *h* to the diopside corner. As melt becomes depleted in this ISC, LC moves directly away down the cotectic toward *a*, as shown by the arrows; an intermediate LC is labeled. As the liquid cools and becomes more sodic, the crystallizing plagioclase becomes more sodic as well, as indicated by the dashed tie lines in Figure 4-12B. When the liquid reaches *a*, the tie line from plagioclase *b* to diopside passes through the bulk composition, indicating that the proportion of liquid has dropped to zero and that the system consists entirely of diopside + plagioclase of composition *b*. The TSC path

begins at diopside and only leaves this composition when the LC reaches *g* and plagioclase crystallizes. The addition of this new solid mix to the already crystallized diopside draws the TSC down toward *h*. The TSC remains on the diopside-plagioclase tie line as LC evolves down the cotectic, but as this tie line sweeps to the left, so must the TSC. This progressive shift causes the TSC to move in a curved path (Figure 4-12B) until it intersects point *f* at the same instant that liquid is at *a* and plagioclase at *b*. When TSC equals the initial liquid composition, the liquid is all used up.

To recapitulate: When the liquid at *f* has evolved down to the boundary curve, the system consists of three phases (liquid + plagioclase of composition *h* + diopside) that constitute the corners of a subtriangle within the overall triangular system. At the instant that the LC reaches the boundary curve and plagioclase first begins to crystallize, the original liquid composition (and bulk composition) *f* lies almost exactly on one edge of the triangle — that from diopside to liquid. As liquid and plagioclase compositions both become more sodic during cooling, the triangle sweeps toward the left, pivoting on the diopside corner of the ternary, and both plagioclase and diopside increase at the expense of liquid. During this whole evolution, the original liquid composition lies within the triangle, this position indicating that all three phases are present (their exact proportions can in fact be calculated by using principles from Figure 4-8). When the liquid reaches *a* and the plagioclase reaches *b*, the plagioclase-diopside leg of the triangle now intersects the bulk composition, thus indicating that liquid is totally exhausted.

In both of the above examples, continuous reaction-equilibration between liquids and plagioclase has been assumed. With incomplete reaction (due to rapid cooling, removal of early-formed crystals, and so on), a variety of plagioclase compositions will result (for example, preserved calcic cores in zoned crystals). The liquid in such a nonequilibrium case will be enriched in sodium and may evolve further along the boundary curve than its predicted composition under equilibrium crystallization. This scenario is examined in Chapter 6.

SUMMARY
• • • • • • • • • • • • • • •

Experimentally determined phase diagrams of binary, ternary, and more complex systems support critical models for magma crystallization processes that cannot be directly observed in nature. They show the sequences of minerals that precipitate from melts of various compositions, information that is quite useful in interpreting the meaning of many igneous textures. The phase rule is a basic tool in describing the behavior of phases within a chemical system. The phase assemblages in one- and two-component systems are determined by the particular combination of temperature and pressure to which the system is subjected. Within systems of two or more components, melting can be congruent or incongruent. Crystallization behavior is strongly influenced by the amount of solid solution between solid phases and by the formation of compounds of intermediate composition.

STUDY EXERCISES
• • • • • • • • • • • • • • •

1. Explain the concept of equilibrium as it applies to igneous processes and igneous rocks.
2. In binary temperature (T)-composition (X) diagrams such as Figure 4-2, why are compositions of solid minerals represented by vertical lines, whereas melts can have compositions across the width of the diagram? Why are the two variables T and X chosen for this and most igneous diagrams?
3. In Figure 4-3 or 4-4, pick a composition that lies above the diopside-liquidus curve and work out the equilibrium crystallization of this melt using both the traditional and the LC-ISC-TSC approaches.
4. In Figure 4-6, for bulk composition at the temperature ($\approx 1420°C$) where LC is at d and TSC is at e, what are the relative proportions of melt and plagioclase crystals? (Use a measuring scale for this lever-rule calculation.)
5. What are the relative proportions of the components diopside, forsterite, and anorthite in composition b

shown in Figure 4-10? (*Hint:* Use Figure 4-8B or 4-8C and measure using a scale.)
6. There are three cotectic lines in the diopside-forsterite-anorthite ternary system (Figure 4-10) but only one in the ternary system albite-anorthite-diopside (Figure 4-12). Why? How many ternary eutectic points are there in each system?

REFERENCES AND ADDITIONAL READINGS
• • • • • • • • • • • • • •

Bowen, N. L. 1915. The crystallization of haplobasaltic, haplodioritic and related magmas. *Am. J. Sci.*, *40*, 161–185.

Ehlers, E. G. 1972. *The Interpretation of Geological Phase Diagrams*. San Francisco: W. H. Freeman.

Griffin, D. T. 1992. *Silicate Crystal Chemistry*. New York: Oxford University Press.

Hall, J. 1805. Experiments on whin stone and lava. *Trans. R. Soc. Edinburgh*, 5, 43–75.

Maaloe, S. 1985. *Principles of Igneous Petrology*. New York: Springer-Verlag.

Morse, S. A. 1980. *Basalts and Phase Diagrams*. New York: Springer-Verlag.

Osborn, E. F., and D. B. Tait. 1952. The system diopside-forsterite-anorthite. *Am. J. Sci.*, *250A*, 413–433.

Philpotts, A. R. 1990. *Principles of Igneous and Metamorphic Petrology*. Englewood Cliffs, NJ: Prentice Hall.

Chapter 5

ORIGIN OF MAGMAS BY MELTING OF THE MANTLE AND CRUST

Evidence indicates that virtually all magmas have been generated within the outermost 250 km of the earth by the melting of solid mineral assemblages in either the crust or the upper mantle. Melting occurs when an assemblage of solid minerals (typically including a fluid phase) reaches a temperature that allows minerals and fluid to form a silicate melt. The temperature required for this melting may be as low as 650°C for feldspathic sandstones in the crust in the presence of water to as high as 1200° or 1300°C or more for melting of dry mantle peridotite. In the crust, melting is typically the end result of very high grade metamorphism during orogeny. Conversely, mantle melting is more likely to be anorogenic and related either to vertical movement and pressure release (for example, during mantle convection) or to involvement of aqueous and carbonic fluids. In Chapters 7–10, the actual scenarios for mantle and crustal melting are explored in detail.

Understanding the processes that generate magmas first requires an analysis of the phase relations involved in melting, to enable investigators to interpret the wide range of magma compositions. Pressure becomes an important physical variable when considering melting, in contast to crystallization. The crystallization of many volcanic and hypabyssal magmas takes place on or near the surface at very low pressure (less than 2 kbar). All but the most deeply exhumed exposed plutonic rocks have generally crystallized within the uppermost 15 to 20 km of the lithosphere, a depth range that corresponds to pressures of 5 to 7 kbar. However, melting and magma formation take place over a much greater range of depths, from as little as 10 km to well over 100 km, and thus at pressures from 3 or 4 kbar to many tens of kilobars. Both the temperatures of melting and the compositions of the magmas derived from melting of the same parental material can vary significantly at different pressures. One of the most important ongoing tasks of igneous petrology is to analyze these magma-forming processes. In this chapter we examine the theoretical phase relations involved in melting in binary and ternary systems and then discuss how these can be used to interpret the origins of natural magmas.

MELTING IN BINARY SYSTEMS

Simple Binary Eutectic Systems

The interpretation of melting in a simple binary eutectic system is relatively straightforward. Some complexity is introduced, however, because of the necessity of treating melting as either an equilibrium or a fractionation phenomenon. In this sense, **equilibrium melting** means the melting of a multiphase solid material through a process that generates a continuous succession of melts ranging from initial, low-temperature, eutectic liquids quite distinct in composition from the parental solid material to a final melt that equals the bulk composition of the parent. (The actual liquid evolution path on a phase diagram is exactly the reverse of that during equilibrium crystallization.) The chemical system during equilibrium melting consists at all times of *both* residual solid phases and a silicate melt, and the sum of the two compositions always equals the original bulk composition. This bulk compositional constraint or balance is referred to as a *mass balance*.

Fractional melting is a process in which melt is assumed to be continuously removed from the remaining solid material, with the result that the chemical system is divided in two: One part consists only of melt and the other consists only of residual solids, and the two are assumed to be separate. Because of this division of the rock into two fractions, bulk composition does not have the same meaning that it had in equilibrium melting, and there is no bulk composition constraint or mass balance. In equilibrium and fractional melting scenarios, only the initial melt compositions coincide; for advanced melting, the melt compositions typically diverge considerably, with great implications for magma origins. A combination of these two melting processes (batch melting) can also occur and is discussed in some detail at the end of the chapter.

In discussing phase relations involved in melting, the same systems that served as examples of equilibrium crystallization in the last chapter will be used again. The first of these is the simple binary eutectic system $CaMgSi_2O_6$–$CaAl_2Si_2O_8$ (diopside-anorthite) (Figure 5-1A). Melting involves paths that go *upward* through the diagram, as temperature is increased. When a mixture of 75% anorthite crystals and 25% diopside crystals is heated, no melting occurs until the eutectic temperature is reached at about 1280°C. One very important rule must be remembered: Melting in *any* system starts at the temperature and composition of the invariant point that represents stable coexistence of the initial solid phase assemblage plus liquid, typically a eutectic but possibly a peritectic (see later). In the present system, this point is the 1280°C eutectic at e, where the first melt will appear. (In a practical sense, this first melt could be viewed as a very thin film of silicate liquid that appears along the grain boundaries separating diopside and anorthite grains.)

At the instant that liquid first appears, a horizontal tie line is established from liquid at e to both diopside and anorthite, reflecting the invariant three-phase assemblage. This tie line coincides with the solidus line. The temperature cannot increase until one solid phase is gone and the system becomes univariant. Which solid phase will be lost can be predicted by noting that the initial solid composition has a lower mole percentage of diopside than the eutectic liquid does. Therefore diopside will be totally depleted from the solids before anorthite is used up. Considering the liquid composition (LC) to be at e and the total solid composition (TSC) at a when melting begins, the formation of melt will drive the TSC (commonly called the *residuum composition*) to the right along the solidus line until it intersects the anorthite composition. The tie line segment that indicates the proportion of diopside (initially ab) has thus shrunk to zero length, indicating total diopside depletion and a new two-phase tie line from liquid to anorthite.

With a two-phase univariant assemblage, the temperature is free to rise as further melting occurs. The LC will thus proceed up the anorthite liquidus as anorthite melts and enriches the liquid in $CaAl_2Si_2O_8$. The TSC moves vertically up the anorthite line so that its temperature matches that of the liquid (as it must do, according to the rules for equilibrium). Eventually the LC approaches the initial solid composition, the tie line segment indicating proportion of anorthite (initially ae) has decreased to zero, and the system is entirely melted. Once the last

crystal of anorthite in the residuum is gone, the liquid cannot change its composition further and, with additional heating, will simply rise in temperature. Note that both the LC and the TSC paths are *exactly* the reverse of these paths during cooling and equilibrium crystallization of liquid x.

For fractional melting, the system must be considered differently because of the absence of what has so far been a fundamental constraint—that the compositions of all phases must add up to the initial bulk composition. If liquid is continuously removed, then the bulk composition *must* change, and the evolution of liquid composition is quite different from the equilibrium case.

Consider fractional melting of the same initial solid as before, with 75% anorthite and 25% diopside (Figure 5-1B). The initiation of melting follows the same sequence as in the equilibrium case, with heating to 1280°C and the appearance of melt at the eutectic. Because liquid cannot continuously equilibrate with solids in this case, it is necessary to introduce the concept of *instantaneous liquid composition* to express the composition of liquid being produced at any instant and simultaneously removed from the system. The initial liquid composition [LC(1)] is at the eutectic (e) and remains there until one of the two solid phases disappears. The TSC or residuum composition is driven along the solidus line toward anorthite by the removal of eutectic liquid until diopside is totally depleted. At this point in fractional melting, all melting stops because all liquid produced so far has been removed from the system, and all that is left are crystals of anorthite. In contrast to the equilibrium melting case, there is no liquid present into which to dissolve the anorthite crystals. No more melting can occur until the melting temperature of pure anorthite is reached, producing LC(2), at about 1550°C (c).

Morse (1980) refers to the jumps in temperature and composition of the LC as liquid "hop" (refer to Figure 5-1B). Note that this liquid hop is from the binary invariant point (eutectic) to another invariant point, the melting point of pure anorthite in the $CaAl_2Si_2O_8$ one-component system. For fractional melting in systems of *any* number of components, liquids are restricted in composition and temperature to invariant points. Also note that each fractional melting event in simple eutectic systems results in the depletion of one solid phase, and any further melting occurs in an appropriate system reduced by one component.

Binary Systems with Peritectics

The next step is to examine equilibrium and fractional melting in a system with an incongruently melting intermediate compound, the system Mg_2SiO_4–SiO_2

(A)

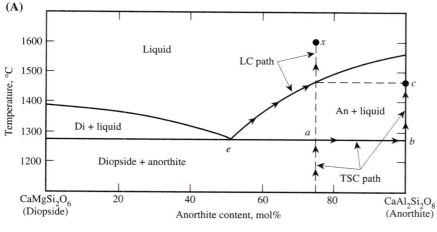

FIGURE 5-1
● ● ● ● ● ● ● ● ● ● ● ● ● ●

Melting relationships in the binary system $CaMgSi_2O_6$–$CaAl_2Si_2O_8$ (diopside-anorthite). **(A)** Equilibrium melting for composition x. The LC and TSC paths are indicated. Point c marks the termination of the TSC path. **(B)** Fractional melting for composition x. The TSC path is much the same as that for equilibrium melting, except that the termination of the path at point c is now at the melting point of pure anorthite. The LC path is a "hop" from the eutectic e to c.

(B)

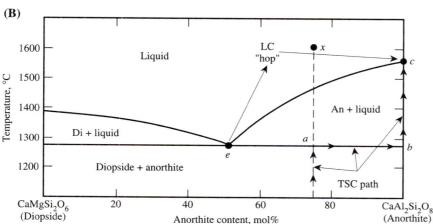

(forsterite-cristobalite, tridymite, or quartz) with enstatite as an intermediate compound (Figure 5-2A). At low pressures, the silica mineral can be either tridymite or cristobalite; at higher pressures (above about 2 kbar; see Figure 4-1), it is normally quartz. This system is a useful one to examine because it is important petrologically and because it constitutes one of the bounding binary systems in a key ternary system (diopside-forsterite-silica) that is discussed later.

Consider a mixture of 10% tridymite and 90% enstatite (the isopleth at x). In low-pressure equilibrium melting, this solid assemblage can be heated and remain unmelted until it reaches 1543°C, at which temperature liquid will appear at the eutectic. Melting to produce this liquid composition from the solids will enrich the residual solids (TSC) in enstatite until tridymite is totally depleted. (Remember that *both* enstatite and tridymite are melting, but tridymite is used up first because there is not as much tridymite as enstatite.) Once tridymite is gone, the liquid is free to rise in temperature and to

dissolve or melt additional enstatite. The TSC therefore rises vertically along the enstatite composition line, and LC follows a path along the liquidus. At 1557°C, LC is at the peritectic point and enstatite has reached the limit of its thermal stability. Any remaining enstatite crystals will react to form peritectic liquid plus forsterite crystals. The TSC is driven toward the forsterite composition by formation of peritectic liquid until all enstatite is gone.

After this, the whole system increases in temperature as forsterite melts and LC becomes more forsterite-rich. Finally, when LC on the liquidus coincides with the compositional isopleth at about 1600°C, all crystals have melted and the liquid can increase its temperature without further composition change. Note that this sequence is *exactly* the reverse of the equilibrium crystallization case, and that LC follows a continuous path from the eutectic to where it leaves the liquidus at about 1600°C. If an initial solid composition that lay between forsterite and enstatite had been chosen, the equilibrium melting sequence would have differed by having the first liquid

FIGURE 5-2

●●●●●●●●●●●●●●●●●

(A) Equilibrium melting within the somewhat simplified binary system Mg_2SiO_4–SiO_2 (forsterite-silica) at low pressure (≈ 1 bar). The temperature axis is not to scale, in order to show relationships more clearly. Equilibrium melting of bulk composition x has the LC path shown. The corresponding equilibrium TSC path starts at a and ends at g.
(B) Fractional melting of bulk composition x within the binary forsterite-silica system. The LC path is discontinuous: Liquid is first produced at the eutectic e, then at the peritectic p, and finally at the melting point of forsterite f. The two discontinuous steps are shown as LC "hop" 1 and LC "hop" 2. The continuous TSC path starts at a and ends at f.

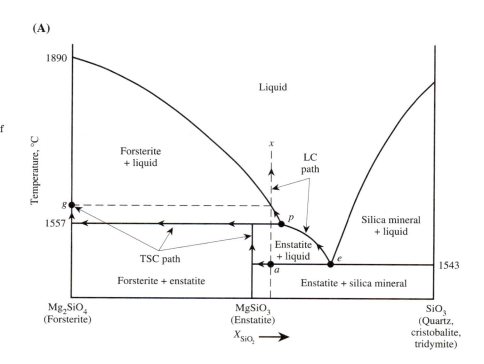

(A)

Binary Systems with Complete Solid Solution

appear at the peritectic rather than the eutectic. Of course, this is because the only liquid that can coexist with forsterite + enstatite is the peritectic liquid.

Fractional melting of tridymite + enstatite begins in the same way (Figure 5-2B) but soon changes dramatically. As the mixture of enstatite and tridymite crystals reaches 1543°C, eutectic liquid appears. The TSC is driven to the left as tridymite is depleted from the solids, until only enstatite remains. Eutectic melting now stops, because the system contains only enstatite crystals. No further melting is possible until the temperature reaches the incongruent melting point of enstatite at 1557°C. At this temperature, liquid will appear at the peritectic point (having "hopped" in temperature and composition from the eutectic to the peritectic) and will drive the TSC toward the forsterite composition. As long as both forsterite and enstatite are present, melting continues at the peritectic and the melt is continuously removed. Once TSC reaches forsterite, enstatite is totally depleted and melting stops. The residuum is monomineralic forsterite, and no further melting can occur until the melting point of pure forsterite is reached at 1890°C. This second liquid "hop" is from 1557°C at the peritectic to 1890°C at forsterite composition. Note that the TSC path in fractional melting is continuous and is similar to that in equilibrium melting. However, the LC path is discontinuous, with liquid appearing only at invariant points and being restricted in composition to these points.

Binary Systems with Complete Solid Solution

For binary systems with complete solid solution, the previous model for this type of system, the $NaAlSi_3O_8$–$CaAl_2Si_2O_8$ (albite-anorthite) system, will be used again (Figure 5-3). Equilibrium melting of a single-phase solid (plagioclase at An_{65}) begins at about 1350°C (point i) when the temperature of the crystals reaches the solidus temperature at An_{65}. The first melt that appears has a composition of about 23% $CaAl_2Si_2O_8$ and 77% $NaAlSi_3O_8$ (point h). As an infinitesimal amount of this sodic liquid is produced, the remaining crystals must become infinitesimally more calcic to provide mass balance. (Note that this type of system is an exception to the rule that first liquids must appear at invariant points.) As temperature increases, the percentage of melt relative to crystals slowly increases also, and the liquid constantly reequilibrates with the remaining solids (TSC) as both become more calcic. Ultimately, at about 1485°C (point f), the LC (65% $CaAl_2Si_2O_8$) equals the initial crystalline composition (An_{65}), and the system is entirely molten.

Fractional melting in the plagioclase system introduces some additional complexity because of the removal of liquid as it is produced. There can be no constant reequilibration between crystals and liquid such as occurred in the equilibrium case, and because the initial liquid that is

(B)

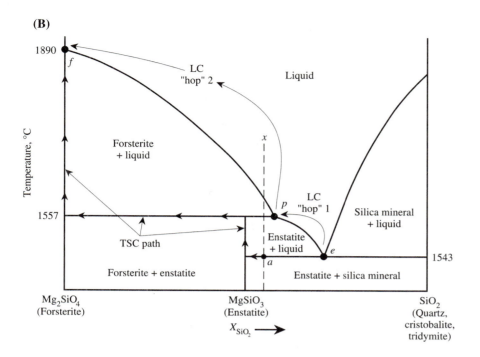

removed is quite rich in $NaAlSi_3O_8$, the remaining plagioclase crystals must be enriched in anorthite relative to their initial composition. The lack of compositional balancing between crystals and liquid effectively drives the residuum composition toward anorthite *more rapidly* than in the equilibrium melting case, and ultimately the residuum becomes essentially pure anorthite before the solids are totally melted. The path of fractional LC must therefore start at about h in Figure 5-3 and move toward pure anorthite, where it converges with TSC for the final

stage of melting. This observation serves to emphasize the point made in the earlier cases—that efficient fractional removal of liquid during melting tends to produce a more refractory (that is, higher melting point) residuum. Fractional melting of *any* plagioclase composition (other than pure albite) produces a final TSC at anorthite, but only for more calcic plagioclase (greater than about An_{50}) will there be more than a tiny fraction of the original crystalline volume remaining at that stage of the melting process.

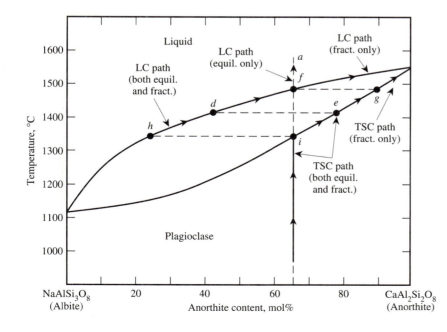

FIGURE 5-3
● ● ● ● ● ● ● ● ● ● ● ● ● ● ●

Binary system $NaAlSi_3O_8$–$CaAl_2Si_2O_8$ (albite-anorthite), illustrating the LC and TSC paths for equilibrium (equil.) and fractional (fract.) melting. The TSC paths are common from i to g. For equilibrium melting, the TSC path ends at g; but for fractional melting, it continues from g to the melting point of pure anorthite. The LC paths are common from h to f. For equilibrium melting, the path goes from f to a; but for fractional melting, it continues past f toward the melting point of pure anorthite.

FIGURE 5-4

• • • • • • • • • • • • • • • •

Ternary system CaMgSi$_2$O$_6$–Mg$_2$SiO$_4$–CaAl$_2$Si$_2$O$_8$ (diopside-forsterite-anorthite), showing the LC and TSC paths for equilibrium melting and the liquid "hops" for fractional melting. The box labeled ISC (instantaneous solid composition) indicates the range of diopside-forsterite mixtures in equilibrium with various liquids along the cotectic. The TSC path for fractional melting is identical to that for equilibrium melting. For details and discussion, see text.

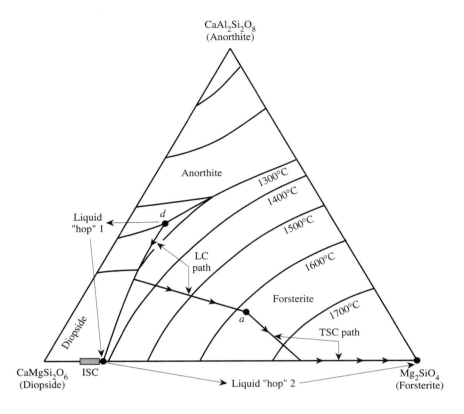

MELTING IN TERNARY SYSTEMS

• • • • • • • • • • • • • • •

Simple Ternary Eutectic Systems

The simplified ternary system CaMgSi$_2$O$_6$–Mg$_2$SiO$_4$–CaAl$_2$Si$_2$O$_8$ (diopside-forsterite-anorthite) provides a good example of melting phase relations in a system with a single ternary eutectic. The basic rule for deciding where the first liquid will appear upon heating was noted earlier: List the solid phases that coexist in the initial crystalline material and locate the one invariant point where liquid can coexist with these solid phases. This rule is very important in ternary systems, which typically contain three solid phases of fixed composition, and especially so in the more complex ternary systems with incongruent melting and thus with peritectic as well as eutectic points. In Figure 5-4, the point of lowest temperature on the liquidus surface where liquid can coexist with diopside + forsterite + anorthite is the ternary eutectic (invariant) point at 1270°C. The initial solid composition *a* indicated on Figure 5-4 contains about 40% diopside, 40% forsterite, and 20% anorthite. When this solid

mixture is heated to 1270°C, melting begins and the composition of the melt is at the eutectic *d*.

In an equilibrium melting sequence, as melting proceeds and a greater proportion of liquid *d* is produced, the TSC must change by moving directly away from the composition of liquid being produced. The TSC path will therefore move along a line directly away from the eutectic toward the diopside-forsterite join. It is clear that anorthite will be the first solid phase to be exhausted during melting; graphically this coincides with the arrival of the TSC at the diopside-forsterite join. Once anorthite is gone, the liquid is no longer constrained to stay at the eutectic. Because the only two remaining solids are diopside and forsterite, the liquid (LC) must increase in temperature along the diopside-forsterite boundary curve. To do so, it must dissolve (melt) both solid phases in a ratio indicated by where the tangent to the boundary curve intersects the diopside-forsterite join (approximately indicated as ISC in Figure 5-4). For every increment of this material removed from the remaining solids, the TSC must be driven an equal increment to the right toward forsterite. Remember that a tie line from TSC to LC at any stage in melting must pass through *a*. When TSC reaches the forsterite corner, diopside is gone, and the system consists of liquid on the boundary curve plus forsterite crystals. Any further melting must be on the forsterite liquidus surface, and therefore as the liquid begins to dissolve the remaining forsterite, LC moves up along the tie line toward a. When LC reaches a, the solids have been completely melted. Note that equilibrium melting in this system has involved *continuous* paths for both LC and TSC.

Fractional melting for composition a in this simple ternary eutectic system begins the same way. Eutectic melt is produced at 1270°C (point d), and the TSC is therefore driven away from the eutectic composition toward the diopside-forsterite join until anorthite is exhausted. Melting must now stop because there are only two solid phases, and the liquid already produced is no longer in the system. In effect, the original ternary system has been reduced to a binary diopside-forsterite system by taking away all the $CaAl_2Si_2O_8$ component in the liquid. Any further melting is therefore restricted to the diopside-forsterite system. The next melt produced is at the binary eutectic and occurs at just below 1400°C. Melting continues at this temperature until all diopside is exhausted and the TSC is at the forsterite corner. With a new residuum consisting only of forsterite, there cannot be any further melting until the system is heated to >1800°C and pure forsterite melts. In the fractional melting sequence, the LC path has been *discontinuous*, with "hops" in both temperature and composition from the ternary eutectic (1270°C) to the binary eutectic (ca. 1390°C) to pure forsterite (>1800°C). Each step in fractional melting produces a single liquid of fixed composition, and complete removal of each of these liquids removes one component at a time from the original system.

Ternary Systems with Solid Solution

Again, consider the system that we discussed earlier: $CaMgSi_2O_6$–$NaAlSi_3O_8$–$CaAl_2Si_2O_8$ (diopside-albite-anorthite) (Figure 5-5). There is only a single boundary curve in this system because of the complete solid solution between anorthite and albite, and there is thus no ternary eutectic. Melting first occurs for any diopside-plagioclase crystalline mixture when the temperature reaches a value sufficient to stabilize a liquid along the boundary curve at a composition in equilibrium with the plagioclase composition. This temperature will be lower for sodic plagioclases and higher for calcic ones. The *particular* liquid composition and temperature cannot be predicted from the phase diagram alone but must be based on experimental information. It may be helpful to think of this in terms of the anorthite-albite melting loop (see Figure 5-3). Melting of plagioclase plus diopside can

be considered to occur along a similar loop, but one that lies directly beneath the boundary curve and is considerably flatter (only about 185°C between the end points of the boundary curve, as opposed to >400°C between the melting temperatures of pure albite and pure anorthite in the binary).

Equilibrium melting for composition a (80% plagioclase of An_{50} and 20% diopside) in Figure 5-5A begins when the temperature reaches that required to melt An_{50} + diopside, that is, point b on the boundary curve at about 1210°C. The first appearance of liquid at b generates a three-phase triangle from An_{50} to diopside to liquid, as shown in the diagram. As the temperature increases, the remaining plagioclase must become more calcic as it equilibrates constantly with the increasing proportion of liquid, and the LC itself must move up the boundary curve. During this process, the three-phase triangle sweeps to the right as it pivots on the diopside corner. Note that as this occurs, the bulk composition a becomes interior to the triangle, then finally is approached by the plagioclase-liquid edge of the triangle. Geometrically, this configuration indicates that the proportion of solid diopside remaining is approaching zero, as also indicated by the curved TSC path. When the tie line from liquid to plagioclase passes through a, no diopside remains (TSC is on the plagioclase join) and the melt is no longer constrained to stay on the boundary curve. As plagioclase melts and the LC moves up across the plagioclase liquidus surface, the LC must follow exactly the same curved path as in the equilibrium crystallization case, because the residual plagioclase (TSC) becomes very calcic as LC approaches a, and the addition of this calcic component to the liquid must draw the LC toward a point close to anorthite. Of course, when LC reaches a, the solids are entirely melted and the TSC path ends at the final plagioclase composition. As an exercise, the reader might try to construct the melting scenario for an initial solid composition along the same tie line (from An_{50} to diopside) but rich enough in diopside to be above the boundary curve. To check the answer, look back to equilibrium crystallization for point f in Figure 4-12B; the equilibrium melting paths for LC and TSC are exactly the reverse of those for equilibrium crystallization.

Fractional melting in this system is considerably trickier to interpret (Figure 5-5B). The onset of melting for composition a is the same as just described in the equilibrium case, with melting beginning at about 1210°C and the first liquid appearing at b. As temperature increases, LC moves up the boundary curve. The constant removal of this liquid has the effect of shifting the residuum composition (TSC) toward the right and downward in a path that is less strongly curved than the equilibrium path. When diopside is totally depleted, the TSC

FIGURE 5-5

●●●●●●●●●●●●●●●●●●

(A) Equilibrium melting in the complex ternary system diopside-anorthite-albite for bulk composition a. The LC path is from b to a; the TSC path is from a to i. The two dashed triangles show the initial and final three-phase triangles. The final plagioclase composition i cannot be determined from the diagram a priori but can only be ascertained from experimental calibration. Labeled points are explained in the text. The portion of the LC path from b to j corresponds to the TSC path from a to k; the j to a portion of the LC path corresponds to the k to i portion of the TSC path. **(B)** Fractional melting of composition a in the diopside-anorthite-albite ternary system. Melt first appears at b, then shifts up along the cotectic to c, at which point all diopside is exhausted from the residuum. The remaining plagioclase has composition d, and when the temperature reaches the solidus for composition d, a new melt appears at e. Melting now proceeds as in Figure 5-3, with final melt at the composition of pure anorthite. The TSC shifts from a to the plagioclase join; the instant it reaches the join, production of melt at c ceases. Once melting begins in the plagioclase join, the TSC shifts toward the composition of anorthite.

(A)

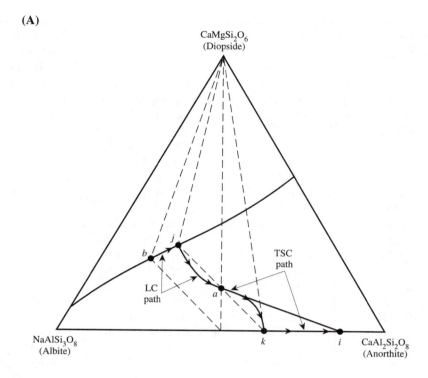

will intersect the plagioclase join and melting on the boundary curve must cease at a temperature <1250°C. Because the residuum now consists only of plagioclase crystals, further melting is restricted to the anorthite-albite binary join (Figure 5-3). The temperature required to melt the calcic plagioclase in the residuum is roughly 1400°C. There is thus a liquid "hop" from LC on the boundary curve at <1250° to LC on the $NaAlSi_3O_8$–$CaAl_2Si_2O_8$ join at >1400°C. Fractional melting in the plagioclase binary then follows the same paths outlined earlier. Both residuum plagioclase composition and liquid composition follow the liquidus and solidus curves, respectively, until they coincide at the melting point of pure anorthite.

Note that the introduction of solid solution into a ternary system has a significant effect on the nature of the LC path. In fractional melting of systems without solid solution, successive liquids are generated at fixed compositions and temperatures, with liquid "hops" between these successive liquids. When there is solid solution in one of the solid phases in the system, fractionally melted liquids can vary in composition over certain intervals but still have one or more liquid "hops" in their evolution.

Ternary Systems with Congruent Intermediate Compounds

Many binary systems have intermediate compounds that lie between the solids that define the compositional limits of the binary. In the simplest of these systems, the intermediate phase melts congruently, splitting the binary into two subsystems, each of which has a eutectic (Figure 5-6A). When a system of this type forms one of the bounding binary systems in a ternary with two simple binaries (Figure 5-6B), the ternary is split into two subsystems, each of which is a simple ternary eutectic system with three boundary curves. This splitting occurs because the tie line from the intermediate compound to the solid phase at the other corner of the ternary forms a true binary eutectic system itself. Each of the two ternary subsystems is independent, and any bulk composition within either side must undergo all melting or crystallization paths entirely within that side without crossing the common binary join. This join is therefore called a **thermal divide** (Figure 5-6B) because temperatures on the liquidus surface decrease away from it toward both sides and crystallizing liquids must move away from it.

(B)

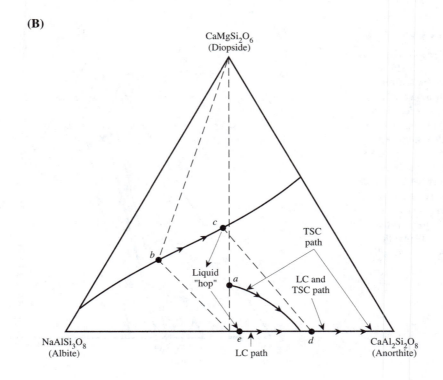

CaMgSi$_2$O$_6$
(Diopside)

TSC
path

Liquid
"hop"

c

b

a

LC and
TSC path

NaAlSi$_3$O$_8$
(Albite)

e

d

CaAl$_2$Si$_2$O$_8$
(Anorthite)

LC path

(A)

Thermal
divide

Liquid

C + L

A + L

C + L

B + L

A + C

C + B

A

C

B

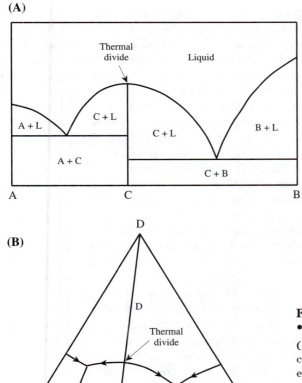

D

(B)

D

Thermal
divide

A

C

B

A

C

B

When the intermediate compound melts incongruently, however, relationships are a little more complex. For example, in the system forsterite-diopside-silica (quartz, cristobalite, tridymite) at low pressure (Figure 5-7), the join enstatite-diopside is not truly binary because there is a stability field along it for a solid phase whose composition lies outside the join. The two ternary subsystems are therefore not independent of each other, and bulk compositions can move back and forth between them during melting or crystallization. This new type of binary join within a ternary is called *pseudobinary,* and it is quite important in advanced phase diagram analysis, which is beyond the scope of this discussion. It must be emphasized that pseudobinary joins are *not* thermal divides, and therefore liquid compositions can move across such joins. The implications of this capability are very important, as shown in the next section.

FIGURE 5-6
● ● ● ● ● ● ● ● ● ● ● ● ● ● ● ● ●

(A) Binary system with a congruently melting intermediate compound. The composition C is called a thermal divide because it separates liquids that will evolve down to the A-C eutectic from liquids that evolve to the C-B eutectic. **(B)** A hypothetical ternary system in which one bounding binary is A-B, as shown in **(A)**. The ternary system is divided into two subsystems, each with its own ternary eutectic. The C-D cotectic line has two segments, separated by the thermal divide: one leads to the left to the ACD eutectic, and the other to the right to the BCD eutectic.

FIGURE 5-7
● ● ● ● ● ● ● ● ● ● ● ● ● ● ● ●

The Mg_2SiO_4–$CaMgSi_2O_6$–SiO_2
(forsterite-diopside-quartz, cristobalite,
tridymite) ternary system at low pressure, showing the LC (heavy solid line)
and TSC (dashed line) paths for equilibrium crystallization of bulk composition
(1). Equilibrium melting of this composition follows the exact reverse paths. For
composition (1), the final equilibrium
melt will be at the ternary peritectic point
b, whereas equilibrium crystallization of
(2) will end at *c*, the ternary eutectic
point. Composition (1) is the approximate location of Ringwood's pyrolite
model for mantle rocks. For a complete
explanation, see text.

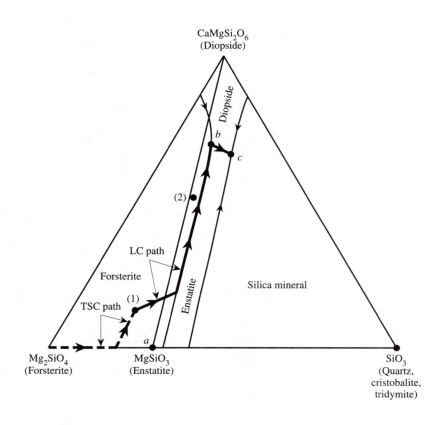

FORSTERITE-DIOPSIDE-SILICA: A MODEL FOR THE ORIGIN OF BASALTIC MAGMAS
● ● ● ● ● ● ● ● ● ● ● ● ● ● ●

Low-Pressure Phase Relations

The Mg_2SiO_4–$CaMgSi_2O_6$–SiO_2 (forsterite-diopside-quartz,
cristobalite, tridymite) ternary system is one of the most
studied of the petrologically important systems because
it provides a useful model for the origin of the different
basaltic magmas, as well as a guide to their crystallization. Although plagioclase is not explicitly represented
within this ternary system, phase relations in forsterite-
diopside-silica are very much a model for relations in
the more complex four-component system with anorthite added. In particular, the implications for silica-
saturated and silica-undersaturated magmas are useful.
The forsterite-diopside-silica system at low pressure
shows some topologic complexity not previously encountered, because of the presence of an incongruently
melting compound (enstatite) in the ternary. Crystallization in this system is examined in some detail in the following chapter, so it is appropriate to introduce it here.

Figure 5-7 shows the one-atmosphere phase relations
in this ternary system. (Experimental studies show that
this system is not actually as simple as portrayed: Both
diopside and enstatite have limited solid solution with

each other.) It has already been shown that enstatite,
the intermediate phase on the Mg_2SiO_4–SiO_2 join,
melts incongruently (see Figure 5-2). What is new and
slightly complicated is the projection of this incongruent melting behavior into the ternary plot as a
pseudobinary join between enstatite and diopside that
splits the ternary into two three-phase subsystems:
Mg_2SiO_4–$CaMgSi_2O_6$–$Mg_2Si_2O_6$ (forsterite-diopside-
enstatite) and $Mg_2Si_2O_6$–$CaMgSi_2O_6$–SiO_2 (enstatite-
diopside-quartz, cristobalite, tridymite).

In Figure 5-7, the incongruent melting behavior of
enstatite affects the topology of the boundary curves and
invariant points within the ternary. For example, the
forsterite-enstatite boundary curve propagates from
the binary peritectic point *a*. Any liquid along this curve
does not crystallize both solid phases simultaneously, as
typically happens on boundary curves, but reacts with
preexisting olivine crystals to form new enstatite crystals. Ternary *peritectic lines*, therefore, like binary peritectic points, represent reaction between liquid and crystals rather than simple crystallization from a melt. Note
in Figure 5-7 that the peritectic line leads to an invariant
point *b*, which represents the liquid composition that
coexists with forsterite, enstatite, and diopside. This
invariant point is a ternary peritectic point: Two boundary curves lead into it, and one (diopside-enstatite) leads
out and ultimately down to the ternary eutectic *c*. Even if
there were no arrows indicating the down-temperature

direction on boundary curves, a ternary peritectic could be distinguished from a ternary eutectic by one simple rule: A peritectic point lies *outside* the triangle connecting the three solids that coexist with the liquid, whereas a eutectic lies *within* such a triangle. For example, note that the peritectic liquid coexists with forsterite, diopside, and enstatite but lies to the right of the triangle connecting these phases. The ternary eutectic liquid coexists with diopside, enstatite, and cristobalite and falls within the triangle of these solids.

First, consider equilibrium crystallization and melting behavior in this system. The crystallization of liquid at bulk composition (1) will commence when the liquid is cooled to the temperature of the forsterite liquidus surface. As forsterite crystallizes, LC leaves (1) and moves directly away from the forsterite composition toward the forsterite-enstatite boundary curve. The LC path extends directly across the diopside-enstatite pseudobinary join without interruption as it follows the forsterite liquidus. When LC reaches the boundary curve, silica-saturated liquid begins to react with already formed forsterite crystals to transform them into enstatite in the reaction forsterite + SiO_2 (in liquid) = enstatite. The LC moves down in temperature along the boundary curve toward the ternary peritectic point *b*. To analyze further progress of LC, the original bulk composition must be considered: Because it falls within the forsterite-diopside-enstatite triangle, crystallization must terminate at a point where liquid can coexist with these three solid phases. The only point that qualifies is the peritectic point *b*, and therefore when LC reaches the peritectic point, forsterite continues to be converted to enstatite while both enstatite and diopside simultaneously crystallize directly from the liquid until liquid is entirely consumed. The TSC path for composition (1) begins at the forsterite corner and stays there until the LC reaches the peritectic forsterite-enstatite boundary curve. As liquid moves down this curve and forsterite is converted to enstatite, TSC moves along the base of the triangle toward enstatite. As long as the LC is on the boundary curve, TSC is restricted to the forsterite-enstatite join. When LC first reaches the peritectic point *b*, TSC lies at the end of a tie line that runs from LC through composition (1) to the base of the diagram. As diopside is added to the solids, TSC moves up along this tie line toward the peritectic point until it coincides with (1), at which time the liquid is all gone.

Initial liquid compositions that lie in the narrow zone between the pyroxene pseudobinary and the forsterite-enstatite boundary curve [for example, composition (2)] undergo similar LC and TSC paths until LC reaches the peritectic point. Because their bulk compositions lie within the diopside-enstatite-quartz triangle, however, these liquids must ultimately evolve down to the ternary eutectic *c*. To do so, they must leave the peritectic point, but can only do so after all the forsterite has been consumed by reaction with liquid. Once LC leaves *b*, it can no longer be in equilibrium with olivine and moves down the enstatite-diopside boundary curve until it reaches the eutectic point and begins crystallizing tridymite as well as diopside and enstatite.

Equilibrium melting of composition (1) follows the exact reverse of the preceding sequence. Composition (1) consists of about 62% enstatite, 25% forsterite, and 13% diopside. Upon heating and melting, the first liquid that appears will be at the temperature and composition of the ternary peritectic *b*, because this is the only liquid that can coexist with the three solid phases forsterite, diopside, and enstatite. The LC will remain at *b* until diopside is completely consumed; in this process the TSC path is directly away from the peritectic point and toward the forsterite-enstatite join. When TSC reaches the join, diopside is consumed and the LC is free to move up the forsterite-enstatite peritectic boundary curve as enstatite incongruently melts to forsterite + liquid. While this incongruent melting is going on, TSC moves away from enstatite and toward forsterite. The TSC ultimately reaches the forsterite corner as the last enstatite is consumed, and the LC can then leave the boundary curve and move across the forsterite liquidus as the only remaining solid, forsterite, is dissolved into the liquid. Finally, LC intersects composition (1) and the residuum is entirely melted. Equilibrium melting of composition (2) can be worked out by using similar reasoning, and the reader is encouraged to do so. One hint for composition (2): Be sure that you pick the correct invariant point at which the first liquid appears. Remember that this is done by first identifying exactly which three solid phases form the triangle that contains a particular bulk composition.

Fractional melting of (1) starts in the same way as equilibrium melting does. When the temperature of the peritectic point *b* is reached, the solid assemblage of forsterite, enstatite, and diopside begins to melt to a liquid of composition *b*. Because there is much more of the component $CaMgSi_2O_6$ in this liquid than in bulk composition (1), extraction of the liquid from the system will deplete the residuum in diopside relative to forsterite and enstatite. The TSC follows the same path as in equilibrium melting, but when TSC reaches the forsterite-enstatite join, melting must stop. The residuum composition, approximately 60% enstatite and 40% forsterite, cannot melt until the temperature reaches that of the *binary* peritectic *a*. At this temperature, liquid of composition *a* starts to form and melting continues at this composition until the TSC has followed a path to the left and reached the forsterite corner. Again, melting stops, this time until the melting temperature of pure forsterite is reached. The fractional melting sequence

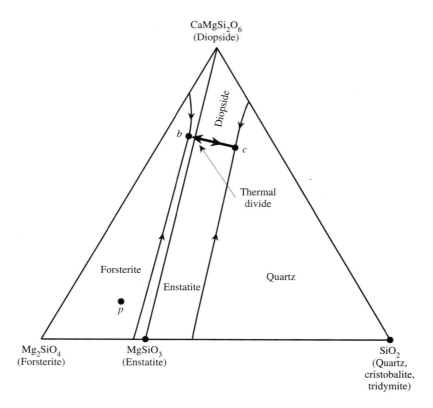

FIGURE 5-8

● ● ● ● ● ● ● ● ● ● ● ● ● ● ●

The forsterite-diopside-silica ternary system at high pressure (above about 8 kbar). Note the shift of the enstatite-forsterite boundary curve from its peritectic location in Figure 5-7 to its co-tectic location in this diagram. The enstatite-diopside join is now a thermal divide separating the forsterite-enstatite-diopside (ternary eutectic *b*) and enstatite-diopside-silica (ternary eutectic *c*) subsystems. Ringwood's pyrolite model for mantle compositions is shown as point *p*. For a complete explanation and discussion, see text.

therefore has two liquid hops: from the ternary peritectic *b* to the binary peritectic *a*, and from *a* to the forsterite composition.

The presence of dunite (essentially pure olivine rock) in the shallow lithospheric mantle has been documented in numerous xenolith suites in basalts and by tectonically emplaced ultramafic bodies. Because forsterite, and thus dunite, melts at about 1800°C, a temperature rarely if ever attained in the lithosphere, the presence of dunite as a refractory residuum from fractional melting and extraction of basalt magmas is perfectly reasonable.

High-Pressure Phase Relations

Research by Kushiro (1969) has shown that the effect of pressure on the diopside-forsterite-silica system is to shift the positions of boundary curves, particularly the forsterite-enstatite peritectic curve. This effect is most likely due to differential compressibility of solids and liquids, which changes the amount of each solid component that can dissolve in any liquid. Kushiro found that increasing pressure expands the liquidus fields of pyroxenes relative to forsterite and silica minerals. The most important effect is to displace the binary and ternary peritectic points, *a* and *b*, respectively, into the forsterite-diopside-enstatite half of the diagram (Figure 5-8) and thus turn the diopside-enstatite join into a true binary

and the forsterite-enstatite boundary curve from a peritectic curve to a eutectic curve. The topology of the system thus becomes like the simple ternary eutectic system illustrated in Figure 5-6B, and at about 10 kbar the diopside-enstatite join becomes a thermal divide that absolutely separates the melting and crystallization behaviors of silica-saturated compositions (to the right) from olivine-saturated compositions (to the left).

The importance of the pressure effect in the forsterite-diopside-silica system lies in its implications for the generation of basaltic magmas within the mantle. Keep in mind that virtually all petrologic and geochemical evidence indicates that the upper mantle is composed of ultramafic rock—in particular peridotite, an olivine-rich rock that also contains calcic clinopyroxene and orthopyroxene. Therefore, mantle peridotite compositions can be modeled by using the forsterite-enstatite-diopside subtriangle within the overall forsterite-diopside-silica system. One proposed upper mantle composition, the so-called **pyrolite** model of Ringwood (1975), is shown in Figures 5-7 and 5-8. Ringwood created this artificial composition as an experimental starting material by mixing one part basalt with four parts dunite. Much of the modeling and experimentation of basaltic magma origins that petrologists have done over the last 20 years is based on derivation of basaltic magmas from pyrolite or something similar. On the basis of the principles discussed in this chapter, fractional melting of an ultramafic composition such as pyrolite at low pressures must produce an initial liquid composition (*b* in Figure 5-7) that actually falls within the silica-saturated half of the system. Removal of this liquid from the source area (as fractional melting requires) and subsequent transport to a surface extrusion as lava will result in crystallization of a silica-saturated basalt, for example, a quartz tholeiite.

Note that initial *low-pressure* fractional melting of *any ultramafic composition* cannot produce an olivine-saturated (or silica-undersaturated) liquid as long as point *b* is peritectic and lies to the right of the pyroxene join. However, as pressure increases beyond about 2 kbar (5–6 km depth), the melting of enstatite becomes congruent. The exact pressure is not known, but Kushiro has established experimentally that melting is clearly congruent at 10 kbar (25–30 km depth). Once the shift to congruent melting occurs and point *b* shifts from being peritectic to being eutectic, the melting of olivine-bearing rocks can *only* produce liquids that lie within the olivine-saturated portion of the system (Figure 5-8). Subsequent removal and equilibrium crystallization of these liquids must produce olivine-bearing basaltic rocks such as olivine tholeiite. On the basis of this observation, petrologists have concluded that quartz tholeiite and other silica-saturated mafic magmas represent melting in the relatively shallow lithosphere, whereas the olivine-bearing basalts probably reflect melting at deeper levels.

Analogous arguments made from phase relations in sodium-bearing systems have been used to support the interpretation that the alkali olivine basalts represent the deepest levels of origin of basaltic magma, perhaps as much as 100 km or more.

ALBITE-ORTHOCLASE-QUARTZ: A MODEL FOR MELTING AND CRYSTALLIZATION OF GRANITES
● ● ● ● ● ● ● ● ● ● ● ● ● ● ●

Low-Pressure Phase Relations

Whereas the diopside-forsterite-silica system provides a useful model for melting and basaltic magma generation within the ultramafic rocks of the mantle, the system $NaAlSi_3O_8$–$KAlSi_3O_8$–SiO_2 (albite-orthoclase-quartz) provides a similar model for granitic magma generation within the continental crust. Like the preceding system, this system introduces another new topologic complexity. It involves the presence of solid solution in a bounding binary; but instead of a continuous liquidus-solidus loop as seen with the plagioclase feldspars, the albite-orthoclase binary has a loop with a minimum temperature in the middle of the system (Figure 5-9). A system with a minimum can be considered to be analogous to a

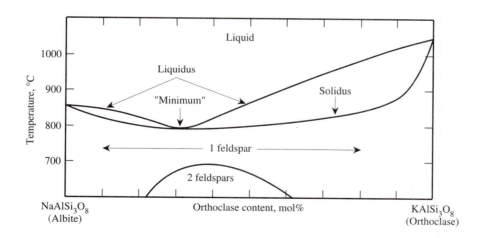

FIGURE 5-9
● ● ● ● ● ● ● ● ● ● ● ● ● ● ● ●

The $NaAlSi_3O_8$–$KAlSi_3O_8$ (albite-orthoclase) binary system at low pressure, with several important features of this type of diagram indicated. The relationships at the orthoclase end of the system have been simplified for clarity of illustration, specifically eliminating pseudobinary behavior involving leucite. The liquidus and solidus curves come together at the "minimum," the composition of the lowest temperature liquid that can exist in the system. A single alkali feldspar phase occurs along the solidus at all compositions. The subsolidus region contains a miscibility gap at lower temperatures, thus causing the solid-state breakdown of the single feldspar into perthites that are intergrowths of potassium-rich and sodium-rich lamellae.

FIGURE 5-10

• • • • • • • • • • • • • • • •

The SiO_2–$NaAlSi_3O_8$–$KAlSi_3O_8$ (quartz-albite-orthoclase) ternary system at low pressure. The binary minimum in the albite-orthoclase system (m1; see Figure 5-9) propagates into the ternary and ultimately to the quartz-saturated cotectic lines as m2, the ternary minimum, at about 775°C.

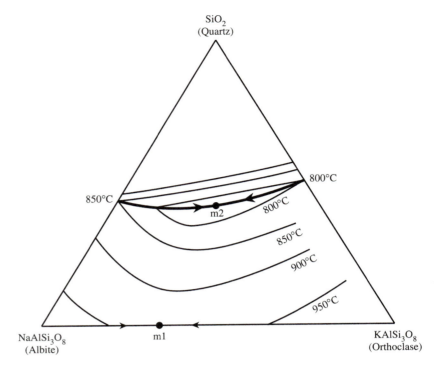

eutectic system because, during crystallization, liquids evolve toward the minimum-temperature point from both sides and cannot leave the minimum once they have reached it. The minimum propagates into the interior of the ternary as a trough in the liquidus surface, much the way that a binary eutectic generates a ternary boundary curve. In fact, a binary minimum can easily become a eutectic if the mutual solid solution of the two bounding compositions is eliminated from the middle of the system and restricted to the two ends. In the albite-orthoclase binary, there are two factors that may accomplish this: increasing pressure *in the presence of water*, and the solution of even very limited amounts of calcium (anorthite component) in the albite. The effects of water pressure are discussed briefly below and more fully in Chapter 10. The phase diagram is also complicated at low pressure by the incongruent melting of orthoclase to leucite + liquid [$KAlSi_3O_8 \rightarrow KAlSi_2O_6$ + silica (in liquid)]. This incongruence results in a pseudobinary orthoclase-albite join at low pressure, exactly like the enstatite-diopside join in the forsterite-diopside-silica system just discussed.

Crystallization and melting in the albite-orthoclase-quartz system at low pressure can be interpreted by using reasoning similar to that employed for the diopside-albite-anorthite diagram, but modified to account for the minimum in the feldspar join. The one-atmosphere phase relations depicted in Figure 5-10 show a superficial similarity to those of the diopside-albite-anorthite system (see Figure 5-5). Note, however, that the alkali feldspar-

quartz boundary curve slopes downward (in temperature) from both sides toward the minimum point. Therefore, silica-poor potassic liquids on the feldspar liquidus will first crystallize potassium-rich feldspar and evolve in a curved path toward the boundary curve (Figure 5-10). After reaching this curve, they will begin to crystallize quartz as well as potassic feldspar and will generate a three-phase triangle with a liquid apex that points toward the minimum. This triangle will sweep leftward until the liquid reaches the minimum composition and temperature, where it stays until crystallization is complete. Sodic liquids take an inverse path: They begin by crystallizing sodium-rich feldspar on the feldspar liquidus and evolve along an inverse curved path toward the boundary curve. After reaching this curve and beginning to crystallize quartz as well as feldspar, the liquid composition forms a rightward-pointing three-phase triangle as liquid moves down the boundary curve toward the minimum. As with the plagioclase system, only experiment can tell which alkali feldspar is in equilibrium with any particular liquid; this cannot be predicted from the geometry of the diagram itself.

Equilibrium melting of any composition within this system will involve the first appearance of liquid at the temperature and composition of the minimum. If the crystalline mixture is potassium-rich, the liquid will move toward the right up the boundary curve with further melting, whereas in sodic compositions, the liquid will move to the left up the boundary curve. In equilibrium melting, the limit on liquid movement up the boundary curves either way is based on intersection of the feldspar-liquid or quartz-liquid legs of the three-phase triangle with the initial composition, much as discussed for the diopside-albite-anorthite system. Following this intersection, the LC path moves up the liquidus surface of either feldspar or quartz, depending on whether the initial composition is feldspar- or quartz-rich, respectively. In fact, the path is exactly the converse of the equilibrium crystallization path, as is always the case in liquidus diagrams. In fractional melting, the LC can evolve all the way to the albite-quartz or orthoclase-quartz binary eutectics in the first episode of melting, depending on whether the initial composition is sodium-rich or potassium-rich.

High-Pressure Phase Relations

The effect of increased water pressure (more than about 3 kbar) is to restrict the solid solution within the alkali feldspar join to albite-rich and orthoclase-rich compositions and to make intermediate compositions unstable. The alkali feldspar minimum is thus turned into a binary eutectic, and the ternary minimum trough is turned into a boundary curve that separates the crystallization fields of orthoclase and albite. In turn, this change generates a ternary eutectic where the three boundary curves intersect, that is, where the liquidus fields of sodium-rich feldspar, potassium-rich feldspar, and quartz meet. Phase relations are therefore those of a simple ternary eutectic system, with the modification that limited solid solutions of albite toward orthoclase and orthoclase toward albite are possible. Experimental studies have shown that the position of the ternary eutectic is dependent on water pressure (as well as on minor solid solution of anorthite in the sodic feldspar) and that it shifts markedly toward the albite corner at higher pressures (Figure 5-11).

Melting behavior at water pressures above about 3.5 kbar thus involves generation of an initial melt at the ternary eutectic, which can be at temperatures as low as 650–700°C. Because of limitations on how high temperatures can get in the crust, rarely is the temperature high enough to generate melts that deviate much from the eutectic. Geologic mapping, geophysical studies, and laboratory work have shown that the continental crust contains a significant volume of rocks with quartz plus two feldspars that are all candidates for fractional melting at the highest grades of regional metamorphism. Some are originally sedimentary, such as feldspathic and lithic sandstones, whereas others have igneous origins such as felsic volcanics or plutonic rocks. In some metamorphic terranes, outcrops called migmatites consist of interlayered dark-colored metamorphic layers and light-colored, apparently igneous, felsic rocks. This observation has prompted the hypothesis that at least some migmatites are the result of fractional melting of feldspathic parental material, and that fractionally melted liquids of eutectic composition are locally derived. Bulk

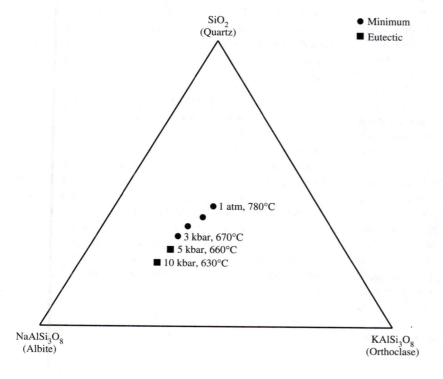

FIGURE 5-11
● ● ● ● ● ● ● ● ● ● ● ● ● ●

The SiO_2–$NaAlSi_3O_8$–$KAlSi_3O_8$ (quartz-albite-orthoclase) ternary system from 1 atmosphere to 10 kbar pressure, showing the shift in composition and temperature of the ternary minimum (L + quartz + alkali feldspar) or ternary eutectic (L + quartz + sodium-rich feldspar + potassium-rich feldspar) as a function of water pressure. [Modified from W. C. Luth, R. H. Jahns, and O. F. Tuttle, 1964, *J. Geophys. Res., 69,* Fig. 5.]

compositions of the light-colored granitic layers are commonly quite similar to the experimentally determined position of the ternary eutectic in the quartz-albite-orthoclase system. The similarity of bulk compositions of many large granitic plutons that have moved far from their source areas suggests that they too may have originated through fractional melting within the crust. In some studies, researchers have even attempted to estimate the depths of magma origin by correlating granite magma compositions with the experimentally calibrated positions of the eutectics in plagioclase-orthoclase-quartz phase diagrams.

EQUILIBRIUM AND FRACTIONAL MELTING: DO THEY OCCUR IN NATURE?

Melting behavior in a number of different systems has been examined by using two idealized concepts: (1) *equilibrium melting*, in which liquid always remains in contact, and in equilibrium, with the melting solids; and (2) *fractional melting*, in which liquid is constantly removed and cannot further interact with remaining solids. These two processes are obvious extreme end members for melt generation, but are they reasonable processes in nature? Within the last 10 years, theoretical and experimental studies of partially molten systems have strongly suggested that melt cannot ordinarily be removed from such systems until a critical minimum volume of melt has been produced. Estimates of such volumes average about 30% for basaltic liquids and somewhat higher for granitic liquids, indicating that some blend of equilibrium and fractional melting behavior may be typical in nature.

Petrologists and geochemists have called this intermediate melting behavior **batch melting.** Batch melting involves the removal of liquid (perhaps by gravitational forces) when the critical melt proportion has been achieved. For solids that have a composition significantly different from that of the eutectic or peritectic liquid that first appears as they melt, the proportion of melt produced in this first stage of melting may not achieve the proportion needed for escape from the source area. Because the liquid is assumed to remain in contact and in equilibrium with the residuum, the equilibrium model for melting is more appropriate than the fractional model. The LC path will therefore be a continuous equilibrium melting path, with liquid following a boundary curve away from an invariant point after the first solid phase is depleted from the residuum. This continuously evolving liquid can escape at any time that a criti-

cal proportion is achieved, and thus a range of magma compositions is possible for different initial bulk compositions. In pure fractional melting, liquids should be restricted in composition because they can be produced only at eutectic or peritectic points.

Figure 5-12 portrays a hypothetical quantitative batch melting scenario in the forsterite-diopside-silica (quartz, cristobalite, tridymite) system at low pressure. Suppose that the olivine-rich parental material that is melting (an approximation of mantle peridotite) is represented in the diagram by the bulk composition 60% forsterite, 32% enstatite, and 8% diopside. The first melt will appear at the peritectic point [LC(1)], and melt of this composition will continue to be produced until diopside is totally depleted from the residuum. At this stage in melting, the TSC will be on the forsterite-enstatite join at the end of the tie line from the peritectic point through the initial bulk composition. Furthermore, the melt produced constitutes 14% of the parental material (tie line length a / $a + b = 0.14$); the other 86% is a mixture of forsterite (72%) and enstatite (28%). If it is assumed that melt must total at least 30% before it can escape, then melting must proceed in an equilibrium model until enough melt has been produced. Melt composition therefore proceeds up in temperature along the forsterite-enstatite boundary curve as enstatite incongruently melts to form forsterite + liquid. When LC reaches the point marked LC(2), 30% melt proportion has been attained (c / $c + d = 0.30$). The TSC has now moved toward the forsterite corner, and the composition of the residuum is 90% forsterite and 10% enstatite. If the melt escapes from the system at this stage, the resulting magma will have the composition shown in the diagram and will be substantially richer in enstatite component and poorer in diopside than magma formed at the peritectic point.

SUMMARY

In this chapter we have developed principles for interpreting equilibrium and fractional melting paths in binary and ternary phase diagrams. It should be reiterated that both types of melting can occur under the defined conditions of *chemical* equilibrium. So-called equilibrium melting involves a melting process in which the melt always remains in contact and in equilibrium with the remaining solid crystals in the residuum, whereas fractional melting requires continual removal of any liquid produced. The principles of melting have been applied to two phase diagrams of petrologic significance: the forsterite-diopside-silica system, which is a model for generation of basaltic magmas in the mantle, and the

FIGURE 5-12

• • • • • • • • • • • • • • •

The $CaMgSi_2O_6$–Mg_2SiO_4–SiO_2 (diopside-forsterite-quartz, cristobalite, tridymite) ternary diagram at low pressure, illustrating the process of batch melting. Phase relations among silica minerals have been simplified. The initial bulk composition is shown, along with the initial melt composition, LC(1), and the melt composition after 30% melting, LC(2). The LC path lies along the forsterite-enstatite phase boundary between LC(1) and LC(2), as shown. Two TSC points are shown: when diopside is depleted from the residuum (72% Fo, 28% En) and at 30% melting (90% Fo, 10% En). For a complete discussion and an explanation of letters, see text.

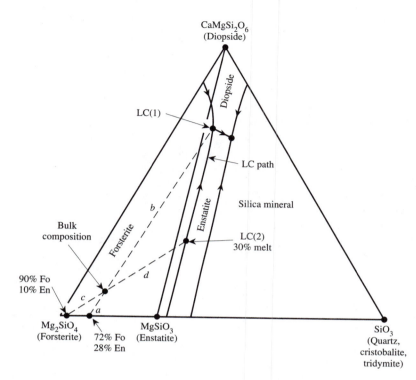

quartz-albite-orthoclase system, which is the basis for interpreting granitic magma formation in the continental crust.

The concluding discussion has focused on the relevance of either pure equilibrium or fractional models to natural melting processes. There are good petrologic reasons to believe that neither process occurs in a pure form. An alternative is batch melting, a blend of equilibrium and fractional melting that is ultimately controlled by magma physics and the gravitational driving force for mobilization.

STUDY EXERCISES

• • • • • • • • • • • • • •

1. Using Figure 5-1, predict what the temperature for beginning of melting would be for a rock consisting of 40% anorthite and 60% diopside. Predict the initial melting temperature for a rock with 80% anorthite and 20% diopside.

2. Construct a melting scenario in the system forsterite-silica (Figure 5-2) for both equilibrium and fractional melting for a rock consisting of 75% enstatite and 25% forsterite. Follow the melting process from inception of melting to disappearance of the last few crystals.

3. Construct a melting scenario in the ternary system diopside-albite-anorthite (Figure 5-5) for a rock composition between the diopside-plagioclase cotectic line and the diopside corner.

4. In Figure 5-6B, trace the crystallization of a variety of melts just on either side of the C-D binary join to prove to yourself that C-D is in fact a thermal divide that cannot be crossed by evolving melts.

5. Why is melting in the ternary system diopside-forsterite-silica (Figure 5-7) a good and relatively simple model for melting of the upper mantle?

REFERENCES AND ADDITIONAL READINGS

• • • • • • • • • • • • • • •

Kushiro, I. 1969. The system diopside-forsterite-silica with and without water at high pressures. *Am. J. Sci.*, *267A*, 269–294.

Luth, W. C., R. H. Jahns, and O. F. Tuttle. 1964. The granite system at pressures of 4 to 10 kilobars. *J. Geophys. Res.*, *69*, 759–773.

Maaloe, S. 1985. *Principles of Igneous Petrology*. New York: Springer-Verlag.

Morse, S. A. 1970. Alkali feldspars with water at 5 kb pressure. *J. Petrol.*, *11*, 221–253.

Morse, S. A. 1980. *Basalts and Phase Diagrams*. New York: Springer-Verlag.

Ringwood, A. E. 1975. *Composition and Petrology of the Earth's Mantle*. New York: McGraw-Hill.

Yoder, H. S., and C. E. Tilley. 1962. Origin of basalt magmas: An experimental study of natural and synthetic rock systems. *J. Petrol.*, *3*, 342–532.

EVOLUTION OF MAGMAS: FRACTIONAL CRYSTALLIZATION AND CONTAMINATION

The preceding chapters covered the principles that govern the origins of magmas by melting and the equilibrium crystallization of such magmas. The remaining important area of igneous petrology that must be explored by using phase diagrams is the modification of magma bulk composition after a magma has left the place in the earth where it originated. Decades of observation and analysis of igneous rocks have revealed enormous diversity, even within limited groups such as basalts. If all melting were purely fractional, less diversity of magma types might be expected. The operation of batch melting adds some diversity to initial magma compositions, but igneous petrologists have generally concluded that although melting processes can exert substantial control over final magma compositions, the principal control occurs during magma evolution.

The processes that govern magma evolution are fractional crystallization and assimilation-contamination. *Fractionation* or *differentiation* is any chemical process in which the chemistry of a system is altered by physical removal of any *fraction* of the system that differs in composition from that of the whole system. The manufacture of maple or sorghum syrup by boiling tree or plant sap is a good example of fractionation: Sap is a mixture of sugar and water, and as it boils, removal of pure water as steam concentrates the sugar in the remaining liquid. *Fractional crystallization* includes processes that produce modifications in the bulk chemical compositions of magmas through physical *removal* of early-crystallizing minerals. These processes thus provide a mechanism that allows magmas to evolve chemically from the eutectic or batch melt compositions that are in equilibrium with multiple solid phases in the residuum toward modified new compositions. *Contamination* involves modification of magmas through *addition* of extraneous material. The most commonly proposed mechanisms for contamination are the melting of wall rocks in situ or disaggregation and melting of wall rock xenoliths by ascending magma. This process can be particularly important in modification of high-temperature, mantle-derived mafic magmas by addition of felsic crustal material. Both fractional crystallization and contamination can be documented by mineralogy, major and trace element geochemistry, and isotopic compositions.

This chapter addresses the principles of fractional crystallization in some binary and ternary systems. Petrologic evidence will be examined in several magmatic environments where either fractional crystallization or magmatic contamination (or both) has acted as an agent of compositional modification of magmas. The layered gabbroic intrusions (Figure 6-1) are the classic examples of fractional crystallization. In these intrusions, the layers have resulted from sequential crystallization of high- to low-temperature minerals, with consequent gravitational settling of dense crystals through less dense liquid. The effects of fractional crystallization on the chemistry of most basaltic magmas during their long journey from the mantle to the surface are more subtle, because the crystalline material removed from the magma has been left behind in the mantle or lower crust.

FRACTIONAL CRYSTALLIZATION

The principles of fractional crystallization in eutectic phase diagrams are similar to those of equilibrium crystallization. The main difference is that idealized fractional crystallization assumes that all crystallized solids

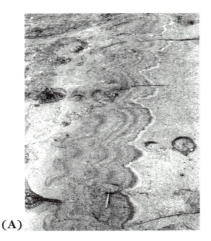

(A)

(B)   (C)

FIGURE 6-1
● ● ● ● ● ● ● ● ● ● ● ● ● ● ● ●

Banding and layered structures in the Skaergaard Intrusion, East Greenland.
(A) Colloform growth structures in the Marginal Border Group, with intrusion
contact to the left. **(B)** Cross-bedded belt in the Lower Zone. **(C)** Rhythmic layering, Middle Zone. [Photographs by Dr. T. N. Irvine, Geophysical Laboratory,
Carnegie Institution of Washington.]

are removed from the magma immediately (by unspecified processes). Thus there is no longer a topologic constraint of a fixed bulk composition and no requirement
for mass balance between crystals and liquid. Crystal
removal can occur by various processes. The most commonly assumed mechanism is gravitational settling,
in which denser crystals sink through less dense magma. High-temperature ferromagnesian minerals such as
olivine and pyroxene have densities of more than
3.0 g/cm^3, whereas typical basaltic melts have densities
of 2.6 to 2.7 g/cm^3. Other proposed mechanisms include
crystal floating, in which less dense crystals (typically
plagioclase, with a density of about 2.6 g/cm^3) rise
through the magma; filter pressing; and flow segregation. These latter processes operate where magma under a pressure gradient flows around or through crystal
aggregates, commonly when crystals stick to the walls

of a magma chamber or feeder dike, or when lath-
shaped crystals weld together into latticework "sieves"
or "filters." "Removal" of crystals can also be due to ar-
moring of high-temperature crystals by coronas or overgrowths of a secondary mineral. This special texture
represents incomplete reaction at peritectic points
because of kinetic factors. As long as an armored crystal
is kept from contacting magma, it is effectively removed
from the reactive magmatic system. Natural examples
are the occurrences of relicts of olivine inside orthopyroxene crystals or nephelines inside feldspars in silica-
saturated plutonic rocks.

The most important implication of removal of a
bulk compositional constraint is to allow all cooling and
compositionally evolving magmas to reach the lowest
temperature invariant point (a eutectic) in the system or
subsystem. The final igneous rock that results from frac-

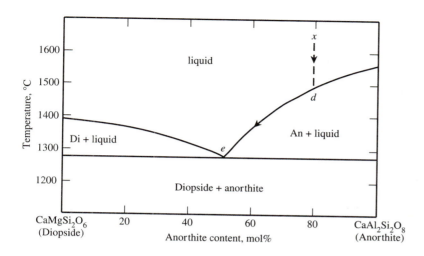

FIGURE 6-2
● ● ● ● ● ● ● ● ● ● ● ● ● ● ● ●

Fractional crystallization in the binary system $CaMgSi_2O_6-CaAl_2Si_2O_8$ (diopside-anorthite). Liquid at composition x follows the indicated liquid descent path from x to d to e (the binary eutectic) and sequentially crystallizes anorthite along the liquidus and then anorthite + diopside at the eutectic point e.

tional crystallization of a magma will therefore have a chemical composition (and mineralogy) identical to that of the eutectic, and quite different from the original magma, if all minerals formed during the crystallization path have been completely removed. It is very important to remember that the only way a magma can reach the eutectic by fractional crystallization is to lose most of its volume through crystal removal. Therefore a rock of eutectic composition must have a volume less than that of the original magma—commonly, much less.

Binary Eutectic and Peritectic Systems

The binary $CaMgSi_2O_6-CaAl_2Si_2O_8$ (diopside-anorthite) system provides a useful first example of fractional crystallization. In Figure 6-2, with cooling of an initial liquid of composition x, crystallization commences slightly below 1500°C, and the crystals are pure anorthite. As anorthite crystals are removed, the liquid composition shifts down the liquidus away from $CaAl_2Si_2O_8$ until the liquid ultimately reaches the eutectic e at about 1275°C. However, in contrast to equilibrium crystallization, all the anorthite crystals have been removed in this process, and therefore the magmatic system at 1275°C consists only of a liquid of composition e. At the eutectic, diopside joins anorthite in crystallizing from the liquid. The liquid composition cannot change further because the amounts of diopside and anorthite crystals that form are identical to the proportions of these two "molecules" in the liquid. Of course, a cooling liquid to the left of the eutectic composition will follow a similar path except that it intersects the diopside liquidus first, then continuously shifts down to the eutectic.

In a slightly more complicated system also discussed in previous chapters, a binary system with incongruent melting of an intermediate compound, the same basic principle applies. As it cools, the crystallizing liquid will always shift continuously down the liquidus to the lowest temperature invariant point, invariably a eutectic point. In the binary system illustrated in Figure 6-3, $Mg_2SiO_4-SiO_2$ (forsterite-silica), the starting liquid is again indicated by x and has a composition between $Mg_2Si_2O_6$ (enstatite) and SiO_2. As it cools, the liquid intersects the forsterite liquidus and begins to crystallize Mg_2SiO_4 (forsterite), as in the equilibrium crystallization case examined earlier (Figure 4-5). The liquid composition shifts down the liquidus curve as forsterite is removed, until the peritectic point p is reached.

Recall that in equilibrium crystallization the liquid must pause at this point to react with the already-formed forsterite, converting it to enstatite, before any further cooling and crystallization can occur. In fractional crystallization, there can be no pause, because in the conceptual model, there has been continual removal of all forsterite crystals formed in this first phase of crystallization. Thus when the liquid reaches the peritectic point, it simply switches over from crystallizing forsterite to crystallizing enstatite with no pause and no liquid-solid reaction. The liquid then follows the enstatite liquidus curve down to the eutectic point e, at which time a silica mineral (tridymite or quartz) also begins to crystallize. No further evolution in liquid composition is possible. If this eutectic liquid is emplaced in a magma chamber or as a dike, sill, or extrusive lava flow, the resulting igneous rock would consist of enstatite and quartz in the same proportions as those of the eutectic composition.

The third type of binary system, complete binary solid solution with liquidus-solidus "loops" (for example, the

FIGURE 6-3

• • • • • • • • • • • • • • •

Fractional crystallization in the binary system Mg_2SiO_4–SiO_2 (forsterite-cristobalite, tridymite, quartz). As it cools, a liquid of composition x begins to crystallize forsterite when it intersects the forsterite liquidus.

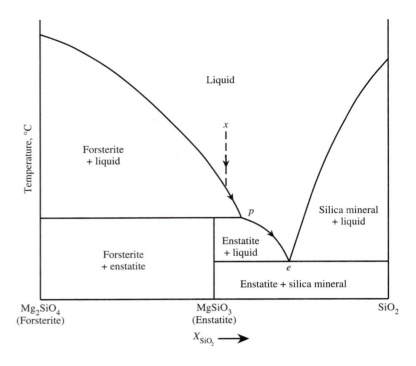

albite-anorthite or forsterite-fayalite systems), will not be discussed in detail except to note that the same basic principles apply for fractional crystallization. The liquid must continually shift down the liquidus to the lowest temperature point in the diagram. In the case of albite-anorthite, this point is not a eutectic but the albite end of the liquidus curve (see Figure 4-6), where the liquidus and solidus come together. All liquids in the albite-anorthite system that fractionally crystallize will therefore end up consisting of a small volume of pure $NaAlSi_3O_8$.

Ternary Eutectic Systems

Like the principles of equilibrium crystallization (Chapter 4), the principles of fractional crystallization of binary eutectic systems can be simply extrapolated to ternary and higher order eutectic systems. The topologic relations are slightly more complicated because of the extra representational dimension required but can be grasped easily with some practice.

The basic ternary eutectic example is once again the system $CaMgSi_2O_6$–$CaAl_2Si_2O_8$–Mg_2SiO_4 (diopside-anorthite-forsterite) shown in Figure 6-4. A liquid at composition a will initially crystallize forsterite when it has cooled to the forsterite liquidus surface. The liquid composition will shift directly away from Mg_2SiO_4 as forsterite is removed until it reaches the forsterite-

diopside cotectic line. Keep in mind that all these forsterite crystals have been removed as they form. Liquid on the cotectic line crystallizes both forsterite and diopside as the liquid shifts down the cotectic toward the ternary eutectic. When liquid composition reaches the eutectic, anorthite joins forsterite and diopside in being crystallized simultaneously. All three minerals are removed in proportions identical to their proportions in the liquid at the eutectic, so the liquid now stays fixed in composition. If this eutectic liquid were emplaced as a magma, the resulting igneous rock would consist of roughly equal amounts of anorthite and diopside (roughly 45% of each) with only a small amount of olivine (about 10%). These proportions are not unlike those in a typical olivine basalt. The reader should note that the initial liquid was much more olivine-rich than this final rock, with very little anorthite component in it. Fractional crystallization has thus allowed an essentially ultramafic magma composition to produce a feldspar-rich olivine basalt. Similar degrees of magma modification have been proposed by petrologists in natural examples.

Volumetric Relationships

It should be apparent to the reader that fractional crystallization is a potentially important mechanism for modifying the compositions of magmas as they make

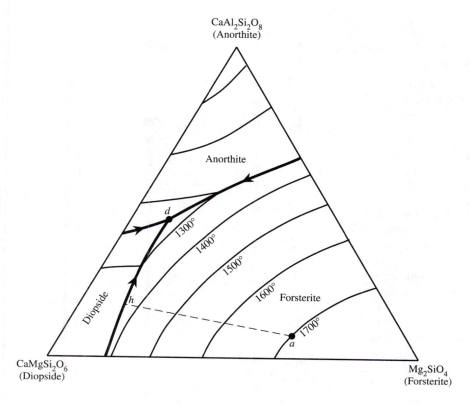

CaAl$_2$Si$_2$O$_8$
(Anorthite)

Anorthite

d

1300°

1400°

1500°

Diopside

h

1600°   Forsterite

1700°

a

CaMgSi$_2$O$_6$
(Diopside)

Mg$_2$SiO$_4$
(Forsterite)

FIGURE 6-4
● ● ● ● ● ● ● ● ● ● ● ● ● ● ●

Fractional crystallization in the simplified ternary system CaMgSi$_2$O$_6$–Mg$_2$SiO$_4$–CaAl$_2$Si$_2$O$_8$ (diopside-forsterite-anorthite). The liquid path follows the dashed line from *a* to *h*, then follows the cotectic line from *h* to *d*. An igneous rock resulting from fractional crystallization of liquid *a* would have a mineralogic composition corresponding to the eutectic point *d*, that is, about 45% diopside, 45% anorthite, and 10% forsterite. The initial liquid at point *a* is much richer than this in forsterite and poorer in anorthite.

their way through the mantle and crust to a site of final emplacement. One point must be emphasized, however: The fractionally crystallized magma will have less (perhaps much less) volume than the initial magma because of the removal of crystalline material along the way.

For example, earlier in the twentieth century it was thought by many petrologists that most granitic magmas were the result of fractional crystallization of basaltic magmas derived from melting in the mantle. The impracticality of this hypothesis was demonstrated by examination of the chemical compositions of basaltic and granitic magmas. The lower temperature chemical (or mineralogic) components that are abundant in granite (especially potassic feldspar and excess silica) are present in low concentrations in typical high-temperature basaltic magma. It has been estimated that it would take approximately nine times the volume of basaltic magma as there is granitic magma in a pluton to produce the granite by fractional crystallization. The enormous volumes of granite in some batholithic intrusions would have required incredible volumes of basaltic magma. There is no evidence either of this volume of basaltic magma or of the removed crystalline fractionation products. This is not to say that felsic magma cannot be produced by fractional crystallization. Persuasive evidence for at least limited fractionation exists in nature, particularly in layered intrusions. However, it is not a viable mechanism for formation of most granite magmas, and

the recognition of this mass balance dilemma forced petrologists to come up with alternate hypotheses for granite genesis, as we discuss in Chapters 9 and 10.

LAYERED GABBROIC INTRUSIONS: A NATURAL EXAMPLE OF FRACTIONAL CRYSTALLIZATION
● ● ● ● ● ● ● ● ● ● ● ● ● ● ● ●

At the beginning of this discussion of fractional crystallization, we noted that it was not necessary to define the specific mechanism for removal of early-formed crystalline material, although gravitational settling has been commonly suggested. Most continents contain small to very large plutons that have features indicative of large-scale chemical fractionation through crystal settling during crystallization. The most famous of these are the Skaergaard Intrusion in East Greenland, the Stillwater Complex in southwestern Montana, the Kiglapait Intrusion in Labrador, the Dufek Intrusion in Antarctica, the Muskox Intrusion in Canada, and the Bushveld Complex and Great Dyke in southern Africa. Layered intrusions span an age range from Archean [older than 2.5 gigans (Ga)] for the Bushveld to Cenozoic [younger than 60 megans (Ma)] for the Skaergaard, although most

FIGURE 6-5

• • • • • • • • • • • • • • •

Bowen's classic experiment illustrating the sinking of olivine crystals within a melt. A melt containing crystals was quenched after 15 minutes. Thin sections made at different levels within the crucible revealed that most of the olivine crystals had sunk 1–2 cm to form an accumulation at the base. [From N. L. Bowen, 1915, *Am. J. Sci.*, 4, 29, Fig. 1.]

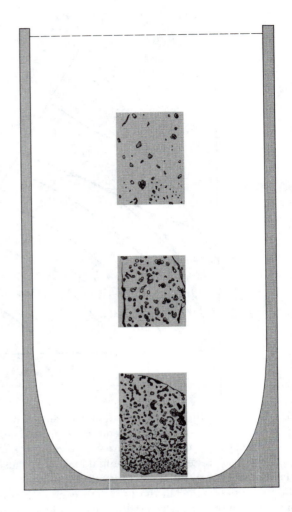

are Precambrian. They show a substantial range in size, from less than 100 km^2 to over 50,000 km^2.

Layered intrusions have many petrologic features in common. All of them have an overall or average basaltic composition, most commonly olivine tholeiite, and all show a prominent structural layering. This layering is a quasi-stratigraphy that subdivides intrusions into various types of olivine- or pyroxene-rich, ultramafic, layered rocks at the base; plagioclase-rich, mafic, plutonic rocks such as gabbros, norites (orthopyroxene-rich gabbros) and anorthosites in the middle levels; and finally highly fractionated felsic rocks such as syenites and granophyres at the top. There are some significant differences between intrusions in the details of sequences and rock types, but a subdivision into layers is universal.

Igneous petrologists have long interpreted the layering as evidence of crystal settling. Pioneering work on this model was done by Harry Hess and others in the 1930s and 1940s in classic studies of the Stillwater Complex in Montana. Hess was able to combine experimental information on the phase relations of igneous minerals, which was being produced in the burgeoning field of experimental petrology, with long-standing empirical observations on the physics of magma chambers. It may surprise some modern readers to learn that Charles Darwin and Leopold von Buch suggested gravitational settling of olivine in thick basalt lava flows in the 1840s! In 1915 N. L. Bowen reported experimental verification of olivine settling in experiments on the crystallization of basaltic magma (Figure 6-5). An essential clue from the bottom layers of layered intrusions was the presence of high concentrations of olivine, orthopyroxene, and chromite, all of which are denser than the magma and are among the earliest crystals to form from basaltic magma.

A comparison and correlation of the layers found in three of the best-studied layered intrusions are shown in Figure 6-6. Note that a composite section has been constructed for the Skaergaard and Stillwater Intrusions because of lack of exposure of the bottom of the Skaergaard and the top of the Stillwater. A feature common to many layered intrusions is the presence of a thin zone that appears along the margins or at the base and has been called a "marginal," "border," or "chilled" zone. Commonly a fine-grained gabbroic or noritic rock, this

chilled margin has been interpreted as representing a sampling of the initial unfractionated magma, which cooled so rapidly against wall rocks that it had no chance to fractionate chemically. Once formed, this thin zone would provide thermal insulation for the magma and would permit it to cool slowly enough to allow gravitational settling to produce monomineralic or bimineralic rock layers. The common occurrence of norite in chilled margins, however, suggests that some caution is required in using chilled margin chemistry to model the parental magma. As discussed later in this chapter, norite is typically not a "primary" igneous rock but is best explained as an olivine basalt magma contaminated with aluminous crustal material.

Crystallization Sequence in Layered Intrusions

An examination of fractional crystallization in appropriate liquidus phase diagrams is required in order to understand the "stratigraphy" of layered intrusions. The

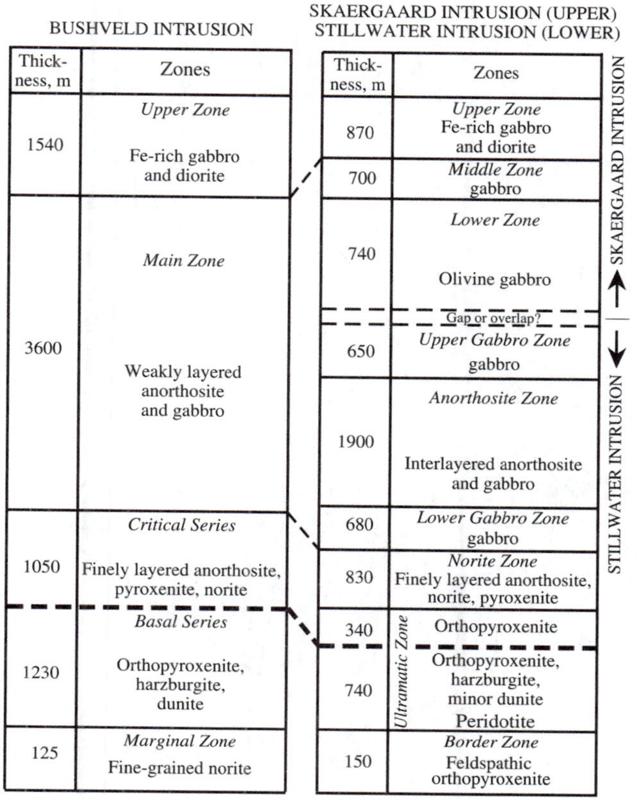

BUSHVELD INTRUSION		SKAERGAARD INTRUSION (UPPER) STILLWATER INTRUSION (LOWER)		
Thickness, m	Zones	Thickness, m	Zones	
1540	*Upper Zone* Fe-rich gabbro and diorite	870	*Upper Zone* Fe-rich gabbro and diorite	
		700	*Middle Zone* gabbro	
3600	*Main Zone* Weakly layered anorthosite and gabbro	740	*Lower Zone* Olivine gabbro	
			Gap or overlap?	
		650	*Upper Gabbro Zone* gabbro	
		1900	*Anorthosite Zone* Interlayered anorthosite and gabbro	
1050	*Critical Series* Finely layered anorthosite, pyroxenite, norite	680	*Lower Gabbro Zone* gabbro	
		830	*Norite Zone* Finely layered anorthosite, norite, pyroxenite	
1230	*Basal Series* Orthopyroxenite, harzburgite, dunite	340	Orthopyroxenite	
		740	Orthopyroxenite, harzburgite, minor dunite Peridotite	
125	*Marginal Zone* Fine-grained norite	150	*Border Zone* Feldspathic orthopyroxenite	

FIGURE 6-6

Composite generalized "stratigraphic" sections and approximate correlations for the Bushveld (left) and Stillwater and Skaergaard (right) Intrusions. The Bushveld Intrusion is mostly exposed, but Stillwater and Skaergaard sections have been combined because the bottom of Skaergaard is unexposed and the top of Stillwater has been removed by erosion. The occurrence of fine-scale rhythmic or cyclic layering within the major zones (for example, Upper, Middle, Lower) is illustrated. The position of economically important chromitite layers ("reefs") is shown by the heavy dashed line near the bases of the intrusions. [Modified after D. W. Hyndman, 1985, *Igneous and Metamorphic Petrology*, 2nd ed. (New York: McGraw-Hill), Fig. 6-27.]

actual chemical composition of the parental basaltic magmas is, of course, more complex than that shown in the simplified systems of three components. Perhaps the most useful idealized system to illustrate basalt fractional crystallization is the four-component (quaternary) system Mg_2SiO_4–$CaMgSi_2O_6$–$CaAl_2Si_2O_8$–SiO_2 (forsterite-diopside-anorthite-quartz). The low-pressure liquidus surfaces and cotectic lines in this system are illustrated in a perspective view in Figure 6-7. However, because it is so difficult to visualize the phase relations in three dimensions, it is more useful to examine separately the subordinate ternary systems forsterite-diopside-SiO_2 (Figure 6-8) and forsterite-anorthite-SiO_2 (Figure 6-9).

A typical olivine tholeiite bulk composition is shown in Figure 6-8. Note that this bulk composition falls within the triangle made by the compositions of the three solid phases olivine-diopside-enstatite. Under rules and conditions of equilibrium crystallization, the crystallization of this liquid composition should terminate with the liquid at the peritectic point *p*, because this point is the only place in the phase diagram where liquid can coexist with forsterite + diopside + enstatite. However, in fractional crystallization, the liquid composition must ultimately end up at the lowest temperature invariant point possible via cotectic or peritectic lines—in this case, the ternary

eutectic *e*. The final small fraction of liquid will therefore be able to crystallize quartz, even though the initial liquid was rich in the forsterite component.

The fractional crystallization sequence in Figure 6-8 is as follows: The liquid will begin to crystallize forsterite when it cools to the liquidus temperature. Liquid composition will shift directly away from the forsterite corner of the diagram until it intersects the olivine-enstatite peritectic curve. With fractional crystallization, it does not stay on the curve until forsterite is totally reacted but simply switches from crystallizing forsterite to crystallizing enstatite. This switch occurs because all of the forsterite that has so far been formed is assumed to have been physically removed. Figure 6.8 shows this change in liquid path direction as the liquid shifts across and down the enstatite liquidus. The next event in the evolution of the liquid is its intersection with the enstatite-diopside cotectic line. The liquid now begins to crystallize diopside as well as enstatite and thus shifts down the cotectic line toward the eutectic. Upon reaching the eutectic, the liquid cannot change further, and it crystallizes a mixture of diopside, enstatite, and quartz.

If fractional crystallization is occurring in a closed magma chamber and dense crystals settle toward the bottom of the chamber by gravitation, what is the sequence of layers formed? In the initial stages of

FIGURE 6-7

• • • • • • • • • • • • • • • •

Phase diagram model for the quaternary (four-component) system olivine-clinopyroxene-plagioclase-quartz (Ol-Cpx-Pl-Qz), illustrating the generalized liquidus relations at low pressures for basalts and related magmas. Note that the cotectic planes in this diagram are three-dimensional equivalents of the cotectic lines in two-dimensional ternary diagrams; that is, they bound the primary crystallization fields of single minerals, as labeled on this diagram. Cotectic planes therefore represent the simultaneous crystallization of two solids from the magma. Planes intersect to form lines that represent the three-dimensional extensions of invariant points (eutectics or peritectics) in a ternary diagram: Three solids coexist with liquid along these lines. Liquids evolve through the volumes to the planes, along the planes to the lines, and finally along the lines to quaternary invariant points. [From T. N. Irvine, 1979, in *Evolution of the Igneous Rock: Fiftieth Anniversary Perspectives*, ed. H. S. Yoder, Jr. (Princeton, NJ: Princeton University Press), Fig. 9-1.]

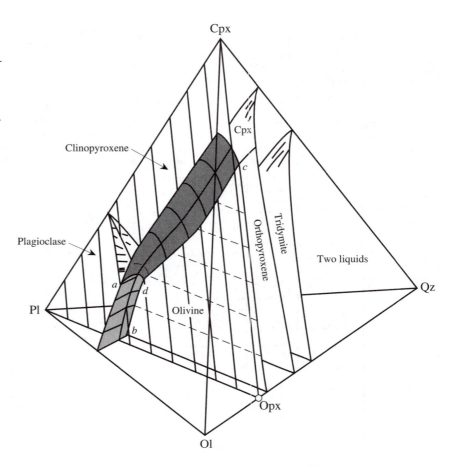

crystallization, only olivine is precipitated from the liquid, and its accumulation on the bottom of the chamber will create a rock layer consisting almost entirely of olivine, that is, dunite. Note in Figure 6-6 that both the Stillwater and Bushveld Intrusions have dunite layers immediately above the bottom chilled zone. Mixed with the dunite layers are very thin bands (sometimes called "reefs") of chromitite, rocks that consist almost entirely of chrome spinel (chromite). Chromitites are economically important because they carry substantial amounts of platinum and related rare metals, as well as chromium. Chromitite layers form when chromite crystallizes simultaneously with olivine; it is not clear why the chromite concentrates into the reefs. Chromite crystallization does not show on the simplified phase diagrams because Cr_2O_3 is not shown as a separate chemical component. Chromite always crystallizes early in the sequence in basaltic magmas and is sometimes referred to as a refractory mineral.

The shift in crystallization from olivine to orthopyroxene appears to be so abrupt that the magma probably contains both types of crystals suspended for some time. They settle together, producing layers of harzburgite, a type of ultramafic rock composed of olivine plus ensta-

tite. After olivine ceases crystallizing and all crystals have settled out, orthopyroxenite layers containing only enstatite and perhaps minor spinel succeed the harzburgite layers. With the onset of crystallization of diopside in addition to enstatite, the accumulated layers become pyroxenite (websterite) with both enstatite and diopside. The presence of these types of layers in layered intrusions is documented in the rock sections of Figure 6-6.

In most layered intrusions, the lowermost ultramafic layers are succeeded by plagioclase-bearing rocks in the middle levels of the intrusions. Because it is impossible to illustrate the formation of plagioclase-bearing layers by using a plagioclase-free diagram (Figure 6-8), it is now necessary to shift to the system forsterite-anorthite-silica shown in Figure 6-9. On the basis of an analysis of the forsterite-diopside-silica system, it is reasonable to assume that diopside appears in the crystallization sequence after olivine and enstatite. Because diopside-bearing ultramafic rocks occur below plagioclase-bearing layers in layered intrusions, we can also assume that the interval of diopside crystallization in natural basaltic liquids falls between, and overlaps, those of both enstatite and plagioclase. In the fractional crystallization path illustrated in Figure 6-9, production of both norites

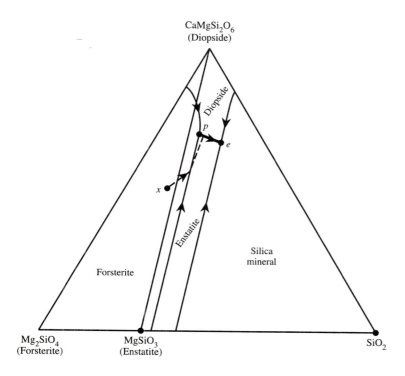

FIGURE 6-8
• • • • • • • • • • • • • • • •

Liquidus phase relations in the $CaMgSi_2O_6$–Mg_2SiO_4–SiO_2 (diopside-forsterite-silica minerals) ternary system (the right front face of the quaternary system in Figure 6-7). The initial liquid composition x indicates where a typical olivine tholeiite magma would plot on this diagram. Fractional crystallization of liquid x results sequentially in crystallization of forsterite, enstatite, enstatite + diopside, and enstatite + diopside + silica along the liquid path shown by the dashed line in the diagram. p indicates the peritectic point and e the ternary eutectic.

and gabbros can be envisioned readily as the liquid path intersects the pyroxene-plagioclase cotectic. Note the presence of these layers in the rock sections (see Figure 6-6).

The common occurrence of anorthosite in the middle levels of layered intrusions presents a problem. Anorthosites are plagioclase-rich rocks (>90% plagioclase), and those in layered intrusions consist of roughly An_{60} plagioclase. Phase diagram interpretation provides no obvious mechanism for generating such nearly monomineralic plagioclase rocks. To explain such thick layers, petrologists have invoked physical mechanisms involving crystal settling and flotation in the magma chamber. The ferromagnesian minerals olivine and pyroxene are denser than magma and therefore sink. Calcic plagioclase (>An_{65}) is also denser than magma and probably sinks. However, more sodic plagioclase (less than about An_{60}) is of equal or lower density than magma and can therefore either have neutral buoyancy and remain fixed in position or even float upward. Thus anorthosites with plagioclase more sodic than about An_{65} are explained as layers that resulted from plagioclase flotation or from preferential loss of the denser ferromagnesian minerals.

Unfortunately, such a simple explanation does not work for the Stillwater Intrusion, because the thick accumulation (over 1000 m) of anorthosite layers contains plagioclase of An_{75-80}. Harry Hess noted the problem with the simple settling versus flotation hypothesis and suggested that large-scale convective overturn of magma or crystal-liquid mushes in the magma chamber could

explain the anorthosites. He argued that contamination of fresher liquid in the center of the magma chamber with more fractionated magma overturned from near the roof could displace the intermixed or contaminated magma over a cotectic curve into the plagioclase primary crystallization field. This displacement could cause an abrupt shift from crystallization of norite or gabbro to crystallization of plagioclase alone.

The "stratigraphic" sections in Figure 6-6 show that the crystallization sequence is also complicated by repeated intervals of gabbro and anorthosite formation in the middle levels of layered intrusions and by formation of olivine gabbro *above* anorthosite in the Stillwater. Studies of the petrology and geochemistry of layered intrusions during the late 1980s have suggested that these repetitions of lithology probably represent periodic injections into the magma chamber of fresh, new unfractionated basaltic magma which mixes with and contaminates the fractionated liquid and causes perturbations and repetitions in the crystallization sequence. This more complicated history of magma injections into a magma chamber can be modeled by using simple phase relations such as those already discussed; but they can be demonstrated in a compelling fashion by documenting subtle shifts in trace element and isotopic chemistry of the layered rocks.

The final stages of fractionation are represented by intermediate and even granitic layers near the roofs of layered intrusions. These layers include diorites and **granophyres** in which even highly fractionated

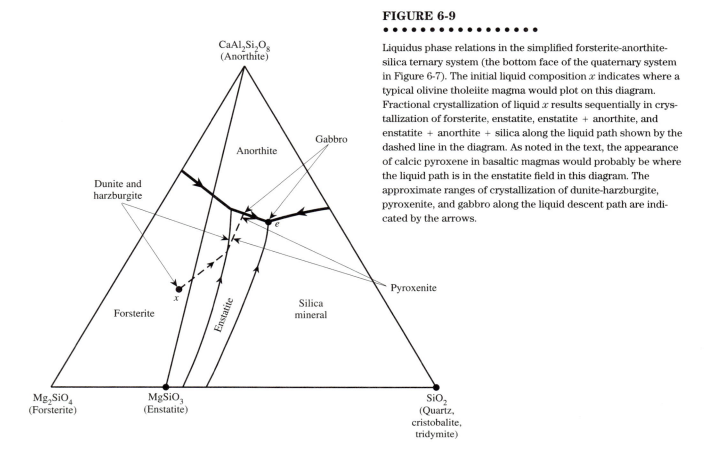

FIGURE 6-9

● ● ● ● ● ● ● ● ● ● ● ● ● ● ● ● ●

Liquidus phase relations in the simplified forsterite-anorthite-silica ternary system (the bottom face of the quaternary system in Figure 6-7). The initial liquid composition x indicates where a typical olivine tholeiite magma would plot on this diagram. Fractional crystallization of liquid x results sequentially in crystallization of forsterite, enstatite, enstatite + anorthite, and enstatite + anorthite + silica along the liquid path shown by the dashed line in the diagram. As noted in the text, the appearance of calcic pyroxene in basaltic magmas would probably be where the liquid path is in the enstatite field in this diagram. The approximate ranges of crystallization of dunite-harzburgite, pyroxenite, and gabbro along the liquid descent path are indicated by the arrows.

minerals such as alkali feldspar and quartz occur. An interesting characteristic of these layers is that most ferromagnesian minerals, for example, olivine and pyroxenes, are considerably more iron-rich than in the gabbroic rocks lower in the section. This composition trend reflects the tendency in the ferromagnesian solid solution series to fractionate toward progressive iron enrichment at lower crystallization temperatures. A classic example of this type of trend is displayed by the liquidus phase relations of olivine in the forsterite-fayalite binary system (Figure 6-10); similar trends are seen in the low-calcium pyroxene (enstatite-ferrosilite and pigeonite) and clinopyroxene (augite-ferroaugite) series, as shown in the calcium-magnesium-iron pyroxene "quadrilateral" (Figure 6-11). The trend has been called the tholeiite trend and has a characteristic pattern in the AFM diagram (Figure 6-12) that is controlled by the fractionation of low-calcium ferromagnesian minerals.

Layered intrusions thus present us with a demonstration of the effect of fractional crystallization on the chemistry of basaltic magmas. Early-crystallizing, high-temperature minerals such as magnesium-rich olivine, pyroxene, and chromite deplete the magma in certain chemical components such as MgO and Cr_2O_3, and enrich it in FeO, Na_2O, K_2O, Al_2O_3, and SiO_2. Investigations of layered intrusions have enabled igneous petrologists and geochemists to establish whether major and trace elements are fractionated into early high-temperature crystals and rock layers or instead are concentrated into the residual magma. Understanding the behavior of elements in such a "closed" system allows us to use the relative enrichment or depletion of any element in a magmatic system as a monitor of degree of fractionation.

Physical Characteristics of Cumulate Layers in Layered Intrusions

The cumulate layers in many layered intrusions, particularly the lowermost ultramafic layers, have physical characteristics that are consistent with an origin through gravitational settling. Evidence includes both microscopic textural features and larger scale structures analogous to sedimentary structures. These layers have been called *cumulates* to reflect their possible origin by *accumulation* at the bottom of a magma chamber. It should be noted that many of these features can also be explained by other growth mechanisms that do not involve gravitational settling. Research into the origin of both layering and cumulate features is currently a very

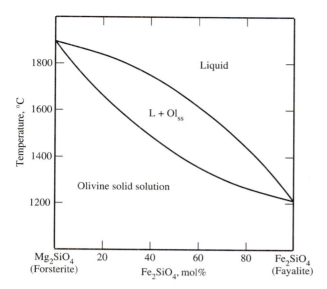

FIGURE 6-10

●●●●●●●●●●●●●●●●

Binary olivine system Mg_2SiO_4–Fe_2SiO_4 (forsterite-fayalite) diagram showing the liquidus-solidus loop and complete solid solution of iron and magnesium in olivine. This system serves as a model for crystallization of many iron-magnesium solid solutions in which the magnesium end member has a higher melting point than the iron end member. Fractional crystallization of olivine from basaltic magmas will therefore produce iron enrichment in evolved liquids.

active area of interest to igneous petrologists, and substantial revisions of the classic explanations of igneous layering and related features are possible.

The nomenclature of chemical and textural features is to a major extent based on classic petrogenetic models for the origin of these features. Both igneous and metamorphic petrologists have been working with varying success for several decades to eliminate genetic implications from descriptive terminology. Note that current cumulate terminology must be used with caution because it contains some possibly misleading genetic implications. **Cumulate texture** (Figure 6-13) consists of an interlocking fabric of subhedral to euhedral medium to large crystals on the order of 5 mm or more in size **(cumulus crystals)** and interstitial patches of other minerals **(postcumulus crystals).** Postcumulus minerals form by crystallization of intercumulus liquid. The

large olivine and pyroxene crystals in bottom layers apparently grew to significant size during their possible descent through the magma. Because of their shapes, there was a limit to the packing of these crystals after settling, and thus small pockets of interstitial magma remained between them. **Adcumulate texture** involves the largest amount of intercumulus liquid, up to 50%. In adcumulates, communication between interstitial liquid and the main body of magma is so efficient that little liquid gets trapped, and adcumulates have less than 10% postcumulus minerals. **Mesocumulates** have a greater proportion of trapped liquid and are generally defined as having 10 to 40% postcumulus crystals.

In practice, the recognition of and distinction among these textures may not be easy. Patches of evolved intercumulus liquid must have been multiply saturated with crystalline phases, including the cumulus crystals themselves. Crystallization of these liquids in a dense crystal mush should involve formation of crystallographically and optically continuous overgrowths on cumulus crystals, thus significantly modifying the original shapes produced by unconstrained growth in liquid. Crystallization of minerals not represented among the cumulus phases produces newly formed interstitial crystals, for example, plagioclase in an ultramafic cumulate dominated by olivine and pyroxene. Another commonly observed textural feature of many cumulates is *poikilitic* texture

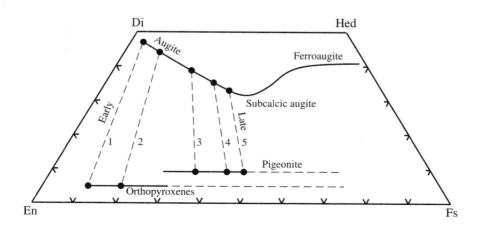

FIGURE 6-11

●●●●●●●●●●●●●●●●

The pyroxene quadrilateral diopside (Di)-hedenbergite (Hed)-enstatite (En)-ferrosilite (Fs), showing crystallization trends of low-calcium and high-calcium pyroxenes in mafic rocks, especially layered intrusions. Five representative equilibrium pyroxene pairs from basaltic rocks are shown, from a typical early augite-bronzite pair (1) to a late subcalcic augite-pigeonite pair (5). Orthopyroxenes are typical in early-crystallizing, magnesium-rich pairs, whereas monoclinic pigeonites occur with augite or subcalcic augite in more fractionated, iron-rich magmas.

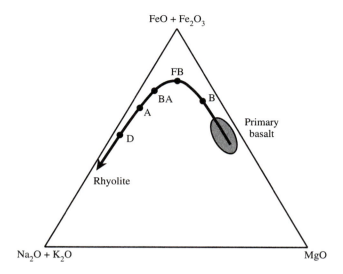

FIGURE 6-12
● ● ● ● ● ● ● ● ● ● ● ● ● ● ● ●

FM diagram showing the tholeiite trend for magma fractionation from basalt to rhyolite. Areas for primary basalt and for rhyolite are shown at the beginning and the end of the trend, respectively. Approximate positions for various fractionated magmas are shown along the trend: B, basalt; FB, ferrobasalt; BA, basaltic andesite; A, andesite; D, dacite. [After M. Wilson, 1989, *Igneous Petrogenesis* (London: Unwin-Hyman), Fig. 1-4.]

(Figure 6-14), in which large, late postcumulus crystals overgrow and enclose smaller cumulus crystals. These large crystals, called **oikocrysts,** can exceed several centimeters in diameter, are optically continuous, and can enclose tens or hundreds of euhedral cumulus crystals, called **chadocrysts.**

An intriguing feature of many layered intrusions is the occurrence of small- and medium-scale structures identical to those found in sedimentary rocks (refer to Figure 6-1B). These structures include ripple marks, cross-bedding of various types, troughs, load casts, and rip-up structures. Some of the best descriptions of such structures come from the Skaergaard Intrusion in East Greenland. It is difficult to envision any explanation for these features other than the existence of currents within the magma chamber that move sedimented crystals on the active chamber floor much as sand and gravel are moved in a stream or ocean current. These currents can be either localized or related to convection of magma on a large scale within the chamber. Such structures remain problematic but are among the most compelling evidence for the gravitational settling model.

The occurrence of cumulate layers defeats bulk chemical analysis, one of an igneous petrologist's favorite analytical tools for reconstructing petrogenetic processes. Compositions of the final rock types do not represent any sensible liquid composition because of the coincidental mixing of cumulus minerals and intercumulus liquids that may originate in different parts of the magma chamber. Further, the intercumulus liquids themselves are probably complex mixtures of different batches of variably evolved magma. Compaction of cumulus crystals in layers occurs through the weight of overlying layers. In this compaction process, intercumulus liquid is squeezed upward toward the interface between cumulate layers and magma in a process called **filter press-**

ing. A complex pattern of diffusion of chemical components through the liquids is coupled with the physical movement of these liquids. As a result, any cumulate rock can have a composition that has been affected dramatically by what is called infiltration metasomatism.

New Ideas Regarding Layered Intrusions

The classic interpretations of the origins of layered intrusions have clearly helped to advance the quantification of igneous processes, but some igneous petrologists now question the classic model and argue that even elaborate modifications of the fundamentally simple models of fractional crystallization and crystal settling cannot explain many features of layered intrusions. They cite several lines of evidence that are inconsistent with gravitational settling, including the occurrence of almost vertical layering on the sides of several intrusions and the inconsistency between small changes in the Fe:Mg ratio of olivines and pyroxenes and the large amount of fractional crystallization required for generating different rock types in a layering sequence.

New models proposed for layered intrusions retain one aspect of the classic models: fractional crystallization of basaltic magma. They differ in the exact mechanism by which chemically evolved liquid is separated from its crystalline products. Irvine (1980) has studied the Muskox Intrusion in Canada for many years, and he has concluded that the magma chamber received at least 20 major pulses of fresh magma during the differentiation process. Cyclic units containing (in upward order) olivine-rich, pyroxene-rich, and plagioclase-rich layers are repeated many times. Irvine proposed that each repetition represents a fresh batch of magma that displaced

FIGURE 6-13
● ● ● ● ● ● ● ● ● ● ● ● ● ● ● ●

Cumulate textures. **(A)** *Plagioclase orthocumulate.* Early calcium-rich cumulate plagioclase is originally surrounded by melt of different composition. The melt has crystallized and no reaction between melt and plagioclase has occurred, resulting in a zoned plagioclase rim. **(B)** *Plagioclase mesocumulate.* An intermediate situation between plagioclase orthocumulate and adcumulate growth. Considerable unzoned adcumulate growth has occurred, but the outer portion of the plagioclase rim shows orthocumulate growth with zoning. Other minerals have also crystallized from the melt. **(C)** *Plagioclase adcumulate.* Plagioclase cumulate surrounded by a melt of different composition has grown by acquisition of material from an external source. The plagioclase forms unzoned crystals as the interstitial liquid is forced out. **(D)** *Polymineralic adcumulate.* The three minerals have grown by adcumulate growth, and all are unzoned. **(E)** *Olivine heteradcumulate.* The cumulate olivine grains show no evidence of growth. The surrounding melt has nucleated and undergone adcumulate growth, forming large, essentially unzoned poikilitic pyroxene and plagioclase. **(F)** *Olivine crescumulate.* Cumulate olivine has grown upward into the melt above, changing composition slightly to correspond to the surrounding melt. All olivine shown is in optical continuity. The surrounding pyroxene and plagioclase are also essentially unzoned and grown as heteradcumulate. [From L. R. Wager, G. M. Brown, and W. J. Wadsworth, 1960, *J. Petrol.*, *1*, Figs. 1, 2.]

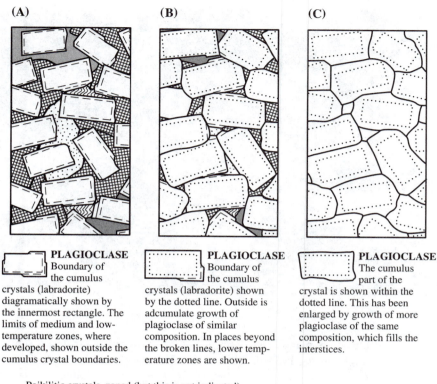

(A) **(B)** **(C)**

PLAGIOCLASE
Boundary of the cumulus crystals (labradorite) diagramatically shown by the innermost rectangle. The limits of medium and low-temperature zones, where developed, shown outside the cumulus crystal boundaries.

PLAGIOCLASE
Boundary of the cumulus crystals (labradorite) shown by the dotted line. Outside is adcumulate growth of plagioclase of similar composition. In places beyond the broken lines, lower temperature zones are shown.

PLAGIOCLASE
The cumulus part of the crystal is shown within the dotted line. This has been enlarged by growth of more plagioclase of the same composition, which fills the interstices.

Poikilitic crystals, zoned (but this is not indicated)

Pyroxene   Olivine   Iron ore   Quartz and orthoclase, locally the final residuum

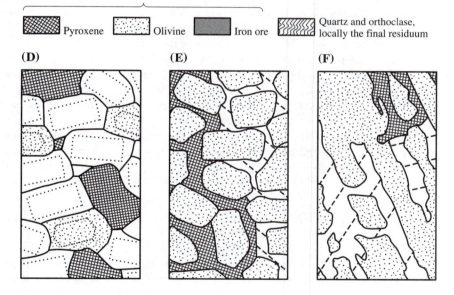

(D) **(E)** **(F)**

differentiated magma and began the fractionation cycle over again. The application of fluid dynamic theory to layered intrusions has raised the possibility of far more complicated convective magma movement and localized diffusive transfer of chemical components to explain much of the fractionation observed (McBirney and Noyes 1979). Layered intrusions, however, remain a use-

ful example of fractionation of basaltic magma and a natural laboratory for testing models for large-scale modification of magma chemistry.

Characterizing processes of magma chemical modification is especially important to understanding how much basaltic magmas are modified in their ascent to the surface. It is now widely accepted by petrologists

FIGURE 6-14
● ● ● ● ● ● ● ● ● ● ● ● ● ●

Poikilitic texture. Large twinned plagioclase grain enclosing several smaller euhedral to subhedral augite grains. Sample is a gabbro from the Bushveld Complex. Crossed polarizers; width of field of view is about 3.5 mm.

that most basaltic magma originates through melting of ultramafic rocks in the lithospheric mantle. Our knowledge of the chemistry and mineralogy of the mantle is based mostly on indirect evidence, for example, seismic information, suites of ultramafic xenoliths (apparent pieces of mantle wall rock brought to the surface by basalts and kimberlites), and basalt magma chemistry. Experimental petrology has calibrated the dependence of magma chemistry on the chemistry and mineralogy of source rocks, and so basalts and related rocks have been used in the last two decades as probes to investigate the nature of their source areas in the mantle. Modification of basalt magma by fractional crystallization during ascent as well as by crustal contamination (see below) must dramatically affect our interpretations.

MAGMA CONTAMINATION
● ● ● ● ● ● ● ● ● ● ● ● ● ● ●

Petrologists now believe that very few magmas move far from their source areas without some degree of **contamination,** or compositional modification through addition of material. Contamination mechanisms taken together can be classified as **assimilation,** and they involve the common process of partial to complete melting of adjacent country rock or xenoliths and thus incorporation of foreign chemical material into the magma. Alternatively, assimilation can occur without melting by entrainment into the magma of solids from single crystals to large xenoliths. On occasion, the actual digestion of country rock and its addition to the magma can be seen in outcrop (Figure 6-15). The most important con-

trols on the degree of contamination are the physical and chemical access of magma to country rocks, the ambient preintrusion temperatures of the country rocks, and the relative values of magma temperatures and country rock melting points. Some magmatic rocks such as norite are considered by petrologists to be useful indicators of assimilation because they rarely occur as primary igneous magmas. (*Primary* magmas are considered to be those that have undergone little fractionation and had no contamination.) On occasion, norites result from crystallization of highly unusual aluminum-rich primary magmas, but isotopic evidence (particularly strontium and oxygen isotopes) suggests that most norites represent contamination of normal basaltic magma by aluminous crustal material.

Physical and Chemical Access

Magmas move through the upper mantle or crust by several mechanisms that depend on the mechanical properties of the magma. Low-viscosity magmas such as basalts are able to exploit, or create, cracks and fractures that act as conduits for flow. The common occurrence of basalts in thin tabular bodies such as dikes or sills attests to this process. High-viscosity magmas such as granites cannot flow through narrow conduits and thus typically form larger rounded or bulbous magma chambers.

The movement of basaltic magma upward in the crust through a series of narrow "braided" conduits presents a maximum of surface area contact between magma and wall rocks. The more irregular the shape of the conduit

FIGURE 6-15
● ● ● ● ● ● ● ● ● ● ● ● ● ● ●

Xenoliths of layered phyllite and fine-grained mafic rock being mechanically disaggregated and partially digested by granitic magma, coastal Maine. The horizontal dimension of the photo is about 1 m. [Photo by Dr. A. K. Sinha, Virginia Tech.]

walls, the greater the contact area. The transfer of magmatic heat to the surroundings, a critical component of assimilation through melting, is most effective with greater contact area and with higher temperature magmas. In addition, the fracturing of wall rock into tabular fragments or slivers further increases the surface area. Large surface area also enhances chemical access by creating a maximum of interface across which movement of chemical species can occur. The tendency of granites to form larger, discrete, rounded bodies reduces the surface contact area relative to magma volume (note that the lowest surface to volume ratio is in a sphere).

Chemical access of magmas to the constituents contained within country rocks is controlled by a complex array of chemical, physical, and thermal factors. One of the most fundamental of these is the difference in bulk chemistry between magma and country rock. Transfer of chemical mass by diffusion, even in the absence of melting of the country rocks, typically occurs *down* chemical concentration gradients in much the same way that thermal energy is transferred from hotter to cooler bodies. Thus a basaltic magma that has intruded a sequence of old basaltic lava flows of similar composition will have little chemical interaction with its surroundings because of a lack of driving force for chemical mass transfer. Even if assimilation did occur, addition of basaltic contaminants to a basaltic magma would have little to no bulk mineralogic or chemical effect, except perhaps in isotopic compositions. Conversely, basaltic magma intruding shale might be significantly altered by addition to the magma of country rock components, such as K_2O, FeO, Al_2O_3, and SiO_2, that are more abundant in the shale than in the magma. These components would tend to migrate into the magma if shale (or schist) minerals containing them broke down under the magmatic heating. (By the same token, some components might migrate out of the magma into the surroundings as metasomatism if they are more abundant in the magma.)

An important factor in both heat transfer and chemical interaction of magma with its surroundings is the speed of magma flow. Because of their low viscosity, basaltic magmas flow fast enough that interaction with the surroundings is limited. Magmas such as these in fact undergo a minimum of either contamination or fractional crystallization and arrive at the surface maintaining a primitive chemistry. When magmas move upward in the upper mantle and crust through a series of holding chambers, which impede velocity, strong interactions are more likely.

Some assimilation undoubtedly takes place through the physical disaggregation of xenoliths or in situ wallrock into individual grains and subsequent dissolution of these crystals by the magma, a process distinct from melting. Thermal shock could significantly aid

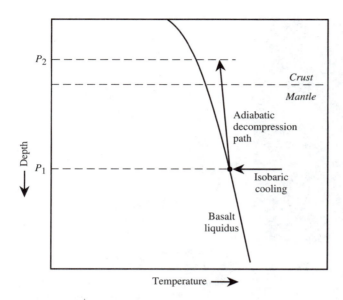

FIGURE 6-16
● ● ● ● ● ● ● ● ● ● ● ● ● ● ●

Temperature-depth diagram showing a typical liquidus curve for a basaltic magma as a function of depth (pressure). If a magma that is cooling isobarically and just beginning to crystallize at P_1 in the mantle undergoes a rapid ascent through the mantle into the crust to P_2 without losing heat to the surroundings ("adiabatic decompression"), the magma can gain superheat by increasing the ΔT between itself and the liquidus at P_2.

disaggregation when cold country rock comes in contact with magma, because very rapid differential thermal expansion of minerals causes breakage along grain boundaries. Dissolution of liberated grains depends on the solubilities of these minerals in the magma. Relatively insoluble crystals could persist as xenocrysts, and either soluble or insoluble species might be overgrown and armored by a magmatic mineral that could nucleate on the xenocrystic surface. Examples of partial dissolution or resorption are commonly seen in thin sections of rocks, and even more complicated scenarios of partial resorption followed by later euhedral overgrowth can be found, particularly with refractory accessory minerals such as zircon.

Thermal and Energy Effects

The thermal or energy budget of magma-country rock interactions may play an important role in the extent of contamination. Whether through melting of country rock or other chemical processes such as diffusion, contamination is typically enhanced by high temperatures. A considerable amount of energy is required to heat the surroundings of a magma body by conduction, and even more is needed to partially melt country rocks. When a magma has a higher temperature than its liquidus temperature (at the beginning of crystallization), it is said to have superheat, or excess thermal energy, that can heat the surroundings, and simultaneously cool the magma, without causing it to begin to crystallize.

In general, however, magmas are assumed to have left their source areas immediately after or even during melting without any further temperature increase and thus

are likely to be at or close to their liquidus temperature (Figure 6-16). In this example of a dry basaltic magma, only with a rapid ascent and purely adiabatic cooling (the cooling that is due *solely* to slight expansion of the magma during pressure release) does the magma get much above its liquidus *P-T* curve. If it is finally emplaced at the depth indicated in the figure, it does have some superheat.

Perhaps a more important source of thermal energy in a magma is the heat (or enthalpy) of crystallization. Significant energy is required to keep mineral molecules such as $CaMgSi_2O_6$ in a disordered liquid rather than in an ordered crystalline state. As these molecules organize themselves into crystals during crystallization, a dramatic amount of energy is given off. It has been estimated that this energy plays the major role in thermal and chemical interactions between magma and country rocks. However, it is typically made available at a stage of magma evolution when mobility is decreasing as the result of an increased effective viscosity of the partially crystallized magma and thus is likely to be more important in affecting magma compositions in the final magma chambers than en route to them.

One other thermal effect that is obvious, but should be mentioned, is the reduced likelihood of assimilation by low-temperature magmas such as granite relative to that of high-temperature magmas such as basalt. Low-temperature magmas do not have the heat content required to melt and assimilate country rocks with high solidus temperatures such as basalt (or its metamorphic equivalents, greenstone or amphibolite). Conversely, basaltic magmas are hot enough to raise the temperatures of metasedimentary and metaigneous rocks of the continental crust above their initial melting points and thus are more likely to have been affected by assimilation.

Effects on Basaltic Magma Chemistry

In the deeper crust, most assimilation probably takes place by partial melting of the rock surrounding a magma conduit or chamber and subsequent mixing of these secondary melts with the main magma. The magma need not melt country rock entirely; magma only has to heat it to its lowest temperature eutectic point. For example, country rock that has a composition equivalent to shale might not *entirely* melt until its temperature reaches 1000°C or higher. However, a small fraction of the rock will melt to form a magma of roughly granitic composition at 650°–750°C. This liquid might physically move across a magma chamber wall and mix with the main magma. A small fraction of this type of granitic magma mixed with a larger volume of basaltic magma could convert basalt into norite. Norite formation occurs with addition of Al_2O_3 and SiO_2 to the basaltic magma; CaO, which would ordinarily go into formation of calcic pyroxene, instead forms calcic plagioclase, thus promoting formation of orthopyroxene. The common occurrence of norite in the marginal or border phases of layered intrusions, as noted earlier, provides an important clue that these rocks are not truly pristine chilled margins and thus do not preserve the chemistry of the original magma. Instead, they represent magma contaminated by addition of aluminous material from surrounding country rocks.

Production of norite is one rather apparent effect of contamination, but basaltic magmas can be contaminated more subtly without producing such obvious major element or mineralogic effects. Because petrologists are concerned with using basaltic magmas as probes of mantle source areas, they must be cautious about the amount of contamination such magmas have undergone. In general, not enough is known about the detailed chemistry of the upper mantle to be able to predict primitive basaltic magma chemistries well enough to be able to discriminate contaminated magmas. It is here that trace elements and particularly isotopes can help with defining subtle contamination effects. The *incompatible* trace elements are those that are typically fractionated into later liquids in fractionally crystallized basaltic magmas. Examples include ions such as barium, cesium, hafnium, and zirconium. These ions tend to be enriched in felsic crustal rocks relative to basaltic magmas or their source areas. Anomalously high concentrations of these ions or unusually high or low ratios of elemental concentrations are important clues to contamination.

Even more useful than trace elements are radiogenic and, to a lesser extent, stable isotopes. Fractionation of certain radioactive elements into crustal rocks over geologic time has generated isotopic ratios in the crust that are quite different from those in the mantle, particularly in ratios between the strontium, samarium, neodymium, rhenium, and osmium isotopes. Relatively young basaltic rocks that have passed through the crust without much interaction or contamination should have distinctive strontium and neodymium isotopic "signatures" typical of the mantle. Research has shown that a relatively small amount of crustal contamination can alter these isotopic ratios dramatically, providing igneous petrologists with a very sensitive tool for studying magma-crust interaction.

MAGMA MIXING

One additional mechanism for production of nonprimary magma compositions is magma mixing. This process was little studied before the advent of trace element and isotopic techniques in igneous petrology because petrographic and mineralogic effects are not particularly diagnostic. As noted by McBirney (1993), magma mixing typically involves the mingling of two somewhat evolved or differentiated magmas to produce a single new magma that appears *less* differentiated than either of the two that were mixed.

In most cases, mixing of magmas is a physically problematic process. Magmas of different densities that are introduced into a single magma chamber should tend to become stratified gravitationally. Some physical mixing mechanism is required to promote local interactions, perhaps turbulence or rapid convection. Magmas with substantial differences in density are also likely to have quite different temperatures. For example, if basaltic magma at 1200°C were injected into a magma chamber containing 800°C granitic magma, the large temperature difference would inhibit mixing because the basalt would chill and crystallize rapidly when it came in contact with the "cool" granitic magma. Examples of such mingling of basaltic and granitic magmas exist, especially along the New England coast from Boston northward to Maine and in the Sierra Nevada of California.

SUMMARY

A review of the principles that govern the generation, crystallization, and possible compositional modification of magmas is now complete. Fractional crystallization is a potent mechanism for either subtle or significant shifts in the chemical compositions of magmas after they leave their source areas. In general, there is not enough evidence to say whether most basaltic magmas have been modified through fractionation, but the large mafic layered intrusions provide unequivocal evidence for the operation of this process on a grand scale. Another major process by which magmas can be modified chemically is contamination through assimilation of foreign chemical material into the magma. In situ melting of country rocks, melting of country rock xenoliths, and incorporation of unmelted solid material are the principal mechanisms of assimilation. Petrologists and geochemists have available analytical tools that can detect even very subtle contamination. If magmas, particularly basalts, are to be used as probes to examine the nature of their lithospheric mantle source areas, petrologists must be certain that the chemistry of the final rocks is truly representative of the primary magma derived from the source and not an artifact of later magmatic evolution.

STUDY EXERCISES
• • • • • • • • • • • • • •

1. In your own words, describe how fractional crystallization differs from equilibrium crystallization.

2. How would you use patterns of igneous rock compositions (specifically, clustering of compositions for specific rock types such as ocean floor basalt or intraplate alkali basalt) to determine whether most magmas undergo substantial fractionation during ascent?

3. Using the ternary systems diopside-forsterite-silica (Figure 6-8) and forsterite-anorthite-silica (Figure 6-9), try to work out fractional crystallization sequences that would give rise to the layering found in layered intrusion (Figure 6-6). Experiment with each diagram to see what compositions of mixed magmas would result from the mixing of fractionated melts and batches of new magma. (*Hint:* You might find some interesting "zigzag" liquid descent lines.)

4. Is there any sensible way to track the continuous evolution in the bulk compositions of fractionated magmas in a layered intrusion when a "stratigraphic" layering develops as a result of such physical processes as gravitational settling? If not, how do we know that fractionation actually occurred?

5. Summarize the ways that trace elements and isotopes can be used to track the processes of magma contamination and magma mixing.

REFERENCES AND ADDITIONAL READINGS
• • • • • • • • • • • • • •

Bowen, N. L. 1928. *The Evolution of the Igneous Rocks.* Princeton, NJ: Princeton University Press.

DePaolo, D. J. 1981. Trace element and isotopic effects of combined wallrock assimilation and fractional crystallization. *Earth Planet. Sci. Lett.,* *53,* 189–202.

Hess, H. H. 1960. *Stillwater Igneous Complex, Montana: A Quantitative Mineralogical Study. Geol. Soc. Am. Mem. 80.*

Irvine, T. N. 1979. Rocks whose composition is determined by crystal accumulation and sorting. In *The Evolution of the Igneous Rocks: Fiftieth Anniversary Perspectives,* ed. H. S. Yoder, Jr. Princeton, NJ: Princeton University Press, pp. 245–306.

Irvine, T. N. 1980. Magmatic infiltration metasomatism, double diffusive fractional crystallization and adcumulus growth in the Muskox Intrusion and other layered intrusions. In *Physics of Magmatic Processes,* ed. R. B. Hargraves. Princeton, NJ: Princeton University Press, pp. 325–384.

McBirney, A. R. 1993. *Igneous Petrology,* 2nd ed. Boston: Jones and Bartlett.

McBirney, A. R., and R. M. Noyes. 1979. Crystallization and layering of the Skaergaard Intrusion. *J. Petrol.,* *20,* 487–454.

Wager, L. R., and G. M. Brown, 1967. *Layered Igneous Rocks.* San Francisco: W. H. Freeman.

Yoder, H. S., Jr., ed. 1979. *The Evolution of the Igneous Rocks: Fiftieth Anniversary Perspectives.* Princeton, NJ: Princeton University Press.

Chapter 7

PETROLOGY OF THE MANTLE

Petrologists generally accept that basaltic and related mafic magmas originated in the earth's mantle. But knowledge of the chemical and physical characteristics of the deep interior of the earth is indirect, because of physical inaccessibility. Nevertheless, a substantial body of theory has developed regarding the composition, mineralogy, density, structures, and thermal patterns of the mantle and is based on geophysical measurements of gravity, magnetism, and seismic velocities and extrapolation of meteorite compositional data. New geophysical techniques developed in the 1980s, especially computer enhancement of the enormous seismic database that has been accumulated over recent decades, have enabled "seismic tomography" of the mantle, which reveals structural complexity heretofore only guessed at.

Direct observation of mantle rocks is limited to rare samples of the mantle that have been brought to the surface as ultramafic xenoliths in mantle-derived magmas. The magmas themselves can provide information about their mantle source areas, but this information must be interpreted in light of the possible modification magmas undergo through fractionation and contamination as they rise through the crust. Models of the deeper mantle become increasingly inferential for depths below several hundred kilometers because there are no known direct samples and models are largely based on geophysical interpretations of temperature-depth profiles, rock density and mineralogy, phase changes with depth, and other parameters. Geophysical models can be constrained by geochemical modeling of the earth's interior that is based mostly on major-element, trace-element, and isotopic analyses of deeply derived magmatic rocks and comparison with samples from other planetary bodies, that is, lunar rocks and meteorites. This chapter presents infor-

mation on the current knowledge of the earth's deep interior and discusses some of the implications of mantle petrology for the melting of mantle peridotite to produce mantle-derived magmas, including basalts, kimberlites, and lamprophyres.

GROSS VERTICAL STRUCTURE OF THE INTERIOR

Most information about the earth's interior comes from interpretation of arrival times and travel paths of seismic waves within the earth. Seismic events include fault-motion earthquakes, nuclear explosions, or any other sudden release of a large amount of energy in the earth. Such events create several types of waves. Some of these waves move along the earth's surface, whereas others penetrate the interior, even traveling radially all the way through the earth. Seismic waves are refracted or reflected by various inhomogeneities or boundaries within the earth, including density gradients and abrupt interfaces between different materials, which produce nonlinear paths and even sudden shifts in propagation direction (Figure 7-1). Geophysical analysis of detailed seismic wave paths has been coupled with constraints from such astrophysical data as the average density of the earth and its moment of inertia to calculate the density distribution within the earth. From these density calculations, the mineralogy of the mantle at various depths can be estimated.

Two types of waves penetrate the interior. *P waves* are longitudinal or compressive waves (similar to ordinary sound waves) that represent the propagation of density or volume variations in the direction of *P* wave

FIGURE 7-1

• • • • • • • • • • • • • • • • •

(A) Cutout showing the pattern of *P*-wave paths through the earth's interior. The numbers show the travel time in minutes for the waves to reach the associated broken line. Note the shadow zone, the region not reached by *P* waves (for this hypothetical earthquake at the North Pole) because they are deflected by the earth's core. [From F. Press and R. Siever, 1986, *Earth*, 4th ed. (New York: W. H. Freeman), Fig. 18-29.] (B) *P* and *S* waves radiate from an earthquake focus in many different directions. Waves reflected from the earth's surface are called *PP* or *SS*. *PcP* is a wave that bounces off the core, and *PKP* is a wave transmitted through the liquid core. *S* waves cannot travel through the core. [From F. Press and R. Siever, 1986, *Earth*, 4th ed. (New York: W. H. Freeman), Fig. 18-30.]

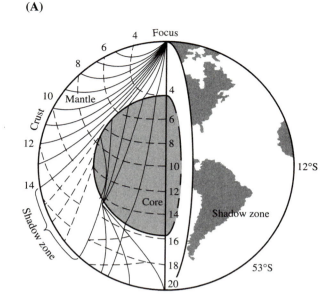

(A)

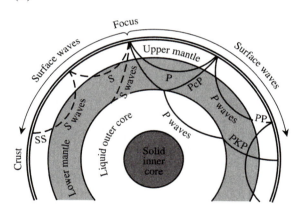

(B)

travel. *S waves* are transverse or shear waves. *P* waves propagate through both solids and fluids, but fluids do not allow the transmission of *S* waves because fluids are not rigid and thus have no shear strength. Although considerable variation exists in wave transmissions near the surface of the earth, the patterns of wave transmission in the deep interior clearly show a concentrically layered earth. On the basis of a plot of *P* and *S* wave velocities, petrologists can subdivide the earth into several major units (see Figure 7-2 for details of the outermost 1000 km). On the largest scale, the commonly agreed subdivisions are crust, mantle, and core. Each of these major units can be further subdivided on the basis of seismic data. A. E. Ringwood (1975) has suggested a more detailed breakdown of the earth into crust, upper mantle, transition zone, lower mantle, outer core, and inner core.

Crust

The crust is a thin layer at the earth's surface that ranges from 6 to 8 km thick beneath the oceans to 30 km or more beneath continents; some continental crust is more than 70 km thick. The average density of the crust is about 2.8 g/cm^3. The most obvious subdivision separates crust into oceanic and continental fractions. Oceanic crust is young (generally less than 200 million years) and composed largely of basalt with minor ultramafic rocks. Continental crust ranges from very young to very old (the oldest known continental rocks are just over 4 billion years old) and is composed of a diverse mix of plutonic and metamorphic rocks, with a thin veneer of sedimentary rocks in many places. Average density near the surface approximates that of granite, whereas the deep continental crust has a higher density more typical of basalt. Seismic velocities in oceanic crust indicate densities consistent with basaltic rocks.

The base of the crust is defined by a very distinct and relatively abrupt increase in seismic velocity called the

Mohorovičić discontinuity—or Moho, for short. Once thought to be a profound boundary of mechanical decoupling at the base of the crust, the Moho is now known to be much less important dynamically than the slightly deeper boundary in the upper mantle between the lithosphere and asthenosphere (see below).

Upper Mantle

The *mantle* extends from the Moho to a depth of about 2900 km. Ringwood defined the *upper mantle* as the layer of the mantle between the Moho and a depth of about 400 km (see Figure 7-2). The upper mantle contains significant substructure, as shown by seismic velocities. The uppermost sublayer of the upper mantle is called the lithospheric mantle. At a typical depth of

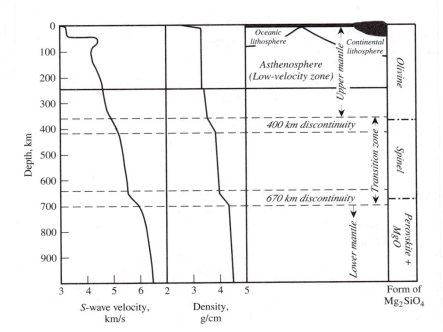

FIGURE 7-2

• • • • • • • • • • • • • • • •

Summary of characteristics of the outer 1000 km of the earth, including the *S*-wave velocities, densities, and form of Mg_2SiO_4 for the upper mantle and the uppermost part of the lower mantle. Note how changes in seismic velocity and density mark transitions from crust to lithosphere to asthenosphere. These changes are especially abrupt over the narrow transition zones where the structure of Mg_2SiO_4 changes from olivine to spinel at about 400 km (the 400 km discontinuity) and from spinel to perovskite plus MgO (the 670 km discontinuity). [After D. P. McKenzie, 1983, The earth's mantle, *Scientific American* (September).]

about 70 km under oceanic crust (somewhat deeper under continental crust), seismic velocities abruptly drop and stay low in a zone called the **asthenosphere** (literally, "sphere of weakness"), or *low-velocity zone*, that is typically 50–100 km thick. The asthenosphere is thought to have lower velocities because of reduced rock densities caused by a few volume percent of interstitial melt. Geophysical modeling suggests that the asthenosphere is less mechanically rigid than either the lithosphere (shallow mantle and crust) or the deeper mantle. The small volume of interstitial magma that causes reduced seismic velocities also appears to produce this mechanical effect. In modern geophysical and plate tectonic theory, it is the asthenosphere rather than the Moho that is the focus of mechanical decoupling in the outer earth. The earth's mobile tectonic plates are composed of lithosphere (crust plus lithospheric mantle), which lies above the ductile asthenosphere.

Transition Zone

The base of the upper mantle at 400 km (a depth corresponding to a pressure of 100–120 kbar) is marked by an important mineralogic transition that causes a notable increase in the density of mantle rocks. This transition is the first of a series of such transitions that occur in the next 500 km or so down to the base of the transition zone at 700–900 km (corresponding to a pressure of 250–300 kbar) (see Figure 7-2). This first of these mineral transformations is the polymorphic shift of Mg_2SiO_4 from the olivine structure to the silicate spinel structure,

and it causes the so-called 400 km discontinuity. The density of Mg_2SiO_4 increases about 8% in this transformation. Note that the mineral spinel ($MgAl_2O_4$) is *not* involved in this transition; the name refers to the spinel *structure*. The second principal mineral reaction in the transition zone occurs somewhat deeper, at a depth between 650 and 750 km, and is the pressure-driven breakdown of Mg_2SiO_4 spinel: Mg_2SiO_4 (spinel) = MgO + $MgSiO_3$. The $MgSiO_3$ crystallizes with the perovskite structure (a very dense crystal structure type named after the mineral perovskite, $CaTiO_3$), and MgO has a cubic (halite) structure. The density increase in this transition is about 10%. Again, note that reference to the "spinel-perovskite transformation" involves *only* the chemical composition Mg_2SiO_4. For a variety of chemical and physical reasons, both transformations are spread out over a range of depths and are not particularly abrupt. The mineralogic reactions in the transition zone are discussed in greater detail in this chapter's final section.

Lower Mantle, Outer Core, and Inner Core

The lower mantle extends from 700 km (210 kbar) to 2900 km (850 kbar) and is bounded below by the outer core. In contrast to shallower parts of the mantle, the lower mantle shows relatively smooth increases in seismic velocity corresponding to density increases with increasing pressure. A sharp increase in rock density and decrease in *P*-wave velocity, along with termination of *S* waves at 2900 km, marks the outer boundary of the *core*, which extends from 2900 km to the earth's center at 6371 km. The termination of *S* waves at the core-mantle boundary indicates that the outer portion of the core is a liquid which has no shear strength and thus will not allow transmission of shear waves. A solid inner core below 5000 km is suggested by refraction of *P* waves at an abrupt density boundary at 5000 km. The lower mantle and core are of only limited interest to most petrologists, so they will not be considered further in this book.

Lithosphere and Asthenosphere

The subdivision of greatest importance for petrology and tectonics divides the outer few hundred kilometers of the earth into *lithosphere* and *asthenosphere*. A substantial body of evidence indicates that virtually all igneous rocks have been derived from melting within the outer 250 km or so, that is, within the crust and upper mantle. Lithosphere combines the uppermost mantle and the crust. Total lithosphere thickness is normally about 70 km in oceanic regions (60 km of which is lithospheric mantle) and about 125 km under continents (80–90 km of mantle), but the lithosphere thins to essentially zero thickness at the mid-ocean ridge spreading centers, where asthenosphere is nearly at the surface. The reasons for this difference between oceanic and continental areas appear to correlate mostly with age. Young lithosphere is thin and old lithosphere is thicker, approaching a long-term constant value of 125 km. Asthenosphere extends from the base of the lithosphere down to about 250 km and has special properties (see later). Readers are strongly encouraged to be conversant with the terminology and the characteristics of the lithosphere and asthenosphere, because both will be widely referred to in discussions of igneous rock petrogenesis and the relationships of igneous suites to plate tectonic regimes.

PHYSICAL CHARACTERISTICS OF THE MANTLE
• • • • • • • • • • • • • • •

Horizontal Density Heterogeneity

Calculated densities of rock at different depths within the mantle, based on seismic velocities, increase with depth (except for the anomalous low-velocity zone) and range from 3.5 to 5.5 g/cm^3. Density might be expected to increase regularly, because rocks are compressible. In fact, density generally does increase in the upper mantle down to about 400 km and in the lower mantle (1000–2900 km). There are several abrupt jumps in this

intermediate or "transition" zone of the upper mantle, which are due to the pressure-driven isochemical phase changes described earlier.

The discussion so far has been confined to variability of seismic velocities with depth in the mantle. The development in the late 1980s of highly sophisticated computer programs that deconvolute (mathematically unscramble) complex seismic data and the increase in computing power have enabled seismologists to uncover major lateral variations in seismic velocities. Compilation of lateral density variations at multiple vertical levels in the mantle has generated a new way of looking at the mantle in three dimensions. This new type of investigation has been called *seismic tomography*, because of its resemblance to the methods that medical science has developed to image the human body.

The causes of variations in lateral seismic velocity (and thus density) are poorly understood. One possible simple cause is lateral compositional variability—namely, that peridotites of subtly different bulk and mineralogic compositions might have different densities. The origins of such compositional variability could be very old and related to the initial segregation of the primordial earth into mantle and core. Alternatively, mantle compositional heterogeneity might be due to ongoing and dynamic processes such as the recycling of old oceanic crust back down into the mantle in subduction zones. Earthquake seismic data suggest that downgoing oceanic lithospheric slabs roughly maintain their mechanical integrity down to 600 km or more. The ultimate fate of such material is not presently known. Some petrologists have suggested that it is converted into eclogite (dense garnet-clinopyroxene rock) and sinks to the mantle-core boundary, where it very slowly melts to release upward-streaming plumes of basaltic magma. Others suggest that recycled oceanic crust is simply stirred into the upper mantle to make an increasingly more complex petrologic "marble cake" over geologic time.

Another explanation is that lateral density variations reflect lateral temperature perturbations. This model implies that geothermal gradients (see the following section) vary slightly over smaller distances than previously assumed. The very slow upward-welling movement that must accompany large mantle convection cells (see later) may lie behind such temperature variations, as may smaller scale **mantle plumes** of hotter lower mantle rock that rise into the upper mantle, as proposed by J. T. Wilson, Jason Morgan, and others. Evidence for mantle plumes is examined in Chapter 8.

Temperature Distribution

The change of temperature with depth in the earth is known as the **geothermal gradient.** There are different

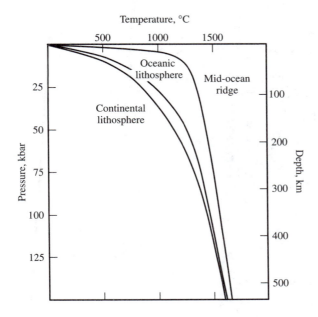

FIGURE 7-3

● ● ● ● ● ● ● ● ● ● ● ● ● ● ● ●

Representative average geothermal gradients for the upper
mantle down to a depth of about 500 km in continental (shield)
lithosphere, for oceanic lithosphere, and for mid-ocean ridge
environments. [From data in G. C. Brown and A. E. Mussett,
1981, *The Inaccessible Earth* (London: Unwin-Hyman).]

gradients for different regions of the earth, particularly
for oceanic versus continental areas. Knowledge of temperature distribution with depth is critical in developing
a theory of formation and ascent of basaltic and related
magmas within the upper mantle. Geothermal gradients
are calculated from temperature measurements made
near the earth's surface in deep wells and drill holes; the
calculations use heat flow models based largely on heat
conduction theory. Because the actual temperature measurements that constrain gradients are taken only within
a few kilometers of the surface (the deepest wells are
about 12 km deep), extrapolation of temperatures to
depths of several hundred kilometers or more is highly
model-dependent.

There are three basic sources of thermal energy
within the earth. Two are of major importance: (1) heat
that is held within the deep interior of the earth and is
left over from the accretion of the planet more than 4.5
billion years ago, and (2) heat released by decay of
radioactive elements. The third, frictional heat generated
along fault planes or shear zones, is minor and local relative to the other two and is probably restricted to near-
surface parts of the earth. The earth has an estimated
temperature of 4000°C at its center. Temperature decreases gradually and roughly linearly with decreasing
depth up to a depth of 500–1000 km; but a more ex-
ponential decrease takes place from there to the surface
(Figure 7-3). Calculated geothermal gradients are highly
speculative at depths below several hundred kilometers,
and models are strongly affected by assumptions regarding processes such as radioactive decay, earth tides or
other gravitational perturbations, and possible convective overturn that resulted in the formation of internal
layering.

Thermal energy due to radioactive decay is thought to
be especially important in the near-surface regime (the
crust and perhaps the uppermost lithospheric mantle)
because of inhomogeneous distribution of the principal
radioactive heat-producing elements uranium, thorium,
and potassium. These elements are described by igneous
geochemists as *large-ion lithophile* (*LIL*) or *incompatible* elements, which means that they are not typical components of early-formed, high-temperature crystals but
instead are concentrated into fractionated magmas. Thus
their concentrations are much higher in granitic and
other felsic rocks than in basaltic or ultramafic rocks.
Crust-forming processes and major fractionation (mostly
through partial melting) over several billion years have
tended to concentrate these elements in the near-surface
part of the earth.

Slow release of heat by radioactive decay accom-
plishes two things: (1) Local heating produces at least
modest temperature increase; and, (2) perhaps more
important for temperature distributions, heating of this
near-surface material creates a blanketing effect (much
like an electric blanket) that retards the flow of heat
from the interior. Reduction of the thermal gradient at
the surface tends to trap heat below, because the rate of
heat flow depends on temperature gradients (the greater
the temperature difference over a given distance, the
more rapid the flow or *flux* of thermal energy). Models
for geothermal gradients such as those in Figure 7-3 are

(A)
(B)

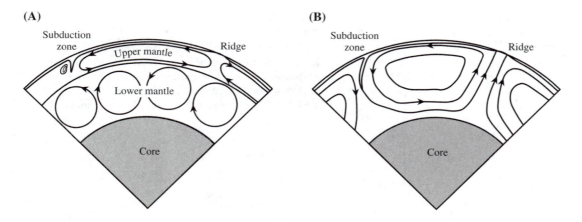

FIGURE 7-4
● ● ● ● ● ● ● ● ● ● ● ● ● ● ● ●

Models for mantle convection. **(A)** Two-layer model, with separate levels of convection cells in the upper and lower mantles. The levels are separated by the 670 km discontinuity (heavy line). **(B)** Mantle-wide model. [From BVSP, 1981, *Basaltic Volcanism on the Terrestrial Planets* (New York: Pergamon), frontispiece to Section 9, p. 1128.]

highly dependent on assumptions about the distribution of radioactive elements in the earth. For example, if it is assumed that the estimated amounts of uranium, thorium, and potassium in the earth are distributed evenly over twice as thick a layer at the surface, then calculations show that temperatures at a depth of 20 km can be increased by 100°C or more over the models shown in Figure 7-3.

Average geothermal gradients for oceanic and continental (shield) lithosphere and for mid-ocean ridges are illustrated in Figure 7-3 (from data in Brown and Mussett 1981), although numerous others have been proposed. These geothermal gradients are referred to as *steady-state geotherms*, meaning that they do not change over time. Tectonic effects, however, can produce significant short-term perturbations in the geotherm, particularly near the surface. It may take tens to hundreds of millions of years for the geothermal gradient to return to a steady-state configuration following a major perturbation. Note that the geothermal gradient shown for oceanic lithosphere is considerably higher than that for continental lithosphere, even though the measured or estimated surface heat flux is similar for both (in this case, "higher" refers to higher temperatures at shallower depths). Because of the concentration of radioactive heat-producing elements in continental crust, the similar heat flux values indicate that there must be a higher gradient at greater depths beneath the oceanic regions to produce a greater flow of heat into the base of oceanic lithosphere. The two gradients are presumed to converge below 1000 km because the lithospheric influences fade

with depth and because of the likelihood of lateral mantle homogenization through convection. In any case, the thermal distributions of the uppermost several hundred kilometers are of the most interest to petrologists, because mantle-derived magmas appear to have come from these shallower regions.

Mantle Convection

The only plausible driving force for horizontal movement of the lithospheric plates is thermal convection (coupled with buoyancy), which cycles deeper, hotter mantle toward the surface as shallow, cooler mantle sinks. Geophysicists and petrologists agree that convection is important, but argument continues over the exact geometry. Movement of solid mantle (perhaps with a small volume of lubricating melt) may not seem feasible, but the process is slow and accomplished by *creep* of mantle materials at high temperatures. Creep is a mechanism that combines a component of grain-boundary sliding with intracrystalline deformation within grains. If movement rates of surface tectonic plates are comparable to the velocity of underlying convecting mantle, the flow rate by mantle creep is several centimeters per year.

The focus of arguments regarding convection geometry is whether the individual cells are mantle-wide or are arrayed at several levels and separated by zones of mechanical decoupling. The two principal arguments were summarized by the Basaltic Volcanism Study Project (1981) and are shown in Figure 7-4. Adherents of a two-layer model argue that the density (that is, seismic velocity) changes over short vertical distances (for example, at the 650–670 km discontinuity) represent boundaries between multiple layers. Geophysicists also now believe that lithospheric plates are not simply passive passengers riding on the convecting mantle beneath. Especially at subduction zones, lithosphere becomes active as old, cold lithospheric plates sink back into the mantle. One useful test of the two models shown in Figure 7-4 is the depth to which there is evidence of subducted lithosphere, specifically whether or not lithosphere can be traced below 670 km. Unfortunately, the current evidence is ambiguous.

CHEMISTRY AND MINERALOGY OF THE MANTLE AND CORE
• • • • • • • • • • • • • • •

The chemical and mineralogic makeup of the lower crust, mantle, and core have been deduced from a variety of lines of evidence, mostly indirect, including meteorites, seismic data, experimental petrology, magma compositions, and xenoliths (both crustal and from the mantle lithosphere and possibly asthenosphere for some kimberlite xenoliths). Extrapolating from the direct evidence of chemical and mineralogic compositions of meteorites and lithospheric xenoliths to the rocks of the earth's interior is less obvious than it may seem and requires numerous assumptions. The indirect lines of evidence rarely produce unique or unequivocal compositional information. The best models of deep crustal and mantle compositions and mineralogy have been derived therefore from the combined application of several techniques, thus providing greater constraints.

Sources of Information

METEORITES.   The use of meteorites as analogs of unexposed deep earth materials is based on the generally accepted assumption that meteorites are fragments of small planets or planetesimals that formed at the same time (and in the same way) as Earth, at about 4.5 Ga. Most meteorites that strike the earth have trajectories indicating that they come from the asteroid belt, a region in the solar system between the orbits of Mars and Jupiter. Meteorites are classified either as falls (observed to fall and immediately collected) or as finds (collected and later identified as meteoritic material). All collected meteorites certainly represent only a very small proportion of material that has actually fallen on Earth. Because meteorites represent a random sampling of space debris swept up by Earth as it moves along its orbit, they

have a wide range of compositions, textures, and mineralogy. In general, however, they fall into two major categories, the stony meteorites and the iron meteorites. The iron meteorites consist largely of iron-rich iron-nickel alloy with minor amounts of other minerals such as silicates and sulfides. When the densities of iron meteorites are extrapolated to the temperature and pressure of the earth's core, they are consistent with core densities estimated from seismic information, data supporting the hypothesis that the core is composed of this type of material.

Similarly, many stony meteorites (or *stones*) have properties consistent with materials of the mantle. They consist mostly of the silicate minerals olivine, augite, and enstatite, with minor oxides, sulfides, and metals. Stony meteorites, similar to lunar rocks and in contrast to earth rocks, are notably devoid of hydrous minerals and excess oxygen; in particular contrast to most known terrestrial rocks, iron occurs mixed in the 0 and +2 oxidation states rather than in the +2 and +3 states. Relative abundances of nonvolatile elements in the stony meteorites (magnesium, iron, calcium, aluminum, and silicon) are similar to those in the sun and other stars. Although there is no guarantee that meteorites represent a statistically valid sampling of planetary material, it is interesting and perhaps more than coincidental that the proportions of stones and irons (roughly 80% and 20%, respectively) are similar to the relative volumes of mantle and core in the earth (84% and 16%, respectively). On the basis of stony meteorites, it has been estimated that the mantle is ultramafic in composition (roughly peridotite) and dominated by FeO, MgO, and SiO_2 (mostly as olivine and orthopyroxene, or their high-pressure equivalents in deep mantle). The rest is mainly CaO, Al_2O_3, and Na_2O (as augite and various accessory minerals). Some stony meteorites have abundant calcic plagioclase and roughly basaltic compositions, and are thought to represent basaltic crustal material. A few, in fact, are now thought to be crustal or upper mantle fragments of the Moon or Mars that were ejected into space by enormous meteorite impacts on these planets.

XENOLITHS AND XENOCRYSTS.   An additional and important source of information about the petrology of the mantle comes from xenoliths and xenocrysts in basalt, kimberlite, and lamprophyres. Basaltic and related magmas that originate in the upper mantle may carry up rock fragments or single large crystals that are pieces of country rock from the source region or fragments eroded from the conduit walls during ascent (Figure 7-5). Interestingly, the occurrence of xenoliths in volcanics and near-surface intrusions is essentially limited to intraplate magmatic environments such as oceanic islands (for example, Hawaii and Tahiti) or continental

FIGURE 7-5

● ● ● ● ● ● ● ● ● ● ● ● ● ● ● ● ●

A block of alkali basalt about 1 m across, which is laden with ultramafic xenoliths, including peridotite, dunite, and pyroxenite, and large xenocrysts of olivine and amphibole. San Carlos Apache Reservation, San Carlos, Arizona.

volcanic provinces or rifts (for example, the southern Colorado Plateau and the Eifel district in Germany). Xenolith occurrences are considerably rarer in interplate environments such as collisional magmatic provinces or island arcs. This absence may be largely due to the necessity for xenolith-bearing magmas to come rapidly to the surface or near-surface, and thus may be related to the simpler magmatic plumbing systems possible in intraplate environments. Ultramafic xenoliths are denser than the transporting magma and tend to sink after incorporation, in a manner similar to sedimentary particles. A slow-moving magma or one that is subject to periodic stops in intermediate magma chambers may lose its xenoliths through settling.

The richest source of a variety of ultramafic xenoliths and xenocrysts is kimberlite, an ultramafic rock rich in potassium and volatiles, apparently because it originates deep in the mantle (estimates range from 150 to 300 km or more) and ascends rapidly to the surface (Mitchell 1986). Some kimberlite magmas may pass through most of the upper mantle on the way to the surface. Occurrences of kimberlites are noteworthy in southern Africa (South Africa, Zimbabwe, Lesotho, and Botswana) and in Siberia, but they occur on virtually all continental **cratons.** Some are rich in carbonate and hydrous minerals, an observation suggesting that the magmas had significant contents of volatile species such as carbon dioxide and water. The occurrence of diamond (a high-pressure mineral) as xenocrysts in kimberlites supports both the deep origin of the magma and the presence of carbon in some mantle rocks at this depth.

Typical mantle peridotites called **lherzolites** constitute a major proportion of xenoliths in kimberlite. Some lherzolites contain small amounts of either magnesium-rich garnet, chromium-aluminum spinel, or calcic plagioclase as a characteristic aluminous mineral; the specific aluminous mineral in any lherzolite is a function of pressure (that is, depth). Plagioclase occurs in lherzolites only at the lowest pressures, corresponding to a depth of less than about 25 km. Spinel is a crystal structure that is more densely packed than feldspar and thus replaces plagioclase as the characteristic aluminum-rich mineral at depths greater than 25 km (about 8 kbar). Even more densely packed magnesium-rich garnet is the stable aluminum-rich mineral at depths below about 60 km (18 kbar) (Figure 7-6). All these lherzolite types occur as xenoliths in kimberlites, along with dunite (olivine-rich rock), harzburgite (an olivine-orthopyroxene rock), eclogite, and a variety of rare rock types.

Eclogite is a high-pressure rock of basaltic rather than ultramafic composition; it is composed of garnet and calcium-sodium clinopyroxene (omphacite). Layers of eclogite are thought to occur in the uppermost lithospheric mantle (just below the Moho) in both subcontinental and suboceanic regions. Such eclogites represent basaltic magmas that were trapped and ponded in the uppermost mantle and then crystallized a high-pressure eclogite mineral assemblage rather than the more typical assemblage of low-pressure pyroxene plus plagioclase. Eclogite may play an especially important role in continental lithosphere as a transitional rock type between felsic gneisses of the lower continental crust and peridotite of the upper mantle. (Eclogite is an important

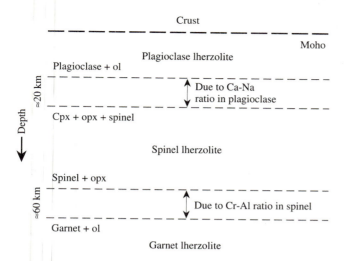

FIGURE 7-6

• • • • • • • • • • • • • • • •

Mineralogic "stratigraphy" of the uppermost 100 km of the lithospheric upper mantle. Note that plagioclase lherzolite does not occur in all lithospheric mantle at shallow levels and is particularly likely to be absent in subcontinental lithosphere. The depth ranges that represent the two mineral reactions are dependent on mineral chemistry in the solid solution minerals, specifically anorthite content of plagioclase and Cr:Al ratio in spinel. Within these bands, all minerals can coexist in a reaction relationship.

metamorphic rock type, and its metamorphic occurrence is discussed in detail in Chapter 22.)

The range of xenoliths found in basalts is considerably more restricted than that in kimberlites, apparently because basalts originate at shallower levels in the mantle (typically at depths no greater than about 120 km) and thus sample less of the upper mantle on their way to the surface. Spinel lherzolites are by far the most common rock type in xenolith suites in alkali basalts (as seen, for example, on Oahu in Hawaii, in the Rio Grande Rift of the southwestern United States, and in the volcanic district of central Europe), with rare plagioclase lherzolites; garnet lherzolites are essentially unknown from basalts. In contrast, garnet lherzolites are relatively abundant in kimberlites, especially those of southern Africa and Siberia. Other ultramafic and mafic rock types that occur as xenoliths in basalt include harzburgite, dunite, and eclogite.

Harzburgite and dunite almost certainly represent residua from fractional melting of mantle peridotite and extraction of basaltic magmas. Refer to the discussion of melting in Chapter 5 (especially see the liquid path in Figure 5-12) and recall that the first solid phase that disappears during melting of typical mantle peridotite (60% olivine, 30% orthopyroxene, and 10% clinopyrox-

ene) is clinopyroxene, thus leaving a harzburgite residue. Higher degrees of melting and extraction deplete the residual harzburgite in orthopyroxene, ultimately leaving dunite. Harzburgite and dunite xenoliths are therefore thought by petrologists to reflect both mineralogically and geochemically depleted regions of the upper mantle from which substantial volumes of basalts have been derived. The variety of xenoliths sampled by basaltic and kimberlitic magmas certainly indicates that the upper 200 km or so of the mantle are far from homogeneous and contain a diversity of ultramafic rock types indicating variable but commonly extensive melt extraction.

EXPERIMENTAL PETROLOGY. Another approach to characterizing mantle mineralogy has been through experimental petrology, which uses high-pressure techniques that approximate the conditions of the mantle. The goals of experimental studies have been diverse, but there have been two principal aims. The first has been to determine the stable compositions and structures of likely mantle minerals over a wide range of pressures and temperatures, in order to calibrate the occurrences of minerals in xenoliths (see Figure 7-6) and the polymorphic transitions that are probable causes of seismic velocity shifts. The second has been to demonstrate the viability of mantle mineral assemblages as potential parental materials for basaltic magmas. This latter branch of experimental petrology has been quite successful over the last several decades in showing how depth and degree of partial melting control the compositions of basaltic liquids.

Estimates of Mantle Chemistry

Petrologists have assembled information from the various data sources noted earlier to construct chemical models for the upper mantle. Meteorites (particularly the carbonaceous chondrites) provide chemical constraints for the total earth. With some assumptions, the material required to segregate and form the core can be subtracted, allowing calculation of the bulk composition of so-called primordial mantle. As noted by Marjorie Wilson (1989, p. 52), three important deductions can be made about primordial mantle: (1) SiO_2, MgO, and FeO constitute more than 90% of the mantle by weight; (2) Al_2O_3, CaO, and Na_2O total 5–8 wt%; (3) these six oxides make up more than 98% of the mantle, and no other oxide has a concentration greater than 0.6 wt%. Present-day mantle has been modified slightly by differentiation, especially segregation of crustal material, but modification can only be minor, because crust constitutes less than 1% of the mass of the earth; the mantle constitutes 68%.

Table 7-1 presents data on compositions of mantle peridotites compiled by Maaløe and Aoki (1977).

MANTLE PETROLOGY

• • • • • • • • • • • • • • •

The development of the various **pyrolite** models of the late A. E. Ringwood of the Australian National University in the 1960s led to a major advance in understanding the petrology of the upper mantle. Ringwood was a pioneer and major figure in high-pressure experimental petrology and modeling of the earth's mantle. The term *pyrolite* refers to a synthetic or model pyroxene-olivine rock that simulates mantle peridotite. Ringwood was trying to create a close approximation of "fertile" (unmelted) primordial mantle to use as the starting material in experiments on mantle phase equilibria, both solid-state mineral transformations and melting. His approach was based on the assumption that dunite and harzburgite were typical residua from mantle melting and that basalt was the material extracted. He therefore experimented with mixtures of various proportions of typical Hawaiian basalt and ultramafic rock, usually about one part basalt to four parts ultramafic material. Over a period of years, Ringwood proposed various pyrolite models; a typical one is given in Table 7-1.

Figure 7-7 shows changes in mantle mineralogy of a pyrolite bulk composition as a function of pressure and temperature, up to pressures of about 50 kbar (corresponding to a depth of about 150–160 km). As the figure shows, the changes in mineral assemblages have relatively shallow slopes; that is, the reactions that cause them are mainly functions of pressure rather than of temperature. Two types of mineralogic changes are illustrated on the phase diagram: (1) the loss or gain of a mineral phase, affecting the mineral assemblage (very similar to metamorphic reactions that are examined in Part III), and (2) the change in solid solution composition of a mineral phase as a function of pressure or temperature. This latter effect is shown by the shift in Al_2O_3 content of orthopyroxene within the garnet peridotite field, indicated by the labeled contours. Note that these contours show a significantly greater dependence on temperature than the assemblage boundaries do. Alumina dissolves into orthopyroxene by entering both octahedrally and tetrahedrally coordinated sites. Because the Al^{3+} cation is larger than Si^{4+}, the substitution is affected by relative compression or expansion of the crystal structure. Of course, a pressure increase compresses the structure, whereas a temperature increase causes it to expand. At constant temperature, increasing pressure causes the solubility of aluminum in orthopyroxene to drop; at constant pressure, increasing temperature causes the Al solubility in pryoxene to increase (Figure 7-7).

The mineralogic changes that yield the field boundaries shown in Figure 7-7 occur for the following reasons. At low pressures (less than about 8 kbar), the Al_2O_3 in peridotite is distributed between pyroxenes and a small amount of calcic plagioclase. The resulting rock, plagioclase peridotite, occurs only rarely, for two rea-

TABLE 7-1   Measured and proposed compositions of mantle ultramafic material

• •

	Range for garnet lherzolite	Average for garnet lherzolite	Range for spinel lherzolite	Average for spinel lherzolite	Ringwood's pyrolite model
SiO_2	43.8–46.6	45.89	42.3–45.3	44.2	45.2
TiO_2	0.07–0.18	0.09	0.05–0.18	0.13	0.71
Al_2O_3	0.82–3.09	1.57	0.43–3.23	2.05	3.54
Cr_2O_3	0.22–0.44	0.32	0.23–0.45	0.44	0.53
FeO	6.44–8.66	6.91	6.52–8.90	8.29	8.48
MnO	0.11–0.14	0.11	0.09–0.14	0.13	0.14
NiO	0.23–0.38	0.29	0.18–0.42	0.28	0.20
MgO	39.4–44.5	43.46	39.5–48.3	42.21	37.48
CaO	0.82–3.06	1.16	0.44–2.70	1.92	3.08
Na_2O	0.10–0.24	0.16	0.08–0.35	0.27	0.57
K_2O	0.03–0.14	0.12	0.01–0.17	0.06	0.13
P_2O_5	0.00–0.08	0.04	0.01–0.06	0.06	—
100Mg/(Mg + Fe)	89–93		89.1–92.6		
100Cr/(Cr + Al)	7.4–18.6		7.0–31.7		

Source: From Maaløe and Aoki (1977).

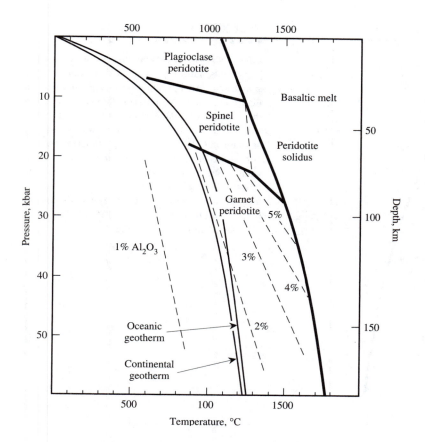

FIGURE 7–7
••••••••••••••••

Stability fields of different subsolidus mineral assemblages in mantle peridotite and a peridotite solidus for Ringwood's pyrolite composition. Dashed lines are contours of Al_2O_3 content (wt%) in orthopyroxene. The small triangular field represents peridotite without spinel: At these high temperatures, clino- and orthopyroxenes can accommodate all the Al_2O_3 in the rock within their structures. Typical continental and oceanic geotherms are shown. Note that they do not intersect the solidus. [After D. H. Green and A. E. Ringwood, 1970, *Phys. Earth Planet. Interiors, 3,* Fig. 2.]

sons: (1) the substantial thickness of continental crust (30 or more km) eliminates any possibility of subcontinental upper mantle occurring in the pressure range (less than 6 kbar) at which plagioclase is stable; and (2) suboceanic upper mantle is so commonly depleted in basaltic components through melt extraction that the ultramafic rocks that occur at shallow mantle levels are Al_2O_3-depleted residual types such as dunite or harzburgite. With increasing pressure in undepleted peridotites, plagioclase is eliminated by reaction and spinel appears through the solid-state reaction plagioclase + olivine = clinopyroxene + orthopyroxene + spinel. The actual position of this reaction on a *P-T* diagram (for example, Figure 7-7) depends on the Mg:Fe ratio of olivine, clinopyroxene, and spinel, the Ca:Na ratio of plagioclase, and the Cr:Al ratio of spinel. These ratios are fixed in a single pyrolite composition, so the phase boundary is a line in Figure 7-7. It should be emphasized that there is considerable speculation involved in applying the results of any single technique (thermal modeling, seismic data, or experimental petrology, for example) to the determination of mantle structures and mineralogy. However, the overall consistency of modern models for the mantle with most or all indirect or inferential observations lends validity to the models. For example, experimental petrology predicts mineral phase changes as a function of temperature and pressure that cross the geothermal gradient at the depths indicated by seismic data to be where the density increases occur. This degree of consistency is unlikely to be coincidental.

Petrologic Significance of the Asthenosphere

Geophysicists were long puzzled by the rather abrupt decrease in seismic velocity (and thus rock density) that provides evidence for the asthenosphere. There are good reasons for relatively abrupt or discontinuous *increases* in velocity or density with increasing depth, such as those based on the solid-state phase changes discussed above. However, there is no simple explanation for a *decrease* in density with increasing pressure. Furthermore, a slight damping-out of the energies of seismic waves accompanies the decrease in seismic velocities and densities. What physical process can be consistent with these observations?

A density decrease could be caused by a local increase or aberration in the average geothermal gradient that was due to thermal expansion of solids. But the small magnitude of such an increase, its local nature, and its probable short duration all appear to preclude this explanation. The alternative explanation, which is now thought to be correct, is that the asthenosphere

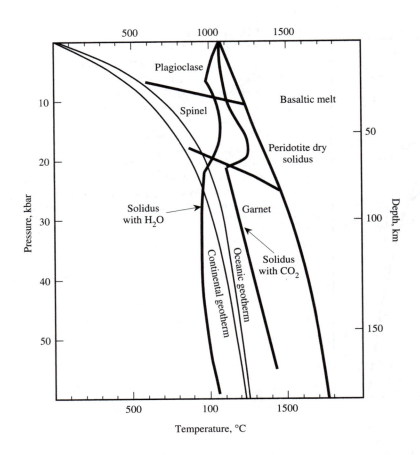

FIGURE 7-8
● ● ● ● ● ● ● ● ● ● ● ● ● ● ● ● ●

Effects of H_2O and CO_2 on the peridotite solidus. Dry solidus, subsolidus boundaries (plagioclase, spinel, and garnet peridotites), and geotherms as in Figure 7-7. Solidus with H_2O is for partial melting in the presence of small amounts of H_2O (<0.4 wt%), and solidus with CO_2 is for melting with small amounts of CO_2 (<5 wt%). For portions of these curves, free fluid is present, and for other parts, all volatiles are in hydrous or carbonate minerals. Note that the depth at which the water-present solidus intersects the geotherms coincides with the upper boundary of the asthenosphere. [After P. J. Wyllie, 1981, *Geol. Rundschau, 70*, Fig. 4.]

represents a zone of incipient melting (to the extent of several volume percent), and that the melt is trapped in unconnected interstitial pockets. Small amounts of trapped melt should satisfactorily explain the physical properties of the asthenosphere: The presence of melt will both decrease seismic velocities and absorb some seismic energy, particularly that associated with S waves.

However, models for incipient melting of the upper mantle have a problem: The anhydrous *peridotite solidus* [the *P-T* curve that represents the initial appearance of melt at a eutectic or peritectic point for the melting of olivine + clinopyroxene + orthopyroxene + garnet (or spinel) (see Figure 7-7)] does not intersect the typical geothermal gradient. Thus mantle peridotite should remain unmelted to great depths, perhaps throughout the mantle. But the discussion in Chapter 5 showed clearly that basaltic melts originate in the upper mantle.

Earlier discussion assumed that melting is "dry," that is, no volatiles are involved. However, the mantle actually contains minor amounts of volatile species such as water and carbon dioxide (and perhaps others) in minerals such as amphibole, phlogopite, and carbonates, which have been observed in some mantle xenoliths. Volatile species have the capability of reducing melting temperatures to an extent proportional to their

solubility in the resulting magma. For example, water is highly soluble in granitic magma, and its presence during melting lowers the granite solidus by several hundred degrees. Water and carbon dioxide are both soluble to a lesser degree in basaltic liquids and have the rough effect on the solidus that is illustrated in Figure 7-8. Small amounts of these volatiles in upper mantle peridotite may be adequate to allow a small amount of melting over a range of depths where the geothermal gradient crosses the modified solidus (Figure 7-8).

The physics of melt formation, coalescence, and extraction is a very active area of petrologic and geophysical research. Very small degrees of melting (less than about 5%) result in distribution of melt in the so-called "triple junctions," where mineral grains (mainly olivine and pyroxenes) come together (Figure 7-9A). There is no physical connection between these pockets of melt and thus no possibility of flow of the melt to allow coalescence into a magma body and ultimate extraction by upward movement. With greater degrees of melting (Figure 7-9B), melts not only occur at multiple grain junctions but also wet the curved surfaces bounding adjacent grains. This wetting effectively creates melt permeability and three-dimensional interconnection, which allow flow and melt extraction. The small amounts of inter-

FIGURE 7-9
● ● ● ● ● ● ● ● ● ● ● ● ● ● ● ●

Schematic illustration of how melt can be distributed among solid grains in a partially molten aggregate such as mantle peridotite in the asthenosphere. (**A**) Small degrees of melting, allowing wetting only of the triple junctions but not of the grain boundaries. (**B**) Higher degrees of melting in which melt has moved out from triple junctions along grain boundaries to such an extent that all triple junctions are connected and a three-dimensional network of melt is created.

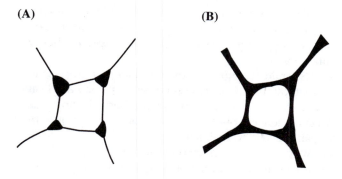

(A)   (B)

stitial melt that apparently give the asthenosphere its special properties can remain trapped for very long times at appropriate temperatures if they never achieve the volume necessary for coalescence, flow, and extraction.

Experiments and calculations have also shown that an odd phenomenon may occur at mantle depths roughly corresponding to the asthenosphere and below. At shallow mantle depths, basaltic magmas are less dense than surrounding peridotite and therefore can rise gravitationally under the right conditions. Like most liquids, however, basaltic magma is more compressible than solid minerals. Thus, with increasing depth in the upper mantle, magma may undergo a more rapid increase in density than peridotite. In fact, it has been calculated that basaltic magma has the same density as the solid rock that confines it at depths of roughly 100–150 km (30–45 kbar). Therefore deep basaltic magmas have no gravitational driving force for upward movement and remain trapped in the mantle no matter what the degree of melting or of coalescence. It has been suggested that the observed lack of basaltic magmas from deeper than 150 km can be explained by this *density inversion*. Kimberlites and related magmas appear to come from depths as great as 300 km, but their chemistries (and particularly their volatile contents) are sufficiently different from basalt to explain the discrepancy.

SUMMARY
● ● ● ● ● ● ● ● ● ● ● ● ● ●

Seismic data indicate a concentrically layered structure for the earth, with relatively abrupt density increases marking the boundaries between crust, mantle, and core. These gross divisions have been subdivided into crust, upper mantle, transition zone, lower mantle, outer core, and inner core. The lithosphere and asthenosphere are more important subdivisions of the outer part of the earth from a dynamic point of view. The lithosphere

varies in thickness from 0 km at mid-ocean ridges to 70 km in oceanic regions, up to 125 km in continental regions. Lithosphere consists of the crust and uppermost part of the mantle and is relatively rigid, making up the mobile tectonic plates that move over geologic time. The 100-km-thick asthenosphere lies beneath the lithosphere and provides the ductile material to make movement of lithospheric plates possible. Most mantle-derived magmas come from the outer 250 km of the earth.

Compositional data from meteorites and xenoliths, along with densities corrected for ambient temperature and pressure within the earth, indicate that the mantle is composed of olivine-rich peridotite and that the core is iron-rich. Experimental petrology and examination of xenoliths reveal the changes in mantle mineralogy, particularly of aluminous accessory minerals, as depth increases. These mineralogic changes are consistent with shallow seismic discontinuities; deeper discontinuities are inferred to be the result of pressure-driven polymorphic changes and mineral reactions. The asthenosphere, or low-velocity zone, appears to be the result of incipient melting at depths ranging from near-surface at the mid-ocean ridges to 70 km under the oceans and as much as 250 km under the continents. Physical properties of melts, particularly compressibility, explain why melts can be denser than peridotite and thus remain trapped in the asthenosphere and below for extended periods of time.

STUDY EXERCISES
● ● ● ● ● ● ● ● ● ● ● ● ● ●

1. Our knowledge of the composition and structure of the mantle is based on both direct and indirect (remotely sensed) observation, combined with modeling and experimental simulation. How deep in the upper mantle does our direct observational database extend? What does this database consist of?

2. The perovskite crystal structure is of great interest to geophysicists. Why is this crystal structure so relevant to magnesium silicate phase transitions in the upper mantle?

3. "Seismic tomography" of the mantle shows significant lateral variations in seismic velocity that have been linked to lateral compositional or thermal variations. What influence would these variations have on generation and ascent of basaltic magmas?

4. Why do xenolith-bearing basalts typically show much less diversity of xenolith rock types than kimberlites show?

5. Discuss petrologic explanations for the physical properties and mechanical behavior of the asthenosphere (low-velocity zone).

REFERENCES AND ADDITIONAL READINGS
• • • • • • • • • • • • • • •

Anderson, D. L. 1970. The earth as a planet: Paradigms and paradoxes. *Science, 223,* 347–355.

Basaltic Volcanism Study Project. 1981. *Basaltic Volcanism on the Terrestrial Planets.* New York: Pergamon.

Brown, G. C., and A. E. Mussett. 1981. *The Inaccessible Earth.* London: Unwin-Hyman.

Dawson, J. B. 1980. *Kimberlites and Their Xenoliths.* New York: Springer-Verlag.

Dodd, R. T. 1981. *Meteorites.* London: Cambridge University Press.

Green, D. H., and A. E. Ringwood. 1970. Mineralogy of peridotite compositions under upper mantle conditions. *Phys. Earth Planet. Interiors, 8,* 359–371.

Jacobs, J. A. 1992. *Deep Interior of the Earth.* London: Chapman and Hall.

Jeanloz, R., and T. Lay. 1993. The core-mantle boundary. *Scientific American* (May), 48–55.

Jeanloz, R., and A. B. Thompson. 1983. Phase transitions and mantle discontinuities. *Rev. Geophys. Space Phys., 21,* 51–74.

Jordan, T. H. 1979. The deep structure of the continents. *Scientific American, 240,* 92–100, 103–107.

Maaløe, S., and K. Aoki. 1977. The major element composition of the upper mantle estimated from the composition of lherzolites. *Contrib. Mineral. Petrol., 63,* 161–173.

MacKenzie, D. P. 1983. The earth's mantle. *Scientific American* (September), 66.

Mitchell, R. H. 1986. *Kimberlites: Mineralogy, Geochemistry and Petrology.* New York: Plenum Press.

Ringwood, A. E. 1975. *Composition and Petrology of the Earth's Mantle.* New York: McGraw-Hill.

Takahashi, E., R. Jeanloz, and D. C. Rubie, eds. 1993. *Evolution of the Earth and Planets. Geophysical Monograph 74.* Washington, DC: American Geophysical Union.

Wilson, M. 1989. *Igneous Petrogenesis: A Global Tectonic Approach.* London: Unwin-Hyman.

Wyllie, P. J. 1971. *The Dynamic Earth.* New York: Wiley.

Wyllie, P. J. 1981. Plate tectonics and magma genesis. *Geol. Rundschau, 70,* 128–153.

Chapter 8

IGNEOUS ROCKS
OF THE OCEANIC LITHOSPHERE

The principal magma type of the oceanic crust is basalt in its various manifestations, including the voluminous mid-ocean ridge basalts **(MORBs)** and ocean island basalts **(OIBs)** with their related alkaline derivatives. The oceanic crust is made up almost entirely of mid-ocean ridge basalts that have been either intruded or extruded at the ridge crest spreading centers and then covered by a thin veneer of sedimentary material. The source of the magmas appears to be the uppermost mantle, either the oceanic lithosphere beneath oceanic crust or the asthenosphere; the magmas are thus characteristic of divergent plate margins in oceanic settings. (Divergent boundaries also occur in continental plates, and these are explored in Chapter 10.) Experimental studies have demonstrated a low-pressure (shallow) mantle origin for most of these magmas. The source rock appears to be spinel or garnet **lherzolite,** which undergoes significant degrees of melting and melt removal (up to 25% or more), a process leaving residual ultramafic rock types such as dunite or harzburgite in the upper mantle. The existence of these residua has been demonstrated by the common presence of dunites and harzburgites as xenoliths in intraplate basalts.

Ocean island basalts (OIBs) are the second major type of magmatic occurrences in oceanic lithosphere. They are far less voluminous than the MORBs but provide very important information on higher pressure melting processes in the mantle. The Hawaiian Islands in the center of the Pacific plate are perhaps the classic example of an occurrence of ocean island basalts, but numerous other occurrences can be found on almost all the oceanic lithospheric plates. The range of major- and trace-element chemistry of the OIBs is significantly greater than that of the MORBs because of the wider range of conditions of magma formation, especially the depth of melting (from 30 or 40 km to 100 km or more) and the proportion of melting of the parental material (typically less than 10%).

THE NATURE OF
THE MID-OCEAN RIDGES

The mid-ocean ridges are one of the more spectacular topographic features of the earth's surface but were virtually unknown prior to World War II. They were revealed by surveys of the seafloor that were done during and after the war years that used remote sensing and depth sounding instruments developed during the war. The length (65,000 km) and continuity of the ridges, as well as their roughly medial positions within each of the ocean basins, struck several geologists, notably Arthur Holmes and Harry Hess, as significant and prompted the development in the late 1940s and 1950s of early theories of seafloor spreading that were the antecedents of modern plate tectonic theory. These theories proposed that new basaltic material was constantly added to the oceanic crust by active magmatism at the ridge and was then transported away from the ridge crest, ultimately to be returned to the mantle through subduction at the oceanic trenches. Later research, including paleomagnetic and heat flow studies and the return of samples from the ridges and from abyssal ocean floor by dredging, drilling, or deep submersible sampling, has persuasively confirmed these ideas and supplied much information on the processes that produce seafloor spreading. Detailed sampling and geophysics have also been done on portions of the ridge that form islands above sea level, such as Iceland and the Azores on the Mid-Atlantic Ridge.

As topographic features, the ridges are typically about 2000 km across, and the highest axial parts are 2 to 3 km below sea level except for an occasional subaerial occurrence as an island (such as Iceland). There may or may not be an axial rift valley, apparently depending on spreading rate. The East Pacific Rise is a relatively fast-spreading ridge and has only a poorly developed axial valley, whereas a slowly spreading ridge such as the Mid-Atlantic Ridge has a prominent rift valley. Active volcanism is restricted to only a few kilometers in the axial zone of a ridge and to isolated volcanic centers along transform faults. Magma does not rise uniformly to the surface all along the axis of a ridge but, instead, rises as diapirs spaced about 10 km apart. Volcanism along the ridge axis is fed by extensive horizontal movement of magmas between these major vertical feeders.

Detailed evidence on the nature of the mid-ocean rift zones has come from the French-American Mid-Ocean Underseas Study of the 1970s (known as Project FAMOUS) and from subsequent rift zone studies that used both manned and robotic deep submersibles. The project area covered a fairly well known portion of the Mid-Atlantic Ridge near the Azores Islands. The studies were both regional in extent (the greater FAMOUS area) and quite detailed in a smaller area (the FAMOUS area), as shown in Figure 8-1. Studies included measurements of bathymetry, magnetics, gravity, seismic refraction, bottom currents, heat flow, and water temperature. Samples were extensively collected from very specific locations by both French and American submersibles.

The Mid-Atlantic Ridge in the FAMOUS area includes a central rift valley 1.5 to 4 km wide and 100 to 400 m deep (Figure 8-2). The valley is essentially symmetrical in cross section and contains a central discontinuous ridge (100 to 240 m high and 800 to 1300 m wide) or in places a central trough (200 to 600 m wide). It has been established by various studies that the central ridge is located along the major rifting line and contains the youngest volcanic rocks, which are fresh, glassy, tholeiitic basalts with pillows and lava tubes, and contain phenocrysts of olivine, clinopyroxene, and plagioclase. After extrusion of these lavas, the central ridge undergoes subsidence, perhaps as a result of both rifting and deflation of the underlying magma chamber. The roof of the magma chamber appears to be floating on molten magma, which comes within 2 km of the seafloor. In the outer

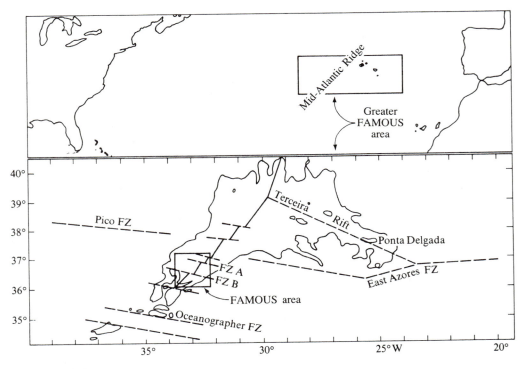

FIGURE 8-1
• • • • • • • • • • • • • •

The FAMOUS and greater FAMOUS areas of the Mid-Atlantic Ridge. A fault zone (FZ) is represented by a dashed line. The 2000-m isobath is represented by a solid line. [From J. R. Heirtzler and T. H. van Andel, 1977, *Geol. Soc. Am. Bull.*, *88*, Fig. 2.]

FIGURE 8-2
● ● ● ● ● ● ● ● ● ● ● ● ● ● ● ●

Orthographic drawing of part of a rift valley floor
and walls. [After R. Hekinian, J. G. Moore, and W. B.
Bryan, 1976, *Contrib. Mineral. Petrol.*, *58*, 107.]

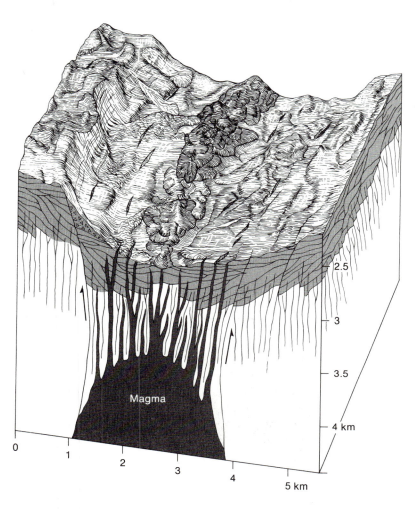

part of the rift valley, the walls are elevated relative to
the rift as a result of normal faulting and subsidence of
the valley floor. Extrusion of the central lava flows is
episodic, and the location of the volcanic axis varies
through time (Figure 8-3). It should be emphasized that
ridge-crest volcanism is far from continuous and that the
volumes of lava extruded in each event are not large
compared with the single basaltic extrusions on oceanic
islands or in continental rifts. However, the process of
episodic basaltic extrusion at the ridges over long periods of time results in the ultimate extrusion of more basaltic magma than in any other terrestrial environment.

Figure 8-4 shows two different diagrammatic views of
the inner rift valley as conceived by Ballard and van
Andel (1977) and by Bryan and Moore (1977). Both views
of the rift valley indicate clearly that magmatism is not
limited to the central volcanic axis. Numerous vertical
dikes are present within the rift valley floor and adjacent
to the rising valley walls (flank lavas and dikes) as well.
The relative ages of various flows have been determined
on the basis of the thicknesses of surface weathering
rind on the basaltic glasses and of iron-manganese oxide

encrustations. Estimated ages indicate that the central
lavas are in general younger than the lavas that lie on the
flanks (sloping sides) of the central axis; the flank lavas
are in turn younger than the basaltic rocks penetrated by
flank feeder dikes.

The FAMOUS study and others have indicated that
the compositions of the various lavas at the ridge axis
are not identical. Regular compositional variation exists
from the center of the rift valley out to the valley walls.
Lavas along the flanks have a lower ratio of olivine
phenocrysts relative to those of plagioclase and clinopyroxene than axial lavas. In flank lavas, volcanic glasses
contain larger concentrations of SiO_2, K_2O, and water
and a higher FeO:MgO ratio. These features might reflect either a greater degree of fractionation of the
magmas before near-surface emplacement or local removal of olivine, plagioclase, and clinopyroxene from
the liquids after emplacement.

FAMOUS investigators have invoked a process of progressive tapping of an actively fractionating, zoned
magma chamber to explain the heterogeneity of magma
compositions. The cooler portions at the edges have

FIGURE 8-3

●●●●●●●●●●●●●●●●

Plan (map) views of an inner rift valley in the FAMOUS area, showing changes in the active volcanic axis (VA) and subsequent rift (SR) over a 180,000-year period. Part H shows all rifts superimposed. Numbers on all diagrams are the dates of the flows (B.P., years before present). [From R. D. Ballard and T. H. van Andel, 1977, *Geol. Soc. Am. Bull.*, *88*, Fig. 17.]

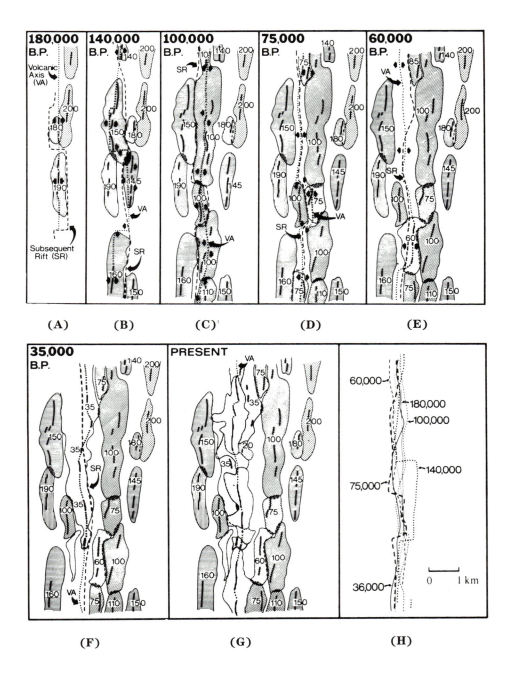

cooled more rapidly than central portions (see Figure 8-4) and thus have changed their compositions through removal of phenocrysts. Fractionation is accomplished by gravitational settling of dense olivine, clinopyroxene, and spinel phenocrysts, which form a cumulate layer on the magma chamber floor. The magma has a low water content and is relatively dense, so the plagioclase phenocrysts float and remain along the walls and roof of the magma chamber, thereby forming a plagioclase-rich cumulate distinct from the plagioclase-poor cumulate on the floor of the chamber. As the magma chamber solidifies laterally, the outer portions form a relatively struc-

tureless gabbro. The process continues during rifting, with episodic formation of unfractionated basaltic feeder dikes in the axial zone, intrusion of fractionated dikes and extrusion of unfractionated flows in the flanks, and subsurface fractionation and solidification of the outer edges of the magma chamber at depth.

As cooling continues at the magma borders, the newly solidifed materials become part of the diverging plate edges. New magma enters the central area, new vertical fractures develop, and the multilayered plates continue to move away from the ridge axis. Reaction of some of the uppermost basaltic material with heated seawater

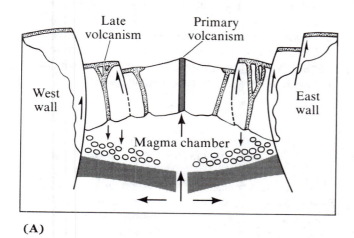

(A)

FIGURE 8-4

• • • • • • • • • • • • • • • •

Two models of the inferred magma chamber beneath the median valley. Lateral magmatic differentiation is indicated by variable thickness of the cumulate layer at the base and by differentiated melt along the relatively cool edges. Half-headed arrows indicate relative fault movement; full-barbed arrows indicate directions of magma flow. Cpx, clinopyroxene; Ol, olivine; Sp, spinel. [Part A from R. D. Ballard and T. H. van Andel, 1977, *Geol. Soc. Am. Bull.*, *88*, Fig. 19; Part B from W. B. Bryan and J. G. Moore, 1977, *Geol. Soc. Am. Bull.*, *88*, Fig. 16.]

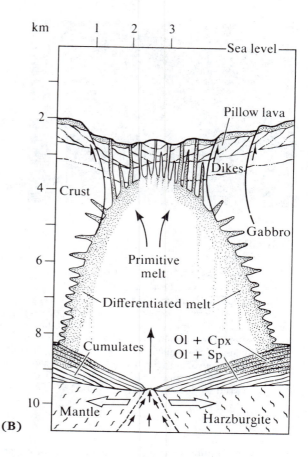

(B)

circulating through fractures (see later) can cause profound changes in bulk chemistry, especially enrichment in Na_2O, which forms the altered rock type called spilite. The general characteristics of oceanic crust and upper mantle that form at the ridges are shown in Table 8-1. The particular "stratigraphic" section consists of a thin surface layer of pelagic sediments underlain by a section of variably altered pillow basalts of tholeiitic and alkalic composition, in turn underlain by a so-called sheeted-dike complex of massive gabbro cut by large numbers of vertical basaltic dikes and finally with a base of cumulate harzburgite grading downward into peridotite. This

TABLE 8-1 Layered structure of upper oceanic lithosphere

• •

Lithospheric portion	Thickness (km)	Materials
Crust		
Layer 1	0.5–1.0	Thin chert, lutite, pelagic limestone.
Layer 2	1.5–2.0	Pillow basalts: tholeiites, alkali basalts, spilites; manganese nodules and encrustations are common.
Layer 3	4.5–5.0	Gabbroic rocks with minor diorite; hydrothermally altered basaltic rocks and serpentinites; numerous vertical dikes in upper portion; ultramafic cumulates present at the base in various thicknesses.
Mantle		
Lithospheric mantle	Up to 60	Peridotite, including depleted lithologies such as harzburgite and dunite, especially in upper part.

FIGURE 8-5

● ● ● ● ● ● ● ● ● ● ● ● ● ● ● ● ●

A plume of hot (>300°C) hydrothermal fluid issues from a hydrothermal vent on the East Pacific Rise near Baja California. These mineral-rich fluids form what are called black or gray smokers, depending on the temperature and the density of mineral precipitation as the hot fluids hit cold seawater. Note the chimney of precipitated minerals (mainly sulfides) that is building up around the vent. Hydrothermal vents of this type are thought to be responsible for many seafloor mineral deposits. [Photo by D. B. Foster, Woods Hole Oceanographic Institution.]

section is very similar to ophiolite sequences found at convergent ocean-continent margins worldwide (see Chapter 9).

Observations by submersibles in the 1980s have provided new information on the process of **spilitization,** or hydrothermal alteration of shallow oceanic crust. This process involves the addition of sodium to originally basaltic rocks, coupled with the nearly total removal of iron and calcium and some depletion of silica. The oxides Al_2O_3 and TiO_2 are neither added nor removed but have greatly increased concentrations because of the massive removal of other elements.

At several locations on the East Pacific Rise west of Central America and northern South America, the submersible *Alvin* found vents near the ridge crest that emit jets or plumes of very hot water (up to 350°C). Upon contact with the cold seawater, large amounts of very small sulfide and oxide crystals precipitate from the hot fluids, thereby giving the plumes a smoky appearance and inspiring the terms *black smoker* and *gray smoker*, depending on the density of the precipitate (Figure 8-5). Stalagmite-like columns of sulfide minerals and colonies of unique marine animals from bacteria to worms, clams, and crabs commonly surround the vents. The smokers have been interpreted as the surface emanations of the seawater that penetrated the surface of relatively recent flows through cracks and fissures, was heated by residual heat of the flows, and caused massive alteration of the material that it passed through. The same type of alteration that appears to be happening with the present-day smokers has been recognized in orogenic belts as the process that formed unusual bulk compositions in what are now metamorphic rocks.

Petrographic and Chemical Characteristics of MORBs

The chemical variability of mid-ocean ridge basalts, although limited compared with that of the basalts of oceanic islands (Table 8-2), encompasses several major recognizable types. The bulk of ocean floor basalts fall into a category that has been called **N-MORB** (or *normal* MORB). An overwhelming majority of these basalts are hypersthene- and olivine-normative (olivine tholeiites), although some are quartz-normative (quartz tholeiites). A second, subordinate type of mid-ocean ridge basalts is **E-MORB** (for *enriched* MORB), also sometimes called plume-type MORB. Alkali basalts (generally combining normative nepheline and lacking normative hypersthene) are rare along the submarine portions of ridges but common on some islands that fall along the ridges (Iceland, for example) or along fracture zones. These occurrences are probably more akin to ocean island basalts (OIBs) than to typical ridge basalts, however.

Trace-element and isotopic chemistry of N-MORBs distinguishes them from the E-MORBs. Both types have essentially total overlaps in major-element chemistry, but N-MORBs are depleted in trace elements, particularly the light rare earth elements (lanthanum through europium), relative to E-MORBs, which are strongly enriched in these elements and others, for example, potassium, barium, rubidium, and zirconium. A relatively minor third type of seafloor basalt is called T-type, for its transitional nature between N- and E-MORBs. The relationship of E-MORBs to N-MORBs has been a subject of debate for some time. Some petrologists have proposed that E-MORBs are derived by fractional crystallization of the same magmas that crystallize as N-MORBs. However, the major-element chemistry of both types overlaps, especially the ratio $Mg/(Mg + Fe)$, or Mg#, which should strongly increase with removal of magnesium-rich olivine and pyroxene crystals, thus precluding a fractionation relationship between the two types of MORB. It is likely that E-MORBs are derived from melting of a compositionally distinct type of parental mantle peridotite.

Within suites of volcanic rocks from small areas, compositional variations can be attributed to several factors, of which fractionation is the most common. For example, a suite of volcanic rocks can represent fractions of

TABLE 8-2   Representative chemical compositions of various basalts

Components	N-MORB[a]	E-MORB[a]	OIT[b]	OIAB[c]	IAT[d]	CFT[e]
SiO_2	48.77	47.74	50.51	47.52	51.90	50.01
TiO_2	1.33	1.59	2.63	3.29	0.80	1.00
Al_2O_3	15.90	15.12	13.45	15.95	16.00	17.08
Fe_2O_3	1.33	2.31	1.78	7.16	—	—
FeO	8.62	9.74	9.59	5.30	9.56	10.01
MnO	0.17	0.20	0.17	0.19	0.17	0.14
MgO	9.67	8.99	7.41	5.18	6.77	7.84
CaO	11.16	11.61	11.18	8.96	11.80	11.01
Na_2O	2.43	2.04	2.28	3.56	2.42	2.44
K_2O	0.08	0.19	0.49	1.29	0.44	0.27
P_2O_5	0.09	0.18	0.28	0.64	0.11	0.19

[a]N-MORB, normal mid-ocean ridge basalt; E-MORB, enriched MORB.
[b]OIT, ocean island tholeiite.
[c]OIAB, ocean island alkali basalt (Kohala, Hawaii).
[d]IAT, island arc tholeiite.
[e]CFT, continental flood tholeiite.

Source: After Wilson (1989), Tables 5.3, 5.4, and 9.6.

liquid tapped from a deep magma chamber over a long period of time, during which the initial magma partially crystallizes within the magma chamber. The individual liquids can be either more primitive or more evolved, depending on when they were removed from the chamber. The exact patterns of compositional variation are functions of which mineral or minerals are crystallizing within the magma chamber (see the earlier discussion of fractional crystallization in Chapter 6). As an example, olivine is a common early-crystallizing mineral in basaltic liquids and typically has a higher Mg# and lower silica content than the liquid; removal of olivine from a basaltic magma causes the remaining liquid to become depleted in magnesium relative to iron and enriched in silica. Calcic plagioclase has much higher CaO and Al_2O_3 contents than typical magmas, and its removal will therefore strongly deplete the liquid in these components. If chemical data for an entire suite are graphically portrayed, the changes in relative concentrations of each element can be correlated with a prediction of how the concentration of that element should change if a particular mineral fractionally crystallizes.

Figure 8-6 shows chemical data for glassy basaltic volcanics from the Amar Valley in the FAMOUS area. Glassy rocks were chosen by the FAMOUS geologists to guarantee rocks that represent actual liquid compositions (that is, are not partially cumulate) and to minimize the possibility of alteration due to seawater contact. In this commonly used variation (Harker) dia-

gram, the weight concentrations of each element are plotted against MgO content (in weight percent), which decreases as a result of olivine crystallization and removal. The data indicate that minerals other than olivine must have crystallized from the initial MORB magma to explain the chemical evolution within the suite. The increase in FeO content is, of course, due to olivine removal, but the decreases in CaO and Al_2O_3 content must be due to plagioclase removal. In addition, CaO content decreases more, relative to Al_2O_3 content, than can be explained with plagioclase removal alone, and requires that a mineral with a high ratio of calcium to aluminum also must have crystallized; this mineral is most likely clinopyroxene. All of the magmas within the FAMOUS area must be related to one another by fractional crystallization of olivine, plagioclase, and augite; and this conclusion is borne out by the presence of these minerals as phenocrysts in the basalts, although clinopyroxene is typically found only in more evolved basalts. An important conclusion is that the MORBs from the FAMOUS area are not primary, mantle-derived liquids but have compositions that are the result of low-pressure fractional crystallization in near-surface magma chambers. This conclusion is very important because it means that most of these MORBs cannot be used to infer anything directly about the nature of the mantle from which they came, nor about the melting process itself. Experimental studies of the crystallization of MORB liquids have confirmed that the order of

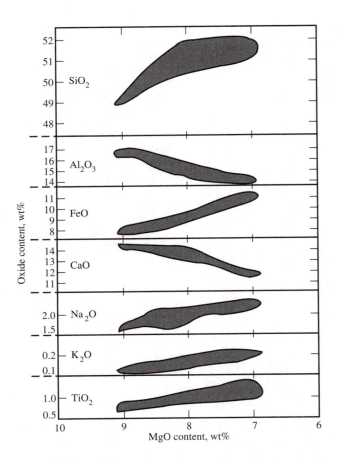

of the elements in question can be characterized geochemically as *incompatible;* that is, they have a moderate to strong preference for the relatively unstructured liquid state as opposed to crystal structures of minerals. As a result, when mantle peridotite begins to melt, much of the content of the incompatible elements in the parental material goes into the first few percent of melt that is formed. As long as none of the original minerals in the peridotite is totally depleted (see Chapter 6), further melting does not change the concentration of major elements in the melt very much but can significantly dilute the concentrations of the trace incompatible elements. It has therefore been proposed that the most primitive E-MORBs represent an approximation of liquids derived from small amounts of melting of mantle peridotite, whereas primitive N-MORBs are melts that come from greater degrees of melting, perhaps two or three times as much. All but the most primitive magmas in each group (those with the highest magnesium:iron ratios and calcium and aluminum contents, and the lowest iron and silicon contents) have undergone variable degrees of near-surface fractional crystallization.

Petrogenesis of Seafloor Basalts

The formation of basaltic magmas in the upper mantle beneath the mid-ocean ridges must be related to the occurrence of rifting and the consequent lithospheric extensional stress that produces a reduction of lithostatic pressure in the shallow mantle. To build a reasonable picture of what is happening at divergent plate boundaries such as the mid-ocean ridges, it is necessary to consider all the relevant information on this environment. The facts are the following:

The rate of heat flow is high at mid-ocean ridges relative to that of the normal seafloor.

Basaltic volcanism is ubiquitous at the mid-ocean ridges.

The basalts extruded at the ridges, and the underlying mantle peridotite, contain very low contents of radioactive heat-producing elements such as uranium, thorium, and potassium.

Experiments have shown that the first melting products of peridotite at moderate to high pressures are basalts with major-element compositions similar to MORBs.

The high heat flow at the ridges probably reflects both the presence of basaltic magmas at shallow levels of the mantle and a general upwelling of hotter mantle toward the surface. The paucity of radioactive heat-producing elements in the ridge environment rules out that heat source as a cause of melting. Melting is now thought to

FIGURE 8-6

• • • • • • • • • • • • • • • •

Harker MgO content variation diagrams, showing the ranges of chemical data for glassy basalts of the Amar Valley of the Mid-Atlantic Ridge. Especially note the patterns of increasing SiO_2 and FeO contents and decreasing Al_2O_3 and CaO contents as MgO content decreases to the right. These patterns indicate that olivine removal alone cannot explain the fractionation trend and suggest removal of plagioclase and clinopyroxene as well. [From D. S. Stakes, J. W. Shervais, and C. A. Hopson, 1984, *J. Geophys. Res.,* *89,* Fig. 8.]

phenocryst crystallization is olivine → plagioclase → clinopyroxene, indicating that, as a late-crystallizing mineral, clinopyroxene should be found only in the more evolved rocks.

If there are essentially no differences between N-MORB and E-MORB in major-element patterns, how are the trace-element differences to be explained? The chemical compositions and evolution of both types of basalt appear to be related to low-pressure fractional crystallization, as discussed above, but the trace-element and isotopic distinctions cannot be explained this way. Rather, they are related either to characteristics of the source area of melting or to the melting process itself. All

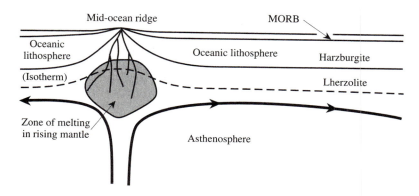

FIGURE 8-7
• • • • • • • • • • • • • • • • •

Schematic cross section of a divergent oceanic plate junction with a spreading
ridge (mid-ocean ridge) and coupled mantle convection cells. Convection flow
lines are shown by heavy lines. Note that upward flow of mantle asthenosphere
beneath the ridge deflects isotherms upward and produces a zone of melting
due to the decompression accompanying mantle uprise. At the ridge itself,
asthenosphere extends virtually to the surface. Typical rock types of oceanic
crust (MORB), depleted oceanic lithosphere (harzburgite), and asthenosphere
(lherzolite) are shown in their appropriate locations.

result from convective processes that explain not only
the upwelling of hot mantle and resultant melting but
also the divergent movement of lithospheric plates.

The idea behind convection ultimately relates to the
cold-accretion theory of the origin of the earth. This
theory states that the earth was formed from the accretion of solid particles of condensate from the solar nebula at temperatures of a few hundred degrees Celsius. The
accretionary process led to heating of the earth and formation of large volumes of melt. Under physical, chemical, and gravitational constraints, a layered body was
created. It is now thought that the differentiation of the
earth was incomplete and that compositional differences
exist within the various layers of the mantle, including
differences in content of radioactive heat-producing elements. The result is that some portions of a particular
level within the earth are hotter than others. As noted in
Chapter 7, calculated values of geotherms under shield
and oceanic areas reveal differences of 100° Celsius or
more at a depth of 100 km. Considering the possible viscosity of the mantle and the unequal distribution of heat
sources, lateral temperature differences should be common within the mantle. Even small temperature differences should be sufficient to cause density differences
between similar rocks at the same level.

Mantle seismic tomography has now revealed the
existence of these density contrasts within the upper
mantle. Less dense rocks migrate upward by plastic flow
while denser ones sink, a process that produces a convection cell with crudely horizontal cylindrical flow. The

depth and lateral extent of convection cells are not well
constrained, and current theories include single cells
that extend downward to the core-mantle boundary, single cells limited to depths below the low-velocity zone,
and multiple levels of convection separated by horizontal
boundary layers. Most models provide for pairs of convection cells with rising parts under the mid-ocean
ridges, horizontally moving currents diverging from the
ridges, and downward portions that sink at or near the
subduction zones at continental margins (Figure 8-7).

It is reasonable to assume the presence of vertically
rising hot mantle peridotite beneath the mid-ocean
ridges (see Figure 8-7). Rising is probably due to convective motions but could also be due to attenuation
of lithosphere and concomitant pressure decrease as
the divergent plates pull apart. This latter explanation
is considered unlikely, because there is no reasonable
physical process for getting the plates to move apart,
in the absence of convection. After vertical rise of mantle peridotite is initiated, the temperature difference
between the hotter rising material and surrounding
cooler rocks increases as the hot material moves upward, but the rising peridotite can be considered to take
a roughly adiabatic path as it moves upward. (Remember
that **adiabatic** means that essentially no heat is lost
from the rising diapir to the surroundings.)

This process of upward heat transfer in the earth
by bulk movement of hot material (either hot solid or
molten rock or hot aqueous fluids) rather than by
conduction is known as *advective* heat transfer and is

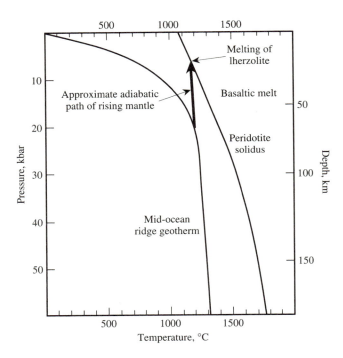

FIGURE 8-8
● ● ● ● ● ● ● ● ● ● ● ● ● ●

Hypothetical *P-T* diagram for melting of mantle peridotite in rising diapirs beneath mid-ocean spreading
centers. Note the advective transfer of anomalously hot
mantle peridotite to shallower depths along a roughly
adiabatic path that intersects the solidus. Even the elevated mid-ocean ridge geotherm shown does not intersect the dry solidus.

exceptionally important in both igneous and metamorphic rocks. Figure 8-8 shows a diagram of depth and
(pressure) as a function of temperature. The rising hot
material will encounter a zone of incipient melting as it
rises to shallower mantle levels at temperatures increasingly above the geotherm. Incipient melting of several
percent of the hot material may aid the rising of the partially molten aggregate by acting as a "lubricant" that
facilitates deformation. When significant melting occurs
(perhaps 20–30%), the melt may be able to escape from
the residual solids and move upward toward the surface
as a magma. There have been intensive study and debate
among petrologists and geophysicists about this melt
escape process.

Advective heating of the ridge zones produces a characteristic pattern of isotherms in the shallow mantle
(Figure 8-9). These diagrams are what geophysicists
refer to as a "half-space": The ridge axis itself is shown

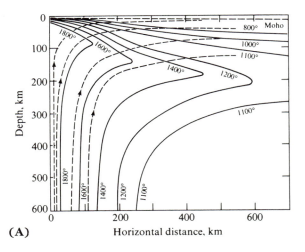

(A)

FIGURE 8-9
● ● ● ● ● ● ● ● ● ● ● ● ● ●

Thermal models for the mantle beneath a mid-ocean ridge. The
dashed lines with arrows show directions of particle motion.
[Part A after E. R. Oxburgh and D. L. Turcotte, 1968, *J. Geophys.
Res.*, *73*, Figs. 11, 12; Part B after K. E. Torrance and D. L.
Turcotte, 1971, *J. Geophys. Res.*, *76*, Fig. 2; and D. W. Forsyth
and F. Press, 1971, *J. Geophys. Res.*, *76*, Fig. 12.]

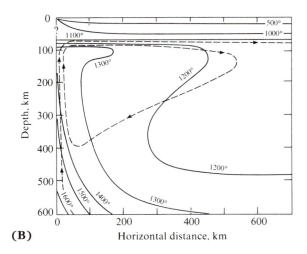

(B)

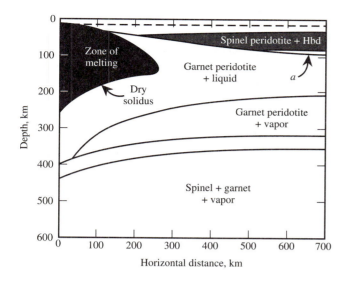

FIGURE 8-10

● ● ● ● ● ● ● ● ● ● ● ● ● ● ● ●

Schematic diagram showing the effect of an ocean ridge thermal regime (from Figure 8-9) on peridotite phase equilibria, including melting. A large area of melting is present under the ridge, and phase changes are vertically depressed. Hbd, hornblende; area *a* consists of garnet peridotite + vapor. [From P. J. Wyllie, 1971, *J. Geophys. Res.*, *76*, Fig. 12.]

as the left edge of each diagram, and the simplified mathematics yields a symmetrical distribution of temperature contours on either side of the rift. On the basis of similar calculations, P. J. Wyllie has developed a speculative schematic section (Figure 8-10) that shows the distribution of rock types and magma at a typical active mid-ocean ridge plate junction. The rise of isotherms at and under the rift itself has increased the depth of phase changes within the mantle (remember that these phase boundaries increase in pressure with increased temperature; that is, they have a *positive slope* on a *P-T* diagram). Note also in Figure 8-10 that there is a significant thickening of the zone of dry peridotite melting under the rift, with the solidus temperature exceeded over a horizontal and vertical distance of more than 200 km.

Regarding petrogenesis of mid-ocean ridge basalts, evidence suggests that very few are primary, that is, have compositions that reflect the phase equilibria of peridotite melting and not low-pressure fractional crystallization. Minimum conditions for candidacy as primary basaltic melts are (1) a glassy character, which indicates a minimum likelihood of fractionation and crystal removal; (2) a high Mg#, (MgO concentration over about 10–11 wt%), which indicates that olivine was probably not removed; and (3) the highest concentrations of calcium and aluminum within the suite, a composition reflecting a lack of plagioclase fractionation. There are only a few analyzed MORBs that fit these qualifications (some from the FAMOUS area), and they have been experimentally tested as primary melts. The experiments can be done in two ways: either checking to see whether the glasses themselves are in chemical equilibrium with presumed mantle source peridotites at high tempera-

tures and pressures, or melting synthetic peridotite (pyrolite) at these high pressures and temperatures to see whether a melt like MORB is generated.

The first type of experiment is illustrated in Figure 8-11. Experiments are done at a series of pressures (commonly 5-kbar intervals) in which a sample of glassy basalt is heated until completely melted, then cooled until the first appearance of each mineral. The temperature of first appearance of each phenocryst mineral typically increases with pressure in crystallization of a dry melt. In the simplified schematic *P-T* diagram of Figure 8-11, the temperature of first appearance of each of the phenocryst minerals olivine, orthopyroxene, and clinopyroxene has been plotted as a function of pressure. Remember that at a single pressure, the boundaries indicate the first appearance of each mineral on a liquidus surface, such as those in the isobaric ternary liquidus diagrams of Chapter 4, and thus reflect progressive intersections of the liquid with a liquidus surface, a cotectic or peritectic line, and a ternary eutectic point. At temperatures at and below the boundary for each mineral, the melt is saturated with that mineral. For example, at 5 kbar, the melt becomes saturated first with olivine, then with orthopyroxene, finally with clinopyroxene. (Note that this figure is only schematic and does not reflect the more complicated actual crystallization sequence at low pressures in MORB liquids, in which plagioclase appears immediately after olivine, and orthopyroxene does not appear at all on the liquidus.)

In Figure 8-11, the convergence of the saturation lines for olivine, orthopyroxene, and clinopyroxene at a single point at high pressure is very important: It means that this particular experimental melt composition can be

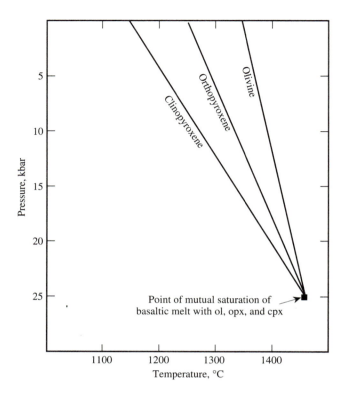

FIGURE 8-11
● ● ● ● ● ● ● ● ● ● ● ● ● ● ● ●

Schematic pressure-temperature (*P-T*) diagram showing the experimental technique for demonstrating that a particular basaltic liquid could be in equilibrium with mantle peridotite at mantle conditions. Lines are experimentally determined *P-T* positions of the initial crystallization of each mineral as the liquid is cooled at various pressures. If the lines intersect at a single point (as shown here at 25 kbar and 1450°C), this basaltic liquid could be multiply saturated with olivine, orthopyroxene, and clinopyroxene *at this P and T*, and thus could have originated by melting of peridotite at 25 kbar.

simultaneously in equilibrium with these three minerals *at a single P-T point.* This point represents the melting point of peridotite to yield this particular melt. Of course, the experiment does not prove that the particular basalt composition is a primary melt but it does indicate that it could be. Cumulative results of numerous experimental studies suggest that primary-appearing MORB compositions may have originated through peridotite melting at temperatures and pressures ranging from 1200°C at 7 kbar to 1300°C at 12 kbar, corresponding to depths of about 20 to 36 km. The degree of melting appears to be moderate (roughly 25%) for N-MORB melts and about a third as much for E-MORB. Alternatively, these melts may themselves be the results of high- pressure fractionation at 20 to 36 km of even more primitive melts from deeper in the mantle. Experiments have definitively

shown, however, that MORBs cannot be the *direct* crystallization products of deeper melts without high-pressure fractionation. These deeper basaltic melts, more alkalic and lower in silica, are characteristic of ocean island occurrences.

OCEAN ISLAND BASALTS AND RELATED ROCKS
● ● ● ● ● ● ● ● ● ● ● ● ● ● ●

Although most of the volcanic activity in the oceans occurs at mid-ocean rift zones or at convergent boundaries between two oceanic plates, there are widely scattered occurrences of so-called intraplate volcanic activity. These locally voluminous extrusions form a number of prominent islands in all the world's oceans (for example, the Hawaiian Islands, the Society Islands, the Canary Islands), as well as the seamounts and guyots that represent either active submarine volcanoes or eroded remnants of former ocean islands. It has been estimated that the volume of basalts extruded in intraplate oceanic volcanism is about one-tenth of that derived at the mid-ocean ridges. The occurrence of basaltic volcanism at plate boundaries is readily explained, but it is more difficult to envision a cause for intraplate volcanism. As early as the 1960s, the importance of such occurrences was recognized and ascribed to point sources of heat in the underlying mantle. The original model has been modified and expanded in the intervening years. In any case, the characteristics of oceanic intraplate volcanic rocks are sufficiently different from those of ridge basalts to indicate a quite different mode of origin.

The most striking physiographic feature of many ocean island volcanic occurrences is the prominent linear trend of the individual volcanic centers. Figure 8-12 shows maps of some of these linear trends: the Hawaiian Islands–Emperor Seamounts and the Society Islands and other south-central Pacific island chains. Research on these trends has indicated a definite connection between the ages of individual centers and their positions within the chain. Figure 8-13 shows results of radiometric dating for the Hawaiian chain. In both cases, the youngest submarine volcano is several tens of kilometers farther to the southeast than the youngest volcanic island and is a seamount that has not yet broken sea level (Loihi in the Hawaiians and Mouaa Pihaa in the Societies). The southeasternmost island in the Hawaiian chain is Hawaii, which is still active; in the Society Islands, Tahiti is farthest southeast, and its most recent volcanism is a little less than 1 million years old. In both chains, the volcanic rocks are older as one progresses from island to island toward the northwest. The most recent activity on Maui is about 1 million years old and on Oahu about

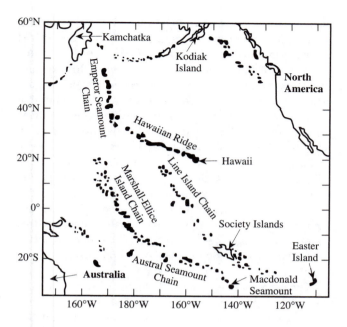

FIGURE 8-12

● ● ● ● ● ● ● ● ● ● ● ● ● ● ● ●

Map of linear island trends in the Pacific.

4.5 million years old. The most northerly seamounts in the Emperor chain are more than 70 million years old.

The linear patterns provide fairly persuasive evidence that such intraplate volcanism is due to "hot spots" in the mantle, as first suggested by J. T. Wilson and amplified by Jason Morgan. **Hot spots** or **mantle plumes** are areally restricted magma sources in the mantle that must be relatively fixed in the sublitho-

spheric mantle. Magmas rising from such hot spots progressively penetrate moving lithospheric plates in a linear pattern, much like the passage of a sheet of paper over a fixed candle flame. The patterns of the Hawaiian and Society Islands in the Pacific plate indicate a northwesterly movement of the plate over the last few tens of millions of years. The bend in the Hawaiian-Emperor trend further indicates an abrupt shift in direction of plate movement. Geophysical data indicate that there is little or no melting actually occurring in the lithosphere immediately beneath oceanic island volcanoes, supporting the idea that the magmas actually come from deeper in the mantle. Although the Wilson-Morgan model for hot-spot magmatism is an attractive explanation for intraplate linear magmatic trends, there are some occurrences that are better explained by other mechanisms such as fracture propagation.

One might ask whether such hot spots and intraplate magmatism are restricted to oceanic plates. It has not yet been proved, but evidence suggests that such hot spots also exist under continental lithosphere. Continental lithosphere is typically about 125 km thick, whereas oceanic lithosphere away from the ridges is much thinner, between 60 and 80 km. It therefore seems much more likely that sublithospheric magmas will pass all the way through oceanic plates to become volcanoes, whereas continental intraplate magmatism is much more likely to result in deep plutonic rocks that are modified by the crust (see Chapter 10).

Most volcanic islands in the oceans are the exposed upper parts of single or coalesced shield volcanoes. Mauna Loa on Hawaii is the largest shield volcano on Earth and, in fact, the tallest mountain on Earth if one considers that its total height from the seafloor to its top (at about 15,000 feet elevation above sea level) is roughly 31,000 feet (9500 meters). The island of Hawaii actually consists of several coalesced shield volcanoes that represent an enormous outpouring of basaltic lavas (perhaps as much as 50,000 km³) whose enormous weight has actually deflected the Pacific lithosphere downward (Macdonald, Abbott, and Peterson 1983).

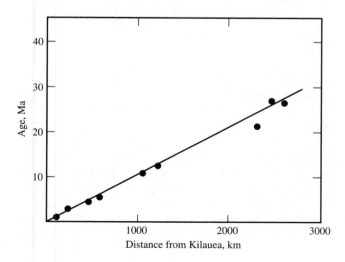

FIGURE 8-13

● ● ● ● ● ● ● ● ● ● ● ● ● ● ● ●

Ages of volcanic activity in the Hawaiian Island–Emperor Seamount linear trend (refer to Figure 8-12), as a function of distance from the currently active volcano Kilauea on the island of Hawaii. Age determinations are generally based on potassium-argon dating. If we assume a fixed hot spot, then the line represents a spreading rate for the Pacific plate of about 10 cm per year. [Hawaiian-Emperor data from H. R. Shaw, E. D. Jackson, and K. E. Baragar, 1980, *Am. J. Sci., 280A.*]

Chemical and Mineralogic Characteristics of Ocean Island Rocks

Various ocean island chains have attracted considerable petrologic interest over the last hundred years or more, especially the Hawaiian Islands, but much work remains to be done in understanding the genesis of these rocks. Single volcanoes undergo a very definite evolution in magma chemistry, but it is unclear what the earlier stages actually are. The reason for this is a practical one: The earliest lavas are buried beneath all the later ones and are not accessible for sampling. Over geologic time, these deeper roots of oceanic volcanoes are not exposed as continental orogenic belts are, because oceanic rocks are at best only imperfectly preserved at ocean-continent plate boundaries and are most likely subducted entirely. Observations have shown that there are basically three types of basaltic magmas that constitute the bulk of oceanic islands: **tholeiite basalts, alkali basalts,** and **nephelinites,** each with progressively lower SiO_2 content relative to total alkali content.

The traditional model for ocean island volcanism proposed that initial magmatism involved olivine tholeiite, which made up the great bulk of the shield volcanoes. A late-stage chemical evolution broken by long dormant intervals eventually produced small volumes of alkali basalt and finally nephelinites and basanites, magmas that are radically different from the initial tholeiites. It is now known, however, that alkali basalts play a larger role in earlier stages of shield building. Critical evidence has come from the seamount Loihi, southeast of Hawaii. Presumably still in its shield-building phase, Loihi has yielded dredge samples consisting of a wide range of basalt types, including many alkali basalts. In addition, virtually no olivine tholeiites have been reported among several hundred analyzed samples from the island of Tahiti, almost all of which are alkalic.

THOLEIITE BASALTS.   Ocean island tholeiites bear a superficial mineralogic resemblance to MORBs. Large phenocrysts of magnesium-rich olivine and chromium-bearing spinel are common. Clinopyroxene and plagioclase are typically not among the phenocrysts, however, but occur instead as fine crystals in the uniform groundmass of the rock, a texture reflecting crystallization of these minerals only during the final quenching event as the magma was extruded. This characteristic means that chemical variability among relatively primitive ocean island tholeiites is mostly due to fractional removal of olivine, in contrast to the fractionation control of both olivine and plagioclase in MORBs. The tholeiites can be either quartz-normative or olivine-normative. Olivine occurs in both types, although it should be noted that its occurrence in quartz-normative rocks is a result of meta-stable persistence during rapid quenching and that neither quartz nor tridymite actually occurs. Some of the quartz-normative tholeiites contain phenocrysts of orthopyroxene, a mineral that is never seen in MORBs, or rims of orthopyroxene surrounding corroded olivine crystals. This latter effect is due to the reaction relationship that accompanies the transition from crystallization of olivine to that of orthopyroxene (refer to the discussions in Chapters 4 and 6 and to Figures 4-6 and 6-8). The crystallization sequence is typically olivine (+ spinel) → clinopyroxene → plagioclase → iron-titanium oxides in olivine-normative magmas, although very subtle differences in chemistry can reverse the order of plagioclase and pyroxene. In the quartz-normative magmas that crystallize orthopyroxene, the typical order is olivine (+ spinel) → orthopyroxene → clinopyroxene and/or plagioclase → iron-titanium oxides.

The range of ocean island tholeiite chemical compositions is virtually as wide as that of MORBs (see Table 8-2), but the patterns differ, particularly between OIB and the MORB glasses that are the best candidates for primary MORB magma. Striking contrasts between OIB tholeiites and MORBs include the following:

1. MgO contents of tholeiites range from a high of 20 wt% to a low of about 5 wt%, although the most magnesium-rich tholeiites appear not to be primary but to have been enriched in MgO through accumulation of olivine crystals; this enrichment does not occur in MORBs.
2. CaO and Al_2O_3 contents increase with decreasing MgO content in OIB, reflecting the olivine-removal control on liquid compositions; in MORB, CaO and Al_2O_3 contents decrease with decreasing MgO content, reflecting fractional removal of clinopyroxene and plagioclase.
3. Trace elements in OIBs are enriched relative to MORB, particularly potassium, thorium, uranium, titanium, and light rare earth elements.
4. The OIBs and MORBs have very different patterns in ratios of neodymium, strontium, and lead isotopes.

Therefore, both mineralogic and chemical data point to a very different process of formation of ocean island tholeiite magmas from that of MORB magmas.

ALKALI BASALTS AND NEPHELINITES.   This class of magmas is considerably more diverse than the tholeiites, both compositionally and mineralogically. It ranges from very magnesium-rich alkali olivine basalts to highly fractionated alkali-rich rocks such as trachytes (fine-grained equivalents of monzonites) and phonolites (fine-grained equivalents of nepheline or leucite syenites). The less

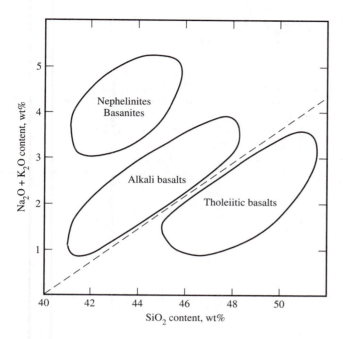

FIGURE 8-14

• • • • • • • • • • • • • • • •

The ratio of $Na_2O + K_2O$ content and SiO_2 content defines important compositional distinctions within the tholeiite-alkali basalt-basanite suite of ocean islands. The dashed dividing line between alkali basalts and tholeiites marks the approximate boundary of nepheline-normative basalts (above) and orthopyroxene-normative basalts (below).

fractionated members of the series typically contain phenocrysts of olivine, and virtually all of these rocks have augite phenocrysts. The augite, however, is quite different chemically from the calcic clinopyroxene of the tholeiite suite, containing significantly higher concentrations of titanium, aluminum, and sodium. Orthopyroxene is unknown in this suite because of the advanced degree of silica undersaturation of the magmas: The presence of normative nepheline precludes normative (or modal) orthopyroxene. This characteristic is readily seen on the margins of mantle harzburgite or lherzolite xenoliths: Orthopyroxene crystals in the xenoliths incongruently melt where they come in contact with the silica-undersaturated host basalt (see Figure 2-13). The alkali-bearing minerals plagioclase, nepheline, melilite, and, more rarely, leucite and sodalite, most typically are in the groundmass.

The bulk chemistry of the alkali and nepheline basalts and related rocks reflects the substantial degree of differentiation within the series. As shown earlier, MgO content can be used as a measure of differentiation, and it ranges from as much as 13–15 wt% in alkali olivine

basalts down to less than 1 wt% in trachytes. The SiO_2 content is typically lower to much lower than in tholeiites and this, along with the high Na_2O content, is the cause for appearance of feldspathoid minerals such as nepheline. The relative ratio of total alkalis to silica has in fact become one of the classic ways of distinguishing tholeiites from alkali basalts (Figure 8-14). The nomenclature of these rocks is complicated, but a simplified scheme is based on the actual mineralogy (mode) rather than normative mineralogy. Thus, if a rock contains plagioclase as its only sodic mineral but falls above the line in Figure 8-14, it is called an *alkali olivine basalt*. If it contains both plagioclase and nepheline, it is called a *basanite*. If it has only nepheline, it is a *nephelinite*, and if it contains melilite as well as nepheline, it is a *nepheline-melilite basalt*. In practice, distinguishing among these rocks is difficult because of the very fine grain size of volcanic rocks.

The series also seems to have anomalously high TiO_2 contents (commonly greater than about 3 wt%) that give rise to minerals such as titanian augite and titanium-rich magnetite. The unusually high concentrations of a number of incompatible elements (the alkalis, titanium, and phosphorus, as well as light rare earth elements) indicate that the magmas either came from very unusual mantle or else resulted from very small degrees of melting (probably 5% or less). This latter interpretation is based on the fact that incompatible elements have a preference for the melt and will be strongly concentrated in the first few percent of melt; any further melting will dilute these elements. Current petrologic thinking is that small degrees of melting best explain the compositional data.

Petrogenesis of Ocean Island Magmas

By far the most studied ocean island rocks are those from the Hawaiian Islands. Early studies indicated that the volume of alkali basalt and related rocks was minuscule relative to the enormous volume of tholeiite that was presumed to make up the mass of the shield volcanoes. An obvious conclusion was that the small volume of late alkalic magmatism was a differentiation product of the tholeiitic parental magma. However, experimental studies, and in particular those of Yoder and Tilley (1962) in a landmark scientific paper, showed that the phase equilibria in relevant ternary systems did not allow fractional crystallization of a tholeiite to yield a nepheline-normative magma (or vice versa) *at low pressures*. They concluded that there could be no common parent to both magma types. But could such a connection of tholeiites and alkali basalts through fractionation be established at higher pressures?

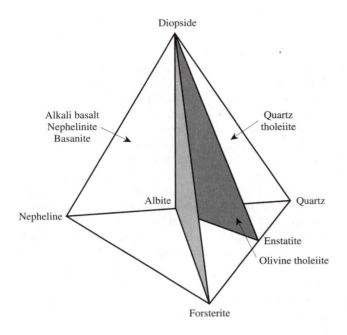

FIGURE 8-15

• • • • • • • • • • • • • •

The basalt tetrahedron, based on the minerals quartz, forsterite, nepheline, and diopside. This tetrahedron is a useful model for simple basalts and alkali basalts. The two crosshatched planes are the enstatite-albite-diopside plane, or *plane of silica undersaturation;* and the forsterite-albite-diopside plane, or *critical plane of silica undersaturation.* The compositional fields for quartz tholeiites, olivine tholeiites, and alkali basalts (and related magmas) are indicated as subvolumes of the tetrahedron. [From H. S. Yoder and C. E. Tilley, 1962, *J. Petrol., 3,* Fig. 1.]

To begin the discussion, consider the chemical system that Yoder and Tilley designed to examine basalt petrogenesis experimentally, the so-called basalt tetrahedron (Figure 8-15). This is a quaternary (four-component) system based on the minerals forsterite, diopside, quartz, and nepheline. There are therefore four ternary systems that form the faces of the tetrahedron and two additional ternary planes within the tetrahedron that are due to the existence of two intermediate minerals, albite on the nepheline-quartz join and enstatite on the forsterite-quartz join. (Clearly albite is not the best approximation of the An_{60} plagioclase that occurs in basalts, but it would require an additional dimension to illustrate phase relations involving anorthite.) Note that the right front face of the tetrahedron is the ternary system forsterite-diopside-silica, which was examined in some detail in Chapters 5 and 6. The two interior planes are important because they divide the overall system into chemical subsystems that relate to basalt classifications. Magma compositions that fall within the right-hand volume of the tetrahedron will crystallize to form rocks containing normative quartz, that is, quartz tholeiite. The central volume contains olivine tholeiite magmas, with normative olivine, plagioclase, calcic clinopyroxene, and orthopyroxene. Finally, the left volume contains nepheline-normative alkali basalts and nepheline basalts. The diopside-enstatite-albite plane is referred to as the *plane of silica undersaturation* because it separates quartz-bearing from quartz-absent compositions. The albite-forsterite-diopside plane is called the *critical plane of silica undersaturation* because it separates orthopyroxene-bearing from nepheline-bearing rocks. This tetrahedron has become the fundamental basis for understanding basalts for a generation of igneous petrologists, but the reader should note an unavoidable deficiency—the absence of two important basalt components, FeO and anorthite ($CaAl_2Si_2O_8$). Portraying them would require more dimensions than graphical analysis permits, but the phase equilibria can still be shown adequately and important points displayed in the simplified system.

As shown in Chapters 5 and 6, the plane of silica undersaturation can be penetrated by crystallizing liquids at low pressure because of the reaction relation between olivine and orthopyroxene. The critical plane, however, is a thermal divide at high pressure, because there is no reaction relation and liquids that start on either side of the plane are required to move away from the plane during crystallization, and thus to stay within their volumes (refer to Figure 5-6B). It has been discovered, however, that the mineral albite becomes unstable at high pressures and is replaced on the phase diagram by jadeite ($NaAlSi_2O_6$). This replacement shifts the position of the critical plane and makes it possible for SiO_2-saturated and mildly undersaturated tholeiites to evolve toward critically undersaturated compositions through high-pressure fractional crystallization. The volatile content of the parental rocks also affects melting behavior.

Although experimental data support fractionation as a mechanism for relating the various alkalic magmas, compositional data essentially preclude this mechanism. The Mg#'s and the trace and incompatible element contents make it impossible to derive alkali basalts from tholeiites, or basanites and nepheline basalts from alkali basalts. Therefore it appears that each of these magma types must arise through a distinct and separate melting event. There are two basic alternatives: different melting processes in essentially similar mantle, or generation of each magma type from fundamentally different mantle parents. Many experimental studies have shown that by varying melting conditions, particularly depth (pressure) and volatile content (dry, water-rich, or carbon dioxide–rich), a variety of silica saturated and undersaturated magmas could be produced from the same peridotite parent. However, isotopic analyses (particularly strontium and neodymium isotopes) of Hawaiian alkali basalts have conclusively demonstrated that different mantle sources must have been involved in generating the different magma types. Compositional heterogeneity appears to characterize mantle beneath single volcanoes as well as larger areas of oceanic crust. Possible magma mixing is a further complication that has arisen in some studies. A few primitive-appearing magmas have both isotopic and major-element characteristics that indicate mixing of several different types of primary magma in deep mantle reservoirs.

Ultimately, petrologic thinking has begun to ascribe ocean island magmatism to the effects of deep mantle plumes (Figure 8-16). Rising columns or diapirs of hot rock within the mantle have become the preferred explanation for the hot spots discussed earlier as a cause for linear ocean island trends such as the Hawaiian-Emperor trend. Moreover, linear magmatic patterns are not restricted to the oceanic lithosphere: Hot-spot tracks have been proposed for numerous linear trends on continents—for example, the Mesozoic White Mountains–Monteregian Hills intrusive series in the northeastern United States and adjacent Canada, which is continuous with a seamount chain extending into the Gulf of Maine. More than 40 active hot spots have been proposed and, with an apparent average duration of about 100 million years, they have probably had a major impact throughout much of geologic history. The origin of mantle plumes is currently an area of intense research and debate among petrologists, geophysicists, and geochemists. One pro-posal suggests that plumes represent old subducted oceanic lithosphere that has been converted into eclogite, a dense metamorphic form of basalt, and has sunk gravitationally to the core-mantle boundary. There it slowly remelts, releasing diapirs or columns of hot magma that rise as thermal anomalies through the mantle and ultimately affect the current lithosphere (see Figure 8-16).

SUMMARY
• • • • • • • • • • • • • • •

Oceanic lithosphere underlies about 70% of the surface of the earth, and its crustal portion consists almost entirely of basalt. The seafloor basalts were generated by melting of upwelling mantle at the divergent plate boundaries that are marked by mid-ocean ridges. The ridges have topographic and geologic features that reflect the magmatic and mantle convection processes that have formed them and have given their names to the dominant rock of the seafloor, mid-ocean ridge basalt (MORB). The MORBs have chemical and mineralogic features that indicate derivation from compositionally heterogeneous mantle and the effects of considerable low-pressure fractional crystallization. The more primary MORBs have chemical features that indicate modest degrees of melting of mantle source material, perhaps 25%.

A second major type of oceanic magmatism occurs entirely within single plates (intraplate volcanism) and produces oceanic island basalt (OIB). Causes for OIB magmatism appear to be quite different from those from MORB magmatism. Compositionally more diverse than MORBs, OIBs range from olivine tholeiites to very magnesium-rich alkali basalts, nepheline, and melilite basalts, and finally to silica-poor differentiates such as trachyte and phonolite on the larger oceanic islands. The OIBs seem to have undergone much less fractionation on their way to the surface than the MORBs have and thus provide more easily yielded information on

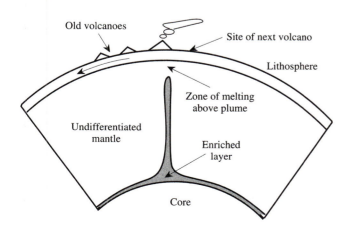

FIGURE 8–16
• • • • • • • • • • • • • • •

Cartoon showing the plume model to explain hot spots and linear volcanic trends. An enriched layer at the core-mantle boundary (possibly of old subducted lithosphere) feeds vertically rising plumes of partially molten, mobile mantle. Plumes serve as "point" heat sources that produce melting in the moving lithosphere above them.

their mantle sources. Although sometimes contradictory, this evidence generally indicates that the suboceanic mantle is quite heterogeneous in both elemental and isotopic chemistry. The common localization of OIBs in linear chains of islands or seamounts is quite compatible with an interpretation that melting was caused by localized thermal perturbations driven by rising plumes in the mantle. These plumes create the "hot spots" of plate tectonic theory and are global in distribution, although their effects are more commonly observed in oceanic lithosphere where surface volcanic features occur. Phase equilibria and chemistry of the OIBs indicate that these magmas result from melting deeper in the mantle (60 to 100 km or more) than the MORBs and at a far smaller degree of melting, probably less than 5%.

STUDY EXERCISES
• • • • • • • • • • • • • • •

1. Is the discovery of "smokers"—that is, vents for mineral-rich hydrothermal waters on mid-ocean ridges—consistent with the incipient alteration of seafloor basalts that is commonly observed in dredge samples? If so, how?

2. Summarize the chemical distinctions between N-MORB and E-MORB, and relate these to the genesis of each type of magma.

3. How does the petrogenesis of MORB magmas differ from that of OIB magmas? Do their differing chemistries relate more to different depths of origin in the mantle or to fundamental chemical contrasts in their mantle source areas?

4. Are hot spots limited to oceanic lithosphere? Is there any petrologic or tectonic reason why they should or should not be?

5. OIB magmas are considerably more diverse chemically than MORB magmas. Is the great compositional range of OIB magmas related more closely to (a) chemical evolution of a common, mantle-derived parental magma for the suite, or (b) complex melting of compositionally heterogeneous mantle over a range of depths?

REFERENCES AND ADDITIONAL READINGS
• • • • • • • • • • • • • •

Ballard, R. D., and T. H. van Andel. 1977. Morphology and tectonics of the inner rift valley at latitude 36 degrees 50 minutes N on the Mid-Atlantic Ridge. *Geol. Soc. Am. Bull.*, *88*, 507–530.

Basaltic Volcanism Study Project. 1981. *Basaltic Volcanism on the Terrestrial Planets.* New York: Pergamon.

Bender, J. F., F. N. Hodges, and A. E. Bence. 1978. Petrogenesis of basalts from the Project FAMOUS area: Experimental study from 0 to 15 kbars. *Earth Planet. Sci. Lett.*, *41*, 277–302.

Bryan, W. B., and J. G. Moore. 1977. Chemical variation of young basalts in the Mid-Atlantic Ridge rift valley near lat. 36 degrees 49 minutes N. *Geol. Soc. Am. Bull.*, *88*, 556–570.

Geological Society of America. 1977. *Bulletin* (April–May). Various articles on Project FAMOUS.

Green, D. H., and A. E. Ringwood. 1969. The origin of basaltic magma. In *The Earth's Crust and Upper Mantle. Geophys. Monogr. 13*, P. J. Hart, ed. American Geophysical Union, pp. 489–495.

Heirtzler, J. R., and T. H. van Andel. 1977. Project FAMOUS: Its origin, programs and setting. *Geol. Soc. Am. Bull.*, *88*, 481–487.

Hess, P. C. 1989. *Origins of Igneous Rocks.* Cambridge, MA: Harvard University Press.

Macdonald, G. A., A. T. Abbott, and F. L. Peterson, eds. 1983. *Volcanoes in the Sea: The Geology of Hawaii*, 2nd ed. Honolulu: University of Hawaii Press.

Morgan, J. P., D. K. Blackman, and J. M. Sinton, eds. 1992. *Mantle Flow and Melt Generation at Mid-Ocean Ridges. Geophys. Monogr. 71.* American Geophysical Union.

Morgan, W. J. 1971. Convection plumes in the lower mantle. *Nature, 230*, 42–43.

Morgan, W. J. 1983. Hotspot tracks and the early rifting of the Atlantic. *Tectonophysics, 94*, 123–139.

Shaw, H. R., E. D. Jackson, and K. E. Baragar. 1980. Volcanic periodicity along the Hawaiian-Emperor chain. *Am. J. Sci., 280A*, 667–708.

Stakes, D. S., J. W. Shervais, and C. A. Hopson. 1984. The volcanic-tectonic cycle of the FAMOUS and AMAR valleys, Mid-Atlantic Ridge (36 degrees 47 minutes N): Evidence from basalt glass and phenocryst compositional variations for a steady-state magma chamber beneath the valley midsections, AMARS. *J. Geophys. Res., 89*, 6995–7028.

Vink, G. E., W. J. Morgan, and P. R. Vogt. 1985. The earth's hot spots. *Scientific American, 252*, 50–57.

Wilson, J. T. 1973. Mantle plumes and plate motion. *Tectonophysics, 19*, 149–164.

Wilson, M. 1989. *Igneous Petrogenesis: A Global Tectonic Approach.* London: Unwin-Hyman.

Wyllie, P. J. 1971. *The Dynamic Earth.* New York: Wiley, pp. 105–137.

Yoder, H. S. 1976. *Generation of Basaltic Magma.* Washington, DC: National Academy of Sciences.

Yoder, H. S., and C. E. Tilley. 1962. Origin of basaltic magmas: An experimental study of natural and synthetic rock systems. *J. Petrol., 3*, 342–532.

Chapter 9

IGNEOUS ROCKS OF CONVERGENT MARGINS

The convergent or active margins where lithospheric plates collide are the most tectonically active parts of the earth. Of the three fundamentally different types, subduction plays the dominant role in magma generation in the first two, but other processes also operate in the third:

1. Convergence of oceanic plates, where the typical result is a deep ocean trench and an offshore island arc such as the Japanese islands, the Indonesian arc, or the Aleutian arc.
2. Convergence of an oceanic and a continental plate, resulting in a deep ocean trench, a metamorphic complex on the continental margin, and a magmatic arc; classic examples include the Andes and the Sierra Nevada.
3. Convergence of two continental plates, where neither slab of thick lithosphere can be subducted and the result is a collisional mountain range with greatly thickened crust. The Himalaya, the Alps, and the Urals are classic examples.

Each of these three types of convergent margins has its own characteristic form and pattern of magma generation, plutonism, and volcanism, although generation of large volumes of andesitic magma is common to the first two. The third type is the termination of convergence of two plates when the continental lithosphere within each of the two plates ultimately converges. In a plate tectonic cycle, however, continent-continent collision with its own unique igneous rocks is typically preceded by episodes of one or both of the first two types as leading edges of oceanic lithosphere in the plates interact. Continental collision zones therefore contain complex, tectonically intermingled igneous rock products of the whole cycle, and unraveling their spatial and chronologic relationships is a major task for geologists.

As discussed in Chapter 8, the plate tectonic cycle involves creation of new basaltic oceanic crust at the divergent plate margins, also referred to as *constructive* margins. Earth has retained a relatively constant radius for most of Earth history, however, so oceanic lithosphere must be consumed at about the same rate that new lithosphere forms at the mid-ocean ridges. Oceanic lithosphere is subducted at convergent margins, also called *destructive* margins, where old, cold oceanic lithosphere sinks gravitationally back into the mantle. The processes of heating and compressing this old (200 million–250 million years) lithosphere are keys to the generation of magmas at most convergent margins. The margins of areas of continental lithosphere within the plates are especially important, because it is in these geologic regimes that most new continental crust has been created during Earth history. New continental crustal material is a mixture of oceanic crust, sedimentary rocks, and newly created magmatic rocks, all of which are reprocessed at the continental margin and welded to the existing continent. Much of the discussion in this chapter will focus on convergent margins of the first two types that border the Pacific plate and related smaller oceanic plates. Figure 9-1 illustrates an interesting concentration of volcanoes around the Pacific that has been recognized by geologists for a long time. This "necklace" of volcanic activity and related earthquakes has been called the "ring of fire" and involves active volcanic island arcs such as the Aleutian arc and those in the western and southwestern Pacific (including Mt. Pinatubo in the Philippines), as well as volcanoes associated with continental magmatic arcs such as the Andes and the Cascades (where Mt. St. Helens is located).

IGNEOUS ROCKS OF CONVERGENT OCEANIC PLATES
• • • • • • • • • • • • • • • •

Convergent plate margins involving oceanic lithosphere are the realm of the volcanic island arcs. Most of the earth's currently active major island arcs are in the Pacific Ocean (see Figure 9-1); the majority are in the western and southwestern Pacific. There are several small arcs in the southern Atlantic (not shown) and Caribbean. The importance of the island arcs of the western Pacific was recognized as long ago as the mid-nineteenth century by Alfred Wallace, a natural scientist and contemporary of Darwin. The common occurrence of the volcanic rock type **andesite** in these curvilinear

arcs prompted the term *andesite line*, which actually marked the southwestern edge of the Pacific plate, although the geologists of that time did not recognize it as such. The volcanoes of island arcs are typically built directly on basaltic oceanic crust, although there are several exceptions, for example, in northern Japan, where apparent slivers of old continental crust lie within an island arc.

Arcs form half of what has been called the *arc-trench association*. As the name implies, arcs always form a curved or arcuate pattern that typically has a deep oceanic trench running along the outer side of the curve. These trenches mark the surface topographic expressions of subduction zones and are the deepest parts of the ocean basins; the bottom of the Mariana Trench is more than 30,000 feet below sea level. The so-called *polarity* of subduction zones is well displayed in map view (Figure 9-2A), because the magma that feeds the volcanoes rises from the subducted slab. Therefore, the subduction zone itself must dip from the trench under the island arc.

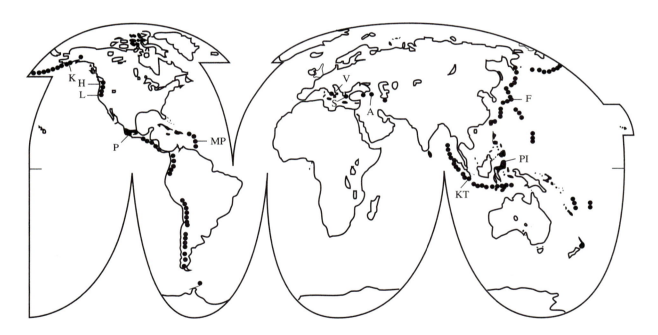

FIGURE 9-1
• • • • • • • • • • • • • • • •

Locations of active and recent volcanoes associated with subduction zones worldwide. Especially note the strong concentration of convergent-margin volcanoes around the edge of the Pacific plate in Alaska, western North and South America, Japan, and Indonesia. Locations of some notable volcanoes are indicated: K, Katmai (Alaska); H, Mt. St. Helens (Washington); L, Lasser Peak (California); P, Paricutin (Mexico); MP, Mont Pelee (Martinique); V, Mt. Vesuvius (Italy); S, Santorini (Greece); A, Mt. Ararat (Turkey); KT, Krakatau (Indonesia); PI, Mt. Pinatubo (Philippines); F, Fujiyama (Japan). [After A. R. Philpotts, 1990, *Principles of Igneous and Metamorphic Petrology* (Englewood Cliffs, NJ: Prentice Hall), Fig. 14-5.]

(A)

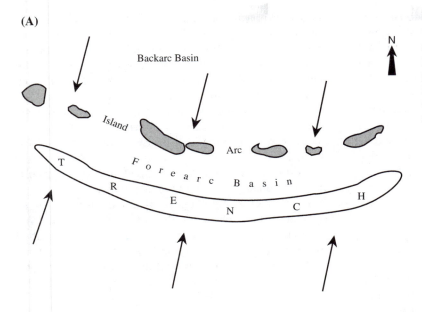

(B)

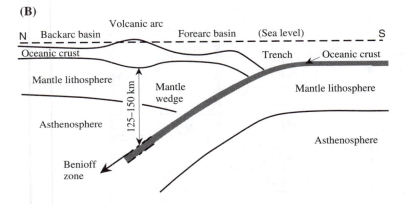

FIGURE 9-2

● ● ● ● ● ● ● ● ● ● ● ● ● ●

(A) An offshore island arc in plan (map) view. The bold arrows show the direction of convergence of the two plates. **(B)** Cross-sectional view of the island arc from north to south, showing the important geologic-petrologic features. For explanation, see text.

Both a schematic map and cross-sectional view of an island arc are shown in Figure 9-2. Although they can be several thousand kilometers long, arcs are fairly narrow, typically 200–300 km wide. Because they are built as thick accumulations of volcanic rocks directly onto the seafloor, the sources of island arc magmas must be within the mantle. The principal geometric features of the arc-trench association are shown in Figure 9-2. The *trench* marks the contact of underthrust and overthrust oceanic lithosphere and the bending of the subducted slab from a horizontal to a dipping position. Subduction zone dip angles vary significantly, as inferred from seismic data on Benioff zone earthquakes. The *Benioff zone* is a planar trend of earthquakes along the upper boundary of the subducted slab, which can extend as deep as 700 km into the mantle. The earthquakes apparently mark the boundary between subducted rigid oceanic lithosphere and more ductile mantle peridotite. The distance from the trench to the axis of the volcanic arc varies as well and seems to show an inverse relation-

ship: The steeper the dip angle, the closer the arc is to the trench (refer also to Figure 9-6). This effect appears to be related to a relatively constant depth of magma generation under the arc, probably 125–150 km (see Figures 9-2 and 9-6). With a shallow dip angle, the top of the subducted slab is farther from the trench before it reaches this depth.

The island arc proper begins on the overriding plate with the *forearc*, which contains both volcanic rocks and sedimentary rocks derived by weathering of the arc. Thicker accumulations of sediment are referred to as a *forearc basin*. Behind the forearc is the main volcanic arc, which runs parallel to the trench at a distance that correlates with the dip angle of the subducting slab: Shallow subduction zones have volcanic arcs farther from the trench than steeply dipping ones. Finally, behind the volcanic arc is a *backarc basin*, which is underlain by basaltic oceanic crust and may contain a secondary spreading ridge. Extensional tectonic features are common in backarc environments. The trench and forearc realms are characterized by anomalously low heat flow because the downgoing slab of cold oceanic lithosphere acts as a heat sink. The volcanic arc and the adjacent backarc basin are areas of unusually high heat flow, partly because of localized magmatic activity but principally because of mantle upwelling; this effect is poorly understood (refer to Figure 19-13 for additional detail). Note the presence of a mantle wedge beneath the arc. Either this wedge or the subducted slab is the likeliest source of arc magmas; the role of the wedge in magma generation is discussed in the section on petrogenesis.

Chemistry and Petrography of Island Arc Volcanics

Andesite is by far the most voluminous rock type in older and larger arcs such as the Indonesian and Philippine arcs. The volcanic rocks of island arcs are a very diverse suite, however, ranging from relatively primitive high-magnesium tholeiitic basalts to strongly differentiated dacites (the fine-grained volcanic equivalent of granodiorite) and rhyolites (the volcanic equivalent of granite). Volcanic rocks of the island arcs are therefore called the **basalt-andesite-rhyolite association.** The tholeiitic basalts show one particular chemical distinction from the tholeiitic basalts of the seafloor and oceanic islands: They are notably higher in Al_2O_3 (typically in excess of 16 wt%), inspiring the name *high-aluminum basalts*. Although arc basalts overlap the MORB suite chemically, this special compositional feature of the island arc tholeiites seems to be virtually ubiquitous and is an important clue to the petrogenesis of the suite.

Andesites are the most distinctive volcanic rock type of convergent margins. The volcanic equivalent of diorite, andesite is distinguished from basalt principally by having somewhat more sodic (less than about An_{50}) and strongly zoned plagioclase, by its commonly porphyritic character, and by the common presence of amphibole or biotite and the rare occurrence of olivine. Chemically, andesites have greater than 52 wt% SiO_2, and basalts have less. Some andesites have exceptionally high Mg#'s (a typical clue to the primary character of mafic magmas) and have been proposed as unmodified mantle melts. In general, however, petrologists have concluded on the basis of major- and trace-element chemistry and isotopic analyses that most andesites cannot be primary mantle melts but are derived from fractional crystallization of a more primitive parent, most likely high-aluminum tholeiite. The more felsic members of the association, dacites and rhyolites, apparently are in turn derived from the andesites by fractionation. As discussed by Ewart (1982), this can be shown persuasively

on a Harker variation diagram (Figure 9-3), similar to the one in Chapter 8 (Figure 8-6). However, because there is considerably more variation in SiO_2 content than in MgO content in the basalt-andesite-rhyolite suite, this diagram uses variation in chemical constituents as a function of SiO_2 content (rather than MgO content), and hence as a more general function of fractionation within a cogenetic suite. Although there is some scatter, the smooth patterns of decreasing Al_2O_3, MgO, FeO, CaO, and TiO_2 content or increasing Na_2O and K_2O content indicate a relatively orderly fractionation relationship among all the members of the suite.

Another powerful chemical technique for relating or distinguishing rock suites that involve fractionation is called the AFM (or FMA) diagram, which was described in Chapter 3. (This is *definitely* not to be confused with a very different AFM diagram that is used by meta-

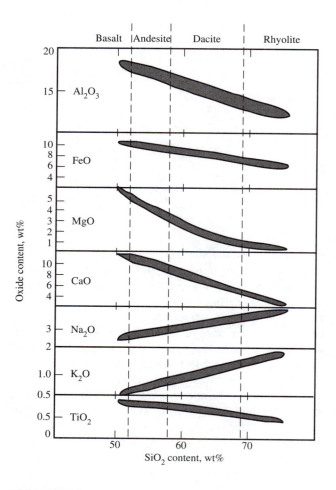

FIGURE 9-3
● ● ● ● ● ● ● ● ● ● ● ● ● ●

Harker SiO_2 content variation diagrams for basalts, andesites, dacites, and rhyolites from island arcs [Data from A. Ewart, 1982, in *Andesites*, ed. R. S. Thorpe (New York: Wiley).]

morphic petrologists and also is used in the metamorphic section of this book.) The AFM diagram (Figure 9-4) is a ternary plot of total content of alkalis (Na_2O + K_2O) versus FeO and MgO contents. Within a rock suite, the trend of rock compositions on this diagram will depend on the mineral or minerals being removed from the initial parental liquid by fractionation. For example, prolonged olivine fractionation alone (as occurs in the ocean island tholeiites) will have a much stronger effect on the Mg# at an early stage than on the proportion of total alkalis (refer back to Figure 6-12), thus causing a fractionation trend that moves from the field of primary basalt on the diagram toward the FeO apex. Only at a fairly late stage in fractionation does alkali content (very low at the start) begin to increase notably, and the trend then turns toward the alkali corner. This AFM trend has been called the **tholeiite trend** and is typical of fractionation in nonorogenic rock suites, both plutonic and volcanic (it is typical within layered intrusions, for example). Alternatively, if fractionation of parental magma involves removal of minerals in addition to, or other than, olivine that cause very little change in the Mg# during an increase in alkali content, the trend will be more nearly radial toward the alkali corner (Figure 9-4). This general trend has been referred to as the **calc-alkaline trend** and is typical of orogenic igneous suites at convergent plate margins. (As an exercise, the reader should think about which minerals might crystallize from basalt to cause the calc-alkaline trend.)

Igneous petrologists have further subdivided the broad calc-alkaline type of trend into individual trends that lie higher or lower in the band in Figure 9-4. Each of these trends contains a complete spectrum of rock types from basalt to rhyolite, but in each trend a different rock type predominates. Distinctions among these trends are considered significant in terms of petrogenetic processes. For example, trends near the upper edge of the calc-alkaline band are called the low-K series. They are dominated by basalts and basaltic andesites and are thought to represent initial stages of mantle melting beneath an island arc, with subsequent fractionation of these initial melts. Paucity of highly differentiated felsic magmas reflects the very primitive nature of the initial basaltic magmas and the very low contents of alkalis and silica. Low-K trends are characteristic of young, immature, and typically smaller arcs. Trends near the lower edge are called the calc-alkaline (in a strict sense) and high-K series. They tend to be dominated by andesite, and even by dacite, and are characteristic of late-stage magmatic activity in larger, older, and more mature arcs like the Japan arc.

Many different minerals occur as phenocrysts in the volcanic rocks of the basalt-andesite-rhyolite association, their diversity reflecting the enormous range of chemical compositions due to fractionation. Except for

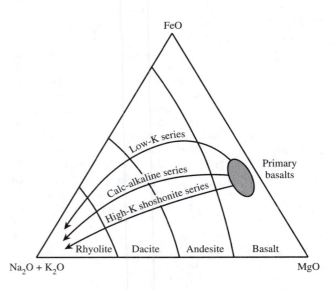

FIGURE 9-4
● ● ● ● ● ● ● ● ● ● ● ● ● ● ● ● ●

AFM diagram showing the calc-alkalic trends, including the low-K series, the calc-alkaline series, and the high-K shoshonite series. Approximate compositional ranges for basalt, andesite, dacite, and rhyolite are shown, as is the field of compositions for primary, mantle-derived basalts.

certain very primitive magnesium-rich tholeiites and the highly fractionated rhyolites, one common feature of rocks of this association is that they are packed with phenocrysts. As discussed in Chapter 8, this is an important clue to whether or not these rocks could be primary mantle-derived magmas. Primary mafic magmatic rocks tend to be fine-grained or glassy and very poor in phenocrysts. Most of the basalt-andesite-rhyolite suite are therefore likely to be evolved magmas that have resulted from fractional crystallization of more primitive parents. Hess (1989) has arranged the information on phenocrysts of the basalt-andesite-rhyolite suite into a useful tabular form that illustrates the evolution of phenocryst assemblages within several of the AFM trends discussed earlier (Table 9-1). As he noted, this table makes an important assumption that there are in fact relationships through liquid lines of descent within each trend, and other data suggest that this is a valid assumption.

As Table 9-1 illustrates, the common phenocryst assemblage in basalts of all trends is olivine + augite + plagioclase + magnetite. Although olivine is not common in andesites, its presence in magnesium-rich varieties supports the idea that andesite is part of a trend from high-aluminum tholeiite to more fractionated magmas caused by olivine removal. Simultaneous fractional removal of the iron-rich phenocryst mineral magnetite is likely to balance the effect of olivine fractionation in

TABLE 9-1   Phenocrysts of island arc volcanic rocks

Rock type	Low-K series	Calc-alkaline series	High-K series
Basalt	Olivine Augite Plagioclase ±Ti-magnetite	Olivine Augite Plagioclase ±Ti-magnetite	Olivine Augite ±Plagioclase
Andesite	Plagioclase Augite Orthopyroxene ±Olivine ±Ti-magnetite	Plagioclase Augite Orthopyroxene ±Hornblende ±Ti-magnetite ±Biotite	Plagioclase Augite Orthopyroxene ±Hornblende ±Ti-magnetite ±Biotite
Dacite/rhyolite	Plagioclase Augite Hypersthene Quartz Fe-Ti oxides ±Fayalite ±Sanidine	Plagioclase Hornblende Biotite Orthopyroxene Quartz ±Augite ±Sanidine	Plagioclase Hornblende Biotite Sanidine Quartz ±Fayalite

Source: Adapted from Hess (1989).

producing the calc-alkaline trends with a relatively constant Mg:Fe ratio. Orthopyroxene is common to ubiquitous in andesite, apparently occurring instead of olivine in these derivative magmas richer in silica. Whereas basalt contains calcic plagioclase (An_{50-70}), plagioclase in andesite is more sodic (An_{30-50}), a composition reflecting fractionation. Low-K andesite contains no hydrous minerals; but in the higher-K series, both hornblende and biotite occur as phenocrysts. This is exceptionally important information because it indicates a role for water in both the generation and evolution of the calc-alkaline magmas. Involvement of water was not seen in any of the oceanic basalts, all of which appear to have been anhydrous magmas throughout their entire evolution. With further fractionation, all of the series can produce dacites or rhyolites that contain both potassium-rich feldspar (sanidine) and sodic plagioclase. By this late stage of fractionation, ferromagnesian minerals in these rocks are relatively iron-rich, and iron-rich olivine (even fayalite) can occur, with or without quartz.

Petrogenesis of Island Arc Magmas

With the advent of plate tectonic theory, attention in the 1960s was focused on two of the most prominent geologic features that underlie the theory, mid-oceanic ridges and arc-trench subduction zone systems. As geophysicists developed thermal models to describe the

effects of subduction on temperature distributions at convergent margins, it became clear that several interesting things were happening thermally. First, the subducting cold slab became heated by an influx of heat from the hotter mantle below. Second, deep mantle isotherms behind the arc and the outer part of the backarc basin were ascending, a result of both the thermal blanketing effect of the pile of arc volcanics and arc-related magmatism (Figure 9-5). These thermal effects inspired the proposition that the downgoing slab was melting when the appropriate depth and temperature were reached. Theory suggested that the hydrated and altered basaltic material of the old oceanic crust would melt more readily than peridotitic mantle. Furthermore, island arc magmas have certain chemical and isotopic characteristics that suggest involvement of either sediments or altered oceanic crust in the source region, including high K_2O, FeO, and $^{87}Sr:^{86}Sr$ ratio.

Subducted lithosphere undergoes a variety of processes as it is heated. Much of the basaltic crust and probably some of the lithospheric upper mantle had been metamorphosed by hydrothermal fluids near the ridge crest and during travel toward the subduction zone. Low-grade diagenetic or low-grade metamorphic minerals such as chlorite, serpentine, carbonates, talc, and clays in these rocks contain significant quantities of water and, to a lesser extent, carbon dioxide. Most of these minerals begin to undergo thermal decomposition at no higher than 250°–400°C, a process releasing significant quantities of fluid. Progressive dehydration within

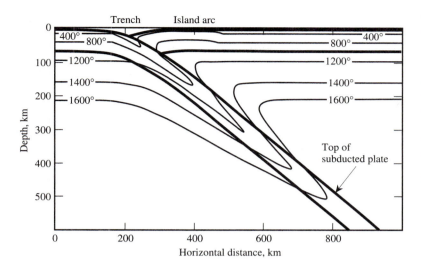

FIGURE 9-5
● ● ● ● ● ● ● ● ● ● ● ● ● ● ● ● ●

Generalized view of thermal structure beneath an offshore island arc, with oceanic lithosphere in both plates. Lithosphere slab margins are shown by heavy lines, and isotherms by fine lines. Numerous modeling studies indicate qualitatively similar results. Note that the subduction rate in this model is sufficiently rapid relative to heat conduction that the downgoing slab acts as a heat sink. Also note the upward deflection of near-surface isotherms under the island arc. Refer to text for detailed explanation.

the subducted slab produces less hydrous, amphibole-bearing assemblages as temperatures and pressures approach those required for ultimate conversion of the basaltic material to anhydrous eclogite, estimated to occur at a depth of about 100 km. The fluid released not only pervades the upper part of the slab but also infiltrates upward into the overlying peridotitic mantle wedge. Experimental studies indicate that at the temperatures estimated to occur at a depth of 125–150 km, where arc magmas originate, either eclogite in the subducted slab (the old oceanic crust) or metasomatized peridotite in the overlying mantle wedge (or both) *could* melt to yield arc magmas.

As attractive as this model for slab melting appears to be, however, further experimental and geochemical studies have shown that it is highly unlikely for the basaltic (or eclogitic) material in the subducted slab to melt to any significant degree to yield arc magmas. Instead, the currently preferred model (although it is *only* a model, and much further work is needed) is that arc magmas are derived from melting of the mantle wedge, a process that can occur because there has been infiltration of sufficient water to reduce its melting temperature (solidus) by several hundred degrees. The geochemical signatures of oceanic crust or sedimentary material that are contained in arc magmas probably come from components that were dissolved in the aqueous fluids generated during heating. These fluids then infiltrated the mantle wedge and metasomatized it to some degree. Finally, being strongly incompatible, these elements were fractionated into the newly formed magmas. The melt that is generated from metasomatized mantle is almost certainly a nearly dry olivine tholeiite magma that closely resembles E-MORB and is derived by substantial degrees of melting. Rarely, however, does any of this primary basaltic arc magma make it to the

surface. The likeliest place to find it is in the suite of low-potassium tholeiites in young and immature arcs (the Caribbean and Aleutian arcs, for example). Here there is relatively thin crust for the magmas to penetrate, and they apparently do so rapidly, with relatively little cooling and fractional crystallization, thus giving rise to an abundance of tholeiite and basaltic andesite.

As arcs mature, the successive primary magmas appear to represent smaller degrees of melting of the mantle wedge and thus contain somewhat larger amounts of water, alkalis, and other elements that tend to concentrate in melts (incompatible elements). As magmas rise, they encounter thicker lithosphere and have more difficulty penetrating to the surface. Significant underplating of the arc may occur as these magmas pond beneath the arc and undergo fractional crystallization. This stage of arc evolution produces voluminous andesites and the first significant volumes of dacites and rhyolites, the more evolved members of the association. These lower-density magmas are more likely to rise through the thickened crust and erupt to form volcanic rocks because of greater buoyancy. There also is a relationship between the depth of melting and the chemistry of the magmas derived. Island arcs commonly have younger and more potassium-rich volcanic rocks farther from the trench. These magmas that occur farther from the trench must come from deeper melting (Figure 9-6).

A good example of the spatial and temporal distribution of arc volcanic rocks can be found in the young volcanic provinces of Japan (Figure 9-7). The westward-dipping active subduction zone and the oceanic trench occur to the east or southeast, bordered immediately to the west by a nonvolcanic arc-trench gap. Beyond the volcanic front marking the beginning of the main volcanic arc, the sequence of progressively younger volcanic zones is (1) tholeiitic and calc-alkaline rocks;

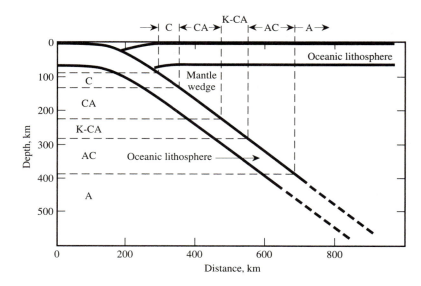

FIGURE 9-6
● ● ● ● ● ● ● ● ● ● ● ● ● ● ● ● ●

The relationship of depth of melting along, or above, a subducting lithospheric plate to the location of various igneous rock types behind the subduction zone. C, calcic series; CA, calc-alkaline series; K-CA, high-potassium calc-alkaline series; AC, alkalic-calcic series; A, alkalic series. The more alkalic rock types form intrusive or extrusive rocks farther from the trench, reflecting greater depth of origin of these magmas. Note that a greater subduction angle would shift all the zones *closer* to the trench. [After S. R. Keith, 1978, *Geology, 6*, Fig. 1.]

(2) calc-alkaline rocks; and (3) calc-alkaline to high-K series rocks. Figure 9-7 also shows the locality of one possible exception to an origin of most andesite through fractionation of basalts. This is an unusual, magnesium-rich orthopyroxene-bearing rock called **boninite,** which occurs in the Izu-Bonin arc, an offshoot of the Japan arc between its northeast and southwest branches. Boninite is thought by some petrologists to be a candidate for a very primitive andesite that was derived directly by melting of metasomatized mantle.

Interestingly, boninite-like plutonic rocks called sanukitoids have now been found in Precambrian orogenic belts of several continents, one good example being the Rainy Lake district of northern Minnesota and southern Ontario along the southern margin of the Canadian Shield. Primary andesite-like magmas may have been more voluminous in the past than they appear to be in modern arc environments.

Knowledge of arc-related magmas is still being assembled, and both experimental and geochemical dis-

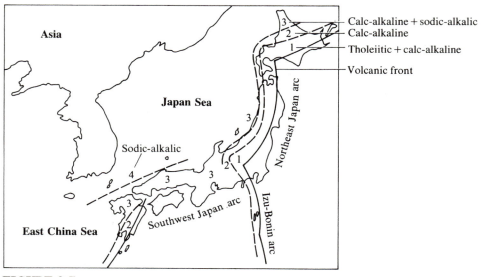

FIGURE 9-7
● ● ● ● ● ● ● ● ● ● ● ● ● ● ● ● ●

Petrographic provinces for the Quaternary volcanic rocks of Japan. The solid lines show the oceanside limits of volcanic rocks. 1, tholeiitic and calc-alkaline rocks; 2, calc-alkaline rocks; 3, calc-alkaline and sodic-alkalic rocks; 4, sodic-alkalic rocks. [From A. Miyashiro, 1972, *Am. J. Sci., 272*, Fig. 4.]

coveries come with considerable frequency. This discussion has obviously been somewhat generalized, without a critical review of much of the experimental and geochemical data that underlie the model. The interested reader is referred to the detailed discussions of this topic by Hess (1989) and Wilson (1989).

IGNEOUS ROCKS OF THE CONTINENTAL MARGINS
• • • • • • • • • • • • • • • •

Continental margins are major sites of sedimentary deposition as well as igneous and metamorphic activity. Throughout Earth history, they have been the places where most continental accretion occurred. Material was carried to them "conveyor-belt" style, subsequently added to the continental edge by deformation associated with metamorphism, and the whole sutured by magmatic rocks. There are several different types of continental margins. *Passive margins*, such as the Atlantic margins of most of North and South America, lie entirely within a lithospheric plate and do not involve interplate tectonic activity. *Active margins*, with currently active subduction, lie on the boundaries between oceanic and continental plates and typically have a well-developed offshore trench; a typical example is the western margin of South America. Finally, some active margins at plate boundaries involve oblique rather than direct convergence, a geometry resulting in transcurrent faulting rather than subduction, as seen in the Pacific margin of North America along the California coast or along the Anatolian Fault in Turkey and adjacent countries. For the purposes of igneous petrology, the most interesting type of continental margin is the active subduction margin, because that is the only one of the three that shows significant contemporary igneous activity.

Interpreting the igneous rocks of an active margin that has been active for a significantly long time requires caution, however. Continued subduction at a continental margin subduction zone may carry multiple island arcs toward the continent, where they become successively accreted because they are too thick and low density to be subducted. This process has apparently occurred along the northwestern margin of North America from northern California northward to Alaska. Both volcanic rocks and island arc plutons exposed by deep erosion can be erroneously interpreted as continental margin rocks even though they are actually island arc rocks; sophisticated techniques such as isotope geochemistry are needed to distinguish them. Although subduction-related continental margin igneous rocks have many important similarities to island arc rocks, there are critical distinctions as well, most of which appear to be due

to interactions of mantle-derived magmas with thick continental crust. Several examples that illustrate important processes in petrogenesis are discussed in the following sections.

The Ophiolite Suite

The term **ophiolite** has been in common geologic usage since the early 1970s and refers to a distinctive rock assemblage containing ultramafic, gabbroic, and basaltic rocks, which are commonly capped by a thin veneer of deep-sea pelagic sediments. The assemblage of rock types and the particular spatial relationships between them bear a striking resemblance to newly created oceanic crust and uppermost mantle near a mid-ocean ridge (refer to Figure 8-4 and Table 8-1). Mineralogic and geochemical signatures of igneous rocks in ophiolites reveal MORB affinities. The presence of ophiolitic suites at continental margins and in orogenic belts has provided powerful evidence in support of plate tectonics, in particular, the transport of what were once mid-ocean rocks to the continental margin where they became accreted to the continent. Relatively complete ophiolite suites are located on the western coast of Newfoundland (particularly the Bay of Islands), in the Troodos Complex on Cyprus, in eastern Papua in New Guinea, and at the Samail Complex in Oman. Most ophiolites, however, are exposed in orogenic belts only as relatively small tectonic slivers of originally more extensive bodies that have been dismembered in the process of emplacement. The Samail ophiolite is much more extensive and better exposed than most others and thus has undergone considerable petrologic and geochemical investigation. Although not directly related to continental margin igneous processes, ophiolitic suite rocks nonetheless have significant petrologic importance.

Emplacement of ophiolites is still not well understood but certainly seems to be the ultimate result of convergence between oceanic and continental plates. Several scenarios for emplacement have been envisioned, some of which involve complicated interactions of lithospheric plates (Figure 9-8). Unfortunately, as noted earlier, the poor state of preservation of most ophiolites makes it difficult to constrain their emplacement unambiguously. In the simplest and most straightforward emplacement model, the beginning of subduction at a newly activated old, passive (Atlantic-type) margin results initially in some of the oceanic plate overriding the edge of the continental plate before the more normal subduction of oceanic lithosphere is established. This process, called *obduction*, can result temporarily in an *oceanward*-dipping subduction zone. The obducted ocean crust is in overthrust contact with

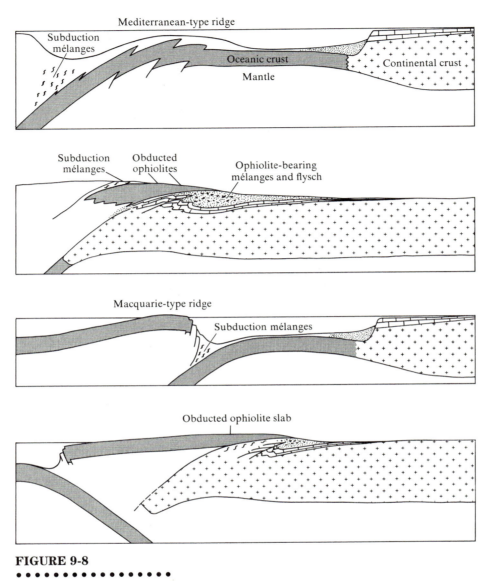

FIGURE 9-8
● ● ● ● ● ● ● ● ● ● ● ● ● ● ● ●

Possible mechanisms for obduction of ophiolite sheets onto continental margins.
[From J. F. Dewey and J. M. Bird, 1971, *J. Geophys. Res.*, *76*, Fig. 6.]

surrounding material (as most ophiolites are) and displays a cross section of the uppermost layers of oceanic crust and shallow mantle when exposed by erosion. An alternative model involves substantial fracturing of oceanic crust during the inception of the normal subduction process. As a mixed zone of fractured and deformed rocks called tectonic *mélange* develops along the contact between continental lithosphere and the downgoing slab, large blocks of oceanic crust with preserved layered structure become incorporated in the mélange and are accreted to the continental margin. Such mélanges have been described from both western Newfoundland and in the Franciscan Formation in

California, where ophiolitic material and serpentinized peridotites are closely associated with areas of intense deformation and shearing.

The Bay of Islands Complex, Newfoundland, is one of the most extensively studied ophiolites (Figure 9-9). This complex extends along the west coast of Newfoundland for about 96 km and is roughly 24 km wide. The entire sequence is structurally allochthonous; that is, it lies above and in thrust-fault contact with the underlying rocks. These consist of Precambrian Grenville gneisses about 1 billion years old, a Cambrian-Ordovician shallow-water sequence of clastic and carbonate rocks representing the early Paleozoic North

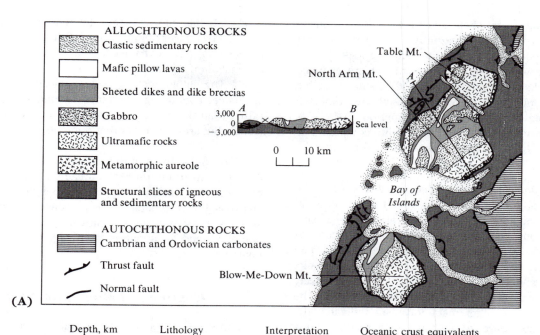

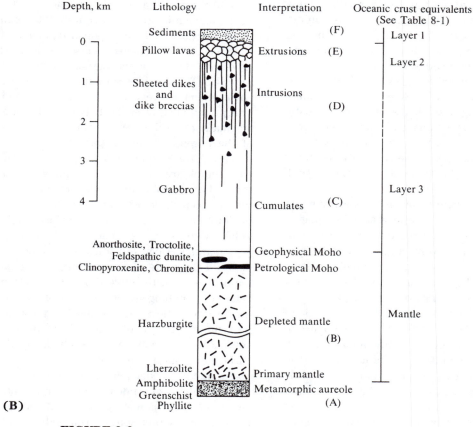

FIGURE 9-9

• • • • • • • • • • • • • • • • •

(A) Geologic map of the Bay of Islands ophiolite, Newfoundland. [From R. G. Coleman, 1977, *Ophiolites* (New York: Springer-Verlag), Fig. 64, compiled from earlier sources.] **(B)** The sequence of rocks found in the Bay of Islands ophiolite, Newfoundland, compared to oceanic crust and mantle. The letter symbols are explained in the text. [From R. G. Coleman, 1977, *Ophiolites* (New York: Springer-Verlag), Fig. 65 and Table 15, compiled from earlier sources.]

American continental shelf, and thrust slices of clastic sedimentary rocks that apparently came from the east and were emplaced onto the shelf rocks before the ophiolite was. Some discontinuous layers of mafic volcanic rocks (now metamorphosed) occur along the base of the ophiolite. The ophiolite itself was emplaced at 460–500 Ma, during the Ordovician period.

The sequence of rocks in the ophiolite complex (see Figure 9-9B), from the base upward, is the following:

A. Metamorphic aureole. Because it lies *below* the ophiolite proper, the aureole is inverted, an unusual feature. The highest grade rocks are at the top, closest to the igneous rocks, where temperatures were as high as 700°–800°C. Both the high-grade metamorphic rocks, which contain pyroxene, plagioclase, amphibole, and garnet, and the adjacent ultramafics are intensely sheared and reflect the thrust nature of the contact. The intensity of the contact metamorphism indicates the very high temperatures of the igneous complex, and it is therefore possible or even likely that juxtaposition occurred near the ridge crest rather than during obduction and that both the ophiolite and its contact aureole were both transported and obducted together.

B. Ultramafic rocks immediately above the metamorphic aureole appear to show several facies of mantle peridotite, including a sequence of dunite (mostly olivine) and lherzolite (containing olivine, orthopyroxene, and clinopyroxene), as well as some harzburgite (olivine and orthopyroxene). These rocks are distinctly layered, and some are highly sheared and serpentinized, particularly at the base.

C. Above the ultramafic zone, there is a transition zone characterized by the first appearances of plagioclase and cumulate textures, apparently as a result of crystal settling. There are layers of anorthosite (rich in calcic plagioclase), dunite, and troctolite (plagioclase-olivine rock). Above the thin cumulate zone, the rock type is a relatively massive and structureless gabbro that shows some vertical change in the Mg# characteristic of crystal fractionation. These rocks, from cumulates to gabbro, have been interpreted as a fractionally crystallized basaltic magma chamber that initially formed in the ridge crest environment. Similar gabbros and ultramafics have been found in the FAMOUS area and other ridges.

D. Above the gabbro is a zone of so-called sheeted dikes, that is, a complex of basaltic dikes so pervasive that there is essentially no interstitial interdike material. Some of these dikes are younger than the gabbro and can be traced down into and through it. Some incipient low-grade metamorphism and hydrothermal alteration of the basalts are present.

E. Moving upward, the sheeted dike complex is overlain by a layer of pillow basalts that marks the original seafloor with extruded basaltic lava flows. Hydrothermal alteration also occurs here.

F. The uppermost layer is a clastic sedimentary sequence containing siliceous sediments and chert rather than the interlayered abyssal sediments and volcanics that are typical of other ophiolites.

All of the features of this ophiolite indicate an origin at or very near a mid-ocean spreading center: the presence of very shallow mantle, the existence of extensive basaltic magmatism, the presence of extensional stresses and fracturing (as shown by the sheeted dike complex), and the apparent near-ridge gabbroic magma chamber.

Continental Magmatic Arcs

Most of the great batholithic provinces of the earth constitute magmatic arcs that occur on the continental side of subduction zones at Andean-type margins. Classic examples are the Andes themselves in western South America and the great batholiths of western North America. The processes of magma generation in this environment are very similar to those in the island arcs. The principal differences between the island arc and magmatic arc suites result from postmelting evolution of the parental magmas, particularly through interactions with thick continental crust in magmatic arcs. The continental arcs contain both plutonic and volcanic rocks, but the plutonic rocks are far more abundant than in even the most mature of island arcs. The most obvious cause for this is the orogenesis and subsequent uplift and erosion that characterize continental convergent margins. Deep erosion over millions of years has exposed the roots of the volcanic arcs that were contemporaneous with active subduction, permitting observation of plumbing systems like those that must underlie modern volcanic arcs. Presumably these modern arcs have unseen deep active plutonism, which involves the solidification of magma in the chambers that feed the surface outpourings of lava.

Subduction of oceanic lithosphere beneath continental lithosphere at Andean-type margins has the same geometry and similar geophysical evidence as the oceanic subduction zones discussed earlier. Above the subduction zone, there is an equally well-developed, though somewhat thinner, mantle wedge, which appears to be the principal source of magma generation. Earthquake foci along a Benioff zone form a plane that dips underneath the continent to a depth of at least 600 km and perhaps as much as 750 km. Recent geophysical studies of the Andean subduction zone for several thousand kilometers north-south along the west coast of South America have shown that it is not a simple plane, nor does it have a constant dip. Seismic data show as much as 10° variation in dip angle along the zone. In one section, flattening of the dip angle to near zero at a depth of several hundred kilometers is then succeeded by a return to relatively steep subduction. The thickness of the continental crust overlying the subduction zone also varies significantly, from only 30 km in the northern and southern Andes to over 70 km in the central portion.

It is unclear whether these changes in subduction geometry or crustal thickness are related to the changes in magmatic patterns that also occur in a north-south sense within the Andean belt. The areas of thinner crust are dominated by surface volcanic rocks such as tholeiite, basaltic andesite, and andesite, whereas the thicker central part is characterized by andesitic to dacitic lavas and even by thick sheets of rhyolitic welded tuff. These patterns are probably a direct result of the relative difficulty of magma penetration through thinner or thicker crust and the subsequent evolution through fractionation that occurs when magma ponds up at the base of the crust.

One relatively common characteristic of volcanism in continental magmatic arcs (although it does occur in island arcs as well) is the mixture of rock types as diverse as basalt and rhyolite, occasionally occurring as nearly contemporaneous eruptions from single volcanoes. This so-called **bimodal magmatism** appears to be a direct result of the nature of magma interactions with thicker continental crust. Most of the primary magmas pond up in chambers at deep levels where they fractionally crystallize to produce the more evolved andesitic-dacitic-rhyolitic magmas. However, a particularly vigorous pulse of relatively primary basaltic magma occasionally makes its way straight through toward the surface, entering the plumbing system normally occupied by more evolved magmas.

An additional notable feature, spectacularly displayed by Mt. St. Helens and other volcanoes in the Cascade Range in Washington and Oregon, is the tendency toward explosive volcanic activity in the continental volcanic arcs (Figure 9-10). This activity is primarily a result of the increase in the silica content of the evolved magmas, a change that leads to a dramatic increase (by orders of magnitude) in the viscosity of the magma. Aiding the process is a concomitant increase in water content of the chemically evolved magmas. Any magmatic water tends to be concentrated in the remaining melt during fractional crystallization of anhydrous minerals. Near-surface boiling of these magmas occurs

FIGURE 9-10
• • • • • • • • • • • • • • • •

Mt. St. Helens in the Cascades, Washington, following the cataclysmic eruption of May 18, 1980 (*right*).

in volcanic plumbing systems when the magma becomes saturated with fluid. Gas expansion during boiling under near-surface low-pressure conditions, coupled with extreme viscosity of magma in the volcanic edifice, commonly produces explosive events such as the one in May 1980 that blew away about the top one-third of Mt. St. Helens. Similar events over millions of years in the northwestern United States have produced thick accumulations of volcanic ash deposits in Washington, Oregon, Idaho, Montana, and Wyoming.

The great batholiths of the earth are also a phenomenon peculiar to the continental margin regime (Figure 9-11). They are enormous in size (tens to hundreds of thousands of square kilometers of exposed area) and extremely complex. Most single large batholiths actually consist of hundreds of individual plutons that intruded one another over a period of millions of years. Internal structures or fabrics such as foliations can develop at the margins of individual plutons and provide mappable markers. Although commonly referred to as the "granite batholiths," these composite plutons do not actually contain much granite but are dominated by granodiorite, tonalite, and quartz diorite, with lesser volumes of the felsic rocks like granite or

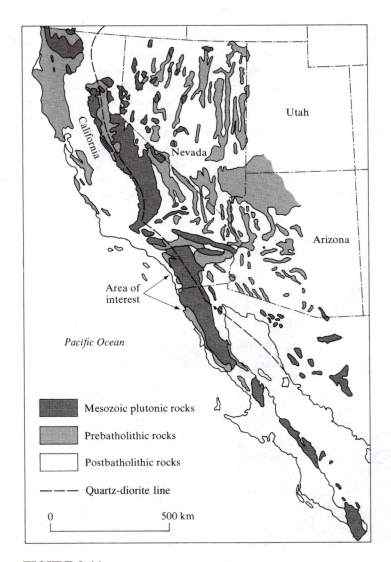

FIGURE 9-11
● ● ● ● ● ● ● ● ● ● ● ● ● ● ● ● ●

The general distribution of Mesozoic plutonic rocks in California and Baja California. The Peninsular Ranges are located in southern and Baja California. The Sierra Nevada are in east-central California. [From E. S. Larsen, 1948, *Geol. Soc. Am. Mem.*, *29*, 198–210.]

monzonite, or intermediate to mafic rocks such as diorite, and rare bodies of gabbro. The term **granitoid**, which is used as a field term for granitic rocks (see Chapter 3), has also been used commonly to describe any lighter colored felsic but not strictly granitic, plutonic rock.

Chemical Compositions and Petrography

The petrography of the continental arc volcanic rocks is essentially similar to lavas from the island arcs. The rare basaltic members have olivine and pyroxene phenocrysts, but typically not plagioclase, a composition reflecting the absence of low-pressure fractional crystallization. Andesites typically contain abundant orthopyroxene, clinopyroxene, and plagioclase phenocrysts, but hornblende is also relatively common and biotite occasionally occurs. The andesites and basaltic andesites are generally phenocryst-rich as a result of their origin as fractionated magmas. Two petrographic features of andesites that are especially notable are strongly zoned plagioclase phenocrysts and partially resorbed hydrous minerals (especially biotite or hornblende).

Larger plagioclase phenocrysts in andesites typically show a wide range of chemical zoning from calcic cores (up to An_{60}) to sodic rims (An_{20-30}). The zoning itself is spectacular in thin section, with abruptly bounded concentric bands in cross-polarized light and occasional multiple reversals of the overall trend to increasing Na:Ca ratio (called **oscillatory zoning**) (refer to Figure 4-7). Although such zoned plagioclase crystals are most common in the volcanic andesites, they also occur in plutonic rocks, particularly tonalites and diorites. The second microscopic feature is the common resorption of the hydrous minerals hornblende and biotite in andesites. These minerals do not crystallize from magmas at low pressure because they require high water pressures to be stable. Hornblende or biotite phenocrysts initially crystallize from water-bearing magmas in deep chambers and then are carried with the magmas toward the surface. If the ascent and eruption process is slow enough, the hydrous minerals decompose to anhydrous breakdown products such as clinopyroxene, orthopyroxene, or potassic feldspar. Hornblende begins to break down at 6 kbar P_{H_2O} (corresponding to a depth of about 20 km) and biotite at about 3 kbar (about 10 km). Breakdown can be interrupted by eruption during relatively rapid ascent, and resorbed crystals with scalloped or irregular margins result. At very shallow levels, oxidation commonly accompanies breakdown, and fine grains of magnetite or even hematite commonly coat the irregular outer margins of the resorbed crystals.

In the plutonic rocks, mineral associations include combinations of plagioclase, potassic feldspar, quartz, biotite, and hornblende. Minor phases include clinopyroxene, orthopyroxene, titanite, ilmenite, apatite, and zircon. A granodiorite typically contains up to 50% plagioclase, 20% quartz, and 20% potassic feldspar (refer to the section on rock classification in Chapter 3), with 10 to 20% hornblende and biotite. Tonalite is similar, but with a higher proportion of plagioclase and less potassic feldspar. Diorites have an even higher proportion of plagioclase and commonly have little or no quartz or potassic feldspar. The more typical mafic mineral in tonalites and diorites is hornblende, and clinopyroxene is not uncommon. Biotite is more abundant in granodiorites. Pyroxenes (especially augite) can occur in unusually water-poor magmas.

Major-element chemistry of the plutonic rocks of continental arcs, like that of the volcanics, shows considerable overlap with the island arc calc-alkaline suite. Most important, the plutonic rock compositions appear to correspond very well to compositions of volcanic rocks that can properly be assumed to represent melt compositions, although probably not primary melts. This fact is important, because the uniformly coarsely crystalline nature of plutonic rocks makes it impossible to use previously discussed criteria to assess whether they represent feasible melts. The occurrence of these rocks in large homogeneous bodies is also good evidence for their origin through crystallization of a single batch of magma. Although the major-element chemistries of magmatic arc rocks are indistinguishable from those of island arc rocks (probably indicating very similar petrogenesis for both types of magma), the trace-element and especially the isotopic compositions are very different. These geochemical features unquestionably reveal much greater degrees of contamination in magmatic arc rocks, that is, chemical interaction between continental arc magmas and the old rocks of the continental crust. In particular, ratios of oxygen isotopes, strontium isotopes, and neodymium isotopes are especially useful for revealing contamination.

Petrogenesis of Continental Arc Magmas

One thing that can be said with certainty about the major continental margin batholiths is that their structural and chemical diversity renders them almost immune to being characterized with simple models. There are, however, some important points of these rocks that serve to illustrate their origins and evolution. The preceding petrogenetic discussion of island arc magmas actually serves quite well to describe the most likely scenario for the origin of continental arc magmas, and little additional detailed discussion is required. Dehydration of the downgoing oceanic lithospheric slab at depths in excess of 125 km produces fluids that permeate and metasomatize the overlying mantle wedge. The hotter parts of the peridotite wedge begin to melt, producing olivine tholeiite or MORB-like magmas that ascend to or into the overlying continental crust. The earliest magmas have difficulty penetrating the crust because it is relatively cool and too much magmatic heat is lost to the surroundings. Later batches of magma that pass through preheated crust can ascend farther before freezing. In general, however, most of the primary basaltic magma must pond up at the base of the crust and undergo fractionation. The more evolved magmas tapped from the subcrustal or deep crustal magma chambers move upward in the crust and either solidify as plutons of diorite, tonalite, or granodiorite or reach the surface as basaltic andesite, andesite, or dacite lavas. Late-stage fractionation can actually produce magmas as silica-rich as rhyolite (granite). In general, the composite batholithic bodies can be viewed as the plumbing system for magmas that vent at the surface from volcanoes. Occasionally, as seen in parts of the Sierra Nevada range of California and in the central Andes, the plutonic rocks actually intrude the lower parts of their own earlier volcanic emanations.

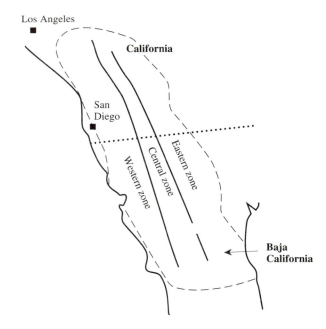

FIGURE 9-12

• • • • • • • • • • • • • • •

Generalized map of the northern part of the Peninsular Ranges batholith, southern California and northern Mexico. The longitudinal zones of the batholith have been drawn by the original authors to reflect consistent isotopic and rare earth element patterns. [After L. P. Gromet and L. T. Silver, 1987, *J. Petrol.*, *28*, Fig. 1.]

As an example of the development of an arc magmatic complex, consider the classic and intensely studied Peninsular Ranges batholith of southern California and Baja California (Figure 9-12). This plutonic complex has been a major focus of study by petrologists and geochemists for many years, summarized in Gromet and Silver (1987). The Peninsular Ranges batholith seems to be typical of such bodies in western North America and is a good example of mineralogic and chemical features common to them. The largest scale and most notable of these features is an asymmetry of pluton age. In this batholith, the batholithic rocks of the continental margin regime appear to share an important, perhaps key, feature with the volcanic rocks of island arcs.

Volcanic arcs show a distinct zonation, from older, less evolved magmas (basalts and basaltic andesites of the low-K series), which lie close to the surface expression of the subduction zone, to younger, more evolved rocks (andesites, dacites, and rhyolites of the high-K series), which lie well behind the subduction zone. The Peninsular Ranges batholith (and other Cordilleran batholiths that have been examined) show this same potassium-enrichment pattern going east from the coast. In addition, there is a major difference in country rocks on the western and eastern sides of the batholith. The west is underlain by rocks of volcanic derivation that probably represent a former offshore island arc, whereas the eastern country rocks are older metamorphic rocks of thick continental crust. On this basis, L. T. Silver and coworkers have suggested that the magmas of the batholith actually span the accreted island arc and the continental margin itself, suturing them together, and should therefore show zoning reflecting the very different types and thicknesses of crust the magmas passed through.

Silver's model describes the various features of the batholith. The ages of the Peninsular Ranges plutons decrease significantly from west to east, spanning several tens of million years, starting at about 120 Ma in the west. The western zone of Gromet and Silver (1987) contains a range of rock types but is dominated by intermediate rocks like tonalite and granodiorite; diorite occurs, although not abundantly. The eastern zone has both tonalite and granodiorite but shows distinctly increased silica and alkalis relative to those in the western zone. Along a roughly north-south line approximately median to the batholith, diorites gradually diminish in volume going eastward. This zonation appears to be a plutonic analog of a zoned volcanic arc such as the Japanese islands.

Accompanying the trend in major-element chemistry are similar trends in trace-element and isotopic composition. The contents of incompatible elements increase from west to east. Recall that this could reflect either decreasing degrees of melting in the source region or increasing fractionation toward the east. The rare earth elements also show increasing fractionation from west to east. Ratios of oxygen and strontium isotopes show strong zonation as well. These ratios are very sensitive to contamination of mantle-derived magmas by old continental crustal material. The oxygen isotopic ratio ($^{18}O:^{16}O$) is strongly increased by surficial processes and is therefore high in sedimentary rocks; it approximately doubles from west to east across the batholith. The strontium ratio ($^{87}Sr:^{86}Sr$) increases with age and ranges from 0.7035 in the west to 0.7075 in the east. An increase in both ratios in plutonic rocks of the Peninsular Ranges batholith from west to east indicates the increasing contamination of magmas by assimilation as they penetrate increasingly thicker and older continental crust to the east.

The Role of Secondary Melting: Introduction to the Granite Problem

It was noted earlier that, although granites do not constitute a major part of the continental margin magmatic suite, they do occur in this environment. Some are undoubtedly the result of extreme fractionation of mantle-derived parental basaltic magma, but it is also possible to derive granite magmas (and, in fact, tonalite and granodiorite magmas as well) from the melting of metasedimentary and metavolcanic rocks in orogenic belts. Metamorphic temperatures in an orogenic belt can reach levels sufficient to cause melting of certain metamorphic rock types, particularly when normal metamorphic heating is amplified by the presence of plutons, as in magmatic arcs. Pelitic schists (rocks of shale composition) can begin melting at about 650°C, and amphibolites (metamorphic equivalents of andesitic volcanics) initially melt at just over 700°C. Two questions represent the crux of "the granite problem" that has been a major focus in igneous petrology: (1) How does one distinguish a granite (or granodiorite or tonalite) derived by fractionation of more mafic melt from one that originated by secondary melting within the continental crust? (2) Which of the two processes is more likely to have generated sizable volumes of truly granitic magma?

The topic is a complicated one that obviously goes into the gray area where igneous and metamorphic petrology overlap. The topic of metamorphic melting is examined in the third part of this book, but for now some initial observations on the topic are appropriate. One of the basic chemical classification schemes for granites is based on molar ratios of Al_2O_3, alkalis, and CaO (refer to Chapter 3). *Peralkaline* granites are those in which $Al_2O_3 < Na_2O + K_2O$. In effect, this criterion means that there is not enough alumina to make feldspar from all the alkalis and that an alkalic mineral other than albite or orthoclase must be calculated in the norm; this mineral is acmite ($NaFe^{3+}Si_2O_6$). Not coincidentally, many peralkaline granites actually contain this mineral or, in the presence of water, a similar sodium-bearing amphibole. *Peraluminous* granites have $Al_2O_3 > Na_2O + K_2O + CaO$. This results in excess normative alumina and thus in normative corundum. Actual (modal) corundum cannot coexist with quartz in a granite, so an aluminous silicate appears instead, typically muscovite, garnet, or cordierite. *Metaluminous* (or normal) granites have $Na_2O + K_2O + CaO > Al_2O_3 > Na_2O + K_2O$. This criterion results in no unusual normative consequences and is typical of simple biotite or hornblende granites.

One of the strongest and most obvious signals of melting of metasedimentary rocks is the occurrence of peraluminous granite. There are possible scenarios for the generation of peraluminous granite through processes other than melting of metamorphosed shaly rocks, but all are considered highly unlikely. Petrologists are so confident about the origin of peraluminous granites that three terms have been coined as shorthand petrogenetic designations for granite magmas, largely on the basis of the above chemical classification and qualified slightly by other data: S-type, I-type, and A-type granites (summarized in Table 9-2).

TABLE 9-2   Characteristics of S-type, I-type, and A-type granites

Granite type	Tectonic environment	Chemical signature	Typical accessory minerals
S-type	Orogenic	Metaluminous to strongly peraluminous, high $^{18}O/^{16}O$, $^{87}Sr/^{86}Sr$	Muscovite, garnet, cordierite, tourmaline
I-type	Orogenic	Metaluminous	Biotite, hornblende
A-type	Anorogenic, rift-related	Metaluminous to mildly peralkaline, Fe-enriched	Fe-biotite, Na-amphibole, Na-pyroxene, hedenbergite, fayalite, titanite

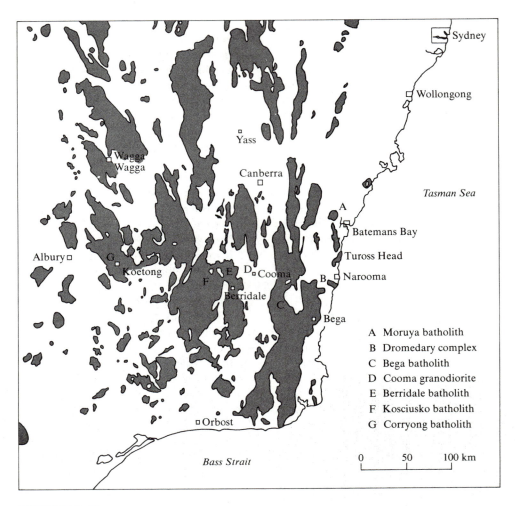

FIGURE 9-13
● ● ● ● ● ● ● ● ● ● ● ● ● ● ● ●

The distribution of granitic rocks (gray) in the Lachlan Mobile Zone of southeastern Australia. [Modified from B. W. Chappell and A. J. R. White, 1976, *Plutonic Rocks of the Lachlan Mobile Zone*, 25th International Geologic Congress, Excursion Guide No. 13, Fig. 1.]

S-type granites are so designated because they are thought to have come from a sedimentary rather than an igneous source, through crustal melting (anatexis). This class of granites undoubtedly includes the peraluminous ones and can include some metaluminous ones as well, which come from the melting of certain less aluminous sedimentary rocks such as feldspathic sandstones or arkoses. The second category, **I-type granite**, is thought to originate through remelting of an igneous source. This process implies ultrametamorphism and melting of an originally solidified igneous rock, but such rocks could be chemically indistinguishable from felsic magmas that fractionated from more mafic parents. In the case of melting of altered felsic metavolcanics, the distinction between S-type and I-type granites may blur somewhat. Classic I-type granites are virtually always metaluminous. Most peralkaline granites also have an apparent igneous source, although probably through fractionation rather than crustal melting. Here a genetic nomenclature system betrays fundamental weakness, because of the potential confusion of I-type and A-type granites, which, according to the definition of the latter, includes most peralkaline granites. **A-type** was coined well after the other two terms and was meant to imply an anorogenic granite, that is, one which did not originate through subduction-related or convergent plate processes. A-type granites are thought to result from remelting of materials that are residua of earlier melting episodes. A-types thus contrast with S- and I-types, which are definitely orogenic granites in the area where they were originally defined (White and Chapell 1983), the Lachlan Mobile Zone of southeastern Australia.

A detailed examination of the Lachlan zone is worthwhile because it illustrates the character of S-type and I-type granites well. Granitoids within the Lachlan zone have been classified into three types: (1) regional-aureole granites, surrounded by schists and gneisses of regional extent, whose metamorphism was due to regional-scale magmatic heating; (2) contact-aureole granites, surrounded by a localized narrow aureole only several hundred meters wide; and (3) subvolcanic granites, surrounded by a very narrow aureole and intimately associated with volcanic and volcaniclastic rocks. Although a depth contrast is implied by these observations, all the plutons appear to have been intruded at about the same depth. Typical granites of the regional aureole series are quartz- and potassic feldspar–rich, with lesser amounts of plagioclase, biotite, and muscovite, a mineralogy indicating that they are peraluminous or S-type. The common presence of migmatites and abundant xenoliths of high-grade metamorphic rocks in the granites suggests that metamorphic conditions were at or near those necessary for melting of the country rocks. It is likely, and supported by strontium isotopic studies, that these granites were derived by melting of a metasedimentary source, probably at slightly deeper levels than currently exposed.

Granites of the contact-aureole type in the Lachlan zone make up most of the composite batholiths consisting of large numbers of smaller individual plutons of circular to elliptical shape (Figure 9-13), in contrast to the more irregular shape of the regional-aureole plutons. Xenoliths are either high-grade aluminous metamorphic rocks, as in the regional-aureole type, or contain hornblende. Both types of xenoliths are thought to have been transported by the magma from a deeper source area. Plutons of the contact-aureole series are typically I-type and do not contain any muscovite. They probably were intruded at higher temperatures than the S-type granites. There is no evidence for a forceful, rapid-injection intrusion mechanism of the I-type or contact-aureole granites because they are accompanied by little deformation. Neither major assimilation of country rocks nor a stoping mechanism of intrusion is indicated. Parallelism of the contacts with structures in the country rocks suggests a mode of emplacement that involved upward streaming of magma in a diapiric fashion. Similar relationships for I-type granites elsewhere have been taken as evidence of secondary magmatic heating and melting through the intrusion of superheated magmas derived either through melting of the base of the crust or by fractionation of mantle-derived magmas.

A-type granites are typically metaluminous to mildly peralkaline—although they can also (rarely) be peraluminous—and are characterized by high $K_2O:Na_2O$ and $Fe:Fe + Mg$ ratios. They are typical of rift zones and the interiors of continental plates, and thus are discussed in Chapter 10. Classic A-type granites occur in the granitic province from coastal New Brunswick and Maine southward to Massachusetts, where they were originally defined.

CONTINENT-CONTINENT COLLISION

Examples of continent-continent collision in recent Earth history, such as the Alps and Himalaya, have mixed igneous suites. Many of the rocks are characteristic of ocean-continent interactions, because oceanic lithosphere of at least limited lateral dimensions must have once intervened between the continental plates that are now in contact and thus must have been involved in subduction. This process has clearly happened in the case of the closing of the Tethys sea in the last 100 Ma. Both the Alps and the Himalaya have resulted from the near-total disappearance of oceanic crust, but magmatic arc rocks related to subduction are preserved, at least imperfectly, within the tectonic collision zones. Based on the Alpine and Himalayan cases, an unusually large degree of crustal thickening is a typical result of continent-continent collision, producing not only impressively high mountains but also the deep crustal roots to balance them isostatically. Crustal thickening is mostly accomplished by thrust faulting, which plays a significant role in both generation and emplacement of granitoids.

The Himalaya are an excellent example of the complex patterns of plutonism that accompany continent-continent collision. French, Swiss, and Chinese researchers made major contributions to an understanding of the Himalaya of Tibet and Nepal in the 1970s and 1980s through a combination of extraordinary high-altitude fieldwork and sophisticated laboratory analysis of samples (see Debon et al. 1985). The high Himalaya and related ranges of southern Asia have resulted from the relatively recent collision of the Indian subcontinent with Asia, thereby completing the closure of the Tethys sea (see the generalized geologic map in Figure 9-14).

Three principal episodes of plutonism have been identified in the Himalaya near Mt. Everest. A lower Paleozoic magmatic event (the *Lesser Himalaya*; roughly 500 Ma), north of the Main Central Thrust, involved shallow emplacement of S-type granitoids and is unrelated to much more recent events of the Himalayan orogeny; these older rocks of the Asian continental margin, however, appear to have been caught in the later collision and tectonically remobilized or remelted by the later orogeny.

(A)

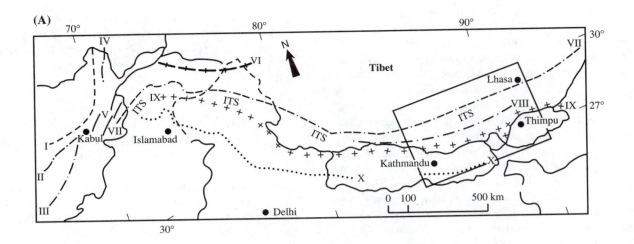

(B)

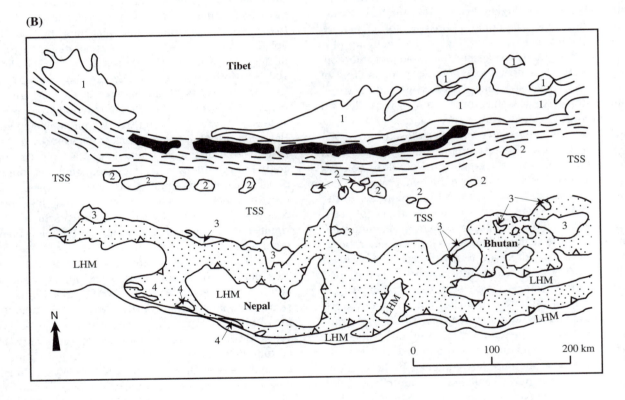

FIGURE 9-14

• • • • • • • • • • • • • •

(A) The ten principal plutonic belts in the India-Eurasia suture zone in the Himalaya, along the boundaries of Tibet (north), Nepal and India (south), Pakistan and Afghanistan (west), and Bhutan (east). The rectangular area corresponds to part B. The Indus-Tsangpo suture (ITS) lies just south of the Transhimalaya belt (VII). [From F. Debon, P. LeFort, S. M. F. Sheppard, and J. Sonet, 1986, *J. Petrol.*, *27*, Fig. 1.] **(B)** Sketch map of the main structural divisions of the Himalaya-Transhimalaya and of the four plutonic belts studied in the Nepal-Bhutan-southern Tibet area. The northern boundary of the Tibetan sedimentary series (TSS) with the Indus-Tsangpo flysch and mélange series (lined pattern with black indicating ophiolite) is not well known and is tectonically complex. The Tibetan crystalline slab is shown by the dotted pattern, and the Lesser Himalaya Midlands are labeled LHM. The four plutonic belts, from north to south, are (1) Transhimalaya belt; (2) North Himalaya belt; (3) High Himalaya belt; and (4) Lesser Himalaya belt.

The second episode, the *Transhimalaya Group*, is of late Cretaceous age and is linked to northward-dipping subduction of Tethys oceanic lithosphere under the main Asian continent. The earliest rocks appear to be associated with an island arc that was accreted to southern Asia during the orogeny, and the later ones show a transition into magmatic arc character. The magmatic arc had a tremendous lateral extent all the way westward into Pakistan. Magmatic activity of the Transhimalaya Group started about 100 Ma and continued until about 50 Ma. This latter age is thought to represent the approximate time of final convergence and initial continent-continent collision and the cessation of subduction. Plutons of this period are mainly quartz diorite, monzodiorite, and granodiorite, with smaller volumes of more evolved rocks like granite and less evolved ones like gabbro. Most granitoids are metaluminous, show quite typical calc-alkaline geochemical patterns, and have distinctive mantle signatures in their trace-element and isotopic chemistry, consistent with their origin as magmatic arc rocks. Their chemistries show some contamination by crustal material, but to a limited extent. The geographic position of the Transhimalaya batholith straddling the collided oceanic arc to the south and the Asian continental margin to the north, combined with its activity spanning the time of accretion of the arc, are also strikingly similar to the characteristics of the Peninsular Ranges batholith.

The third group, the Cenozoic (mainly Eocene and younger) *Himalaya Group*, which occurs in the high Himalaya and part of the North Himalaya belts, is very different in character from the Transhimalaya plutons. Himalaya Group plutons range in age from Eocene (about 50 Ma) to as young as 6 Ma. (The most recent data suggest ages perhaps as young as 1 Ma in Pakistan!) These latter ages are startlingly young and emphasize just how tectonically active the Himalaya still are. Petrographically, the Himalaya Group plutons are fundamentally different from the Transhimalaya plutons: They are almost all granite or quartz monzonite and all are peraluminous. All of them contain at least muscovite, and many contain other aluminum-rich minerals such as garnet, cordierite, or even aluminum-silicates. They are high to very high in oxygen and strontium isotope ratios, confirming that they are unequivocally S-type granitoids and indicating a probable origin through melting of aluminous metasedimentary rocks. Geographically, these plutons are linked to the Main Central Thrust and many have tabular, dipping forms suggesting that they may have been "smeared out" along the thrust surfaces during active faulting. Much of the dramatic postcollisional thickening of the crust north of the Main Central Thrust was certainly accomplished by north-directed thrusting. The crustal thickening led directly to heating of the crust and large-scale melting or anatexis of metasedimentary rocks.

SUMMARY

The zones of convergence of various combinations of oceanic and continental plates are sites of generation of enormous volumes of igneous rocks and important indicators of the petrogenesis of magmas. Convergence of oceanic lithosphere and the subduction of one oceanic plate beneath another produce island arcs above the subduction zone. These arcs begin as small accumulations of relatively primitive basalt and basaltic andesite derived at depths of 125–200 km from hydrothermally metasomatized mantle peridotite immediately above the downgoing slab. Altered oceanic crust in the subducted slab dewaters as it is heated, causing metasomatism of the overlying mantle wedge. The slab also contributes key trace elements and isotopes to the overlying melting mantle through fluid infiltration but does not otherwise directly produce magma for the arc. As the arc builds and matures, the ascending magmas begin to fractionate more extensively in deep magma chambers below the arc and produce increasingly more evolved andesitic to dacitic and even rhyolitic magmas that find their way to the surface. Deeper and less extensive melting in the mantle above an increasingly dehydrated subduction slab eventually contributes magmas to the surface along a band in the back of the arc and farther from the oceanic trench. Richer in incompatible elements, particularly alkalis, these magmas fractionate to form the potassium-rich volcanic series.

Where the oceanic trench is immediately offshore from continental crust and the subduction zone dips beneath the continent (an Andean margin), the process of melting seems much the same as that for island arcs. The surface expression of magmatism is substantially different, however, because of the thickness of continental crust that the magmas must pass through. Much of the rising magma solidifies within the continental crust as plutons of tonalite, diorite, and granodiorite, with lesser volumes of more mafic and more felsic rocks. At least some of the plutons represent magma chambers that served as feeders for surface volcanism, and some shallow ones can actually intrude their own early volcanic debris. Chemical contamination of magmas by continental rocks is common in this environment. Heating of the crust by magmatic activity in a continental margin orogenic belt can lead to anatexis, the secondary melting of metasedimentary or older igneous rocks within the crust.

The third major type of convergent margin occurs between two continental plates, but collision must be preceded by the closing of an oceanic basin between the two continents. This regime therefore can contain a complex mix of early subduction-related volcanic arc and magmatic arc rocks and later anatectic magmas that originated following the collision. The Himalayan plutonic belts of Tibet and Nepal are good examples of the magmatic effects of continent-continent collision.

STUDY EXERCISES

1. How many different types of convergent plate tectonic regimes are there? In general terms, how do their igneous patterns differ?

2. What are the components of a typical island arc complex, and which igneous rock types are characteristic of each?

3. What are the most important chemical and mineralogic contrasts between the two most voluminous island arc magma types, tholeiites and andesites?

4. Suppose you are a field geologist mapping a deformed terrane in an orogenic belt. Describe the petrologic characteristics you would use to recognize an ophiolite suite. What does the presence of an ophiolite suite tell you about the tectonic process?

5. What are the petrologic characteristics of the great granitoid batholiths of the earth, and why are these rocks restricted to continental margins?

6. What are the petrologic and tectonic implications of the so-called alphabet granites—S-type, I-type, and A-type?

REFERENCES AND ADDITIONAL READINGS

Arculus, R. J., and R. W. Johnson. 1978. Criticism of generalized models for the magmatic evolution of arc-trench systems. *Earth Planet. Sci. Lett.*, *39*, 118–126.

Chappell, B. W., and A. J. R. White. 1974. Two contrasting granite types. *Pacific Geol.*, *8*, 173–174.

Coleman, R. G. 1977. *Ophiolites*. New York: Springer-Verlag.

Debon, F., P. LeFort, S. M. F. Sheppard, and J. Sonet. 1985. The four plutonic belts of the Transhimalaya-Himalaya: A chemical, mineralogical, isotopic and chronological synthesis along the Tibet-Nepal section. *J. Petrol.*, *27*, 219–250.

Dewey, J. F., and J. M. Bird. 1971. Origin and emplacement of the ophiolite suite: Appalachian ophiolites in Newfoundland. *J. Geophys. Res.*, *76*, 3179–3206.

Ewart, A. 1982. The mineralogy and petrology of Tertiary-Recent orogenic volcanic rocks: With special reference to the andesite-basalt compositional range. In *Andesites*, ed. R. S. Thorpe. New York: Wiley, pp. 25–95.

Gromet, L. P., and L. T. Silver. 1987. REE variations across the Peninsular Ranges batholith: Implications for batholithic petrogenesis and crustal growth in magmatic arcs. *J. Petrol.*, *28*, 75–125.

Hess, P. C. 1989. *Origins of Igneous Rocks*. Cambridge, MA: Harvard University Press.

Kieth, S. R. 1978. Paleosubduction geometries inferred from Cretaceous and Tertiary magmatic patterns in southwestern South America. *Geology*, *6*, 516–521.

MacDonald, G. A., A. T. Abbott, and F. L. Peterson. 1983. *Volcanoes in the Sea*. Honolulu: University of Hawaii Press.

Miyashiro, A. 1972. Metamorphism and related magmatism in plate tectonics. *Am. J. Sci.*, *272*, 629–656.

Pitcher, W. S. 1978. The anatomy of a batholith. *J. Geol. Soc. London*, *135*, 157–182.

Ringwood, A. E. 1975. *Composition and Petrology of the Earth's Mantle*. New York: McGraw-Hill.

White, A. J. R., and B. W. Chappell. 1983. Granitoid types and their distribution in the Lachlan fold belt, southeastern Australia. *Geol. Soc. Am. Mem. 159*, 21–34.

Wilson, M. 1989. *Igneous Petrogenesis: A Global Tectonic Approach*. London: Unwin-Hyman.

Yoder, H. S., Jr., ed. 1979. *The Evolution of the Igneous Rocks: Fiftieth Anniversary Perspectives*. Princeton, NJ: Princeton University Press.

Chapter 10

IGNEOUS ROCKS
OF CONTINENTAL LITHOSPHERE

This chapter completes an overview of magmatism associated with plate tectonic regimes by examining continental igneous rocks that do not directly involve plate interactions. These igneous rocks form either in continental shield areas or in major continental fracture or extensional rift zones long after the cessation of major orogenic periods. They are an extraordinarily diverse group that includes tholeiitic basalts and their coarser grained plutonic equivalents in layered mafic intrusions; rocks of the continental rift suite such as tholeiite, alkali basalt, carbonatite, nephelinite, phonolite, and rhyolite; anorthosite; anorogenic granitoids; and the alkali-rich mafic to ultramafic rocks such as lamproites and kimberlites.

Of the various continental lavas, tholeiitic basalts are by far the most common and voluminous. Originating in continental rifts, they have flooded large areas with hundreds of thousands of cubic kilometers of lava. Examples include the Columbia River basalts of the Pacific Northwest, the Karroo basalts of South Africa, and the Deccan basalts of India. The major layered mafic intrusions already described in Chapter 6, for example, the Bushveld, Stillwater, and Skaergaard Complexes, are probably plutonic expressions of this type of voluminous, rift-related basaltic magmatism. Other volcanic expressions of continental magmatism include the alkalic provinces, which are also generally localized in rifts, and the rare Precambrian ultramafic lavas called komatiites.

Continental plutonic rocks include the mafic layered complexes already mentioned, as well as enigmatic rocks such as anorogenic peralkaline granitoids and anorthosite complexes. Both of these igneous rock types have been subjects of intense investigation. All continental magmatic rocks are the subject of investigation by igneous petrologists, not only because these rocks are easier to study in the field than oceanic rocks but also because of the evidence they can provide about the subcontinental mantle. Research suggests that the mantle lithosphere beneath continental crust is substantially more heterogeneous chemically and mineralogically than oceanic lithosphere. Petrologic and geochemical examination of some of the earth's oldest exposed rocks in the continental cratons, as well as some of the youngest in the continental rifts, enables geologists to learn more about the nature of the subcontinental mantle.

CONTINENTAL
BASALT PROVINCES
● ● ● ● ● ● ● ● ● ● ● ●

Flood Basalts

From the Precambrian to the mid-Cenozoic, enormous outpourings of tholeiitic basalt in extensional rifts on continents have marked short stretches of geologic time. Some episodes have been remarkably brief, for example, the Miocene Columbia River Plateau basalts: Over 100,000 km^3 of basaltic flows were extruded in a little over 3 million years. Other major flood basalt provinces are the Proterozoic Keweenawan lavas of the Lake Superior region and upper Midwest; the Jurassic-Cretaceous Paraná basalts of South America and Etendeka basalts of Namibia; the Permian basalts of the Siberian Platform; the Cretaceous Deccan traps of southern India; and the Jurassic Karroo basalts of South Africa and Ferrar dolerites of Antarctica. Although most of these basalts underwent substantial chemical evolution on their way to the surface, some are relatively primitive in character and yield information about the subcontinental mantle. Much of the discussion in the preceding two chapters has involved use of oceanic and arc basaltic magmas as tracers to learn something about the character of the oceanic mantle. Can the voluminous continental basalts serve a similar purpose for subcontinental mantle?

Petrologic argument abounds on this point. A brief look at the character of the flood basalts may help to illustrate the debate, and a good place to start is with the basalts of the Columbia River Plateau. Figure 10-1A shows a generalized map of the western United States, giving the location of the Columbia River basalts of Oregon and Washington and the nearby Snake River province in Idaho, Montana, and Wyoming. Note that the location of the Columbia River Plateau is to the continental side of the Cascades magmatic arc. This region, along with much of the western margin of North America, was very active magmatically from the late Cretaceous through the Cenozoic, but the peak of regional activity was from 20 to 14 Ma. This period includes the extrusion of the bulk (over 90%) of the Columbia River basalts from 17 to 14 Ma, although eruptive activity continued until about 5 Ma. The Columbia River Plateau occupies what was effectively a backarc position relative to the continental margin subduction zone and was affected by extensional stresses common to backarc environments.

The Columbia River basalts have a total thickness of about 1000 m and have been subdivided into five units or formations (Figure 10-1B), but by far the greatest volume of basalt (in excess of 75%) is in the Grand Ronde Formation. The next most voluminous unit, the Wanapum Formation, contains about 15% of the total volume, and the other three, the Saddle Mountains, Picture Gorge, and Imnaha basalts, together contain less than 10%. About 95% of the total volume was extruded in about 3 million years, from 17 to 14 Ma. Perhaps the best known of these units among petrologists is the Picture Gorge because of the detailed examination it has received. Quartz tholeiite is the dominant rock type among the Columbia River basalts, although more and less evolved types also occur. There is no particular pattern to compositional variations within the plateau that could be ascribed to a consistent pattern of fractionation, and relatively primitive basalts occur throughout the thickness of the volcanic pile. Some flow units show evidence for fractionation over short intervals.

Most of the Columbia River lavas contain few if any phenocrysts, which suggests that they could represent primary liquids. But their bulk and trace-element chemistries are not consistent with an origin in typical mantle. In particular, the most primitive of these olivine tholeiite basalts have Mg:Fe ratios too low for credible "normal" mantle melts. The patterns shown both by the rare earth elements and by various isotopic systems indicate the likelihood of contamination by continental crust, not a surprising possibility given the thick crust through which these magmas must have passed. However, these patterns also could have been caused by origin of the basalts from *atypical* mantle, particularly mantle that was heterogeneous chemically and also

(A)

FIGURE 10-1
● ● ● ● ● ● ● ● ● ● ● ● ● ● ● ●

(A) Map of the western United States showing the location of the Columbia River Plateau (CRP) and Snake River Plain (SRP) continental flood basalt provinces, relative to the axis of the Cascades volcanic/magmatic arc. The proposed location of the Yellowstone Hot Spot is shown at the eastern end of the SRP trend. Also note the location of the Rio Grande Rift, south and east of the Colorado Plateau, and the East Pacific Rise. **(B)** Schematic, roughly proportional stratigraphic section of the Columbia River Plateau basalts, showing ages and volumetric percentages of total basalt volume. [Adapted from data in BVTP, 1981.]

enriched in components normally found in continental crust. There are two potential complications to the use of the Columbia River basalts as probes of subcontinental lithosphere. First, the east-dipping subduction zone under the Cascades must also have passed under the location of the Columbia River Plateau and could have

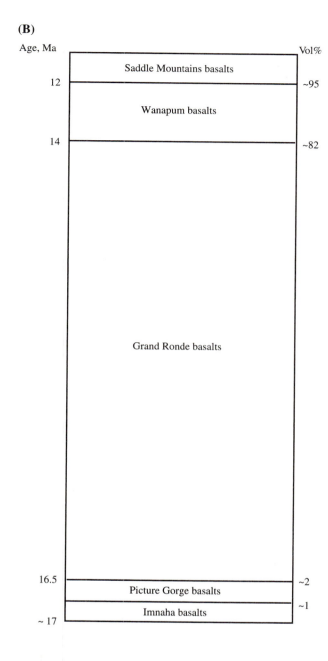

(B)

Age, Ma

Saddle Mountains basalts

Wanapum basalts

Grand Ronde basalts

Picture Gorge basalts

Imnaha basalts

12

14

16.5

~ 17

Vol%

~95

~82

~2

~1

apparent hot-spot track, however, lies the interesting magmatic province underlying Yellowstone National Park (Figure 10-1A). Well known for its abundant geothermal effects, including geysers, boiling mud springs, and fumaroles, the Yellowstone region is clearly underlain by unusually hot rocks at unusually shallow depths. The immediate area has been volcanically active for several million years, and very thick piles of rhyolitic and dacitic ash deposits are exposed in the walls of the Yellowstone River canyon and are punctuated by an occasional columnar-jointed basalt flow. One of the most interesting volcanic phenomena displayed at Yellowstone is a thick layer of obsidian at Obsidian Cliffs. This very dark glassy rock of rhyolitic composition appears to be a flow but is actually at the bottom of a thick section of rhyolitic tuffs. With their very high viscosities, rhyolites almost always occur as pyroclastic rocks (ignimbrites, for example) rather than as flows. In this and other occurrences of obsidian, the weight of overlying tuffs has compressed the welded glassy shards into a dense compact mass in which all interstitial pore spaces have been eliminated.

Some other continental flood basalt provinces show a diversity of rock types similar to that of the Columbia River Plateau-Snake River Plain province. Basalts of the Paraná province of Brazil and the Etendeka province of Namibia in southwestern Africa show a substantial range from relatively primitive magnesium-rich olivine tholeiites to much more evolved silica-enriched lavas. The Keweenawan basalts of the northern midcontinental United States and the Deccan traps (*trap* is an old quarrying term for a fine-grained rock, particularly basalt) are overwhelmingly olivine- or quartz-tholeiitic by volume. The Deccan traps have a much smaller range of volcanic rock types (typically nothing more felsic than andesite), but there are some major rhyolite occurrences associated with the Keweenawan basalts in Minnesota. A major important feature of the Keweenawan basalts is the contemporaneous and apparently comagmatic occurrence of the Duluth mafic complex on the north shore of Lake Superior. Although not a classic layered intrusion, this large body has enough similarities to emphasize a link between flood basalts and layered intrusions.

The current petrologic consensus, although with much remaining debate, is that the subcontinental mantle, including both lithosphere and asthenosphere, is indeed quite heterogeneous, with depleted and enriched portions. Ironically, this character may ultimately derive from its long-term stability. Oceanic lithosphere comes and goes according to convection in the upper mantle and the plate tectonic cycle, and thus ultimately becomes mixed in and reprocessed with other mantle by subduction. Subcontinental lithospheric mantle appears to become welded or attached to the overlying

contributed subduction-zone magmas to the backarc province. Second, the crudely linear belt of volcanics in the Snake River province just to the east (see Figure 10-1A) has been ascribed by some workers to the passage of a hot spot beneath this part of North America, the so-called Yellowstone Hot Spot. The subcontinental mantle under the northwest at this time must have seemed like a petrologic Times Square!

The flood basalts of the Snake River Plain in southern Idaho are very similar in character to those of the Columbia River Plateau. At the eastern end of the

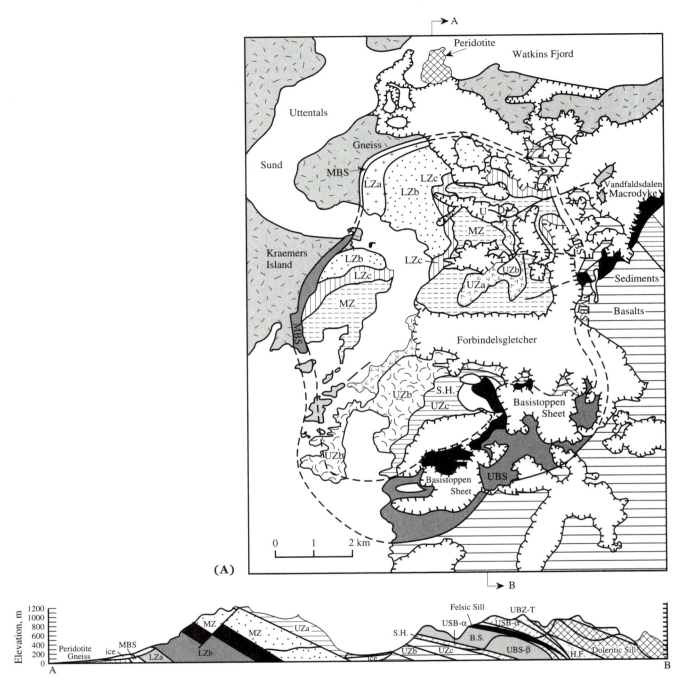

(A)

FIGURE 10-2

●●●●●●●●●●●●●●●

(A) Simplified geologic map and cross section of the Skaergaard Intrusion of East Greenland. Units of the layered series are Lower Zone (LZ) a, b, and c; Middle Zone (MZ); and Upper Zone (UZ) a, b, and c. The marginal border series (MBS) and upper border series (UBS) merge in the southern part of the map and are not shown separately. The cross section is drawn along the line A–B, normal to the southward dip of the intrusion. [From A. R. McBirney, 1993, *Igneous Petrology*, 2nd ed. (Boston: Jones and Bartlett), Fig. 6-21.] (B) Cross section of the Skaergaard Intrusion, restored to original orientation according to the Wager-Brown interpretation. The solid irregular line indicates the present erosion surface. The intrusion is considered to be funnel-shaped, with a large unexposed hidden zone at its base. [From L. R. Wager and G. M. Brown, 1967, *Layered Igneous Rocks* (San Francisco: W. H. Freeman), Fig. 8.] (C) Cross section of the Skaergaard Intrusion, restored according to the McBirney interpretation. The present level of erosion is indicated by a solid line. Much of the hidden zone has been eliminated, and the extent of the middle and lower zones has been considerably increased. [From A. R. McBirney, 1975, *Nature*, *253*, Fig. 7.]

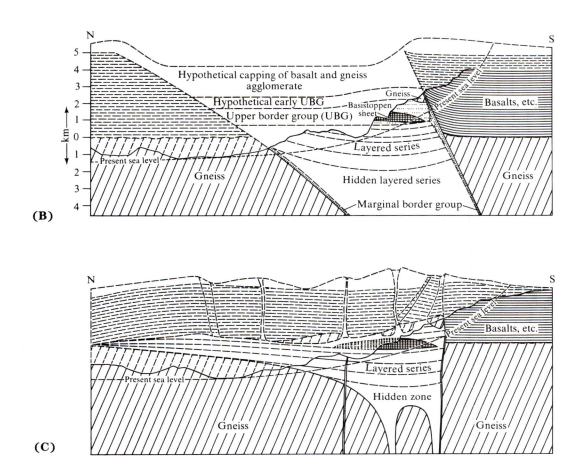

continental crust quite early, however, and thus is saved from being reprocessed in the deeper mantle. Continents grow by slow accretion over geologic time, and subduction zones carry material beneath continental margins that may variably affect the overlying mantle (**mantle wedge**) through metasomatism by fluids or by trapped magmas. Perturbation of long-term stable thermal patterns in this subcontinental mantle, particularly by tectonic extension and thinning of both lithospheric mantle and overlying continental crust, can have dramatic effects on magma production. It has been proposed that continental flood basalts, and perhaps other continental rift-related magmas, originate from melting of enriched sublithospheric (asthenospheric) mantle of the type that also produces E-MORBs and OIBs.

Layered Mafic Intrusions

Chapter 6 explained how phase diagrams can be used to understand important aspects of the formation of layering in mafic intrusions. In this section, we present further information on their nature and occurrence. Layered mafic intrusions have been included in the dis-

cussion because there is increasing tectonic evidence that they are related to periods of major crustal extension and rifting in continental lithosphere. This is not particularly surprising, given the enormous volumes of basaltic magma that were apparently injected rapidly into some of them. Attention has been focused on a possible role of layered intrusions as deep crustal feeders for the continental flood basalts that are also associated with continental rifting.

A number of classic twentieth-century petrologic studies have examined the layered mafic intrusions. For example, the Stillwater Complex of Montana was described by Hess (1960) and the Skaergaard Intrusion of East Greenland by Wager and Brown (1967) and McBirney (1975 and 1993) (Figure 10-2). Layered mafic complexes occur on most continents, but the most well known range from the two giants, Bushveld (South Africa) and Dufek (Antarctica), both of which have over 50,000 km^2 of exposed rock, to more typical 3000- to 5000-km^2 bodies such as Stillwater (Montana), Muskox (Northwest Territories, Canada), and the Great Dyke (Zimbabwe), and finally down to the tiny but classic Skaergaard (East Greenland) of only about 100 km^2.

All of these intrusions share two common aspects: They are of an overall basaltic composition (tholeiitic to olivine tholeiitic), and they are subdivided into thick layers or units that range from ultramafic (at the bottom) to felsic (at the top). These compositions are more extreme than the *average* compositions of the intrusions as a whole. Smaller scale layers of centimeters to meters in thickness occur within the gross layered units (see Figure 6-1). Many models for the origin of the layering have been proposed over the years, but the classic consensus was that layering resulted from low-pressure fractional crystallization of a basaltic parent and subsequent gravitational crystal settling and accumulation, as discussed in Chapter 6. This long-standing model has been questioned by some petrologists who are applying principles of fluid dynamics and fine-scale liquid convection to the problem, and the debate is far from resolved. It has been argued that the relatively abrupt emplacement of most of the parental basaltic magma in one batch into a shallow crustal chamber could have allowed the layering to develop, and that repeated injections of large volumes of fresh magma and the resultant stirring would probably have disrupted both the fine-scale and larger-scale chemical and physical layers. Conversely, numerous repeated-injection mechanisms have been proposed and backed up with substantial physical and geochemical evidence. The layered intrusions thus remain a focus for intense petrologic study, as they have been for much of the twentieth century.

The inferred stress regimes for magma emplacement and the large volumes of basaltic magma required, at least in the cases of Bushveld and Dufek, bear striking resemblance to the scenario for the origin of continental flood basalts. Investigators of layered intrusions have used two different models to constrain the compositions of the parental liquids, but both have problems and complications. In the first technique, average compositions can be estimated by an algebraic process that "adds up" the different layers. However, in layered intrusions, the lower or upper parts are typically missing, either unexposed below the surface or removed by erosion, and approximations must be made. In the second technique, fine-grained chilled rocks at the margin of the intrusion (the so-called *chilled margin facies*), which are assumed to represent parental melts before any fractionation could occur, have been used to estimate compositions. With this technique, however, the possibility of country rock contamination limits the accuracy. Therefore, precise comparisons of chemical compositions of parental liquids for layered intrusions and continental flood basalts may not be practical. It seems likely, however, that the mafic layered intrusions are the eroded remnants of magmatic plumbing systems that supplied magma to old, and now eroded, flood basalt provinces.

Komatiites

Komatiites are rare rocks believed to represent ultramafic lava flows. They occur in layered stratigraphic patterns and commonly have pillow morphologies at the tops of individual flows. A brief discussion of this petrologic oddity is presented here because igneous petrologists have debated whether it is possible for ultramafic magmas to exist in a sufficiently molten state to form lavas. N. L. Bowen established early in the twentieth century that ultramafic rocks have liquidus temperatures of 1400°–1600°C or more at low pressures and thus are unlikely to occur as liquids even in the earth's interior and extremely unlikely to occur at the surface. In 1969, Viljoen and Viljoen first reported Archean lava flows of ultramafic composition in the greenstone belt of the Barberton Mountainland, South Africa, near the Komati River (hence the name of this rock type). Since then, similar rocks, almost all Archean in age and in greenstone belts, have been recognized elsewhere, notably in western Australia, Africa, and Canada.

Komatiites form fairly thick and extensive lava flows that are mixed with olivine tholeiites and have highly characteristic petrographic textures, the most notable of which is called **spinifex,** after the large, native, dry-country grasses in the type locality. This texture is composed of highly elongated bladelike phenocrysts of olivine (or pseudomorphs after olivine; see later) that can be many centimeters in length and are set in a finer matrix of skeletal calcic clinopyroxene and other minerals. The spinifex-texture blades are either random or oriented and typically occur in the middle to upper parts of the flows. Flow bottoms commonly show cumulate textures of olivine, pyroxenes (both ortho- and clino-), and chromite, textures that reflect at least some crystal settling. Komatiite samples rarely retain original igneous mineralogy. Most have undergone at least low-grade metamorphism along with other lithologies in greenstone belts. Preservation of original volcanic textures is commonly remarkably good, but the igneous minerals have been passively replaced in most samples by metamorphic or alteration minerals such as chlorite, serpentine, calcite, and actinolite.

Spinifex is a volcanic texture caused by extremely rapid cooling, even quenching, of the original magma. Similar, though much smaller, skeletal crystals of olivine and pyroxene have been produced in the laboratory in quenched mafic and ultramafic runs. Some experimental data, however, suggest that the key factor is rapid *crystallization* rather than necessarily rapid cooling. It is well known that liquids can be supercooled (chilled to temperatures well below their ordinary crystallization temperatures) by suppressing crystal nucleation (this can easily be done in the case of water). The cooling rate can actually be slow, but when crystallization of supercooled liquids does begin, it proceeds more rapidly than normal, and large crystals with skeletal morphologies commonly result. In general, though, the bulk of the evidence indicates a volcanic origin of komatiites.

The implications of ultramafic volcanism in the Archean are potentially profound. To generate ultramafic melts, the mantle must have been much hotter at shallow depths than today's mantle, and probably closer to the surface; that is, the crust was thinner. It is certainly impractical to derive ultramafic magma from the mantle now, and special thermal conditions are required to produce even olivine-rich basaltic liquids. In addition, ultramafic magma would approximate the composition of parental mantle peridotite itself, thus requiring degrees of melting that could approach or exceed 80%. To melt this much peridotite would require anomalous amounts of heat. Nisbet and Walker (1982) have speculated that much higher mantle temperatures and possibly more extensive melting of the upper parts of the Archean mantle (oceanic as well as continental) could have significantly affected Precambrian plate tectonics. Seafloor spreading and subduction may actually have operated more effectively and rapidly in the Archean than today, thus allowing for very efficient recycling of the early products of crust building—and explaining why rocks from this earliest part of Earth's history are so rarely seen. In any case, komatiites have been known and studied for such a short time that much is yet to be learned from them.

CONTINENTAL RIFTS

Large-scale continental extensional environments *probably* represent the initial rifting stages of single continental plates as they begin to break up and drift apart to create new oceanic crust, much as happened when the current Atlantic basin was created some 120 million to 200 million years ago. "Probably" is emphasized, because there are many "failed" rifts around the globe that never matured into full-scale new oceanic crust. For example, the Triassic-Jurassic rift grabens of eastern North America (from Georgia to eastern Canada) are thought by many geologists to be a failed first attempt to break up part of the Pangaea supercontinent. Successful continental rifting finally came more than 50 million years later, and slightly to the east, when the present Atlantic opened. Other notable ancient and modern rifts include the Dead Sea and East African Rifts, the late Paleozoic Oslo Graben of Scandinavia, the modern Rhine Graben between Germany and France, the Rio

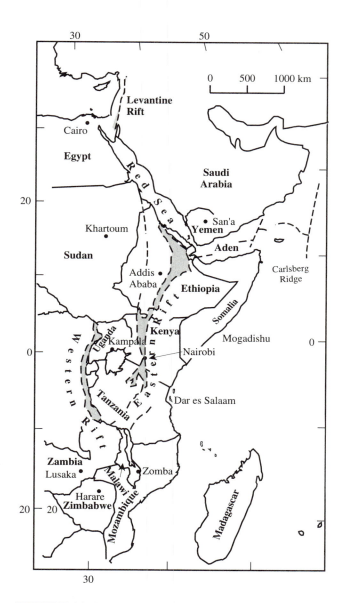

FIGURE 10-3

The East African Rift–Red Sea Graben–Dead Sea Graben (Levantine Rift) extensional system. [From B. H. Baker, P. H. Mohr, and L. A. J. Williams, 1972, *Geology of the Eastern Rift System of Africa, Spec. Paper 136* (Boulder, CO: Geological Society of America), Fig. 1.]

Grande Rift in the southwestern United States, and the Baikal Rift in Siberia.

One modern environment where many tectonicists and petrologists confidently believe that extensional stresses are pulling a continent apart is the East African Rift system, which runs from Syria in the north to Mozambique in the south (Figure 10-3). Normal faulting is typical of extensional regimes, and a sequence of

FIGURE 10-4

●●●●●●●●●●●●●●●

Volcanism map of the East African Rift in Ethiopia, Uganda, Kenya, and Tanzania, showing the distribution of volcanics of different age and type. [From B. H. Baker, P. H. Mohr, and L. A. J. Williams, 1972, *Geology of the Eastern Rift System of Africa, Spec. Paper 136* (Boulder, CO: Geological Society of America), Fig. 6.]

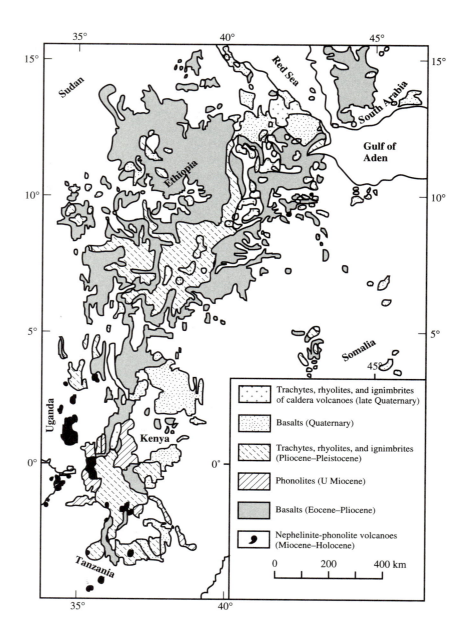

spectacular rift valleys or grabens has been created along the whole length of this zone by such faulting. The prominent valley system on the African continent is continuous with the Red Sea Rift Graben to the north and this in turn with the Dead Sea Rift, which extends as far north as the Israel-Jordan border. The rift valleys in Kenya, notably Olduvai Gorge, are famous as the locales where many of the most important fossil finds in primate paleontology were made. Early hominids coexisted with the vigorous volcanic activity in the rift, which continues even today, and their remains were preserved in volcaniclastic deposits as old as 5 million years.

The East African Rift is volcanically active (Figure 10-4), like other continental rifts. This activity is proba-

bly due to the large-scale extensional stress regime in the crust, which has caused normal faulting and provided pathways for mantle-derived magmas to reach the surface quickly. The rift in east-central Africa has experienced essentially continuous volcanic activity over the last 30 million years. In common with other rifts, volcanic rocks of the East African Rift are dominantly alkalic. Nephelinites and alkali basalts occur early in the sequence and are followed by phonolites, trachytes, and rhyolites. There is an overall trend toward increasing silica content in later magmas, and eventually silica saturation is reached at a late stage. This trend and another toward increasing Fe:Mg ratio in phenocrystic ferromagnesian minerals are key indicators of fractionation processes at work in deep magma chambers. Many

FIGURE 10-5
● ● ● ● ● ● ● ● ● ● ● ● ● ● ● ● ●

Some of the volcanoes in the San Francisco volcanic field west of the Rio Grande
Rift and south of the Colorado Plateau, near Flagstaff, Arizona. Part of Sunset
Crater (about 900 years old) and the Bonito lava flow are in the foreground; Leary
Peak is in the background. Sunset Crater consists of alkali olivine basalt, whereas
Leary Peak and adjacent San Francisco Mountain are built of intermediate to silicic
rocks. [From C. A. Wood and J. Kienle, eds., 1990, *Volcanoes of North America:
United States and Canada* (Cambridge: Cambridge University Press), p. 279.
Photo by E. W. Wolfe, U.S. Geological Survey Cascades Volcano Observatory.]

of the volcanoes are small cinder cones, but very large
composite and even shield volcanoes occur (for example, Mount Kenya and Mount Kilimanjaro). Plateau
lavas of basaltic to trachytic or phonolitic composition
also have erupted from time to time from fissures or
from shield volcanoes and have flooded large areas
within the rift. Some late volcanoes are characterized by
highly alkalic and silica-deficient lavas, and even by
carbonatites (like those found at the volcano Oldoinyo
Lengai in Uganda). Later extensional normal faulting
of the thick, horizontal lava piles has created high
scarps that expose the flow stratigraphy.

The Rio Grande Rift in the southwestern United
States (see Figure 10-1A) is another notable continental
rift. It extends from northern Mexico along the southern
border of Texas, forms the southern edge of the
Colorado Plateau province through west Texas and New
Mexico, and ends in central Colorado. Like the East
African Rift, it is characterized by high to very high
continental heat flow values, has recent volcanic and
plutonic activity along much of its length, and shows

magma evolution from mafic to intermediate and even
silicic volcanics. Fresh cinder cones dot the landscape
west and north of Albuquerque and Santa Fe in New
Mexico. Although not strictly part of the rift, the area
west of Albuquerque toward Flagstaff, Arizona, is also a
recent volcanic terrane. The San Francisco Volcanic
Field lies south of the Grand Canyon, near Flagstaff, and
has numerous volcanic features, including the San
Francisco Peaks, an eroded Late Cenozoic volcanic center. One small alkali olivine basalt cinder cone near
Flagstaff, Sunset Crater (Figure 10-5), is estimated to
have last erupted only about 900 years ago, between A.D.
1064 and 1068.

The Rio Grande Rift shares another petrologic characteristic with the East African Rift: the scattered occurrence of alkalic rocks mixed with the dominant
basalt-to-rhyolite volcanics. Alkali olivine basalts form
a number of the smaller cinder cones throughout
the rift, whereas larger volcanic centers, such as the
San Francisco Peaks in Arizona and the Valles Caldera
near Taos and Santa Fe, New Mexico, contain diverse

FIGURE 10-6

• • • • • • • • • • • • • • •

Wall of a small canyon in a xenolith-bearing alkali olivine basalt flow from the Peridot Mesa diatreme vent in the San Carlos volcanic field, near Globe, Arizona. The wall of alkali basalt contains many coarse ultramafic mantle xenoliths (peridotite, lherzolite, pyroxenite) and large olivine megacrysts.

volcanic rock types from basalt to dacite, trachyte, and rhyolite. Rhyolite and other fractionated rock types commonly form thick sheets of welded tuff and other explosively erupted pyroclastic rocks. Mantle derivation of the most mafic volcanic rocks (alkali basalts and basanites) and a direct, rapid ascent to the surface are amply demonstrated by the number of localities in the region where mantle xenoliths are found. Notable xenolith localities include Kilbourne Hole and related basaltic diatremes or maars (explosive volcanic craters) in southern New Mexico north of El Paso, as well as San Carlos, Williams, and Chino Valley in Arizona. A wide range of spinel-bearing lherzolite, harzburgite, and dunite xenoliths in these occurrences indicates heterogeneous mantle beneath this part of the Colorado Plateau (Figure 10-6). Some basaltic lavas (for example, Williams, Arizona, and Kilbourne Hole) also contain significant numbers of lower crustal xenoliths (mainly granitic gneisses), which provide unusually detailed information about the nature of the deep continental crust.

The Rio Grande Rift does not appear to be as likely a continent-splitting feature (that is, potential divergent plate boundary) as the East African Rift. Crustal extension undoubtedly plays a major role in the Rio Grande Rift, as it does in the Basin and Range province to the north, but the extension has been attributed to subduction of the East Pacific Rise spreading center under the southwestern United States and is probably less extreme than in eastern Africa. Note that the East Pacific Rise extends up the Gulf of California (see Figure 10-1A), causing Baja California to rift away from mainland Mexico, before the Rise plunges under the continent. The Rio Grande Rift is thought to reflect the thermal and extensional consequences of a subducted spreading ridge under the continent itself.

Chemistry, Petrography, and Petrogenesis of Continental Rift Magmas

The East African Rift contains a remarkable variety of volcanic rocks. These range from very primitive, silica-undersaturated basanites, nephelinites, alkali basalts, and even carbonatites to silica-rich rocks such as peralkaline rhyolites, with virtually every intermediate member present as well. Many petrologic questions arise with such a diverse suite. How are these magmas related to one another? How many, if any, are mantle-derived, and how many are products of fractionation in mantle or deep crustal magma chambers? Has crustal melting or large-scale contamination played a role in magmatic evolution?

The chemistry of the silica-undersaturated alkali basaltic, basanitic, and nephelinitic volcanic rocks is similar to that of the same rock types on oceanic islands, a chemical signature suggesting a similar origin by deep melting of enriched mantle, possibly asthenosphere.

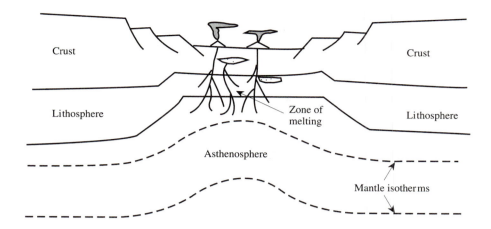

FIGURE 10-7

• • • • • • • • • • • • • • • •

Schematic crust-mantle cross section for the East African Rift. Crustal thickness is
about 20 km in the rift and approximately 50 km outside the rift. Speculative thinning of
subcontinental mantle lithosphere under the rift is illustrated. Upward deflection of
mantle isotherms raises temperatures above the melting point in shallow mantle (both
lithosphere and asthenosphere) and produces a zone of melting (shown), which is the
source for rift magmas. Some magmas penetrate the crust and form volcanoes, whereas
others pond at the base of the crust or solidify within the crust to form plutons.

Some of the more interesting rocks of the continental
rifts, however, are the more evolved ones, for example, phonolites, trachytes, and rhyolites among the
volcanics, and their respective intrusive equivalents,
nepheline syenites, syenites, and granites. All these
rocks appear to be the products of fractionation.
Trachytes and phonolites are actually very similar to
each other in composition, and both types typically
contain phenocrysts of a single alkali feldspar, either
a sodic sanidine or anorthoclase. Trachytes lie just on
the border of silica saturation, whereas phonolites are
silica undersaturated and typically contain nepheline.
Plutonic granitoids crystallized only a single alkali
feldspar as well, now an exsolved perthite, indicating
a shallow emplacement depth and hypersolvus crystallization (see the section on anorogenic granitoids).
Most important, an overwhelming majority of the felsic
rocks, both volcanic and plutonic, are peralkaline, some
strongly so. The occurrence of sodic pyroxenes and
amphiboles such as aegerine, aegerine-augite, riebeckite, and other, rarer minerals, as well as high Fe:Mg
ratios in ferromagnesian minerals and the presence
of fluorite, reflect this chemical character. In general,
the magmatic association of basalt-trachyte-phonolite-
rhyolite, with a tendency to peralkalinity, is probably a
key indicator of the association of continental rifting
and continental magmatic provinces.

There seems to be little doubt that the initial stages
of magmatic activity in the East African Rift represent
mantle-derived melts. The highly alkalic nature of the
magmas and high concentrations of incompatible elements (like those in OIBs) indicate small degrees of
melting of enriched mantle at great depths (60–100 km
or more). A variety of geophysical data suggest that the
mantle beneath the East African Rift is not normal subcontinental lithospheric mantle but is upward-bulging
asthenosphere (Figure 10-7). The lithosphere apparently ruptures as part of the initiation of rifting and
asthenospheric mantle comes into direct contact with
the lower continental crust (refer to Figure 8-7 for comparison with the analogous process in an oceanic setting, that is, a mid-ocean ridge). The continental crust
itself is thinned to about 20 km within the rift by ductile
extension processes but remains a normal thickness
outside the rift.

When rifting is initiated, the first melting event in
the ascending asthenosphere generates very small
volumes of basanitic-alkali basaltic-nephelinitic melts
and also carbonatites, which move upward through
the thinned crust. The presence of carbon dioxide–
rich fluids during melting has been inferred from experiments and indicated by carbon dioxide fluid inclusions
in xenolithic olivine. As further and more extensive
melting occurs, transitional basalts and more MORB-
like compositions are produced in substantial volume.
Accumulation of this magma in chambers both below
and within the continental crust allows for relatively
low-pressure fractional crystallization and development

of the later, evolved, more silicic magmas that form the trachyte-phonolite-rhyolite series. This petrogenetic model appears to work reasonably well for the East African Rift, but does it apply as well to the Rio Grande Rift? The answer is not entirely clear but is probably no. The Rio Grande Rift does not display the range of lavas found in East Africa, particularly the highly alkalic, very silica-undersaturated early ones. For the Rio Grande Rift, a better model is that lithospheric extension has caused some asthenosphere upwelling and has allowed some deep, xenolith-bearing alkali basalt magmas through, but the bulk of magmatic activity has been of a more normal basalt to rhyolite type. Crustal thinning has not been of nearly the same magnitude as in Africa, and the lithosphere is probably less thinned as well. Magmatic activity has therefore been in the form of volcanism derived from basaltic magma fractionation in upper mantle or deep crustal magma chambers.

Carbonatites

Carbonatites deserve special mention here. They are rare igneous rocks that contain at least 50% carbonate minerals, mainly calcite but also dolomite, magnesite, and sodium carbonate. Their igneous origin has been debated, but in the East African Rift they clearly occur in volcanic cinder cones and flows and in shallow intrusive bodies. (Other well-known carbonatite localities at Oka in Quebec, on the Kola peninsula in northern Russia, and in southern Norway are undoubtedly intrusive.) They are actually high in sodium as well as in calcium, and always contain sodium-rich pyroxenes and amphiboles. Other typical major minerals include apatite, phlogopite, and magnetite, with the accessory minerals fluorite, perovskite ($CaTiO_3$), monazite, pyrochlore (an exotic niobium- and tantalum-rich oxide mineral), and barite. Carbonatite magmas very commonly have anomalously high contents of various exotic incompatible elements and crystallize equally rare and exotic minerals.

Carbonatites almost invariably occur in small, shallow intrusive complexes accompanied by alkali-rich rocks such as nepheline syenite; in modern settings like the East African Rift, they can also show a surface volcanic expression. They appear to require a continental environment of at least extensional stress, if not actual rifting, to make it to high levels in the crust. An idealized geologic map of a typical carbonatite occurrence is shown in Figure 10-8. The general pattern is that of a group of more or less vertical and apparently concentric rings, although some complexes are elongated along

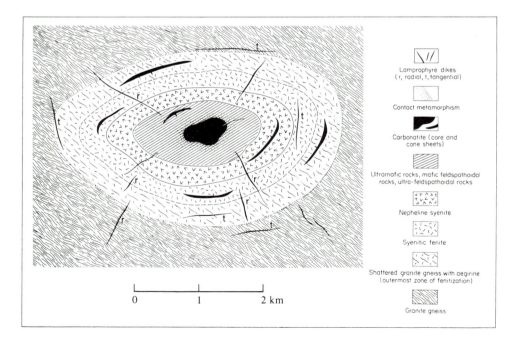

FIGURE 10-8

• • • • • • • • • • • • • • •

Idealized geologic map of an alkalic-ultramafic carbonatite complex. Carbonatites form a central core and cone sheets. Lamprophyre dikes are both radial and tangential. The complex is surrounded by a "fenitized" (metasomatized) halo and shattered metamorphic rocks.

fracture zones. Within these complexes, carbonatites themselves make up a small proportion of the total and apparently intrude the core of the complex at a late stage: The initial magmatic intrusion is of nepheline syenite in an intermediate ring. The shallow nature of the intrusion, combined with likely emanations of hot, expanding gases, leads to abundant fracturing of the surroundings, penetration by hot alkali-rich fluids, and **metasomatism** involving intense chemical alteration, addition of sodium, and oxidation. This process, called **fenitization,** was first described in the Oslo Graben in Norway, which is effectively the type locality for metasomatism.

Carbonatites and some kimberlites (see later) both have an unusual feature—they contain magmatic carbonate minerals. Both magmas apparently originate by small degrees of fractional melting of mantle peridotite, which contains either carbonate minerals or a carbon dioxide–rich fluid phase. However, petrologic evidence indicates little other linkage between them. Experimental work has demonstrated that carbon dioxide plays an important role in peridotite melting, and the presence of diamonds in some igneous rocks produced by mantle-derived magmas is persuasive evidence of the existence of carbon in the earth's deep interior. However, the details of the origin of carbonatite magmas remain a largely unexplored area of igneous petrology.

ANOROGENIC GRANITOIDS

Anorogenic granitoids are granites and related plutonic rocks that occur in tectonic environments where no obvious relationship to subduction or plate-convergence processes is obvious. The granite chemical and tectonic classification scheme discussed in Chapter 9 (S-type, I-type, A-type) has been modified to include these rocks as A-type but not entirely successfully, because of a problem common to genetic classification schemes. There is inevitable substantial overlap in chemistry for granitoids from different tectonic environments. The so-called anorogenic granitoids that form the basis for the A-type definition are virtually all either metaluminous or peralkaline. In addition, granitoids of this suite typically have chemical characteristics indicating significant fractionation, especially high to very high Fe:Mg ratio in ferromagnesian minerals. They therefore may be confused with the I-type granites that are very definitely related to orogenic processes, like those in southeastern Australia, or with the magmas related to continental rifting. This last point is critical, because it is likely that many of the so-called anorogenic granitoids are simply the deep plutonic expressions of rift-related (or possibly hot-spot) magmatism that have been exposed by erosion. It may well be that the extensive rhyolitic volcanism at Yellowstone indicates the formation of anorogenic granite at depth.

One classic anorogenic granitoid province extends from the northeastern United States into adjacent Canada (Figure 10-9). In fact, rocks from this province inspired the definition of A-type granite. Two belts of granitoid plutons cross this province about at right angles to each other. A broad, roughly linear belt of Mesozoic (Jurassic-Cretaceous) plutons runs from the eastern coast of New England (Boston to southern Maine) to the northwest through New Hampshire, Vermont, and Quebec. Some plutons have associated volcanic rocks (for example, Moat Mountain, New Hampshire, near Mt. Washington) or hypabyssal dikes that are probably former volcanic feeder dikes. The belt is roughly continuous to the southeast, with a linear trend of seamounts across the continental shelf in the Gulf of Maine. The White Mountains of New Hampshire are part of this belt, as are the Monteregian Hills intrusions in Quebec near Montreal. The linear nature of the belt and age relations indicating older plutons in the northwest and younger ones in the southeast have prompted the proposal that the belt represents an old hot-spot track. This track is essentially similar to the Hawaiian-Emperor trend in the Pacific but contrasts with it in having both plutonic and volcanic expression and extending across both continental and oceanic lithosphere. In detail, however, age relations are considerably more complicated than in other proposed hot-spot tracks, and the hot-spot model does not stand up well under detailed scrutiny.

The White Mountains–Monteregian Hills belt crosses an older group of anorogenic granitoids along the Atlantic coast. This second belt, referred to as the Coastal Maine Magmatic Province (CMMP), extends from northeastern Massachusetts (Peabody, Cape Ann) to New Brunswick, Canada, and also has some associated volcanic rocks in addition to plutons. In contrast to the younger belt, the CMMP appears definitely to be related to crustal extension or rifting rather than to a hot spot. The ages of plutons (mostly Ordovician to Silurian) do not show the progressive trend along the belt that characterizes hot-spot tracks. Reconstruction of the paleotectonic setting of the CMMP belt, however, is complicated by the accretion of the province to North America in the late Paleozoic and subsequent tectonic modification.

Both magmatic belts contain a variety of plutonic rock types ranging from rare alkalic gabbros to intermediate granodiorites and monzonites, and finally to abundant syenites and granites. Silica-undersaturated rocks are more common in the White Mountains–Monteregian

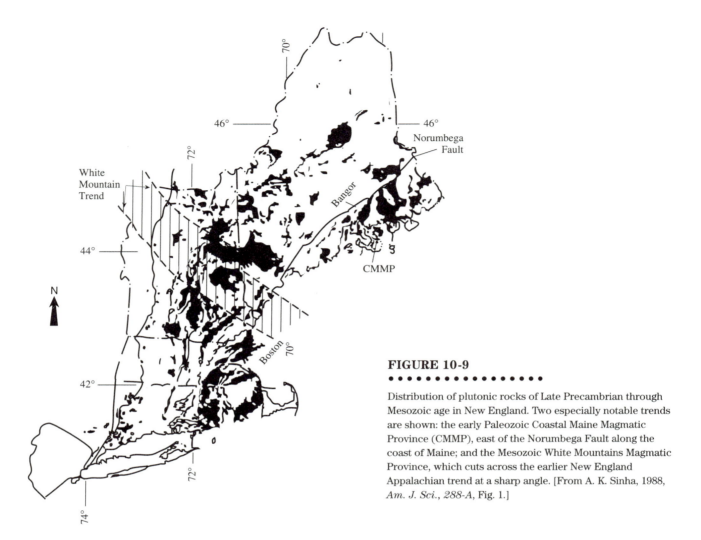

FIGURE 10-9

• • • • • • • • • • • • • • • •

Distribution of plutonic rocks of Late Precambrian through Mesozoic age in New England. Two especially notable trends are shown: the early Paleozoic Coastal Maine Magmatic Province (CMMP), east of the Norumbega Fault along the coast of Maine; and the Mesozoic White Mountains Magmatic Province, which cuts across the earlier New England Appalachian trend at a sharp angle. [From A. K. Sinha, 1988, *Am. J. Sci.*, *288-A*, Fig. 1.]

belt, which includes rock types such as nepheline syenite and nepheline-sodalite syenite at Red Hill in New Hampshire and other locations, and even rare carbonatite and related silica-poor rocks along the St. Lawrence River near Montreal in Quebec. The mafic rocks of both belts are dominantly tholeiitic to alkalic and the granitoids are metaluminous to dominantly peralkaline, and all are characterized by high to very high Fe:Mg ratios in ferromagnesian minerals, compositions indicating fractionation. Many of the granites and syenites from both belts contain fayalite, hedenbergite, riebeckite, and other iron-rich minerals. In fact, the Cape Ann Granite of the CMMP in northeastern Massachusetts has given its name to the iron-rich end member of the biotite solid solution—*annite*, which occurs on Cape Ann in pegmatites.

One common characteristic of many of the anorogenic granitoids is the occurrence on the liquidus of only a single alkali feldspar, rather than the separate crystallization of a potassium-rich alkali feldspar and a sodium-rich plagioclase, as is typical of most orogenic

granitoids. Upon cooling, this single feldspar decomposes (exsolves) into the lamellar intergrowth called perthite with potassium-rich and sodium-rich lamellae (Figure 10-10). This process is referred to as **hypersolvus** crystallization and is an important petrogenetic clue to relatively low total pressure or water pressure (indicating shallow depth) during crystallization of these magmas. Simultaneous crystallization of two feldspars is called **subsolvus** texture. The effects of pressure, particularly water pressure, on the liquidus phase relations of alkali feldspar solid solutions determine whether hypersolvus or subsolvus crystallization occurs.

Figure 10-11 shows the effects of P_{H_2O} on granite crystallization in a simplified P-T diagram. In Chapter 5 we noted that increased "dry" pressure typically causes eutectic temperatures to increase (as shown by the "dry" melting curve in Figure 10-11) because liquids, being less dense, are more compressible than solids of identical composition. Increased pressure at constant temperature therefore favors the solid, and the solidus

FIGURE 10-10
● ● ● ● ● ● ● ● ● ● ● ● ● ● ● ●

Perthitic feldspar in granite, as seen in thin section. Perthite typically consists of a potassic feldspar (orthoclase or microcline) host with lamellae of sodium-rich plagioclase (albite or oligoclase). Plagioclase lamellae are commonly recognized by the presence of albite twinning, which shows clearly in cross-polarized light. Photo taken with crossed polarizers; width of field is about 0.8 mm.

becomes a line of positive slope on the *P-T* diagram. However, when a melt can dissolve large amounts of water, as granite magmas do, the corresponding compositions above and below the solidus become melt with dissolved water (above the solidus) versus solid minerals *plus water vapor* (below the solidus). The separate water vapor phase is much more compressible (at least at low pressures) than the same amount of water *dissolved in melt*, and pressure thus favors the melt. This effect causes the "wet" melting curve to have a negative slope (see Figure 10-11). At high pressures, the effect of relative compressibility lessens, and the "wet" melting curve becomes essentially vertical on the diagram and ultimately parallels the "dry" curve. The presence of water during the melting or crystallization of granite can thus lower melting or solidification temperatures by hundreds of degrees.

Figure 10-12 is a binary *T-X* phase diagram (refer to Figure 5-9) that shows the liquidus-solidus phase relations in the KAlSi$_3$O$_8$–NaAlSi$_3$O$_8$ (orthoclase-albite) binary system at various water pressures. The behavior of alkali feldspars was introduced in Chapter 5 but will be reviewed briefly here. The liquidus-solidus loop reflects solid solution effects similar to the plagioclase system (albite-anorthite), but now the lowest temperature feldspar is in the middle of the composition range rather than at one end. Melts at either end therefore migrate toward the middle as they crystallize, much as in a binary eutectic system. The lowest temperature point (*minimum*) acts much like a eutectic point, except that melt at this point crystallizes only a single

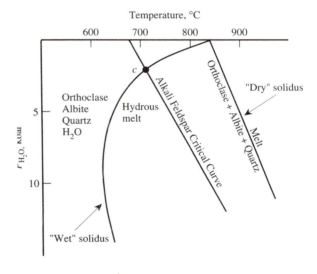

FIGURE 10-11
● ● ● ● ● ● ● ● ● ● ● ● ● ● ● ●

P$_{fluid}$-*T* diagram for the simplified granite system, illustrating the shift in the granite solidus caused by the presence or absence of water. The alkali feldspar critical curve is also shown. Hydrous granite magmas that have a high Na:Ca ratio and crystallize at pressures above *c* will display subsolvus texture with discrete alkali and plagioclase feldspars (that is, crystallization occurs at a temperature lower than the critical point). Those crystallizing at pressures lower than *c* will be hypersolvus (containing a single perthitic feldspar) because crystallization occurs at a temperature higher than the critical point. For details, see text. [Compiled from data in O. F. Tuttle and N. L. Bowen (1958), W. C. Luth, R. H. Jahns, and O. F. Tuttle (1964), and S. A. Morse (1980).]

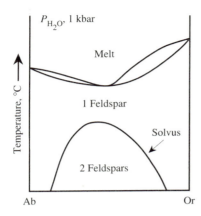
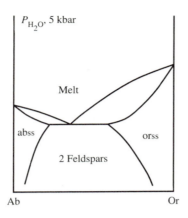

FIGURE 10-12
••••••••••••••••

T-X diagrams for the albite (Ab)-orthoclase (Or) binary system at 1 kbar P_{H_2O} and 5 kbar P_{H_2O}. Note the direct crystallization of a single feldspar from melt at 1 kbar and subsequent exsolution to perthite upon cooling below the alkali feldspar solvus. At 5 kbar, melt crystallizes either orthoclase-rich solid solution or albite-rich solid solution until it reaches the eutectic; then it simultaneously crystallizes two feldspars. The solidus and solvus curves intersect at about 2.5 kbar P_{H_2O}.

solid rather than two. Note in Figure 10-12A the inverted parabola (solvus) labeled "2 feldspars" and recall from mineralogy that if a single feldspar cools into this region, it will break down into the two feldspars that form perthite. Increased pressure (wet *or* dry) causes the top of the solvus (critical point) to move upward in the diagram, thus widening the gap between coexisting feldspar compositions. This effect exists simply because increased pressure causes too much strain in the feldspar crystal structure (and thus excess strain energy) when large potassium and much smaller sodium atoms are all mixed together randomly. To reduce energy, the ions will therefore tend to cluster into potassium-rich regions and sodium-rich lamellar regions—hence perthite. The increase in temperature of the critical point with increased pressure is shown on the *P-T* diagram (see Figure 10-11) as the "critical curve."

As increasing water pressure causes the solvus to move upward in temperature, at the same time it causes the liquidus-solidus loop to migrate downward, for the reasons noted above and illustrated in the *P-T* diagram (see Figure 10-11). At 2–3 kilobars, the solidus and the critical curve meet (point *c* in Figure 10-11), and the continuous one-feldspar region below the solidus curve (across the whole width of the diagram) vanishes. When magmas crystallize above the pressure where critical curve and solidus meet, two feldspars (one potassium-rich and the other sodium-rich) will crystallize separately and be recognizable petrographically in the rock (each may individually exsolve slightly, but they will remain recognizably potassium-rich and sodium-rich). This subsolvus crystallization behavior indicates relatively deep, moderate to high pressure crystallization. Crystallization of a magma at lower water pressures results in only a single feldspar, which then becomes perthitic upon further cooling—hypersolvus crystallization. One strong qualification of the distinction between subsolvus and hypersolvus behavior as a depth indica-

tor is that it applies only for granites with relatively high Na:Ca ratios. The introduction of even small amounts of calcium into feldspar solid solutions causes the solvus to widen dramatically, and essentially all feldspar crystallization becomes subsolvus. Granite petrologists have long debated the utility of this approximate scheme of pressure estimation, but it is a useful petrographic technique that has been widely used, and the student of petrology should be aware of its significance.

ANORTHOSITES
•••••••••••••••

On the basis of the Streckeisen classification, **anorthosites** are plutonic rocks consisting of greater than 90% plagioclase, with the remainder consisting of anhydrous ferromagnesian silicates and accessory minerals. The definition is probably the only simple thing about anorthosites, however. This unusual rock type has widely varied occurrences that range from the highlands of the lunar surface to centimeter-thick layers within terrestrial layered mafic complexes to large independent massifs. The most extensively studied anorthosites are those that occur in independent bodies ranging from smaller lensoid shapes to very large batholith-scale complexes (the anorthosite *massifs*). Extensively studied anorthosite massifs include the Adirondacks (New York), numerous bodies in Quebec and Labrador (for example, Nain, Harp Lake, St. Urbain), the Laramie Complex (Wyoming), and the Rogaland and Hidra Massifs (Norway). Almost all anorthosites are of Precambrian age. The greatest majority are large massifs of Proterozoic age that were intruded between about 1.6 and 1.2 Ga, but smaller Archean (>2.6 Ga) bodies are known on all continents except Antarctica (for example, Fiskenaesset Complex in Greenland, Bad Vermilion Lake Complex in Ontario, and Limpopo Complex in South Africa). A few Paleozoic "anorthosites" are known (for example, the Betts Cove in Newfoundland), but they tend to be plagioclase-enriched gabbroic rocks (commonly called "gabbroic anorthosite") and are much more commonly associated with voluminous mafic igneous rocks than Precambrian anorthosites are.

Lunar Anorthosites

The Apollo program of lunar exploration made six landings on the moon's surface between 1969 and 1972, and astronauts returned with over 700 kg of lunar rocks. Although the majority of the samples are basalt, regolith (unconsolidated shocked particulate matter), and breccia, some plutonic rock samples were collected from lunar highlands areas, particularly by *Apollo 17* astronauts. The relative abundance of anorthosite samples surprised most petrologists and sparked the development of novel theories for the development of the early lunar crust.

Like other lunar rocks, lunar anorthosites are sodium-poor and contain much more calcic plagioclase than most terrestrial anorthosites ($>An_{90}$). Minor phases include magnesium-rich olivine and pyroxene, as well as chromium-rich spinel. Radiometric dating of anorthosite, troctolite (olivine-plagioclase cumulate rock), and norite samples revealed ages very nearly as old as the moon itself (4.3 to 4.5 Ga). Lunar petrologists hypothesized that immediately after its formation, the early moon had an outer molten layer of basaltic composition that they called the Lunar Magma Ocean (LMO). As it cooled, the LMO began to fractionally crystallize calcic plagioclase, which, being less dense than the liquid, clotted together and floated. Minor amounts of other silicates and oxides were trapped in these "rockbergs," which coalesced to form a largely anorthositic crust a few kilometers thick, surrounding a partially molten lunar mantle. Intense bombardment by meteorites and even larger planetesimals in the first few hundred million years of lunar history caused breaching of the anorthositic crust and outpourings of enormous volumes of basaltic magmas to form the dark-colored circular lunar *mare*. Anorthosites and related rocks remained exposed in the light-colored highland areas between impact basins.

Although it was regarded as revolutionary when first proposed, this theory is now widely accepted. Its possible application to very early terrestrial history is unclear but fascinating. Continuing geologic activity on the earth, in contrast to the moon, has obscured virtually all rocks and events of the first billion years. Could the Archean anorthosites on the earth (see the following section) be the reworked and tectonically modified remnants of an original widespread early anorthositic crust on the earth?

Archean Megacrystic Anorthosites

The Archean anorthosites (>2.6 Ga) are primarily of interest because their closer chemical similarity to highland anorthosites on the moon than to the younger Proterozoic terrestrial massif anorthosites suggests that they may hold important information about early crust-forming processes on the earth. They commonly occur in lensoid or layered bodies associated with mafic rocks such as gabbro, norite, and ultramafic rock types, all of which are typically cumulate. The country rocks are so-called supracrustals or **greenstone belts** common in Archean tectonic settings and include shallow water sediments and mafic volcanics metamorphosed to low grade. Note that this is also the locale of the komatiites.

The two striking aspects of these anorthosites are the large size and well-formed crystal shape of the plagioclase **megacrysts** (up to 1 m across) and the extraordinarily calcic compositions of the unzoned crystals ($>An_{80}$ and up to An_{95}), considerably more calcic than in typical basalts. These anorthosites have bulk compositions that are too rich in Al_2O_3 to represent possible liquids, even when they are averaged volumetrically with other mafic and ultramafic rocks that occur with them in complexes. As noted earlier in the komatiite section, it has been suggested that Archean mantle or thermal structure was sufficiently different that a unique type of basaltic magma could have been generated then. However, for a variety of reasons, including the absence of compelling supporting evidence, this explanation has not been widely accepted.

One model for the origin of Archean anorthosites has focused on the presence in the greenstone belts of basaltic dikes, which also contain unusually calcic plagioclase megacrysts (Ashwal 1993). These dikes, however, are of more normal or typical high-aluminum basalt composition. Archean anorthosites may represent trapping of such melts in fairly large magma chambers deep in the crust or at the crust-mantle boundary, thus allowing very slow cooling that could produce large crystals. Sudden tapping of the chambers would allow a crystal-rich mush to rise and be emplaced at shallower levels, thereby producing a magmatic rock anomalously rich in plagioclase.

Massif Anorthosites

The massif anorthosites are younger and have a tectonic setting different from that of the Archean anorthosites, as well as significantly different chemistry. They are virtually all of Proterozoic age (2.5 to 0.6 Ga) and have been emplaced into Precambrian shield areas. Most are associated with highly fractionated granitoids (very high Na:Ca and Fe:Mg ratios), including granites, syenites,

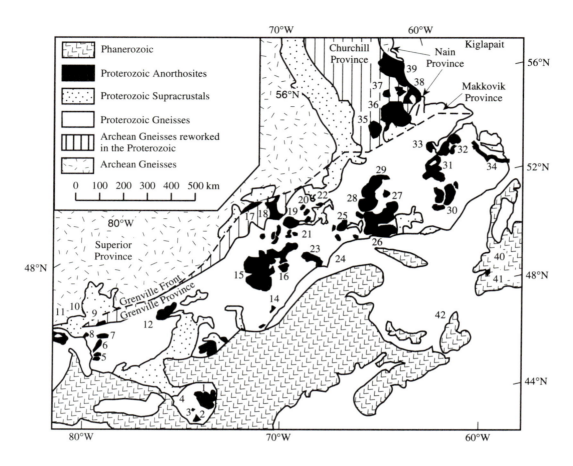

FIGURE 10-13
• • • • • • • • • • • • • • • •

General geology of the eastern Canadian Shield (northeastern United States and eastern Canada), showing distribution of massif-type anorthosites. Map numbers refer to individual intrusions, of which the more notable are the Marcy and Oregon Dome Massifs in the Adirondacks (1–2); Morin (13), Lac St. Jean-Labrieville (15–16), Lac Allard-Romaine R. (26–27), and Lac Fournier-Atitonak R. (28–29) in Quebec; and Harp Lake (36) and Nain (39) in Newfoundland. [From L. D. Ashwal, 1993, *Anorthosites* (New York: Springer-Verlag), Fig. 3.2. (Compiled from numerous sources.)]

and monzonites of the so-called **charnockite** series (charnockite is a name used for orthopyroxene granite). Granitoids of the charnockite series commonly contain iron-rich minerals such as fayalite and hedenbergite. Whereas Archean anorthosites have unusually calcic plagioclase, this mineral is actually slightly more sodic in the massif anorthosites (An_{40-60}) than in typical basalts.

A major belt of massif anorthosites occurs in northeastern North America from Labrador south through Quebec to the Adirondacks of New York and westward to Ontario (Figure 10-13). Other notable anorthosites of similar age in North America occur in the Wolf River Complex (Wisconsin), the Duluth Complex (Minnesota), the Laramie Complex (Wyoming), the San Gabriel Mountains (near Los Angeles), and Oaxaca,

Mexico. Variably tectonized fragments of anorthosites and their associated charnockites have even been found in the Blue Ridge and Piedmont of the southern and central Appalachians from Pennsylvania to Virginia. Notable occurrences outside North America include Rogaland, Bergen, and Jotun (Norway), numerous occurrences in European and Asian Russia, Upanga and Kunene in Africa, and Eliseev in Antarctica. Important questions about the massif anorthosites involve the origin of the magmas that formed them, the reasons for plagioclase concentration, and any possible genetic connections between the anorthosites and their nearly ubiquitous granitoid companions. A comprehensive review can be found in Ashwal (1993).

Anorthosites were long thought to be deep intrusions, but petrologic evidence increasingly suggests a shallow intrusive depth for most of the massif anorthosites. In Labrador, the country rocks are of very low metamorphic grade (greenschist), and anorthosite bodies are surrounded by contact aureoles with low-pressure metamorphic minerals. The anorthosites themselves are typically unmetamorphosed and undeformed—their own intrusion was the most recent petrologic event in the area. Similarly, the Laramie Anorthosite has imposed a low-pressure contact aureole on its country rocks. South of the Grenville Front (see Figure 10-13), which marks the northern boundary of a zone of major crustal deformation and metamorphism at roughly 1050 Ma in Quebec and New York, both anorthosites and country rocks have undergone deformation and very high grade metamorphism at great crustal depths (>20 km). Petrologists formerly thought that anorthosite intrusion and regional metamorphism in the Adirondacks were contemporaneous, thus indicating a deep intrusive origin for the anorthosite. Discovery of preserved low-pressure contact metamorphic rocks in the Adirondacks now confirms that intrusion was essentially anorogenic and at shallow depths long before the high-pressure regional metamorphism. This finding supports a consistent shallow intrusion depth model for all the massif anorthosites of eastern North America.

At present, the exact nature of the parental liquid for massif anorthosites is unknown. A variety of geochemical and mineralogic data, however, including the sodic character of the plagioclase and high Fe:Mg ratios of mafic silicate minerals such as pyroxenes and olivines, indicates that the parent magma was almost certainly not derived directly from normal mantle. An origin of parental magma either by fractionation of a more primitive basaltic liquid or by melting of anomalous mantle are both possibilities. In any case, some degree of residence in a deep magma chamber is necessary to allow crystallization of plagioclase and its concentration in derivative melts. Many massif anorthosites contain minor volumes of a variety of transitional rock types with various proportions of plagioclase and mafic minerals, compositions suggesting variable amounts of fractionation or crystal enrichment. These have been called **gabbroic anorthosite** (somewhat richer in mafic minerals than typical anorthosite), **anorthositic gabbro** (somewhat richer in plagioclase than typical gabbro), and **troctolite** (olivine-plagioclase rock). As in other continental magmatism, there is a possibility of an unusual degree of mantle heterogeneity in subcontinental mantle, in this case heterogeneity that must be at least 1.5 billion years old.

One further question posed by massif anorthosites involves the essentially ubiquitous association between the anorthosite and charnockite series. Were these magmas both derived from the same source, either a direct mantle source or a fractionating more primitive liquid, or did anorthosite emplacement cause secondary crustal melting? The highly fractionated nature of the charnockite suite rocks is *consistent* with the first model but is otherwise not of much help in discriminating possibilities. The granitoids of the charnockite suite are always intrusive into anorthosites where relative age relationships are displayed in outcrop. Determination of ages in various massifs has shown that the granitoids may range from virtually as old as the anorthosite to considerably younger.

The clue to the relationship between anorthosite and the charnockite series comes from isotopic analyses. Cogenetic magmas (that is, magmas derived from the same source) must share certain isotopic ratios, in particular, the ratio of strontium isotopes, $^{87}Sr:^{86}Sr$, which are unlikely to be altered by any subsequent degree of crystal fractionation after the magmas diverge from the parental line. The strontium isotopic analyses of the anorthosites and charnockites unequivocally show that although an ultimate mantle source is likely for the precursor of the anorthosite magma, the granitoids cannot share this source. Instead, they appear to have been derived by anatexis or secondary melting of the continental crust. Their metaluminous to peralkaline character means the source rocks are themselves of probable igneous rather than sedimentary origin. It is easy to visualize the large amounts of melting of lower crustal material that must have accompanied the long residence of the voluminous anorthosite parental magmas in deep magma chambers.

Extraordinary questions remain, however. For example, why do these unusual rocks occur only in this relatively narrow window of geologic history? But remember that spans of time in the Precambrian are longer than they at first seem. Consider that the span of time for massif anorthosite intrusion is almost as long as all of the Phanerozoic, from the Cambrian to the present! The fact that anorthosites are demonstrably not unique to the earth but are *planetary* geologic features has sparked much research interest in comparing terrestrial anorthosites with those of lunar and meteoritic origin.

KIMBERLITES AND LAMPROITES
• • • • • • • • • • • • • • •

Geologists have expended an inordinate amount of effort in examining these rocks, which are volumetrically minuscule. One significant reason, of course, is economic: Some of these rocks are the major ores of diamonds. They are of more than simply commercial interest, however. Indeed, the presence in these rocks of diamond, a mineral that forms only at very great pressure, is an indication that these magmas are probes of deeper parts of the mantle that can be directly sampled in no other way. In this section we summarize the extraordinary amount of knowledge that has been gained by igneous petrologists about these fascinating rocks.

Kimberlites are potassic ultramafic rocks that generally occur as small plugs or pipes. Lamproites are ultramafic members of a family of highly porphyritic ultrapotassic rocks that occur as shallow dikes or small extrusions. Early petrology texts called kimberlites "mica peridotites," and indeed they typically contain both phlogopite and magnesian olivine, along with pyroxenes, potassic amphiboles, garnets, carbonates, oxides, and, rarely, diamonds. Lamproites are distinct from kimberlites in that lamproites contain preserved glass and lack carbonate; there is otherwise a substantial mineralogic overlap. Lamprophyres constitute another group of potassic porphyritic dike rocks that are related to lamproites but are rarely ultramafic.

One major problem with kimberlites lies in deciphering just which minerals belong to the "kimberlite" magma and which minerals are part of the abundant xenolithic and xenocrystic population. Typical kimberlite pipes may not represent magma at all. The apparently rough-and-tumble process of movement of kimberlite magmas toward the surface has yielded a composition with far more xenolithic and xenocrystic material than other magmas—estimated to be as much as 75% or more of some kimberlite intrusion masses by volume. Larger xenoliths have become granulated or comminuted during emplacement, and their individual grains (megacrysts) can be mistaken for phenocrysts. It is a daunting task for kimberlite petrologists to try to deduce a true magmatic composition. Complicating the

task is the presence in the magma of abundant volatiles. In contrast to other mafic and ultramafic rocks that crystallized from essentially dry magmas, kimberlites and related rocks come from magmas that were apparently rich in both water and carbon dioxide. The rapid ascent velocity was probably the result of violent boiling of the magma during ascent as decompression caused gases to escape from the liquid. Final emplacement at shallow crustal levels was apparently explosive. Most kimberlite samples are richer in carbonate minerals than true kimberlite magma should produce, and it has been suggested that emplacement in pipes involved a relatively cool mush of residual kimberlite liquid mixed with gas and entrained mineral and rock fragments. Fortunately, some kimberlite pipes grade downward into thin dikes and sills of much more igneous-looking material that was apparently emplaced as a hot igneous melt and can be sampled by drilling. Sampling of this more pristine igneous material has provided the best information regarding kimberlite magma chemistry.

The most famous kimberlites are those of the African cratons in South Africa, Botswana, Lesotho, Angola, and West Africa (Figure 10-14) and of Siberia, but kimberlites have been found on virtually all the continents. Some well-known North American kimberlites occur in northern Colorado and southern Wyoming (near Laramie), in the southern Colorado Plateau in Utah, in upstate New York near Ithaca, and in the southern Appalachians; many of these are diamond-bearing. Worldwide, kimberlites appear to favor occurrences in old, stable, thick continental crust. No particular age for kimberlite intrusion is dominant; in Africa, kimberlite intrusive ages range from Precambrian to Cretaceous, and similar age ranges occur for Siberian kimberlites.

Diamonds are quite rare even in relatively diamond-rich kimberlite localities like those of southern Africa: It typically takes several tons of ore to produce a diamond (an average figure is 1 part diamond to 20 million parts kimberlite). Diamond has also been found in Africa and elsewhere in placer gravels, where diamonds have been concentrated in coarse sediments derived from the weathering of kimberlite. Because of its high density, diamond acts as a heavy mineral and thus is concentrated in fluvial and marine environments. It was long debated whether diamond was a xenocryst mineral accidentally brought to the surface by kimberlite or was in fact a liquidus mineral in kimberlite magma. The consensus is that diamond is xenocrystic and reaches the surface in kimberlites because of the extremely rapid ascent and cooling. Diamonds require very high pressure to form (in excess of 40 kbar—equivalent to a depth greater than 120 km—at about 1200°C). At high temperature and lower pressure, they either recrystallize to graphite, the stable low-pressure form of carbon, or oxidize to form carbon dioxide.

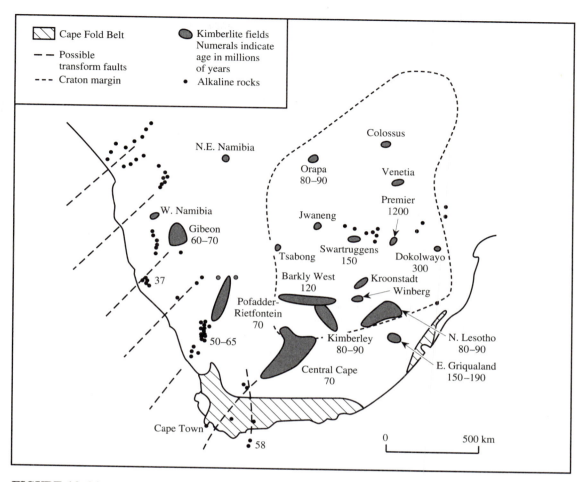

FIGURE 10-14

• • • • • • • • • • • • • • • •

Map of southern Africa showing distribution of the kimberlite fields in the kimberlite province. The outline of the South Africa craton is shown, as well as the Cape Fold Belt and possible transform faults with associated deep crustal fracturing. All unpatterned rocks outside the craton are accreted belts. [From R. H. Mitchell, 1986, *Kimberlites* (New York: Plenum), Fig. 5.7.]

The presence of diamond xenocrysts in kimberlites is persuasive evidence that the magmas must have originated at least as deep in the mantle as the depths required to stabilize diamonds, so that diamonds could be sampled by the magmas as they passed upward (Figure 10-15). Quantitative pressure estimates have been made by using coexisting minerals in garnet-bearing peridotite xenoliths, and these confirm that kimberlite magmas must have come from a depth of at least 150 km. But these are *minimum* depth constraints. Is there any way to constrain actual depth of origin more precisely?

Experimental petrology provides some answers. Melting experiments show that a melt of presumed kimberlite magma composition is in simultaneous equilibrium with the mantle minerals olivine, clinopyroxene, orthopyroxene, and garnet at about 60 kbar

(roughly equivalent to a depth of 180 km), and this has been widely accepted by petrologists as the depth of kimberlite genesis. But intriguing research by A. E. Ringwood and colleagues in 1993 has suggested that melting depths may be much greater. Diamond xenocrysts have been documented to have tiny inclusions of an unusual mineral called *majorite*. This mineral is an ultrahigh-pressure variant of garnet and has some silicon in octahedral coordination (as seen in the silica mineral stishovite). Melting experiments suggest that kimberlite magma can coexist simultaneously with majorite, olivine, and pyroxenes only at about 160 kbar (or roughly a depth of 480 km!). On this basis, Ringwood and coworkers have constructed a model for the origin of kimberlite magmas and for mantle plumes that involves melting at the 650 km discontinuity in the upper mantle. This work is provocative and highly

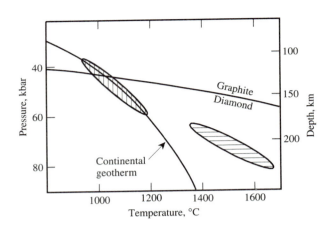

FIGURE 10-15
• • • • • • • • • • • • • •

Temperature-depth diagram showing the stability of diamond versus graphite and the estimated continental geotherm for southern Africa. Pressure-temperature estimates made from peridotite xenoliths in kimberlites are shown in the two patterned areas. Pressure-temperature estimates from diamond-bearing xenoliths are shown in the horizontally ruled field; estimates from xenoliths without diamond lie within the vertically ruled field. [From A. R. Finnerty and F. R. Boyd, 1984, *Geochim. Cosmochim. Acta*, Fig. 4.]

controversial, but it indicates how the techniques of igneous petrology are forging new (and deep) frontiers for this old discipline.

SUMMARY
• • • • • • • • • • • • •

Continental lithosphere away from plate margins is a magmatically active part of the earth. The continental terranes are not dominated by magmatic rocks as are the oceanic plates, but continental magmatism is nonetheless important for much of Earth's history since early in the Precambrian. These continental magmatic occurrences can be regarded as anorogenic because they do not actively involve plate interactions, but they all seem to require extensional tectonics—either incipient or advanced rifting—for magmas to ascend through the thick continental crust.

Although continental volcanic rocks span an enormous compositional range from silica-poor carbonatites and nephelinites to silica-rich rhyolites, this suite is dominated volumetrically by tholeiitic continental flood basalts. These great outpourings of lava have occurred throughout Earth's history, although each represents only a very short interval of active volcanism. The reasons for the sudden generation of such great volumes of basaltic magma in subcontinental mantle remain poorly understood. Layered mafic intrusions are probably plutonic expressions of the same magma generation events. Related to flood basalts are the continental rift igneous provinces, of which the East African Rift is one of the best examples. Magmatism here is both volcanic and plutonic, and ranges from early alkalic mafic rocks to late, highly fractionated dacites and rhyolites. There

are clearly some failed continental rifts, which were briefly tectonically active and then died, but the East African Rift appears to involve genuine continental breakup with eventual creation of new oceanic lithosphere. Upwelling of mantle under extensionally thinned continental crust has produced the thermal perturbation necessary to provoke widespread melting in the upper mantle.

Continental plutonic rocks are an equally diverse group, probably because these rocks represent the plumbing systems for the surface volcanics. Compositionally unusual rocks like anorthosites, carbonatites, and kimberlites may provide special clues to processes of magma generation and evolution, especially in the early earth, but the petrogenesis of these rocks remains a highly controversial subject among igneous petrologists and geochemists.

STUDY EXERCISES
• • • • • • • • • • • • •

1. Are continental flood basalts a recent phenomenon in Earth history, or do they extend back into Precambrian time? What reasons can you propose for the greater abundance of relatively young flood basalts? Contrast the petrology and geology of flood basalts with those of komatiites.

2. What is the likely genetic relationship between flood basalt provinces and the great mafic layered intrusions of the earth?

3. What types of magmatic rocks are most common in continental rifts? Are these relatively primitive, mantle-derived magmas or more chemically evolved ones? Explain.

4. Why are A-type granites properly characterized as anorogenic? Are they related to continental rifting or some other tectonic process?

5. Contrast the different types of anorthosites, focusing on chemistry, mineralogy, and field occurrence. Are there any young anorthosites? Explain.

REFERENCES AND ADDITIONAL READINGS
• • • • • • • • • • • • • •

Arndt, N. T., and R. W. Nisbet. 1982. What is a komatiite? In *Komatiites*, ed. N. T. Arndt and E. G. Nisbet. London: Allen & Unwin, pp. 19–28.

Ashwal, L. D. 1993. *Anorthosites*. New York: Springer-Verlag.

Baker, B. H., P. A. Mohr, and L. A. J. Williams. 1972. *Geology of the Eastern Rift System of Africa. Geol. Soc. Am. Spec. Paper 136.*

Basaltic Volcanism Study Project. 1981. *Basaltic Volcanism on the Terrestrial Planets.* New York: Pergamon.

Bergman, S. C. 1987. Lamproites and other potassium-rich igneous rocks: a review of their occurrence, mineralogy and geochemistry. In *Alkaline Igneous Rocks*, ed. J. G. Fitton and B. J. G. Upton. *Geol. Soc. London Spec. Pub. 30*, pp. 103–189.

Dawson, J. B. 1980. *Kimberlites and Their Xenoliths.* Berlin: Springer-Verlag.

Hess, H. H. 1960. *Stillwater Igneous Complex, Montana: A Quantitative Mineralogical Study. Geol. Soc. Am. Mem. 80.*

Luth, W. C., R. H. Jahns, and O. F. Tuttle. 1964. The granite system at pressures of 4 to 10 kilobars. *J. Geophys. Res., 69*, 759–773.

McBirney, A. R. 1975. Differentiation of the Skaergaard Intrusion. *Nature, 253*, 691–694.

McBirney, A. R. 1993. *Igneous Petrology*, 2nd ed. Boston: Jones and Bartlett.

Meyer, H. O. A. 1979. Kimberlites and the mantle. *Rev. Geophys. Space Phys., 17*, 776–788.

Mitchell, R. H. 1986. *Kimberlites: Mineralogy, Geochemistry and Petrology.* New York: Plenum.

Mitchell, R. H., and S. C. Bergman. 1991. *Petrology of Lamproites.* New York: Plenum.

Morse, S. A. 1980. A partisan review of Proterozoic anorthosites. *Am. Mineral., 67*, 1087–1100.

Ringwood, A. E., S. E. Kesson, W. Hibberson, and N. Ware. 1993. Origin of kimberlites and related magmas. *Earth Planet. Sci. Lett., 113*, 521–538.

Tuttle, O. F., and N. L. Bowen. 1958. *Origin of Granite in Light of Experimental Studies in the System $NaAlSi_3O_8–KAlSi_3O_8–SiO_2–H_2O$. Geol. Soc. Am. Mem. 74.*

Tuttle, O. F., and J. Gittins, eds. 1966. *Carbonatites.* New York: Wiley.

Viljoen, M. J., and R. P. Viljoen. 1969. Evidence for the existence of a mobile extrusive peridotitic magma from the Komati Formation of the Onverwacht Group. *Geol. Soc. South Africa Spec. Pub. 2*, 87–112.

Wager, L. R., and G. M. Brown. 1968. *Layered Igneous Rocks.* San Francisco: W. H. Freeman.

Williams, L. A. J. 1982. Physical aspects of magmatism in continental rifts. In *Continental and Oceanic Rifts*, ed. G. Palmason. Washington, DC: American Geophysical Union, pp. 193–222.

Wilson, M. 1989. *Igneous Petrogenesis: A Global Tectonic Approach.* London: Unwin-Hyman.

Wyllie, P. J. 1979. Kimberlite magmas from the system peridotite–CO_2–H_2O. In *Kimberlites, Diatremes and Diamonds: Their Geology, Petrology and Geochemistry*, Vol. 1, ed. F. R. Boyd and H. O. A. Meyer. Washington, DC: American Geophysical Union, pp. 319–329.

Yoder, H. S., Jr., ed. 1979. *The Evolution of the Igneous Rocks: Fiftieth Anniversary Perspectives.* Princeton, NJ: Princeton University Press.

·······································

PART II

Sedimentary Rocks

·······································

· ·

Chapter 11

THE OCCURRENCE
OF SEDIMENTARY ROCKS

Pre-Holocene sediments and sedimentary rocks cover 66% of the continental surfaces and probably most of the ocean floor. The main reason for this wide areal extent is the chemical instability of igneous and metamorphic rocks under atmospheric conditions. Rocks and minerals are in equilibrium only under the set of physical and chemical conditions in which they form. Under different conditions, they tend to move toward a new equilibrium state. Igneous and metamorphic rocks form at temperatures and pressures much higher than those at the earth's surface and in environments containing less water, less oxygen, less carbon dioxide, and very little organic matter. It is to be expected that such rocks will be unstable and will undergo chemical and physical changes when brought to the surface by tectonic, erosional, or isostatic forces. These changes in rocks constitute the process we call weathering. Based on the Le Châtelier principle, we would expect the products of this chemical change to contain more water, more oxygen, more carbon dioxide, and more organic matter than the unweathered rocks.

Sedimentary rocks consist almost entirely (>95%) of three types: **sandstones, mudrocks,** and **carbonate rocks. Sand** is defined as fragmental sediment 2.0 to 0.06 mm in diameter; **mud,** smaller than 0.06 mm. *Carbonate rocks* are composed largely of $CaCO_3$ (calcite or aragonite) or $CaMg(CO_3)_2$ (dolomite), with other carbonates (such as siderite and magnesite) being rare. As expected, there are transitional rocks that do not fit neatly in the three pigeonholes. For example, **coquina** is a fragmental rock composed of sand-size (or fine gravel-size) fragments of fossil shells. It is both a sandstone (or a conglomerate) or a limestone, but it is usually included in the limestones. How should we classify a rock composed of subequal amounts of **clay** minerals and microcrystalline carbonate material **(marl)**? There are no perfect answers to such questions, only agreed-on compromises—and sometimes, not even those. Rocks termed *detrital* are those composed of sediment that has undergone significant transport. The term *clastic* is a synonym of fragmental, and does not necessarily imply that the sediment has been transported.

The most abundant sedimentary rocks are the mudrocks, which form 65% of all sedimentary rocks. A moment's reflection about the mineralogy of igneous and metamorphic rocks suggests why mudrocks dominate the average stratigraphic section. Igneous and metamorphic rocks are composed of approximately 20% quartz and 80% other silicate minerals; only the quartz is chemically stable under most surface conditions. The other minerals are unstable when exposed at the surface and are altered to a variety of substances, but mostly to clay minerals. Clay minerals are mud size; hence, mudrocks are the dominant sedimentary rock. The quartz in crystalline rocks is very resistant to chemical attack and occurs chemically unchanged in both mudrocks and sandstones. Many thousands of analyses by X-ray, polarizing microscope, and chemical techniques have shown mudrocks and sandstones to have the average detrital mineral compositions listed in Table 11-1. A weighted average shows that the detrital sediment in the sedimentary column consists of 45% clay minerals, 40% quartz, 6% feldspar, 5% undisaggregated rock fragments, and 4% others. About 85% is either clay minerals or quartz, the most stable minerals under most surface conditions. Clearly, weathering has been a very effective process through geologic time.

TABLE 11-1 Average detrital mineral composition of mudrocks and sandstones

Mineral composition	Mudrocks, %	Sandstones, %
Clay minerals	60	5
Quartz	30	65
Feldspar	4	10–15
Rock fragments	<5	15
Carbonate	3	<1
Organic matter, hematite, and other minerals	<3	<1

The average thickness of sedimentary rocks on the continents is about 1800 m but is quite variable, ranging from 0 over extensive areas such as the Canadian Shield to more than 20,000 m in some basinal areas such as the Louisiana-Texas Gulf coastal region. The maximum possible thickness is determined by the amount and rate of subsidence, by the **geothermal gradient** in the area, by fluid chemistry, and by the chemical reactivity of the detrital particles. The average temperature at a given depth can vary greatly among geographic-tectonic areas. For example, at a depth of 10,000 m under Pittsburgh (cratonic, dormant), it is about 150°C; under New Orleans (cratonic, salt flowage), 200°C; under Las Vegas (cratonic, active), 260°C; and under Los Angeles (convergent, strike-slip margin), greater than 300°C. The mineral content of the sandstones in the sedimentary section can vary, from those composed entirely of quartz grains, which are relatively resistant to destruction or recrystallization, to sandstones composed mostly of calcic plagioclase grains and basaltic rock fragments, which readily alter chemically at very low temperatures.

DESTRUCTION OF THE ROCK RECORD

Examination of geologic maps of the continental areas of the earth reveals that 66% of the areas are underlain by sedimentary rocks. The maps also reveal that rocks of more recent geologic periods are more abundant in outcrop than those of older periods and that the decrease in abundance with increasing age follows the same logarithmic law that describes the radioactive decay of elements (Figure 11-1). The data indicate that half of all outcropping sedimentary rocks are younger than 130 Ma, that is, Cretaceous or younger. As we examine older sedimentary rocks, we have less rock to examine, so our interpretations must be more general-

FIGURE 11-1

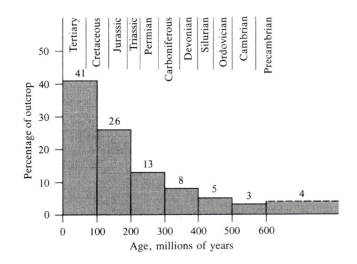

Relationship between sedimentary rock age and amount of area. [From H. Blatt and R. L. Jones, 1975, *Geol. Soc. Am. Bull.*, *86*, 1087.]

ized. The world of Cambrian time will forever remain more speculative than the world of more recent times.

TYPES OF SEDIMENTARY ROCKS
• • • • • • • • • • • • • • •

The major types of sedimentary rocks are mudrocks (65%), sandstones (20–25%), and carbonate rocks (10–15%), with all other rocks totaling less than 5%. However, these world averages vary widely on both a regional and a local scale, with some sedimentary basins filled largely with detrital sediments in various proportions and others dominated by nonclastics. For example, the thick Cenozoic section of the northern part of the Gulf of Mexico basin contains about 90% mudrocks and 10% sandstones, with only minor amounts of nonclastic rocks. The Michigan Basin, in contrast, contains only 18% mudrocks and 23% sandstones, but it has 47% carbonate rocks and 12% evaporite beds. Within a single stratigraphic section, still greater extremes occur. In later chapters we discuss each of the abundant types of sedimentary rocks (and some of the less abundant ones) in detail, but it is useful to introduce their general characteristics here.

Mudrocks

Mud particles are defined as grains less than 62 μm (10^{-6} m) in size (silt size is 62-4 μm; clay size is <4 μm); most are less than 5 μm. Because of this, the particles or aggregates of them (**flocculs**) are easily kept in suspension by even the weakest of currents and can settle and accumulate only in relatively still waters. Many such quiet-water environments exist, in both nonmarine and marine settings—for example, floodplains, deltas, and lakes on the continents and lagoons and areas below wave base in the marine environment. The thickest accumulations of mudrocks occur adjacent to ancient continental margins, with mudrock thicknesses ranging up to at least 2000 m in the central Appalachians of Pennsylvania and in the Ouachita belt in Arkansas.

Sandstones

Sand accumulates in areas characterized by relatively high kinetic energies, that is, environments of rapidly moving fluids. Examples of these environments include desert dunes, beaches, marine sandbars, river channels, and alluvial and submarine fans. However, some sands deposited on the shallow seafloor are subsequently carried down to great depth in the sediment-water mixtures called **turbidity currents.** As a result, coarse-grained sediments can occur in quiet water.

Typical sites of sand accumulation are elongate, such as beaches and rivers, but in the geologic record the sands deposited in these environments are commonly sheetlike in character. This difference results from the displacement of the depositional site through time; for example, a beach migrates inland during a marine transgression, the migration resulting in a slight increase in the thickness of the sand body but an extreme increase in its width. It is possible, however, for a sand-dominated beach-dune complex to exist at the same geographic locality for a long period of time. This stationary location could occur in the tectonic setting of a slowly subsiding basin, resulting in pure sand deposits that are hundreds of meters thick but relatively narrow in areal extent. The Cambro-Ordovician quartz sands of the western United States may be an example of this.

As with mudrocks, the environment of deposition can commonly be related to mineral composition. Sands deposited in loci of highest kinetic energy, such as beaches and desert dunes, tend to be more quartz-rich than the sandbars of sluggish rivers. This enrichment in quartz occurs because of the relative ease of breakage and elimination of cleavable minerals, such as feldspar, or of foliated fragments, such as shale or schist. However, it is not a good idea to base an environmental interpretation on detrital mineral composition. Some fluvial sands contain more than 90% quartz, and many modern beaches contain high percentages of feldspar and rock particles (see Chapter 13).

Carbonate Rocks

Modern carbonate sediments are rich in the hard parts of marine organisms, and there is every reason to believe this has been true of carbonates throughout Phanerozoic time. Because of the great chemical reactivity of calcium carbonate, however, most carbonate particles recrystallize sometime after deposition; thus, their organic (biochemical) origin is not always evident. This is particularly true of the microcrystalline particles that form the bulk of ancient limestones. Some of these particles may be inorganic chemical precipitates rather than biochemical fragments.

Because a large proportion of carbonate particles is organic in origin, the abundance of limestones is tied to the occurrence of **phytoplankton,** which are at the base of the food chain; the phytoplankton, in turn, are tied to the depth of penetration of light into seawater. If there is no light, there can be no photosynthesis and no phytoplankton. The depth of penetration of light in seawater is shown in Table 11-2, and it is apparent that most light is absorbed at very shallow depths. Because

TABLE 11-2 Percentage of incident light penetrating to specific depths in seawaters

Depth, m	Clearest ocean water, %	Turbid coastal water, %
0	100	100
1	45	18
2	39	8
10	22	0
50	5	0
100	0.5	0

Source: G. L. Picard and W. J. Emery, 1990, *Descriptive Physical Oceanography*, 5th ed. (New York: Pergamon), p. 30.

of this, most organisms live within 10 m of the sea surface **(photic zone).** However, the carbonate material generated in these shallow waters need not be deposited there. **Planktonic,** carbonate-shelled organisms such as *Globigerina*, are abundant in the open ocean and can settle to the deeper ocean floor. Few of these deep-sea carbonates appear in the stratigraphic record because the shells dissolve in the cold waters of the deep ocean. As a result, most carbonate rocks we see in the stratigraphic record are of shallow marine origin and originated in relatively warm waters.

Sand-size fragments of carbonate-shelled organisms are not difficult to recognize and identify in thin sections of limestones, at least to the level of phylum and class (see Chapter 16). Sometimes even the genus and species can be specified, and rather detailed reconstructions of depositional environments can be made on the basis of these and other data. For example, analyses of the relative amounts of the various isotopes of oxygen that are present in the calcite of unrecrystallized shell material can be used to determine the temperature of the water in which the organism lived. Because organisms are very sensitive to their environment, limestones can be gold mines of information about the shallow marine waters of the geologic past.

DEPOSITIONAL BASINS AND PLATE TECTONICS

From analysis of stratigraphic data accumulated during the past 100 years, it is clear that preserved sedimentary rocks are dominantly marine. There are several reasons for the dominance of marine sediments:

1. The light sialic material that forms a large proportion of continental masses is limited in volume, with the result that continental areas constitute only about 30% of the earth's surface. The marine areas are sediment traps that cover 70% of the earth's surface.

2. Because of the movement and subduction of oceanic lithosphere at the edges of some continental blocks, topographically depressed areas **(trenches)** with adjacent easily eroded highland areas exist at active continental margins.

3. There has been a pronounced tendency through time for broad areas of the continental blocks to be invaded by shallow marine waters **(epicontinental seas).** The resulting shallow-water marine sediments are laterally extensive on the craton, although typically they are thin. Figure 11-2 shows the number and extent of the major global transgressions and regressions of these seas, as determined from seismic records; but many more minor transgressive-regressive sequences are known to be present. The tectonic explanation for at least the major cycles is the changing volume of the oceanic rift system. An increase in volume of the ridges results in a decrease in volume of the ocean basins and subsequent flooding of the low-lying parts of continents. It is clear that there have been two first-order transgressions, the first extending from Early Cambrian to Late Cambrian time and the second from Early Cretaceous to Late Cretaceous time. Extensive continental flooding also was prominent from Cambrian through Mississippian time. The causes of higher order cycles might be either the waxing and waning of ice sheets or regional tectonic movements. Phanerozoic continental glaciations are known to have occurred in every geologic period except Cambrian, Triassic, Jurassic, and Cretaceous.

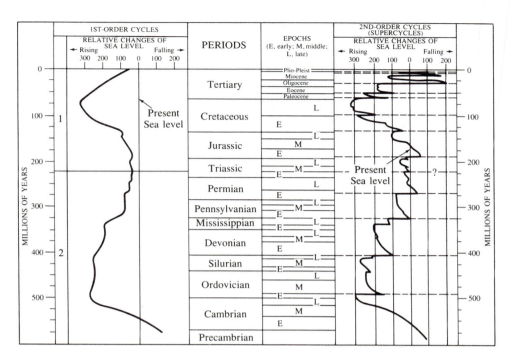

FIGURE 11-2
• • • • • • • • • • • • • • • • • •

First- and second-order global cycles of relative change of sea level during
Phanerozoic time. [From P. R. Vail et al., in C. E. Payton, 1977, *Am. Assoc.
Petrol. Geol. Mem. 26*, p. 84; metric calibration after W. C. Pitman, Jr., 1978,
Geol. Soc. Am. Bull., *89*, 1389–1403.]

4. Continental deposits are, by definition, formed above
sea level and hence are subject to removal should the
rate of accumulation fall behind the erosion rate. It is
no accident that stratigraphic sections on the continental blocks contain many more large unconformities than deeper marine or oceanic sections.

It is clear from studies of sediment distribution in the
oceans that the thickest accumulations of Cenozoic sediment occur adjacent to the continental margins; and
paleogeographic reconstructions based on stratigraphy,
paleontology, and facies analysis indicate that the same
pattern has existed throughout geologic history. Figure
11-3 shows the extent of accretion onto the Precambrian nucleus of the United States during Phanerozoic
time. Also shown is the position of the United States
with respect to the equator during each geologic period;
it is apparent that the area was located within 30° of the
equator from Cambrian through Triassic time. This position implies a continual tropical to subtropical climate,
and climate is an important control of the mineral composition of sediments (see Chapter 12).

A universally accepted classification scheme for sedimentary basins has yet to appear, but those proposed
so far are based on three factors: whether the basin is
underlain by continental or oceanic crust; whether the
basin was formed along a divergent, convergent, or
strike-slip plate margin; and the position of the basin on
the plate. The geologic history of each basin may then
be subdivided into cycles on the basis of three parameters: basin-forming tectonics, depositional sequences,
and basin-modifying tectonics. All types of complexities
can be accommodated within this framework.

In the present discussion we group the major types of
sedimentary basins into six categories, each of which
has distinctive structural and sedimentary petrologic
characteristics.

1. *Oceanic basins* are areas of deposition underlain
by oceanic lithosphere, for example, the bulk of the
Atlantic and Pacific Ocean basins. Mesozoic examples of oceanic deposits now accreted to the continental blocks are well known in the circum-Pacific
region. Paleozoic examples occur in eastern Canada.

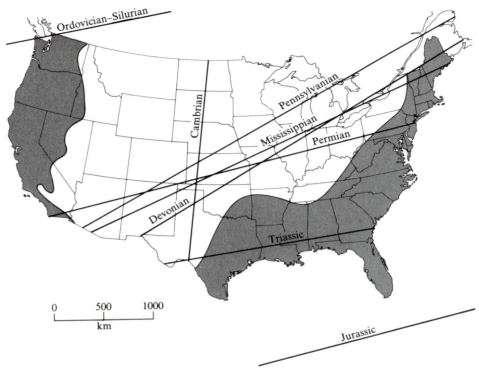

FIGURE 11-3

• • • • • • • • • • • • • • • •

Present outline of conterminous United States, showing accretion (shaded areas) to Precambrian craton during Phanerozoic time. Also shown is the current best estimate of the location of conterminous United States in relation to the equator (labeled lines) during each Phanerozoic period. During Cretaceous and Tertiary times, the United States was located more than 2000 km north of the equator because of a consistent drift northward of the North American plate (1100 km = about 10° of latitude).

2. **Arc-trench system** *basins*, the most complex zone of basin development, form in convergent-margin areas and pose very difficult problems of interpretation because of the intense tectonism that accompanies subduction. Examples have been described from around the Pacific rim.

3. *Continental-collision basins* develop along zones where continental plates collide and are different from the collision zone of an oceanic plate and a continental plate. Examples of such basins have been described from the Ouachita (Carboniferous) and Himalaya belts.

4. *Basins in displaced terranes* occur along continental margins. Geologists sometimes visualize the oceanic plate that is subducted during convergence as a relatively flat surface composed of biogenic oozes and abyssal clay. Commonly, however, the plate also contains old **island arcs** and other oceanic terranes; and as the plate is subducted, these arcs and their included basins are plastered onto the side of the continental plate. These oceanic plate fragments are termed exotic terranes and occur along the western margin of the United States from Alaska southward to southern California (Figure 11-4). They are recognized in many other regions as well. Such terranes are structurally, stratigraphically, and paleontologically discordant with the adjacent continental block and may contain oceanic sediment, forearc basins, intraarc basins, or backarc basins.

5. *Grabens along continental margins* are depocenters formed by rifting along the margins of stable continental crust in association with the formation of an oceanic basin. Most such grabens are oriented parallel to the sides of the oceanic basin, but others can be oriented at a large angle to the coastline, in a reentrant (an **aulacogen**).

6. **Intracratonic basins** are depocenters formed independently of tectonic activity at plate margins and at unpredictable locations within the continents. Their locations are thought to reflect buried ancient

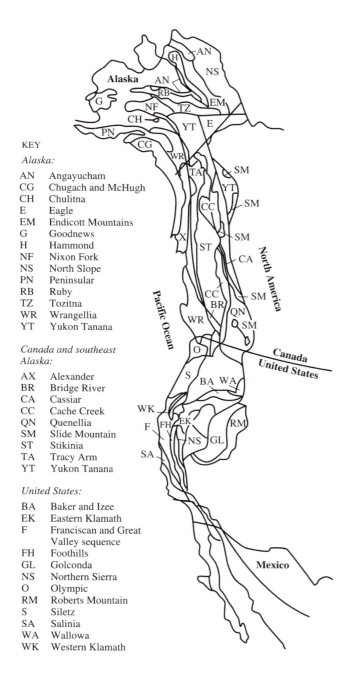

KEY

Alaska:

AN   Angayucham
CG   Chugach and McHugh
CH   Chulitna
E   Eagle
EM   Endicott Mountains
G   Goodnews
H   Hammond
NF   Nixon Fork
NS   North Slope
PN   Peninsular
RB   Ruby
TZ   Tozitna
WR   Wrangellia
YT   Yukon Tanana

Canada and southeast Alaska:

AX   Alexander
BR   Bridge River
CA   Cassiar
CC   Cache Creek
QN   Quenellia
SM   Slide Mountain
ST   Stikinia
TA   Tracy Arm
YT   Yukon Tanana

United States:

BA   Baker and Izee
EK   Eastern Klamath
F   Franciscan and Great
      Valley sequence
FH   Foothills
GL   Golconda
NS   Northern Sierra
O   Olympic
RM   Roberts Mountain
S   Siletz
SA   Salinia
WA   Wallowa
WK   Western Klamath

FIGURE 11-4

• • • • • • • • • • • • • • • • •

Permian through Cretaceous accretion of terranes to the western edge of the North American craton. Rocks in each area differ sharply in geology, paleontology, and paleomagnetic properties from one another and from rocks in the adjacent ancient craton. Many of the terranes are composed of rocks that originally formed on the ocean floor. Some of them originated thousands of kilometers to the south of their present position; this conclusion is based on paleomagnetic evidence. Different terranes are indicated by different letters. [From Jones, in Dewey et al., 1991, *Allochthonous Terranes* (New York: Cambridge University Press), p. 25.]

Oceanic Basins

The principal settings of oceanic facies controlled by tectonic relations are shown in Figure 11-5. They are (1) bathymetric highs of ridge crests at divergent plate junctions, where igneous rocks are formed along the trends of the spreading centers; (2) the zone where the oceanic substratum gradually subsides as it cools in moving away from spreading centers; and (3) deep basins beneath which the thermal contraction of the lithosphere is essentially complete.

The pelagic sediment that covers the basaltic ocean crust has a stratigraphy and a facies relationship that reflect changing water depths. Near the upper part of the ridge, the water depth is commonly less than about 4000 m; here **pelagic** shells of calcium carbonate can accumulate. Below this depth—called the **carbonate-compensation depth (CCD)**—the degree of undersaturation of the water with respect to calcite or aragonite is so great that shell accumulation is not possible (see Chapter 16). Lower on the rise flanks, in the deep basins and in cold waters, brown clay or siliceous shells accumulate. If the depositional site is sufficiently near a landmass, then continental sediment is carried into the deep oceanic basin by turbidity currents (see Chapter 13) and is interbedded with the calcareous and siliceous oceanic deposits.

Arc-Trench System Basins

Arc-trench systems occur along convergent plate margins and include a variety of **morphotectonic** elements (Figure 11-6), each of which can accumulate a distinctive sedimentary assemblage:

The *trench*, a bathymetric deep, floored by oceanic crust

The *subduction zone*

The *arc-trench gap*, a belt within which a *forearc basin* may occur between the trench and the magmatic arc

rift zones. Sediment thicknesses in the basins are typically 1000 to 5000 m greater than those in the geographic surroundings; examples include the Illinois, Michigan, and Williston Basins. The sedimentary fill in these basins is dominated by carbonates, shales, evaporites, and some quartz-rich sandstones. Rocks such as conglomerates, lithic or feldspathic sandstones, turbidites, and bedded cherts are uncommon to rare because the tectonic setting surrounding intracratonic basins is nonorogenic and the depositional environments are invariably located in shallow waters.

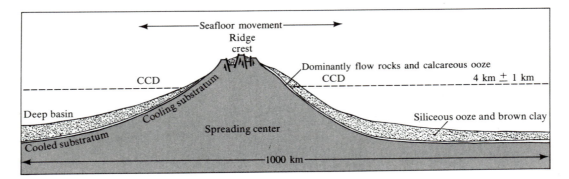

FIGURE 11-5

• • • • • • • • • • • • • • • •

Idealized cross section of an oceanic spreading center, showing accumulations of ponded and peripheral pelagic sediment and volcanic rocks in extensional fault basins at ridge crest and on flanks above carbonate-compensation depth (CCD) and siliceous ooze and brown clay overlying mafic rocks to the sides of the spreading center. Turbidites may be intercalated with siliceous ooze and brown clay near continental edges. Total sediment thickness far from continental margins can be a few hundred meters.

The *magmatic arc*, within which *intraarc basins* may occur

The *backarc area*, within which may lie either an **interarc basin** floored by oceanic crust and separated from the rear of the arc by a normal fault system or a **retroarc** (or **backarc**) basin floored by continental basement and separated from the rear of the arc by a thrust fault system. (A retroarc basin is retro with respect to the arc. The basin can also be called a **foreland basin** because of its position with respect to the continent.)

In each of these morphotectonic settings, sedimentation, volcanism, and plutonic intrusions occur contemporaneously, although not necessarily at the same site.

TRENCH SEDIMENT. On the trench floor, variable thicknesses of turbidites are deposited above the oceanic sediment layers. Transport by turbidity currents within a trench is mainly longitudinal along the trench axis, although the initial entry of sediment into the trench can occur along the inner wall as well as from the ends of the trench. The volume of sediment within the trench at any given time reflects the balance between the rate of supply and the rate of plate consumption into the subduction zone. The trench is one of the possible sites of formation of **mélanges,** which are chaotic mixtures of very large fragments of older sedimentary and crystalline rocks in a pelitic matrix. Mélanges are known from every continent except Antarctica.

FOREARC BASINS. Immediately landward of the top of the trench lie forearc basins, which overlie older, deformed orogenic belts or perhaps oceanic or transitional crustal material. Forearc basins receive sediment mainly

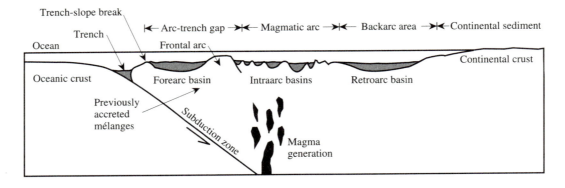

FIGURE 11-6

• • • • • • • • • • • • • • • • •

Generalized sketch of an arc-trench system along a convergent continental margin, showing the spatial relationships and nomenclature of plate tectonics and related sedimentary basins. Marine sediment accumulations in basins are stippled.

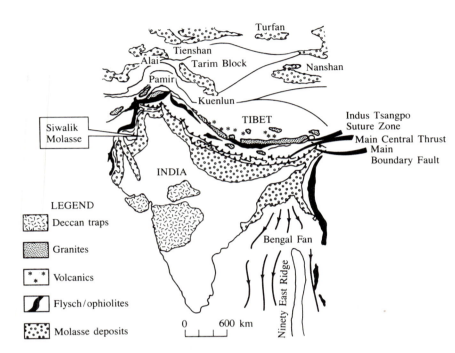

FIGURE 11-7

• • • • • • • • • • • • • • • •

The distribution of the late orogenic deposits in the Himalaya belt. [From R. Kumar and S. K. Tandon, 1985, *Sediment. Geol., 42*, 107.]

from the extensive nearby arc structures, where not only volcanic rocks but also plutonic and metamorphic rocks exposed by uplift and erosion serve as sources. Sources can also include local uplands along the trench-slope break or within the arc-trench gap itself. The sandstones are characteristically rich in volcanic rock fragments and calcic plagioclase grains. There may be little transfer of sediment into the subduction zone from the forearc basins; frequently, the forearc basins seem to completely override the subduction zone. The Great Valley sequence of California is an example of a forearc basin deposit.

By inference from the bathymetry of modern forearc basins and from the sedimentology of older sequences assumed to have been deposited in similar settings, a forearc basin can contain a variety of **facies.** Shelf and deltaic or terrestrial sediments, as well as turbidites, occur in different examples. The local bathymetry is controlled by the elevation of the trench-slope break, the rate of sediment delivery to the forearc basin, and the rate of basin subsidence. Different facies patterns occur in various basins.

INTRAARC BASINS.   The sedimentary strata in modern intraarc basins include turbidite aprons of volcanic debris shed backward from magmatic arcs. These turbidite wedges rest almost directly on the basaltic oceanic crust with few or no intervening pelagic deposits. Landward from the intraarc spreading centers, sedimentation varies markedly.

RETROARC BASINS.   The sedimentary record of retroarc basins includes fluvial, deltaic, and marine strata as much as 5 km thick deposited in terrestrial lowlands and epicontinental seas along elongate, cratonic belts between continental margin arcs and cratons. Sediment dispersal into and across retroarc basins is from highlands on the side toward the magmatic arc and from the craton toward the continental side. The Sea of Japan is a modern example of an extensional retroarc basin. Ancient examples of compressional (thrust-faulted) retroarc basins are the Upper Cretaceous basins of the Western Interior and Rocky Mountain region of North America.

Continental-Collision Basins

Continental collision occurs when two plates carrying continental crust, and possibly associated island arcs, converge either by subduction or by strike-slip motion that has a component of convergence. Examples include the Late Paleozoic convergence of the northeastern part of the South American plate with the southeastern part of the North American plate, and the convergence of the Indian plate with the Asian plate during Tertiary time. Because convergence is seldom orthogonal, remnant ocean basins occur within the collision belt, and enormous quantities of sediment are available to fill basins formed during convergence. The collision of India with Asia has resulted in the accumulation of as much as 17 km of detrital sediment in Tertiary subduction-related troughs in Burma. The collision is also responsible for the formation of major **molasse** basins parallel to the Himalaya belt (Figure 11-7) and their filling with a fluviodeltaic facies of Late Tertiary age, one well-studied part of which is termed the Siwalik Series.

FIGURE 11-8
• • • • • • • • • • • • • • • •

Locations of major rift grabens (shaded) produced in Early Mesozoic time along the eastern coast of the United States and Canada.

Basins in Displaced Terranes

Basins within tectonically displaced terranes are now recognized in many parts of the world. More than 100 highly diverse fragments of exotic terranes have been identified along the eastern margin of the North Pacific Ocean. Although some are similar in geologic aspect, most record a geologic history significantly different from that of their neighbors. In most cases the differences are so pronounced that it is inconceivable for the rocks of neighboring terranes to have formed in close proximity.

Grabens Along Continental Margins

Prominent along the eastern coast of the United States are grabens created during Triassic-Jurassic time as an early stage of failed rifting as the North American plate separated from a larger plate (Pangaea) that included South America, Africa, and Europe. The axis along which separation occurred between North America-South America and Africa-Europe was the site of domal uplift and intense tensional forces that produced a large

number of rift grabens. Those on the western side of the Atlantic Ocean are shown in Figure 11-8; an equivalent set occurs in North Africa. The grabens develop initially within the domal uplifts but later extend as an essentially continuous branching network along the full trend of the rift belt. In the rift valleys, feldspathic continental redbeds of alluvial fan facies are interbedded with volcanics that continued to erupt through the growing system of crustal fractures. Broad regions to either side of the eventual zone of rupture between the separating continents can be affected by the extensional faulting. For example, the Triassic basins of the Appalachian region, which are filled with nonmarine sedimentary and volcanic rocks, lie 250–500 km inland from the present continental slope; the slope can be taken as marking roughly the line of Jurassic continental separation.

As continued crustal separation induces subsidence along the zone of incipient continental rupture, the floors of the main rift valleys become partially or intermittently flooded to form protooceanic gulfs. Restricted conditions in these basins may promote the deposition of evaporites in suitable climates. For example, layers of evaporites 5–7.5 km thick are present in the subsurface beneath parts of the Red Sea. Extensive evaporites sev-

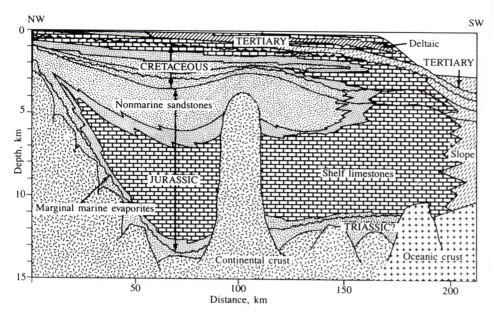

FIGURE 11-9

• • • • • • • • • • • • • • • •

Schematic cross section (northwest to southeast) through the Baltimore Canyon trough, which underlies the present coastal plain, continental shelf, and upper continental slope off the eastern United States. The trough is about 500 km long, parallel with the present shoreline, and 100–200 km wide, with up to 14 km of mainly Mesozoic sediments. Triassic continental rift valley sandstones are followed by partially synchronous marginal marine evaporites and then very thick transgressive Jurassic shelf limestones with carbonate reef buildups over highs. Shoreward there is passage into marginal marine sandstones. A major Early Cretaceous prograding wedge is followed by a Late Cretaceous transgression. The Tertiary configuration is one of delta progradation with minor transgression, passing seaward into slope deposits of uncertain age and thickness. [From H. G. Reading, 1982, *Proc. Geol. Assoc.*, *93*, 327.]

eral thousand meters thick are known also from coastal basins on both sides of the Atlantic Ocean, where they apparently represent dismembered portions of the same elongate trend of evaporite basins.

When filling of the grabens was completed, which along the eastern seaboard of the United States required about 60 million years, a sequence of marine and nonmarine shallow-water clastics 100–250 km in width was deposited on the continental terrace (shelf), as a wedge-shaped accumulation thickening offshore toward the continental margin. From the shelf margin, the continental slope leads down at perhaps a 2° slope to deep water, where turbidites accumulate along the edge of oceanic crust. Shelf, slope, and basinal turbidite sediments are Cretaceous and Cenozoic in age. At present, the Triassic grabens and their arkosic fill are mostly buried under this post-Jurassic cover (Figure 11-9), but they crop out particularly well in Massachusetts and Connecticut. Along most of the east coast, the grabens are easily recognized on seismic records.

Intracontinental Basins

The Michigan Basin, which includes an area of about 320,000 km^2 (Figure 11-10), is an example of a major intracontinental depocenter. The basin was initiated in Cambrian time, and before it was finally filled had accumulated more than 5000 m of sediment—about 47% carbonates, 23% sandstones, 18% shales, and 12% evaporites. As is characteristic of intracontinental basins, all sediments have a shallow-water origin, and many stratigraphically important unconformities are present; 12 have been recognized within the Cambrian to Pennsylvanian section. Because the Pennsylvanian rocks are overlain by Pleistocene glacial deposits, post-Carboniferous geologic history in the area is unknown.

Intracontinental depocenters such as the Michigan Basin can be identified in the geologic record by virtue of their paleogeographic location, abundance of unconformities, and preponderance of nearshore sedimentation. Particularly noteworthy is an abundance of carbonates and evaporites.

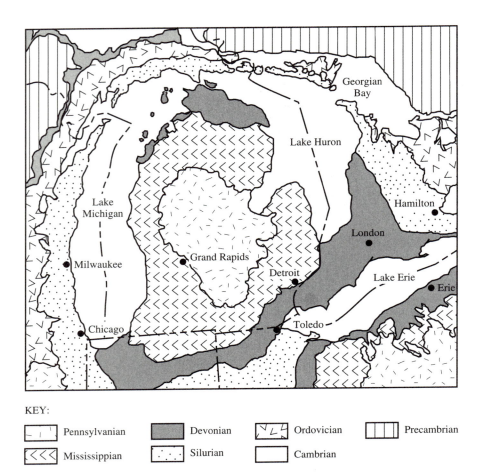

FIGURE 11-10

• • • • • • • • • • • • • • • •

Bedrock geology map of the Michigan Basin region.

KEY:

Pennsylvanian	Devonian	Ordovician	Precambrian
Mississippian	Silurian	Cambrian	

CLIMATE

• • • • • • • • • • • • • • •

The effects of climate on the amounts and types of sedimentary accumulations are not always easy to document except in extreme cases. For example, bauxite and ferruginous laterite are clear indications of moist tropical conditions, and tillite indicates a glacial regime. But regional patterns of temperature and precipitation have always been present on Earth and have affected sedimentary patterns. For example, the formation of extensive marine limestones requires warm temperatures throughout most or all of the year. What are the controls of regional climatic patterns and how have they varied through geologic time?

One important control of regional climatic variations results from changes in insolation with latitude. These changes occur because the sun's radiation strikes the earth's surface closer to the vertical at low latitudes than at higher ones. The lower angle of incidence at higher latitudes means that temperatures there will always be lower than at the equator, although it is not clear that the difference is always large enough to promote ice formation and glaciation. There are several modifying factors:

1. Probable variations in the amount of carbon dioxide in the atmosphere affect climate and sedimentation, with more carbon dioxide causing higher temperatures through a "greenhouse effect."
2. Changes in the relative proportions of land and sea. Water has a much lower albedo (reflectivity) and higher heat capacity than rocks and vegetation on the land surface. As a result, transgressions of the sea cause both regional warming and a moderation of extremes in temperature variation. The major factor controlling sea level probably is the formation of mid-ocean ridges, and **hypsometric** studies indicate

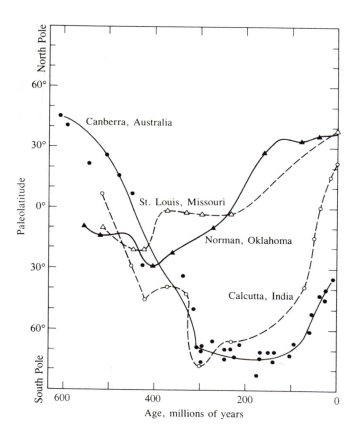

FIGURE 11-11

● ● ● ● ● ● ● ● ● ● ● ● ● ● ● ●

Measurements of paleomagnetic latitudes at various sites indicate the extent of continental drift. Phanerozoic paleomagnetic latitudes of Canberra, Australia; Calcutta, India; Norman and St. Louis, United States, are shown. Solid and open circles and triangles are data points for the four best-fit curves shown. [From W. B. Hamilton, 1979, *U.S. Geol. Surv. Prof. Paper No. 1078;* and data from C. Scotese et al., 1979, *J. Geol., 87,* and P. Tapponnier, 1977, *Scientific American, 236.*]

that at least 10% of the area of exposed continents is flooded for each 100 m of sea level rise. Orogenic movements on the continents are a secondary control of temperature distribution.

3. Continental fragmentation and drift through tens of degrees of latitude (Figure 11-11) can result in vast changes in sediment distribution patterns. Note that between Late Cambrian and Late Ordovician time, the site of Calcutta, India, moved from 8°N to 30°S latitude; Norman, Oklahoma, a more conservative area, hardly moved at all. But even an apparently small latitudinal difference can mean the difference between tropical rain forest and arid desert, as can be seen today in northern Africa. Continental drift can also alter the path of oceanic currents and consequently the temperatures on land areas adjacent to the currents. For example, both central England and southern Labrador lie at 53°N latitude, but only England is easily habitable, a result of the warming effect of the Gulf Stream.

Geologic data are always insufficient for paleoclimatic reconstruction. The problem may be the absence of strata because of erosion, **diagenetic** alteration, or dissolution of minerals and organic matter, or it may be that many climate parameters leave no discernable mark in the rocks. The most useful information comes from environmentally sensitive fossils, which can yield information about temperature, precipitation, and elevation or bathymetry. Textures and structures of sediments can be informative as well, examples being eolian cross-bedding in sandstones (Chapter 13) and bird's-eye structures in limestones (Chapter 16).

Perhaps the greatest insufficiency in reconstructing paleoclimates is the lack of adequate numerical dating. Although Neogene rocks can sometimes be dated to $\pm 10^4$ years, the precision of radiometric dating of pre-Miocene rocks is no better than 1 million years; and the precision becomes progressively worse, in terms of years, in more ancient rocks. The Pleistocene record bears witness to the magnitude of climate changes possible in only 100,000 years. And, of course, the geologic materials required for radiometric dating may be absent in an area of particular interest. Biostratigraphically dated paleomagnetic reversals can sometimes be helpful but their use for rocks older than about 10 million years is not yet possible because the lengths of the recognized polarity cycles are of the same magnitude as the dating errors by the potassium-argon method.

SUMMARY
• • • • • • • • • • • • • • • •

Sedimentary rocks are composed almost entirely of mudrocks (65%), sandstones (20–25%), and carbonate rocks (10–15%). Clay minerals and quartz grains form about 85% of the mineral grains in detrital rocks. The thickness of sedimentary rocks on the continents ranges from 0 over extensive areas such as the Canadian Shield and the Siberian Shield to more than 20,000 m in the deepest parts of some basinal areas; the lower limit is set by the local geothermal gradient and the susceptibility of the minerals to metamorphism.

Mudrocks are composed of 60% clay minerals, which, because of their small grain size, can accumulate only in areas of low kinetic energy. Sandstones dominate in areas of high kinetic energy such as beaches and desert dunes. The occurrence of carbonate rocks is controlled primarily by the depth of penetration of light into the sea; thus, most carbonate rocks accumulate within a few tens of meters of the sea surface.

The location and size of depositional basins are controlled by continental drift and plate tectonics. Five distinct areas of structurally controlled accumulation of sediments can be recognized: oceanic basins, arc-trench systems, continental-collision basins, rifted continental margins, and intracontinental basins. Numerous examples of each type of basin are known. The mineral composition of the sediments in each type is determined by its location with respect to a continental margin, the nature of the underlying crustal material, the types of plate boundaries nearest the basin, and climate.

STUDY EXERCISES
• • • • • • • • • • • • • •

1. The average thickness of sedimentary rocks on the continental surfaces is only 1800 m, an accumulation representing more than 4 billion years. List and discuss the reasons why so little sediment took so long to accumulate.
2. Why do you think that the decrease in abundance of sedimentary rocks with increasing age follows the same logarithmic law that describes the radioactive decay of chemical elements? Do you think all types of sedimentary rocks have the same "half-life"? Explain why.
3. Describe a way in which the geographic relationship among the various types of arc-trench system basins might be used to reconstruct the paleogeography in a tectonically disturbed area.
4. Suppose you wanted to determine the relative amounts of sandstone, mudrock, and carbonate rock in the earth's crust. How might you go about sampling sedimentary accumulations, and what tectonic or sedimentary factors would you have to consider?
5. Some scientists who study global climate change believe that the average temperature in the lower part of the earth's atmosphere is increasing at a rate greater than at any time in the geologic past. The cause is thought to be an increasing "greenhouse effect" resulting from industrial emissions of carbon dioxide during the past 100 years. What effects do you think such a change might have on the accumulation of the different types of sediments at the earth's surface?

REFERENCES AND ADDITIONAL READINGS
• • • • • • • • • • • • • • •

Algeo, T. J., and B. H. Wilkinson. 1991. Modern and ancient hypsometries. *J. Geol. Soc., 148,* 643–653.
Dallmeyer, R. D., ed. 1989. *Terranes in the Circum-Atlantic Paleozoic Orogens. Geol. Soc. Am. Spec. Paper 230.*
Dewey, J. F., I. G. Gass, G. B. Curry, N. B. W. Harris, and A. M. C. Sengör, eds. 1991. *Allochthonous Terranes.* New York: Cambridge University Press.
Firstbrook, P. L., B. M. Funnell, A. M. Hurley, and A. G. Smith. n.d. (probably 1980). *Paleooceanic Reconstructions, 160–0 Ma.* Washington, DC: National Science Foundation.
Hallam, A. 1992. *Phanerozoic Sea-Level Changes.* New York: Columbia University Press.
Kleinspehn, K. L., and C. Paola, eds. 1988. *New Perspectives in Basin Analysis.* New York: Springer-Verlag.
Sloss, L. L., ed. 1988. *Sedimentary Cover—North American Craton: United States. Geology of North America,* Vol. D-2. Boulder, CO: Geological Society of America.

Chapter 12

WEATHERING AND SOILS

Weathering and soil development are the oldest sedimentary processes on the continents, and the fact that only 10^2–10^3 years are needed for a soil profile to develop suggests that soils should be common lithologic units in outcrops of ancient continental rocks. Yet it is only within the past decade that sedimentologists and stratigraphers have searched intensively for **paleosols.** The reasons for this neglect are that many types of ancient soil zones are difficult to recognize in outcrop and that few students of sedimentary rocks have had training in soil science. But paleosols are indeed present in nonmarine stratigraphic sections. For example, more than 1000 paleosols were formed during a 3.5-million-year period represented by 770 m of the Willwood Formation in Wyoming. A large number of paleosols have been recognized in continental rocks as old as Archean. The importance of soil horizons to sedimentary petrologic interpretation is made evident by the observation that sand and mud particles spend most of their surface sedimentary history stationary in soils rather than in transport. Most ancient alluvial deposits contain numerous superposed ancient soil profiles, and geologists have identified thick alluvial sequences that consist largely of stacked paleosols. Are the minerals in rock samples from such sequences likely to be representative of upstream paleogeology?

REACTIONS AND PRODUCTS

The rate of alteration of a fresh mineral grain in the sedimentary environment is determined by four factors: chemical composition, structural integrity, crystallinity, and the chemical character of the environment.

CHEMICAL COMPOSITION.   Bonding forces between cations and anions in silicate minerals are generally either ionic or covalent. Bonds that are dominantly ionic in character are formed between elements in widely separated columns in the periodic table of the elements, for example, between K^+ (column 1) and Cl^{1-} (column 7) or between Ca^{2+} (column 2) and O^{2-} (column 6). Bonds of mostly covalent character occur between elements in nearly adjacent columns in the periodic table, for example, between Si^{4+} (column 4) and O^{2-} (column 6) or between Al^{3+} (column 3) and O^{2-}. For considerations of weathering, the most significant difference between ionic and covalent bonds is their susceptibility to breakage by water dipoles. Ionic bonds are easily disrupted (example, NaCl, table salt) but covalent bonds are not (example, quartz). Silicate and aluminosilicate structures are relatively insoluble in water because both the Al–O and Si–O bonds are dominantly covalent. A relatively large proportion of the sodium or potassium ions must be leached from albite or orthoclase before the affected part of the structure will decompose into $H_4SiO_4^0$ and either $Al(OH)_3^0$ or $Al(OH)_4^-$ and be carried away in groundwater or rainwater.

STRUCTURAL INTEGRITY.   Because chemical attack occurs along surfaces, crystals containing fractures, cleavages, phase boundaries, or twin surfaces will decompose more rapidly than crystals that lack these features. A corollary of this is that smaller crystals, which have relatively high surface:volume ratios, will decompose faster than larger crystals. Polycrystalline fragments will be altered along crystal boundaries and be split into their constituent minerals.

CRYSTALLINITY.   An ideal stoichiometric crystal that lacks defects is the most resistant state of a mineral during weathering. With element substitution and lattice defects, mineral stability decreases. Thus, a potassic feldspar containing some sodium in solid solution is not as stable as a pure potassic feldspar; and magnesian calcite is less stable than pure calcite. Amorphous solids such as opal or volcanic glass are the most disorganized and hence most easily dissolved by rainwater or percolating groundwater.

CHEMICAL CHARACTER OF THE ENVIRONMENT. Silicate minerals are less stable in acidic environments than in basic ones and less stable at higher surface temperatures than at lower ones. Soils in semiarid climates tend to be basic, with calcite accumulation in the B horizon; those in more humid climates are acidic because of the growth and decay of organic matter (organic acids) and because of carbon dioxide from the atmosphere that is dissolved in soil water. Hence, soils in humid areas are more mineralogically mature.

When we examine the types of mineral alterations in soil profiles in different stages of development, we find abundant examples of incomplete alteration, pseudomorphing of an original grain by an alteration product, cavities within original grains, etching of original grain surfaces, and perhaps most important, the production of large amounts of clay minerals. From these observations, supplemented by laboratory experiments, we find the following common mineral reactions among silicates during weathering:

$$3\ KAlSi_3O_8 + 2\ H^+ + 12\ H_2O \rightarrow \qquad (1)$$
$$\underset{\text{orthoclase}}{}$$
$$\underset{\text{muscovite}}{KAl_3Si_3O_{10}(OH)_2} + \underset{\substack{\text{dissolved} \\ \text{silica}}}{6\ H_4SiO_4{}^0} + 2\ K^+$$

$$2\ \underset{\text{muscovite}}{KAl_3Si_3O_{10}(OH)_2} + 2\ H^+ + 3\ H_2O \rightarrow \qquad (2)$$
$$3\ \underset{\text{kaolinite}}{Al_2Si_2O_5(OH)_4} + 2\ K^+$$

$$\underset{\text{kaolinite}}{Al_2Si_2O_5(OH)_4} + 5\ H_2O \rightarrow \qquad (3)$$
$$2\ \underset{\text{gibbsite}}{Al(OH)_3} + \underset{\substack{\text{dissolved} \\ \text{silica}}}{2\ H_4SiO_4{}^0}$$

$$2\ \underset{\text{orthoclase}}{KAlSi_3O_8} + 2\ H^+ + 9\ H_2O \rightarrow \qquad (4)$$
$$\underset{\text{kaolinite}}{Al_2Si_2O_5(OH)_4} + \underset{\substack{\text{dissolved} \\ \text{silica}}}{4\ H_4SiO_4{}^0} + 2\ K^+$$

$$\underset{\text{muscovite}}{KAl_3Si_3O_{10}(OH)_2} + H^+ + 9\ H_2O \rightarrow \qquad (5)$$
$$3\ \underset{\text{gibbsite}}{Al(OH)_3} + \underset{\substack{\text{dissolved} \\ \text{silica}}}{3\ H_4SiO_4{}^0} + K^+$$

$$NaAlSi_3O_8 \cdot CaAlSi_2O_8 + H_2O + H^+ \rightarrow \qquad (6)$$
$$\underset{\text{Ca-montmorillonite}}{\overset{\text{Na-montmorillonite}}{}} + H_4SiO_4{}^0 + \underset{Ca^{2+}}{\overset{Na^+}{}}$$

$$\text{ferromagnesian minerals} + H_2O + H^+ \rightarrow \qquad (7)$$
Na-montmorillonite
Ca-montmorillonite
muscovite
kaolinite
gibbsite $\qquad + \underset{\substack{\text{dissolved} \\ \text{silica}}}{H_4SiO_4{}^0} + \underset{\substack{\text{amorphous} \\ \text{ferric} \\ \text{hydroxide}}}{Fe(OH)_3}$

$$2\ Fe(OH)_3 \rightarrow \underset{\text{hematite}}{Fe_2O_3} + 3\ H_2O \quad \text{or} \qquad (8)$$

$$Fe(OH)_3 \rightarrow \underset{\text{goethite}}{FeOOH} + H_2O$$

$$KAlSi_3O_8 = H_2C_2O_4 + 8\ H_2O + 2\ H^+ \rightleftharpoons$$
$$K^+ + (AlC_2O_4 \cdot 4\ H_2O) + 3\ H_4SiO_4$$

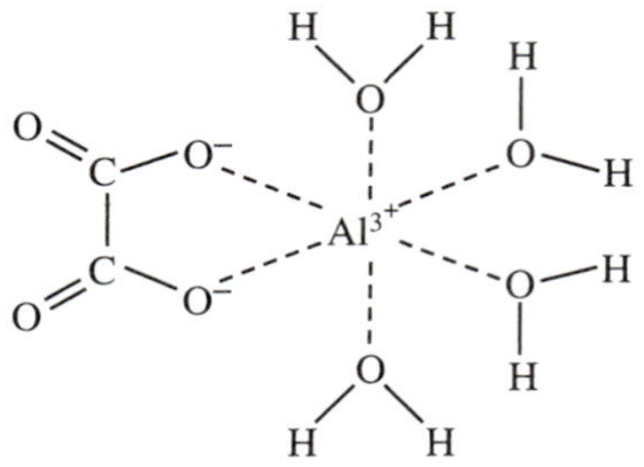

FIGURE 12-1
● ● ● ● ● ● ● ● ● ● ● ● ● ● ●

Equation for the total dissolution of potassic feldspar into soluble constituents. The schematic diagram shows the complexation of aluminum by a carboxylic acid (oxalic acid).

Although these reactions appear clean and precise when expressed in the customary forms shown, in fact they are less so when viewed closely. In reaction (1), for example, it is not muscovite that forms from alteration of potassic feldspar but illite, a chemically less well defined clay mineral that contains less potassium and more iron and magnesium than muscovite. In reactions (1) and (4), the "orthoclase" may be microcline or sanidine and very commonly contains some sodium substituting for potassium in the crystal structure. The gibbsite in reactions (3), (5), and (7) may actually be diaspore (AlOOH) plus water [Al(OH)$_3$ = AlOOH + H$_2$O]. Most important, perhaps, the equations assume that aluminum is conserved and is not solubilized relative to silica. This assumption is known to be inaccurate in aqueous systems containing complex organic compounds, as is the case in humid temperate and tropical climates, at least. In the acidic soils (**podsols**) that characterize such climates, carboxylic and phenolic organic acids **chelate** much of the aluminum from the primary aluminous mineral (Figure 12-1). The aluminum is then transported to other locations in the soil profile or to the soil surface, where it is carried away in streams. All cations can be chelated by organic compounds, but the effect is most intense for small, highly charged ions such as silicon and aluminum.

Despite the qualifications concerning the accuracy of chemical equations that illustrate the interaction between soil minerals and shallow groundwaters, two generalizations hold true: (1) Reaction products contain less silica and a smaller proportion of metallic cations than do the input minerals. (2) As temperature and precipitation increase, the stable reaction products become increasingly less complex; under the most intense humid tropical conditions, simple aluminous and ferric compounds (crystalline or amorphous) are the most stable.

THE STRUCTURE OF CLAY MINERALS

The major clay minerals are kaolinite, montmorillonite, and illite. Nearly all clay-bearing sedimentary rocks contain more than one type of clay mineral; this fact, coupled with the very small size of all clay particles (length 0.1–1.0 μm; thickness 0.001–0.01 μm) and the similar optical properties of most clay minerals, requires the use of X-ray diffraction or other techniques to analyze these rocks. The techniques are standardized and provide unequivocal identification of the various kinds of clay minerals. In clay mixtures, however, the estimation of relative percentages of each clay mineral is only semiquantitative. Estimates to within 10% are acceptable for most purposes.

The crystal structure of clay minerals is the same as that of micas: sheeted-layer structures with strong bonding (covalent) within each sheet and among the sheets, but weak bonding (hydrogen bonds) between the adjacent two- or three-sheet layers (the clay flakes). The weak bonding between layers permits not only the excellent cleavage of clays but also the adsorption of metallic cations and organic substances on clay mineral surfaces; this latter factor is important in chemical reactions that occur in weathering horizons.

The clay mineral structure contains two types of sheets. The first type is a *tetrahedral sheet* composed of tetrahedrally coordinated Si–O and Al–O groups in which three of the four oxygen atoms of each group are

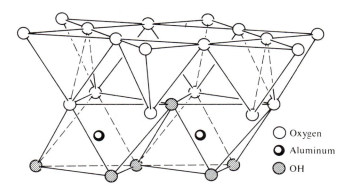

FIGURE 12-2

The structure of kaolinite. A tetrahedral sheet (above) bonded on one side to an octahedral sheet (below).

shared with adjoining groups. The cations in this sheet are at least 50% silicon atoms; the proportion of aluminum is greatest in illite and least in kaolinite. The second sheet type is an *octahedral sheet* composed of Al–OH or Mg–OH groups in octahedral coordination, with ferric iron sometimes substituting for aluminum and ferrous iron sometimes substituting for magnesium.

A single unit of kaolinite consists of one Si–O tetrahedral sheet and one Al–OH octahedral sheet (Figure 12-2), with essentially no cation substitution in either sheet. All other clays (and micas) are composed of three sheets: an octahedral sheet sandwiched between two tetrahedral sheets (Figure 12-3). There is abundant substitution within all three sheets: aluminum for silicon

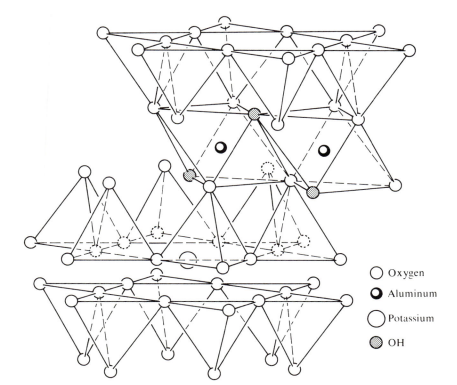

○ Oxygen
◑ Aluminum
○ Potassium
◍ OH

FIGURE 12-3

The structure of muscovite (illite), showing the position of the required interlayer cation.

in the tetrahedral sheets, magnesium and ferrous iron for aluminum in the octahedral sheet. These substitutions cause charge imbalances within the sheets (Al^{3+} versus Si^{4+}; Mg^{2+} and Fe^{2+} versus Al^{3+}) that are rebalanced by the adsorption of metallic cations on the surfaces of each clay flake (see Figure 12-3). The cations adsorbed are those available in soil waters: the K^+ that was released from orthoclase, the Na^+ and Ca^{2+} that were released from plagioclase and ferromagnesian minerals. Illite prefers potassium; montmorillonite prefers sodium and calcium. These relationships are summarized in Table 12-1.

SOILS
• • • • • • • • • • • • • • •

The result of mineral alteration in exposed rocks is the development of a soil, a granular aggregate capable of supporting plant life. On the basis of studies of modern soils, soil scientists have defined a large number of soil groups, subgroups, and so on. In pre-Quaternary rock sequences, however, erosion and superposition of soils formed penecontemporaneously (in the geologic time sense) make detailed classification impossible. A "geologic instant" may encompass tens of thousands of years in an ancient rock sequence, enough time for large changes in climate to occur with attendant changes in soil character. Even without climatic changes, the normal changes in depositional environment through time at a locality can cause complexities in soil development. For example, in a deltaic setting, a soil formed on a well-drained floodplain can be overprinted by a boggy swamp soil in which organic acids are an important modifying agent.

Paleosols

Paleosols are ancient profiles of weathering developed on rocks or sediments that were exposed long enough to be modified by soil-forming processes. These processes modify or destroy textures and structures in the parent rock, support growth and subsequent oxidation of organic matter, lower pH, control variations in **Eh** (oxidation potential), dissolve minerals and remove soluble cations (Table 12-2), produce clay minerals and move clay into the B horizon, and concentrate insoluble ferric and aluminum oxides and hydroxides in the B horizon of the soil profile. The result is a lithologic unit that has characteristics distinct from ordinary sand-

TABLE 12-1 Abundant phyllosilicates in rocks and weathering horizons, showing their idealized composition[a]

• •

Name	Chemical formula	Comment
Two-Sheet Structure		
Kaolinite	$Al_2Si_2O_5(OH)_4$	Almost no substitution
Antigorite	$Mg_3Si_2O_5(OH)_4$; platy	Forms from serpentine; no Al is present
Chrysotile	$Mg_3Si_2O_5(OH)_4$; fibrous	Forms from serpentine; no Al is present
Three-Sheet Structure		
Pyrophyllite	$Al_2Si_4O_{10}(OH)_4$	Almost no substitution
Montmorillonite	$Al_2Si_4O_{10}(OH)_2 \cdot nH_2O$	Mg may partly replace Al; interlayer Na and Ca present
Muscovite (illite)	$KAl_2(AlSi_3O_{10})(OH)_2$	In illite, Mg, Fe partly replace octahedral Al; interlayer K present
Talc	$Mg_3Si_4O_{10}(OH)_2$	
Vermiculite	$Mg_3Si_4O_{10}(OH)_2 \cdot nH_2O$	
Phlogopite	$KMg_3(AlSi_3O_{10})(OH)_2$	
Biotite	$K(Mg,Fe)_3(AlSi_3O_{10})(OH)_2$	
Chlorite	$Mg_5Al(AlSi_3O_{10})(OH)_8$	

[a] The Si_2O_5 or Si_4O_{10} grouping ($\pm Al$ substituting for Si) is the tetrahedral sheet. The other cations and hydroxyl groups are the octahedral sheet.

TABLE 12-2   Properties of major cations in aqueous solutions

Cation	Radius, Å	Ionic potential, Z/r^a	Dominant species[b]	Dehydrated or crystalline form[b]
K^+	1.33	0.75	K^+	
Na^+	0.97	1.0	Na^+	
Ca^{2+}	0.99	2.0	Ca^{2+}	Soluble ions
Mn^{2+}	0.80	2.5	Mn^{2+}	
Fe^{2+}	0.74	2.7	Fe^{2+}	
Mg^{2+}	0.66	3.0	Mg^{2+}	
Fe^{3+}	0.64	4.7	$Fe(OH)_3$	Fe_2O_3 (hematite)
Al^{3+}	0.51	5.9	$Al(OH)_3$	$Al(OH)_3$ (gibbsite)
Mn^{4+}	0.60	6.7	$Mn(OH)_4$	MnO_2 (pyrolusite)
Si^{4+}	0.42	9.5	$Si(OH)_4$	SiO_2 (quartz)
B^{3+}	0.23	13.0	BO_3^{3-}	
P^{5+}	0.35	14.3	PO_4^{3-}	Soluble ions
S^{6+}	0.30	20.0	SO_4^{2-}	
C^{4+}	0.16	25.0	PO_4^{3-}	

[a]Ions with ionic potentials between 3 and 12 are very insoluble and are almost totally precipitated at moderate pH values. Z, ionic charge; r, ionic radius.

[b]At the earth's surface, in oxidizing conditions and near-neutral pH values.

stones, mudrocks, and carbonate units (Table 12-3). However, because the degree of distinctness and ease of recognition of paleosols depend on the climate and stage of development of the soil profile, the paleosol literature contains many more descriptions of end member or peculiar soils than of the much more abundant but less extreme types.

Duricrusts

Duricrusts are hard, generally impermeable crusts on the surface or in the upper horizons of a soil. They are characteristically formed in the more extreme climates, either semiarid or humid tropical, and include varieties termed **calcrete** (or **caliche**), **ferricrete** (or ferruginous laterite), **aluminocrete** (bauxite), and **silcrete.**

CALCRETE.   Accumulation of calcium carbonate at and near the ground surface can occur in regions with rainfall as high as 100 cm/yr but is best developed in dryer areas. When the mean annual precipitation is less than about 50 cm/yr, an essentially continuous layer of

soil calcite (caliche) is formed within a few tens of centimeters of the ground surface. As rainfall increases, however, the depth of carbonate formation increases to perhaps 1 m, and the caliche horizon becomes discontinuous, consisting of irregularly spaced nodules. The restriction of major calcium carbonate precipitation to areas of low mean annual rainfall results from two factors. First, calcium ion has a high solubility (ionic potential, 2.0). Thus, whereas some precipitation is required to leach the calcium from parent silicate minerals, excessive rainfall removes all the calcium in groundwater flow. Second, in areas of high rainfall, vegetation is very abundant. Thus, the soil water is rich in carbon dioxide and organic acids from plant growth and decay, and carbonate ion is converted to HCO_3^-.

Important physicochemical processes in calcrete formation include physical disintegration and decomposition of rocks and minerals, precipitation of calcium carbonate, recrystallization, and metasomatic replacements. Wetting and drying cycles cause development of shrinkage cracks and channels. **Pedogenic** and pedodiagenetic processes occur together and are inseparable in a paleosol.

TABLE 12-3   Useful criteria for recognition of paleosols

• •

- Stratiform and relatively thin (usually <20 m). Nonstratiform paleosols occur on steep paleosurfaces as a result of either slumping or drainage variations. Soil thickness varies with type of parent rock, climate, and time.

- Transitional lower boundary but sharp upper boundary. The lower boundary is gradational because of the decrease in weathering intensity with depth. The sharp upper boundary results from the unconformity between the soil and overlying sedimentary or volcanic rocks. Gradational upper boundaries occur in exceptional circumstances, as when an arkose overlies a soil developed on granite.

- Characteristic textures and structures. Residual soils lack current structures such as cross-bedding and ripple marks, are poorly "sorted," and contain much clay. In thin section, illuviation and clay (plasma) flowage may be present, as may numerous other microtextures characteristic of soil. Clay coatings (cutans) on grains are common.

- Fossils. In addition to in situ organic growths such as tree trunks or root casts, there may be trace fossil assemblages characteristic of soils.

- Color variations; color mottling (gleying). The common colors generated during pedification are red, which results from movement of ferric iron downward from the A to the B horizon, and white, a result of precipitation of calcite in the B horizon in semiarid regions. The hematite normally appears to be a continuous color zone; the calcite is often patchy and nodular.

- Mineralogic variations within the soil. If enough of the vertical profile of the paleosol is preserved, the upper part tends to be enriched in stable constituents such as quartz, kaolinite, zircon, and tourmaline.

- Element variations. Alkali and alkaline earth cations are depleted (except in calcretes), as are other soluble elements with ionic potentials less than 3 or greater than 12. Elements with ionic potentials between 3 and 12 are enriched in soils; for example, silicon, aluminum, titanium, ferric iron, and manganic ion. The heavier elements of the rare earth group are depleted.

- Rip-up clasts in overlying sediments and/or dikes of material from overlying sediments washed down into desiccation cracks in the soil. Patterned shrinkage-swelling features (gilgai).

Most research concerned with calcrete development has tended to concentrate on inorganic processes because the role of biologic processes has been underestimated. These processes include mineral dissolution by organic acids and chelating compounds, control of $CO_2 - H_2CO_3 - HCO_3^- - CO_3^{2-}$ equilibria by bacterial and plant growth and decay, selective uptake of ions by plants, soil disruption and brecciation by roots and burrowing organisms, and calcification of organic matter to form biogenic carbonate structures.

Calcrete profiles are exceedingly complex, not only because of the high solubility of the microcrystalline calcite crystals that compose the primary calcrete precipitate, but also because of the complex interrelationships among the factors of climate, bedrock, topography, hydrology, organisms, and time. All models of calcrete formation must of necessity be generalized. Boundaries between designated horizons tend to be ir-

regular and diffuse, and the horizons themselves may be repeated several times or be absent altogether.

ALUMINOCRETE. Aluminocrete (bauxite, aluminous laterite) is a soil composed largely or entirely of amorphous or crystalline aluminum hydroxide (gibbsite, diaspore, boehmite); it is usually **pisolitic** (Figure 12-4). Such soils form only in humid tropical to subtropical regions such as the Amazon and Congo River basins. Bauxites can form on any aluminum-bearing parent rock but normally will be thicker and better developed on a more aluminous substrate. The key to alumina concentration in the humid tropics lies in removal of most of the "contaminant" ferric iron and silica. Removal can be accomplished in two ways. If the environment is reducing (low Eh), the ferric iron will gain an electron and be converted to the ferrous state, in which iron is very soluble (see Table 12-2) and can be carried away in

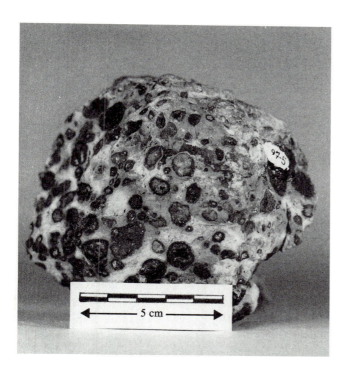

FIGURE 12-4

● ● ● ● ● ● ● ● ● ● ● ● ● ● ● ●

Typical appearance of pisolitic laterites, either aluminous or ferruginous. [From J. M. Guilbert and C. F. Park, Jr., 1986, *The Geology of Ore Deposits* (New York: W. H. Freeman), p. 782. Photo by W. K. Bilodeau.]

acteristic of iron enrichment remains. Well-developed iron-laterite contains at least 50% ferric oxide and hydroxide. These compounds are much less soluble in inorganic solutions than their aluminous counterparts (Figure 12-5) and, like them, must be chelated by organic compounds in order to be mobilized in significant quantities. In the presence of organic acids the solubilities of these normally insoluble, small, highly charged ions can be increased by 10^2–10^3.

SILCRETE.   Aluminocrete and ferricrete are residual soils, deposits formed by removal of all other constituents. Silcrete, in contrast, is essentially a secondary siliceous deposit and is composed of chemically precipitated opal, chalcedony, chert, and quartz in variable

near-surface water flow. Alternatively, the Fe^{3+} ion can be chelated and removed if suitable chelating agents are available. The various organic species that chelate cations have preferences for some cations over others, so iron can be removed while aluminum is largely unaffected. The silica is probably removed from aluminocretes by chelation. Complete separation of iron, aluminum, and silica is usually not achieved in soils and interfingering of aluminous and ferruginous laterite is common; minor quartz also occurs. In addition, gibbsite typically contains some iron in its crystal structure; hematite contains aluminum substitution.

Aluminous soils are chemically unstable in the presence of dissolved silica. Six parts per million of silica in solution in contact with $Al(OH)_3$ is sufficient to resilicate the soil, converting the aluminum hydroxide to kaolinite, as in reaction (3) (p. 232). Nearly complete absence of silica from near-surface waters is needed if an aluminocrete is to form and be preserved during diagenesis.

FERRICRETE.   Ferruginous laterite is perhaps the most commonly recognized paleoduricrust because of its color, and the most thoroughly studied modern crust because of its abundance and importance in resource-poor countries. Like the other duricrusts, ferricretes have a field appearance complicated by erosion and redeposition of the original crust, but the essential char-

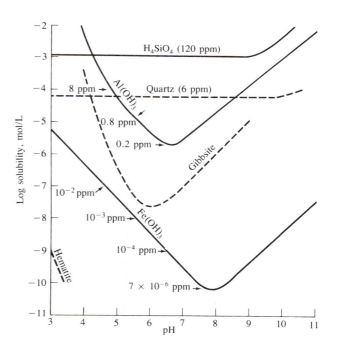

FIGURE 12-5

● ● ● ● ● ● ● ● ● ● ● ● ● ● ● ●

Solubility of crystalline and amorphous forms of silica, aluminum hydroxide, and ferric hydroxide as a function of pH.

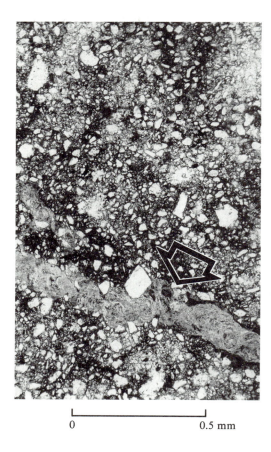

|—————————————————————|
| 0   0.5 mm |

FIGURE 12-6
● ● ● ● ● ● ● ● ● ● ● ● ● ● ● ●

Thin section of iron oxide–stained massive soil silcrete, showing contorted lamellar fabric (arrow); plane light. [From M. A. Summerfield, 1983, *J. Sediment. Petrol.*, *53*, 905.]

proportions. Detrital quartz may also be present. Formation of silcrete involves precipitation from surface or groundwater solutions, either directly through crystal growth, or through dehydration of silica gels followed by crystallization to crystalline silica. Silicification of clays and organic debris in silcretes is typical, as illustrated by silicification of wood and calcareous fossils in many deposits.

The occurrence of silcretes of Tertiary age (Figure 12-6) is commonly associated with laterite occurrences of apparently contemporaneous age, suggesting that the silica needed for silcrete formation was derived from lateritization of nearby soils. This inference seems quite reasonable. As shown in Figure 12-5, the silica content of waters from which opal can precipitate must exceed 120 ppm silica; and, because near-surface waters normally contain less than 20 ppm, an extraordinary source of dissolved silica is required to form sil-

crete. Given the normally high quartz content of most rocks and the near-absence of quartz in lateritic soils, developing laterites are an excellent source of dissolved silica. The chelated silicon in solution need only encounter a more oxidizing environment for the organic chelate to be destroyed and the silicon atoms released. This condition may occur seasonally or during a change from a humid tropical to a semiarid climate over a longer time period. It is no doubt significant that these two types of climate are juxtaposed between the equator and 30° latitude.

SUMMARY
● ● ● ● ● ● ● ● ● ● ● ● ● ● ● ●

Weathering is the attempt of minerals and rocks formed at high temperatures and pressures to reach chemical equilibrium at surface conditions. Because the lower temperatures and pressure in the surface environment are complemented by an excess of water and oxygen, sedimentary residues tend to be enriched in hydrous minerals and to contain polyvalent elements in their most oxidized state. Clay minerals and ferric iron are examples of this phenomenon.

The large-scale result of weathering is soil development, a process that requires times on the order of 10^2–10^3 years to occur, depending on climate and type of bedrock. Tens to hundreds of distinct soils can be contained within a few hundred meters of continental stratigraphic section and can provide very detailed data about the paleoclimatic conditions at the site. Paleosol recognition is handicapped by the general lack of training of geologists in soil science and by the inability to distinguish in ancient rocks periods of time on the order of 10^2–10^3 years. A result of the poor time resolution is the amalgamation of soils formed under different climatic and hydrologic regimes.

STUDY EXERCISES
● ● ● ● ● ● ● ● ● ● ● ● ● ● ● ●

1. What are the important climatic and geographic factors that control the rate of soil formation? Does an answer to this question require you also to consider tectonic variables? Why?
2. Many geologists believe that the atmosphere during the first 2 billion years of Earth history contained a much higher ratio of carbon dioxide to oxygen than that which exists today. Assuming they are correct, in

what ways might you expect the higher ratio to affect weathering?

3. Describe the ways by which the presence of plants on the land surface might affect rock weathering and soil development. Land plants did not become abundant on the land surface until Devonian time. Might it be possible to detect this change in paleosols? How?

4. Ferricrete is used as a building material by poor people in tropical countries. What can you infer from this about the physical and chemical characteristics of ferricrete? Would it be possible for those who live in drier climates to use calcrete in the same way? Explain.

5. List the common types of igneous and metamorphic rocks in order of their susceptibility to weathering in a humid temperate climate. What textural or mineralogic factors of the rocks must be considered in organizing such a list?

REFERENCES AND ADDITIONAL READINGS
• • • • • • • • • • • • • • •

Crowley, T. J., and G. R. North, eds. 1991. *Paleoclimatology.* New York: Oxford University Press.

Mack, G. H., W. C. James, and H. C. Monger. 1993. Classification of paleosols. *Geol. Soc. Am. Bull., 105,* 129–136.

Martini, I. P., and W. Chesworth, eds. 1992. *Weathering, Soils, and Paleosols.* New York: Elsevier.

Nahon, D. B. 1991. *Introduction to the Petrology of Soils and Chemical Weathering.* New York: Wiley.

Reinhardt, J., and W. R. Sigleo, eds. 1988. *Paleosols and Weathering Through Geologic Time. Geol. Soc. Am. Spec. Paper 216.*

Retallack, G. J. 1990. *Soils of the Past.* Boston: Unwin Hyman.

Chapter 13

CONGLOMERATES AND SANDSTONES

Conglomerates and sandstones form 20–25% of the stratigraphic record but have received much more than 25% of the attention of sedimentary petrologists. This imbalance has occurred for several reasons. The grains in sandstones are coarse enough to be seen easily, and these rocks typically contain **textures** and structures that can be described and photographed, such as pebble shape, grain rounding, **cross-bedding, ripple marks,** and **graded bedding.** These can be diagnostic of transport mechanism or depositional environment. Also, sandstones supply about half of the world's production of petroleum and natural gas and most of the groundwater we drink.

FIELD OBSERVATIONS

Grain Size

The most important textural feature of conglomerates and sandstones is grain size (Table 13-1), both the average size and the distribution of sizes in the rock. Most grain-size distributions have a single dominant size **(unimodal),** but detrital rocks composed of two abundant sizes with little sediment of intermediate size are also common, such as a rock composed of 20% pebbles and 80% fine sand. This pebbly fine-grained sandstone (or fine sandy pebble **conglomerate**) has two modes **(bimodal)** and might have been formed by a mixing of sediments from different environments, for example, **sand** blown by the wind onto a **gravel** beach. It might, however, represent a single environment such as a river channel in which the fine sand was deposited on top of the gravel and subsequently infiltrated into the gravel's pore spaces.

The distribution of sizes is normally given as the range, in ϕ **(phi) units,** that includes approximately two-thirds of the grains. This range is twice the **standard deviation.** For example, if the mean size is 2.0 ϕ and two-thirds of the grains have sizes between 1.5 ϕ and 2.5 ϕ, then the standard deviation is 0.5 ϕ. *Standard deviation* is the accepted measure of the **"sorting"** of the sediment (Figure 13-1), and sorting values tend to differ among different sedimentary environments.

After the average grain size and the range of sizes have been estimated, the sediment must be named. There is little agreement among geologists concerning a method of naming mixtures of different sizes. For example, some field geologists are likely to call a deposit a conglomerate even if gravel forms less than 50% of the rock. The reason for this is that it requires only 10% or 20% pebbles or cobbles to create a strong impression in an outcrop, with the result that many "conglomerates" contain less than 50% gravel (matrix-supported conglomerates).

Dispersal Pattern and Transport Distance

The factors that control the amount of sediment generated from land areas are relief and climate, with relief dominating; adjacent to active faults, erosion rates can be as high as 1000 cm/10^3 years. Over large areas, however, rates are always lower, for example, a maximum of 100 cm/10^3 years in parts of the Alps and Himalaya. Areas of high relief worldwide have an average erosion rate of about 50 cm/10^3 years; and lowland areas,

TABLE 13-1 The standard grain-size scale for clastic sediments[a]

Name	Millimeters	Micrometers	ϕ
	4096		−12
Boulder			
	256		−8
Cobble			
	64		−6
Pebble			
	4		−2
Granule			
	2	———	——— −1
Very coarse sand			
	1		0
Coarse sand			
	0.5	500	1
Medium sand			
	0.25	250	2
Fine sand			
	0.125	125	3
Very fine sand			
	0.062	———	——— 4
Coarse silt			
	0.031	31	5
Medium silt			
	0.016	16	6
Fine silt			
	0.008	8	7
Very fine silt			
	0.004	———	——— 8
Clay			
	↓	↓	↓

GRAVEL — Boulder, Cobble, Pebble, Granule
SAND — Very coarse sand, Coarse sand, Medium sand, Fine sand, Very fine sand
SILT — Coarse silt, Medium silt, Fine silt, Very fine silt

[a] As devised by J. A. Udden and C. K. Wentworth. The ϕ scale was devised to facilitate statistical manipulation of grain-size data and is commonly used. $\phi = -\log_2$ mm.

5 cm/10^3 years. Clearly, most detrital particles originate in mountainous regions.

Because of the great relief in mountainous regions, stream gradients there are relatively high (sometimes as high as 1°, 17.4 m/km); the ability of the stream to transport large particles (**stream competence**) is correspondingly high, resulting in many coarse particles being moved downstream. As the stream leaves the mountains, however, the stress of the water on the streambed decreases with the slope, competence decreases, and coarse particles can no longer be moved. Many studies of conglomerates have used this concept to infer distance of transport from a mountain front (Figure 13-2). In detrital sequences lacking cobbles and pebbles, a better way to detect an areal decrease in grain size is through sand to mudrock ratio maps or isopleth maps.

Textural Maturity

Following the release of grains from a source rock and their chemical alteration in the soil, the texture of a detrital sediment is modified in stages during transport in a stream. Within a few minutes, the clay and most of the silt are washed out and carried downstream. Within a few hours or days, the stream currents separate different sizes of sand and gravel, and the best sorting possible in the environment is achieved. As the sand is transported from the upstream areas, the grains are abraded and rounded. This process can be rapid or slow, depending on the size and mechanical resistance of the grains. For example, carbonate sand rounds rapidly; feldspar more slowly; and quartz even more slowly or not at all. This sequence of clay removal, sorting, and rounding is termed *textural maturity* (Figure 13-3); it is a key concept in the analysis of sandstones. There are three problems in its application, however. First, the roundness of quartz sand is controlled in part by depositional environment. Environments of high kinetic energy such as beaches and sand dunes will round quartz sand, but those of lower kinetic energy, such as rivers, will not. The second problem occurs when there is mixing of sediment from different environments, for example, when rounded grains from a marine sand bar are blown by the wind into a clay-rich lagoonal environment to create a **textural inversion**. The third problem is that clay **matrix** can be produced after burial in sandstones that contain chemically unstable grains.

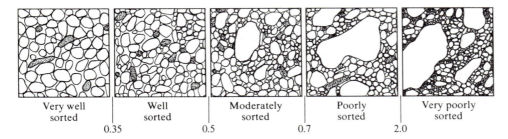

Very well sorted	Well sorted	Moderately sorted	Poorly sorted	Very poorly sorted
0.35	0.5	0.7	2.0	

FIGURE 13-1

● ● ● ● ● ● ● ● ● ● ● ● ● ● ●

Classification of degrees of sorting as seen through a square hand lens. Silt- and clay-size sediments are indicated by fine stipple. Values of standard deviation that divide each class of sorting are also shown. [From R. R. Compton, 1962, *Manual of Field Geology* (New York: Wiley), p. 214.]

Mineral Composition

The majority of the **detrital** fragments in sandstones and conglomerates are coarser than 0.06 mm in diameter, but accurate identification of the mineralogy of the fragments can be difficult. Quartz is easy to identify, as is pink potassic feldspar; but chert can look like rhyolite, phyllite can look like shale, and those little black things you see through the hand lens may be basalt, magnetite, or black **chert.** Considerable experience is needed to achieve a reasonable degree of accuracy in the identification of sand-size particles in hand specimens.

CONGLOMERATES

● ● ● ● ● ● ● ● ● ● ● ● ● ●

Nearly all mineral composition studies of conglomerates are field studies (of necessity!). The sampling technique is to select at random (meaning that every grain has an equal chance of being selected) 100 gravel-size grains, crack them open with a geology pick, and identify them with a hand lens. The fundamental categories for classification are extrusive igneous, plutonic igneous, metamorphic, and sedimentary. Each of these four groups should be further subdivided to an extent that seems useful.

FIGURE 13-2

● ● ● ● ● ● ● ● ● ● ● ● ● ●

Areal distribution of maximum pebble size in the Pocono Formation (Mississippian), Pennsylvania and Maryland. Measurement of cross-bedding directions also indicates current flow from southeast to northwest. [From B. R. Pelletier, 1958, *Geol. Soc. Am. Bull.*, *69*, 1054.]

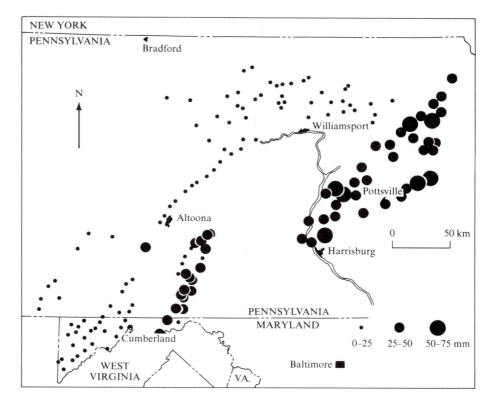

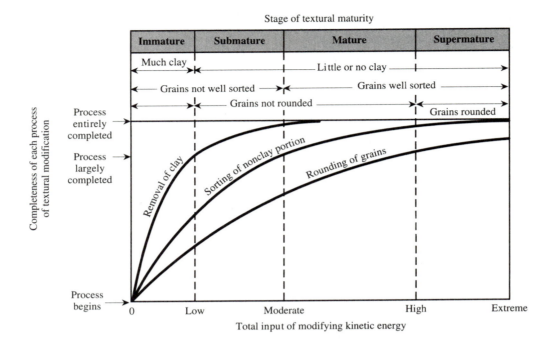

FIGURE 13-3

• • • • • • • • • • • • • • • • •

Textural maturity of sands as a function of input of kinetic energy.
[From R. L. Folk, 1951, *J. Sediment. Petrol.*, *21*, 128.]

A conglomerate is termed **oligomictic** if it is composed almost entirely of a single type of fragment (for example, granite, rhyolite, or quartz); it is **polymictic** if it is composed of a variety of fragment types. The composition of a gravel or conglomerate depends on several factors:

Lithology and climate of the source area
Initial size of the fragments
Transport distance
Grain size of the gravel particles in the conglomerate

The significance of the source area is obvious. If a rock type is not exposed upcurrent, then it cannot occur in a downcurrent deposit. However, weathering in the source area may cause the proportions of the fragment types in the sediment to differ from the proportions of source rocks. Normally, the object of a petrologic study of a conglomerate is to determine the nature of the source area.

The initial size of fragments released from the up-current sources differs for rocks of different lithology because of the different boulder-forming capabilities of rock types. For example, metaquartzite fragments initially are quite large because of their resistance to chemical weathering. Their initial size is determined by the

thickness of bedding and the spacing of joints. **Shale** fragments typically begin sedimentary transport in the finer pebble sizes because of their fissile nature and closely spaced joint pattern in outcrop. Further, fragments have various resistances to size reduction during transport. Metaquartzite, chert, and rhyolite fragments are quite durable during transport, whereas limestone, schist, and shale are not.

Because of these factors, the interpretation of the petrology of conglomerates is not so straightforward as it seems initially. Nevertheless, conglomerates are particularly valuable deposits for the determination of **provenance** (source area). Unlike most sandstones, conglomerates are typically rich in fragments of the undisaggregated source rock rather than being composed largely of individual mineral grains, which might have originated in a variety of rock types. In addition, most accumulations of gravel-size fragments are located much nearer the source terrane than are sand deposits. As a result, the average gravel-size fragment has undergone less modification since being released from its parent rock; so interpretation of gravel in terms of paleogeology is more reliable.

The detritus from areas with high relief and rapid rates of erosion often contains abundant gravel of low resistance to weathering and transportation. Examples of such detritus include schists, granites, and lime-stones. These polymictic conglomerates are characteristic of orogenic areas, such as the Alps during the Tertiary Period and the Rocky Mountains during the Pennsylvanian Period. The conglomerates can be alluvial fan deposits, gravel bars in high-gradient streams, beach deposits, offshore bars, or matrix-rich conglomerates emplaced in the deep ocean basin by turbidity currents.

POWAY CONGLOMERATE (CRETACEOUS), SOUTHERN CALIFORNIA. The usefulness of conglomerate petrology for tectonic reconstructions is shown by the Poway Conglomerate and coeval submarine fan conglomerates of the Jolla Vieja Formation on Santa Cruz Island in California (Figure 13-4). The unit is composed largely of rhyolite and dacite porphyry clasts, rock types not found as bedrock in surrounding mountainous areas of southern California; the clasts apparently are derived from far to the east of the present Poway outcrop. The search for a source led to outcrops of Upper Jurassic volcanic rocks near El Plomo in Sonora, Mexico, an area suggested by tectonic considerations and the known northward displacement of southern California with respect to Mexico during Tertiary time.

Two methods were used to determine whether the El Plomo volcanics were the source for the Poway and Jolla Vieja gravel particles. First, samples of both the possible source rock and clasts from the Eocene conglomerates were dated using radiometric techniques; both groups of samples gave statistically identical ages, early to middle Jurassic. Second, a group of 16 trace elements, selected to include a wide range of behaviors during magmatic and alteration processes, were analyzed. The Eocene sediments and Jurassic volcanics have virtually identical trace-element means (Figure 13-5). The concentrations of the 16 trace elements have been controlled by either magmatic or postmagmatic processes and, therefore, contain a record of the evolutionary history of the rocks—from fusion, through fractional crystallization, to postmagmatic alteration. It is difficult to believe that the two suites of rocks could have such similar trace-element distributions (evolutionary history) unless they were drawn from the same population. Thus, it was concluded that the rhyolitic bedrock west of El Plomo is a remnant of the source terrane for the Eocene Poway Conglomerate in San Diego (500 km to the northwest) and also for the coeval submarine fan deposits on Santa Cruz Island (300 km to the northwest of the Poway outcrop). This depositional

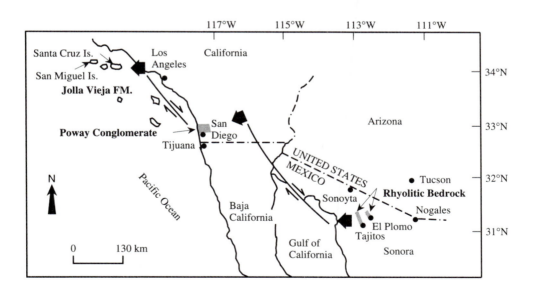

FIGURE 13-4
• • • • • • • • • • • • • • • • •

Location map of the Poway Conglomerate and the Jolla Vieja Formation. Half-headed arrows indicate direction of movement on opposite sides of a fault (solid line). The three large arrows represent dismembered segments of an east–west–oriented piercing point created by a fluvial and alluvial fan–submarine canyon–submarine fan depositional system. [From P. L. Abbott and T. E. Smith, 1989, *Geology, 17*, 329.]

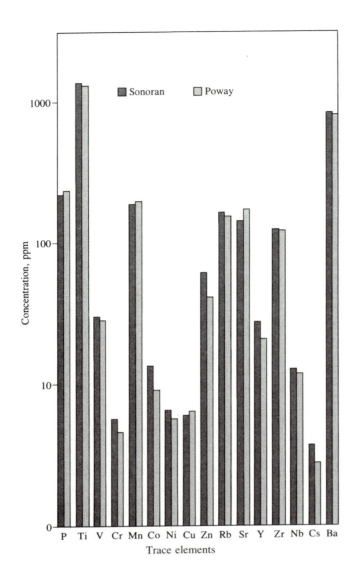

FIGURE 13-5
• • • • • • • • • • • • • • •

Histograms showing trace-element means for Sonoran bedrock rhyolite (black) and Poway Conglomerate rhyolite casts (gray). [From P. L. Abbott and T. E. Smith, 1989, *Geology*, *17*, 331.]

pattern indicates a minimum average northwestward movement of 4–5 mm/yr along the strike-slip faults involved.

Most conglomerates contain abundant finer sediment between the gravel particles. Commonly, the sand is equal in abundance to the gravel. But whether or not this occurs, the matrix material should be sampled for later examination in thin section. Frequently, the mineral composition of the sandy or muddy matrix is differ-

ent from that of the gravel. Perhaps the difference is simply due to disaggregation: The gravel is composed of granite pebbles, and the sand is a mixture of feldspar, quartz, and mica. Conversely, it may be that the gravel was locally derived and the sand originated at some distance from the depositional site. The mud fraction in a conglomerate may have been derived from the same place as the gravel and transported and deposited with the gravel, as in a turbidity-current deposit on a deep-sea fan. However, it may be that the mud entered the gravel through infiltration downward into an originally permeable gravel; for example, a mud layer deposited on the surface of a stream gravel filters downward. Examination of the matrix can resolve these uncertainties.

SANDSTONES
• • • • • • • • • • • • • • •

Sandstones are at least an order of magnitude more abundant than conglomerates and are, therefore, the rocks that are most often used for determining the source area (provenance) of detrital grains. The objective in studying detrital mineral grains in sandstones is to make possible the construction of paleogeologic maps for the geologic past that are as accurate and detailed as the geologic maps of present outcrop patterns. This goal can never be attained because of the information loss between the time of formation of the detrital particles and the present day, but the goal is clear.

Quartz

Two-thirds of the detrital fraction of the average sandstone is quartz. Quartz is the most abundant mineral in most sandstones for several reasons: It is abundant in the most common crystalline rocks—granitoid igneous rocks, gneisses, and schists. It is mechanically very durable, with a hardness of 7 on the Mohs scale and a poor cleavage. It is very resistant to chemical attack because the bonds between silicon and the shared oxygen ions are strong.

In thin section studies, detrital quartz grains are usually classified as either **monocrystalline** (grain at least 90% composed of a single quartz crystal) or **polycrystalline** (composed of two or more crystals). Thus, a polycrystalline quartz grain is in fact a rock fragment; it is treated as such in some sandstone classification schemes.

The crystalline rocks that are the source of nearly all silicate mineral grains in sandstones reach the surface by means of tectonic uplift. During this process, most

FIGURE 13-6
• • • • • • • • • • • • • • • •

Photomicrograph of fine-grained St. Peter Sandstone (Ordovician) near Foley, Missouri, showing several quartz grains with undulatory extinction and overgrowths of secondary quartz. The sharp boundary between some extinction zones suggests formation by rotation of broken grain fragments rather than by plastic deformation of an unbroken grain.

quartz crystals are plastically deformed. The crystal deformation is seen in thin sections as **undulatory extinction** (Figure 13-6); that is, the crystal does not extinguish as a single unit on the slightest rotation of the microscope stage but instead extinguishes in sectors through a rotation of several degrees. The amount of stage rotation required for extinction is not a very effective criterion to use in distinguishing between igneous and metamorphic origins for a quartz crystal. Undulatory extinction can also be produced in a sedimentary rock during folding and faulting.

Polycrystalline quartz grains can have several types of internal structures:

Elongate structure. If the individual quartz crystals within the grain are elongate (Figure 13-7), they have been deformed in a nonhydrostatic stress field. Such stretched quartz crystals are commonly found in foliated metamorphic rocks such as schists and gneisses. Hence, detrital grains that contain these crystals are mostly of metamorphic derivation. Stretched quartz crystals are also found along and adjacent to fault surfaces, where extreme stretching can be accompanied by granulation and recrystallization to produce the rocks **mylonite** and **phyllonite.**

Bimodal distribution of sizes. Metamorphic rocks are, by definition, recrystallized rocks. Recrystallization begins at points of stress concentration within the rock rather than simultaneously at all locations. Thus, when recrystallization ends, some quartz crystals are in a dif-

ferent stage of the process than others. This is reflected by a great variation in crystal size within a detrital quartz fragment, commonly a bimodal distribution of crystal sizes illustrating a recrystallization "caught in the act" (Figure 13-8). The smaller crystals are the newly developing ones that have not yet grown to equilibrium size.

Efficient packing. Many times in metamorphic rocks it happens that a thermally induced recrystallization occurs as the last recrystallization event in the history of the rock. In such cases, the intercrystalline boundaries among quartz crystals are rather straight and intersect at an angle of 120°, a geometry reflecting a minimization of surface energy and the most efficient use of space among crystals that form simultaneously.

Grain size. Granitoid igneous rocks are coarse grained, with the quartz crystals in them typically exceeding 0.5 mm (sizes of several millimeters are common). Quartz crystals, however, are fine grained in many metamorphic rocks, such as phyllites, most schists, and some gneisses. Therefore, the more quartz crystals in a detrital polycrystalline grain of a given size, the more likely the grain is to be of metamorphic derivation. A sand-size quartz grain composed of more than five separate crystals is probably of metamorphic derivation.

Sutures. Intercrystalline suturing among quartz crystals is common in both igneous and metamorphic rocks, although it may be more intense in polycrystalline grains of metamorphic origin. It is an unreliable criterion to use for provenance.

FIGURE 13-7

Photomicrographs of coarse sand-size quartz grains (crossed nicols) from **(A)** igneous rock and **(B)** metamorphic rock. Granitic quartz is composed of fewer crystals of more equant shape and has little or no intercrystalline suturing. Metamorphic grain consists of perhaps 10 times as many crystals, which are stretched and elongate and have intensely sutured contacts. These grains are extreme examples of polycrystalline grains from the two classes of crystalline rocks.

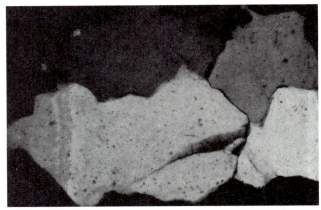

(A)

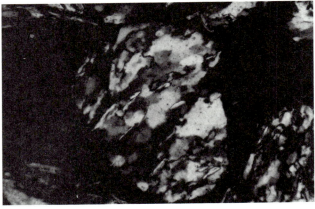

(B)

Granitoid plutonic rocks tend to be coarser grained than metamorphic rocks. Therefore, monocrystalline quartz grains of medium sand size and coarser that appear in sandstones are likely to have been derived from granites. Fine and very fine sand size monocrystalline quartz grains can be produced by several processes: release in these sizes from fine-grained metamorphic rocks such as slates, phyllites, and some schists; breakage and chipping of larger quartz grains of any provenance; and disaggregation of polycrystalline quartz grains during soil-forming processes.

Quartz grains in silicic volcanic rocks and tuffs typically have euhedral crystal outlines of β-quartz (high-temperature quartz) and have **nonundulatory extinction,** but quartz crystals from these sources are trivial in abundance relative to the amount of quartz derived from the more abundant igneous and metamorphic rocks.

Quartz veins are common in most areas of crystalline rocks and, as is the case for metaquartzites, typically release coarse gravel-size grains into the sedimentary

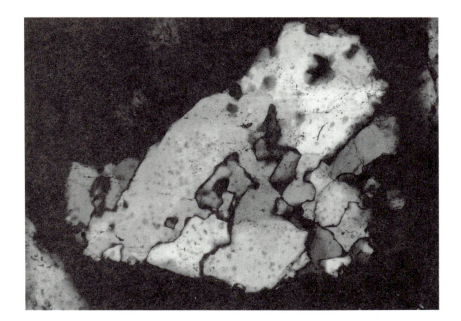

FIGURE 13-8

Polycrystalline quartz grain from a metamorphic rock caught in the act of recrystallization.

environment. Veins form from solutions that are more analogous to aqueous subsurface brines than to silicate magmas. As a result, vein quartz can contain unusually large volumes of water-filled cavities, which give the quartz a milky color. Not all quartz veins are milky, however; thus, much vein quartz in sediments no doubt is unrecognized. Also, like quartz from rhyolites, vein quartz tends to be swamped by the more abundant grains from granites, gneisses, and schists.

The most durable variety of quartz in the sedimentary environment is nonundulatory monocrystalline quartz that does not contain inclusions. Polycrystalline grains are weaker because of their internal discontinuity surfaces (crystal boundaries). Grains with undulatory extinction are weaker because they have been plastically deformed. Grains with **inclusions** are weak because they are composed of two distinct phases, either two solid phases or a solid and a liquid (or gaseous) phase; as with polycrystalline grains, discontinuity surfaces are present within these grains.

Because of these differences in durability, the concept of survival of the fittest can be applied to quartz in the sedimentary environment. Assemblages of quartz grains that have spent more time in the sedimentary environment should be relatively enriched in nonun-

dulatory monocrystalline grains and depleted in undulatory polycrystalline grains. Examination of thin sections of sandstones reveals that this is indeed the case (Figure 13-9). The average crystalline rock contains less than 20% nonundulatory quartz crystals in its quartz population, as does relatively unabraded sand-size detritus. Sandstones composed mostly of feldspar and lithic fragments and less than 40% quartz still contain about 20% nonundulatory grains in the quartz population. Sandstones that contain more than 90% quartz average 40–45% nonundulatory grains. These nearly pure quartz sandstones contain less than 1% polycrystalline quartz. Quartz grains that contain mineral inclusions are also rare among grains in pure quartz sandstones.

Feldspar

Feldspars are the most abundant group of minerals in crystalline rocks, forming 60% of the mineral grains in igneous rocks and perhaps a similar percentage in metamorphic rocks. Feldspars are unstable in the sedimentary environment, relative to quartz. Hence, although the feldspar to quartz ratio in crystalline

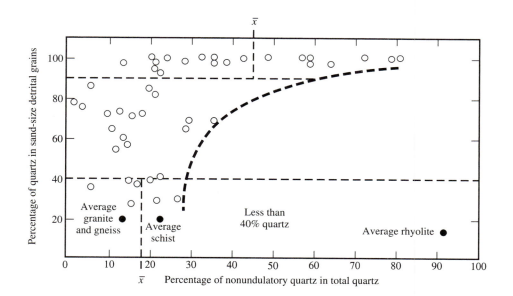

FIGURE 13-9

• • • • • • • • • • • • • • • •

Relationship between percentage of quartz in crystalline rocks and sandstones and percentage of that quartz that has nonundulatory extinction. The dashed, curved line includes all sandstone sample points. Undulatory extinction is harder to see in smaller crystals; so schists, which typically contain quartz crystals of fine sand size, appear to have a higher proportion of nonundulatory quartz grains. [From H. Blatt and J. M. Christie, 1963, *J. Sediment. Petrol.*, *33*, 569, Fig. 4.]

FIGURE 13-10
• • • • • • • • • • • • • •

Photomicrographs of medium sand-size feldspar grains (crossed nicols). **(A)** Orthoclase, with alteration parallel to right-angle cleavages. **(B)** Plagioclase, showing polysynthetic albite twinning. **(C)** Microcline, showing characteristic grid twinning. **(D)** Microperthite, showing spindle-shaped, exsolution lamellae of sodic and potassic feldspar.

(A)

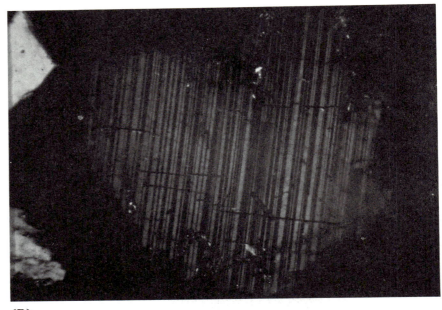

(B)

source rocks of sandstones is 3:1, in the sandstones themselves the ratio is about 1:6. Feldspars form only 10–15% of the detrital fraction of the average sandstone.

Because of the complex nature of the feldspar minerals, they have been subdivided into many categories on the basis of their chemical, physical, and structural characteristics. In routine thin section work, however, many of these characteristics are not determined. The categories of feldspar normally used by sedimentary petrologists are the following (Figure 13-10):

Potassic feldspars: orthoclase, microcline, and sanidine

Plagioclases: albite through anorthite

Microperthite: an intergrowth of sodic feldspar and potassic feldspar

The methods used to recognize each of these types of feldspar in thin section are given in Table 13-2. However, it is often difficult to distinguish detrital quartz from untwinned feldspar, and even more difficult to distin-

(C)

(D)

guish orthoclase from untwinned albite, which may be abundant in fine-grained sandstones. For this reason, thin sections of sandstones commonly are stained with organic dyes so that total feldspar and feldspar varieties can be recognized more precisely.

TYPES OF ALTERATIONS.   Several types of alterations of feldspars are seen commonly in thin sections of sandstones:

Vacuolization. Water percolates into the feldspar grain along cleavage planes, fractures, and twin composition surfaces, producing no visible chemical change in the mineral structure. The grain appears turbid in transmitted light but has a whitish, cloudy appearance in reflected light. The whitish color can be seen by switching off the light under the stage of the microscope and shining a beam on the surface of the thin section.

Illitization. In a chemical process that is not well understood, the tectosilicate crystal structure of a

TABLE 13-2   Thin section characteristics of feldspars

• •

Feldspar	Characteristics
Potassic feldspars	*Orthoclase*[a]: Untwinned or with Carlsbad twinning; refractive indices below Lakeside or quartz; birefringence less than that of quartz; large 2V; optically negative; internal alteration to illite flakes, commonly along cleavage planes.
	Microcline: Unique and distinctive grid twinning pattern.
	Sanidine: Same as orthoclase except that the 2V is very small, 5°–10°.
Plagioclase feldspars	Typically polysynthetically twinned (albite twinning); internal alteration to montmorillonite flakes, commonly along cleavage planes; can be concentrically compositionally zoned, reflecting change in Ab:An ratio; two directions of twinning at a large angle to each other (albite and pericline twinning); Ab:An ratio determined by Michel-Lévy method. Albite has indices of refraction lower than quartz, oligoclase is about the same as quartz in indices and birefringence, and andesine through anorthite have indices higher than quartz.
Microperthite	Parallel intergrowth of two phases with contrasting indices of refraction and birefringence.

[a] Potassium- and calcium-sensitive stains are commonly used to distinguish orthoclase from untwinned plagioclase, which is common in metamorphic rocks and sediments derived from them.

potassium-rich feldspar grain is converted in places to the phyllosilicate structure of illite. Typically, the illite flakes appear first along surfaces where water can permeate, such as cleavages, fractures, and twin surfaces. The flakes are 5–10 μm in length and have a straw-yellow interference color.

Montmorillonitization. A type of alteration directly analogous to illitization, except that it occurs in sodic-calcic feldspars. Montmorillonite has the same birefringence as illite but lower indices of refraction than quartz.

Kaolinization. The chemical transformation of either potassic, sodic, or calcic feldspar to kaolinite reflects a more intense and/or prolonged process of alteration than either illitization or montmorillonitization. Kaolinite has a very low birefringence (0.005) and is dark gray in thin section. If the flakes are very small, kaolinite can look like chert.

In describing these types of chemical changes, we use the term *alteration* rather than the more specific term *weathering*, because there is substantial alteration of feldspar grains after burial. Internal chemical alteration of feldspar grains can occur as easily and rapidly during diagenesis as during surface weathering. Therefore, altered feldspars in sedimentary rocks cannot be used as an indicator of climate at the outcrop where the grain originated.

There are, however, many sandstones in which nearly all the feldspar grains are fresh and unaltered. This con-

dition indicates an arid or glacial climate, the freshness of feldspars reflecting a lack of sufficient water to accomplish chemical alteration. Of course, freshness indicates a lack of diagenetic alteration as well.

PROVENANCE.   Feldspar occurs in nearly all types of crystalline rocks. Zoned plagioclase feldspars occur only in magmatic rocks and more commonly in volcanic than in plutonic rocks.

The percentage and type of feldspars in a sandstone depend on the rate and type of tectonic activities and on climate. In a tectonic setting characterized by block-faulted and uplifted crust (such as occurred within the craton in Colorado during the Pennsylvanian Period), uplift, erosion, and burial are rapid and the resulting sands contain up to 50% feldspar. What is meant by the expression *rapid uplift*? In the Pennsylvanian Period in Colorado at least 1500 m of uplift occurred in less than 40 million years (4 cm/10^3 yr), resulting in a wedge-shaped deposit of granitic conglomerate and feldspathic sandstone that is now 1500 m thick at the base of the Front Range (Fountain Formation), 60 km in width (east-west), and 350 km in length (north-south). The climate in the area was probably semiarid. The combination of high topographic relief and low intensity of chemical weathering is ideal for the accumulation of a thick sedimentary sequence.

In a quiescent cratonic setting like that during the early Paleozoic Era in central North America, very little sediment can be produced from the low-lying granitoid crust. The small amount that is produced is reworked repeatedly by waves and currents in the beach-dune complexes that characterize this tectonic setting. Low topographic relief permits extensive transgressions and regressions of epicontinental seas; thus, environments of high kinetic energy dominate the geographic setting and the resulting abrasion of sand grains removes nearly all the feldspar. Also, central North America during early Paleozoic time was located within 20° of the equator (see Figure 11-5), so the climate was hot and possibly humid.

The variety of feldspar that is most abundant in a sandstone also depends to an important degree on the tectonic setting in which the sandstone forms. Plagioclase dominates when erosion and burial are rapid and the granitoid rocks exposed are granodiorites and quartz diorites, conditions that occur along convergent plate margins (for example, Tertiary sandstones in California). Potassic feldspars (orthoclase and microcline) dominate the feldspar suite of sandstones formed in rifted intracratonic settings (for example, Mesozoic basins of the Appalachians and Pennsylvanian sandstones in Colorado). In the California sandstones the plagioclase to potassic feldspar ratio is typically 2:1, with abundant unstable calcic feldspar; in the Colorado sandstones there are only trace amounts of plagioclase of any composition. Quiescent cratonic settings also result in only very small amounts of plagioclase feldspar among the total feldspar grains.

The proportion of total feldspar in a sandstone (and perhaps the type of feldspar) is also affected by the grain-size distribution in the rock. Detrital feldspars, because of their relative instability, are normally finer grained than associated quartz grains. As a result, it is not uncommon to find, for example, that the medium sand fraction contains 90% quartz and 10% feldspar; in the fine sand fraction, 80% and 20%; and in the coarse silt fraction, 70% and 30%. The percentage of feldspar then decreases to about 5% in the medium silt fraction as the rapidly increasing surface to volume ratio of the smaller grains causes a rapid increase in the dissolution rate of the feldspars. The average mudrock contains about 5% feldspar. The clay-size fraction contains almost no feldspar.

Lithic Fragments

Pieces of polymineralic source rock (called *lithic fragments*) form 15–20% of the average sandstone (Figure 13-11). These fragments not only indicate whether the source rock was igneous or metamorphic but also reveal other things, such as the silica content of the magma, its rate of crystallization, and the character of the premeta-morphic sedimentary rocks from which the metamorphic rock was formed.

Although any type of rock fragments can be found in a sandstone, some types are much more common than others. The types that will occur are determined by the following factors:

1. Areal abundance in the drainage basin
2. Location in the drainage basin: whether in the highland or lowland areas
3. Susceptibility of the rock fragments to chemical and mechanical destruction by sedimentary processes
4. Size of the crystals in the fragments

Obviously, the greater the areal extent of the source rock (factor 1), the better the chance of finding pieces of it downstream; and we have previously noted the greater rates of erosion from areas of high relief (factor 2).

Factors 3 and 4 determine the **survival potential** of the fragment. Their importance can be illustrated by consideration of mudrock fragments. Mudrocks form two-thirds of the stratigraphic column, and probably most of the mudrock is shale. Therefore, we might anticipate that most lithic fragments in sandstone would be fragments of shale. But this is the opposite of what is found: Shale fragments are quite uncommon in ancient rocks. The explanation is that shale fragments are practically untransportable because they are very soft and also split rapidly along fissile surfaces (factor 3); that is, they have a low survival potential. Extension of this principle leads to the expectation that fragments of gabbro will be poorly represented because of their chemical instability relative to granite; fragments of older sandstones will be rare because the common cements calcite and hematite break easily during rock transport. Most fragments of older sandstone that survive will be either pieces of quartz-cemented quartz sandstone or chert fragments.

The crystal size in the lithic fragment (factor 4) determines the minimum size of fragment necessary for the fragment to exist. For example, a fragment of granite is unlikely to occur in a fine-grained sandstone because fragments tend to break along crystal boundaries, and the crystals in a granite are coarser than fine sand. Fragments such as rhyolite or chert, however, can occur with equal ease in sand of any size. Neglect of this factor can lead to erroneous paleogeologic inferences. It is clear that the interpretation of upstream paleogeology from sandstone petrology is not straightforward, even when the sandstone contains pieces of the source rock itself.

In many rocks that contain silicic volcanic fragments, it is difficult to distinguish some of them from detrital chert. In such cases the following criteria are useful:

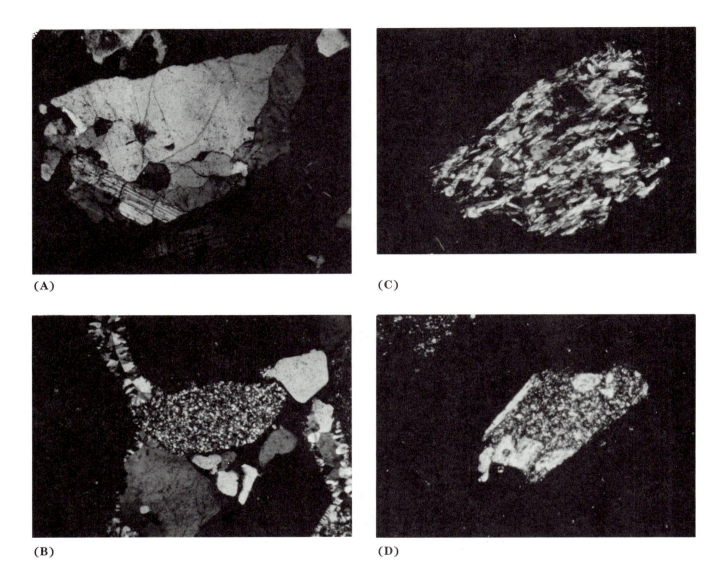

(A)

(B)

(C)

(D)

1. Felsite contains microcrystalline feldspar, and these crystals are commonly cloudy because of fluid inclusions or chemical alteration. Quartz microcrystals are not cloudy.
2. Feldspars may be lath-shaped (elongate or tabular) rather than subequant, as is always true of associated quartz microcrystals.
3. Felsite may contain glass, which is isotropic.

Accessory Minerals

The **accessory minerals** in sandstones include all detrital minerals except quartz and feldspar, although micas are typically excluded from the accessory group because of their extremely platy shape and resulting anomalous behavior during transport. Any mineral that

FIGURE 13-11
● ● ● ● ● ● ● ● ● ● ● ● ● ● ● ●

Photomicrographs of coarse sand-size lithic fragments (crossed nicols), showing characteristic appearances. **(A)** Granite fragment composed of large quartz crystal, orthoclase crystal (right, dark gray), and twinned plagioclase crystal (lower left). **(B)** Chert fragment (center) surrounded by three finely microcrystalline chert fragments (black with gray speckles) and several quartz fragments (white and gray). Elongated crystals of quartz cement have grown from chert grain surfaces to bind the grains. **(C)** Mica-quartz schist fragment containing a few opaque mineral crystals. **(D)** Volcanic rock fragment containing three large and altered potassic feldspar crystals (one of which is euhedral) and large mica flake (upper left edge of grain) set in groundmass of felsitic crystals of low birefringence, probably quartz and potassic feldspar.

occurs in igneous and metamorphic rocks can occur in sandstones. The relative amounts of accessory minerals in a sandstone depend on the abundance of each mineral in the source rock; its survival potential during weathering, transport, and diagenesis; and its specific gravity. Because of the wide range in specific gravities of the common accessory minerals, there may be significant segregation among them during transport (placer deposits; Figure 13-12). The range in specific gravity among the common accessories is 3.0–5.2. In contrast, the range among quartz and feldspars is only 2.56–2.76.

Other than micas, no common detrital minerals occur with specific gravities in the range 2.8–3.0. This fact is the basis for the usual method for separating quartz plus feldspar from the accessory minerals: The loose sediment (or disaggregated sandstone) is dropped into a liquid with a specific gravity in the 2.8–3.0 range, with the result that the quartz and feldspar float while the accessories sink. For this reason, the accessory minerals are termed **heavy minerals.** The heavy minerals typically form less than 1% of a sandstone. The percentage is commonly related to the proportion of lithic fragments in the light-mineral fraction of the rock, particularly the proportion of metamorphic lithic fragments. A high percentage of metamorphic fragments suggests

that a high percentage of heavy minerals may be present (perhaps 3%) because most metamorphic rocks contain more nonmicaceous accessory minerals than plutonic igneous rocks do. In addition, most species of heavy minerals in sandstones originate in metamorphic rocks. This is true because metamorphic rocks form in a much wider range of temperatures and pressures than do igneous rocks, thereby allowing a larger number of species to crystallize. For the purposes of provenance determination, it is useful to group accessory minerals according to the type of crystalline rock in which they usually form (Table 13-3). Unfortunately, many of the more common accessories in sandstones, such as zircon, tourmaline, and magnetite, form in abundance in both igneous and metamorphic rocks. Some minerals, such as tourmaline, occur in a variety of colors, and color variation may be related to provenance. For example, brown tourmaline is believed to be diagnostic of metamorphic rocks.

The fact that many of the more common heavy minerals in sandstones occur in both igneous and metamorphic rocks means that it is necessary to employ techniques more sophisticated than simple mineral identification to determine provenance (Figure 13-13). It is clear that two distinctly different sources supplied

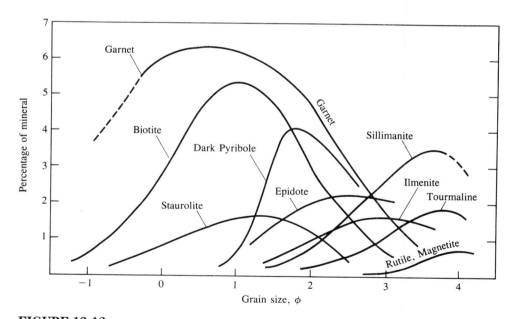

FIGURE 13-12

• • • • • • • • • • • • • • •

Size distributions of heavy minerals in the Kitt Brook delta (Pleistocene), Connecticut. The distributions are the combined effect of the sizes of these minerals in the source terrane and the sorting processes during transport and deposition. [From E. R. Force and B. D. Stone, 1990, *U.S. Geol. Surv. Bull. 1874,* 19.]

TABLE 13-3   Provenance of common accessory minerals in sandstones

• •

Igneous rocks	Metamorphic rocks	Indeterminate[a]
Aegerine	Actinolite	Enstatite
Augite	Andalusite	Hornblende
Chromite	Chloritoid	Hypersthene
Ilmenite	Cordierite	Magnetite
Topaz	Diopside	Sphene
	Epidote	Tourmaline
	Garnet	Zircon
	Glaucophane	
	Kyanite	
	Rutile	
	Sillimanite	
	Staurolite	
	Tremolite	

[a] Common in both igneous and metamorphic rocks.

garnets to the Etive Formation in the Murchison Field and that there may be a difference in garnet populations in the two portions of the Ness Formation, those in the Murchison Field and those in Tern Field.

Nearly all studies of heavy minerals in sedimentary rocks have been made of the nonopaque fraction, although it commonly forms only a minor part of the heavy-mineral crop. Most **opaque minerals** in sedimentary rocks are described by the three-component system $FeO–Fe_2O_3–TiO_2$ and include phases such as magnetite and ilmenite, with a variety of intergrowth structures and trace-element variations, some of which can be diagnostic of provenance. Perhaps the most surprising result from an opaque mineral study of sandstones was the finding of 18% chromite in the fine sand-size heavy-mineral fraction of a Lower Ordovician sandstone in Quebec. Chromite in crystalline rocks is essentially restricted to ultramafic rocks such as peridotites, dunites, and serpentinites. On the basis of the abundance of chromite in the sandstone, the emplacement of a large sheet of ophiolite during Ordovician time in Quebec was inferred, although very little ophiolite is now present in the area.

Mica

Detrital micas of sand size are a minor constituent of most sandstones. They are most abundant in fine-grained sandstones that contain abundant micaceous metamorphic lithic fragments. The relative abundances of biotite, chlorite, and muscovite in sandstones are unknown. In thin section, we can see that green or brown biotite has changed to pale green chlorite with the development of an anomalous interference color, a bluish color different from the blue color of normal polarized light. In a few sandstones, brown biotite occurs in euhedral hexagonal flakes, a perfection of crystal habit that reflects crystallization in a fluid (probably lava) and an absence of grain-to-grain impacts in the sedimentary environment. We infer that the biotite flakes were blown out of a volcano (pyroclastic).

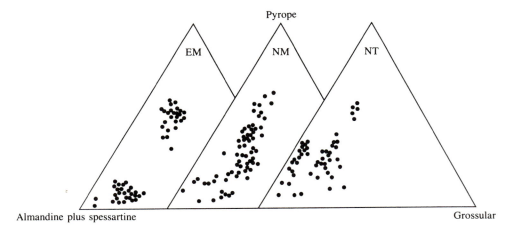

FIGURE 13-13

• • • • • • • • • • • • • • • •

Comparison of electron microprobe analyses of garnet grains. Sample EM is from Etive Formation, Murchison Field; NM, from Ness Formation, Murchison Field; NT, from Ness Formation, Tern Field, which is about 50 km from Murchison. The Ness Formation immediately overlies the Etive Formation. [Modified from A. C. Morton, 1985, *Sedimentology, 32,* 556, Fig. 1.]

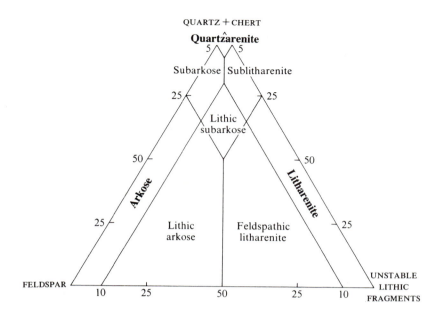

QUARTZ + CHERT

Quartzarenite

Subarkose | Sublitharenite

Lithic subarkose

Arkose

Litharenite

Lithic arkose

Feldspathic litharenite

FELDSPAR

UNSTABLE LITHIC FRAGMENTS

FIGURE 13-14

• • • • • • • • • • • • • • • •

One of many mineralogic classifications of sandstones in common use. Numbers give percentages. [From E. F. McBride, 1963, *J. Sediment. Petrol.*, *33*, 667, Fig. 1.]

Typically, such flakes occur in sandstones that contain other evidence of volcanic provenance, such as rhyolite fragments, sanidine grains, and quartz with β outlines.

Muscovite is volumetrically uncommon in granitoid rocks because of aluminum deficiency; most of the potassium needed to form the mineral goes instead into potassic feldspars during crystallization. Abundant muscovite in a sandstone suggests derivation from metamorphic rocks. Biotite, in contrast, occurs in a wide variety of crystalline rocks, and the proportions of iron and magnesium in the biotite reflect the composition of the rock in which the biotite formed.

Glauconite

Glauconite is a distinctive sand-size, granular material that occurs in thin sections as green or brown (oxidized), structureless near-spheres **(peloids)**, most of which are believed to have originated as fecal pellets. The peloids are exclusively marine in origin. Mineralogically, the peloids range from green smectite to a 10-Å dioctahedral mica (glauconite) and from a 7-Å trioctahedral serpentine (berthierine) to a 14-Å chlorite (chamosite). Ancient chamositic granules typically have oolitic coatings.

Green peloids are not restricted to a particular type of sandstone, although they are most common in essentially pure quartz sandstones of shallow marine origin. Typically, phosphatic debris is present as well, either as ovoid peloids and **ooids** or as shell fragments of the brachiopod *Lingula*. Glauconitic greensands and chamositic ironstones commonly occur above a coarsening- or

shoaling-upward facies sequence. Many of them are cross-bedded and burrowed, and some are interbedded with a ferruginized or phosphatized hardground, indicating a **diastem** and submarine cementation. Glauconitic and chamositic beds are most common at times when the cratonic blocks were widely dispersed and sea level was high, in Cambro-Ordovician and Cretaceous times, but the reason for this association is not known.

CLASSIFICATION
• • • • • • • • • • • • • •

The first objective of a petrologic project is a thorough description of the rocks, which is then summarized. The object of a summary is to convey the most important information as briefly as possible. If we are to summarize by means of a classification scheme, we are limited to consideration of only a few variables—those that we believe give the most insight into the genesis of the rocks. For nearly all rocks (igneous, metamorphic, and sedimentary), texture and mineral composition are the most important variables. A large number of suggestions has been published concerning the best way to combine texture and mineral composition in a classification scheme.

The easiest mineralogic separation to make in the field is among quartz, feldspar, and lithic fragments. These can form the poles of a classification triangle, and the inside of the triangle can be subdivided in any convenient manner. The triangle shown in Figure 13-14 is one of many in common use among sandstone petrologists.

TABLE 13-4 Descriptions of thin sections and the summary rock names of four detrital rocks

Mineral composition, %	Texture	Diagenetic effects	Complete rock name[a]
50 garnet schist fragments 15 amphibolite fragments 15 hornblende gneiss fragments 10 polycrystalline quartz 5 untwinned feldspar 5 metasiltstone fragments	No clay matrix; grains poorly sorted; grains well rounded; mean grain size 5 mm	Dolomite	Submature, pebbly, dolomite-cemented lithic conglomerate
40 quartz, mostly polycrystalline 25 twinned plagioclase 10 orthoclase 25 granodiorite	No clay matrix; grains well sorted; grains subrounded; mean grain size 1.2 mm	Calcite cement	Mature, very coarse-grained, calcite-cemented lithic arkose
10 quartz, some with β-habit 20 feldspar, mostly sanidine 70 rhyolite fragments	No clay matrix; grains well sorted; grains subangular; mean grain size 0.6 mm	Chalcedony	Mature, coarse-grained, chalcedony-cemented, volcanic feldspathic litharenite
60 monocrystalline quartz 25 chert 10 mudstone fragments 5 microcline	10% mud matrix; grains well sorted; grains rounded; mean grain size 0.2 mm	Squashed mudstone fragments	Immature, fine-grained mud-cemented, chert-bearing sublitharenite

[a]Uses the textural maturity concept of R. L. Folk and the mineralogic triangle of E. F. McBride.

The texture of the sandstone is indicated by the use of the concept of textural maturity. It is also useful to indicate in the rock name the type of cement that holds the detrital grains together. The presence in the sandstone of unusual constituents may also be worth noting if they are present in amounts greater than 5%. Examples of nomenclature that result from the use of the triangle in Figure 13-14 and the Folk scheme of textural maturity in Figure 13-3 are given in Table 13-4.

INTERPRETIVE PETROLOGY

After a conglomerate or sandstone has been described and perhaps classified, it must be interpreted. Four factors are decisive in determining mineral composition during deposition, and the success of interpretive clastic petrology hinges on the investigator's ability to recognize the relative importance of each factor. The four factors are tectonics, climate, depositional environment, and diagenesis.

Tectonics and Detrital Mineral Composition

The persistent movement of plates across the earth since at least Proterozoic time is the main cause of mountain building on the continents. It is also the main cause of the major depocenters that have come and gone during the past 2 billion years of geologic time. When continental plates collide, mountain belts are formed parallel to the collision surfaces, as occurred along the Atlantic coastline of the United States when the Iapetus Ocean closed during Paleozoic time. Mountain belts also form when an oceanic plate moves under a continental plate, as is now occurring along the western edge of the South American plate (Andes Mountains). When a plate splits to create a new ocean basin, as occurred starting in Triassic time at the site of the Mid-Atlantic Ridge, graben and horst mountains are formed parallel to the major rift.

Thus, different types of plate interactions result in different types of structural deformation, which in turn results in exposure of different types of sedimentary and crystalline rocks by erosion. By studying the mineral

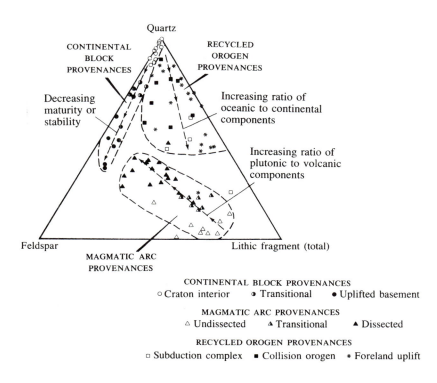

FIGURE 13-15

• • • • • • • • • • • • • • • •

Triangular quartz-feldspar-lithic fragment plot, showing the relationship between plate tectonics and sandstone composition. [From W. R. Dickinson and C. A. Suczec, 1979, *Am. Assoc. Petrol. Geol. Bull.*, *63*, 2171.]

composition of conglomerates and sandstones, the sedimentary petrologist hopes to determine the types of source rocks and, continuing the journey backward from effect to cause, the type of plate interaction that initiated the entire process (Figure 13-15).

Sediments associated with undissected magmatic arcs will tend to contain high percentages of mafic volcanic fragments and high plagioclase to potassic feldspar ratios. Quartz content will be low, commonly less than 10%. Examples include many Jurassic sandstones in Oregon and Washington. If erosional unroofing has occurred, cogenetic silicic plutons are exposed and the derived sandstones will contain more quartz and potassic feldspar. Examples include the Great Valley sequence (Cretaceous) in California.

Sands derived from sources on continental blocks are of two types: those from rifted areas associated with continental separations and those derived from broad positive areas in the interiors of stable cratons. The first group of sandstones will contain richly quartzofeldspathic detritus with high ratios of potassic feldspar to plagioclase. Typically, these sandstones form wedge-shaped accumulations that pinch out within 100–200 km from the granitic source. Examples include the Fountain Formation (Pennsylvanian) in Colorado and the many arkoses in Triassic basins along the eastern coast of the United States. In contrast, sands derived from broadly positive cratonic areas will be highly quartzose and contain less than 10% feldspar, nearly all of it potassic.

Sandstones derived from recycled orogen provenances tend to be the most complex. Three subcategories of provenance are distinguished: subduction complex provenance, collision orogen provenance, and foreland uplift provenance. Subduction complexes are composed of deformed ophiolitic and other oceanic materials and, therefore, supply sediment with various proportions of chert, argillite, graywacke, and serpentine fragments. Examples have been described from the Paleozoic and Mesozoic of Alaska.

Orogens formed by continent-continent collisions elevate the sedimentary and metasedimentary sequences present along each continental margin prior to collision. In addition, some oceanic and arc slices may occur as trapped fragments in the suture zone. Sands derived from these collisions are composed mostly of recycled sedimentary materials, have intermediate quartz contents, and have an abundance of sedimentary-metasedimentary lithic fragments. Examples include the Devonian sandstones of Pennsylvania, Carboniferous flysch in Arkansas and Oklahoma, and the Bengal Fan in the Indian Ocean, which resulted from the collision of India with the southern edge of Asia.

Foreland fold-thrust belts shed sediment into adjacent foreland basins and supply mostly sands recycled from sediments within the belts. Quartz, chert from replaced carbonate rocks, and sedimentary lithic fragments dominate in sandstones derived from this provenance. Feldspar contents commonly are nearly zero.

FIGURE 13-16
• • • • • • • • • • • • • • • •

Hypothetical climatic map of Pangaea, plotted with respect to a Jurassic paleomagnetic reference frame. Lambert Equal Area Projection with 30° and 60° paleomagnetic lines indicated. Note the complexity of the climatic pattern. [From W. W. Hay et al., 1981, *Geol. Rundschau, 70*, 310.]

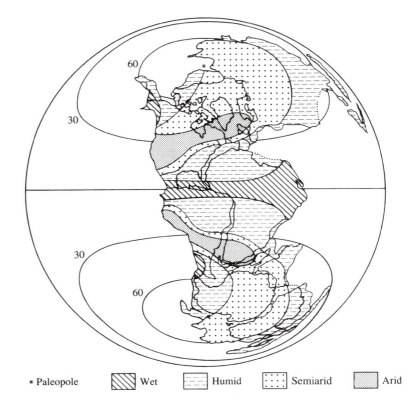

* Paleopole   ▨ Wet   ▤ Humid   ▨ Semiarid   ▨ Arid

Climate and Detrital Mineral Composition

As discussed in Chapter 12, weathering and soil formation result in the formation of sedimentary lithic fragments and the release of individual mineral grains from source rocks. However, the same processes also result in the disaggregation of the lithic fragments and decomposition of the mineral grains, with resultant loss of provenance and tectonic information. Probably more information is lost during weathering processes than during succeeding processes of transportation, deposition, and diagenesis combined.

The main control of climate is latitude; as a result, knowledge of the latitudinal position of plates during geologic time is critical to provenance studies. The smaller the interval of time for which the position is known exactly, the better. Knowledge of the sizes, shapes, and latitudinal positions of Earth's plates is essential for accurate provenance studies (Figure 13-16).

One way to approach the question of the effect of climate on sand compositions is to examine modern fluvial sands derived from the same types of crystalline rocks in different climates. Comparison of sands in semiarid and humid climates (Figure 13-17) indicates that an increase in annual precipitation from 37 cm

(semiarid) to 120 cm (humid) results in a loss of about two-thirds of the rock fragments and an increase of about 140% in the percentage of quartz. Granites, which contain 20–25% quartz, yield sand in humid climates that contains 60% quartz even before the sand has suffered significant stream transport. The effect in foliated metamorphic terranes is even more severe.

The great increase in the quartz percentage from crystalline source rock to first-cycle sediment in humid temperate climates leads to the possibility that a pure quartz sand might be generated as first-cycle detritus in humid tropical climates. In the humid tropical Orinoco basin of South America, first-cycle quartz arenites are produced in areas of diverse tectonic setting and relief. In the lowland of the basin, gentle slopes, tropical climate, and low erosion rate combine to produce the pure quartz sands. In the highland areas of the basin, the sands are stored on alluvial plains for extended periods of time, thereby allowing chemical weathering to destroy all rock fragments and minerals other than quartz. (The quartz grains will not be rounded, however.) Thus, it is clear that climatic effects have the capacity to obliterate tectonic signals. Without tectonic signals, plate regimes cannot be deciphered. Fortunately, the effects of weathering are incomplete in most sandstones, because an average sandstone contains 25–30% feldspars and lithic fragments.

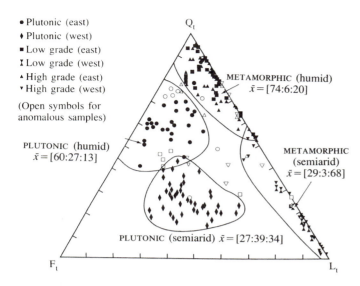

- Plutonic (east)
- Plutonic (west)
- Low grade (east)
- Low grade (west)
- High grade (east)
- High grade (west)

(Open symbols for anomalous samples)

METAMORPHIC (humid)
$\bar{x} = [74:6:20]$

PLUTONIC (humid)
$\bar{x} = [60:27:13]$

METAMORPHIC (semiarid)
$\bar{x} = [29:3:68]$

PLUTONIC (semiarid) $\bar{x} = [27:39:34]$

Transporting Agent, Depositional Environment, and Detrital Mineral Composition

Suppose a sand is generated from crystalline rocks in a temperate climate and then is transported by streams toward the coastline. What types of changes in fragment composition are to be expected as a function of transport rigor and distance? The answer to this question is not known quantitatively, although it is generally agreed that the effects are greater in high-gradient streams and that rock fragments will tend to disintegrate into their constituent minerals. Much change in detrital mineral composition occurs during periods when the grains are stranded on sandbars and floodplains and subjected to the effects of soil formation for extended periods of time.

The effect of depositional environment on sand mineralogy is better known than the effect of stream transport. It would be expected that the higher kinetic energies of the beach environment would make the sand more mineralogically mature (that is, more quartzose), and that is what many studies of both modern and ancient sediments have found. Data for 11 river-beach pairs from the U.S. Gulf Coast and from the coastline of

FIGURE 13-18

• • • • • • • • • • • • • • • •

Compositional differences between modern beach sands (triangles) and their associated fluvial sands (circles). Arrows connect pairs and also indicate amount and kind of compositional change; numbers refer to different sample sites in various parts of the world. [From Sedimentation Seminar, 1988, *J. Geol. Educ.*, *36*, 81.]

FIGURE 13-17

• • • • • • • • • • • • • • • •

$Q_tF_tL_t$ ternary plot of the composition of the medium sand size of Holocene fluvial sand of known first-cycle parentage shows the effect of climate on sand composition. [$\bar{x}$ = average percentages.] Anomalous samples are those that plot outside the inferred boundaries of their fields. [From L. J. Suttner et al., 1981, *J. Sediment. Petrol.*, *51*, 1236.]

South America (Figure 13-18) show that loss of lithic fragments is more pronounced than loss of feldspar in the samples. Loss of feldspar can be important, however. For example, after the depositional environments of the Lyons Formation in Colorado (Permian) were defined by sedimentary structures, it was found that the fluvial facies of the formation averaged 28% feldspar, whereas the beach-dune facies averaged only 8% feldspar. The variation occurred frequently within a short stratigraphic interval, illustrating the speed and effectiveness of high kinetic energy as a disintegrator of feldspar grains.

Many other studies of sandstone sequences deposited in different environments fail to find significant differences in mineral and rock fragment composition, however. The explanation for the different results is uncertain but may relate to the relative rates of fluvial sediment input, shoreline wave abrasion of the sand, and burial of the sand below the depth of wave effectiveness.

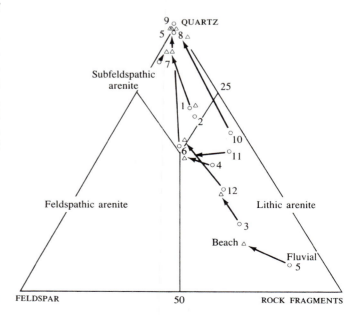

QUARTZ

Subfeldspathic arenite

Feldspathic arenite

Lithic arenite

Beach

Fluvial

FELDSPAR   50   ROCK FRAGMENTS

RECYCLING OF SEDIMENT
• • • • • • • • • • • • • • •

Our consideration of sand mineralogy thus far has concentrated on provenance determination and its implications for plate movements. To accomplish these objectives, the discussion has concentrated on three questions: What are the abundant minerals (and rock fragments) in sandstones? Did these minerals originate in metamorphic or in igneous rocks? What does the presence of these source rocks tell the sedimentary petrologist about the character of the plate and location of the sandstone? We have been concerned with the ultimate sources of the grains; however, two-thirds of the continental surface is covered by sedimentary rocks, not by metamorphic and igneous rocks. If we want to construct an accurate paleogeographic map for the Silurian or Cretaceous period, we have to determine which of the sand grains came directly from igneous or metamorphic sources and which were released from sedimentary rocks. We must distinguish between **ultimate sources** and **proximate sources.** Perhaps the quartz or garnet grain in a Jurassic sandstone emanated last from a Triassic sandstone, and before that resided in a Permian mudrock and an Ordovician conglomerate, since being released from a Proterozoic gneiss.

Four approaches are currently used for distinguishing ultimate from proximate sources:

1. *The percentage of quartz among the detrital grains.* The principle involved is that repeated reworking over long periods of time is required to remove completely all the feldspars and lithic fragments from an assemblage of sand grains. Therefore, if a sandstone is composed entirely of quartz, the grains were derived from older sandstones rather than directly from an igneous or a metamorphic rock. The difficulty with this criterion of **recycling** is that it ignores the effect of climate as a purifier of sand composition. Some workers believe that most pure quartz sandstones are produced by climatic effects, with abrasion of grains during transport of secondary importance.

2. *The percentage of superstable accessory minerals in the nonopaque heavy-mineral assemblage.* Because the most resistant minerals are zircon, tourmaline, and rutile, the *ZTR index* is a commonly used criterion of the importance of recycling. The principle is the same as for quartz and the limitation is also the same. What about climate?

3. *The degree of rounding of the quartz grains.* It requires repeated abrasion over long periods of time in a beach or dune environment to produce a well-rounded quartz grain from the angular grains released by crystalline rocks. Therefore, an assemblage of well-rounded grains indicates not only environments of deposition but recycling as well. Most pure quartz sands in the geologic column consist almost entirely of well-rounded grains. This suggests that they were not produced by climatic effects alone. A relatively small percentage of rounded grains may have been produced by solution of quartz in humid subtropical and tropical climates.

4. *The presence of abraded secondary growths on quartz grains.* It is common to find secondary quartz deposited from underground waters on the surfaces of detrital quartz grains. Subsequently, the rock may be disaggregated and the enlarged quartz grains released and abraded. The abraded overgrowths can be seen in thin sections of the later sandstone deposit (Figure 13-19) that includes the overgrown grains, and the overgrowths constitute excellent evidence of recycling. Unfortunately, quartz grains with recycled overgrowths are uncommon in sandstones, perhaps because most sandstones are cemented by calcite rather than quartz or perhaps because the overgrowths are detached from their host grain before redeposition.

SUMMARY
• • • • • • • • • • • • • •

Many petrologic characteristics of conglomerate and sandstone sequences are best seen in the field. For this reason, they should be described while the investigator is standing at the outcrop. Back in the laboratory, it is difficult to remember whether the conglomerates (which were too coarse-grained to sample adequately) contained a variety of types of rock fragments or only quartz pebbles or whether the conglomerates were more poorly cemented than the interbedded sandstones (which were sampled). Perhaps pebble compositions varied geographically or stratigraphically or with pebble size. Was there a relationship between the mean size of the gravel bed and that of the underlying or overlying sandstone bed? Each of these observations has the potential to contribute to the petrologic interpretation of the rock sequence.

Sandstone petrology is best studied in the laboratory. The most generally useful technique is thin section analysis, but scanning electron microscopy and various types of microchemical analyses are important tools in some investigations. The central control of sandstone mineralogy is plate tectonics, and this should be the framework into which mineralogic analyses are cast.

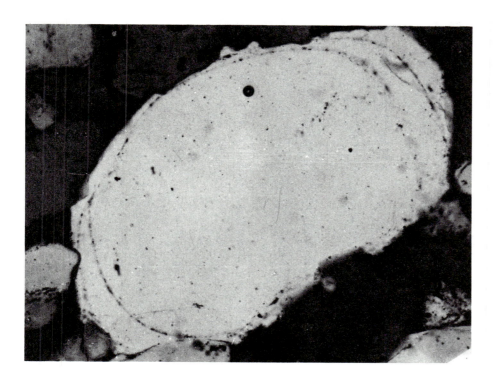

FIGURE 13-19
• • • • • • • • • • • • • • •

Photomicrograph of medium sand-size detrital quartz grain (crossed nicols), showing multiple rounded overgrowths, Weber Sandstone (Pennsylvanian-Permian), Utah. Border of detrital part of grain is marked by inner oval ring of water-filled vacuoles. [From I. E. Odom et al., 1976, *J. Sediment. Petrol.*, *46*, 867. Photo courtesy I. E. Odom.]

The mineral composition of the detrital rock is the best indication of the types of source rocks that supplied sediment to the rock being studied. Most provenance investigations require a sound understanding of igneous and metamorphic petrogenesis to be successful because mineralogic variations in sandstones are frequently quite subtle. Quartz grains may be plastically deformed or not, monocrystalline or polycrystalline; and polycrystalline grains may show a variety of types of internal structures characteristic of an igneous or a metamorphic parentage. Feldspar grains may be orthoclase, microcline, sanidine; a perthitic intergrowth; or plagioclase of variable composition. Different feldspar types and compositions are characteristic of different origins; but even with these grains, problems in interpretation exist. Minor accessory minerals are valuable sources of information, but many of the ones most abundant in sandstones occur in most types of crystalline source rocks.

The mineral composition of sandstones can be used to infer the tectonic setting in which the rocks formed. Cratonic sands are rich in quartz and chert. Sandstones formed along divergent continental margins are somewhat less rich in quartz and chert and typically contain 20–30% of foliated sedimentary or metamorphic lithic fragments. In intracontinental rift zones, richly feldspathic sands are formed (dominantly potassic feldspars) as granitic rocks are exposed by tensional forces. Along convergent continental margins, volcanic lithic fragments tend to be abundant and are accompanied by a feldspar suite with a high proportion of fairly calcic plagioclase grains (An content >30%). Within a convergent margin setting, many smaller tectonic units are present, and it may be possible to distinguish among them mineralogically.

STUDY EXERCISES
• • • • • • • • • • • • • • •

1. Explain why grain size is an important factor in the textural analysis of sandstones. What information can it provide about the origin of the rock?

2. Suppose you wanted to do a study of dispersal pattern and transport distance from a mountain front. What hydrologic and rock-mineral factors would you need to consider to do a good study?

3. In what ways is the provenance information obtained from study of conglomerates better than the information from associated sandstones? What types of tectonic, geographic, or hydrologic information might you infer from the ratio of sandstone to conglomerate in a stratigraphic section?

4. Categorize the common types of igneous and metamorphic rocks in terms of their survival potential. Make two lists, one in order of physical stability, the second in terms of chemical stability. How would you combine these lists to arrive at an "overall stability"?

5. In what ways do plate movements affect the mineral composition of sandstones? Does it make any difference whether the movements are latitudinal or longitudinal? Explain.

REFERENCES AND ADDITIONAL READINGS
● ● ● ● ● ● ● ● ● ● ● ● ● ●

Basu, A., and E. Molinaroli. 1989. Provenance characteristics of detrital opaque Fe-Ti oxide minerals. *J. Sediment. Petrol.*, *59*, 922–934.

Bilodeau, W. L., ed. 1986. Plate tectonics and petrologic suites. *Sediment. Geol.*, *51*, 1–135.

Chandler, F. W. 1988. Quartz arenites: Review and interpretation. *Sediment. Geol.*, *58*, 105–126.

Ibbeken, H., and R. Schleyer. 1991. *Source and Sediment*. New York: Springer-Verlag.

Johnsson, M. J. 1990. Tectonic versus chemical-weathering controls on the composition of fluvial sands in tropical environments. *Sedimentology*, *37*, 713–726.

Kleinspehn, K. L., and C. Paola, eds. 1988. *New Perspectives in Basin Analysis*. New York: Springer-Verlag.

Morton, A. C. 1985. Heavy minerals in provenance studies. In *Provenance of Arenites*, ed. G. G. Zuffa. Boston: D. Reidel, pp. 249–277.

Owen, H. G. 1983. *Atlas of Continental Displacement, 200 Million Years to the Present*. New York: Cambridge University Press.

Pettijohn, F. J., P. E. Potter, and R. Siever. 1987. *Sand and Sandstone*, 2nd ed. New York: Springer-Verlag.

Smith, A. G., A. M. Hurley, and J. C. Briden. 1981. *Phanerozoic Paleocontinental World Maps*. New York: Cambridge University Press.

Weijermars, R. 1989. Global tectonics since the breakup of Pangaea 180 million years ago: Evolution maps and lithospheric budget. *Earth-Science Rev.*, *26*, 113–162.

Chapter 14

DIAGENESIS OF SANDSTONES

Diagenesis is defined as all the physical, chemical, and biological changes that a sediment is subjected to (excluding folding and fracturing) after the grains are deposited but before they are metamorphosed. Some of these changes occur at the seawater-sediment interface, but the bulk of diagenetic activity takes place after burial. During burial, the main diagenetic processes are **compaction** and **lithification.**

COMPACTION AND CEMENTATION

Measurements of modern muds reveal that a mixture of mud and water on the seafloor is typically at least 60% water, and values as high as 80% have been recorded. The "sediment" has a "porosity" of up to 80%! The water squeezes out easily because the clay minerals in the mud are ductile (flexible) and platy and so can be compacted very tightly at relatively low pressures.

A marked contrast in compactibility is seen if one steps on a beach composed of quartz sand. The crunching sound produced reflects the friction of grain against grain as the pressure causes an increase in the tightness of packing. Very little water spews out of the sand around one's feet. Quartz grains are subspherical to ellipsoidal in shape, relatively closely packed when deposited, and are not ductile at sedimentary pressures and temperatures.

The pressure exerted on a layer of sedimentary rock at depth is equal to ρgh (ρ, density of water; g, accelera-

tion of gravity; h, height of water column); and if we assume that the sediment is permeable from its burial depth to the surface, ρgh is equal to the weight of a column of water of that height, approximately 10 bar/100 m. In some areas of rapid sedimentation, such as the Gulf Coast, many impermeable clay seals are formed in the stratigraphic column. In such cases the sediment pile is not permeable above certain layers of sand and on up to the surface; water in the sand supports some of the weight of the overburden (both rock and water), and the pressure on the water in the sand layer is greater than 10 bar/100 m. Fluid pressures may be as high as 25 bar/100 m, 2.5 times the pressure exerted by a column of water at that depth. The upper limit is set by the density of the overlying column of rock.

Pure quartz sandstones compact from initial depositional porosities of about 45% to values of 25% ±5%, depending on sorting. But many types of nonquartz grains occur in sands, and many of them are quite ductile. Fluvial sands may contain sand-size aggregates of floodplain mud, and other sands may contain clay matrix and appreciable percentages of ductile lithic fragments such as shale, slate, phyllite, and schist. All of these types of fragments deform easily at relatively shallow depths. They bend around the more rigid quartz and feldspar grains to fill pores and lithify the rock (Figure 14-1). Porosity can be reduced from the original 45% to nearly zero simply by squeezing ductile grains into pores. In such cases the original thickness of the sand must have been reduced by nearly 50%.

In summary, the amount of compaction of a freshly deposited sand primarily depends on four factors: clay content, percentage of ductile sand grains, sorting of the sand grains, and burial depth or effective stress.

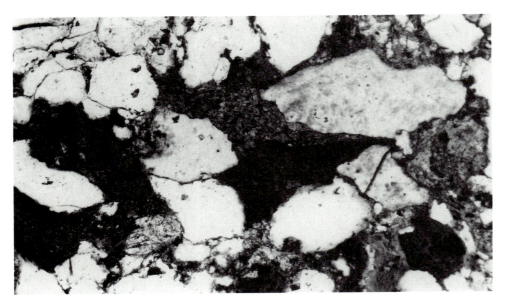

FIGURE 14-1
● ● ● ● ● ● ● ● ● ● ● ● ● ● ● ●

Center of photomicrograph shows two pelitic lithic fragments squeezed into conformable contact with each other and with surrounding quartz grains. Paleocene sandstone at 2380 m, Pinedale anticline, Green River Basin, Wyoming. Width of photo 1.3 mm. [Photo courtesy D. W. Houseknecht.]

The average sandstone contains only 15% lithic fragments, many of which are not ductile (for example, rhyolite, gneiss, and granite), and the amount of detrital clay in the average sandstone is probably about 5%. Therefore, the intimate grain-to-grain contact in most sandstones that results in lithification must be achieved largely by the introduction of chemical precipitates—that is, cements. The growth of new mineral matter into the depositional pores creates the intimate surface-surface contact needed for lithification. The degree of lithification depends on the amount of cement-to-grain contact produced. If only a small amount of secondary mineral matter is precipitated in the pores, the rock can be disaggregated into individual grains by finger pressure; such rocks are termed **friable.** An increased amount of pore filling produces a rigid but still porous sandstone, the type in which much of the world's petroleum, natural gas, and uranium are located. In the extreme, all porosity is lost and the sandstone is truly "hard as a rock."

QUARTZ CEMENT
● ● ● ● ● ● ● ● ● ● ● ● ● ● ● ●

Significant amounts of pore-filling quartz are, with few exceptions, restricted to sandstones whose detrital grains are nearly all quartz. Such sandstones occur mostly in intracratonic, foreland, and passive-margin basins. They are rare in rift basins, which typically contain arkoses, and in collision-margin basins, which typically contain lithic sandstones. Pure quartz sandstones are most common in depositional environments of high kinetic energy; therefore, quartz cement occurs most commonly in ancient beaches, marine bars, desert dunes, and some fluvial sandbars. It is rare in alluvial fans and turbidite sandstones.

Pore spaces that remain after compaction of the sediment may subsequently be partially or completely filled by quartz. This secondary quartz in detrital quartz sandstones grows as a coating on the detrital grains and in crystallographic continuity with them (Figure 14-2). The growths can take root at several locations on each grain but never at a place where the boundary between zones of undulatory extinction (a surface of discontinuity) intersects the grain surface. The **overgrowths** nucleate on both sides of this line of intersection and join during growth directly above the line, thus propagating the discontinuity surface into the overgrowth.

The boundary between the detrital grain and the overgrowth may or may not be visible with standard petrographic techniques. When visible, the boundary is marked by substances different in petrographic character from quartz. Sometimes the substance is the red mineral hematite, sometimes dark-colored organic-rich material, sometimes clay minerals, and sometimes petrographically irresolvable material (Figure 14-2). In any event, the substances that coat the detrital grain must be discontinuous; otherwise, secondary quartz cannot nucleate on the detrital grain. Secondary quartz in a quartz sandstone will not nucleate on the surface of a clay flake or bit of organic matter, but it will nucleate on the exposed surface of the detrital grain, grow laterally, and perhaps cover the bits of extraneous material. Growths of secondary quartz are initiated as rhombohedra or prisms and grow to coalesce, a growth pattern resulting in large planar surfaces that abut overgrowths from other grains somewhere in the pore space. When the surfaces abut, they generally lose their planar character and form crystallographically irrational compromise boundaries.

In many pure quartz sandstones the grains seem to interlock and adhere, but there appear to be no secondary growths. The perfectly contoured fit of grain against grain indicates that the texture could not have been produced by depositional processes. There are at least two alternative explanations: (1) The detrital grain had no extraneous substances on its surface at the time that secondary quartz was precipitated, so the host–new growth contact is petrographically invisible. (2) No secondary quartz is present, and the detrital grains were welded together by **pressure solution,** which is intergranular dissolution of adjacent grains as a result of nonhydrostatic stress.

Distinguishing between these two methods of lithification can be very difficult in routine petrographic work, but the distinction is important. If the first alternative is correct, the sandstone is a sink for silica in solution; so we have to find a large, outside source of silica to lithify the quartz sandstone. If the second alternative is correct, the sandstone is a source of silica in solution; so a pore solution capable of dissolving quartz and carrying the silica in solution to an area outside the area covered by the thin section is needed.

The special technique that resolves this problem is **luminescence petrography.** In this technique, a thin section is placed on a microscope stage and bombarded by electrons from below, a bombardment that causes certain parts of minerals to luminesce. The parts that luminesce are those that contain either **"activator elements"** as trace impurities (commonly transition elements or rare earth elements) or certain types or amounts of crystal defects. In the case of quartz, detrital grains nearly always luminesce, but secondary growths do not (Figure 14-3). This difference in response to electron bombardment can be related to the presence of impurity elements in the detrital quartz, which formed at temperatures above 300°C, and the lack of such impurities in the secondary growths, which formed at temperatures below 150°C.

Silica for quartz cement can originate in either shale or sandstone beds in the stratigraphic section, possibly including deeply buried rocks undergoing low-grade metamorphism. Many silica-generating mechanisms are known to operate during diagenesis, but their relative importance is usually uncertain. The most likely sources of silica for sandstone cementation are pressure solution, feldspar alteration and dissolution, and perhaps carbonate replacement of silicate minerals. It has been suggested that dissolved silica is transported vertically

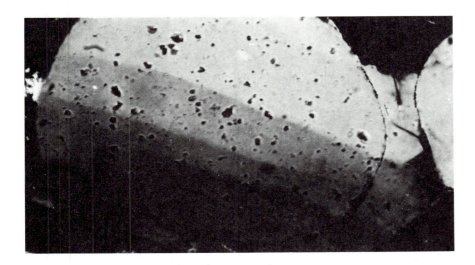

0   50 μm

FIGURE 14-2

• • • • • • • • • • • • • • • •

Thin section of a detrital quartz grain with rhombohedral and prismatic overgrowths (crossed nicols). The outline of the detrital grain is marked by a discontinuous coating of "dirt." Bands of undulatory extinction pass through both detrital grain and overgrowth. [From E. D. Pittman, 1972, *J. Sediment. Petrol.*, *42*, 513.]

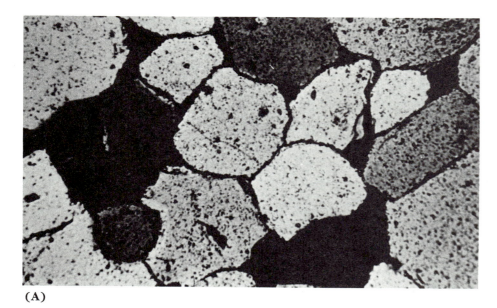

(A)

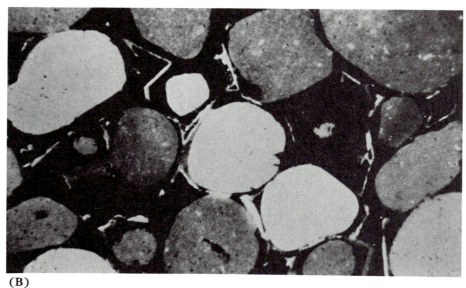

(B)

FIGURE 14-3
● ● ● ● ● ● ● ● ● ● ● ● ● ● ● ● ●

Photomicrographs of medium sand-size quartz grains in Hoing Sandstone (Devonian), Illinois. **(A)** Crossed nicols. **(B)** Cathodoluminescence. Apparent pressure-solution contacts are shown by luminescence petrography actually to be secondary quartz formed in the original pore space. [From P. A. Scholle, 1979, *Am. Assoc. Pet. Geol. Mem. No. 28*, p. 186. Photos courtesy R. F. Sippel.]

from deeper waters at higher temperatures to shallower depths, with the silica precipitating as the waters cool. However, data supporting this hypothesis are lacking. The source of most of the silica seen in silica-cemented quartz sandstones is unknown.

The solution chemistry of silica is relatively simple. Only two solid forms need to be considered for most purposes: **amorphous** silica (opal) and the crystalline equivalent (quartz). The solubility of amorphous silica at 20° is 100–140 ppm, depending on impurities; quartz has a constant composition and, hence, a constant solubility—6 ppm. The chief control of silica solubility in underground waters is temperature. The solubility of both the amorphous and the crystalline solid increases significantly as temperature increases. On the basis of

the relationships shown in Figure 14-4, we can make several inferences:

1. Amorphous silica is much more soluble than quartz at all temperatures and, consequently, at all depths of burial.

2. The solubility of amorphous silica is increased greatly by even small increases in temperature, that is, by only shallow burial. Thus, for practical considerations, it is impossible to precipitate opal as a cement at depth. There is no source of silica adequate to maintain supersaturation with respect to opal when values of several hundred parts per million are required simply to saturate the solution. (Normal surface waters—streams—average only 13 ppm.) Opal cement in a sandstone implies cementation within a few tens of meters of the surface.

3. Because surface waters and shallow underground waters both contain less than a few tens of parts per million of silica in solution, an extraordinary source of silica is required even near the surface to generate a solution containing more than 120 ppm. The source of this needed silica is revealed by examination of opal-cemented sandstones. In every case, the sandstone contains altered volcanic fragments such as basalt or glass shards (fragments). The alteration product is montmorillonite clay. Calculations reveal that the formation of the clay from the volcanic fragments does not require all the silica that is present in the fragments; some is released to the pore waters to increase the amount of silica in solution above the saturation level with respect to opal. The opal precipitates, filling the pore spaces and lithifying the sandstone.

4. The solubility of quartz rises much more slowly than that of amorphous silica, so saturation of pore solutions with respect to quartz may be very common at temperatures less than 100°–150°C. Geochemical data suggest that the content of silica in subsurface pore waters is determined largely by the depth (temperature) at which the water sample is taken. This is equivalent to saying that the solubility of quartz is the major buffer for the silica content of subsurface waters.

At low temperatures, the precipitation of quartz from a supersaturated solution may be delayed or effectively prevented because of an extremely slow rate of nucleation of quartz. For example, average river water contains 13 ppm silica, more than double the amount needed for saturation. Yet quartz never precipitates from rivers, because of nucleation problems.

FIGURE 14-4

• • • • • • • • • • • • • • • •

Solubilities of amorphous silica and quartz as a function of temperature. Amorphous silica becomes so unstable at temperatures greater than about 100°C that it crystallizes very rapidly to quartz, commonly in the form of chert.

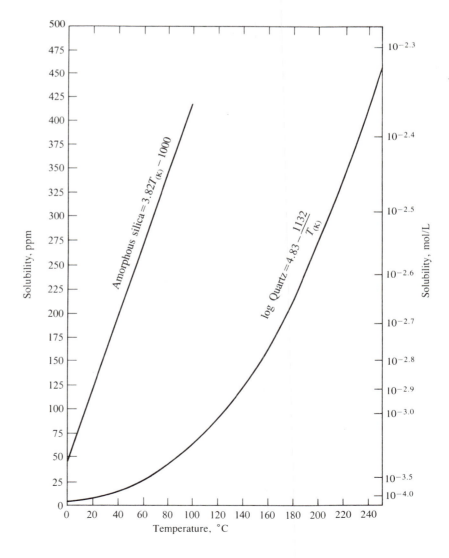

The solubility of silica is increased significantly at high values of pH, but the effect is insignificant at pH values below about 9 for amorphous silica and 10 for quartz. It is, therefore, unlikely that much quartz cement is precipitated as a result of pH changes in the subsurface. However, organic molecules commonly present in subsurface waters chelate silica. The organics cause accelerated dissolution and increased solubility of quartz compared with these processes in inorganic solutions.

CALCITE CEMENT
• • • • • • • • • • • • • •

Microcrystalline calcite with crystal diameters greater than 20 μm is the most common and widely distributed cement in sandstones. It is abundant in rocks deposited in all tectonic settings and depositional environments. Its continuity, however, is quite different from that of quartz cement. Quartz cement, where it occurs, is generally extensive and continuous on the scales of both outcrop and thin section. Calcite, in contrast, tends to have a patchy and irregular distribution; in thin section, the patchiness can occur on a scale of centimeters or less. Also, it commonly is found that calcite cement is absent in surface or near-surface exposures of a sandstone but becomes more abundant in the subsurface, an effect clearly resulting from weathering on the outcrop and dissolution of diagenetic calcite. In thin section, however, the question of whether a patchy distribution of calcite is due to incomplete original cementation or to subsequent dissolution of original cement is not always answerable.

Removal of calcite cement is probably the most common cause of secondary porosity in sandstones, although dissolution of silicate materials can also be important. In thin section, secondary pores can be identified by several criteria:

1. Corroded, irregular boundaries on calcite cement crystals or disruptions in growth zonations in the cement crystals.
2. Inhomogeneity of packing, with areas of apparently unsupported (floating) or loosely packed grains and high intergranular porosity interspersed with areas of tightly packed grains with little or no intergranular porosity.
3. Elongate pores in sandstones composed of subequant grains, a structure that reflects dissolution parallel to lamination and maximum permeability zones. If the detrital grains are elongate, however, such as schist fragments, pores that never were filled with cement will be naturally elongate without secondary dissolution.

4. Pores larger than the largest detrital grains, that is, microcaverns in the rock.
5. Molds of detrital grains (Figure 14-5); that is, the pore is the size and shape of the detrital grains in the rock. Because detrital limestone grains are relatively uncommon in sandstones, this feature usually indicates dissolution of a silicate grain such as a feldspar or a lithic fragment. In some sandstones the loss of such grains is so pervasive that a diagenetic quartz arenite is produced, with a corresponding loss of provenance information.

Geochemistry of Calcite Cement

The chemistry of calcium carbonate in water involves a dissolved gas, carbon dioxide. Because of this, the solubility relationships are considerably different from those of silica. Carbon dioxide is relevant because the solubility of calcium carbonate is severely affected by changes in pH; and the pH of pore waters is in part controlled by the partial pressure of carbon dioxide gas. The essential relationships among CO_2, H_2O, $CaCO_3$, and pH can be summarized by two chemical equations:

$$H_2O + CO_2 \rightleftharpoons H^+ + HCO_3^- \tag{1}$$

$$Ca^{2+} + 2\,HCO_3^- \rightleftharpoons CaCO_3 + H_2O + CO_2 \tag{2}$$

The first equation shows that carbon dioxide dissolved in water generates hydrogen ions and increases the acidity of the water; the increased hydrogen ion concentration causes any calcium carbonate present to dissolve, or makes it more difficult to precipitate calcium carbonate if none of this solid is present. If carbon dioxide escapes from the system, perhaps because of an increase in temperature, the reaction is forced to the left, hydrogen ions are eliminated, and the water becomes more basic. An increase of one pH unit—say, from 6.5 to 7.5—decreases the solubility of calcite in seawater from 500 to 100 ppm.

The second equation shows that the elimination of CO_2 causes more $CaCO_3$ to form because the reaction is forced continually to the right. Thus, both equations show that, other things being constant, it is easier to precipitate calcite cement at higher temperatures (increased depth) than at lower temperatures (shallower depth).

To cement a sandstone with any type of pore filling, a very large amount of water must circulate through the rock. For example, consider calcite cementation, as described by the simple reaction

$$Ca^{2+} + CO_3^{2-} \rightarrow CaCO_3$$

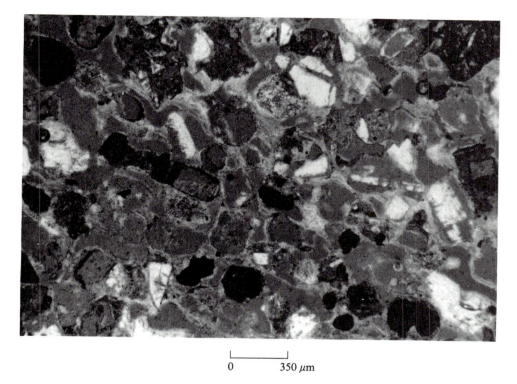

$$0 \qquad\qquad 350\ \mu m$$

FIGURE 14-5

• • • • • • • • • • • • • • • •

Thin section showing almost total dissolution of unstable framework grains, primarily plagioclase, volcanic rock fragments, and augite (Ilagan Formation, Plio-Pleistocene, Philippines). Original grain shape revealed by remnant authigenic clays. [From M. E. Mathisen, 1984, *Am. Assoc. Pet. Geol. Mem. 37*, 186.]

The reaction indicates that one mole of calcium ions will react to form one mole of calcium carbonate, assumed to be calcite. Seawater contains 400 ppm Ca^{2+} or 0.01 mol/L. Thus, one liter of seawater can produce a maximum of 0.01 mol of calcite, assuming that CO_3^{2-} is available as needed; that is, the limiting factor is calcium ion. One mole of calcite contains 100.09 g, and calcite has a density of 2.72 g/cm^3. Thus, 1 mol has a molar volume of (100.09 g)/(2.72 g/cm^3) or 36.79 cm^3, and 0.01 mol has a volume of 0.37 cm^3. If 1000 cm^3 of seawater can produce only 0.37 cm^3 of calcite, filling the pores in a sandstone with calcite would require 1000/0.37 or 2700 pore volumes of water to pass through the pore space. This is a substantial volume of water and requires either a high permeability, or a steep hydraulic gradient, or many millions of years, or some appropriate combination of these three variables.

The relationship between depth of burial of a sediment and temperature is straightforward: greater depth means higher temperatures, the exact temperature depending on the geothermal gradient, transfer of heat by circulating water, and other, more subtle factors. The amount of carbon dioxide dissolved in subsurface water is not as easily predicted. One important source of carbon dioxide in the subsurface is the partial decomposition of organic matter, particularly petroleum located in the pore spaces of detrital rocks. Petroleum consists of a complex mixture of naphthenes, paraffins, aromatic hydrocarbons, and olefins, and all of these carbon-hydrogen compounds can be oxidized. For example, the reaction between the aromatic hydrocarbon cyclohexane and oxygen gas is

$$C_6H_{12} + 9\,O_2 \;\rightarrow\; 6\,CO_2 + 6\,H_2O + heat$$

One mole of cyclohexane yields six moles of carbon dioxide. The amount of carbon dioxide generated in the subsurface by such reactions can be quite large. For example, the partial pressure of CO_2 gas (P_{CO_2}) at the earth's surface is $10^{-3.5}$ bar. But values at least as high as $10^{+1.5}$ (30 bars) have been reported from analyses of gases in subsurface waters from oil-producing basins. The increase of five orders of magnitude in P_{CO_2} from surface to subsurface will usually significantly decrease

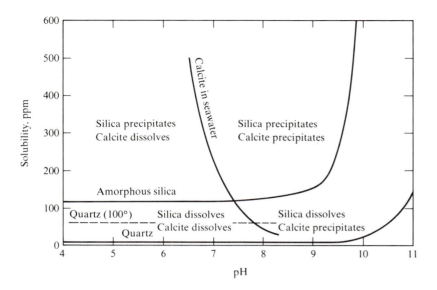

FIGURE 14-6

Relationships between pH and solubilities of amorphous silica, quartz, and calcite in seawater at 20°C. Intersections of curves divide graph into areas in which water is supersaturated or undersaturated with respect to each of the three solid phases. [From H. Blatt et al., 1980, *Origin of Sedimentary Rocks*, 2nd ed., p. 348. Reprinted by permission of Prentice Hall, Inc., Englewood Cliffs, New Jersey.]

the pH in the subsurface water and cause calcite dissolution. This is not always the case, however, because another by-product of petroleum decomposition or kerogen maturation is organic acids; these can buffer the pH of the water so that increasing the P_{CO_2} results in increased carbonate ion and precipitation of calcite rather than dissolution.

Comparative Geochemistry of Quartz and Calcite

In thin section studies of sandstones, it is common to find that detrital quartz grains have been etched by calcite cement or that silica cement contains inclusions of calcite. The frequency with which such relationships are found suggests that the chemical conditions that cause them must be common in subsurface waters. On the basis of the previous consideration of the effects of pH, temperature, and CO_2 pressure on calcium carbonate and silica, it is likely that changes in one or more of these variables is the cause of replacement reactions between calcite and quartz (Figures 14-6 and 14-7). Two important inferences can be made from these graphs:

1. The critical range of pH in which silica-calcite replacement reactions occur is 7 to 9, precisely the range of the vast majority of subsurface waters. At a diagenetic temperature of 100°C, and a burial depth of 2000–3000 m, the point of calcite-quartz intersection is at a pH of about 7.8.
2. At CO_2 pressures that are likely to be present during diagenesis, the calcite-quartz intersection lies at 40°–130°C, a burial depth of 1500–3000 m.

HEMATITE CEMENT

Three diagenetic iron oxides occur in sedimentary rocks, magnetite (Fe_3O_4), goethite ($FeO \cdot OH$), and hematite (Fe_2O_3). Of these, hematite is by far the most common in sandstones. Magnetite is stable only in the absence of gaseous oxygen, and goethite (which is brown) tends to dehydrate to hematite. Limonite consists of poorly crystalline hydrated ferric oxides and is yellowish in color.

The iron atoms that form hematite are derived from alteration of ferromagnesian minerals. Most of these iron atoms are originally in the reduced state, but the iron immediately oxidizes in contact with water and gaseous oxygen. The iron that is oxidized during surface weathering is subsequently transported, either adsorbed on clay minerals or as colloidal ferric oxide or hydroxide in surface or near-surface waters, and is deposited with muds or muddy sands. Some iron oxide may travel as surface coatings on coarser materials such as quartz, feldspars, or lithic fragments. In thin section the occurrence of red pigment along grain contacts is taken to indicate predepositional hematite coatings (Figure 14-8).

Iron-bearing detrital minerals can alter after deposition and burial (Figures 14-9 and 14-10) as well as in soils before burial, and this process occurs readily in permeable, aerated sandstones. Most of the red coloration produced in this way is formed during very early diagenesis; but irrespective of the time at which the hematite is formed, it can be used to obtain a numerical date of the event. Such paleomagnetic dating is based on the premise that as authigenic magnetic

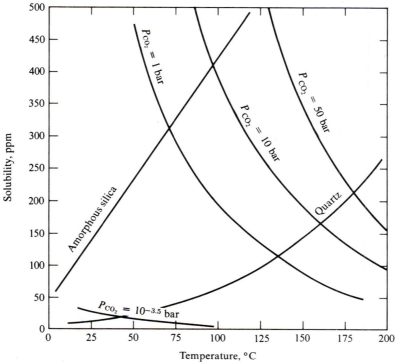

FIGURE 14-7

• • • • • • • • • • • • • • • •

Relationships among solubilities of quartz, amorphous silica, and calcite in seawater solution at different partial pressures of carbon dioxide gas and temperature. Silica solubility is unaffected by carbon dioxide pressure. Areas of dissolution and precipitation of each of the three solid phases can be designated as in
Figure 14-6, depending on the assumed partial pressure of carbon dioxide in subsurface water. Lowest carbon dioxide pressure curve represents the pressure of
carbon dioxide gas in seawater in equilibrium with the atmosphere. Note that the
curve intersections of quartz with calcite at the four carbon dioxide pressures
common during diagenesis occur at low temperatures.

minerals grow, they acquire a stable magnetization
aligned with the earth's magnetic field. If this magnetization can be isolated, we can compare the corresponding pole position with the Apparent Polar Wander Path
and thus can determine the time at which the magnetic
mineral formed. For Phanerozoic rocks, the precision
of these dates is about ±20 million years. In some
hematitic sandstones, several episodes of hematite precipitation have occurred, and in such cases each period
of precipitation can be instrumentally separated and
dated numerically. This technique is one of the few
methods available to obtain the numerical date of a diagenetic event.

The chemistry of iron in sedimentary environments
is complex because iron occurs at the earth's surface in
two oxidation states; the relative amounts of each
are determined by the balance between the amount
of free oxygen available to oxidize organic matter and
the amount of organic matter present to be oxidized.
If there is a surplus of oxygen, the organic matter will
be destroyed completely, and the remaining oxygen will
keep the environment in an oxidizing condition. A relative deficiency of oxygen gas results in partly decomposed and degraded organic compounds that accumulate in the reducing environment. For materials that
are immersed in water at and near the earth's surface,
water that contains dissolved oxygen must circulate
from the surface downward to maintain oxidizing conditions; otherwise, the supply of oxygen below the
water-sediment interface will be quickly exhausted.

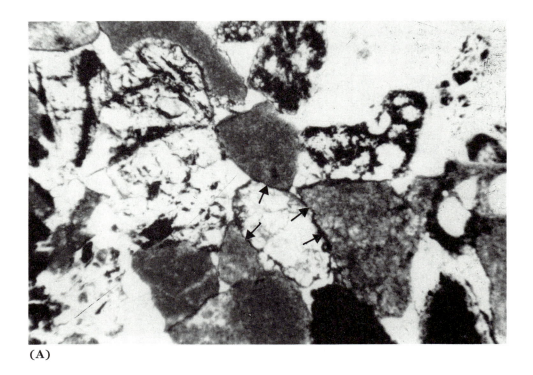

(A)

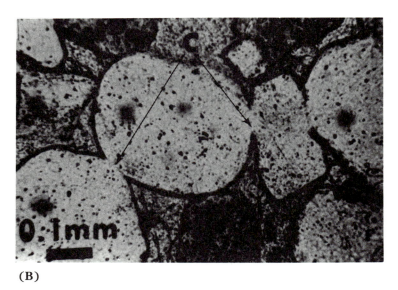

(B)

FIGURE 14-8
• • • • • • • • • • • • • • • •

Photomicrographs of two sandstones illustrating **(A)** predepositional and
(B) postdepositional precipitation of cement. In lower center of part (A) is a
carbonate lithic fragment in contact with four chert grains (arrows); hematite
is continuous along grain-grain contacts. [From Y. Al-Rawi, 1983, *Sediment.
Geol.*, *35*, 184.] In part (B) the gray quartz grains lack authigenic clay at grain-
grain contacts (arrows), an observation indicating that the clay is diagenetic.
[From G. A. Flesch and M. D. Wilson, 1974, *New Mexico Geology Society
Guidebook*, 25th Field Conference, p. 203. Photo courtesy C. T. Siemers.]

FIGURE 14-9
● ● ● ● ● ● ● ● ● ● ● ● ● ● ● ●

Hand specimen of sandstone of Neogene age in arid Baja California, Mexico, showing diagenetic red halo (dark area in center) of hematite spreading downward from zone of concentration of ferromagnesian minerals. [From T. R. Walker, 1967, *Geol. Soc. Am. Bull.*, *78*, plate 2.]

Figure 14-11 shows the relationship between pH, Eh, and the stability fields of amorphous ferric hydroxide and hematite. Both of these ferric compounds have large fields of stability, that is, they are very insoluble at nearly all naturally occurring values of pH and Eh. Either $Fe(OH)_3$ or hematite may be the first ferric-iron precipitate from saturated solution, but the amorphous hydroxide quickly dehydrates to hematite:

$$2 Fe(OH)_3 \rightarrow Fe_2O_3 + 3 H_2O$$

FIGURE 14-10
● ● ● ● ● ● ● ● ● ● ● ● ● ● ● ●

Thin section of intensely etched and partly dissolved hornblende grain, Hayner Ranch redbeds (Miocene), north of Las Cruces, New Mexico. [From T. R. Walker, 1976, *The Continental Permian in Central, West, and South Europe* (Dordrecht, Holland: D. Reidel), p. 261.]

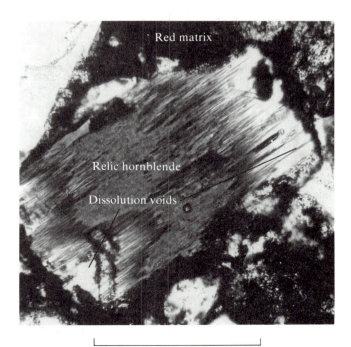

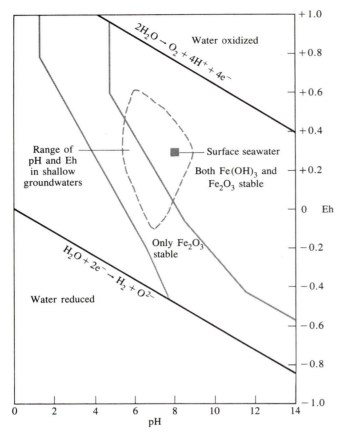

FIGURE 14-11

• • • • • • • • • • • • • • • •

Stability fields of amorphous ferric hydroxide (smaller field) and hematite (larger field) in the system Fe–O₂–H₂O as functions of pH and Eh at 25°C. Outside the stability fields, iron exists as soluble ferric or ferrous species. Among sedimentary environments, only some peat bogs, waterlogged soils, and a few deep diagenetic settings have pH-Eh values outside the hematite stability field.

Further, once hematite is formed, it is very resistant to transforming into more soluble iron species. Most near-surface environments are oxidizing and have pH values greater than 5, so the red color caused by hematite is very common in the sedimentary environment near the earth's surface.

CLAY MINERAL CEMENT
• • • • • • • • • • • • • • •

Individual clay flakes are, of course, too small to be seen with a hand lens, but sometimes it is possible to detect the presence of clay in a hand specimen. If present in amounts of at least 5–10%, clay minerals coat the sand grains to the extent that the boundaries of the grains

appear fuzzy or indistinct. Even when clay is detected, however, it is not possible in the field to determine whether the clay is detrital or secondary. In either instance, it can serve as a binding agent.

Ancient volcaniclastic sandstones very commonly contain abundant diagenetic **matrix,** usually composed of chlorite, sericite, and quartz in older rocks, zeolites and montmorillonite in younger ones. Boundaries between recognizable detrital grains and the irresolvable clay paste are typically fuzzy and indistinct—a textural transition indicating a mineralogic transition between the two materials. It is impossible to make either an accurate or a precise point count in thin sections of such rocks.

In sandstones whose detrital sand grains are all quartz, any clay minerals that grow during diagenesis can be recognized and identified, either in thin section or with a scanning electron microscope, because of their uniform coating thickness around the grains, their orientation normal to the grain surface, and their euhedral shape (Figure 14-12). This is also true for many arkoses. Clay minerals can also enter sandstones by mechanical infiltration from overlying soils. Such clays adhere to detrital sand surfaces and are oriented parallel to these surfaces (Figure 14-13).

In most sandstones the volume of **authigenic** clay minerals is minor, no more than 1–2% of the rock. The prominent exception is volcaniclastic sandstones, rocks that contain large percentages of very unstable lithic fragments that are deposited near convergent plate margins in areas of high heat flow and geothermal gradient. All major clay minerals can form: kaolinite, montmorillonite, and illite, as well as chlorite and various zeolite minerals. Each of these minerals has a field of stability determined by temperature and the activities (effective thermodynamic concentrations) of the elements from which the minerals are formed. For the zeolite minerals, the fugacity (effective partial pressure) of water is also important. A typical chemical reaction that occurs at near-neutral pH conditions and can result in the precipitation of authigenic clay is

$$2\,Al(OH)_3{}^0 + 2\,H_4SiO_4 \rightarrow Al_2Si_2O_5(OH)_4 + 5\,H_2O$$

aluminum   silica   kaolinite   water
in solution   in solution

Addition of appropriate amounts of metallic cations can result in the precipitation of the other clay minerals and chlorite: K^+ for illite, Na^+ and/or Ca^{2+} for montmorillonite, and Mg^{2+} and/or Fe^{2+} for chlorite. It is assumed that adequate amounts of alumina and silica are available. All reactions in the sedimentary environment are solution-precipitation reactions; solid-solid transformations can occur only at metamorphic temperatures and pressures.

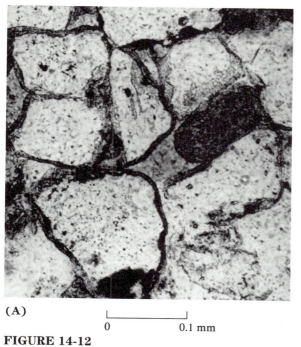

(A)

├──────────┤
0   0.1 mm

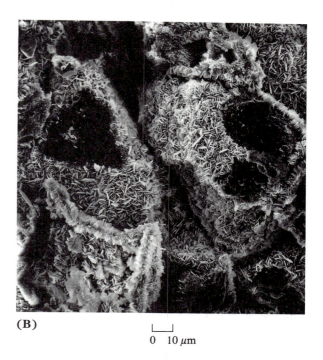

(B)

├──────┤
0   10 μm

FIGURE 14-12
● ● ● ● ● ● ● ● ● ● ● ● ● ● ● ● ●

Authigenic pore-lining chlorite, Tuscaloosa Formation (Cretaceous), Louisiana, as seen in **(A)** thin section (polarized light microscopy) and **(B)** by scanning electron microscopy (SEM). The chlorite rims are green and consist of euhedral, pseudohexagonal crystals 2–5 μm in size. Crystals are on edge and oriented perpendicular to detrital quartz-grain surfaces. Bald areas in the SEM are areas of former detrital grain contact and consequently lack the secondary chlorite. Scale bar in SEM represents 10 μm. [From J. E. Welton, 1984, *SEPM Petrology Atlas* (Tulsa: American Association of Petroleum Geologists), p. 38.]

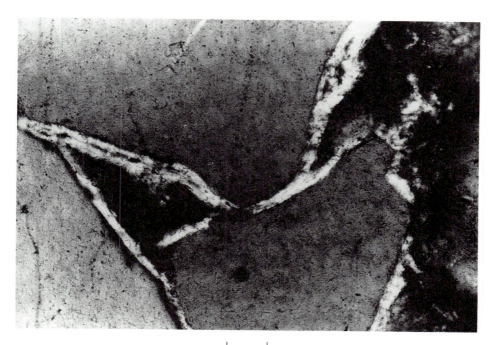

FIGURE 14-13
● ● ● ● ● ● ● ● ● ● ● ● ● ● ● ● ●

Infiltrated soil clay (Jurassic, Aracas oil field, Brazil). Clay platelets are arranged tangentially around the detrital quartz grains. There is some shrinkage of the clay because of dehydration, evidenced by separation of clay layers from each other and from the quartz. [From M. A. S. Moraes and L. F. DeRos, 1990, *J. Sediment. Petrol.*, *60*, 813.]

├──────┤
0   35 μm

The stability fields of the various clay minerals may either enlarge or shrink with respect to other mineral phases as temperature increases with increasing depth of burial.

ZEOLITE CEMENT

Authigenic zeolites are common in altered volcaniclastic sandstones; and, because of the variable chemical compositions of this group of minerals, about 15 different species have been reported from diagenetically altered volcaniclastic rocks (sandstones that are rich in volcanic fragments). The most common species are analcite ($NaAlSi_2O_6 \cdot H_2O$), heulandite [$(Ca,Na_2)Al_2Si_7O_{18} \cdot 6 H_2O$], and laumontite ($CaAl_2Si_4O_{12} \cdot 4 H_2O$), their compositions reflecting the dominance of sodium in seawater and of calcium in the basaltic fragments from which the zeolites form. The zeolite facies can be considered as either high-grade diagenesis or low-grade metamorphism. Its occurrence is confined to sediments rich in volcanic detritus and lacking carbonate rocks. Typical reactions of the facies include the reaction of heulandite to laumontite:

$$CaAl_2Si_7O_{18} \cdot 6 H_2O \rightarrow$$
$$\text{heulandite}$$
$$CaAl_2Si_4O_{12} \cdot 4 H_2O + 3 SiO_2 + 2 H_2O$$
$$\text{laumontite} \quad\quad \text{quartz}$$

and the direct conversion of plagioclase to laumontite:

$$CaAl_2Si_2O_8 + 2 SiO_2 + 4 H_2O \rightarrow CaAl_2Si_4O_{12} \cdot 4 H_2O$$
$$\text{anorthite} \quad\quad \text{quartz} \quad\quad\quad \text{laumontite}$$

ALBITIZATION

In sandstones buried to depths of more than 2000 m and subjected to temperatures of about 100°C, both potassic and calcic feldspars are commonly at least partially albitized. When albitization is complete, the resulting feldspar contains at least 95% albite. Well-described examples occur in the Harz Mountains of northwestern Germany and in numerous Tertiary sandstones in the subsurface of the Gulf Coast. In thin section it may be difficult to detect albitization other than by the possible presence of patchy birefringence or sharply bounded areas of different content of water

vacuoles in partially albitized grains. Partial albitization shows quite clearly in backscattered electron images, however (Figure 14-14).

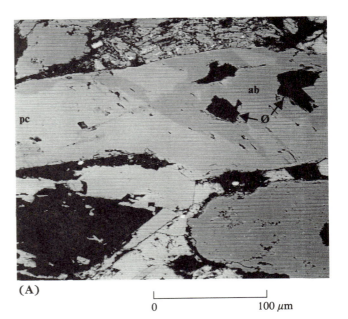

(A)

0   100 μm

(B)

0   100 μm

FIGURE 14-14

Backscattered electron images showing albitization of **(A)** plagioclase and **(B)** potassic feldspar in the Frio Formation (Oligocene) in southern Texas. pc, plagioclase; K, potassic feldspar; ab, albite; ϕ, secondary porosity. [From K. L. Milliken, 1989, *J. Sediment. Petrol.*, *59*, 370.]

CALCITIZATION
• • • • • • • • • • • • • • •

Another commonly observed type of feldspar alteration in many sandstones is calcitization, which may be produced as a by-product of the albitization process:

$$NaAlSi_3O_8 \cdot CaAl_2Si_2O_8 + 4\,SiO_2 + 2\,Na^+ + CO_3^{2-} \rightarrow$$
$$3\,NaAlSi_3O_8 + CaCO_3$$

Calcite can also be generated by the alteration of the calcic zeolite laumontite.

Earlier in this chapter we noted that the light-mineral fraction of sandstones can be changed during diagenesis to the extent that a diagenetic quartz arenite can be produced from a depositional feldspathic or lithic sandstone. This alteration can occur because materials crystallized at very high temperatures are unlikely to be stable during either weathering at 25°C or diagenesis at 50°–250°C. This principle can be applied to the heavy-mineral assemblage of sandstones as well as to the light assemblage. For example, in Holocene sediments of appropriate provenance, detrital augite, hypersthene, and olivine can be abundant and dominate the heavy-mineral suite. These minerals are rarely, if ever, found in pre-Tertiary sandstones. Paleozoic sandstones tend to be selectively enriched in the heavy minerals that are the most chemically stable, such as zircon, tourmaline, and rutile (ZTR). Stereotypically, high content of detrital quartz in the light-mineral fraction is associated with a high ZTR index in the heavy-mineral fraction. There are, however, many exceptions to this generalization, and it is not well documented. The relationship, if any, between opaque minerals and either quartz percentage or ZTR index is unknown.

PLATE TECTONIC CONTROLS
OF DIAGENESIS
• • • • • • • • • • • • • • •

The five principal controls of diagenetic reactions in sandstones are temperature (*geothermal gradient*), rate of sediment burial, mineral and rock fragment composition of the sandstone, pore water composition, and water circulation (porosity and permeability). Each of these variables is affected in some manner by the plate tectonic setting in which diagenetic reactions occur. Heat flow and geothermal gradients are higher in areas where magma is relatively close to the surface, such as in pelagic sediments along mid-ocean rifts, in intraarc basins, and in rift sediments along continental margins. Rates of sediment accumulation and burial are highest along trailing edges (for example, Gulf Coast), subduction zones (for example, Aleutian trench), and areas of continent-continent collision (for example, Himalaya drainage to the Bengal submarine fan). High percentages of unstable and very reactive grains occur in trenches and in forearc and intraarc settings, where volcaniclastic sediments are abundant (for example, California). Pore waters are richest in dissolved constituents near subduction zones because of the ready availability of seawater and glassy volcanic sediment, although the abundant salt that forms the near-basal sediment of trailing margin accumulations can supply high amounts of ions such as Na^+, Ca^{2+}, and K^+ (Gulf Coast). Water circulation is a function of good permeability and high temperature gradient, factors that are not normally combined. Good permeability occurs in depositional environments of high kinetic energy, such as beaches and bars, environments that are not abundant in areas of high heat flow. The best permeabilities generally occur in intraplate sediments such as those associated with epicontinental seas, tectonic settings of low heat flow and geothermal gradient. Clearly, the interrelationship between plate setting and diagenesis is complex and requires considerable knowledge of geology to decipher.

SUMMARY
• • • • • • • • • • • • • •

The main diagenetic processes in sandstones are compaction and the formation of new minerals, either as pore-filling cements or as replacements of the original detrital fragments. The effectiveness of compaction depends on the ductility of the detrital grains and on ductility produced as part of the chemical diagenesis of the rock. Originally ductile fragments include grains of shale and foliated metamorphic rock; secondary ductility can be induced in originally rigid grains by the transformation of these grains into clay matrix. With increased depth of burial, ductile sand grains and clay are squeezed into adjacent primary or secondary pore spaces, thinning the stratigraphic section and diminishing the porosity and permeability of the sandstone.

Porous and permeable sandstones are lithified either by precipitation in the pores of secondary quartz, calcite, or other minerals in the pores or, in the case of quartz arenites, by pressure solution. The growth of secondary minerals in a sandstone requires that the pore fluid be oversaturated with respect to the mineral being precipitated. It is also necessary for very large amounts of pore fluid to pass through the rock. Quartz and calcite form the bulk of authigenic mineral growths in sandstones, but others, such as hematite and clay minerals, can dominate in some sandstones. Much hematite seems to be produced very rapidly by the leaching and oxidation of ferrous iron from ferromagnesian minerals in soils. However, continued destruction of ferromagnesian minerals much later in the geologic history of the rock can cause hematite to be formed throughout the life of a sandstone.

Lithic sandstones have a complex diagenetic history because of the great variety of fragment types they contain. Because of the combined effects of compaction and chemical alteration, it is commonly not possible to distinguish between detrital grains and primary or secondary matrix in these rocks. The problem is particularly severe in volcaniclastic sandstones deposited near convergent plate margins, because of the unstable nature of the fragments and the exceptionally high heat flow in such areas. Adequate evaluation of diagenetic history in volcaniclastic sandstones depends in part on experience and on an understanding of physical chemistry.

STUDY EXERCISES
● ● ● ● ● ● ● ● ● ● ● ● ● ● ●

1. Discuss the factors that affect the closeness of packing in a sandstone. In what way is packing related to tectonic setting, for example, a cratonic setting as contrasted to a continental collision setting?

2. Suggest some reasons why quartz cement is absent from sandstones whose detrital grains are not at least 90% quartz. In what tectonic settings would you expect quartz-cemented sandstones to be most common? Least common?

3. Calcite is the most common cement in sandstones. Suggest reasons why this is the case. Do you think the difference in partial pressure of carbon dioxide in the atmosphere during the first 2 billion years of Earth history might have affected the commonness of calcite cement during that period? Why?

4. In what way might the higher ratio of carbon dioxide to oxygen during early Precambrian time have affected the abundance of hematite cement in sandstones during this period? How might you test your hypothesis?

5. How might albitization of feldspars in some stratigraphic sections affect provenance interpretations and tectonic inferences? In what tectonic settings might you expect albitization of detrital feldspars to be most common?

REFERENCES AND ADDITIONAL READINGS
● ● ● ● ● ● ● ● ● ● ● ● ● ●

Dutton, S. P. 1993. Influence of provenance and burial history on diagenesis of Lower Cretaceous Frontier Formation sandstones, Green River Basin, Wyoming. *J. Sediment. Petrol.*, *63*, 665–677.

Hutcheon, I. E., ed. 1989. *Short Course in Burial Diagenesis*. Toronto: Mineralogical Association of Canada.

Lundegard, P. D. 1992. Sandstone porosity loss— Relative importance of compaction and cementation. *J. Sediment. Petrol.*, *62*, 250–260.

McBride, E. F. 1989. Quartz cement in sandstones: A review. *Earth-Science Rev.*, *26*, 69–112.

Pettijohn, F. J., P. E. Potter, and R. Siever. 1987. *Sand and Sandstone*, 2nd ed. New York: Springer-Verlag.

Scholle, P. A. 1979. *A Color-Illustrated Guide to Constituents, Textures, Cements, and Porosities of Sandstones and Related Rocks. Am. Assoc. Pet. Geol. Mem. No. 28.*

Surdam, R. C., L. J. Crossey, E. S. Hagen, and H. P. Heasler. 1989. Organic-inorganic interactions and sandstone diagenesis. *Am. Assoc. Pet. Geol. Bull.*, *73*, 1–23.

Tada, R., and R. Siever. 1989. Pressure solution during diagenesis. *Ann. Rev. Earth Planet. Sci.*, *17*, 89–118.

Wolf, K. H., and G. V. Chilingarian, eds. 1992. *Diagenesis, III*. New York: Elsevier.

Chapter 15

MUDROCKS

Mudrocks are the most abundant type of sedimentary rock, forming about 65% of the stratigraphic column. They are common in sediments from Archean to Holocene, at elevations from oceanic trenches to Himalayan lakes, and in climates from the Antarctic to tropical jungles. They are the source rock for the petroleum, natural gas, and many of the metals on which our industrial civilization depends.

Mudrocks can be used to help solve numerous global sedimentary problems as well as more local ones. The amount and grain-size distribution of eolian dust in the deep ocean basin are used to unravel changes in atmospheric circulation, and the occurrence of black shales has become a key element in paleoceanographic investigations and the reconstruction of plate motions.

Mudrocks are the **protoliths** for pelitic schists, key rocks for the interpretation of the metamorphic history of complex deformed terranes. It is not possible to understand such terranes without an understanding of the areal distribution, mineral composition, and chemical composition of mudrocks.

FIELD OBSERVATIONS

As is the case for all rocks, an adequate field description of a mudrock should include information about the texture, structure, and mineral composition. It is more difficult to accomplish this for mudrocks than for sandstones or carbonate rocks because of the fine grain size of mudrocks.

Textures

Texture is defined as the size, shape, and arrangement of the grains or crystals in a rock. Individual grains in a mudrock cannot always be seen with a hand lens, but despite this constraint it is still possible to make a semiquantitative estimate of the ratio of **silt** to clay. The method is to nibble a bit of the rock and grind it between the teeth to determine whether the rock is gritty. Mudrocks are formed almost entirely of quartz and clay. Quartz is gritty; clay is slimy. If you sense no abrasion of your teeth, the rock contains more than two-thirds clay minerals; and clay minerals are the bulk of the clay-size particles. If grit is sensed, the rock probably contains between two-thirds and one-third clay minerals (Table 15-1). If it contains less than one-third clay minerals, enough quartz silt is present to be seen with a hand lens.

The shape of the grains in a mudrock cannot be determined in outcrop because of their small size, but thin section observations reveal that the silt-size quartz grains are angular and commonly platy. Particles with diameters less than about 60 μm travel almost entirely in suspension, being small enough to be suspended even by weak currents. Hence they suffer few grain-to-grain impacts and do not round. They can, however, be rounded by chemical weathering during soil-forming processes.

Clay minerals have a shape determined by their crystal structure. Most commonly they are shaped like a sheet of paper, which is one of the reasons that many mudrocks are fissile.

Structures

Sedimentary structures are generally due to differences in rock **fabric,** the spatial arrangement of the minerals and other materials of which the rock is composed. The degree of closeness of the grains **(packing),** the presence or absence of grain orientation, and various types of mineral segregations all contribute to fabric development.

FISSILITY. The major sedimentary structures of mudrocks that are visible in outcrop are **fissility** and **lamination.** Fissility is a property of a mudrock that causes it to break along thinly spaced planes parallel to the bedding and to the orientation of the sheetlike clay flakes (Figure 15-1). The existence of fissility depends on many factors, only one of which is the abundance of clay minerals. Mudrocks with identical percentages of clay can differ greatly in fissility because of the following factors:

TABLE 15-1   Classification of mudrocks

Ideal size definition	Field criteria	Fissile mudrock	Nonfissile mudrock
>2/3 silt	Abundant silt visible with hand lens	Silt-shale	Siltstone
>1/3, <2/3 silt	Feels gritty when chewed	Mud-shale	Mudstone
>2/3 clay	Feels smooth when chewed	Clay-shale	Claystone

Source: From H. Blatt et al., 1980, *Origin of Sedimentary Rocks*, 2nd ed. (Englewood Cliffs, NJ: Prentice Hall), p. 382, Table 11-1.

1. *Differences in degree of preferred orientation of the clay flakes during deposition.* All small particles tend to adhere when they collide during transport, and this tendency is increased by salinity and by the presence of organic matter or other materials in the water. The process of adhesion is termed **flocculation.** Maximum adhesion is achieved when salinity has increased to about 2000 ppm. The higher the proportion of floccules in a mudrock, the less likely it is that the rock will be fissile.

2. *Bioturbation.* Bottom-dwelling organisms also affect the development of fissility (Figure 15-2), as does escape of water during sediment compaction (Figure 15-3). As organisms scavenge through the fresh mud for organic matter, they disrupt the depositional fabric, in-gest sediment, and excrete pellets. In well-aerated environments the rate of fabric disruption can exceed the rate of sediment deposition. The extent to which physical structure is altered depends on the ratio of biological mixing rate to sediment accumulation rate. Ratios exceeding 10 result in complete homogenization of the sediment.

3. *Diagenetic effects.* When clay minerals are buried and subjected to increased temperatures and nonhydrostatic pressures, they recrystallize (see later discussion), a process that can result in the development of fissility. The frequency with which fissibility is produced by recrystallization is not known.

LAMINATION.   *Lamination* refers to parallel layering within a bed. By definition, a *bed* is thicker than 1 cm, and a *lamina* is thinner than 1 cm. Lamination can have many origins related to variations in current strength during deposition and to changes in composition of the sediment deposited. For example, a bed of mudrock may contain laminae of black organic matter, zones of green color within a dominantly red unit, or placering of quartz silt within an otherwise clayey mudrock. Each of these features can supply unique information about oxidation or reduction conditions at the site of deposition, changes in these conditions after burial, or variations in current strength with time during deposition of the bed.

FIGURE 15-1

Well-developed fissility in Antrim Shale (Mississippian), Alpena County, Michigan. Hand lens is about 2 cm in diameter. [From R. V. Dietrich and B. J. Skinner, 1979, *Rocks and Rock Minerals* (New York: Wiley), p. 201. Photo courtesy R. V. Dietrich.]

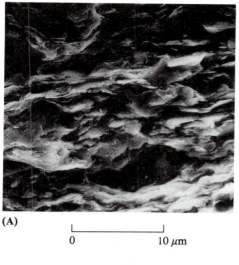

(A)

|_____|
0   10 μm

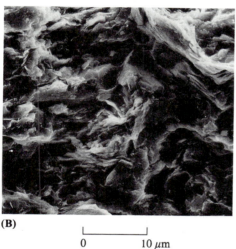

(B)

|_____|
0   10 μm

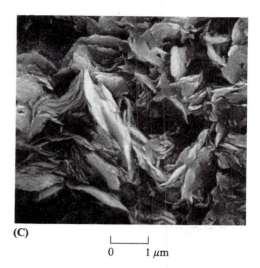

(C)

|_____|
0   1 μm

FIGURE 15-2
● ● ● ● ● ● ● ● ● ● ● ● ● ● ● ● ●

Scanning electron micrographs of mudrocks.
(A) Nonbioturbated shale, showing parallel
particle orientation. **(B)** Random fabric of
highly bioturbated shale. **(C)** Random clay
fabric caused by flocculation rather than bioturbation, as shown by domains of face-to-face
oriented flakes in this mudstone. Note absence
of domains in part (B). [From N. R. O'Brien,
1987, *J. Sediment. Petrol., 57*, 451.]

Colors

The colors of mudrocks fall almost entirely into two
groups: gray-black and red-brown-yellow-green. The
gray-black shades reflect the presence of 1% or more of
free carbonaceous material, usually reflecting deposition in an oxygen-deficient environment. In well-
oxygenated water the concentration of free oxygen is
10^{-4} mol/L. To obtain conditions sufficiently reducing
for black free carbon to accumulate requires a decrease
to about 10^{-6} mol/L. This condition implies a lack of
circulation of aerated water in the depositional environment to prevent complete oxidation of the organic
tissues to carbon dioxide plus water. It is important
to note that lack of circulation is not related to depth
of water. Most of the ocean floor at depths of many
thousands of meters is kept well oxygenated by the
cold bottom currents that originate at the surface in
polar regions and circulate throughout the oceans. And
many shallow areas of the ocean floor are stagnant, for
example, the lagoons between the Texas Gulf Coast
and the offshore barrier bars. Some of the geographically most extensive black mudrocks are known to have
been deposited in shallow waters on the continental
blocks.

The red-brown-yellow-green color grouping reflects
the presence or absence of ferric oxide (red), hydroxide (brown), or limonite (yellow) as colloidal particles
among the clay-mineral flakes. Hematite (Fe_2O_3) is red;
goethite [$FeO(OH)$] is brown. Only a few percent of
hematite is sufficient to produce a deep-red mudrock. If
these compounds are absent, the true green color of
most clay minerals shows through; illite, chlorite, and
biotite are green. In many red mudrocks, ovoid or tubular green spots are present, the green color reflecting the
reduction of ferric iron adjacent to a plant root or other
bit of organic matter and the removal of the ferrous ions
in groundwater.

FIGURE 15-3
● ● ● ● ● ● ● ● ● ● ● ● ● ●

Laminated dark gray to black mudstone consisting of silt interstratified with dark shale. Prominent water-escape structure extends upward through many laminae. Many desiccation cracks are evident along the tops of laminae. Nonesuch Formation (Precambrian), Michigan. [From R. D. Elmore et al., 1989, *Precambrian Research*, *43*, 196.]

LABORATORY STUDIES
● ● ● ● ● ● ● ● ● ● ● ● ●

Both texture and structure can be studied in the laboratory as well as in the field, and laboratory study is essential for almost all mineralogic work with mudrocks. For most studies of mudrocks, the most valuable techniques and instruments are, in order of increasing fineness of the features seen, radiography of rock slabs, sodium bisulfate fusion, polarizing microscopy, scanning and transmission electron microscopy, X-ray diffraction, and the electron microprobe. Figure 15-4 shows the textural and compositional variations visible in the polarizing microscope.

COMPOSITION
● ● ● ● ● ● ● ● ● ● ● ● ● ● ●

Sedimentary petrologists study the detrital mineral composition of rocks to determine source areas (provenance) and to gain insights into tectonic styles and plate movements. Sandstones have been used for these purposes since the late nineteenth century, and mudrocks have the same potential—although they are rarely used. In mudrocks that are not megascopically silty, the abundant minerals are clay minerals (60%), quartz and

chert (30%), feldspar (4%), and carbonate minerals (3%); other substances make up the remaining 3%. In general, the higher the content of silt, the higher the content of quartz and feldspar and the lower the clay-mineral content.

Clay Minerals

The most abundant mineral group in mudrocks is the clays (principally illite, montmorillonite, and kaolinite), because of the great chemical stability of a sheeted Si–O, Al–OH crystal structure. Although the abundance of clays in the stratigraphic column is not surprising, the relative abundances of the main types are. Late Tertiary clay minerals are dominantly of the **expandable** type, either pure smectite or interlayered smectite-illite. In older mudrocks, illite with few or no smectite interlayers becomes more common; and in lower to middle Paleozoic rocks, expandable clays average less than 10% of all clay minerals. This change has been documented worldwide in many stratigraphic sections and results from a diagenetic alteration usually symbolized as

$$smectite + Al(OH)_3^0 + K^+ \rightarrow$$

$$illite + H_4SiO_4^0 + Na^+ + Ca^{2+} + Fe^{2+, 3+} + Mg^{2+} + H_2O$$

The diagenetic change proceeds progressively from a randomly interlayered smectite and illite, to an ordered

FIGURE 15-4
• • • • • • • • • • • • • • • • •

Photomicrograph of striped shale, showing several beds of carbonaceous shale alternating with graded silt-mud couplets. Arrow A points to wavy-crinkly carbonaceous silty laminae that contain carbonaceous matter, clay minerals, tiny pyrite crystals, and scattered silt grains. Arrow B points to drapes of dolomitic clayey shale in carbonaceous silty shale. Arrow C points to tiny silt-rich lenses. Arrow D points to tiny carbonaceous flakes in dolomitic clayey shale. Note the sharp contacts between carbonaceous silty shale beds and silt-mud couplets. Dolomitic clayey shale beds show preferred extinction parallel to bedding because of aligned clay minerals. Photo is 3.7 mm high. [From J. Schieber, 1989, *Sedimentology, 36,* 208.]

interstratification, to discrete illite. As a result of water loss, the illite then undergoes progressive changes in its crystal structure from 1Md to 1M to 2M, which is the mineral muscovite. The 2M mica does not form until the temperature exceeds 250°C for a significant period of time.

The conversion from smectite to illite can begin at a temperature as low as 50°C and is largely completed at 120°C, but the correlation between the existence of smectite sheets and temperature is far from perfect. It depends on many variable factors in the diagenetic environment, including the availability of potassium from breakdown of feldspar, mica, or a less stable smectite-illite phase; mobility of aluminum, which is very insoluble in inorganic solutions (see Figure 12-5); composition of the smectite sheets; subsurface fluid pressure, with higher pressure favoring persistence of smectite; presence of calcite, which may act as a cement and retard fluid movement or may release Ca^{2+} to the fluid and stabilize smectite sheets; and presence of salt **diapirs,** which can have several, possibly opposing, effects: (a) the high thermal conductivity of salt may increase the rate of smectite-illite diagenesis in adjacent sediments; (b) the sodium from the salt may stabilize smectite; and (c) potassium from posthalite evaporites may enlarge the stability field of illite. Clearly, interpretation

of the transformation from smectite-illite to pure illite is not a simple matter.

Mudrocks form about two-thirds of all sedimentary rocks; 60% of the average mudrock is clay minerals, and at least half of all clays are expandable, with 50% or more smectite sheets. Simple multiplication of these percentages indicates that a minimum of 10% of the stratigraphic column consists of smectite sheets, each of which releases silica, sodium, calcium, magnesium, iron, and water during the conversion to illite. Nearly all the materials released during the clay transformation from smectite to illite appear to remain in the mudrock. The released silica may crystallize as quartz or chert; calcium may precipitate in calcite, dolomite, and ankerite; the magnesium, in dolomite; the iron, in ankerite, dolomite, siderite, or hematite; the sodium, in albite. All these minerals occur in mudrocks and some of the grains may be of diagenetic origin.

Quartz

More than 95% of the quartz in mudrocks is composed of single crystals, the remainder being polycrystalline aggregates that may have originated as detrital chert, detrital metaquartz, crystallized diatoms or radiolaria, or silica produced as part of the smectite to illite transformation. The size of the quartz is one-eighth fine and very fine sand, six-eighths silt, and one-eighth clay size. There is a positive correlation between the size of the quartz in a mudrock and the amount of quartz in the rock. Very little quartz occurs in grain sizes smaller than 2 μm.

Nearly half of the quartz in mudrocks appears to originate in slates, phyllites, and fine-grained schists; this estimation is based on stable oxygen isotopic data. Most of the remaining quartz probably originates in low-grade quartzose schists, whose quartz aggregates are disintegrated during soil-forming processes into single crystals of quartz.

Feldspar

The percentage of feldspar in the nonclay fraction of the suspended load of large rivers may be large: for example, 10–15% in the nontropical sections of the Amazon River system and 45% in the Mississippi River at New Orleans. But as a result of destruction in the surf zone and removal during diagenesis, feldspars occur in small amounts in most mudrocks and must be concentrated by sodium bisulfate fusion to be examined. Such studies are uncommon. The few studies available indicate that feldspars in mudrocks can be used for provenance studies, just as feldspars in sandstones are used.

Carbonate Minerals

Carbonate minerals are about as abundant in mudrocks as feldspars are. It is assumed that most of the carbonate is calcite because that is the abundant carbonate in both sandstones and carbonate rocks. The carbonate in mudrocks may consist of whole ostracods or microfossils, or it may originate by seafloor predation of shell material or inorganic precipitation during compaction or diagenesis. Dolomite, ankerite, and siderite are more common in black shales because of the reducing environment in which such shales form.

Organic Matter

Of the minor constituents of mudrocks, organic matter is probably most abundant and certainly is the most important. Mudrocks contain about 95% of the organic matter in sedimentary rocks, although the amount by weight in the average mudrock is probably less than 1%. The occurrence of black organic matter in mudrocks depends on the volume of organic material produced in surface waters, on the sedimentation rate, and on the amount of free oxygen present to decompose the organic matter at the basin floor. Anoxia is not a requirement for the preservation of organic matter if the rate of burial is sufficiently rapid.

The environmental conditions that cause the reduction of the organic matter toward free carbon also cause the reduction of other substances in the water. For example, ferric iron is changed to the ferrous form, and sulfur is reduced from a valence of $+6$ in sulfate ion (SO_4^{2-}) to -2 (S^{2-}). The reduced iron and sulfur then combine to produce amorphous iron(II) sulfide, FeS, and pyrite, FeS_2. Many black mudrocks contain pyrite.

Bentonite

Bentonite is a special and important variety of mudrock that is defined by the type of material from which it formed. It is composed almost entirely of smectite and colloidal silica produced as the alteration product of glassy volcanic debris, generally a tuff or volcanic ash. In a pure bentonite the only other minerals present are clearly volcanic in origin, such as euhedral brown biotite, idomorphic zircon, or relict glass shards (fragments). The montmorillonite is normally the calcic variety, reflecting the fact that the parent material is basaltic or andesitic. Sodic and potassic bentonites are known, however, and contain euhedral sanidine and quartz grains with the high-temperature crystal habit (β-quartz outlines), in addition to the essential montmorillonite. Bentonites frequently are interbedded with impure tuffs.

A bed of bentonite is the result either of a single eruption or of several eruptions within a very brief period, perhaps a few years. Consequently, bentonite beds normally are less than 50 cm thick. The thickness of the bed and the size of the unaltered fragments in it decrease logarithmically with distance from the volcanic vent. These fragments also tend to be graded within the bed, with the larger and/or denser ones near the base. Bentonite beds are widespread, reveal paleowind directions, and can be correlated over many hundreds of kilometers on the basis of mineralogic or chemical variations.

ANCIENT MUDROCKS
• • • • • • • • • • • • • • •

Paleocurrent Indicators

It is more difficult to determine transport directions in mudrocks than in sandstones. The coarser grained rocks commonly contain obvious current indicators such as cross-bedding, ripple marks, or flute casts (see Chapter 13). But mudrocks usually lack these features, except in the coarser siltstones. The currents that

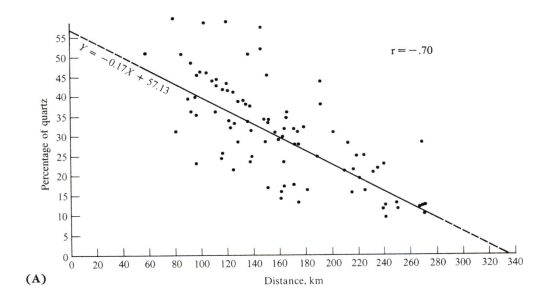

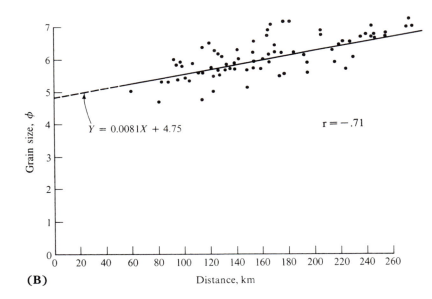

FIGURE 15-5

● ● ● ● ● ● ● ● ● ● ● ● ● ●

Decrease in **(A)** percentage and **(B)** mean grain size of quartz with distance from the sand-mud line offshore of the beach in the Blaine Formation (Permian) epicontinental sea. [From H. Blatt and M. W. Totten, 1981, *J. Sediment. Petrol.*, *51*, 1262 for part (A) and 1264 for part (B).]

deposit muds are weak, and directional structures are correspondingly obscure or absent. With effort, however, directional indicators can be found. Three types have been uncovered:

1. Progressive decrease in quartz grain size and percentage of quartz offshore in marine rocks (Figure 15-5).
2. Alignment of inequant quartz grains or fossils parallel to the depositing current.
3. Magnetic susceptibility fabrics, which typically are nearly parallel to current-controlled grain lineations;

that is, the magnetic lineation of magnetite grains typically is parallel to grain elongation.

Petrology

The most widespread mudrocks exposed on the continental blocks are those deposited in the epicontinental seas that flooded the continents repeatedly during Phanerozoic time. One example is the Pierre Shale (Upper Cretaceous) in western North America (Figure 15-6). The Pierre seaway extended from the Arctic Ocean to the Gulf of Mexico, had a width exceeding 1200 km, and

FIGURE 15-6
● ● ● ● ● ● ● ● ● ● ● ● ● ●

Probable distribution of land (shaded) and sea (unshaded) in North America during late Campanian time, showing the geographic distribution of the seaway that divided the continent into eastern and western parts. Samples were taken over the entire area of outcrop (crosshatched). [From R. L. Jones and H. Blatt, 1984, *J. Sediment. Petrol., 54,* 18.]

existed for nearly 10 million years. It was bounded on both east and west by topographically low areas underlain by sedimentary rocks, although active volcanic areas were present in the area to the west of the seaway, the direction from which nearly all of the Pierre sediment originated. Pierre Shale reaches a maximum thickness of more than 7000 m at its western edge in Idaho (longitude 112°W, 43°N latitude) and thins to a wedge-edge in eastern North and South Dakota (longitude 97°W).

The amount of quartz in the Pierre Shale averages 23% (σ = 9%) and decreases from more than 30% at the western edge of the outcrop area to less than 5% at its feather edge in North Dakota. The mean grain size of the quartz also decreases from west to east, and the per-

centage of clay minerals increases. Not all the Pierre sediment was derived from the western sedimentary sources via fluvial-deltaic transport, as indicated by the presence in the formation of some bentonite layers and large amounts of cristobalite in some samples. These data, coupled with the extremely fine-grained character of the Pierre (Figure 15-7), strongly indicate a significant airborne volcanic contribution into the Pierre epicontinental sea. One way to evaluate the relative importance of airborne sediment in the Pierre is by relating the percentage of feldspar in the formation to distance from the major volcanic center in southwestern Montana (Figure 15-8). World-average mudrock contains 4.5% feldspar and a quartz (Q) plus feldspar (F) content of 35.3%. Therefore, the Q + F fraction of an average mudrock contains 12.7% feldspar and 87.3% quartz. As shown in Figure 15-8, the Q + F fraction of Pierre Shale samples is clearly divisible into two groups, one whose content of feldspar is constant at 13% and a second whose feldspar content depends on the distance from the known volcanic source. About two-thirds of the samples show a clear volcanic influence.

Mudrocks deposited in continental environments can also be used to unravel geologic history, as shown by the Vanoss Formation (Pennsylvanian) in Oklahoma

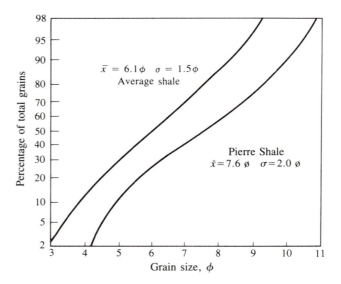

FIGURE 15-7
● ● ● ● ● ● ● ● ● ● ● ● ● ●

Comparison between size distributions of the quartz + feldspar fraction in an average shale and the Pierre Shale. [From R. L. Jones and H. Blatt, 1984, *J. Sediment. Petrol., 54,* 26.]

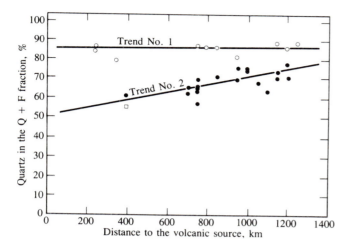

FIGURE 15-8

● ● ● ● ● ● ● ● ● ● ● ● ● ● ●

Relationship between percentage of quartz in the Q + F fraction and distance to the nearest volcanic source area in west-central Montana. Open square represents bentonite sample. Trend No. 1 is a horizontal line at 13% feldspar and 87% quartz, a world-average mudrock. Trend No. 2 is the best-fit line through points significantly distant from the horizontal line. [From R. L. Jones and H. Blatt, 1984, *J. Sediment. Petrol.*, 54, 27.]

(Figures 15-9 and 15-10). The amount of feldspar decreases with distance in a regular manner, consistent with results obtained in associated sandstones in the Vanoss. Analogous plots for normative orthoclase, perthite, and albite revealed that plagioclase was selectively removed during diagenesis, a result consistent with scanning electron microscopic observations of the feldspars (Figure 15-11).

Black Shales

Black shales are a distinctive and paleogeographically important type of mudrock that occurs sporadically throughout the stratigraphic column. The cause of the black color is accumulation of partially decomposed organic matter on the basin floor and/or the presence of finely disseminated pyrite, occurrences that depend on biological productivity, sedimentation rate, and water circulation. In most modern marine settings, the balance among these variables is such that nearly all organic matter is completely oxidized before accompanying inorganic sediment is buried; so only the inorganic hard parts of the organisms remain (calcite, aragonite, opal). But it is not uncommon in the marine environment to find that either surface productivity is increased (for example, by upwelling of nutrient-rich waters), inorganic sedimentation rate is decreased (for example, by the formation of a topographic barrier on the seafloor), or water circulation is decreased (for example,

because of plate movements). When these events occur, organic-rich muds may accumulate and ultimately become the black shales seen in the geologic record.

In an atmosphere containing 21% oxygen, the oxygen content of standard surface seawater varies from about 8 mL/L at 0°C to approximately 2 mL/L in halite-saturated brine at 25°– 30°C. Below the sea surface, oxygen is removed by the decomposition of organic matter that settles from the sea surface; but as long as water circulation is sufficient to keep the dissolved oxygen content above about 1 mL/L, normal aerobic decomposition of organic matter will occur. Organic matter will be completely converted to carbon dioxide and water, leaving no residue. However, as the amount of oxygen decreases below 1 mL/L, the water becomes dysaerobic (suboxic), aerobic decomposition

FIGURE 15-9

● ● ● ● ● ● ● ● ● ● ● ● ● ● ●

Arbuckle Mountain area in south-central Oklahoma, showing outcrop of the Vanoss Formation and its granite source. [From H. Blatt and J. R. Caprara, 1985, *J. Sediment. Petrol.*, 55, 549.]

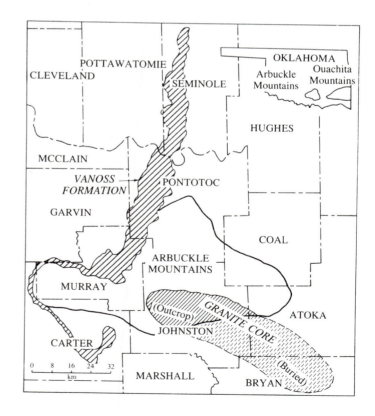

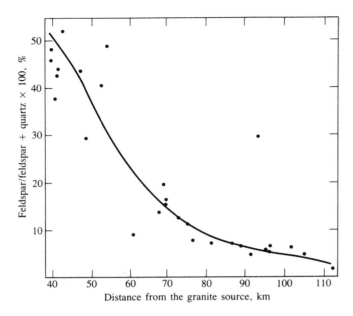

is reduced, benthic burrowing organisms decrease in abundance, and the muds pass from red, brown, or colorless to gray. At oxygen levels below 0.1 mL/L, anaerobic bacteria and metabolism prevail, benthic life is extinguished, and fine-grained black muds accumulate. Diagenetic change subsequently produces the finely fissile black shales that occur in the geologic record.

Black Shales, Paleoceanography, and Plate Motions

Beginning about 150 million years ago in late Jurassic time and continuing until late Cretaceous time about 85 million years ago, large areas of the Atlantic Ocean floor were either dysoxic or anoxic, as evidenced by extensive areas floored by black shale. The stratigraphic occurrence of these black rocks is intermittent throughout this period of 65 million years, and their lateral extent appears to be discontinuous, a conclusion based on results of the Deep Sea Drilling Project and its successor, the Ocean Drilling Project. Thus, although oxygen deficiency was quite common, there was no time during which the entire Atlantic Ocean was dysoxic or anoxic. The causes of the persistent oxygen deficiency were the lack of open circulation associated with the narrowness of the early Atlantic; the existence of many

FIGURE 15-11
● ● ● ● ● ● ● ● ● ● ● ● ● ● ● ● ●

Feldspar grains from the Vanoss Formation showing characteristically fresh orthoclase (A) and altered plagioclase (B). Length of plagioclase grain is 0.25 mm. [From H. Blatt and J. R. Caprara, 1985, *J. Sediment. Petrol.*, 55, 551.]

FIGURE 15-10
● ● ● ● ● ● ● ● ● ● ● ● ● ● ● ● ●

Percentage of normative feldspar in the quartz + feldspar (3.5ϕ–4.0ϕ size) fraction of Vanoss Formation shales. [From H. Blatt and J. R. Caprara, 1985, *J. Sediment. Petrol.*, 55, 549.]

fault-bounded silled basins generated by the creation of new ocean; the upwelling water stimulated by the existence of the topographic barriers on the seafloor; and the restriction or near-absence of cold bottom waters because of the warm, equable Cretaceous climate and the shrinkage of polar ice caps. The warm sea surface extended into relatively high latitudes and led to increased productivity of plankton. As the plankton died and settled to the seafloor, their tissues decayed, a process that used up the oxygen dissolved in the water and created an oxygen-minimum zone. With high productivity and the lack of a well-defined boundary between warm surface water and cold polar bottom water, the oxygen-minimum water layer expanded, with its base frequently extending below the sill depth of the restricted basins on the ocean floor. Further, the creation of the Mid-Atlantic Ridge as the ocean formed decreased the volume of the world ocean basin and caused extensive transgressions onto the continental blocks during Cretaceous time. Plankton productivity in these newly generated shallow sea waters further stimulated the accumulation of black muds in the marine environment. The combination of these factors led to one of the most important black shale periods in Earth history.

In area, the most extensive black shale complex on the North American plate is the early Devonian-Early Mississippian accumulation in eastern and midcontinental North America (Figure 15-12). These black

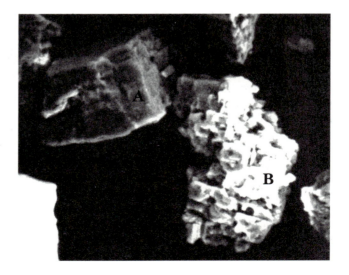

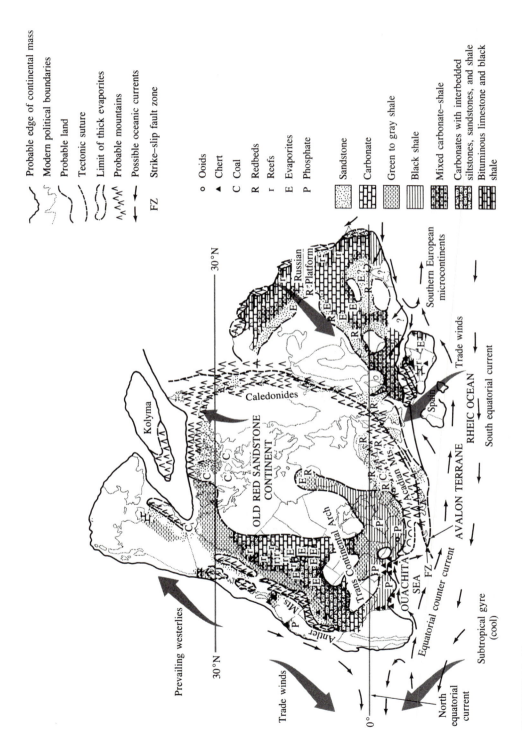

FIGURE 15-12
● ● ● ● ● ● ● ● ● ● ● ● ● ● ● ●

Generalized late Devonian paleogeography showing lithofacies for the Laurasia landmass. The black shales have maximum thicknesses of 200 m and organic matter contents of 30–35%. Recurrent upwelling resulted in periods of siliceous plankton blooms and resultant chert–black shale couplets in some areas. Depth of black shale deposition is estimated to be 100–200 m. [From F. R. Ettensohn, 1985, *Geol. Soc. Am. Spec. Paper 201*, 67.]

shales were deposited during a period of about 50 million years and, like their Jurassic-Cretaceous counterparts in the Atlantic Ocean, are widely distributed in space and time.

The basinal Appalachian deltaic black shales are laminated, commonly on a scale of 1 mm or less, and may contain nodules of early diagenetic pyrite if adequate amounts of iron and sulfur were available. Bioturbation is absent from the blackest shales but begins to occur as the shales pass from black to gray, from the paleoanoxic to the paleodysaerobic zone above the Devonian seafloor. The content of quartz in the black shales is unusually low, averaging 15–20%, and much of it may have been wind deposited. The anoxia that appears to have been widespread in the immense Devonian inland sea was caused largely by thermal density stratification. In the equatorial belt in which the sea occurred, the surface waters are nearly always warmer and less dense than those lower in the water column, so vertical overturn and oxygenation of bottom waters cannot occur. Also, the warm surface waters are favorable for abundant growth of soft-bodied phytoplankton (calcareous plankton had not yet evolved in Devonian time), thus assuring a high volume of organic "rain" and a thick oxygen-minimum zone. Radiolaria proliferated in some areas of the sea, particularly in the Woodford black shale in southern Oklahoma, resulting in unusually siliceous deposits.

SUMMARY

Mudrocks are the most abundant type of sedimentary rock, although their apparent abundance in outcrop may be less than that of sandstones. Mudrocks are not well understood because of their very fine grain size and the difficulty of disaggregating them.

The relative proportion of fissile to nonfissile mudrocks is not known. Fissility can be produced during deposition, but the rapid rate at which burrowing organisms on the seafloor destroy clay-flake parallelism suggests that a large proportion of the fissility seen in ancient nonblack mudrocks has been produced during diagenesis.

Mudrocks contain most of the earth's buried organic matter, so gray-black mudrocks are common. Black organic matter is preserved in oxygen-deficient environments; and in such an environment, aerobic burrowing organisms cannot exist. As a result, black mudrocks are more likely to be fissile than red ones, which form and are preserved under oxygenated conditions.

Clay minerals form 60% of muds and mudrocks. Illite forms about 25% of modern clay-mineral suites but increases in abundance with increasing age to perhaps 80% in Lower Paleozoic mudrocks. Both field observations and geochemical experiments indicate that the cause of the change is the progressive change of montmorillonite and mixed-layer montmorillonite-illite clay to pure illite during diagenesis. The reaction is initiated at temperatures as low as 50°C but is not completed until temperatures above 200°C are attained. Detrital silt-size quartz forms most of the remainder of the typical mudrock.

Mudrocks can be used in provenance studies of modern muds; but the mineralogic character of the clay-mineral suite changes during diagenesis. Trace elements may be relatively immobile in these rocks, however, and may be useful in provenance studies. Quartz, feldspar, and accessory minerals present in mudrocks can also be used for this purpose.

STUDY EXERCISES

1. Most shales in the stratigraphic column are marine and, therefore, the clays in them were flocculated. Floccules are typically of silt size and thus might be expected to produce current-generated sedimentary structures when deposited. Yet features such as ripple marks and cross-bedding are only rarely reported from mudrocks. Suggest reasons why.

2. When smectite is changed into illite during diagenesis, five different ionic species are produced as "byproducts." They are silica, sodium, calcium, iron, and magnesium (see page 285). Suppose you wanted to determine what happens to each of these substances after they are released into the pore waters of mudrock. How might you do this? (*Hint:* Consider minerals that might be formed from the ions and where in the adjacent part of the stratigraphic section you might search for them.)

3. Would you expect the abundance of black shales to be different in Precambrian time relative to Phanerozoic time? Is the composition of the atmosphere an important factor? Does the near-absence of fossils in Precambrian rocks imply a difference in the abundance of organic matter at that time? Explain your reasoning.

4. Describe how plate movements, plate orientations, and global wind patterns affect the dispersal of volcanic ash. What can you infer about these three variables from a distribution of ash you have mapped?

5. Mudrocks contain only small amounts of feldspars. Suggest explanations for this observation.

REFERENCES AND ADDITIONAL READINGS
● ● ● ● ● ● ● ● ● ● ● ● ● ● ● ●

Awwiller, D. N. 1993. Illite/smectite formation and potassium mass transfer during burial diagenesis of mudrocks: A study from the Texas Gulf Coast Paleocene-Eocene. *J. Sediment. Petrol., 63,* 501–512.

Blatt, H. 1987. Oxygen isotopes and the origin of quartz. *J. Sediment. Petrol., 57,* 373–377.

Elder, W. P. 1988. Geometry of Upper Cretaceous bentonite beds: Implications about volcanic source areas and paleowind patterns, Western Interior, United States. *Geology, 16,* 835–838.

Eslinger, E., and D. Pevear. 1988. Shale diagenesis. In *Clay Minerals for Petroleum Geologists and Engineers. SEPM Short Course No. 22,* 5-1–5-46.

Ettensohn, F. R. 1987. Rates of relative plate motion during the Acadian orogeny based on the spatial distribution of black shales. *J. Geol., 95,* 572–582.

Milliken, K. L., and L. S. Land. 1993. The origin and fate of silt-sized carbonate in subsurface Miocene-Oligocene mudstones, south Texas Gulf Coast. *Sedimentology, 40,* 107–124.

O'Brien, N. R., and R. M. Slatt. 1990. *Argillaceous Rock Atlas.* New York: Springer-Verlag.

Pedersen, T. F., and S. E. Calvert. 1990. Anoxia vs. productivity: What controls the formation of organic-carbon-rich sediments and sedimentary rocks? *Am. Assoc. Pet. Geol. Bull., 74,* 454–466.

Schieber, J. 1989. Facies and origin of shales from the mid-Proterozoic Newland Formation, Belt Basin, Montana, USA. *Sedimentology, 36,* 203–219.

Chapter 16

LIMESTONES
AND DOLOSTONES

Limestones and **dolostones** are carbonate rocks that make up 10–15% of the sedimentary rocks, with limestones being considerably more abundant than dolostones in most stratigraphic sections. Carbonate rocks are normally quite free of impurities, which total less than 5% of an average limestone and consist of clay minerals and fine-grained quartz. The lack of significant mineralogic variation in limestones and dolostones and their great solubility relative to that of sandstones mean that textures and structures assume greater importance in carbonate rocks and also that geochemistry becomes essential to the interpretation of petrologic variability. Grain types and mud content are generally of greater importance in interpretation than are grain sorting and rounding and hydraulic sedimentary structures because of the biotic and in situ origin of most clastic carbonate grains. In addition, limestones contain important diagnostic structures that, when coupled with grain types and mud content, are extremely important for environmental interpretations.

Limestone is recognized in outcrop by its softness and by the bubbly evolution of carbon dioxide gas when a few drops of cold, dilute hydrochloric acid are dropped on it:

$$CaCO_3 + 2\,HCl \rightarrow Ca^{2+} + 2\,Cl^- + CO_2 + H_2O$$

Limestone is distinguished from dolostone by the fact that dolomite does not react visibly to dilute hydrochloric acid unless the mineral is powdered. Also, dolostone commonly weathers with a dull brownish yellow appearance (buff color) because it usually contains some ferrous iron as a substitute for magnesium in the crystal structure. The iron is released from the carbonate during weathering and oxidizes, causing the color.

The relative proportions of calcite, dolomite, and quartz (or other silicate minerals) in a limestone can be estimated by etching the rock surface at the outcrop. Dilute hydrochloric acid is dripped onto a clean, flat surface of a hand specimen until the calcite is dissolved in sufficient amounts to allow the less soluble materials to stand out in positive relief. Quartz is identified by its translucency and glassy appearance; chert by its opacity and hardness (if the grains are large enough to be scratched and examined with a hand lens); and dolomite by its white color, softness, and rhombohedral crystal outlines. Calcite only rarely occurs as scalenohedra or rhombohedra, but dolomite in limestone often occurs as scattered rhombs. The etching technique also reveals the distribution in the rock of these relatively insoluble constituents. For example, chert is often located along bedding planes; fossils may have been converted into chert, but the remainder of the rock is still limestone; or the limestone-dolostone contact in the outcrop may be at an angle to the bedding of the rocks, thus proving a replacement origin for the dolomite.

TEXTURES

The textures of limestones are extremely variable because of the complex origins of carbonate rocks. Limestones can have textures that are identical to those of detrital rocks (for example, grain rounding and sorting;

Figure 16-1) or to those of chemical precipitates (for example, equigranular, interlocking crystals, and "porphyritic"). Many carbonate rocks display both types of textures (Figure 16-2). In addition, limestones commonly have biologically produced textures characteristic of the growth habits of living organisms, such as algae (Figure 16-3) or corals. Some of these biological textures are so intricate that they defy adequate written description (Figure 16-4). However, the texture of most limestones can be described adequately by determining the types of clastic grains, the presence or absence of calcium carbonate mud matrix, and the presence or absence of coarsely crystalline calcite cement (visible with a hand lens).

GRAINS
● ● ● ● ● ● ● ● ● ● ● ● ● ● ● ●

Grains (allochemical particles, often shortened to *allochems*) are the gravel-, sand-, and coarse silt-size car-

bonate particles (>30 μm) that typically form the framework in mechanically deposited limestones. They are the equivalent of quartz, feldspar, and lithic fragments in sandstones. Four types of grains are common: fossils, ooids, peloids, and limeclasts.

Fossils

The clearly distinguishable fossils in limestones in outcrop are those that have inequant, biologically determined shapes, such as the concavoconvex outline of pelecypods, brachiopods, and ostracods; the segmented, cylindrical shape of crinoid stems; or the leafy pattern of a bryozoan. Because these fossils are identified by their shapes, however, they are more difficult to recognize if they are broken into small pieces before burial—a common phenomenon in the shallow marine environment in which most limestones are deposited. High current velocities, predators, and scavengers can cause the fragmentation. The minimal size required for recogni-

FIGURE 16-1
● ● ● ● ● ● ● ● ● ● ● ● ● ● ● ●

Well-rounded, poorly sorted pebbles of algal micritic limestone in a matrix of black (organic matter) microcrystalline limestone, Cool Creek Limestone (Ordovician), Arbuckle Mountains, Oklahoma. Algal nature of pebbles is visible only in thin section.

FIGURE 16-2
● ● ● ● ● ● ● ● ● ● ● ● ● ● ● ●

Laminated limestone composed of interlocking microcrystalline calcite crystals a few micrometers in diameter, Kindblade Formation (Ordovician), Arbuckle Mountains, Oklahoma, Laminations indicate the original clastic nature of crystals that has been obliterated during recrystallization from aragonite to calcite.

tion depends on the microstructure of the organism. For example, a fragment of a coral 10 mm in diameter in a hand specimen might be indistinguishable from a bryozoan or an algal fragment, whereas a similar-size piece of crinoid column would be easily identifiable. An echinoid fragment of this size might not be recognized as a fossil fragment at all because echinoderm skeletons disaggregate into sand-size crystals of calcite and might look like cement crystals in limestone hand specimens.

The outer form and internal microstructure of the shells of organisms are varied and extremely complex, but typically the taxa can be identified in thin section. An extensive description with photomicrographs of many types of microstructures is given by Scholle (1978).

Ooids

Ooids are nearly spherical, polycrystalline carbonate grains of sand size that have a concentric or radial internal structure (Figure 16-5). The ooids always contain nuclei of quartz grains or carbonate fragments, around which the oolitic coating has formed; these nuclei can be seen with a hand lens. Some particles barely qualify as ooids, having only one or two thin layers of calcium carbonate surrounding the nucleus. Other ooids have thick coatings. Presumably, the difference reflects the length of time during which the ooid was forming, although the possibility of variations in current strength or change in water chemistry should not be neglected. Because the coating on the nucleus can be precipitated

FIGURE 16-3

• • • • • • • • • • • • • • •

Limestone composed of fragments of calcified algal mat, showing platy limestone fragments characteristic of such algal mats, McLish Limestone (Ordovician), Arbuckle Mountains, Oklahoma.

FIGURE 16-4

• • • • • • • • • • • • • • •

Polished slab of reef rock, Capitan Reef (Permian), New Mexico, composed of coelenterate (?) *Tubiphytes* (*t*), sponges (*s*), and bryozoa (*b*) and alga *Archaeolithoporella* (*a*). Cements are shades of gray bands. [From J. A. Babcock, 1977, SEPM Permian Basin Section, *Field Trip Guidebook to Guadalupe Mountains*, 25. Photo courtesy J. A. Babcock.]

0   1 cm

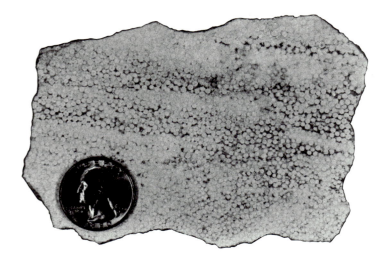

FIGURE 16-5
• • • • • • • • • • • • • • • • •

Ooids cemented by translucent, coarse calcite crystals, Chimney Hill Limestone (Silurian), Arbuckle Mountains, Oklahoma. The upper part of the slab shows horizontal bedding; at the lower right, oolitic layers are cross-bedded, dipping to left.

easily only from highly agitated water, most ooids form in shallow marine waters such as the Bahama platform rather than in lakes or streams. The presence of ooids is evidence that the particle has been transported by strong currents. Oolitic limestones commonly are cross-bedded, a reflection of the high kinetic energies in their environment of deposition.

Particles similar in appearance to ooids include pisoliths and **oncoliths.** *Pisoliths* are concentrically laminated bodies of inorganic origin that are larger than 2 mm in diameter (Figure 16-6A). Although the distinction between ooids and pisoliths depends only on size, pisoliths generally have rather irregular laminae and less smooth surfaces than ooids. Pisoliths are formed most commonly in calcrete soil horizons but they are also known from caves, freshwater and hypersaline lakes, and normal marine settings, in which case they are simply large ooids that differ in no other way from their smaller companions.

Oncoliths are similar to pisoliths in hand specimen. In thin section, however, the oncoliths reveal a filamentous structure characteristic of calcite algal encrustations (Figure 16-6B). That is, oncoliths have an organic origin and form in marine waters of normal salinity.

Peloids

Peloids are aggregates of microcrystalline calcium carbonate that lack internal structure. Their origin is usually uncertain. Most peloids have an ellipsoidal to roughly spherical shape (Figure 16-7), and many are believed to be fecal pellets; they contain organic matter and assorted detritus normally ingested by organisms during feeding. In many limestones the peloids are of rather uniform coarse-silt to fine-sand size, presumably reflecting the anal dimensions of the organisms that pro-

duce them. Studies of the feeding habits of vagrant benthic marine organisms reveal that they burrow through carbonate sediments on the shallow seafloor, swallowing anything that is small enough to ingest and contains nourishing organic matter. The ingested sediment passes through the alimentary tract of the organism and finally is excreted as the microcrystalline aggregates we call fecal pellets.

Peloids can also be produced by other mechanisms. In many modern carbonate enviroments, carbonate sand and silt grains of various kinds are micritized by endolithic (boring) algae, with the destruction of the original fabric of the grain. The grains then appear as particles of structureless microcrystalline calcite.

On the basis of the abundance of peloids in modern marine areas, peloidal limestones seem to be underrepresented among ancient limestones. Probably this underrepresentation results from the initially soft state of fecal pellets, which causes them to merge when compacted and to be indistinguishable from pure mud in ancient limestone. Indeed, mud is the most abundant type of carbonate in ancient limestones, a fact that may be closely related to the abundance of fecal pellets in modern carbonate areas.

Peloids are more easily seen in thin section than in hand specimen, where they are usually difficult to distinguish from the micrite matrix. Even in thin section there exist all gradations, from clear and sharply bounded peloids, to those with diffuse and vague borders, to mud in which the previous existence of peloids can be suspected but not proved.

Limeclasts

Limeclasts are fragments of earlier formed limestone or partially lithified carbonate sediment. They can origi-

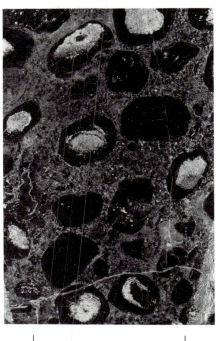

FIGURE 16-6
● ● ● ● ● ● ● ● ● ● ● ● ● ● ● ●

(A) Pisolite, containing numerous pisoliths, Tansill Formation (Permian), southeastern New Mexico. Scale is about 7 cm long. **(B)** Polished slab showing oncoliths in a matrix of packstone and grainstone, Whipple Cave Formation (Cambro-Ordovician), Nevada. Concentric algal laminae are barely visible in some oncoliths. [From P. J. Cook and M. E. Taylor, in P. J. Cook and P. Enos, 1977, *SEPM Spec. Pub. No. 25*, 59.]

(B)   0   2 cm

nate in a number of ways. Most are **intraclasts,** pieces of penecontemporaneous partially lithified carbonate sediment from within the basin of deposition (see Figures 16-1 and 16-3). They may be pieces of semiconsolidated carbonate mud torn from the seafloor by a winter storm on January 3, 100 million years ago. Others are aggregates of peloids (*grapestone*) that stuck together because of mucilaginous organic coatings or were cemented together. Still others are fragments produced by drying and cracking of intertidal mud from

FIGURE 16-7
● ● ● ● ● ● ● ● ● ● ● ● ● ● ●

Peloids (structureless micritic intraclasts) at left center, Arbuckle Formation (Ordovician), southern Oklahoma. Also visible are stylolite seams (just above the matchstick) and several large fragments of clastic limestone, the one at top center being highly peloidal. Scale bar is about 2.5 cm long.

carbonate sediment on the margin of the basin. Any of these origins qualifies the particles as intraclasts.

Limeclasts may be carried into the basin of deposition from the surrounding area, for example, a piece of Mississippian fossiliferous limestone in a Cretaceous limestone. Most limestones do not contain such fragments, and those that do typically also contain other evidence of externally derived (*terrigenous*) detritus, such as noticeable percentages of detrital quartz, feldspar, or silicate lithic fragments. Commonly, the limestone that contains such fragments was deposited in a tectonically active area such as the Alpine region during the closing of the Tethys seaway or in the Marathon region of western Texas during its Pennsylvanian orogenic episodes.

MATRIX

The *matrix* in limestones is the calcium carbonate mud that binds the allochemical grains to lithify the sediment. In hand specimen, matrix carbonate is dull and opaque, much like a piece of rhyolite, because of the extremely small grain size of the crystals (1–5 µm). When this microcrystalline carbonate (micrite) occurs, it usually forms the bulk of the rock, as do hydraulically equivalent clay minerals when they occur in a detrital rock. Carbonate mudstones, like silicate mudstones, imply a depositional environment of low kinetic energy.

Much micrite is a chemical precipitate from supersaturated solution. Chemical data prove conclusively that seawater is such a solution and that it is supersaturated with respect to both calcite and aragonite. However, some micrite is produced by the decay of algae, whose hard parts consist of aragonite needles.

Irrespective of whether the original micritic sediment is a direct chemical precipitate, consists of cryptomicrofossils, or consists partly of broken shell debris, the original sediment is mostly aragonite needles 1–5 µm in length, not the subsequent blocks of calcite seen in ancient micritic limestones. The dissolution-precipitation recrystallization that converts the micritic aragonite to calcite occurs rather rapidly, usually within a few million years, although nonmicritic sedimentary aragonite of Paleozoic age is known.

Most calcite that is 5–15 µm in size (termed **microspar**) is thought to form by recrystallization and size increase of calcite micrite. Microspar occurs largely as envelopes fringing allochems, as matrix in limestones containing few allochems (so the microspar could not be secondary cement), and as irregular patches grading downward in size into normal micrite. In hand specimen, microspar cannot be differentiated from micrite; but in thin section, the microspar appears transluscent like the coarser secondary sparry cement (Figure 16-8).

Secondary calcium carbonate cement in pre-Quaternary limestones has crystal sizes generally larger than 20 µm and is calcite. Because of their relatively coarse size, the crystals are translucent in hand specimen as well as in thin section and differ in no significant way from sparry calcite cement in sandstones.

0   0.4 mm

FIGURE 16-8

Photomicrograph (crossed nicols) of Solenhofen Limestone (Jurassic), western Germany. In hand specimen this rock is so finely crystalline that individual crystals cannot be resolved and it appears to be a homogeneous micrite. Thin section reveals that much microspar is present (clear, translucent areas) among micrite crystals (darker, opaque areas). The micrite seems vaguely peloidal in places. [From P. A. Scholle, 1978, *Am. Assoc. Pet. Geol. Mem. No. 27*, 175.]

FIGURE 16-9
• • • • • • • • • • • • • • • •

Chert nodules and lenses developed
along surfaces parallel to bedding,
Onondaga Limestone (Middle
Devonian), Albany, New York.
Length of ruler at lower left is 15 cm.
[Photo courtesy R. C. Lindholm.]

GRAIN SIZE, SORTING, AND ROUNDING
• • • • • • • • • • • • • •

The interpretation of these textural features in limestones is more difficult than in sandstones, largely because of the biological character of fossils and peloids. For example, the fossils in a limestone can be whole ostracods of a particular species that were buried in the carbonate mud in which they lived. The fact that these allochems are of a certain size and are "very well sorted" is not directly related to current strengths in the depositional environment. Both the mean size and the sorting may be biological in origin rather than hydrodynamic.

We noted earlier the biologically determined, excellent "sorting" of fecal pellets. Rounding can be similarly biologically determined. Crinoid columns and fecal pellets are always round, whether the depositional environment is of high or low kinetic energy. The variety of shapes and sizes of biological particles makes "hydraulic parity" very difficult to determine.

The kinetic energy level in a carbonate environment is evaluated mostly from the presence or absence of carbonate mud in the rock. Geochemical data indicate that seawater has been supersaturated with respect to calcium carbonate for at least the past 600 million years. Consequently, if a limestone lacks mud-size particles, we can conclude that current strengths were high enough to remove them. If the limestone is rich in microcrystalline carbonate, we conclude that the depositional environment was one of low kinetic energy. It must be remembered, however, that the "source area" of most carbonate particles is very close to the site of final deposition; allochems may be produced rapidly and in large numbers, thus forming a rock with many allochems and little mud in an environment of low kinetic energy.

Other factors used to evaluate the kinetic energy level of the environment are evidence of mechanical abrasion during transport; presence of ooids; hydraulically produced sorting; and current structures such as cross-bedding.

INSOLUBLE RESIDUES
• • • • • • • • • • • • • • •

Nearly all the noncarbonate sediment in limestones is either quartz, chert, or clay minerals. Chert in carbonate rocks has two origins. A minor amount is extrabasinal, originating in either older carbonate rocks or in bedded cherts; but nearly all chert in carbonates is intrabasinal and forms by crystallization of the amorphous silica from shells of siliceous organisms. In the modern ocean basin, diatom shells are most abundant, but in Paleozoic oceans only radiolaria and siliceous sponge spicules were present. The shells of diatoms and radiolarians are mostly of silt size and, therefore, tend to accumulate in quiet-water areas with texturally immature sediment.

The intrabasinal chert in limestone occurs mostly as very finely crystalline to powdery quartz or as nodules centimeters to meters in length (Figure 16-9). When in the nodular form, it is concentrated along visible bedding planes, reflecting the migration paths of the silica dissolved from the siliceous shells. The reason for migration of dissolved silica to the centers of crystallization we see as nodules is uncertain.

CLASSIFICATION
• • • • • • • • • • • • •

The classification of limestones is based almost exclusively on textural variations because of the lack of mineralogic variations. But despite the constant mineralogy of ancient carbonates, a useful classification scheme can be constructed by analogy with sandstones. The allochemical grains, calcium carbonate mud, and sparry calcite cement are the analogs of sand grains, clay matrix, and cement in sandstones, respectively. The closeness of the analogy between classification schemes for sandstones and limestones is shown in Figure 16-10.

In the construction of descriptive names for limestones, the key terms are modified so that fossil becomes *bio-*, peloid becomes *pel-*, ooid is shortened to *oo-*, intraclast is *intra-*, microcrystalline calcite is *mic-*, and sparry is *spar-*; thus, we have biosparites, pelmicrites, and oosparites. For gravel-size allochems, the suffix *-rudite* is appended (for example, intrasparudite).

The main part of the name is based on the major allochem and the cement or matrix; appropriate modifiers may precede the main part of the name.

An alternative and widely used system of limestone classification is based on the concept of grain support (Figure 16-11). When deposited, did the sediment consist of a self-supporting framework of allochems, or do the allochems float in a micrite matrix? Application of this system is not so simple as it first appears because of the exotic and extremely irregular shape of some allochems, particularly fossils. Solid spheres, such as those approximated by quartz grains or ooids, form a self-supporting pack with about 60% of the volume being grains; that is, the porosity of packed spheres is about 40%. Arcuate shells, however, form a self-supporting framework with only 20–30% grains; and many limestones contain allochems of a variety of irregular shapes. In practice, one examines the hand specimen and thin section of the limestone and estimates whether or not a grain-supported framework exists. A point-count to determine the grain to mud ratio may not be conclusive. The concept of a grain-supported fabric can be very useful because such fabrics commonly facilitate important diagenetic modifications due to their initially high porosities and permeabilities relative to those of nongrain-supported, micrite-rich limestones. Table 16-1 describes typical limestone compositions and their descriptive names.

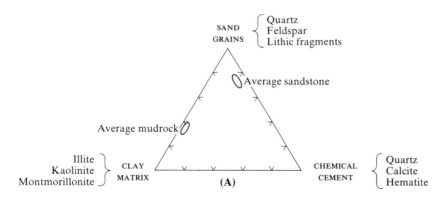

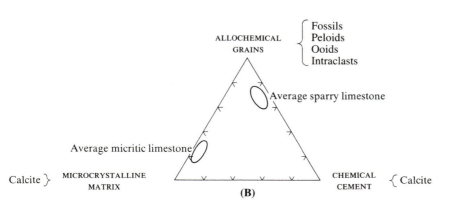

FIGURE 16-10
• • • • • • • • • • • • • •

Triangles showing analogy between components of (**A**) sandstones and (**B**) limestones. [From R. L. Folk, 1959, *Am. Assoc. Pet. Geol. Bull.*, *43*, 11.]

Original components not bound together during deposition				Original components were bound together during deposition ... as shown by intergrown skeletal matter, lamination contrary to gravity, or sediment-floored cavities that are roofed over by organic or questionably organic matter and are too large to be interstices.
Contains mud (particles of clay and fine silt size)		Lacks mud		
Mud-supported		Grain-supported		
Less than 10% grains	More than 10% grains			
Mudstone	*Wackestone*	*Packstone*	*Grainstone*	*Boundstone*

FIGURE 16-11

• • • • • • • • • • • • • • • •

Classification of limestones according to R. J. Dunham (1962, *Am. Assoc. Pet. Geol. Mem. No. 1*, 117).

STRUCTURES

• • • • • • • • • • • • • •

A large variety of sedimentary structures occurs in limestones, many of which reflect their biological origins. Other structures reflect the fact that most limestones are formed of clastic particles that are subject to the same forces that act on silicate particles. Like silicate rocks, all sedimentary structures are subject to destruction by burrowing organisms on the seafloor before the structures are buried below the penetration depth of burrowing organisms.

Current-Generated Structures

The variety of current-generated sedimentary structures in limestones is no different from those in sandstones and mudrocks, although the structures usually are not so evident in the carbonate rocks. The structures are obscure because limestones normally lack the color

TABLE 16-1   Names and compositions of various limestones

• •

Composition of limestone

Allochems		Orthochems	Folk name	Dunham name
70%	Pelecypods	Sparry cement	Oolitic pelecypod biosparite	Oolitic pelecypod grainstone
30%	Ooids			
80%	Ooids	Sparry cement	Glauconitic oosparite[a]	Glauconitic oolitic grainstone
5%	Fossil fragments			
15%	Glauconite			
60%	Peloids	Microcrystalline carbonate matrix	Fossiliferous pelmicrite	Fossiliferous peloidal wackestone or packstone
30%	Fossil fragments			
10%	Intraclasts			
70%	Intraclasts (gravel size)	Sparry cement	Trilobite intrasparudite[b]	Intraclastic grainstone
25%	Trilobites			
5%	Peloids			
40%	Crinoid fragments	Microcrystalline carbonate matrix	Clayey crinoid-brachiopod biomicrite	Crinoid-brachiopod wackestone or packstone
40%	Brachiopods			
20%	Clay minerals			

[a] If terrigenous detritus forms 10% or more of the sand- and gravel-size debris, it is added to the rock name as a modifier.
[b] If the allochems are gravel size (>2 mm), *-rudite* is added to the Folk name.

FIGURE 16-12

●●●●●●●●●●●●●●●●●●

Photograph shows 1140 orthoceratid shells, ranging from a few millimeters to 70 centimeters in length, that are accumulated on the bedding plane of a 1 × 1.4-m slab of black cephalopod limestone from the uppermost Silurian–lowermost Devonian near Tazzarine, eastern Anti-Atlas, Morocco. Orientation patterns indicate a pronounced current direction from the lower left. Most sharp ends of the fossils point upcurrent, which is the rest position that offers the least resistance to a moving current. [Photo courtesy J. Wendt.]

contrast between adjacent laminae that is common in sandstones, for example, the laminae of dark magnetite-ilmenite on crossbed surfaces. But where current structures are visible in outcrop, the most common is cross-bedding, just as it is for shallow marine sandstones. In cross-bedded limestones, the thicknesses of cross-bed sets range from centimeters to several meters, the larger ones reminiscent of the structures in the outcrops of some ancient desert sand dunes. In limestones the very large cross-bed sets may reflect eolian dunes, large submarine dunes, or ancient carbonate tidal deltas. Commonly, the azimuths of cross-bedding in marine limestones show a bimodal distribution, an expression of reversing tidal currents like those described from intertidal sandstones.

Some carbonates are transported and deposited at great depth by turbidity currents and show many of the other sedimentary features characteristic of sandstone beds formed by this mechanism. Limeclasts of extra-basinal (terrigenous) origin are common in such limestones.

Grains are typically very inequant and give rise to directional indicators such as lineation and **imbrication.** The types of grains most likely to show these features are intraclasts and fossils (Figure 16-12). Preferred orientations of allochems result from storm activity, wave action, or tidal effects, and the orientation is either parallel to or normal to the current direction. Both the size and shape of the grains influence the orientation, which must be interpreted in conjunction with other sedimentologic, stratigraphic, and facies data.

Geopetal Structures

Many organisms that become fossils have curved outlines (pelecypods, brachiopods, ostracods, trilobites), and after death the hydrodynamically stable resting position for them is concave downward. Thus, they form a bridge over the underlying sediment that subsequently can be partially or completely filled with internal sediment or sparry calcite. Those subbridge areas that are only partially spar-filled commonly serve as **geopetal structures** (Figure 16-13), that is, structures that permit the up-direction of the bed to be determined in highly deformed stratigraphic sections.

Other geopetal structures that occur in limestones include several types of asymmetrical trace fossils such as vertical U-shaped burrows and graded bedding.

Lamination

Lamination in carbonate rocks can be either inorganic or organic in origin; and in many limestones, both types of processes are important. Laminated carbonate sediment is commonly produced by organic processes in the supratidal (above normal high tide), intertidal, and subtidal shallow marine environment. Probably the most common type of lamination is that produced by soft-bodied, filamentous blue-green algae that grow as mats. The filaments are mucilaginous, and as they are repeatedly swept over by wave- and tide-generated currents, they trap and bind microcrystalline carbonate particles in the water, a process leading to the formation of laminated layers consisting of organic tissue and mud. In ancient limestones the organic tissue is normally absent because of decay that is penecontemporaneous with mat accumulation, and only layered cavities and other textural variations remain in the cryptalgal fabric (Figure 16-14). The cavities are termed **fenestrae**

FIGURE 16-13
● ● ● ● ● ● ● ● ● ● ● ● ● ● ● ●

Geopetal structure inside gastropod shells, Winfield Limestone (Pennsylvanian), Kay County, Oklahoma. Large oval shell in lower right is floored by micrite, with upper part of shell filled with later, coarse sparry calcite. Boundary between the two types of calcite is perpendicular to gravity. Diameter of shell is 3 mm. [Photo courtesy C. Gasteiger.]

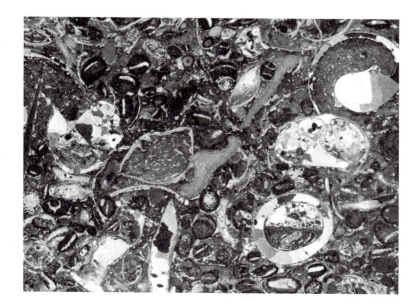

and are defined as gaps in a carbonate rock larger than grain-supported cavities. Planar-laminated cryptalgal fabrics form mostly in the supratidal zone and commonly are dolomitized.

Many algal-laminated fabrics are not planar but appear as bulbous structures termed **stromatolites** (Figure 16-15). They occur in limestones as old as 3.5 billion years and are the oldest megafossils known.

The morphologies of the stromatolites represent not only the microbathymetries at past instants in time but also the record of successive surfaces of equilibrium between interacting physical, chemical, and biological factors in the environment. Geometric analysis of the bulbous structures can indicate current strengths and tidal ranges as well as evolutionary changes in algal populations.

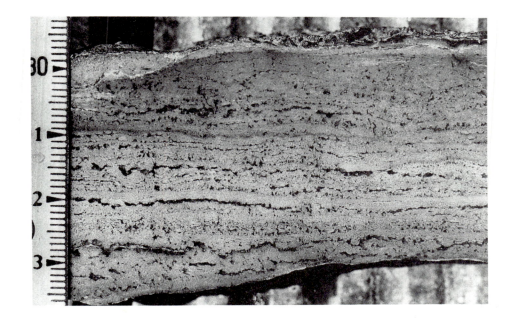

FIGURE 16-14
● ● ● ● ● ● ● ● ● ● ● ● ● ● ● ●

Laminated fenestral cryptalgal fabric in the Pillara Formation (Devonian), western Australia. Scale in millimeters and centimeters. [From J. F. Read, 1973, *Bull. Can. Pet. Geol.*, *21*, 369.]

FIGURE 16-15
• • • • • • • • • • • • • • • •

Plan view of eroded surface of limestone, showing cross sections of algal stromatolites, Hoyt Limestone (Cambrian), Saratoga Springs, New York. White scale is 10 cm in length. [From J. L. Wilson, 1975, *Carbonate Facies in Geologic History* (New York: Springer-Verlag), p. 438.]

CALCIUM CARBONATE DEPOSITIONAL SITES
• • • • • • • • • • • • • •

The occurrence of calcium carbonate sediment depends primarily on two factors: (1) warm temperatures to stimulate supersaturation of the water and rapid formation of $CaCO_3$, and (2) optimum abundance and growth of calcareous-shelled organisms. Calcareous-shelled organisms are overwhelmingly marine and thrive best in areas of abundant light, constant salinity, and clear, warm water. Thus, the water must be very shallow and far from large rivers, which lower the salinity and carry abundant mud and sand. Modern reef-building organisms (corals) have an additional reason for needing abundant light. Most corals contain symbiotic blue-green algae, which grow within the coral tissues (*Zooxanthellae*) and need light.

Although abundant carbonate sediment is produced only near the sea surface, it need not be deposited there. Calcareous plankton such as *Globigerina* have been abundant in the sea since Cretaceous time, and the skeletons of deceased plankton are typically deposited in deep water. Such deposits are identified in ancient rocks by means of a combination of stratigraphic, paleogeographic, and petrographic criteria.

Calcareous sediment formed and deposited in shallow water can also be carried into deep water by various types of mass movements and turbidity currents, caused by both oversteepened slopes and earthquakes.

Platform and Shelf Carbonates

Most **platform limestones** are simultaneously shelf limestones, in that the platform is the shallow-water marine extension of the continental block (Figure 16-16). A modern exception is the Bahama Banks, an isolated fragment of continental crust left behind as the African plate separated from the North American plate in Mesozoic time. Platform limestones also may extend into the continental block as epeiric sea limestones, which are usually indistinguishable from platform shelf limestones in petrologic characteristics. The two types are distinguished more on the basis of paleogeographic location than on sedimentology or petrology.

The rate and type of carbonate formation vary considerably across a carbonate shelf or platform, with the locus of skeletal production or physicochemical fixation being concentrated along the seaward margin. The seaward margin is favored because of the greater water turbulence and the abundance of nutrients upwelling from deeper waters. Reefs commonly grow at the platform margin for this reason. In the absence of reefs, ooid shoals are common; these become oosparite when lithified. Landward of the reef or ooid shoals lies a subtidal area in which carbonate production may be nearly as great as at the margin but where preservation is poor because of storm activity. These storms suspend carbonate sediment and transport some of it seaward into the deeper water offshore. Some sediment is also

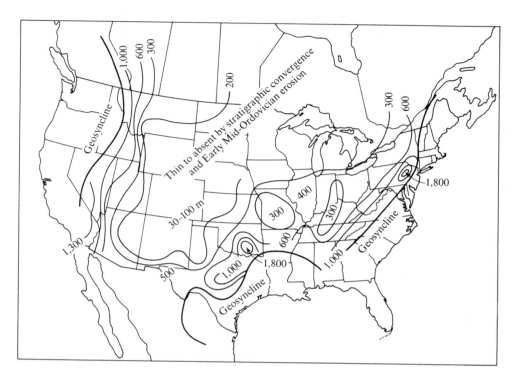

FIGURE 16-16

• • • • • • • • • • • • • • • • •

Thickness (meters) of Lower Ordovician shallow-water carbonate sediments on both sides of the Transcontinental Arch on the North American craton. [From J. L. Wilson, 1975, *Carbonate Facies in Geologic History* (New York: Springer-Verlag), p. 30.]

transported landward onto tidal flats, tidal channels, and shoreline complexes. In nearshore areas protected from strong tidal effects, fine-grained carbonate sediment (micrite, biomicrite, pelmicrite) may accumulate in bays and lagoons, as in Florida Bay. Further landward, above normal high tide (the supratidal area), dolomite typically is formed.

Micritic Mud Mounds

Carbonate mud mounds are circular to oval dome-shaped accumulations of microcrystalline calcite, in Florida Bay and elsewhere. They have thicknesses up to several hundred meters and diameters up to about 1 km and are accretional in origin rather than erosional, as indicated by steep depositional slopes of up to 50°, with fenestral layers and occasional bioclastic layers paralleling their external configuration. Mud mounds form in both deep and shallow water and most commonly have no internal structure. The mounds have uncertain origins but may have accumulated as collections of peloids

swept together by current activity and later compressed into a homogeneous mass by overburden pressure. Occasional cryptalgal structures suggest stabilization of the clastic sediment by plant activity such as *Thallasia* grass or mangrove tree growth. Disintegration of algal particles may also contribute to micritic mound sediment. Nonalgal sediment-trapping organisms such as bryozoa may be important in some mud mounds.

Organic Reefs

Reefs originate almost entirely in low latitudes in shallow marine waters, at depths of a few meters. The carbonate sediment is produced by a variety of frame-building calcareous organisms whose relative abundances have changed greatly through time; these include archaeocyathids, sponges, stromatoporoids, corals, algae, bryozoa, and even rudistid pelecypods. The internal structure of these limestone accumulations is extremely complex, reflecting the biological shapes and ecologies of the organisms involved (Figure 16-17). Whole, undisturbed fossils are commonly abundant and are interwoven with mud and grains that settled into crevices among the growing organisms. Ooids and intraclasts are rare within the reef itself, but intraclasts are abundant in the talus deposits that fringe the reef. The reef core is typically massive and unbedded.

FIGURE 16-17

● ● ● ● ● ● ● ● ● ● ● ● ● ● ● ●

Outcrop photo of an Upper Devonian reef, Blue Fiord Formation, Ellesmere Island, Canada. Visible are domal stromatoporoids and domal to branching tabulate corals. [From N. P. James, in R. G. Walker (ed.), 1984, *Facies Models*, 2nd ed. (Toronto: Geological Association of Canada), p. 237.]

Surrounding the reef core, and typically exceeding the core in volume, lies an apron of reef talus that grades outward into planktonic, fine-grained carbonate sediment and noncarbonate mud. Like the reef core, the talus apron is fundamentally a lithostratigraphic feature that is more easily recognized and defined in the field rather than in thin section.

Detrital reef-derived limestones, such as the talus apron surrounding a reef, typically are very coarse grained and are classified as either *floatstone* or *rudstone*. In a floatstone, more than 10% of the rock particles are larger than 2 mm and the rock is matrix supported. In a rudstone, the rock is clast supported.

Deep-Sea Carbonates

The planktonic carbonate sediment characteristic of the modern deep ocean basin occurs outward from the seaward limit of terrigenous sediment and from carbonate platform talus to maximum depths of about 5 km. At greater depths, only brown clay or siliceous ooze occurs. The lower depth limit of carbonate accumulation has varied through geologic time and is determined by the balance between carbonate production near the sea surface, rate of sediment settling through the water column, and the degree of saturation with $CaCO_3$ of the various water masses through which the sediment passes. Seawater is colder at depth than at the surface, and colder waters contain more dissolved carbon dioxide than do warm waters. The increased carbon dioxide causes an increase in carbonic acid (H_2CO_3) in the water and results in the dissolution of the calcitic and aragonitic shells as they settle toward the seafloor after the death of the organism. Few shells survive below about 5000 m.

The depth below which no carbonate sediment can accumulate is termed the carbonate compensation depth (CCD). The carbonate compensation depth is deeper in equatorial regions largely because calcium carbonate formation is greater in warmer waters. The increasing hydrostatic pressure with increasing depth of water in the ocean also increases the solubility of calcium carbonate, but this effect is only 25% as great as the effect of decreasing temperature. The variation in abundance of deep-sea bioclastic limestone in post-

Jurassic oceanic sediments has been interpreted in terms of changing oceanic current systems and changing depth of the seafloor as carbonate sediment that was deposited near oceanic ridges sinks to greater depth as it moves away from ridge crests. Not all carbonate sediment need be dissolved at great depth, however. Brown clay, siliceous deep-sea sediment, or turbidites may cover the carbonate sediment and protect it from the corrosive seafloor waters.

Deep-sea limestones in ancient rocks are identified in outcrop chiefly by the stratigraphic and tectonic settings in which they occur and by their microcrystalline character. Deep-sea micrites are known from the Franciscan Formation (Jurassic) in California and from several units in the Alps; both are locations where subduction of oceanic crust has occurred, apparently with the incorporation of some of the seafloor deposits (the carbonate sediment) into the continental plate. In these limestones, the most abundant fossils typically are planktonic foraminifera and the silt-size plates of planktonic algae called coccoliths, which can be seen with an electron microscope (Figure 16-18). As with the *Globigerina*

shells, the plates settled to the deep seafloor following the death of the organism. Carbonate-secreting planktonic organisms did not evolve until the Jurassic Period, so there are no pre-Jurassic "microbiomicrites." Triassic and older deep-sea limestones consist of sand- and gravel-size limeclasts carried into deep oceanic waters by turbidity currents. The detritus may have come from an actively rising highland or from sediment swept off continental platforms or shelves.

Lacustrine Carbonates

Ancient lacustrine carbonates are recognized largely on the basis of stratigraphic, facies, and biological criteria. The limestones occur within nonmarine rock sequences in association with either evaporites or fine-grained fluvial sediments and typically are micritic. As is true for marine carbonates, most precipitation of calcium carbonate is thought to result from biological activity, although this is commonly difficult to prove. In many lakes, the lake surface turns white in late spring and

FIGURE 16-18

• • • • • • • • • • • • • • • •

Transmission electron micrograph of Laytonville Limestone (Cretaceous), Franciscan Formation, Sonoma County, California, a coccolith coquina showing whole, segmented, oval plates and a hash of separate segments. Length of chipped plate in left center is 6 μm. [Photo courtesy R. E. Garrison.]

summer **(whitings)** as temperature increases and removal of carbon dioxide from lake waters is at a maximum as a result of active photosynthesis by microscopic plants called charophytes. Charophyte remains are found in many ancient lake deposits and occur as both calcified reproductive structures and as plant-stem encrustations. Like sea grasses, charophytes also baffle and trap fine carbonate mud as well as producing considerable quantities of carbonate.

CALCIUM CARBONATE EQUILIBRIA
• • • • • • • • • • • • • • •

The chief control of the solubility of calcium carbonate at the earth's surface is the hydrogen ion concentration (pH), which is determined by the partial pressure of carbon dioxide according to the linked series of reactions:

$$CO_2 + H_2O \rightleftharpoons H_2CO_3 \qquad K_{25°} = 10^{-1.47} \qquad (1)$$

$$H_2CO_3 \rightleftharpoons H^+ + HCO_3^- \qquad K_{25°} = 10^{-6.40} \qquad (2)$$

$$HCO_3^- \rightleftharpoons H^+ + CO_3^{2-} \qquad K_{25°} = 10^{-10.33} \qquad (3)$$

$$CaCO_3 \rightleftharpoons Ca^{2+} + CO_3^{2-} \quad \begin{matrix} K_{25°} = 10^{-8.48} \text{ (calcite)} \\ K_{25°} = 10^{-8.34} \text{ (aragonite)} \end{matrix} \quad (4)$$

The net result of these interactions can be summarized as

$$CaCO_3 + CO_2 + H_2O \rightleftharpoons Ca^{2+} + 2\,H^+ + 2\,CO_3^{2-}$$

This summary equation makes it clear that carbon dioxide gas dissolved in water is the key substance responsible for both the pH and the CO_3^{2-} content of distilled or fresh waters.

Is river water undersaturated, approximately saturated, or supersaturated with respect to calcite or aragonite? To answer this question, we must compare the solubility product of $(Ca^{2+})(CO_3^{2-})$ that is determined by using deionized water in the laboratory with the ion product $(Ca^{2+})(CO_3^{2-})$ in the river water. The laboratory result at 25°C is $10^{-8.48}$ (Equation 4); the ion product in average river water is between 10^{-9} and $10^{-8.5}$, depending on the assumptions made about the composition of the river water. That is, river water is undersaturated with respect to calcite. The degree of undersaturation is greater with respect to aragonite because aragonite is more soluble than calcite (Equation 4).

What about seawater, which (unlike river water) has a constant composition? The calculation of saturation state for seawater is not as straightforward as for river water because of the high salinity, which causes cation-water and cation-anion interactions that lower the effective concentrations of Ca^{2+} and CO_3^{2-} to values much below those of the true concentration. For example, there is so much magnesium in seawater that the neutral ion pair $MgCO_3^0$ is abundant, a condition that reduces the amount of carbonate ion available to contribute toward saturation of the water with respect to $(Ca^{2+})(CO_3^{2-})$. This reduced amount of carbonate ion available to interact with calcium ion is termed the *activity* or effective concentration of carbonate ion. This terminology distinguishes it from the total amount of carbonate—the concentration—which by definition includes all carbonate whether free or bound as $MgCO_3^0$ or some other ion pair species. The ion activity product of $(Ca^{2+})(CO_3^{2-})$ in shallow seawater at low latitudes is $(1.7 \times 10^{-3})(1.0 \times 10^{-5})$ or 1.7×10^{-8}. This number is more than five times larger than $10^{-8.48}$, so it is clear that normal seawater is supersaturated with respect to calcite. The degree of supersaturation with respect to aragonite is less because of the greater solubility of aragonite.

Because seawater is perpetually supersaturated with respect to calcium carbonate, it is to be expected that either calcite or aragonite or both should be continuously forming in the sea. Is this occurring? The continual new growth of calcareous-shelled organisms in the sea suggests that it is. It is generally accepted that the activities of calcium ion and carbonate ion in seawater are maintained at a constant level by the balance between input of calcium from rivers, the outgo by shell formation and subduction back into the mantle, and the constant partial pressure of carbon dioxide in the atmosphere. In Precambrian time, when the amount of carbon dioxide in the atmosphere may have been greater and the earth's surface temperature warmer, a different equilibrium may have been present. A similar statement can be made for subsurface formation of calcium carbonate. In the subsurface, temperatures are always higher than at the surface and the effective partial pressure of carbon dioxide commonly is greater as well. (The effective partial pressure of a gaseous species is termed *fugacity*; it is the equivalent of activity for ions in water.)

Although seawater is always supersaturated with respect to either phase of calcium carbonate, more abundant and more rapid formation of calcite and aragonite can be stimulated by several natural processes:

increases in temperature and salinity; agitation of the water; and biological activity, including production of carbon dioxide in the soil zone.

1. *Increased temperature.* Higher temperature decreases the solubility of calcium carbonate by driving off carbon dioxide gas, increasing the pH, and increasing the CO_3^{2-}:HCO_3^- ratio in the water. This is one reason reefs concentrate in tropical areas at low latitudes.

2. *Agitation of the water.* As anyone who has shaken a bottle of soft drink or beer is aware, the carbon dioxide dissolved in the liquid at the factory or brewery is released when the container is shaken. The rapid evolution of the gas causes the liquid to overflow the container. An analogous process operates in the sea, where water is blown by the wind over shallow substrates such as cratonic margins and volcanic peaks.

3. *Increased salinity.* Carbon dioxide is less soluble in more saline waters, so an increase in salinity causes carbon dioxide to be released. That is, for two waters at the same temperature, the more saline water will contain less carbon dioxide.

4. *Organic activity.* A reef is a symbiotic community of plants (algae) and animals (mostly corals in modern seas but bryozoa, sponges, or other animals in ancient seas). Plants and animals have contrasting metabolisms in that plants ingest carbon dioxide (during photosynthesis) and emit oxygen, whereas animals (such as humans) ingest oxygen gas and emit carbon dioxide. An additional difference is that plants have a peak in metabolic activity during the daylight hours; animal metabolism is less affected by sunlight. Thus, during the daylight hours, the net effect of the gas exchange involving the reef community is the ingestion of carbon dioxide from the seawater by plant tissues and the precipitation of $CaCO_3$ (formation of shell material); that is, the reef grows. During the night, plant metabolism is depressed and the effect of the animal community dominates; the carbon dioxide content of the water increases and the reef dissolves. As a result, the pH of seawater in solution basins within the reef is about 0.3 pH unit higher in the afternoon than it is 12 hours later because changes in the carbon dioxide content of the water are reflected in its hydrogen ion concentration (Equations 1–3).

5. *Organic carbon dioxide production in the soil zone.* Rainwater contains an amount of carbon dioxide that is tied to the amount in the air mass through which the water falls. The rainwater must then pass throughout the soil zone, in which the partial pressure of carbon dioxide is much greater than in the atmosphere because of the decay of plant tissues. As a result, soil water is markedly enriched in carbon dioxide relative to water in the air. If the soil water then enters a cave, in which the P_{CO_2} resembles that of normal air, carbon dioxide will be released from the water, resulting in the precipitation of calcium carbonate as stalactites and stalagmites.

DIAGENESIS

Diagenesis of carbonate materials can begin almost as soon as the skeletal material is precipitated in a shallow-water setting. A large variety of boring organisms may be involved, including algae, sponges, worms, and arthropods, rasping echinoids and fish, and boring and rasping mollusks. The type and extent of this alteration are environmentally controlled, and both bathymetric and latitudinal zonations are evident. Bathymetric patterns are related to light penetration; latitudinal patterns to temperature variation. The boring and scraping performed by the organisms as part of their search for food produce micritic sediment, but it is normally assumed that the amounts are insignificant in comparison to the amounts produced by inorganic precipitation and algal disintegration.

Cementation

Cementation of carbonate sediment can occur very early in the history of the sediment, either on the shallow seafloor, with the formation of surfaces termed **hardgrounds** (Figure 16-19), or within a few thousand years after formation as the sediment is exposed to meteoric water by a lowering of sea level (for example, during Pleistocene glacial episodes). In modern seas, hardgrounds are known from the Bahama Banks, the Persian Gulf, and Shark Bay, western Australia. Lithification is most intense at the surface and decreases in intensity downward, generally ceasing within tens of centimeters of the sediment-water interface. Cemented crusts occur on carbonate slopes to depths of at least 300 m. Many hardgrounds have been described from ancient limestones. They signify interruptions in sedimentation, hiatuses that typically have durations of several hundred thousand years, as estimated on the basis of faunal discontinuities.

Early carbonate cementation of allochem-rich calcium carbonate also occurs in the deep sea. Deep-sea hardgrounds occur mostly in areas where there is good bottom current movement, such as on the flanks of carbonate platforms and the margins of submarine channels and canyons.

FIGURE 16-19
● ● ● ● ● ● ● ● ● ● ● ● ● ● ● ●

Burrowed and mineralized hardground in shallow-water skeletal wackestone to mudstone of the Deschambault Formation near Quebec City, Canada. The sinuous, tubular features are typical of the ichnogenus *Thalassinoides* and are formed of sediment that differs from that immediately above and below. The overlying unit is the Neuville Formation, also of Middle Ordovician age, which is abiotic and a deeper water facies. [Photo courtesy S. C. Ruppel.]

STRANDLINE CEMENTATION. The strandline environment is more complex than that of the shallow-water seafloor because the diagenetic waters are marine, brackish, and meteoric, and the strandline is intermittently exposed to the atmosphere. Strandline cementation produces **beachrock,** which consists of layers of cemented sand- or gravel-size carbonate particles (calcarenite). Strata generally are localized from the intertidal zone up into the zone of wave splash and disappear seaward, landward, and with depth into unconsolidated sediment. Erosion of rock slabs during storms results in conglomeratic units of widely varying clast size. These units are one of the most diagnostic features of ancient beachrock in outcrop.

Shoreward from the strandline beach lies the supratidal zone, an area a few centimeters above normal high-tide level. This zone is flooded only a few times each month, during full moon or new moon, or when storms or hurricanes cause flooding. Once carbonate sediments are deposited well above normal high tide by storms, they are exposed for long periods and subjected to diagenesis by waters that range in salinity from hypersaline (evaporation of shallow, ponded water) to fresh (meteoric water). The effect of hypersaline diagenesis typically is the formation of gypsum and dolomite.

METEORIC WATER CEMENTATION. The most extensive areas of early carbonate diagenesis by meteoric waters are produced by lowering of sea level, a common occurrence during Quaternary time. The direction of meteoric diagenesis is to produce a mineralogically stable low-magnesian calcite rock from a sediment composed predominantly of less stable aragonite and high-magnesium calcite. This process is generally accomplished within a geologically brief period, commonly less than 1 million years.

Diagenesis by meteoric water is not restricted to carbonate sediments above the normal high-tide line. Given sufficient hydrostatic head, fresh water can migrate laterally seaward for more than 100 km through the shallow sedimentary pile under the continental shelf, even while the shelf is flooded with seawater.

Later Diagenesis

Early diagenesis is defined functionally in terms of processes visible in carbonate sediments less than about 10^6 years old, that is, those carbonates that have been formed very recently and whose history has been controlled by Quaternary glacial-interglacial

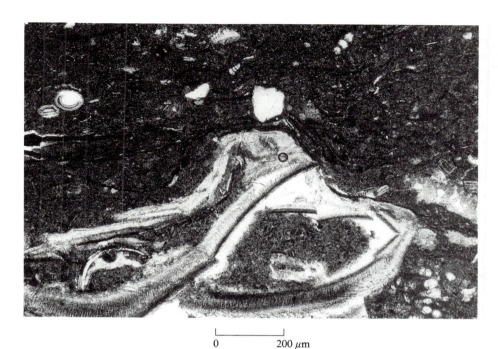

FIGURE 16-20

• • • • • • • • • • • • • • •

Thin section from a core of artificially compacted subtidal micrite, Biscayne Bay, Florida. The wavy organic stringers resemble stylolites and drape over a crushed gastropod shell. The two clear grains above the gastropod are quartz. [From E. A. Shinn and D. M. Robbin, 1983, *J. Sediment. Petrol.*, *53*, 602.]

0   200 μm

episodes. Subsequent diagenetic changes, commonly termed *burial diagenesis*, include compaction, stylolitization, continued cementation and decementation, and recrystallization.

MECHANICAL COMPACTION. Mechanical compaction occurs in unlithified carbonates as it does in unlithified silicate sediments. In carbonates, the effect is most evident in grainstones. Concavoconvex shells such as pelecypods, brachiopods, and ostracods cannot support great weights and will fracture if not protected by very early pore-filling cement. As a result, deformed and broken allochems such as ooliths and fossils are not uncommon in ancient grainstones (biosparites and oosparites).

Petrographic evidence of mechanical compaction in mud-supported limestones (mudstones, wackestones, and micrites) is typically scarce or nonexistent. Micritic sediments can compact readily to half their depositional thickness under overburdens of as little as 100 m, with the production of flattened burrow fillings, mashed or obliterated peloids and fenestral voids, reorientation of fossils to closer parallelism to bedding, and the formation of microstylolite-like layers, which often drape over fossils and other less compactible materials (Figure 16-20).

CHEMICAL COMPACTION. Chemical compaction is the process of porosity loss in carbonate rocks due to pressure solution; it may or may not include cementation. Rocks are thought to have lost material by pressure solution where the surface that separates two objects cuts across the internal fabrics of one or both of them (Figure 16-21). For example, ooids may interpenetrate or stylolites cut across fossil debris. Where truncation is not obvious, pressure solution can sometimes be inferred from the presence of a film of noncarbonate material (clay or organic matter) that can be regarded as an insoluble residue. The stress that causes the dissolution is transmitted by a thin film of water between the two rigid allochems; the film also serves as the transporting pathway for the dissolved Ca^{2+} and CO_3^{2-} (or HCO_3^-) to diffuse outward into the main pore system. The diffusing ions may be precipitated nearby to initiate cementation or may be transported a considerable distance by moving water before precipitating.

Depending on the purity of the rock and the lateral extent of dissolution, pressure solution in carbonate materials may result in the formation of a **stylolite,** an irregular surface within a bed characterized by mutual interpenetration of the two sides, the columns, pits, and toothlike projections on one side fitting into their counterparts on the other (Figure 16-22). Quantitative analyses of stratigraphic sections of limestones in which stylolites are abundant typically reveal that more than 25% of the section has been removed by stylolitization. In one study it was estimated that as much as 90% of the original carbonate deposit had been dissolved, the

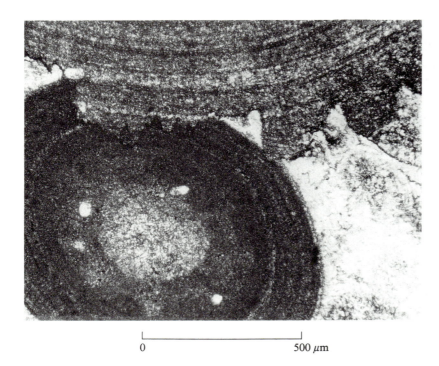

FIGURE 16-21
● ● ● ● ● ● ● ● ● ● ● ● ● ● ● ●

Thin section showing interpenetration of pisolith, ooid, and spar cement, Conococheague Limestone (Cambrian), Maryland. [From H. R. Wanless, 1979, *J. Sediment. Petrol.*, *49*, 440.]

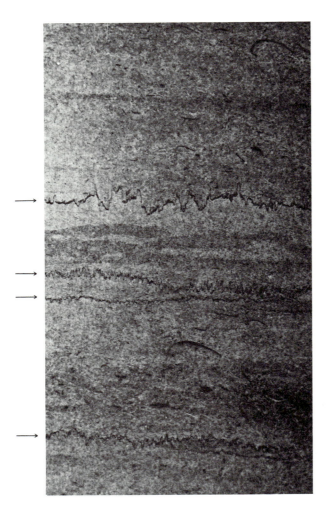

ions having migrated elsewhere. In the absence of seams of insoluble residue, either as gently undulating dark laminae or as fitted tooth-and-socket features, surfaces formed by pressure solution can easily be mistaken for normal bedding planes (Figure 16-23). These pseudobedding surfaces may be much more common than is now recognized and may be an important source of calcium and carbonate ions for subsurface cementation.

BURIAL CEMENTATION. Calcium carbonate, although very soluble in water relative to silicate minerals, is quite insoluble by most other measures. The amount of calcium carbonate in subsurface waters typically is less than 1 g/L; if 10% of this were precipitated, it would require more than 27 L of water to precipitate only 1 cm^3 of calcite. Thus, the amount of water that would have to flow through a carbonate sediment to fill the pores with calcite is prohibitive except very near the surface where pores are large, permeability is great, and water can move vertically and rapidly. On the basis of this reason-

FIGURE 16-22
● ● ● ● ● ● ● ● ● ● ● ● ● ● ● ●

Stylolites in Mississippian brachiopod-crinoid biosparite, formerly used as dividing panel between toilet stalls in Oklahoma Memorial Union, Norman, Oklahoma. Stylolite seams (arrows) are formed of material less soluble than limestone, a mixture of clay and carbonaceous material in this example. Width of photo is 10 cm.

FIGURE 16-23
● ● ● ● ● ● ● ● ● ● ● ● ● ● ● ●

Prominent weathered surface (arrow) passing laterally (to left)
into anastomosing stylolites, which might be mistaken for bedding
surfaces, High Tor Limestone (Mississippian), Wales, Great Britain.
Scale is 1 m long. [From J. Simpson, 1985, *Sedimentology*, *32*, 501.]

ing, we might assume that nearly all cementation of carbonate sediment occurs within perhaps 100 m of the
surface. But porosity in limestones decreases continually with increasing depth. Cementation must continue
progressively to great depth. Such cementation cannot
be the result of near-lateral flow of water through continually less permeable rocks because the time required
for essentially complete lithification is prohibitive. This
conclusion leads to the inference that a large and perhaps major source for the calcium and carbonate ions in
subsurface waters must be internal to the sediment
itself. Cannibalization is required, and the fact that most
limestone beds do not contain stylolitic seams suggests
that nonstylolitic pressure solution is an important
source for calcite cement. Because pressure solution
would increase in abundance and intensity with increasing effective stress, nonstylolitic pressure solution
seams may be very common at depths of perhaps 10^3 m
or more.

Many limestones contain several generations of cements. Fluctuating oxidation states and changing manganese and iron ratios in subsurface fluids cause some
calcite cements to exhibit marked zonation when examined by means of luminescence petrography (Figure
16-24). Although these zones can form at any depth,

they can sometimes be mapped and correlated with
regional stratigraphic discontinuities, enabling the age
and approximate depths of burial to be estimated.

Primary fluid inclusions in carbonate cements have
been analyzed in an attempt to determine the temperature of cement formation. By reference to a reasonable
geothermal gradient, the temperature leads to a depth of
burial at the time of cement precipitation. These inclusion analyses are based on the principle that at the time
of inclusion formation, only liquid was trapped in the
cavity. As temperature decreased with uplift, a gas bubble exsolved to create the two-phase liquid-gas inclusion seen in the cement in thin section. Heating on a
microscope stage causes homogenization of the two
phases, and the temperature of homogenization gives
the minimum temperature of formation of the inclusion
and cement. Homogenization temperatures are affected
by the salinity of the liquid, so that microchemical analysis of the liquid is required for best results.

Secondary Porosity

Diagenetic processes are reversible. If calcite cement
can be precipitated in the subsurface, it can also be
redissolved as chemical conditions change in time and

(A)

0   0.6 mm

FIGURE 16-24

• • • • • • • • • • • • • • • •

Crinoidal biosparite from Lake Valley Limestone (Mississippian), New Mexico. **(A)** Normal thin section view (uncrossed nicols) shows crinoid plate (left center) surrounded by apparently uniform, one-stage growth of calcite cement. [From P. A. Scholle, 1978, *Am. Assoc. Pet. Geol. Mem. No. 27*, 230. Photo courtesy P. A. Scholle.] **(B)** Approximately same area viewed under cathodoluminescence shows at least five generations of cement that can be correlated from sample to sample and related to a variety of tectonic and erosional events. [From W. J. Meyers, 1974, *J. Sediment. Petrol.*, *44*, 844.]

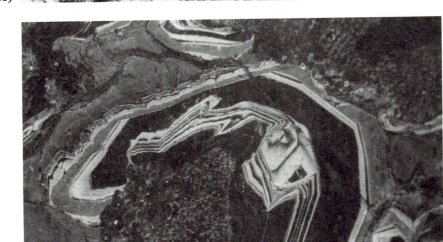

(B)

0   0.6 mm

space, leading to the creation of secondary porosity and perhaps solution breccias. Because of the relatively high solubility of calcite among sedimentary materials in the subsurface, secondary porosity is common in carbonates (and in calcite-cemented sandstones) and may be essential for the creation of pores that are sufficiently numerous for oil fields to form.

The solution pores may be fabric-selective or not and may include molds, **vugs,** and solution-enlarged intraparticle voids (Figure 16-25). Pores commonly cut across both allochems and matrix or cement, as well as stylolites, fractures, or other earlier diagenetic features.

The chemistry of secondary porosity formed by dissolution of calcite in limestones is the same as the chemistry of secondary porosity formed by dissolution of calcite cement in sandstones. The acidity of the pore waters must increase. The most common way for this to occur is through an increase in carbon dioxide fugacity caused by the degradation of organic compounds, particularly petroleum. The relationships among carbon dioxide pressure, temperature, and calcite solubility are shown in Figure 16-26, and it is clear that the greatest effect of increasing carbon dioxide pressure occurs at low pressures. The increase in calcite solubility at any

FIGURE 16-25
• • • • • • • • • • • • • • •

Secondary porosity in limestone as seen in thin section. Phylloid algal plate in an algal mound, Straw Formation (Pennsylvanian), southeastern New Mexico, showing pore space created by dissolution of coarse calcite crystals within plate. Micrite envelopes outline plate. Length of plate is 1.6 mm.

diagenetic temperature is much greater between 0 and 10 bar than between 20 and 30 bar. The greatest effect of temperature on calcite solubility is at the lower diagenetic temperatures, those below 100°C.

Calcite is more soluble in saline water than in distilled or fresh waters, a result of ion pair formation. For example, in seawater a significant proportion of dissolved CO_3^{2-} is tied up as the neutral dissolved species $MgCO_3^0$, and these carbonate molecules are not available to help saturate the solution with respect to $CaCO_3$. Calculations indicate that only 15% of the CO_3^{2-} in seawater is free and available for calcite formation. In an analogous manner, not all the calcium in seawater is available for calcite formation. Calculations indicate that about 12% of it is tied up as ion pairs, mostly as $CaSO_4^0$. The more saline the subsurface water, the less of the dissolved calcium and carbonate in it is likely to be available for calcite precipitation. Some subsurface brines have salinities of several hundred thousand parts per million; normal seawater has 35,000 ppm.

FIGURE 16-26
• • • • • • • • • • • • • • • •

Solubility of calcite in the system $CaCO_3$–CO_2–H_2O in distilled water as functions of temperature and partial pressure of carbon dioxide. Note that increasing carbon dioxide pressure and increasing temperature have opposite effects on calcite solubility. Carbon dioxide is less soluble in the increasingly saline waters that are usually encountered with increasing depth during diagenesis. [From K. Magara, 1981, *Am. Assoc. Pet. Geol. Bull.*, 65, 1341.]

DOLOSTONES
• • • • • • • • • • • • • • • •

Dolostone is a rock composed largely or entirely of the mineral dolomite, $CaMg(CO_3)_2$. Dolostones occur in approximately the same tectonic and physiographic settings as limestones: at low latitudes on the shallow shelves of low-lying continents, most commonly far from the nearest convergent plate margin.

There is no consistent relationship between dolostone abundance and geologic age (Given and Wilkinson

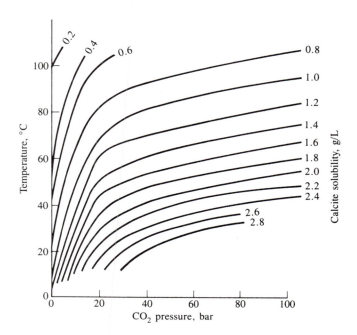

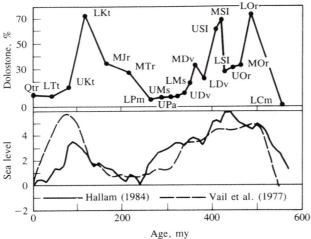

FIGURE 16-27
● ● ● ● ● ● ● ● ● ● ● ● ● ● ● ●

Comparison of global sea level curves (by A. Hallam, 1984, *Ann. Rev. Earth Planet. Sci.*, *12*, 205–243, and by P. R. Vail et al., 1977, *Am. Assoc. Pet. Geol. Mem. No. 26*, 49-212) and the abundance of dolostone (compiled from many sources). The general correlation between high stands of sea level and dolostone abundance is evident. [From R. K. Given and B. H. Wilkinson, 1987, *J. Sediment. Petrol.*, *57*, 1071.]

1987). There does, however, appear to be a positive correlation between dolostone abundance and high stands of sea level (Figure 16-27). Although the correlation is not perfect, perhaps because of imprecisions in the dolostone data collection, it is obvious that first-order eustatic changes in sea level correlate systematically with dolostone abundance through Phanerozoic time.

Field studies also reveal that carbonate rocks tend to be composed of either all calcite or all dolomite. Apparently, whatever the conditions are that produce limestones and dolostones, a marked tendency exists to form one or the other but not subequal mixtures of the two. Why is this so? Does it reveal something about the environments of deposition of the two rock types? Or does it tell us something about the diagenetic history of carbonate accumulations? As usual, the place to begin the search for answers to such questions is in the field.

Field Observations

As was true of limestones, the bulk of dolostone occurs in marine rocks, as indicated by stratigraphic relationships and the presence of fossils, although fossils are noticeably less common in dolostone. Most dolostone units occur as distinct beds or formations interlayered with other nondetrital sedimentary rocks of similar thickness. The contacts with limestone units above and below are usually sharp, but a dolostone bed may grade laterally into another rock type, typically limestone or evaporite. In most cases, the evaporite is gypsum or anhydrite. In some limestone-dolostone sequences, the contact between the two types of rocks is irregular or cuts across bedding planes at a high angle (Figure 16-28). This relationship clearly indicates that the dolostone has formed by replacement of preexisting lime-

FIGURE 16-28
● ● ● ● ● ● ● ● ● ● ● ● ● ● ● ●

Contact between dolostone (light, right) and limestone (dark, left) at high angle to bedding (upper left to lower right), Lost Burro Formation (Devonian), California. The dolostone-limestone contact is sharp but irregular. [From G. V. Chilingar et al., 1979, *Diagenesis in Sediments and Sedimentary Rocks*, *Vol. 1* (New York: Elsevier), p. 495. Photo courtesy D. H. Zenger.]

0   400 μm

FIGURE 16-29
● ● ● ● ● ● ● ● ● ● ● ● ● ● ● ●

Photomicrograph of coarsely crystalline and nonplanar dolomite crystals
with minor interstitial illite (dark), Lost Burro Formation (Devonian),
California. [From D. H. Zenger, 1983, *Geology*, *11*, 520.]

stone. Etching a hand specimen of the limestone associated with the dolostone may reveal selective dolomitization between fossils and matrix. Usually in such cases, the matrix is dolomitized and the fossils are still calcitic, although the reverse can occur. Sometimes it is possible with a hand lens to see that rhombs of dolomite partly transect some of the fossils, although this is more easily seen in thin section in the laboratory. There are, however, beds of pure dolostone that appear to have conformable, sharp contacts with overlying and underlying limestones and, in addition, are microcrystalline (dolomicrite) and contain no visible allochems. In these rocks it is not possible to determine from field observations whether the dolostone bed was originally a limestone or is a primary precipitate.

Laboratory Studies

Calcite and dolomite have very similar optical properties in thin section, so the best method for distinguishing between them is staining. The most commonly used stain, alizarin red S, turns calcite pink but leaves dolomite unaffected. In unstained slides, other criteria are useful although not always certain: euhedral crystal habit, zoning, and twinning.

Many carbonate rocks contain abundant euhedral and subhedral crystals, and staining with alizarin red S reveals that such crystals are nearly always dolomite. Dolomite precipitated at temperatures below 50°–100°C has a very strong tendency to crystallize with a euhedral habit, a feature almost unknown in either primary calcite crystals or in calcite that has replaced earlier calcite or aragonite.

Some coarse-grained dolostone is composed of anhedral crystals (Figure 16-29) that commonly have undulatory extinction. Such dolostone forms at temperatures greater than 50°–100°C, typically as a replacement of an earlier dolostone.

Dolomite crystals (and calcitized dolomite crystals), particularly those of sand size or larger, commonly contain concentric, alternating zones of iron-rich (red) and iron-poor (clear) dolomite that mark stages of growth of

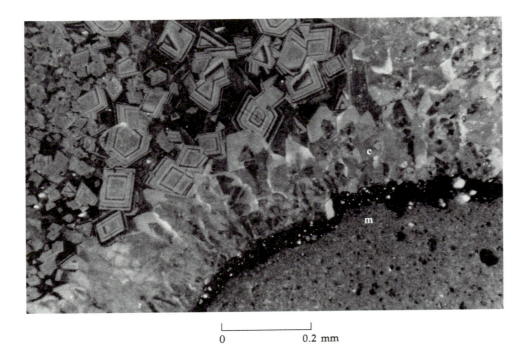

0   0.2 mm

FIGURE 16-30
● ● ● ● ● ● ● ● ● ● ● ● ● ● ● ● ●

Cathodoluminescence photomicrograph showing mudstone pebble (m) in stylolitic contact with weakly to dully luminescent fringe of columnar calcite (c). Rhombs of dolomite have replaced the matrix of a conglomerate. The dolomite crystals share a common microstratigraphy. Dark crystal zones contain at least 0.5% FeO; bright zones contain less and are orange-red in color. [From M. Coniglio and N. P. James, 1988, *J. Sediment. Petrol.*, *58*, 1039.]

the rhomb (Figure 16-30). The ferrous iron must be oxidized to the ferric form (and precipitate as hematite) to be visible with standard petrographic techniques. On oxidation, the iron atoms acquire a magnetic orientation, which can be measured. The orientation can then be compared with the apparent polar wander path during Earth history to establish a numerical date for the diagenetic event, a date accurate to ±20 million years.

Dolomite crystals are twinned only rarely in sedimentary rocks, although calcite is commonly twinned. This difference in frequency may result from the fact that calcite has a plane of easy slip parallel to {0001}; dolomite has no plane of comparably easy slip.

Aggregate Dolostone Textures

Any texture that occurs in a limestone can occur in a dolostone. In most dolostones the replacement origin of

the mineral is clearly evident in the rock texture in thin section. Dolomite rhombs cut across the boundaries of allochems (Figure 16-31) and may even preserve delicate structures within the allochem during the replacement process.

Dolostones that contain replacement textures are, without doubt, formed after deposition of a sediment composed of calcite or aragonite. However, many dolostones lack clear evidence of replacement origin. Perhaps dolomitization completely destroyed the outlines and internal structures of preexisting allochems; perhaps the original sediment was a homogeneous calcitic or aragonitic micrite that did not contain diagnostic features; or, possibly, the dolomite formed as an original chemical precipitate from modified seawater. For many dolostones, none of these three possibilities can be eliminated on the basis of either field or thin section observations.

FIGURE 16-31
● ● ● ● ● ● ● ● ● ● ● ● ● ● ● ● ●

Dolomite rhombs that have replaced ooids, preserving the fabric of the ooids as "ghosts." Diameter of large rhomb is 0.3 mm. [Photo courtesy R. C. Murray.]

ENVIRONMENTS OF FORMATION
● ● ● ● ● ● ● ● ● ● ● ● ● ●

Because the occurrence of Phanerozoic dolostone appears to depend heavily on the existence of high stands of sea level and because the Cenozoic Era was a time of low stand (see Figure 16-27), there are few Cenozoic dolostones. Nevertheless, in both Cenozoic and pre-Cenozoic rocks, dolostones are associated almost exclusively with either limestones or evaporites, and these associations supply the clues to explain the bulk of dolostones in the geologic record.

Most dolostones form by one of two mechanisms, evaporative **reflux** or by mixing of fresh and marine waters. In evaporative reflux, isolated or near-isolated areas of the sea underlain by limestone are evaporated until the concentration of the water is about 100,000 ppm. At this concentration, gypsum precipitates in the brine. The remaining water is now denser than the water in the pores of the underlying limestone (which is normal seawater) and also has an increased Mg:Ca ratio because of the calcium removed by gypsum precipitation. The brine sinks into the limestone, displaces the lighter pore water, and dolomitizes the limestone. The classic example of such an occurrence is the Permian reef complex in west Texas and southeastern New Mexico (Figure 16-32).

Gypsum is a very soluble mineral. Although it may have formed in association with dolomicrites, it may subsequently have been dissolved. Sometimes casts or molds of gypsum are preserved and can be recognized by its crystal habit. Other times, the dolostone stands alone, and its environment of formation must be deciphered by the sedimentary textures and structures it contains. Features that, as a group, are diagnostic of a supratidal environment are desiccation cracks; dolomicrite without fossils; laminations suggestive of replaced algal mats, such as wavy laminations and carbonaceous films; thin bedding; nontectonic, soft-sediment folding and brecciation (flat-pebble conglomerates); and very finely interlaminated dolomicrite and micrite.

For evaporative reflux to be a possible explanation of dolostone occurrences, it is required that significant thicknesses of gypsum and anhydrite (or pseudomorphs of these minerals) directly overlie the dolostone. The volume of calcium sulfate formed must be approximately the same as the volume of dolomite. Many dolostone sections, however, are not associated with evaporites, evaporite mineral pseudomorphs, or solution-collapse structures that would indicate the former presence of evaporites. How are these nonevaporite dolostones to be interpreted?

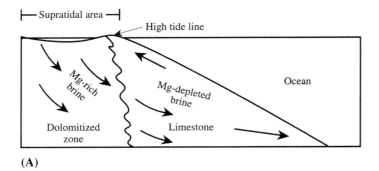

(A)

FIGURE 16-32

● ● ● ● ● ● ● ● ● ● ● ● ● ● ● ●

(A) Schematic cross section showing paths of water flow in the reflux mechanism of dolomitization.
(B) Cross section of Permian reef complex, west Texas and southeastern New Mexico, Guadalupian shelf to Delaware Basin, showing the field relationships in which reflux dolomitization occurred. The time during which dolomitization might have occurred is approximately 5–10 million years. [From West Texas Geological Society, 1982, *Field Trip Guidebook to the Delaware Basin*, 126.]

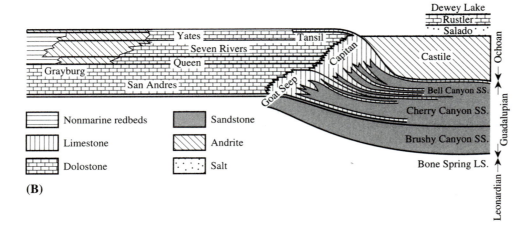

(B)

The answer was provided in 1973 by a study of Pleistocene reefs in Jamaica (Figure 16-33). The dolomite forms below the water table in the zone of brackish water, a setting that is universal along coastlines. It is the zone of contact between fresh, essentially meteoric, waters a few meters below the ground surface and the seawater present below the freshwater lens. Many ancient dolostones lacking evaporites have since been interpreted in terms of this mixed-water model, termed **Dorag dolomitization.**

If the model of dolomite formation by the mixing of seawater and meteoric waters is correct, several inferences can be made:

1. The mineralogic transition from limestone in a seaward direction to dolomite in a landward direction does not coincide with smaller scale changes in depositional environment within the limestone section. It may, however, be closely related to differences in the predolomitization permeability of the limestone.

2. The position of the mineralogic transition is controlled by the position of exposed tracts of the carbonate sequence. The position of exposed tracts, and thus of freshwater lenses, is paleogeographically controlled.

3. Changes in regional paleogeography should produce changes in the position of the limestone-dolostone boundary. Regional regressions of the sea result in a seaward shift of the limestone-dolostone boundary because the position of the subaerially exposed carbonate terranes moves seaward. Conversely, regional transgressions would flood previously exposed tracts; this would push areas of freshwater recharge, and thus the limestone-dolostone boundary, landward.

4. If a close correlation exists in a stratigraphic section between transgressions and regressions and a shift in the limestone-dolostone boundary, it must be that dolomite formation occurred in the shallow subsurface soon after burial of the carbonate units. If dolomitization had occurred after deep burial of the limestone, the effect of surface paleogeography would not be present. Excellent correlation has been found between onlap-offlap and the movement of the limestone-dolostone boundary in many areas of Paleozoic rocks in the United States.

FIGURE 16-33
● ● ● ● ● ● ● ● ● ● ● ● ● ● ● ● ●

Cross section showing contact relations between Hope Gate Formation and Falmouth Formation (Pleistocene), on northern side of Jamaica. Truncated marine terrace underlain by Falmouth Formation is capped by well-lithified, caliche-like caprock (dotted pattern), through which very little water percolates from vadose zone to phreatic zone. Most of the meteoric water in the Falmouth is derived from the phreatic-zone aquifer of the island. [From L. S. Land, 1973, *Sedimentology, 20*, 413.]

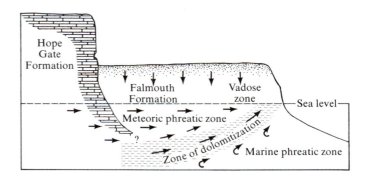

Deep-Burial Dolostone

Both the reflux and mixed-water models for the generation of extensive dolostone units rely on replacement reactions occurring at shallow depths, perhaps within 100 m of the surface. The advantages of very shallow depth are that an inexhaustible supply of magnesium is available from circulating seawater (1290 ppm Mg^{2+}) and that permeabilities of the carbonate sediments tend to be high. The combination of abundant magnesium and high rates of fluid flow is optimal for dolostone formation from preexisting limestone. However, some dolostone is composed of anhedral and/or undulatory dolomite crystals, features characteristic of formation at high temperatures. The source of the magnesium for these dolostones is uncertain, and they may simply be shallow dolostones that have recrystallized at depth.

GEOCHEMISTRY OF DOLOMITE
● ● ● ● ● ● ● ● ● ● ● ● ● ● ● ●

The formation of dolomite crystals can be described chemically by

$$CaMg(CO_3)_2 \rightleftharpoons Ca^{2+} + Mg^{2+} + 2\,CO_3^{2-}$$

The equilibrium constant is

$$K_{25°} = [Ca^{2+}][Mg^{2+}][CO_3^{2-}]^2 = 10^{-18.06}$$

River water is undersaturated with respect to dolomite, as it is with calcite and aragonite. The ion activity product for dolomite in seawater is $10^{-14.6}$, a value indicating that seawater is supersaturated with dolomite. Many subsurface waters probably are also supersaturated with dolomite, although this is less certain because of the variable compositions of these waters and the uncertain effects of very high salinities on the values of activity coefficients. Dolomite does not normally precipitate from unmodified seawater, despite the

apparent supersaturation. The explanation for the lack of primary dolomite precipitates is thought to be the difficulty of stripping from the magnesium ion the water molecules that are bound to it in aqueous solution. Magnesium is a small ion and holds its bound water tightly. At the higher temperatures that are present in the subsurface, however, stripping of the bound water is more easily accomplished; so precipitation should occur if the pore solution is supersaturated.

For dolomite to precipitate at 25°C, the $Mg^{2+}:Ca^{2+}$ ratio in the dolomitizing water must be at least 1:1, but the thermodynamically required ratio *decreases* as temperature increases. At 90°C, dolomitization is possible in solutions that contain only one-fourth as much magnesium as calcium; at 190°C, only one-tenth as much. Combining these data with the relative ease of stripping water molecules from ions at high temperatures, we can infer that deep-burial dolomitization is a likely occurrence in limestones if sufficient magnesium is available.

SUMMARY
● ● ● ● ● ● ● ● ● ● ● ● ● ● ● ●

Nonreefal limestones are composed of grains, matrix, and cement and can be described as calcium carbonate analogs of sandstones. Grains are either fossils, peloids, ooids, or limeclasts; matrix is calcitic mud; and cement is coarse, clear calcite, as in sandstones. Reefal limestones have textures and structures determined by the biology and growth habits of the animal and plant communities that form the reef. All ancient limestones have undergone significant recrystallization because of the relatively high solubility of calcium carbonate, but many important textural and structural characteristics are normally preserved.

Most dolostones originate through replacement of limestones with preservation of many of the original depositional features. Probably most dolostones form within 100 m of the surface, but further burial and higher temperature may result in recrystallization.

STUDY EXERCISES
• • • • • • • • • • • • • • • •

1. Some scientists believe that the temperature of the lower atmosphere is increasing because of the continual input of carbon dioxide into the air by the combustion of fossil fuels. Describe how this increasing temperature and increasing content of carbon dioxide might affect limestone formation.

2. Explain why limestones normally contain only very small percentages of terrigenous detritus. (*Hint:* What is the origin of most particles of calcium carbonate?)

3. List and discuss the various physical, chemical, and sedimentary variables that might affect the formation of stylolites. Why do you think stylolites do not seem to be present in all limestones, given the increased solubility of calcite with increased pressure (burial depth)?

4. List features you might see in a thin section of a porous limestone that would suggest that the porosity is secondary. Suppose these features are absent. Does this necessarily indicate that the visible porosity is all primary? Explain.

5. Fossils, peloids, ooids, and limeclasts are all less abundant in dolostones than in limestones. What are possible explanations for this observation?

REFERENCES AND ADDITIONAL READINGS
• • • • • • • • • • • • • • •

Boss, S. K., and B. H. Wilkinson. 1991. Planktogenic/eustatic control on cratonic/oceanic carbonate accumulation. *J. Geol.*, *99*, 497–514.

Carozzi, A. V. 1989. *Carbonate Rock Depositional Models*. Englewood Cliffs, NJ: Prentice Hall.

Crevello, P. D., J. L. Wilson, J. F. Sarg, and J. F. Read. 1989. *Controls on Carbonate Platform and Basin Development. SEPM Spec. Pub. No. 44.*

Geldsetzer, H. H. J., N. P. James, and G. E. Tebbutt, eds. 1988. *Reefs. Can. Soc. Pet. Geol. Mem. 13.*

Moore, C. H. 1989. *Carbonate Diagenesis and Porosity.* New York: Elsevier.

Nelson, C. S., ed. 1988. Non-tropical shelf carbonates—Modern and ancient. *Sediment. Geol.*, *60*, 3–367.

Scholle, P. A. 1978. *Carbonate Rock Constituents, Textures, Cements, and Porosities. Am. Assoc. Pet. Geol. Mem. No. 27.*

Sellwood, B. W., ed. 1992. *Sediment. Geol.*, *79* (special issue on ramps and reefs).

Shukla, V., and P. A. Baker, eds. 1988. *Sedimentology and Geochemistry of Dolostones. SEPM Spec. Pub. No. 43.*

Tucker, M. E., and V. P. Wright. 1990. *Carbonate Sedimentology.* Boston: Blackwell.

Chapter 17

OTHER TYPES
OF SEDIMENTARY ROCKS

EVAPORITES

*E*vaporite minerals are defined as those minerals produced from a saline solution as a result of extensive or total evaporation of the water. Significant deposits of evaporites are irregularly distributed throughout geologic time and total about 1% of Phanerozoic rocks, the Permian Period being the time of greatest abundance. Presumably, the times of great abundance of evaporites reflect the times when drifting continents had become concentrated into the belt of descending, warm, dry air, at present centered at about 30°N and 30°S latitudes. The latitudinal location of this belt is affected by slight changes in solar radiation, the extent of polar glaciers, the relative amounts of land and sea, and other factors; so the location of this belt has varied throughout geologic time. An additional control on the location of dry areas is the existence of topographic barriers to air flow in the troposphere. These barriers—mountain ranges—have been present at various locations on the continental masses throughout geologic time. Evaporite beds are rare in Precambrian rocks, although casts of evaporite minerals have been found in numerous sequences as old as 3.5 billion years. Presumably, the major reason for the absence of evaporite units in Precambrian rocks is that they have been dissolved during later diagenesis.

Abundance

Evaporite minerals are very soluble and rarely occur in outcrop except in arid regions. In these areas, gypsum is the most common and abundant mineral because (a) gypsum is the first mineral (after calcite) to precipitate from evaporating seawater at temperatures below 40°C, and (b) the anhydrite that forms from gypsum on burial to a few hundred meters is rapidly rehydrated to gypsum when uplifted. Halite in outcrop is a poor second in abundance to gypsum because it is orders of magnitude more soluble than gypsum.

Despite the infrequent occurrence of evaporite beds in outcrop, they are common in the subsurface. Drilling for petroleum and natural gas has revealed that beds of evaporites underlie about one-third of the land surface of the United States (Figure 17-1). Further, halite is more abundant in the subsurface than gypsum, as might be expected because of the dominance of sodium and chloride ions in seawater.

Evaporite rocks are most commonly interbedded with dolostone, limestone, and fine-grained detrital rocks, particularly thin, red shales.

Mineralogy

The variety and abundance of minerals produced from an evaporating body of water depend on its initial composition; and because the range in possible parent waters is very large, so is the variety of evaporite minerals. Approximately 70 evaporite minerals are known, of which 27 are sulfates, 27 are borates, and 13 are halides. A single bed of evaporite is usually composed of more than one mineral, particularly when the deposit consists of the less common and more soluble evaporite minerals.

The bulk of evaporite deposits has seawater as its parent solution, and the most abundant anions in seawater

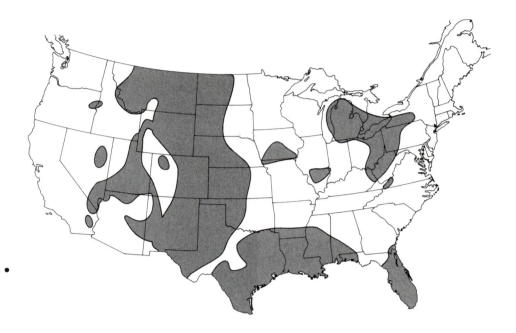

FIGURE 17-1

● ● ● ● ● ● ● ● ● ● ● ● ● ● ●

Areas of conterminous United States underlain by evaporites (shaded area).

are Cl^- (94.5% of anion molarity) and SO_4^{2-} (4.9% of anion molarity). As a result, the abundant evaporite minerals in these deposits are chlorides (halite) and sulfates (gypsum and anhydrite). Table 17-1 lists the major minerals in marine evaporite deposits. Nonmarine saline lake and playa evaporite waters commonly have initial compositions that differ significantly from the composition of seawater; as a result, they may give birth to minerals rarely formed from seawater, including many complex carbonates, sulfates other than gypsum and anhydrite, and many borates.

Distinguishing Marine from Nonmarine Evaporites

A large proportion of modern saline lake concentrated brines are, like seawater, rich in Na^+ and Cl^-, and commonly are rich in sulfate as well. Hence, the most common saline minerals in nonmarine evaporite deposits are, like those of marine evaporites, halite and gypsum and/or anhydrite. Some nonmarine salt pans are very large, with areas in excess of 10,000 km^2 (Lake Uyuni, Bolivia); thicknesses of Cenozoic nonmarine evaporites can exceed 600 m. So we conclude, based on thickness and area, that a nonmarine evaporite deposit may be difficult to distinguish from a marine deposit. For many evaporite rocks, there may have been significant inflow of nonmarine waters to a surface water that was initially an isolated arm of the sea; that is, the water is neither exclusively marine nor exclusively nonmarine.

How can the continentality of an evaporite deposit be determined? Because we know that some modern evaporative saline waters have not been influenced by seawater, minerals diagnostic for nonmarine evaporites have been identified. The most common of these are complex carbonate minerals such as trona ($NaHCO_3 \cdot Na_2CO_3 \cdot 2\,H_2O$), natron ($Na_2CO_3 \cdot 10\,H_2O$), nahcolite ($NaHCO_3$), and gaylussite ($CaCO_3 \cdot Na_2CO_3 \cdot 5\,H_2O$); sulfate minerals such as mirabilite ($Na_2SO_4 \cdot 10\,H_2O$); and borates such as borax [$Na_2B_4O_5(OH)_4 \cdot 8\,H_2O$], kernite [$Na_2B_4O_6(OH)_2 \cdot 3\,H_2O$], and colemanite [$CaB_3O_4(OH)_3 \cdot H_2O$]. However, the

TABLE 17-1 Major minerals in marine evaporite deposits

● ●

Mineral	Chemical formula
Chlorides	
Halite	$NaCl$
Sylvite	KCl
Carnallite	$KMgCl_3 \cdot 6\,H_2O$
Sulfates	
Anhydrite	$CaSO_4$
Langbeinite	$K_2Mg_2(SO_4)_3$
Polyhalite	$K_2Ca_2Mg(SO_4)_4 \cdot 2\,H_2O$
Kieserite	$MgSO_4 \cdot H_2O$
Gypsum	$CaSO_4 \cdot 2\,H_2O$
Kainite	$KMg(SO_4)Cl \cdot 3\,H_2O$

TABLE 17-2   Criteria for distinguishing between primary and secondary features in evaporites
● ●

Syndepositional Features

1. Sedimentary structures such as ripple marks, lamination, and graded bedding
2. Detrital framework textures such as abrasion features, "rafts" of halite crystals formed by settling from the air-water interface
3. Crystalline framework fabrics such as vertically oriented gypsum prisms with euhedral terminations, upward coarsening of crystals, and crystals originating on a common substrate
4. Dissolution-reprecipitation features such as sharp and smooth truncations of vertically oriented crystal frameworks overlain by detrital materials
5. Absence of high-temperature salts such as langbeinite

Postburial Features

1. Massive crystalline mosaics without bedding
2. Mosaic patches that cut across bedding
3. Pressure solution suturing
4. Equigranular mosaics with crystals intersecting at 120° angles
5. Concentrations of former solid inclusions at crystal boundaries, reflecting purging during recrystallization
6. Deformation features such as folds and veins

Ambiguous Features

Minerals with wide temperature stability ranges that occur as
    Pseudomorphous replacements
    Intrasediment growths as euhedra or nodules
    Void-filling cements

Source: From L. A. Hardie et al., in B. C. Schreiber and H. L. Harner (eds.), 1985, *Sixth International Symposium on Salt* (Alexandria, VA: The Salt Institute).

standard sedimentologic and/or stratigraphic criteria, such as fossil content and the characteristics of the enclosing strata, are frequently more diagnostic than mineralogy. Does the evaporite deposit contain land plant remains or freshwater ostracods, gastropods, or diatoms? Are planktonic marine organisms such as foraminifers or radiolaria present? Are the enclosing beds fluvial, as determined from sedimentary structures and textures, or are shallow marine environments indicated? Always keep in mind that in strandline settings there may be sedimentologic and aqueous influences from both marine and nonmarine environments.

Distinguishing Primary from Secondary Features

The extreme solubilities of evaporite minerals cause saline minerals to be altered even during shallow burial. Consequently, features such as dissolution, pseudomorphing, recrystallization, and overgrowth formation are much more common in evaporite rocks than in silicate or carbonate rocks. The ease and frequency of reactions during very early diagenesis, essentially penecontemporaneously with deposition and geologically instantaneous, means that it can be difficult, either in the field or in thin section, to distinguish between original and diagenetic features in an ancient evaporite deposit (Table 17-2).

Sedimentary Structures

Primary sedimentary structures described from evaporite units include lamination, cross-bedding, graded bedding, ripple marks, mud cracks, syndepositional folds and slumps, and other, more unusual clastic and/or chemical structures. The most common structure is lamination (Figure 17-2). A typical evaporite lamination

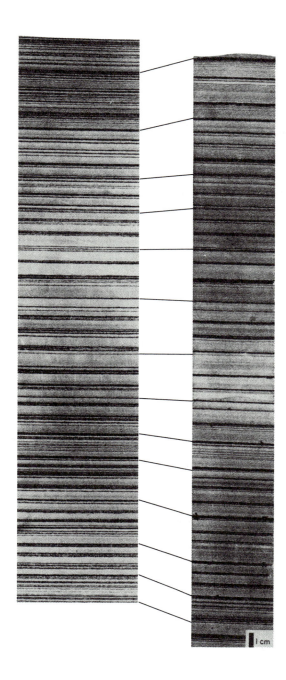

pattern consists of alternate laminae of white, nearly pure calcium sulfate mineral and gray-black lamellae rich in calcite and/or dolomite and organic carbon. The lamellar pair is typically 0.2–2 mm thick but may be as thick as 10 mm, and the organic-rich laminae are generally only 10–20% of the total thickness of the couplet. The evenness and continuity of laminae are interpreted by some sedimentologists as indicative of deposition in "deep" water.

The majority of sedimentary structures seen in outcrops of evaporite beds are of diagenetic origin. This is to be expected because of the extreme solubilities of evaporite minerals. Figure 17-3 illustrates nodular or **chicken-wire structure,** a striking structure formed of elongate nodules of anhydrite set in a darker matrix of microcrystalline sediment, usually anhydrite or carbonate. This anhydrite mosaic can form from dehydrated aggregates of gypsum, usually at shallow depths.

Laboratory Studies

In thin sections of ancient evaporite beds, apparently primary and clearly secondary textures can both be

FIGURE 17-2

• • • • • • • • • • • • • • •

Comparison of presumably correlative laminae of calcite (dark) and anhydrite (light) in subsurface cores from sites about 95 km apart, Castile Formation (Permian), west Texas. [From Dean and Anderson, 1982, in C. R. Handford et al. (eds.), *Depositional and Diagenetic Spectra of Evaporites—A Core Workshop*, p. 340.]

FIGURE 17-3

• • • • • • • • • • • • • • •

Mosaic anhydrite (chicken-wire structure). Actual size. [From W. R. Maiklem et al., 1969, *Bull. Can. Pet. Geol., 17*. Photo courtesy R. P. Glaister.]

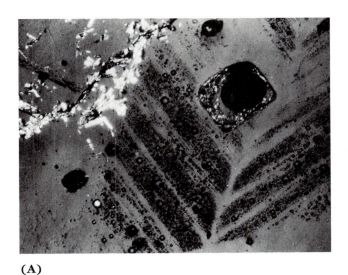

(A)

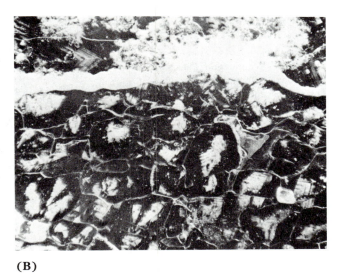

(B)

FIGURE 17-4

●●●●●●●●●●●●●●●●●●

Thin section showing chevron structure within halite crystals.
(A) Close-up, with white anhydrite along crystal boundaries.
Width of photo is 3.4 mm. **(B)** Broader field of view, showing
chevrons within crystals surrounded by clear halite overgrowths
that meet along subplanar boundaries. Width of photo is 4.2 mm.
[From K. A. Holdaway, 1978, *Kansas Geol. Surv. Bull.*, *215*, 25, for
part (A) and 26 for part (B). Photos courtesy Kansas Geological
Survey.]

$$CaSO_4 \cdot 2\,H_2O \; \rightleftharpoons \; CaSO_4 + 2\,H_2O$$
$$\text{gypsum} \qquad\qquad \text{anhydrite}$$

$$2\,CaSO_4 + 2\,K^+ + Mg^{2+} + 2\,SO_4^{2-} + 2\,H_2O$$
$$\text{anhydrite}$$
$$\rightleftharpoons K_2Ca_2Mg(SO_4)_4 \cdot 2\,H_2O$$
$$\text{polyhalite}$$

$$KMgCl_3 \cdot 6\,H_2O \; \rightleftharpoons \; KCl + Mg^{2+} + 2\,Cl^- + 6\,H_2O$$
$$\text{carnallite} \qquad\qquad \text{sylvite}$$

$$K_2Mg_2(SO_4)_3 + Mg^{2+} + 2\,Cl^- + 3\,H_2O$$
$$\text{langbeinite}$$
$$\rightleftharpoons 3\,MgSO_4 \cdot H_2O + 2\,KCl$$
$$\text{kieserite} \qquad\qquad \text{sylvite}$$

observed, although the distinction between the two
is not always clear-cut because of frequent diagenetic
modifications. For example, thin sections of halite beds
commonly reveal the crystals to contain an internal
chevron structure formed by zones of fluid inclusions
(Figure 17-4), reflecting changing growth rates of halite
at the sediment-brine interface. If the halite recrystal-
lizes, fluid and solid inclusions are purged to grain
boundaries, and the depositional texture is completely
eradicated (Figure 17-5).

Evidence of replacement textures is common in evaporite rocks. Pseudomorphs are widespread and reaction
rims are common, as are relics of an earlier evaporite
mineral enclosed within a mineral that replaced it. Well-
known examples are pseudomorphs of anhydrite after
swallowtail twins of gypsum, reaction rims of polyhalite
around earlier anhydrite, relics of carnallite in secondary
sylvite, and pseudomorphs of sylvite plus kieserite after
earlier langbeinite. The following equations describe
these types of transformations recognized in thin section:

Ochoan Series, Delaware Basin

The most spectacular evaporite deposit in North America
probably is the gypsum-anhydrite-halite sequence of the
Ochoan Series, composed of the Castile and Salado Formations (Permian), in the Delaware Basin of west Texas
and southeastern New Mexico. The Castile has a small
area of outcrop, but the Salado occurs only in the subsurface, where it underlies an area more than 150,000 km^2 in
extent (Figure 17-6). The entire evaporite deposit has a
maximum thickness of 1300 m and a volume of approximately 65,000 km^3. The original depositional volume of
the evaporite beds was even greater than at present, as
evidenced by erosional contacts laterally and many anhydrite mudstone solution breccias caused by the removal
of underlying halite beds.

FIGURE 17-5

• • • • • • • • • • • • • • • • •

Photomicrograph of recrystallized halite, Salado Formation (Permian), New Mexico. Texture is an equigranular mosaic in which crystal boundaries meet at triple junctions with angles of approximately 120°, characteristic of all fabrics recrystallized under hydrostatic stress conditions. Most mud impurities have been purged to crystal boundaries. Scattered small bubbles result from slide preparation. [From T. K. Lowenstein and L. A. Hardie, 1985, *Sedimentology, 32,* 640.]

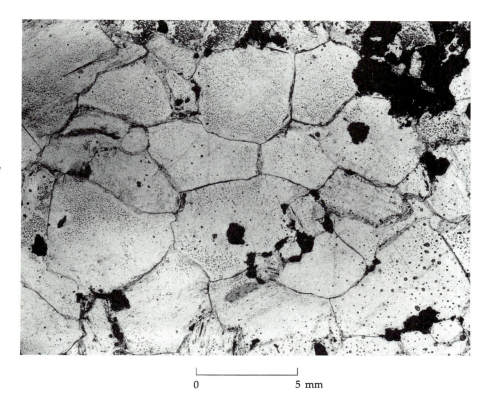

```
0                    5 mm
```

The depositional environment of this giant evaporite deposit was a steep-sided, deep basin bordered by a reef that stood more than 500 m above the basin floor at the onset of Castile Formation deposition (Figure 17-7). The lower portion of the Castile is composed of alternating thin laminae of calcite, anhydrite, and organic matter. The anhydrite layers are two to three times thicker than the calcite layers. In addition to many variations in the character of these primary sedimentary structures, many diagenetic features are present. Calcite laminae disappear into nodular masses of anhydrite. Concretionary anhydrite lenses grow between the laminae and disrupt them. Bands of crinkled laminae occur between undisturbed beds. Sometimes the anhydrite is interlaminated with dolomite rather than calcite.

In the upper part of the Castile, halite becomes more abundant, and the onset of halite precipitation seems to have been synchronous throughout the basin. Halite laminae are up to 10 cm thick, are thoroughly recrystallized, and are interbedded with anhydrite laminae. Toward the top of the Castile, the halite laminae thicken and display vertically aligned pseudomorphs of shallow-water gypsum. At the same time, the beds become less regular. Finally, the Castile passes up into the bedded halite (700 m thick) of the Salado Formation, which contains extensive deposits of sylvite, carnallite, langbeinite, and kainite. In contrast to the calcite mineralogy in the carbonates of Castile time in the basin, abundant dolo-

mite and magnesite are found in the Salado. Because of its shelf location, a strong continental component probably influenced the chemistry of Salado evaporite waters.

The distribution of sedimentary rock types in the Delaware Basin shows a crude concentric zonation (Figure 17-8), characteristic of a desiccating basin. Along the outer fringe are either fine-grained clastics or carbonate rocks, depending on the location of nearby land areas. Within this outer fringe are gypsum and/or anhydrite, followed by halite, and finally by the more soluble salts in the center of the desiccated area—salts such as polyhalite, langbeinite, carnallite, and sylvite. Polyhalite is the least soluble of the posthalite salts; it occurs throughout a greater stratigraphic interval; and it is more widespread in distribution than the more soluble potassium salts. The very soluble potassium salts persist into the lower part of the Rustler Formation (see Figure 17-7), where they are finally terminated by the influx of red muds.

Devonian Duperow Formation, Williston Basin

A different type of evaporite depositional environment is the shallow-basin, shallow-water setting, in which the basin is shallow initially and remains shallow throughout the period of evaporite accumulation. The evaporitic

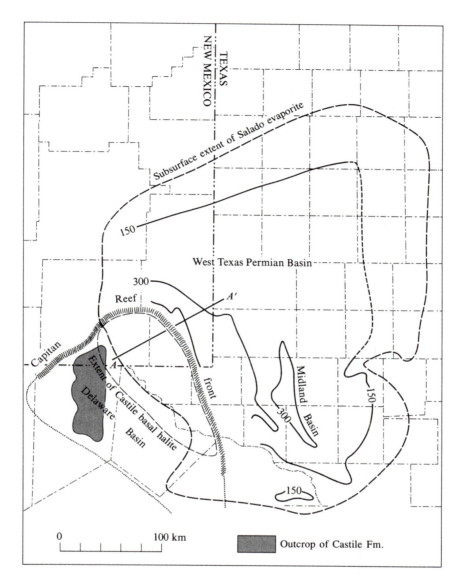

FIGURE 17-6

● ● ● ● ● ● ● ● ● ● ● ● ● ● ● ● ●

Map of west Texas and southeastern New Mexico, showing location and thickness of Salado evaporite (in meters). *A–A'* is line of cross section in Figure 17-7. [From G. A. Kroenlein, 1939, *Am. Assoc. Pet. Geol. Bull.*, *23*, 1683.]

sequence typically consists of interfingering evaporite and mud-flat units often interrupted by karst and erosional surfaces. The overall sediment accumulation consists of a number of shoaling-upward cycles, commonly with intercalated shallow-water carbonates and siliciclastics. Shallow-water, shallow-basin evaporites are often found as rift valley, or aulacogen, fill, where the basinal succession in the rift is continental and passes up-section into marine evaporites. Examples include the sequence beneath Atlantic continental margins and the

sequence currently forming in the Ethiopian sector of the East African Rift valley. Well-studied examples in North America are the Michigan and Williston Basins.

The Duperow Formation in the Williston Basin (Figure 17-9) is part of a great sheet of Upper Devonian strata that extend beneath the Prairie Provinces of Canada southeastward from the Canadian Rockies and arctic Canada into the United States as far south as the transcontinental arch of South Dakota. Its maximum thickness is about 2500 m in southern Saskatchewan.

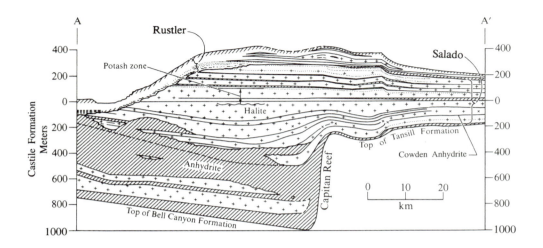

FIGURE 17-7

●●●●●●●●●●●●●●●●●●

Generalized cross section of Ochoan Series normal to reef front (see Figure 17-6), showing changes in evaporite composition and thickness between Delaware Basin and backreef area. The dashed line marks the top of the Castile Formation. [From G. A. Kroenlein, 1939, *Am. Assoc. Pet. Geol. Bull.*, *23*, 1687.]

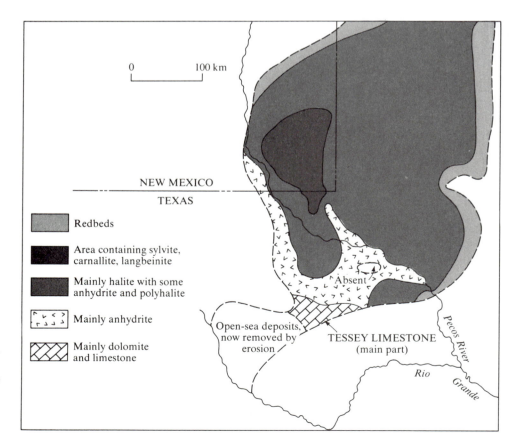

FIGURE 17-8

●●●●●●●●●●●●●●●●●

Map showing distribution of sedimentary rock types during deposition of the Salado Formation in the Delaware Basin, Texas and New Mexico. [F. H. Stewart, 1963, *U.S. Geol. Surv. Prof. Paper 440-Y*, 28.]

The formation is composed of about 12 regular carbonate-evaporite cycles. Each cycle consists of a lower member of burrowed, fossiliferous micrite containing a normal marine, shallow-water fauna of brachiopods, crinoids, and stromatoporoids (a group of extinct calcareous coelenterates); a middle member of brown micrite with a restricted microfauna of ostracods interbedded with unfossiliferous, peloidal beds or laminated micrite; and an upper member of bedded anhydrite with gray-green, silty dolomicrite displaying intertidal and supratidal sedimentary structures such as mud cracks and microbreccias (Figure 17-10). The anhydrite member is typically the thinnest of the three members. Duperow cycles are exceedingly widespread, and constituent beds only 3–5 m thick can be traced for several hundred kilometers across the Williston Basin.

Deposition occurred in a vast backreef lagoon south of a reef belt in Alberta and stretched to a sandy shore in South Dakota and Wyoming. This very shallow basin was periodically and apparently rapidly flooded with marine water, a condition permitting certain benthic organisms to flourish and even small reefs to grow at times. Gradual shallowing as sedimentation filled the basin resulted in extensive evaporitic tidal flats; extensive dolomitization occurred on the peripheral shelves. The time required for the deposition of each cycle is estimated to be between 500,000 and a million or so years, assuming a constant rate of sedimentation through the late Devonian.

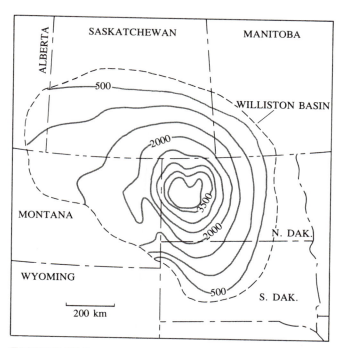

FIGURE 17-9

• • • • • • • • • • • • • • • • •

Outline of the Williston Basin, showing thickness (in meters) of Phanerozoic sediments. [From J. K. Warren, 1989, *Evaporite Sedimentology*, p. 75.]

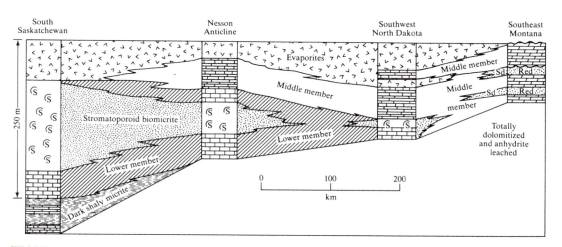

FIGURE 17-10

• • • • • • • • • • • • • • • • •

Idealized stratigraphic diagram showing interpreted Duperow Formation (Devonian) facies belts from north to south across the Williston Basin. Normal marine, burrowed biomicrite and stromatoporoid reefs grade into restricted marine, laminated micrite with microfauna (white area, middle member). Microfaunal micrite grades upward into dolomicrites and evaporites. [From J. L. Wilson, 1967, *Bull. Can. Pet. Geol.*, *15*, 246.]

Origin of Giant Marine Deposits

Giant evaporite deposits such as those in the Prairie Provinces of Canada and the Castile and Salado Formations appear to have formed in standing bodies of water that were isolated or nearly isolated from the sea by reef growth, tectonic activity, volcanic eruptions, or some other cause. Therefore, the simplest model to consider for the origin of thick, pure evaporite deposits is a straightforward process of evaporation to dryness of a standing body of seawater. If the rate of evaporation exceeds the rate of inflow of water over a prolonged period of time, the concentration of dissolved salts in the water increases and an evaporite mineral is precipitated.

Laboratory experiments performed by numerous scientists during the past 130 years have revealed the order in which we can expect the evaporite minerals to precipitate. If we start with normal seawater, gypsum will precipitate when the water has been reduced to one-fifth its original volume; halite, one-tenth the original volume. Other evaporite minerals (sulfates and halides) do not appear until the volume is less than one-twentieth the original.

What thickness of evaporite rock might we expect from the complete evaporation of seawater? Suppose we assume a column of water 1000 m deep, with 3.5% dissolved salts. Complete evaporation would yield a nonporous precipitate of sediment with a thickness of $(0.035 \times 1000)/$(specific gravity of the mineral precipitated). The average evaporite mineral has a specific gravity of 2.2, so the total thickness of the deposit will be 15.9 m. If we assume that all the sodium in seawater is precipitated as NaCl (specific gravity, 2.16), then Table 17-3 indicates that 41.80×2 or 83.6% of the evaporite will be halite; this is 13.5 m [$(35/2.16) \times 0.836$]. If all the calcium comes out as gypsum (specific gravity, 2.32), Table 17-3 indicates that 0.91×2 or 1.82% of the evaporite will be gypsum; this is 0.27 m [$(35/2.32) \times 0.018$]. The remaining 2.1 m of evaporite will be composed of a variety of other evaporite minerals in appropriate proportions.

Our calculation indicates that the evaporation of a standing body of water as much as 1000 m deep can yield only 15.9 m of evaporite deposit. As we have seen in the Castile Formation, however, the thicknesses of anhydrite-halite sequences can be several hundreds of meters. Most evaporite deposits are several meters to several tens of meters thick. The amount of evaporation this thickness requires seems staggering (think of the humidity!), but that is only because of the brevity of the human lifespan relative to the thousands of years during which uninterrupted evaporation can proceed in an isolated body of water.

The simple evaporation-to-dryness model we tested with our calculation has defects in addition to the fact

TABLE 17-3 Relative amounts of dissolved chemical constituents in seawater

Dissolved species	Molarity	Percentage
Cl^-	0.535	48.72
Na^+	0.459	41.80
Mg^{2+}	0.052	4.74
SO_4^{2-}	0.028	2.50
Ca^{2+}	0.010	0.91
All others	0.014	1.33
Total	1.098	100.00

that it could not supply the great thicknesses of evaporites seen in the field. For example, our model produced 50 times as much halite as gypsum because the dissolved material in seawater is mostly sodium and chloride ions. How are we to explain halite to gypsum ratios that are greatly different from 50:1, such as the Castile and Salado Formations in which the ratio is about 1:1? A study of ancient evaporites reveals that there is no norm for this ratio. Some deposits are all gypsum; others, all halite.

One mechanism that allows thick deposits of a single mineral to occur is *reflux*. The requirements for reflux are

1. Influx of seawater
2. Evaporation within a restricted basin
3. Loss of the heavy brine produced, either by seepage down through the underlying sediments (subsurface reflux) or by flow of the brine out of the basin beneath the inflow of new seawater

Under these conditions, a constant salinity can be maintained and a single evaporite mineral deposited. The main factor controlling what the constant salinity will be is the rate of brine loss by reflux. Greater reflux produces lower salinity; lesser reflux, high salinity.

Applying this model to the Castile Formation, we infer that the salinity achieved during evaporation and reflux was sufficient initially to cause the precipitation of calcium sulfate (lower Castile); but as reflux flow decreased, halite was precipitated. These evaporite accumulations will be somewhat younger in age than the rocks of the basin walls and will not grade laterally into contemporaneous sedimentary facies.

If the rate of evaporation and reflux is greater than the rate of inflow, the water level drops (evaporative drawdown). The Great Salt Lake is a familiar example, a remnant of the much larger, glacial Lake Bonneville;

Death Valley in southern California was the site of glacial Lake Manly, as evidenced by the present salt flat and bordering wave-cut benches marking the levels of former shorelines.

Summary of Evaporites

Evaporite rocks are not widely exposed on the earth's surface because of their high solubility, but they underlie about one-third of the land surface and perhaps one-half of the land area covered by sedimentary rocks. In surface outcrops, gypsum is most abundant; in the subsurface, halite probably dominates. Because they are so soluble, many evaporites are recrystallized at least once during their geologic history.

Many ancient, thick evaporite deposits appear to have formed by reflux of shallow, hypersaline brine in basins largely surrounded by considerable topographic relief. The relative rates of evaporation and reflux determine the salinity of the brine and the types of evaporites that will be precipitated. Most other marine evaporites are formed in supratidal (sabkha) settings and are interbedded with dolostone as a result. Nonmarine, lacustrine evaporites are less abundant than marine types and contain many minerals not found in marine types.

CHERTS
• • • • • • • • • • • • • • • •

Chert is a microcrystalline sedimentary rock that is composed largely or entirely of quartz and is precipitated from an aqueous solution. The quartz crystals are typically 5–20 μm, but cherts that consist of coarser or finer crystals are not uncommon. Most chert beds are quite pure, in concert with other chemically formed sedimentary rocks, such as evaporites, limestones, and dolomites, and for many of the same reasons. Rock units composed of chemical or biochemical precipitates cannot form in areas of significant influx of detrital sediment. Impurities in chert normally total less than 5% and consist of calcite, dolomite, clay minerals, hematite, and organic matter.

Field Observations

Chert occurs in three different types of stratigraphic and tectonic settings:

1. As nodules and silt-size grains in cratonic carbonate rocks
2. As pure, bedded chert adjacent to tectonically active plate margins
3. In association with hypersaline lacustrine deposits

Almost all occurrences of chert are in settings 1 and 2.

Chert Nodules

Chert nodules are irregularly shaped, usually structureless, dense masses of microcrystalline quartz; they occur most commonly in carbonate rocks, although they have also been reported in mudrocks and sandstones. The nodules range in size and shape from egg-shaped spheroids a few centimeters in length to large, highly irregular, tuberous bodies. The larger ones typically have warty or knobby exteriors. The outer few centimeters of a nodule may show desiccation cracks filled with secondary chert, probably formed penecontemporaneously with the formation of the nodule. Nodule exteriors are commonly bleached, soft, and porous because of alteration at some stage in the history of the nodule. The interior of the nodule may show concentric color zoning of translucent shades of gray, caused either by carbonaceous matter or by zones of different water content. Occasionally, the bedding, texture, or color shading in the host carbonate rock is seen to pass without interruption through the chert nodule, but this is uncommon. Some nodules contain visible calcareous fossils that are partially or completely silicified, as well as the remains of siliceous organisms such as sponges, radiolarians, and diatoms. But in most nodules no siliceous remains can be seen. Some nodules were precipitated originally as opal and subsequently crystallized to chert. Other nodules were directly precipitated as quartz, probably because the pore waters were undersaturated with respect to opal.

Most chert in limestones occurs in micritic facies because the source of the silica—siliceous sponge spicules (Cambrian-Holocene), radiolaria (Ordovician-Holocene), and diatom tests (Jurassic-Holocene)—are of silt size and are deposited in quiet-water environments. The chert occurs over a wide size range, from powdery quartz seen only in insoluble residues of limestones to nodules several meters in length. Nodular chert is normally concentrated along bedding or condensation surfaces, reflecting episodic accumulation of siliceous fossils and subsequent migration of dissolved silica along surfaces of greater permeability. In some cases, however, nodules cut across bedding, reflecting either burrow filling, gas and fluid escape pathways, or fracture fillings.

Bedded Cherts at Plate Margins

Stratigraphic sections that contain bedded cherts up to several hundred meters thick are commonly found along tectonically active plate margins, in association with graded turbidites, ophiolites, and mélanges. Examples include the Franciscan Formation (Jurassic) in California, many Mesozoic occurrences in the Mediterranean-Himalayan (Tethys) region, Ordovician

FIGURE 17-11

● ● ● ● ● ● ● ● ● ● ● ● ● ● ● ● ●

Bedded chert in Caballos Formation (Devonian-Mississippian), Texas, showing
pinch-and-swell bedding, typical bed thickness, and minor partings of shale.
[From E. F. McBride and A. Thomson, 1969, in *Dallas Geological Society
Guidebook to Marathon Basin, Texas*, p. 59. Photo courtesy E. F. McBride.]

cherts in Scotland, and Jurassic cherts in New Zealand.
In each of these examples, the associated rocks clearly
indicate a deep-water oceanic depositional setting not
too far from a continental margin. Some bedded cherts,
however, are not associated stratigraphically with ophiolites and mélanges; and for these bedded cherts, the
depositional setting may be less certain. In the Caballos
Novaculite of western Texas, for example, structures are
found that can be interpreted as replaced carbonate fecal
pellets, replaced podsolic soils, and evaporite pseudomorphs.

Could some bedded cherts be very shallow-water
deposits? This possibility may not be as outlandish as we
might think, because it appears that ancient radiolarian
cherts were not deposited in wide-open basins or on
continental slopes but instead were deposited near shore
in deep, elongate basins or small basins with restricted
circulation. These relatively confined basins were gulfs
rich in organic matter, modern examples being the Gulf
of California, the Red Sea, and marginal seas such as the
Sea of Okhotsk and the Bering Sea. Such areas of intense
nutrient upwelling enhance the deposition and preservation of radiolarian-rich sediments.

Individual chert layers in bedded cherts range in
thickness from a few centimeters to a meter or more

(Figure 17-11); they are evenly bedded, thinly laminated
to massive, and typically green or black in color, although white bedded cherts termed *novaculite* are common in some areas. The upper and lower surfaces of
chert layers can be either smooth or wavy, and many layers pinch and swell while others bifurcate. Ripple marks,
cross-bedding, and stylolites are occasionally seen in
bedded cherts; but at most outcrops, no sedimentary
structures other than bedding are present. Lithologies
interbedded with the chert layers are typically thin green
or dark-colored siliceous shales.

Cherts in Saline, Alkaline Lakes

Thin beds and nodules of chert occur in many nonmarine
sequences that have been interpreted as lacustrine on
the basis of criteria such as facies relationships, the
absence of marine fossils, and the presence of evaporite
minerals or their pseudomorphs. The chert occurs
chiefly as thin, discontinuous beds or as plates or nodules of irregular shape formed at or slightly below the
water-sediment interface. Typically, the chert is dense,
homogeneous, and translucent and has a thin, soft,
white, opaque rind. The chert beds are characterized by

soft-sediment deformational features such as slump structures, intraformational breccias, various types of surface ornamentation (Figure 17-12), and casts of mud cracks resulting from the 25% shrinkage during the transformation from a primary sodium silicate precipitate to chert. Most of the chert is chalcedonic and commonly contains length-slow fibers, a feature that often indicates the former presence of evaporites.

Lacustrine cherts are forming at present in sodium carbonate alkaline lakes in Oregon and Kenya, the original African discovery being at Lake Magadi. In these lakes, pH values as high as 11 have been measured, with dissolved silica contents as high as 2700 ppm. The silica is precipitated as the sodium silicate mineral magadiite as the pH is sporadically lowered by influxes of freshwater, evaporative concentration, or biogenic production of carbon dioxide. Magadiite layers and laminae dehydrate to chert within a few thousand years by a reaction such as

$$NaSi_7O_{13}(OH)_3 \cdot 3\ H_2O$$
$$\text{magadiite}$$
$$\rightarrow 7\ SiO_2 + 4\ H_2O + Na^+ + OH^-$$
$$\text{chert}$$

Laboratory Studies

Chert in thin section appears as a colorless, microcrystalline aggregate (Figure 17-13), texturally similar in appearance to micrite and microsparite in carbonate rocks. In some bedded cherts, circular or elliptical clear areas are seen in plane-polarized light and are interpreted to be "ghosts" of diatoms, radiolarians, or siliceous sponge spicules (Figure 17-14). Cherts of Tertiary age may contain isotropic (amorphous) material among the microquartz crystals, remnants of an opaline precursor of the chert.

Many cherty rocks contain not only the microcrystalline equant quartz crystals but also an elongate, fibrous variety of microquartz called *chalcedony*. When both chert and chalcedony are present (for example, in chert-cemented sandstones), the chert is always closest to the detrital grain surface, with the chalcedonic quartz filling cavities among neighboring detrital grains. The reason a silica-saturated solution sometimes precipitates fibrous quartz rather than the equant variety is not known but is thought to result either from a difference in

FIGURE 17-12

• • • • • • • • • • • • • • • •

Surface reticulation on lacustrine chert of Gila Conglomerate (Neogene) near Buckhorn, New Mexico. Fragment is 0.5 cm long. [From R. A. Sheppard and A. J. Gude, 1986, *U.S. Geol. Surv. Bull.*, *1578*, 339.]

FIGURE 17-13

• • • • • • • • • • • • • • • • •

Typical appearance of chert in thin section, showing microcrystallinity and black-gray speckled character. Rhombs are manganco-lacite. [E. F. McBride and A. Thomson, 1970, *Geol. Soc. Am. Spec. Paper 122*, Plate 11B. Photo courtesy E. F. McBride.]

```
0                          0.5 mm
```

rate of crystallization or from the presence of other ions in the solution.

Many cherts, particularly those in cratonic carbonate rocks, contain shreds of calcite, which indicates incomplete replacement of limestone by the chert. Rhombic molds may also occur and are interpreted as evidence of partial dolomitization of the preexisting limestone, the dolomite having weathered out after chertification.

FIGURE 17-14

• • • • • • • • • • • • • • • • •

Photomicrograph of novaculite in Caballos Formation in plane-polarized light, showing "ghosts" of sponge spicules about 0.03 mm in diameter. [From E. F. McBride and A. Thomson, 1969, *Geol. Soc. Am. Spec. Paper 122*, Plate 8A. Photo courtesy E. F. McBride.]

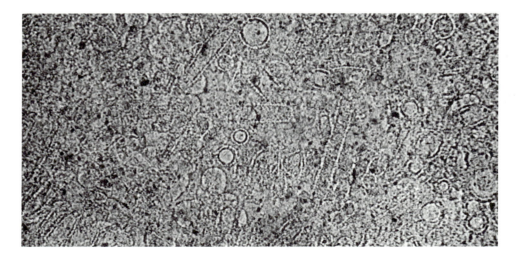

Chemical Considerations

Nearly all cherts are marine, even though the amount of silica in surface seawater is only 1 ppm or less. How can chert form from a solution so undersaturated with SiO_2? As was the case for calcium carbonate, the answer is found in biological extraction. Living organisms can accomplish geochemical feats not possible in inorganic solutions, because living creatures have an energy supply, ultimately obtained from the sun via photosynthesis, and can use this energy to counter the inorganic thermodynamic gradient against precipitation. While they are alive, the silica-secreting organisms in the sea extract silica from undersaturated seawater, bind it within their organic matrix as opal, and keep it from redissolving into the sea. About 12 km^3 of opal are produced annually in this way. When the organism dies and the energy supply to the tissues is stopped, the tissues decompose, the specific gravity of the organism becomes that of the opaline skeleton (2.0–2.2), and the skeleton dissolves as it settles to the seafloor. Only a few percent of the annual production of solid silica reaches the seafloor thousands of meters below. Further, most of the shell fragments that reach the seafloor dissolve before being buried; thus, less than 1% of the surface production of opal makes it into the stratigraphic record. Within a few million years after deposition, on burial to tens or hundreds of meters and at temperatures of 50°–75°C, the opal crystallizes to chert, normally passing through an intermediate stage of opal-CT, that is, opal containing some cristobalite or trydimite crystallites.

As reasonable as this scenario seems for the origin of bedded cherts, it has yet to be documented below the modern ocean floor. All of the many layers of bedded chert sampled during oceanic drilling are replaced carbonate oozes, not in situ crystallized siliceous oozes. The explanation for this difference in mode of origin of oceanic cherts probably relates to the appearance of planktonic calcareous organisms in Jurassic time. The classic rhythmically bedded "ribbon cherts" in the stratigraphic record are nearly all of Paleozoic or early Mesozoic age, before calcareous planktonic organisms evolved. Periodic rise and fall of the carbonate compensation depth may also influence the occurrence of replacement versus nonreplacement origin of bedded cherts.

The fact that the origin of Phanerozoic cherts depends on the existence of siliceous organisms raises the question of Precambrian cherts, formed at a time when there seem to have been few, if any, siliceous organisms, although a significant biota of other types is known. In middle Precambrian rocks, the preserved stratigraphic column contains about 15% chert, an order of magnitude more than that in Phanerozoic rocks. What would happen to the silica brought in by rivers if there were no siliceous organisms? Some investigators believe the dissolved silica would accumulate to many tens of parts per million in the ocean; kinetic problems with quartz nucleation would thus be overcome, and chert would precipitate at the seafloor. However, virtually all Precambrian cherts are of replacement origin, excepting cherts associated with banded iron formations (see below).

Summary of Cherts

Chert is a mineralogically simple rock whose origin can be quite complex. Most Phanerozoic cherts were formed in two stages: (1) removal of dissolved silica from undersaturated seawater by siliceous sponges, radiolarians,

and diatoms and its solidification as opal; and (2) dissolution of the opaline skeletons during shallow burial after death of the organisms, the silica in solution subsequently crystallizing as microcrystalline quartz (chert). In carbonate environments and at sites where the number of opaline shells per unit volume is low, the crystallization results in scattered nodules of chert. At sites where the number per unit volume is large, as occurs in some areas in the deep oceanic basins at present, crystallization results in beds of chert that can be hundreds of meters thick. Cherts in both types of settings can show replacement features, such as undigested remnants of carbonate material.

Many nonmarine cherts may have been formed in hypersaline lakes by the leaching and dehydration of a sodium silicate precursor. For pre-Mesozoic times, no nonmarine siliceous organisms are known, so all nonmarine cherts must have been formed by this or a similar mechanism.

The origin of Precambrian cherts is uncertain. No siliceous organisms are known to have existed at that time, and the majority opinion among petrologists favors an origin for Precambrian marine cherts by precipitation from normal seawater, which would have been supersaturated with respect to quartz.

IRON-RICH ROCKS

Bedded iron-rich rocks are defined by their chemical composition rather than by mineral composition, texture, or structure. Iron-rich deposits are defined as those that contain at least 15% Fe. This is either 21.4% Fe_2O_3, 19.3% FeO, or some combination of the two totaling 15% Fe. These amounts are much greater than the average for mudrocks (4.8% Fe), sandstones (2.4% Fe), or limestones (0.4% Fe). All three groups of rocks can, however, grade laterally into an iron-rich deposit. Two types of iron-rich strata exist: (1) Precambrian cherty **iron formations** and (2) **ironstones,** which are mostly Phanerozoic in age. In the Lake Superior Precambrian province, the cherty iron ores are termed *taconite*.

Cherty Iron Formations

The initial investigations of iron-rich rocks were made about 100 years ago by American and Canadian geologists in the Lake Superior region. They described linear belts of banded, cherty, iron-rich Precambrian rock extending for hundreds to thousands of kilometers, with widths of several kilometers and thicknesses as great as 600 m (Figure 17-15). Many of these iron-bearing

FIGURE 17-15

• • • • • • • • • • • • • • • •

Outcrop of iron-rich Precambrian rock in upper part of Negaunee
Formation, Marquette District, Michigan. Red (dark) bands are
hematitic chert (jasper); gray (light) bands are specular hematite.
[From *Econ. Geol.*, 1973, *68*, 913. Photo courtesy H. L. James.]

rocks are either unmetamorphosed or only weakly
metamorphosed to perhaps greenschist facies, with
preservation of most depositional sedimentary structures and textures. Radiometric dating has established
that such cherty iron formations are mostly older than
2000 ± 200 million years, with the only younger deposits
being 800–500 million years old. Because of their restriction to Precambrian time and their peculiar mineral
composition, iron formations have long been an enigma
to sedimentologists.

MINERAL COMPOSITION. Cherty iron formations
are composed primarily of quartz; iron oxides (hematite
and magnetite); iron silicates such as greenalite
$[(Fe,Mg)_3Si_2O_5(OH)_4]$, minnesotaite $[Fe_3Si_4O_{10}(OH)_2]$,
stilpnomelane (hydrated Mg,Fe,Mn,Al silicate of uncertain composition), and riebeckite $[Na_2Fe_5Si_8O_{22}(OH)_2]$;

iron chlorite; and/or iron carbonates (siderite and
ankerite). Most of these minerals, particularly the oxides
and iron silicates, are probably not primary but were
formed during diagenesis and low-grade metamorphism
of the iron formations. Some of the iron formations are
impure, with the impurities being mostly volcaniclastic
materials related to the volcanic units that are associated
with many iron formations.

TEXTURES AND STRUCTURES. The variety of textures and structures in cherty iron formations is much
like those in Precambrian carbonates. The allochemical
iron-rich rocks contain intraclasts, peloids, ooids, and
pisoliths, and are termed *granular* (not used to indicate
grain size). They also display shrinkage cracks, cross-
bedding, graded bedding, ripple marks, erosion channels,
and flat-pebble conglomerates. Among organically pro-

duced features, stromatolites occur, as do ovoid, gravel-size grains that may be oncoliths.

The nongranular iron-rich units are termed *femicrites* (or felutites) and are texturally analogous to micrites or carbonate mudstones. Femicrite is always banded, in part because of the absence of a burrowing infauna during Precambrian time. In some oxide-rich iron formations, the bands are strikingly colored (red and gray) in outcrop; and as a result, cherty iron formations are typically called banded iron formations (BIF), despite the fact that it is only the originally microcrystalline beds (those that lack sandy textures) that are banded. Among the few other textural and structural features of femicrite is the occasional presence of matrix-rich intraformational breccias that probably reflect slumps and mass movements. Specialists in iron-rich rocks commonly refer to femicrites as *lutitic*; but this usage can be confusing, because the term *lutitic* normally implies a rock dominated by clay minerals.

The banding in femicrites may be lenticular, anastomosing, or nodular; and the appearance and thickness of bands can change over short distances because of the solution and reprecipitation of silica during diagenesis and low-grade metamorphism. The original banding, however, is of depositional origin. It is present in all of the unmetamorphosed femicrite units and occurs as penecontemporaneously reworked, already banded boulders in some beds. Some of the bands have been precisely correlated laterally over hundreds of kilometers and may represent annual layers (varves).

As with limestones, the kinetic energy at the various depositional sites in the basin was the critical factor in determining whether banded femicrite or granular iron formation formed. Thin laminae formed in quiet, undisturbed waters. In agitated, turbulent waters, allochemical (granular) iron formations formed, either by precipitation and accretion of iron and silica around nuclei to form ooids, or by the formation of intraclasts of seafloor femicrite. Many iron formations show an interbedding of these two sediment types, a structure indicating an alternation of quiet and turbulent conditions through time, as is typical of many shallow marine environments.

FIELD RELATIONSHIPS.   Cherty iron formations older than 2600 million years generally differ in tectonic setting from those of younger age; they consist almost entirely of cherty femicrite beds and have been termed the *Algoma* type. Characteristically, they are thinly banded or laminated with interbands of ferruginous gray or red chert (jasper) and hematite or magnetite. The magnetite may be either an original precipitate from anoxic Archean waters or a product of diagenesis.

Siderite and iron sulfide or silicate minerals are present locally or in small amounts. Occasionally, massive siderite and pyrite-pyrrhotite beds form part of the formation. Single iron formations range from millimeters to perhaps 100 m thick and are rarely more than a few kilometers long. Usually, a number of lenses of iron formation are distributed en échelon within a volcanic belt. They are intimately associated with various volcanic rocks, including pillowed andesites, tuffs, pyroclastic rocks, or rhyolitic flows, and with graded graywacke, gray-green slate, or black carbonaceous slate. The stratigraphic and lithologic association, fine grain size, and presence in some outcrops of turbidite structures, such as graded bedding, suggest deposition in a distal turbidite in a marine basinal setting. The source of the iron is thought to be hydrothermal effusive activity associated with fracture systems in the deep ocean.

Cherty iron formations 2600–1800 million years old form the bulk of the Precambrian iron-rich beds. They differ from Algoma formations in lithologic association and in the presence of abundant granular units that accompany the banded femicrite beds. These younger iron formations are referred to as the *Superior* type. Superior iron formations are usually (but not always) larger than the Algoma type; this characteristic, coupled with the relative abundance of granular beds in Superior formations, suggests that shallow-water platforms were more extensive during early Proterozoic time than in Archean time. This inference is consistent with conclusions reached from regional tectonic studies in many terrains that lack iron formations. Although intercalations of volcaniclastic sediment and volcanic rocks may occur in the shallow-water iron-rich beds, they are far less common than in Algoma iron formations and not so intimately associated with the iron-rich rocks. The contrasts between Superior and Algoma iron formations are summarized in Table 17-4.

Iron Formations and Free Oxygen in the Precambrian

Cherty iron formations occupy a central location in debates about free oxygen on the early Earth because of their abundance and because they contain large amounts of a readily oxidizable element, iron. Although a consensus among researchers of this topic has not yet been reached, a synthesis of current viewpoints suggests the following scenario for free oxygen during Precambrian time.

1.   Cherty iron formations are essentially devoid of clay minerals and other detrital debris, suggesting that

TABLE 17-4 Characteristic features of Algoma and Superior banded iron formations
• •

Feature	Algoma type	Superior type
Age	Mostly older than 2600 my	Mostly older than 1800 my
Sedimentary environment	Oceanic tectonic basins of several 100 km diameter; related to green-stone belts	Margins of stable continental shelves; shallow-water; restricted intracra-tonic basins
Extent	Commonly lenticular bodies of a few kilometers	Extensive formations persistent over hundreds of kilometers
Thickness	0.1–10 m	Several meters to 1000 m
Location in sedimentary sequence	Irregular, lenticular bodies within Archean basement rocks	Sheet deposits, transgressive on older basement rocks
Associated rocks	Graywackes and shales; carbona-ceous slates; mafic and rhyolitic vol-canics; pillowed lavas	Coarse clastics; quartzites, conglom-erate, dolomites, black shales (graphitic)
Volcanics	Close association to volcanism in time and space	Volcanics normally absent
Sedimentary facies	Oxide facies predominant; carbonate and sulfide facies thin and discontin-uous; silicate facies; all facies fre-quently closely associated	Oxide facies most abundant; silicate and carbonate facies frequently intergradational
	Sulfide and carbonate facies near the center of volcanism; oxide facies on the margins	Sulfide facies insignificant or absent
	Heterogeneous lithologic assem-blages with fine-grained clastic beds	More homogeneous (especially oxide facies); little or no detritus

From J. Eichler, 1976, in K. H. Wolf (ed.), *Handbook of Strata-Bound and Stratiform Ore Deposits*, Vol. 7, p. 173.

the source of the iron atoms in the formations was not the continental landmasses. The only other reasonable source for the iron is the hydrothermal vents that occur throughout the present world ocean and were probably even more abundant very early in Earth history. The source of the elements in this hydro-thermal fluid emanating from the cracks on the ocean floor is the alteration of seafloor olivine basalt, and studies of the composition of these hydrothermal flu-ids reveal that iron is enriched by greater than 10^5 rela-tive to seawater.

2. The texture of femicrite indicates that it was a chem-ical precipitate rather than a detrital accumulation; and for iron atoms to occur in significant amounts in solu-tion, they must be in the ferrous state. Ferric iron is too insoluble in water to be able to circulate widely and pro-vide iron atoms for a widespread precipitate of iron-rich minerals in a femicrite lamina. The deep ocean waters during precipitation of Algoma femicrites (Archean time) must have been anoxic to permit wide circulation of iron atoms. The oxygen required to oxidize the iron was produced by blue-green algae living in shallow

water. Stromatolitic blue-green algae are present in rocks at least as old as 3.5 billion years. The dissolved oxygen circulated to the ocean floor, and in areas where ferrous-rich effusions and top-to-bottom circulation coincided, femicrite laminae formed.

3. The reason there are few granular iron-formation beds prior to 2600 million years ago is that continental shelves apparently were not yet extensive. With few shallow marine depositional sites available, iron formations must generally be deep-water deposits. This is consistent with the fact that Algoma iron formations are femicrites. Also, the femicrites are formed of undisturbed layers that are only a few millimeters thick and are continuous over areas at least 50,000 km² in extent, an areal extent consistent with the inference of deep water. Archean sedimentary rocks that lack iron formations are dominantly deep-water turbidites, reinforcing the hypothesis that the landmasses of this period lacked continental shelves.

4. So long as the algal biomass was small and the accompanying amount of oxygen produced small in relation to the amount of ferrous iron dissolved in the ocean water, both the ocean and the atmosphere remained anoxic. (*Anoxic* in this context means a level of oxygen less than 1% of the present amount.) This condition was maintained until about 2000±200 million years ago, the time when the major pulse of iron-formation deposition ceased. At this time, the generation of oxygen increased to the level at which general oxygenation of the ocean occurred. Consequently, the ferrous atoms emanating from the seafloor hydrothermal vents oxidized immediately on reaching the ocean floor. Because long-distance transport of ferrous iron through the seawater was no longer possible, deposition of iron formation on continental margins ended.

It is noteworthy that about 1800 million years is also approximately the age of the oldest red soils, hematite-cemented sandstones, and the youngest rocks containing detrital uraninite and pyrite. (Pyrite is the most abundant detrital heavy mineral in many Archean conglomerates in South Africa.) Both of these minerals are unstable in an oxygen-rich atmosphere.

5. The scenario just described leaves one major question unanswered. What caused the burst of iron-formation deposition 800 million–500 million years ago? The key to an answer probably lies in a lowered level of oxygen in the atmosphere and hydrosphere during this period. As of 1995, no good reason for such a lowering has been proposed, and no method of determining whether a lowering actually occurred has been devised.

The level of oxygen in the atmosphere and hydrosphere is, of course, important for the evolution of a complex metazoan fauna, and it is worth noting that there is a clear overlap in time between the formation of the youngest banded iron formations (800 million–500 million years ago) and the existence of complex metazoan faunal assemblages. The Ediacaran fauna has been dated at 590–540 million years and occurs on many Proterozoic shelves. Modern representatives of the coelenterates in this fauna require a level of atmospheric oxygen at perhaps 10% of the amount now present.

Ironstones

Ironstones are mostly unbanded oolitic rocks, of Phanerozoic age (Figures 17-16 and 17-17), composed of goethite, hematite, and chamosite (iron-rich chlorite); siderite is not uncommon. Locally, magnetite or pyrite may predominate. Common nonferriferous minerals include detrital quartz, calcite, dolomite, and nondetrital phosphorite. Authigenic chert is commonly present but is never equal in abundance to the ferriferous minerals.

The areal distribution of the ironstones is much more restricted than that of the Precambrian banded iron formations, with individual basins not exceeding 150 km in length, and the thickness of the units ranges from less than 1 m to a few tens of meters. Interbedded and interfingering sedimentary rocks are carbonates, mudrocks, and fine-grained sandstones of shelf to shallow marine origin.

FIGURE 17-16
● ● ● ● ● ● ● ● ● ● ● ● ● ● ● ● ●

Fractured surface of fresh oolitic ironstone (Cretaceous), Nigeria. Ooids and matrix consist of siderite, magnetite, and chlorite, with subordinate limonite and clay minerals. Width of photo is 30 mm. [From H. A. Jones, 1965, *J. Sediment. Petrol.*, 35, 840.]

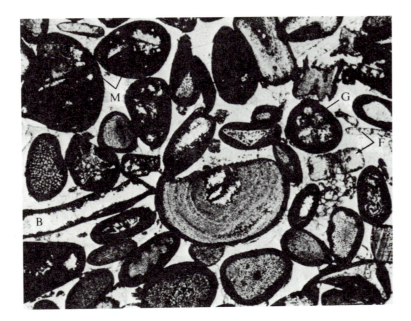

FIGURE 17-17

• • • • • • • • • • • • • • • • • •

Photomicrograph of oolitic hematite, Keefer Sandstone (Silurian), Pennsylvania. Most ooid nuclei are crinoid columnals. Others are brachiopod (*B*) shell fragments, small gastropods (*G*), and intraclasts composed of small fossil fragments in hematite matrix (*M*). Hematite permeates fossil nuclei and their pore systems, but a few fossils (*F*) scattered among the ooids are only slightly replaced by hematite and lack oolitic coatings. Cement is calcite. The large, broken, coated crinoid in the center is 1 mm in long dimension. [Photo courtesy R. E. Hunter.]

Sedimentary structures in oolitic ironstones include large cross-beds, ripples, scour-and-fill structures, ripped-up clasts of penecontemporaneously lithified oolitic sediment, and worm burrows. In thin section, fossils of shallow-water origin are as common as they are in oolitic limestones. Many of the fossils are abraded. Commonly, the fossils in ironstones are partly or completely replaced by ferriferous minerals. The irregularly shaped burrows that occur in many ironstones reveal the presence of abundant shallow-water life forms even where shells are lacking. Tertiary ironstones locally contain abundant wood and tree seeds. All the evidence indicates that the oolitic ironstones are nearshore, shallow-water deposits.

The nuclei of the ooids may be quartz grains (angular or rounded), fossil fragments, peloids, intraclasts, or other sediment locally available. The ooid coatings may be monomineralic or not. For example, some consist of alternating rings of hematite and chamosite. Commonly, individual ooids are broken and regrown. Sometimes, delicate, well-preserved algal borings cut across some of the ooid rings.

The mode of origin of most Phanerozoic oolitic ironstones has not been resolved. The environment of deposition was the subtidal part of a marine shelf at depths up to perhaps 100 m, and the ubiquitous evidence of strong current activity requires that the depositional environment be oxidizing. But it is not clear whether the ooids are depositional, diagenetic, or both.

Analysis of the stratigraphic occurrence of Phanerozoic ironstones indicates that ironstone formation is favored during major transgressive episodes (Figure 17-18) in warm climatic regimes, the same setting that stimulates the formation of marine black shales. However, there is a need for minor regressive episodes within the transgressive periods to stimulate iron mobility. During regressions, a mass of iron-rich muddy sediments of continental origin would cover the calcareous ooids, and ferruginization would occur from downward-migrating, oxygen-deficient solutions. However, evidence for the continental covering is not always present in stratigraphic sections.

Summary of Iron-Rich Rocks

Bedded iron ores are enigmatic rocks of two types. The more abundant type is the Precambrian bedded iron ore or cherty iron formation. These rocks are typically composed mostly of chert but contain at least 15% iron in the form of hematite, magnetite, siderite, or other minerals. These minerals are concentrated in varvelike layers within the chert. The iron-bearing minerals were precipitated from marine water, lake water, or pore solutions during very early diagenesis. No modern counterparts of these rocks are known, so their origin remains uncertain.

The second type of iron-rich rock is the Phanerozoic oolitic iron ore or ironstone. These units are much thinner than their Precambrian relatives and much less extensive. The ooids are composed mostly of goethite and contain a nucleus of detrital quartz or a carbonate shell fragment; the units grade laterally into shallow-water limestone deposits. Some ironstones were formed by replacement of calcareous ooids, but other origins occur as well.

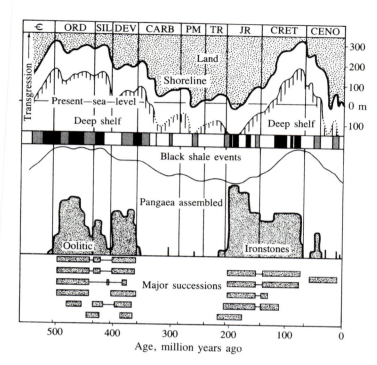

FIGURE 17-18

• • • • • • • • • • • • • •

Occurrence of Phanerozoic ironstones and black
shales in relation to sea level. [From F. B. Van Houten
and M. A. Arthur, in T. P. Young and W. E. G. Taylor,
1989, *Phanerozoic Ironstones*, p. 34.]

phosphatic sequences that may be hundreds of meters
thick. The beds occur in outcrop as black layers
commonly interbedded with carbonates, dark-colored
chert (phosphatic or organic matter), and carbonaceous
mudstone. Coarse detrital sediments are rare. Sedimentary structures include bedding, lamination, scours,
and cross-bedding (Figure 17-19). Megascopically visible
textural features include grain rounding and ooids. Calcareous fossils include mollusks, brachiopods, sponges,
and foraminifera; phosphatic fragments of fish bones and
teeth can also be seen with a hand lens.

Laboratory Observations

Except for the difference in mineral composition (apatite
rather than calcite), thin sections of phosphorites look
very much like thin sections of limestones (Figure
17-20), and the same type of nomenclature can be used
for both groups of rocks. Phosphomicrites, phospho-
sparites, and varieties of both contain variable percentages of fossils, peloids, ooids, and phosphoclasts.

PHOSPHORITES

• • • • • • • • • • • • • • • •

Phosphorites are nondetrital sedimentary rocks that
contain at least 20% P_2O_5, an enormous enrichment over
the amounts in mudrocks (0.17%) and limestones (0.04%).
The phosphorous occurs in ancient phosphorites as
cryptocrystalline fluorapatite, $Ca_5(PO_4)_3F$, typically with
several percent of CO_3^{2-} substituting for PO_4^{3-} in the
crystal structure. The term *collophane* is used commonly
to describe cryptocrystalline to X-ray–amorphous phosphatic material.

The best-known phosphorite deposits are in Morocco
(Upper Cretaceous-Eocene) and the United States
(Permian, Neogene) but phosphorites occur on every
continent and range in age from Proterozoic to Holocene.

Field Observations

Phosphorites most commonly occur as beds ranging in
thickness from a few centimeters to tens of meters in

FIGURE 17-19

• • • • • • • • • • • • • • •

Sedimentary structural and textural features in phosphorite. Dark-
colored, ovoid phosphate grains cemented by light-colored dolomite, Meade Peak Member of Phosphoria Formation (Permian),
Idaho. Note cross-bedding and parallelism of long axis of grains to
bedding. [From G. D. Emigh, 1967, in L. A. Hale (ed.), *Inter-
mountain Assoc. Geol. 15th Annual Field Conference Guidebook*,
p. 111. Photo courtesy G. D. Emigh.]

0   5 cm

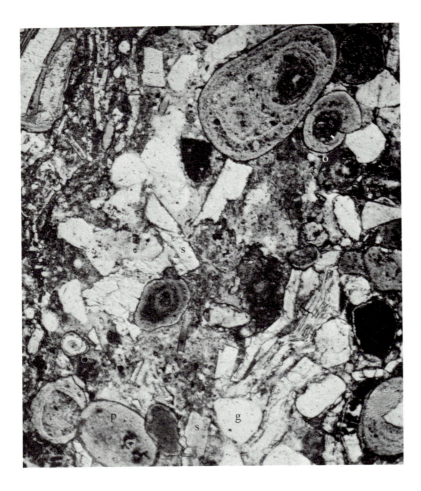

FIGURE 17-20

• • • • • • • • • • • • • • • •

Photomicrograph of phosphorite in Phosphoria Formation, containing quartz sand grains (*g*) and apatite ooids (*o*), "pellets" (*p*), and skeletal (*s*) fragments. Note ooid nuclei consisting of quartz sand grains and apatite coating. Uncrossed polarizers. [From E. R. Cressman and R. W. Swanson, 1964, *U.S. Geol. Surv. Prof. Paper 313C*, 311. Photo courtesy E. R. Cressman, U.S. Geological Survey.]

0   0.5 mm

(Specialists in phosphorites typically do not use carbonate terminology and term all gravel-size phosphoclasts *nodules*; sand-size grains are termed *pellets*.)

The bulk of phosphorite grains and matrix is cryptocrystalline (crystal diameters smaller than 1 μm) rather than microcrystalline; as a result, phosphorite appears as a brownish, isotropic material in thin section. Within many particles, more coarsely crystalline areas show low first-order interference colors. Typically, these areas are not random within the particle but parallel internal structure, such as concentric zones in ooids or thin zones paralleling the shell outline in linguloid brachiopods.

Chemical Considerations

The oldest known phosphorites occur in banded iron formations that are 2600–2000 million years old (Australia); but deposits are not common in Precambrian rocks until Eocambrian time. A marked increase in abundance appears in basins less than 900 million years old, possibly as a result of evolutionary advances in biota, such as the development of phosphatic exoskeletons. A peak in phosphorite abundance occurs in Cambro-Ordovician rocks, and this correlates with the appearance of calcareous skeletons, which are easily replaced and pseudomorphed by apatite when suitable chemical conditions exist at the seafloor. In addition to the Cambro-Ordovician occurrences, the major deposits are Permian, late Cretaceous, Eocene, and Miocene.

On the basis of fossils and sedimentary structures such as cross-bedding, petrologists have concluded that nearly all ancient phosphorite deposits are of shallow marine origin. However, at depths less than about 30 m, the P_2O_5 content of ocean water is low because of consumption in the photic zone by phytoplankton. Hence, the depth range 30–100 m is the likely zone of phosphorite formation. The precipitation of pure fluorapatite is controlled by the reaction

$$Ca_5(PO_4)_3F \rightleftharpoons 5\ Ca^{2+} + 3\ PO_4^{3-} + F^- \qquad K = \approx 10^{-60}$$

so the solubility product is $[Ca^{2+}]^5[PO_4^{3-}]^3[F^-]$. Available data suggest that seawater is near saturation with respect to fluorapatite, and because both calcium and fluoride ions are conservative species in seawater, it is traditional to assume that variations in abundance of phosphate ion control the degree of supersaturation. The availability of PO_4^{3-} is determined by several variables.

1. Phosphate-rich tissues of marine organisms accumulate on the seafloor, and phosphorus released by decay can be raised into overlying waters by upwelling currents. Such vertically moving currents are generated particularly on the western margins of continents in tropical or subtropical latitudes, in the trade wind zone. Quaternary phosphate accumulations in such areas are well documented (Figure 17-21), and paleomagnetic data indicate that more than 90% of major ancient deposits occur at paleolatitudes less than 40°.

2. The PO_4^{3-} concentration in seawater is related to pH by the reaction sequence

$$H_3PO_4 \rightleftharpoons H_2PO_4^- + H^+ \qquad K = 7.5 \times 10^{-3}$$

$$H_2PO_4^- \rightleftharpoons HPO_4^{2-} + H^+ \qquad K = 6.2 \times 10^{-8}$$

$$HPO_4^{2-} \rightleftharpoons PO_4^{3-} + H^+ \qquad K = 4.8 \times 10^{-13}$$

so higher pH (fewer hydrogen ions) favors PO_4^{3-} formation. Higher pH is most easily attained by loss of carbon dioxide stimulated by increased water temperature. Thus, low latitudes and shallower waters are favored locations for phosphate ion to increase in abundance relative to other phosphate species. The effect of temperature change is quite large; apatite is 1000 times less soluble at 25°C than at 0°C.

It is apparent that the variables that control saturation with respect to apatite are the same ones that control calcium carbonate saturation. The ascension of cold, carbon dioxide–enriched waters into a warmer shallow area, with subsequent loss of carbon dioxide and rise in pH, would favor both calcium carbonate and apatite precipitation. Because dissolved carbonate species in seawater have activities several orders of magnitude higher than those of phosphate species, a coprecipitation of apatite and calcite would surely result in the apatite being completely overwhelmed by the calcium carbonate. For large quantities of essentially pure apatite to form, it is necessary that the dissolved phosphate activity be raised high enough so that calcium ion activity is controlled by apatite formation rather than by carbonate equilibria.

3. A large proportion of phosphatic particles is formed during very early diagenesis immediately below the sediment-water interface, possibly through bacterial activity. Many phosphorites are cemented by cryptocrystalline phosphate, which also fills voids of fossils such as bryozoa, gastropods, and foraminifera. The apatite also

FIGURE 17-21

• • • • • • • • • • • • • • • •

Grains of phosphorite of fine sand size from continental shelf, Baja California, Mexico. [From B. F. D'Anglejan, 1967, *Marine Geol.*, *5*, 22. Photo courtesy B. F. D'Anglejan.]

replaces stromatolites, originally aragonitic and calcitic shells, and other detrital grains; entire calcitic beds may be phosphatized. Because of the decay of organic matter and sulfate reduction, concentrations of phosphorus near the seafloor typically reach 1–2 mg/L and have been recorded as high as 8–9 mg/L. At such high concentrations, supersaturation of the water with respect to apatite apparently is great enough to overcome kinetic inhibitions so that apatite nucleates on grains of varied types, such as siliceous tests and carbonate materials. Phosphate ion concentrations as low as 0.1 mg/L are sufficient to initiate replacement and pseudomorphing of calcite by apatite according to

$$5\,CaCO_3 + 3\,PO_4^{3-} + F^- \rightleftharpoons Ca_5(PO_4)_3F + 5\,CO_3^{2-}$$
$$\text{calcite} \qquad\qquad\qquad \text{apatite}$$

It may be significant that almost all known Precambrian phosphorites are cryptocrystalline collophane mudstones analogous to femicrites, whereas Phanerozoic deposits commonly contain sand-size phosphatic particles. Might this fact relate to the absence of calcareous skeletons in Precambrian time?

Ancient Phosphorites

The Phosphoria Formation (Permian) has a maximum thickness of 420 m and was deposited during a period of 15 million years over an area of about 350,000 km². Phosphorites occur in both the platform and the deeper

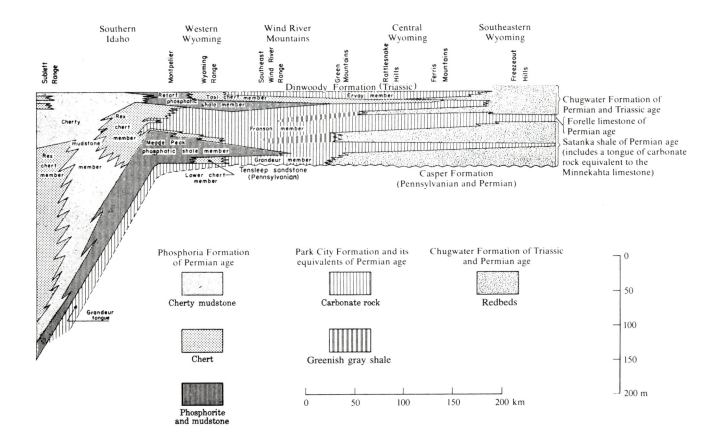

FIGURE 17-22
● ● ● ● ● ● ● ● ● ● ● ● ● ● ● ●

Stratigraphic relations of Phosphoria Formation (Permian), Park City Formation (Permian), and Chugwater Formation (Permian) in Idaho and Wyoming. [From V. E. McKelvey et al., 1959, *U.S. Geol. Surv. Prof. Paper 313A*, 3.]

parts of the Paleozoic Cordilleran structural belt (Figure 17-22). The regional stratigraphy of the Phosphoria reflects its structural setting not only by the great increase in thickness toward the basin but also by lithologic changes. Spiculiferous chert and cherty, carbonaceous mudstone dominate the stratigraphic section in the deeper parts of the depositional basin; they are succeeded successively eastward by dolomitic limestones and calcareous sandstones and ultimately by continental red shales and sandstones. Highly phosphatic beds occur

in both the basinal and the platform facies and appear to reach a maximum thickness near the "hinge line" separating the two facies.

In the vicinity of its type locality in southeastern Idaho, the formation has a thickness of 80–150 m and consists mainly of dark chert, phosphatic and carbonaceous mudstone, phosphorite, cherty mudstone, and minor amounts of dark carbonate rock.

The thickest and most extensive phosphorite unit in the Phosphoria Formation is the Meade Peak Member, about 10 m thick, composed of phosphatic shale, mudstone, and carbonate rocks. The phosphorites are mostly thin-bedded, laminated, dark-colored, well-sorted, granular rocks containing about 50% apatite. The grains include peloids (possibly fecal pellets), ooids, and intraclasts. Linguloid brachiopods, phosphatic fish scales, and phosphatized gastropods, cephalopods, bryozoans, echinoid spines, and sponge spicules also occur in the allochemical fraction.

In the nearshore area, well-aerated waters are suggested by the noncarbonaceous, cross-bedded, well-sorted, granular phosphorites that contain abraded fossils and lineated, elliptical grains. Contrasting with this lithology are the offshore carbonaceous, pyritic, phosphatic mudstones, whose few fossils show no evidence of current transport. The change in lithologic character is strikingly similar to that encountered in limestones as nearshore rocks are followed offshore to a quiet-water micritic area, but a replacement origin for the bulk of the Phosphoria is not evident from field and laboratory studies.

Summary of Phosphorites

Phosphorites are rocks whose occurrence appears to have pronounced spatial, climatic, and temporal controls. They occur mainly in locations of oceanic upwelling within 40° of the equator (or paleoequator) and in waters shallower than 100 m. They are composed largely of cryptocrystalline to microcrystalline apatite, are black in color, and have structures and textures similar to those found in limestones. Phosphorites form in many of the same depositional environments as limestones, and it is possible that many phosphorites were formed by penecontemporaneous phosphatization of carbonate sediment immediately below the water-sediment interface. Primary precipitation may occur, mediated by bacterial activity and organic matter at the seafloor.

STUDY EXERCISES
• • • • • • • • • • • • • • •

1. Describe criteria that you might observe, either in the field or in thin section, that would suggest the former presence of evaporite beds or minerals in areas where none are now present.

2. How might the fluctuating depth of the CCD during geologic time affect the abundance of chert deposits in the deep-sea stratigraphic record? Do you think the depth of the CCD might have fluctuated in pre-Mesozoic times, when there were no planktonic carbonate-shelled marine organisms? Explain.

3. The thicknesses of laminae and beds are variable in banded iron formations. What explanations might there be for this observation?

4. Phosphate deposits are an important source of commercial fertilizer. Suppose you were an economic geologist for a fertilizer company and were asked to locate new phosphate deposits for your company. What types of tectonic, stratigraphic, or geochemical considerations might be important for you to consider before going into the field setting?

REFERENCES AND ADDITIONAL READINGS
• • • • • • • • • • • • • • •

Bentor, Y. K., ed. 1980. *Marine Phosphorites. SEPM Spec. Pub. No. 29.*

Casas, E., and T. K. Lowenstein. 1989. Diagenesis of saline pan halite: Comparison of petrographic features of modern, Quaternary and Permian halites. *J. Sediment. Petrol., 59,* 724–739.

Hardie, L. A. 1984. Evaporites: Marine or non-marine? *Am. J. Sci., 284,* 193–240.

Hesse, R. 1989. Silica diagenesis: Origin of inorganic and replacement cherts. *Earth-Sci. Rev., 26,* 253–284.

James, H. L. 1992. Precambrian iron-formations: Nature, origin, and mineralogic evolution from sedimentation to metamorphism. In K. H. Wolf and G. V. Chilingarian, eds., *Diagenesis III.* New York: Elsevier.

Jarvis, I. 1992. Sedimentology, geochemistry and origin of phosphatic chalks: The Upper Cretaceous deposits of NW Europe. *Sedimentology, 39,* 55–97.

Jones, D. L., and B. Murchey. 1986. Geologic significance of Paleozoic and Mesozoic radiolarian cherts. *Ann. Rev. Earth Plant. Sci., 14,* 455–492.

Lamboy, M. 1993. Phosphatization of calcium carbonate in phosphorites: Microstructure and importance. *Sedimentology, 40,* 53–62.

Lowenstein, T. K., and L. A. Hardie. 1985. Criteria for the recognition of salt-pan evaporites. *Sedimentology, 32,* 627–644.

Maliva, R. G., A. H. Knoll, and R. Siever. 1989. Secular change in chert distribution: A reflection of evolving biological participation in the silica cycle. *Palaios, 4,* 519–532.

Maliva, R. G., and R. Siever. 1989. Nodular chert formation in carbonate rocks. *J. Geol., 97,* 421–433.

Notholt, A. J. G., and I. Jarvis, eds. 1990. *Phosphorite Research and Development.* London: The Geological Society.

Peryt, T. M., ed. 1987. *Evaporite Basins.* New York: Springer-Verlag.

Presley, M. W. 1987. Evolution of Permian evaporite basin in Texas panhandle. *Am. Assoc. Pet. Geol. Bull.*, *71*, 167–190.

Renaut, R. W., ed. 1989. Sedimentology and diagenesis of evaporites. *Sediment. Geol.*, *64*, 207–298.

Schubel, K. A., and B. M. Simonson. 1990. Petrography and diagenesis of cherts from Lake Magadi, Kenya. *J. Sediment. Petrol.*, *60*, 761–776.

Simonson, B. M. 1985. Sedimentological constraints on the origins of Precambrian iron formations. *Geol. Soc. Am. Bull.*, *96*, 244–252.

Van Houten, F. B. 1992. Review of Cenozoic ooidal ironstones. *Sediment. Geol.*, *78*, 101–110.

Van Kauwenbergh, S. J., J. B. Cathcart, and G. H. McClellan. 1990. Mineralogy and alteration of the phosphate deposits of Florida. *U.S. Geol. Surv. Bull. 1914*.

Warren, J. K. 1989. *Evaporite Sedimentology*. Englewood Cliffs, NJ: Prentice Hall.

Young, T. P., and W. E. G. Taylor, eds. 1989. *Phanerozoic Ironstones*. London: The Geological Society.

....

PART XXX

Metamorphic Rocks

Chapter **18**

....

METAMORPHISM
AND METAMORPHIC ROCKS

I n the preceding text we examined the formation of igneous rocks as a function of operation of physical and chemical processes within the earth's mantle or crust, and of sedimentary rocks as a function of interactions of chemical, physical, and biological processes at the surface or shallowly buried. Both rock types appear to be more or less in chemical equilibrium in their sites of origin, although chemical equilibration tends to be more complete in igneous than in sedimentary rocks. The earth is a dynamic system, however, and, once formed, both igneous and sedimentary rocks commonly are tectonically transported or emplaced into new environments where the physical and chemical conditions are quite different from the original environments. Extensive changes may then occur in both texture and mineralogy as rocks and their constituent minerals seek a new equilibrium state, a process broadly defined as **metamorphism.** In general, equilibration processes and rates (characterized as *kinetics*) vary significantly as a function of temperature, the availability of catalyzing fluids, and pressure, especially directed stress. Most environmental changes that lead to metamorphism involve burial of sedimentary or volcanic rocks under younger layered rocks or under tectonic cover; the burial process produces elevated temperatures and pressures. Heating rates and mineral transformations are typically slow in this type of regional-scale metamorphism. In unusual cases, more rapid changes are caused by events such as igneous intrusions, which produce contact aureoles of high-temperature minerals, or even by rare meteorite impacts, which almost instantaneously create ultrahigh-pressure minerals and shocked textures. Metamorphism thus involves a wide range of mineralogic or textural changes above the temperatures of diagenesis and lithification and up to, and including, the temperatures of crustal melting.

DEFINITION
OF METAMORPHISM

• • • • • • • • • • • • •

Metamorphism traditionally has been defined as mineralogic or textural change brought about in a rock *in the solid state* as a result of increase in temperature or pressure. Although this simple definition has merit, it is always advantageous for the student of petrology, either beginning or advanced, to think in terms of initial and final equilibrium states in looking at the processes of textural and chemical change in metamorphic rocks. Which chemical or physical variables must change (and how) to provide a driving force for reequilibration? Are there alternative pathways from the starting state to the final one? It must always be remembered that the two principal environmental variables in metamorphism—pressure and temperature—are to some extent independent of each other. Although pressure and temperature are linked through the average geothermal gradient, local gradients can vary significantly in different tectonic environments and over time. Anomalously high temperatures can be produced at shallow depths through heat input accompanying intrusions of magma. And because metamorphism commonly involves rocks with hydrous or carbonate minerals, other variables such as partial pressures (or fugacities) of water and carbon dioxide can also play an important role.

As a simple introductory example of the way to think about metamorphic processes, consider the incipient metamorphism of mudrock (Figure 18-1A). As discussed in Part II of this book, this sedimentary rock is a physical mixture of very small sedimentary particles, principally quartz and clay minerals. In contrast to the orderly chemical processes that lead to the mineralogy of an igneous rock, the mudrock is a purely accidental assemblage of mineral grains that simply happened to arrive in the same depositional site at the same time. Overall, however, the minerals are more or less chemically stable under depositional conditions. Because this is a sedimentary *rock*, the process of diagenesis during initial burial has expelled most of the pore water and closed most of the pore spaces, aligned the flakes of clay, and cemented the grains together. But metamorphism involves change beyond diagenesis. Given the stabilities of the minerals in a mudrock, what changes can occur? If the mudrock is heated, it is unlikely that the quartz will break down because quartz both crystallizes directly from magma and is stable in sandy sediments, and thus has perhaps the widest thermal stability of any commonly occurring mineral. However, it is likely that heat will affect the clay minerals, which are known to be low-temperature minerals. The temperature limits for the

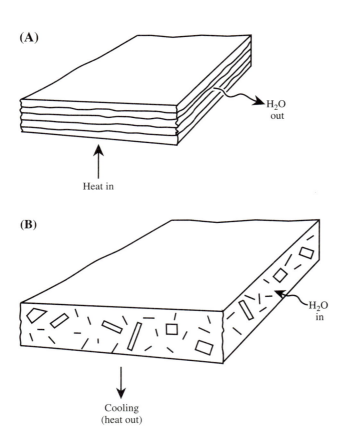

(A)

H₂O
out

Heat in

(B)

H₂O
in

Cooling
(heat out)

FIGURE 18-1

● ● ● ● ● ● ● ● ● ● ● ● ● ● ● ●

(A) Incipient metamorphism of mudrock, in which input of heat causes thermal decomposition of clay minerals and *loss* of water. **(B)** Incipient metamorphism of young basalt, in which cooling in the presence of water allows anhydrous basalt minerals (for example, olivine, pyroxene, plagioclase) to *become* hydrated to metamorphic minerals by addition of water.

stability of clays can be exceeded in either of two ways: A clay mineral can simply decompose by itself during heating to form one or more new minerals containing less water (perhaps a mica or chlorite). Alternatively, a clay mineral can react with quartz or some other mineral in the mudrock to generate new metamorphic minerals. Both are examples of metamorphic reactions, and substantial amounts of water are produced by both. Metamorphism has commenced, as a direct result of heating.

Can these dehydration effects be caused in some way other than by heating? Increasing the pressure alone causes only some minor recrystallization and coarsening of grains. But suppose the mudrock is subjected to an environment that is exceedingly low in water (for example, in a desiccator with very low partial pressure of water vapor). Although the kinetics are very much slower without an increase in thermal energy, the clays will change into less water-rich minerals over a very long time through dehydration *without* heating. This is not likely to happen in nature, but it illustrates a very important point to remember. Different combinations of changes in total pressure, water pressure, and temperature during metamorphism can lead to similar mineralogic changes in rocks from two different areas. Because such mineralogic changes can be produced by changes in *different* variables, several different paths can lead from one starting point (initial mineral assemblage) to one end point (final mineral assemblage).

As a second example, consider a pillow basalt erupted onto the ocean floor at a mid-ocean ridge (Figure 18-1B). The minerals in the basalt are in equilibrium in a dry environment at about 1200°C or more (the solidus of the basalt). If the basalt were maintained indefinitely at 20°C in a *completely dry* environment, essentially nothing would happen to the constituent minerals. *In the presence of water*, however, substantial reaction can occur relatively rapidly at low temperatures (like those on the seafloor), and hydrous metamorphic minerals form because they represent the new, more thermodynamically stable configuration of this chemical composition under the new conditions. Water serves two roles here: as a catalyst to enhance the rates of reaction and as an essential constituent of the new hydrated minerals. In contrast to the first example, it is the decrease in temperature in this case (in the presence of water) that has allowed or caused the metamorphism, because the original mineralogy of the rock represented a stable equilibrium only at high temperature. This second example is an important one, because it is the basis for seafloor metamorphism of the oceanic crust that was referred to in Part I. Interestingly enough, metamorphism in this case involves adding water to the rock, whereas in the mudrock, incipient metamorphism involves dehydration, or subtraction of water.

The lower limit of metamorphism is a little hard to pin down. Temperature increase, typically due to burial, is the most important cause of low-grade metamorphic mineral changes. The accompanying pressure increase has little effect, except perhaps texturally. Aside from expulsion of pore fluids and whatever dissolved components they carry, bulk rock chemistry remains essentially unchanged. Chemical changes within the rock simply involve redistribution of components among the most reactive minerals, including cements. The traditional approach has been to define the boundary between diagenesis and metamorphism by noting the first occurrence of a mineral that normally does not occur as a detrital or diagenetic mineral in surface sediments, for example, chlorite or epidote in mafic rocks or muscovite in mudrocks. Other minerals that have been used as markers of the onset of burial metamorphism are laumontite, lawsonite, albite or other zeolites or framework silicates, and sheet silicates that are transitional between clay minerals and micas. With some variation, formation of all these minerals requires a temperature of at least 150°–200°C, corresponding to a lithostatic pressure of about 1500 bars or a depth of about 5 km under normal geothermal conditions, but the onset of metamorphism can occur at temperatures as high as 350°C for certain rock types. The burial depth at which metamorphic temperatures are achieved depends on the geothermal gradient. In geothermal areas (such as Rotorua, New Zealand, or the Salton Sea in southern California), this depth may be as little as 1000–2000 m, whereas in stable interior basins, it may be deeper than 8000 m.

As a practical matter, defining the upper limit of metamorphism as the beginning of appreciable melting (based on a definition of metamorphism as a solid-state process) makes it relatively easy to distinguish this upper limit in the field and in hand samples. It must be emphasized, however, that this defined limit does not represent a single temperature or even a narrow range of temperatures. Whereas the onset of substantial melting in mudrock compositions can occur at temperatures as low as 650°C, many mafic and even some aluminous quartzofeldspathic rocks continue to undergo solid-state reactions, without melting, at temperatures up to 800°C or higher. Indeed, many petrologists regard solid-state reactions in upper mantle peridotite (for example, the reaction of spinel + pyroxene to form olivine + garnet) as classic solid-state metamorphism even though it occurs at temperatures in excess of 1000°C. Although the standard definition of metamorphism as a solid-state process excludes formation of silicate melts, it certainly does not preclude the involvement of aqueous or carbonic fluids. Such fluids play a crucial role in metamorphism as transport media to remove water and carbon dioxide from sites of reaction and as agents to transport other chemical components, particularly silica and alkalis. Such transport modifies bulk compositions in a special process referred to as **metasomatism.**

TYPES OF METAMORPHISM

Historically, four principal types of metamorphism have been recognized by metamorphic petrologists: **burial metamorphism, regional metamorphism, high-pressure, low-temperature metamorphism,** and **contact metamorphism.** The simplest level of distinction among them is areal: The first three tend to be regional, covering areas of hundreds to thousands of square kilometers, whereas the fourth is more limited in scale and clearly associated with intrusive igneous contacts. Burial and regional metamorphism are distinguished from each other by the restriction of burial metamorphism to relatively low metamorphic grades and to P-T conditions that lie on a normal geothermal gradient (rarely exceeding 2 kbar and 250°–300°C). In contrast, regional metamorphism commonly involves temperatures well above the geothermal gradient and can occur at all crustal levels. High-pressure, low-temperature metamorphism is generally restricted to narrow linear belts and forms at P-T conditions below normal geothermal gradients.

In reality, the distinctions become more subtle when igneous rocks do not crop out at the surface. For example, the heat from large, mushroom-shaped intrusions at depth can affect surrounding country rocks beside and above the pluton on a regional scale. Alternatively, localized heating (and resultant metamorphism) can be produced by passage of magmas through the crust, a passage leaving behind no igneous rocks and little or no obvious evidence of intrusive activity other than thermal metamorphic effects. Many metamorphic petrologists have thus adopted the philosophy that regional metamorphism should be defined as any metamorphism for which there is no obvious and identifiable local heat source such as a pluton.

The importance of distinguishing contact from regional metamorphism has diminished somewhat as petrologists have begun to understand better the thermal processes that cause metamorphism. From the point of view of a geologist mapping a metamorphic terrane, it clearly remains important to determine whether mappable changes in metamorphic grade (called *isograds;* see below) are related to intrusions. In some cases this is trivial, but in surprisingly many instances it is not at all obvious. However, it has become increasingly clear from a process-oriented metamorphic perspective that there can be considerable overlap between contact and regional metamorphic processes. This is not particularly surprising because physicochemical principles (in particular, thermodynamics) suggest that all heated rocks of similar composition should show similar mineralogic responses, regardless of the ultimate source of the heat. Intrusive igneous rocks are emplaced at virtually all levels within the crust; thus contact metamorphism should be possible at virtually any depth, but recognizing it may be a problem. Key distinctions lie in overall confining pressure, heating rate, and cooling rate. At shallow crustal levels, small intrusions can chill so rapidly that they produce virtually no thermal aureole. At deep crustal levels, contact thermal metamorphic effects require emplacement of plutons into country rocks no hotter than the magma. For example, granite magmas at 750°C emplaced in granulite-facies country rocks at 750°–800°C will obviously produce no thermal aureole.

In the late 1970s and 1980s, metamorphic geologists attained a more sophisticated understanding of heat sources for all types of metamorphism, especially the regional variety. It had earlier been widely assumed that regional metamorphism occurred at P-T conditions along more or less normal geotherms. Precise pressure and temperature determinations now make it clear that many if not most regional metamorphic rocks formed at temperatures well above normal geothermal gradients. Because regional metamorphism is typical of convergent-margin orogenic belts where magmatic activity is characteristic, some petrologists have proposed that most regional metamorphism requires at least some input of extra heat directly from magmas or from magmatically derived hydrothermal fluids.

Low-pressure contact metamorphism associated with shallow intrusions is worthy of special attention because there is a class of metamorphic minerals and mineral assemblages produced in no other way. Many of these minerals are unstable at the higher pressures (greater than 2–3 kbar) that are characteristic of regional metamorphism. Furthermore, the highest temperatures of regional metamorphism almost invariably occur at moderate to high pressures; and in these deeper crustal environments, heating and cooling rates are slower. There are special problems concerning the kinetics of metamorphic processes in shallow contact environments. Because heat is gained or lost from rocks more rapidly near the earth's surface, both heating and cooling rates in these shallow environments are especially high and affect both textural and chemical equilibration. The common result is a fine-grained rock (hornfels), with incompletely equilibrated mineral assemblages.

Although some regional metamorphism is aided by magmatic heat input, two types of regional-scale metamorphism clearly do not require excess heat: burial metamorphism and high-pressure, low-temperature metamorphism (sometimes called **blueschist-facies metamorphism**). This latter type is especially problematic because it represents equilibration at temperatures *below* those of a normal geothermal gradient. The P-T conditions of burial metamorphism (typically of low to medium-low grade) essentially lie along a normal geotherm. So much progress has been made recently in understanding the thermal evolution of all these types of metamorphism that a later chapter is devoted largely to this topic.

Shock or **impact metamorphism** is a recently recognized and studied phenomenon. This process involves the production of ultrahigh-pressure (and sometimes high-temperature) phases through two processes: the impact of an extraterrestrial body (meteorite or comet) with the earth's surface or a highly explosive volcanic eruption. Expansion of interest in this field has paralleled the development of lunar and planetary geology, as well as the highly publicized debate over the feasibility of meteorite impacts as the explanation for biological extinction events. Although the origin of the spectacular Meteor Crater in Arizona has long been known, several other impact features were not recognized until recently because of the camouflaging effect of surficial processes on ancient terrestrial impact craters. At suspected impact sites, the recognition of ultrahigh-pressure phases such as the silica minerals coesite and stishovite, microcrystalline shock lamellae in mineral grains, and glassy rocks representing impact melts has confirmed an impact origin for these features.

ORIGIN OF METAMORPHIC PETROLOGY
• • • • • • • • • • • • • •

It is useful for the student of any scientific field to have some knowledge of the evolution of ideas in the field, so we devote a few words to the development of modern metamorphic theory. As far back as the publication of Hutton's *Theory of the Earth* in 1795, it is clear that geologists appreciated that some rocks in continental areas were sufficiently different in mineralogy and texture from sedimentary rocks, while still retaining some characteristics of these rocks, to indicate an origin by heating of original sedimentary material. Throughout the nineteenth century, the development of qualitative ideas about the causes of metamorphism continued, both in Europe and in America, due to the efforts of such famous geologists as Lyell, Grubenmann, von Rosenbusch, Sorby, Silliman, and Dana. The recognition of contact metamorphism as an immediate thermal effect of nearby magmatic rocks and of regional metamorphism as characteristic of deeply exposed and highly deformed axial regions of mountain belts were especially important.

The culmination of this early phase of development of metamorphic theory came with the publication in the 1890s of George Barrow's studies on metamorphic zonation in the eastern Highlands of Scotland. Barrow mapped a number of continuous metamorphic zones that roughly paralleled the Highland Boundary Fault. Zone boundaries (equivalent to isograds) were based on the first appearance of key indicator minerals in progressively metamorphosed mudrocks, and Barrow related the zones to progressive temperature increase. His work was among the first to recognize a decipherable and a mappable order in metamorphic terranes and laid the groundwork for later metamorphic advances.

The true modern phase of metamorphic petrology was simultaneously sparked by the development and systematization of thermodynamics by J. W. Gibbs and others (1870s to 1900s) and by the recognition of an orderly sequence of mineral formation in metamorphism. The real "founding father" of modern metamorphic petrology was V. M. Goldschmidt, who first used the techniques of thermodynamics to estimate the actual pressure and temperature conditions of formation of contact metamorphosed calcareous rocks near Kristiania (now Oslo), Norway, around 1910. At about the same time, Pentti Eskola, a Finnish contemporary of Goldschmidt's, studied the metamorphic rocks of southwestern Finland and rigorously applied the principles of chemical equilibrium to such rocks, developing techniques for graphical portrayal of mineral equilibria. Perhaps most important, he established with his principle of metamorphic facies (see Chapter 19) a "uniformitarian" principle of metamorphism. Between them, Goldschmidt and Eskola placed the study of metamorphic petrology on a firm physicochemical foundation.

Widespread use of the principles they pioneered actually took about half a century, largely because of the lack of high-quality experimental and thermochemical data on stabilities of metamorphic minerals. The development in the post–World War II era (1950s on) of techniques for studying rocks, minerals, and fluids at elevated temperatures and pressures led to the creation of a database on mineral properties that was large enough to allow the rigorous application of thermodynamic principles. Refinement of experimental techniques for study of mineral-fluid equilibria and the development of microanalytical instruments, particularly the electron microprobe, allowed estimation of temperature and pressure to within 1–2 kbar and 25°–50°C. Isotopic techniques for age determination of rocks, minerals, and even discrete thermal events made it possible to track the thermal evolution of metamorphic terranes. These data can now be used in conjunction with modern plate tectonic theory to constrain mountain-building processes spatially, temporally, and thermally with detail never before possible.

METAMORPHIC TEXTURES AND STRUCTURES

Virtually any rock—igneous, sedimentary, or metamorphic—can undergo metamorphism, so a very wide range of compositional and mineralogic types of metamorphic rocks exists. Similarly, metamorphic rocks can form in many different stress environments and with different heating rates and therefore can have textures ranging from massive to highly foliated and grain sizes ranging from extremely fine to very coarse. Some metamorphic rocks contain extremely large and well-formed crystals (Figure 18-2), but others are uniformly fine grained and unexceptional looking. Recognition of metamorphism in either a hand sample or an outcrop, therefore, may require the geologist to apply as many discriminative tests as possible, starting with hand specimen mineral identification and recognition of the mineral textures and planar or linear structures typical of metamorphic rocks.

Recrystallization or mineral growth during metamorphism typically produces textures that are diagnostically important in recognizing metamorphic rocks (Figure 18-3). Metamorphic minerals are called *blasts* (from a Greek root meaning "lump"), and different texture types have -*blastic* as a suffix. For example, where one or more minerals occur as subhedral to euhedral crystals distinctly larger than the matrix (a metamorphic equivalent of igneous porphyritic texture), they are called **porphyroblasts** (Figure 18-3A). In some situations, particularly in metamorphosed volcanic rocks, porphyroblasts can be confused with preserved or passively replaced (**pseudomorphed**) original igneous phenocrysts. Preservation of igneous phenocrysts in anything but very low-grade burial metamorphism is quite rare, and microscopic examination usually reveals the presence of numerous small crystals occupying the original shape of the phenocryst as a pseudomorph. For example, metamorphosed basalts commonly show actinolite and/or chlorite pseudomorphs after original clinopyroxene or serpentine pseudomorphs after olivine. Where there is an absence of porphyroblasts, a metamorphic rock composed of equigranular crystals is called **granoblastic** (Figure 18-3B).

Porphyroblasts that contain numerous small inclusions of matrix minerals (quartz and iron-titanium oxides are very common as inclusions) are called **poikiloblasts** (Figure 18-3C). The process by which smaller mineral

FIGURE 18-2
● ● ● ● ● ● ● ● ● ● ● ● ● ● ● ●

Outcrop photograph of large (up to 50 cm) kyanite porphyroblasts
in a chlorite-albite-rutile matrix, western Connecticut. Note the hand
lens for scale. Horizontal distance across the photo is about 1 m.

grains become included within larger grains is poorly
understood, but it is generally assumed that porphyroblast minerals grow faster than surrounding grains and
thus overgrow them and isolate them from the matrix. In
some cases, minerals that have disappeared entirely from
the matrix through reaction during progressive metamorphism are preserved as relicts in the interiors of
poikiloblasts, thus providing valuable clues to the reaction history of the rock. Poikiloblasts may also overgrow
and preserve earlier microstructures in the rock such as
foliations or microfolds.

The growth of new minerals or modification of old
ones during metamorphism results in crystals that have a
range of development of crystal faces. Descriptive textural terms that have been used are **idioblastic** (well-
formed crystals), **hypidioblastic** (medium crystal development), and **xenoblastic** (anhedral crystals). F. Becke,
the founder of metamorphic petrography, described a so-
called idioblastic series that categorized the tendency
of important minerals to develop well-formed, or idio-

blastic, surfaces against other minerals lower in the
series. Minerals such as rutile, titanite, magnetite, garnet,
aluminum-silicates, and tourmaline are high on the list,
and minerals such as quartz, cordierite, feldspars, and
carbonates are low. Like the corresponding igneous textural terms (for example, hypidiomorphic-granular), the
preceding terms have been more widely used in older
literature.

In the last decade or two, metamorphic petrologists
and structural geologists have increasingly realized the
importance of shear stress in the development of textures and structures in metamorphic rocks. Many finer
grained or inequigranular rocks are now recognized to be
mylonites, that is, rocks in which pervasive ductile
deformation has caused grain-size reduction at elevated
temperature and pressure. The overall effect is similar to
cataclasis or granulation, which are brittle processes that
occur at near-surface conditions. With either brittle or
ductile granulation, larger grains may be left surrounded
by finer grained crushed material in a texture called

(A)   (B)   (C)   (D)

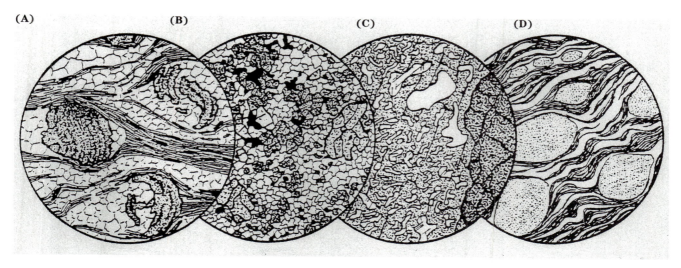

FIGURE 18-3

• • • • • • • • • • • • • • • • •

Drawings of typical metamorphic textures: **(A)** porphyroblastic; **(B)** grano-
blastic; **(C)** poikiloblastic; **(D)** porphyroclastic. For explanation, see text.
[From H. Williams, F. J. Turner, and C. M. Gilbert, 1982, *Petrography*, 2nd ed.
(San Francisco: W. H. Freeman), Figs. 16-1 and 18-1C.]

blastomylonitic. Larger shear-bounded grains are called **porphyroclasts** and are commonly lenticular or eye-shaped (or, in German, **augen**) because of the branching of the microshears around them (Figure 18-3D). The most common example of blastomylonitic texture is the rock type **augen gneiss,** which has alkali feldspar or plagioclase porphyroclasts.

Although some metamorphic rocks have recrystallized passively through heating in a plutonic thermal aureole or simply through burial, the great majority form in orogenic belts where unequal application of stress (nonhydrostatic stress) and resulting deformation are typical. Many of the more obvious features of these rocks in outcrop are planar structures, such as rock cleavage, foliation, schistosity, shear zones, and layering, that are either the results of deformation or represent original features modified by deformation. The recognition and study of metamorphic rocks therefore require at least a basic understanding of how rocks deform. Many of the planar structures in metamorphic rocks develop perpendicular to the direction of maximum compressive stress (Figure 18-4) because they are defined by the parallel alignment of platy minerals. That this is the physically most stable position (that is, lowest energy position) for a platy mineral can easily be demonstrated by pushing on the edges of any platy object, such as a coin or an index card. It will invariably rotate into a position

where its two longer dimensions are perpendicular to the pushing direction, and its shortest dimension is parallel to the stress. In a rock, this rotation is accomplished *both* by physical rotation and active recrystallization of platy minerals and their matrix into the new, more energetically stable, configuration. (The reader should be aware that foliations also commonly form as a result of applied shear stress, and in this case the final planar orientation is not perpendicular to the stress direction.)

Foliation is the general term for the pervasive planar structure found in most metamorphic rocks. It is typically due to a combination of fine-scale compositional layering and parallel alignment of platy minerals. Some foliations may represent features that have been inherited from their sedimentary or volcanic precursors (for example, bedding), but metamorphic geologists now suspect that most sedimentary structures are eliminated during metamorphism and deformation. Even such a prominent outcrop-scale feature as centimeter- to meter-thick compositional layering, commonly assumed to be original bedding, has been shown to form readily during deformation. *Rock cleavage* is the form of foliation that is most typical in fine-grained or low-grade rocks. It is usually manifested by closely spaced planes along which the rock breaks cleanly, as in slate (where it is called **slaty cleavage**). The cleavage planes may parallel original bedding, but quite commonly cut across them at a high

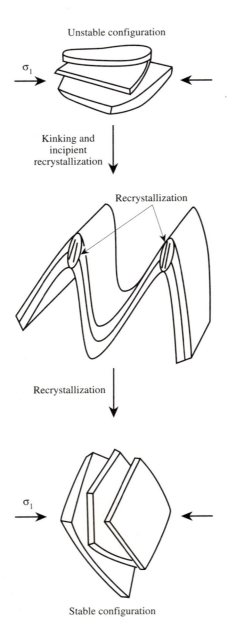

Unstable configuration

σ_1

Kinking and
incipient
recrystallization

Recrystallization

Recrystallization

σ_1

Stable configuration

FIGURE 18-4
● ● ● ● ● ● ● ● ● ● ● ● ● ● ● ● ●

Stable and unstable configurations of platy minerals under simple compressive stress. Increased energy in the top orientation causes rotation or recrystallization to the lower orientation. This phenomenon can cause parallel alignment of platy minerals (for example, chlorite, micas) and development or enhancement of metamorphic foliations. Note that parallel alignment of platy minerals can also be due to a shear stress couple and, in this case, forms at an angle less than 90° to the stress direction.

higher grades of metamorphism, although deformed and metamorphosed igneous rocks may be gneissic at any grade of metamorphism.

By its nature, deformation that results from unequal application of stresses tends to produce linear as well as planar structures. Linear structure is produced because ductile or plastic material that is compressed tends to flow in the direction of *least* compressive stress. If a rock contains appropriate prismatic or needle-shaped crystals (commonly amphiboles, sillimanite, or tourmaline), they will become aligned in this direction and will form what is called a *mineral lineation*. Mineral lineations almost always form within foliation planes, except for a rare structure called *rodding* in which the rock is dominated by linear structures but no identifiable planar structure exists. Linear structures can form in other ways as well, particularly *intersection lineations*, which are produced by the intersection of two sets of closely spaced, non-parallel planar structures or foliations. Recognition and measurement of lineations in outcrop is a normal and important part of field work in metamorphic rocks because such measurements enable the geologist to reconstruct stress directions during metamorphism.

angle. Cleavage planes characteristically are dull-looking surfaces in which individual mineral grains are hard to see or identify. **Schistosity** is another type of foliation characteristic of more intensely metamorphosed rocks in which medium to coarse grains of platy minerals such as mica or chlorite can be easily identified with the naked eye or a 10× hand lens. Schistose rocks commonly break irregularly along the planes of schistosity. When the foliation of a rock consists of millimeter- to centimeter-scale layers in which mineral proportions, colors, or textures vary, but along which there is no particularly strong tendency to break, the rock is described as **gneissic.** In general, gneissic rocks are poor in platy minerals such as micas and represent

METAMORPHIC
RECRYSTALLIZATION
● ● ● ● ● ● ● ● ● ● ● ● ● ● ● ●

Initiation of Metamorphism

As noted earlier, the minimum temperature at which typical regional metamorphic processes begin in sediments is about 150°–200°C, at pressures of 0.5 to 1.5 kbar (depths of 2–5 km), depending on thermal gradient. At this stage in the evolution of the rock, diagenetic processes are complete (although some geologists regard diagenesis and metamorphism as a continuum of processes). The rock has been compacted during burial, and most of the pore spaces have been eliminated either by pressure solution and reprecipitation or by growth of secondary minerals or cements from pore solutions. Most of the water in the rock is confined to hydrous minerals, chiefly clays; clay minerals are the major part of the clay-size fraction (some is quartz and feldspars), and clay minerals contain 5–15 wt% H_2O, depending on temperature. A small amount of saline fluid may also be present in residual pore spaces.

The beginning of metamorphism, as indicated by the formation of nonsedimentary minerals or the development of observable metamorphic recrystallization, depends on rock type. Quartz-rich arenites commonly show few changes until metamorphism is well along in other rock types, because quartzites are relatively nonreactive. Highly reactive metastable materials such as tuffs, volcanic glasses, high-temperature volcanic minerals and sedimentary zeolites respond easily and quickly to modest increases in temperature and pressure. The concept of burial metamorphism is based essentially on observed reactions and assemblages in these highly reactive rock types, particularly volcanics. Glassy materials rich in aluminum and silica readily recrystallize to form zeolites, a large family of hydrated framework silicates whose members typically contain calcium, sodium, or potassium in aluminosilicate frameworks similar to feldspars. Very low grade reactions include the formation of zeolites from volcanic glasses and the dehydration of zeolites to form feldspars. One of the unique aspects of this type of ultralow-grade metamorphism is that neither large volume change nor deformation occurs, so mineralogic transformations can proceed while the relatively delicate sedimentary structures and textures are preserved.

Two areas where low-grade metamorphism in volcanic and sedimentary rocks is actively being studied are northern New Zealand, near Rotorua, and the Salton Sea region in southern California. Both areas share the advantage of having anomalously high heat flow (both are, in fact, geothermal areas), and thus active metamorphism is going on at shallow depths relatively easily accessible to drilling and coring. These areas serve as natural experimental laboratories for low-grade metamorphic processes because in situ temperature measurement can be made to calibrate ongoing active recrystallization.

Increase in Grain Size

One of the principal effects of temperature increase in metamorphism is to cause an increase in the grain size of most rocks. This process is especially important at moderate pressures, in the absence of shear stresses that may work to reduce grain size, and in the presence of flu-

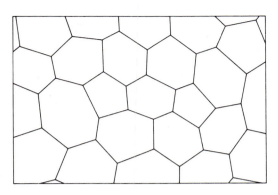

FIGURE 18-5
● ● ● ● ● ● ● ● ● ● ● ● ● ● ● ●

Polygonal texture. Most of the grains have five or six relatively straight sides, and intersections are usually made between three grains, with boundaries intersecting at about 120° (so-called triple junctions). If the grains had been xenoblastic as well as equidimensional, the texture would also be granoblastic.

ids. Metamorphic petrologists have applied the theory and observations of ceramics and metallurgy to understanding the processes and driving forces for recrystallization. Many of these petrologic processes become fairly readily understandable when viewed as energy-reduction processes that serve to minimize energy and bring the rock toward a state of textural equilibrium.

The phenomenon of grain-size increase in many rocks can generally be categorized as **annealing,** a widely observed process in ceramics and metals subjected to elevated temperatures for a long time. Annealing involves both an increase in average grain size and an approach to a *polygonal texture* in nearly monomineralic rocks such as quartzite, marble, or metamorphic dunites (Figure 18-5). Polygonal texture is characterized by the presence of five- and six-sided grains and abundant *triple junctions*, where three grains come together and the three grain boundaries intersect at approximately 120° angles. The grain boundaries typically have arbitrary crystallographic orientations and bear little relation to normal crystal faces, even in minerals that are normally idioblastic to subidioblastic. The restriction of this texture to monomineralic rocks suggests that where grain boundaries coincide with chemical boundaries (as occurs in most polymineralic rocks), there are more complicated energetic relationships that stabilize normal crystal faces and more complex textures. Angular relationships in polygonal texture are actually quite similar to those in aggregates of soap bubbles, widely cited by physicists as an example of energy minimization. In metamorphic rocks poor in platy or prismatic minerals, annealing produces *granoblastic* texture, with most grains equidimensional and of similar size. In rocks

(A)

(B)

FIGURE 18-6
● ● ● ● ● ● ● ● ● ● ● ● ● ● ● ● ●

(A) Decussate texture. Crystals are hypidioblastic, prismatic, and randomly oriented.
(B) Idiotropic texture. Crystals are generally idioblastic and randomly oriented.

containing minerals that have pronounced inequigranular crystal shapes, such as amphibolites or mica-rich rocks, annealing leads to **decussate** textures (Figure 18-6). Triple junction intersections may still be quite common, but grain boundary intersection angles are commonly not 120°. Because rocks are polymineralic as a general rule, several minerals usually recrystallize simultaneously. As each mineral with a pronounced shape tendency interacts with other similar mineral grains, the result is a complex texture, with some minerals being subhedral and others anhedral.

As noted earlier, the driving force for annealing is the tendency of the rock to reduce its internal energy. Much of the energy in the rock is related to the identity and proportions of its minerals, temperature, and pressure, but a small though significant energy contribution also comes from the mineral surfaces; and this surface energy is a function of surface *area*. As a general rule, one large crystal tends to have less surface area per total mass than a number of smaller crystals of the same total mass. A rock can thus lower its energy without changing pressure, temperature, or chemical composition simply by making the average size of the grains larger. The coarsening processes themselves are actually rather complicated, but a simplified view is that the input of thermal energy into a metamorphic rock causes some of the smaller grains to dissolve, and the material from them is added in crystallographic continuity to other grains, thereby making them larger. These crystallographically continuous overgrowths are called **epitaxial overgrowths,** and they are relatively common in recrystallized metamorphic rocks.

Growth of Porphyroblasts

One of the more interesting and obvious things that happens during recrystallization in many, but not all, metamorphic rocks is the development of porphyroblasts. These conspicuously large crystals (see Figure 18-3A) are the metamorphic equivalents of phenocrysts in igneous rocks. Despite the normal tendency of metamorphic rocks to develop roughly equigranular textures, porphyroblasts are relatively common, especially in metapelites. Porphyroblasts are rare, however, in monomineralic rocks and, if present, are always crystals of a minor mineral derived from trace detrital constituents, for example, porphyroblasts of garnet or kyanite in quartzites. Minor silicate minerals in marbles actually tend to form smaller crystals than the matrix carbonates. Some minerals almost never are found as porphyroblasts, for example, quartz, whereas others such as staurolite and garnet occur almost exclusively as porphyroblasts. Because porphyroblasts are only required to be large in relation to other minerals in a rock, some porphyroblasts may be pinhead size (1–2 mm); good examples of this are garnet porphyroblasts in low-grade chlorite phyllites (Figure 18-7). In other cases, rocks with very large general grain size (1 cm or more) may have spectacularly large porphyroblasts, for example, garnets up to 30 cm in size in coarse amphibolite from the Adirondacks (Figure 18-8), and kyanite blades up to 50 cm long in chlorite schist in Connecticut (see Figure 18-2).

The mechanism of porphyroblast formation has received considerable attention from petrologists for many years. Both spatial and size distributions of garnets in garnet-mica schists have been carefully investigated by using techniques as straightforward as breaking the rock, hand-picking garnets, and measuring them, or as sophisticated as X-ray tomography (equivalent to a rock CAT scan). As in magmatic crystallization, the principal requirement for making a few large crystals is a low rate of crystal nucleation. If many crystal nuclei initially form, the result will probably be many small crystals. Relatively rapid diffusion through the rock of the material needed to grow a porphyroblast is coupled with low

FIGURE 18-7
● ● ● ● ● ● ● ● ● ● ● ● ● ● ● ●

Fine-grained chlorite phyllite with abundant, uniform, small (roughly 1 mm) euhedral garnet porphyroblasts (dark spots). Note the coin for scale. Although the porphyroblasts are small, the even smaller grain size of the matrix makes them stand out.

FIGURE 18-8
● ● ● ● ● ● ● ● ● ● ● ● ● ● ● ●

Photograph of quarry wall from the former Barton Garnet Mine, Gore Mountain, North Creek, New York. Large (up to 25–30 cm) garnets are surrounded by a rim of hornblendite and are set in a granulite-grade metagabbro matrix consisting of ortho- and clinopyroxenes and plagioclase.

nucleation rate. Because the chemical composition of a porphyroblast mineral is different from the rock matrix, this material may have to come from a substantial distance. Ultimately, the spacing between the porphyroblasts and the size to which they grow will be controlled by a delicate balance between nucleation rate and growth rate. Research has now focused on attempting to define the actual causes for nucleation in particular sites. Some sites of crystal growth are clearly determined by concentrations of particular chemical constituents that have limited migration rates, for example, alumina needed for aluminum-rich minerals such as garnet, staurolite, or aluminum-silicates. In other cases, nucleation is favored in parts of the rock that have been highly strained or mechanically disturbed by deformation; these places have high energy and thus are relatively unstable. Nucleation and growth of garnet and staurolite porphyroblasts in the cores of microfolds have been proposed to explain the common observation of folded inclusion patterns in these minerals. Some research has begun to focus on chaotic perturbations of either composition or energy as likely nucleation sites.

COMMON METAMORPHIC ROCK TYPES
• • • • • • • • • • • • • • •

Metamorphic petrologists must communicate their observations to others, and this requires that metamorphic rocks have names that transmit the intended information. As noted earlier for igneous rocks, nomenclature can become a problem because of the proliferation of nonsystematic names, many of which are based on geographic localities. Igneous rock nomenclature was revised and systematized by Streckeisen and the International Union of Geological Sciences Commission (as noted in Chapter 3), but metamorphic rocks have not received the same formal attention. Fortunately, the names of most metamorphic rocks are based primarily on textural, structural, and mineralogic features and thus are not totally random. Widely used general rock names such as schist or gneiss are usually modified by a prefix or term indicating a conspicuous textural feature, if present (for example, *augen* gneiss), or by the names of key minerals (*garnet-biotite* schist). If a rock is almost monomineralic, it may be named for the dominant mineral (for example, marble, quartzite, or serpentinite). Other terms may invoke genetic assumptions and thus are inherently hazardous. For example, some petrologists have used the prefixes *ortho-* and *para-* to indicate igneous or sedimentary precursors (often called **pro-**

toliths), respectively, as in orthogneiss and paragneiss. When the protolith is obvious, a rock may be named genetically, for example, *granitic* gneiss. It is rarely straightforward, however, and commonly impossible to distinguish between the two possibilities in a thoroughly recrystallized or deformed high-grade metamorphic rock.

The basis of metamorphic nomenclature springs from mineralogic and textural characteristics inherent to rocks of different bulk chemistry. The great majority of metamorphic rocks fall into three broad categories, based on protolith chemistry: (1) aluminous clastic sedimentary rocks, including the mudrocks and the variably aluminous sandstones like greywackes and arkoses; (2) calcareous rocks, such as limestones, dolostones, and marls; and (3) mafic or intermediate volcanic or pyroclastic rocks, including basalts, andesites, and dacites, and some ultramafic rocks. These rock types are dominant in metamorphic belts because they dominate the sedimentary and volcanic tectonic environments that become incorporated into metamorphic and orogenic belts. Other, more uncommon rock types also have their metamorphic equivalents, for example, relatively pure quartz sandstones (quartzites), ultramafic rocks, and iron formations (ironstones). These are volumetrically insignificant but may otherwise hold great petrologic interest because of the unusual mineral assemblages in them.

Mudrocks

As an example of nomenclature, we will examine the progressive regional metamorphism of mudrocks (refer to Table 18-1). These rocks are commonly called **metapelites,** from the relatively uncommon sedimentologic term **pelite,** which is synonymous with mudrock. Metamorphism of these rocks results initially in recrystallization and modification of the clay minerals present. Grain sizes remain microscopic, but a new direction of preferred orientation may develop in response to the stress. The resultant fine-grained rock typically shows excellent rock cleavage and is called a **slate.** Submicroscopically, crystallographic changes that can be detected only by X-ray analysis occur in the sheet silicates as they gradually change from clay minerals to muscovite and chlorite. Continued metamorphism commonly causes an increase in grain size while maintaining the preferred orientation of the platy minerals. Modest grain size increase produces a fine-grained rock called **phyllite,** which is similar to slate but has silky or shiny, rather than dull, cleavage surfaces. At this stage of metamorphism, small porphyroblasts may start to develop and are observed as small bumps on otherwise smooth cleavage surfaces. In rocks from higher grades of meta-

TABLE 18-1   General characteristics of progressive regional metamorphism of mudrocks

Grade	Rock name	Grain size	Key minerals	Notable textures
Premetamorphic	Mudrock	Very fine; clay-size particles	Clays, quartz	Fissile shale bedding or massive mudrock texture
Very low	Slate	Very fine; recrystallized particles	Clays, chlorite, quartz, Fe oxides	Bedding, slaty cleavage
Low	Phyllite	Fine	Micas, chlorite, quartz, Fe oxides	Bedding, phyllitic foliation
Moderate	Schist	Fine to moderate, with random isolated large crystals	Micas, chlorite, quartz, plagioclase, garnet, staurolite, kyanite, Fe-Ti oxides	Schistosity, porphyroblasts, rare bedding
High	Schist/gneiss	Moderate to coarse	Micas (mainly biotite), quartz, plagioclase, garnet, cordierite, sillimanite, Fe-Ti oxides	Schistosity or gneissosity, segregation layering, porphyroblasts
Very high	Gneiss	Moderate to coarse	Biotite, quartz, plagioclase, orthoclase, sillimanite, garnet, cordierite, Fe-Ti oxides	Gneissosity, segregation layering, porphyroblasts, migmatitic layering

morphism, many of the crystals are visible without a hand lens, and the platy or elongate minerals have a distinct preferred orientation forming a conspicuous foliation (schistosity) along which the *schist* breaks readily into irregular flakes or slabs. The development of porphyroblasts or poikiloblasts in schists is common. Many are idioblastic, or nearly so, and can be easily identified as the characteristic metamorphic minerals garnet, staurolite, kyanite, sillimanite, and cordierite.

At still higher grades of metamorphism (temperatures in excess of roughly 600°C), the schistosity may become less pronounced as hydrous micas are gradually replaced by less platy anhydrous minerals such as alkali feldspars, sillimanite, garnet, and pyroxenes. As the schist undergoes transition into a *gneiss*, it becomes more massive and commonly also develops gneissic banding or layering through poorly understood solid-state segregation processes. This segregation tends to develop alternating lighter layers rich in quartz and feldspar and darker layers enriched in ferromagnesian and aluminum-rich minerals. The mineralogy at these high metamorphic grades can be quite similar to that of igneous rocks, but a metamorphic origin of gneisses is typically indicated by the gneissic banding as well as by preferred orientation of inequigranular mineral grains. The highest grades of

metamorphism commonly produce *migmatites* (from a Greek root meaning "mixed rock") in which layers or patches of light-colored material have igneous textures and probably represent crystallized locally derived melts of granitic to dioritic composition (Figure 18-9).

Calcareous Rocks

Similar processes of progressive mineral reaction involving devolatilization and coarsening occur in other protolith compositions as well. Among the calcareous (calcium-rich) rocks, **marble** is a carbonate-rich rock consisting mostly of calcite and dolomite, but few marbles are purely carbonate. Most contain minor amounts of a variety of silicate and oxide minerals that have been derived by reaction from the original detrital minerals of the limestone or dolostone. At low metamorphic grades, the noncarbonate minerals are dominated by hydrous species such as brucite, phlogopite, chlorite, and tremolite, which are replaced at higher grades by the anhydrous minerals diopside, forsterite, wollastonite, grossular, and calcic plagioclase. Marbles are typically granoblastic at all grades and in hand specimen reflect increasing metamorphic grade principally through grain

FIGURE 18-9

• • • • • • • • • • • • • • • • •

Migmatitic schist outcrop on Parker Island, Quabbin Reservoir, Massachusetts.
Folded, coarse leucocratic layers consist of quartz, orthoclase, sodic plagioclase,
with minor amounts of muscovite, garnet, and sillimanite. Schistose melanocratic
layers are biotite-, sillimanite-, and garnet-rich and probably represent the residuum
following extraction of the leucocratic melt from normal pelitic schist.

size increase. Marbles are not common in metamorphic
terranes, however, and the **calc-silicate** rocks are much
more typical of the calcareous rock family. These rocks
are metamorphic equivalents of impure carbonates or
marls that contained abundant detrital silicate material.
Their character in progressive metamorphism resembles
that of the mudrocks: They commonly exhibit relict bedding or other foliation at low grades and become progressively more massive with increasing grade. Carbonate minerals typically constitute less than half the
volume of these rocks at low grade and in many cases
are completely reacted away before the medium and
higher grades are achieved. At middle to high grades,
many calc-silicate rocks closely resemble amphibolites
that are the metamorphic equivalents of mafic volcanics,
and care is required to discriminate between the two
possibilities. Characteristic low-grade calc-silicate minerals are chlorite, epidote, and actinolitic amphibole. At
higher grades, these are typically replaced by augite,
hornblende, garnet, and calcic plagioclase.

Mafic and Ultramafic Rocks

The third principal compositional type of metamorphic
rocks—mafic and ultramafic schists and gneisses—are
largely derived from metamorphism of intermediate
through mafic volcanics (both flows and pyroclastics)
and shallow intrusions, including the ultramafic cumulates such as those that form in mid-ocean ridge magma
chambers. Many of these mafic protoliths are the classic
igneous suite associated with convergent plate margins,
particularly the island arc suite, mixed with a component
of basaltic oceanic crust that has been added tectonically. Igneous and sedimentary rocks produced at active
convergent margins are typically the most common protolith rock types in metamorphic belts because it is the
convergent margin environments that evolve into deformed and metamorphosed orogenic belts accreted to
the continental edges. The most volumetrically important protolith volcanics are andesitic and basaltic in composition, with lesser volumes of more felsic volcanics

such as dacite or rhyolite (refer to Chapter 9 for a discussion of the island arc suite). Thus the bulk compositions are relatively rich in iron, magnesium, and calcium and poor in alkalis and silica. Low-grade mafic schists contain characteristic minerals that reflect these bulk compositional characteristics, including chlorite, epidote, actinolite and sodic plagioclase. Hydrous ferromagnesian minerals at this grade are typically green and influence the color of these rocks so strongly that low-grade mafic schists are commonly called *greenschists* or **greenstones.** Extensive Precambrian greenstone belts like those in southern Africa, Australia, and the Canadian Shield are thought to represent major preserved and metamorphosed island arc terranes of great age, potent evidence of plate tectonic processes at least as far back as 3 Ga.

Assemblages of very hydrous minerals evolve into less hydrous ones at higher metamorphic grades. Because it is chemically complex, hornblende can take the place of *several* lower grade minerals, and typical mafic rocks quickly become dominated by the two minerals hornblende and plagioclase in the rock type **amphibolite,** which exists over a wide range of metamorphic grades. Again, as noted earlier, the name *amphibolite* is descriptive and is simply used to indicate a hornblende-plagioclase rock. It should not be taken to imply a volcanic origin for the protolith. Certain nonvolcanic calcareous rocks or finely interbedded layers of shale and limestone may coincidentally have bulk compositions similar to andesites or basalts and will therefore also form metamorphic amphibolites. It should always be kept in mind that metamorphic mineral assemblages are a function of both metamorphic grade and original bulk chemical composition.

At the highest metamorphic grades (above about 700°C), even hornblende becomes unstable and reacts to form anhydrous mineral assemblages, including minerals such as orthopyroxene, augite, garnet, calcic plagioclase, and olivine. In this process, the mineralogy of high-grade mafic gneisses begins to resemble the original mafic igneous protolith mineralogy, and the practical distinction between igneous and metamorphic rocks in high-grade terranes is difficult and may be based solely on the presence or absence of metamorphic textures.

Other Rock Types

The preceding discussion of the three dominant protolith classes has provided definitions of many metamorphic rock names, but a few others remain.

Quartzite. The metamorphic equivalent of a relatively pure quartz sandstone. Quartzite is commonly massive,

shows granoblastic or annealed metamorphic textures, and typically contains trace amounts of aluminous or ferromagnesian minerals that have resulted from reactions involving the original trace detrital constituents.

Eclogite. A rare rock that is chemically equivalent to basalt and has recrystallized under high pressure–high temperature conditions to an assemblage of omphacite (a sodium- and aluminum-rich augite) and magnesium-rich garnet. Grain size ranges from fine to coarse, but most eclogites are medium to coarse grained, are granoblastic, and have a notable red and green appearance produced by the bright green pyroxene and red garnet.

Granulite (granofels). A granoblastic rock that is rich in quartz and feldspars and has only minor amounts of platy or prismatic minerals. Lenticular to tabular compositional layering may be present. Granulites are the metamorphic equivalent of a clay-poor sandstone or a fine- to medium-grained granitic rock. This term is avoided by many petrologists because of its possible confusion with the same word used to express the highest grades of metamorphism (as in granulite facies).

Hornfels. A very fine-grained, granoblastic or porphyroblastic rock, which has a baked appearance similar to a kiln-fired ceramic. These rocks are dense and compact, and commonly splinter when broken. Hornfels forms at low pressure by contact metamorphism, typically of a mudrock protolith.

Serpentinite. A massive to slightly schistose, dark green rock that is composed of platy or fibrous serpentine, talc, chlorite, and minor carbonates. It forms by metamorphism and hydration of ultramafic rocks and is commonly thought to represent metamorphosed slivers of upper mantle rock tectonically incorporated into crustal rocks or in ophiolite complexes.

Skarn. A granoblastic rock that is rich in calcium silicate minerals such as calcium-rich garnet (andradite or grossular), epidote, vesuvianite, diopside, and wollastonite. The small amount of carbonate distinguishes skarn from marble. Skarn forms by silica metasomatism at igneous contacts. Rare and unusual minerals that are rich in economically important trace elements (for example, tungstates and molybdates) are common constituents of skarns.

Iron formations and ironstones. At low grade, these magnetite-rich quartzites (taconites) commonly contain iron-rich hydrous minerals such as iron-serpentine (greenalite), iron-talc (minnesotaite), and iron-rich amphiboles (ferroactinolite, ferrocummingtonite, ferrogedrite). High-grade assemblages include quartz, magnetite, fayalite, iron-rich orthopyroxene, hedenbergite, and almandine garnet. Weathered and oxidized occurrences may be very hematite-rich. Iron-rich metamorphic lithologies represent metamorphosed middle Precambrian layered iron formations such as the famous taconite ores of the Lake Superior region in Minnesota and Michigan or the rarer Proterozoic or Phanerozoic ironstones at Sanford Lake or Benson Mines in the Adirondacks (see Chapter 17).

Mylonite. A very fine-grained, to locally glassy, commonly streaky or layered rock that is produced by ductile deformation and granulation along deep shear zones. Mylonites are essentially the higher temperature, deeper equivalents of cataclasites that develop along near-surface fault planes. Depending on the amount of shearing, mylonites may have a variable amount of residual unbroken fragments.

Figures 18-10 through 18-19 illustrate metamorphic textures, structures, and rock types. These examples are only a supplement to laboratory exercises, for there is no substitute for a careful study of a well-labeled set of metamorphic rocks and minerals.

FIGURE 18-10
● ● ● ● ● ● ● ● ● ● ● ● ● ● ●

A muscovite-biotite-garnet-staurolite schist in Daggett County, Utah. Larger dark angular crystals are mainly staurolite and rounded ones are garnet. Note the irregular crosslike twinning in staurolite. Abundant dark specks near the hammer point and butt of the handle are biotite. A weakly developed schistosity is evident from upper right to lower left. [From W. R. Hansen, 1965, *U.S. Geol. Surv. Prof. Paper 490*, Fig. 6.]

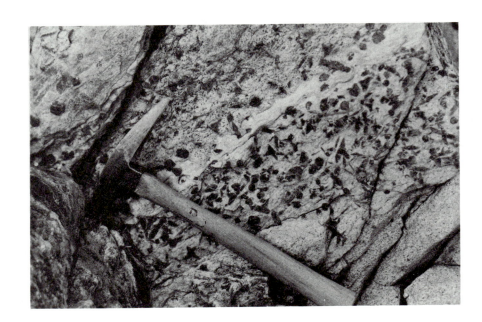

FIGURE 18-11
● ● ● ● ● ● ● ● ● ● ● ● ● ●

Interlayered biotiote (light) and hornblende-biotite (dark gray) gneisses, Caldwell County, North Carolina. White layers are rich in potassic feldspar. The variable amount of mafic versus felsic minerals in the different layers results in an unusually strong foliation. Scale (center) is about 18 cm long. [From B. H. Bryant, 1970, *U.S. Geol. Surv. Prof. Paper 615*, Fig. 68.]

FIGURE 18-12

• • • • • • • • • • • • • • • • •

Slate, held in a position that shows the excellent
development of the slaty cleavage.

FIGURE 18-13

• • • • • • • • • • • • • • • •

The same rock oriented to provide a view of the cleavage surface. This dark gray rock has several dark bands ("intersection" lineations) that represent original bedding in a premetamorphic mudrock. Slates commonly show this effect, with very
different orientations of slaty cleavage planes and original bedding planes. Typically, slaty cleavage develops as a roughly
axial-planar foliation in folded bedding in slates.

FIGURE 18-14

• • • • • • • • • • • • • • • • •

A gneiss containing irregular dark bands of
biotite- and hornblende-rich rock separated
by lighter areas consisting mainly of quartz
and feldspars.

FIGURE 18-15
• • • • • • • • • • • • • • • • •

Kyanite-biotite-muscovite-quartz schist. The photo is taken perpendicular to the plane of schistosity. Pale blue kyanite porphyroblasts (gray in photo) and black biotite grains lie within the schistosity planes but show no dominant lineation.

FIGURE 18-16
• • • • • • • • • • • • • • • • •

Contact-metamorphic marble with a typical granoblastic texture. The minerals are gray to tan calcite (light gray in the photo) and green diopside (dark gray).

FIGURE 18-17
• • • • • • • • • • • • • • • • •

Very coarse talc-actinolite rock. Pale green, platy talc crystals (light in photo) surround large bladed or prismatic, dark green actinolite crystals (dark). This metamorphic rock typically represents mafic volcanic protoliths.

5 cm

FIGURE 18-18
● ● ● ● ● ● ● ● ● ● ● ● ● ● ● ●

Granoblastic calcite marble with scattered diopside
crystals. Calcite is typically pale orange in this sample,
except adjacent to diopside crystals, where it is white.

FIGURE 18-19
● ● ● ● ● ● ● ● ● ● ● ● ● ● ● ●

Polished slab of veined serpentinite. The
rock is irregularly colored from dark green
to black. Lighter veins contain carbonates,
and magnesite-bearing areas to the left are
pale gray. Such rocks commonly contain
small euhedral grains of either magnetite or
chromite, or both, and are highly magnetic.

0   5 cm

FIELD OBSERVATIONS
OF METAMORPHIC ROCKS
● ● ● ● ● ● ● ● ● ● ● ● ● ● ● ●

Observations of metamorphic rocks in the field have
revealed that they can be grouped into several categories
as a function of their regional extent and circumstances
of origin. As already discussed, the broadest of these categories are regional and contact metamorphism. It is typically not possible (and certainly not wise!) to attempt to
distinguish contact from regional metamorphic rocks in
hand specimen or thin section alone, although some rare
high-temperature, low-pressure mineral assemblages (for

example, wollastonite- or monticellite-bearing ones in
calcareous rocks; mullite- or cordierite + andalusite +
potassic feldspar-bearing ones in aluminous rocks) are
almost invariably of contact metamorphic origin. Mapping and field study of metamorphic rocks provide the
firmest basis for their categorization.

Regional Metamorphic Rocks

The most common metamorphic rocks are those of
regional extent that occur over large areas, typically
within orogenic belts. A map of the intensity of metamorphism commonly forms an elongated concentric pattern

along an orogen, with higher grade zones in the center, surrounded by lower grade ones. Igneous activity typically accompanies the metamorphism, particularly in the higher grade zones, but it is sometimes unclear whether this is genetically important or simply coincidental. Dikes or small stocks with contact aureoles are characteristic of lower grade zones, whereas the high-grade zones contain larger and more irregular bodies or highly deformed sheets. In extremely elongated linear metamorphic belts, the high-grade metamorphic core regions may correspond to the deeper parts of the magmatic arcs of convergent margins; that is, they represent the walls of the conduits through which most of the magmas passed upward to shallower levels.

Regional metamorphic rocks are found in all the major orogenic regions of the world. These typically occur on former convergent plate margins where active erosion and sedimentation, volcanic and intrusive igneous processes, and both vertical and horizontal tectonic movements have significantly perturbed the equilibrium thermal structure of the crust. During burial and exhumation in these tectonically active zones, protoliths have become metamorphosed, recrystallized, and deformed before their eventual exposure at the surface through erosion. Because nonhydrostatic stress is typical of orogenic belts, development of preferred orientation of platy or elongate minerals is typical. Measurement of directions and orientations of structures such as lineations, foliations, and rock cleavages is a major and important part of mapping and field investigation in regional metamorphic terranes. Analysis of these deformational structures begins in the field but can be pursued in the laboratory through examination of thin sections taken from oriented samples. Modern microstructural techniques for such analysis include characterization of grain sizes and shapes, as well as measurement of crystal orientations, especially in sheared rocks. The combined use of such techniques can yield much information about the deformational history of a metamorphic rock.

Contact Metamorphic Rocks

Contact metamorphism occurs where heat has been directly transferred from an igneous intrusion to the surrounding country rocks by conduction or convection (by heated fluids). Metamorphic effects range from the baking of thin zones (up to tens of centimeters wide) along the contacts of dikes and sills, producing hornfels, to kilometers-wide **contact aureoles** around stocks or batholiths. Contact aureoles occur around both lower temperature (700°–850°C) granitic plutons and higher temperature (900°–1100°C) gabbroic and dioritic plu-

tons. Despite their lower temperatures, water-bearing granitic plutons commonly have wider thermal aureoles than "dry" mafic plutons do, an observation that emphasizes the potential importance of fluids as heat transport agents in metamorphism. Different mineralogic zones are common in wider aureoles, with higher grade minerals closer to the igneous contact and lower grade zones farther away. Texturally massive rocks are produced because thermal metamorphism is generally a static event that operates in the absence of nonhydrostatic stress. In contact metamorphism at shallow crustal levels (less than 6–7 km depth), where heat can be rapidly dissipated to the surface, fine-grained rocks (for example, hornfels) typically form because the short duration of the heating event does not permit substantial grain growth. Deeper crustal intrusions commonly are surrounded by aureoles of coarser contact metamorphic rocks. Metasomatism (compositional modification) of country rocks (particularly calcareous rocks) is a typical feature of contact aureoles surrounding granitoid intrusions. Aluminous calcareous rocks, such as calcareous shales, can develop both porosity and permeability because of mineralogic reactions during contact metamorphism and thus may act as metamorphic aquifers or conduits for aqueous fluids. These hot fluids either originate from the crystallizing magma or represent convection of pore fluids in the country rocks. In either case, these fluids carry dissolved components (especially silica) and can deposit them (and remove others). Abundant veins in a contact aureole are important clues to intense fluid activity.

Identification of Rock Types

Mapping and fieldwork in metamorphic rocks begin with identification of rock types. As previously discussed, names of metamorphic rocks are generally based on descriptive features, specifically texture, structure, and mineralogy. The texture (including grain size) is often an important clue to the intensity (grade) and type of metamorphism, whereas mineralogy tells more about the nature of the protoliths (and perhaps the premetamorphic tectonic environment) than about the metamorphism itself. Extremely fine-grained metamorphic rocks are typically the result of either low-grade regional metamorphism or shallow contact metamorphism. Increases in grain size in regional metamorphic rocks are generally a useful indicator of increasing intensity of metamorphism; but the effect can vary from one compositional type to another, so care must be taken. Increasing metamorphic grade typically produces a progressive obliteration of original sedimentary or volcanic features as well. In general, the type of foliation also provides a clue to metamorphic grade. Slaty and phyllitic structures indicate low grade, schistosity is typical of medium-grade metamorphism, and gneissosity is characteristic of high-grade metamorphism.

In addition to identifying and mapping rock types, it is also critical to determine the metamorphic mineralogy as completely as possible. This can be done successfully in the field even with relatively fine-grained rocks by using a 10× hand lens. Especially fine-grained matrix minerals may require later microscopic identification. Hand specimen mineral identification, like thin section petrography, is an endeavor in which the brain is as important as the eye. Experience with metamorphic and igneous rocks gradually impresses upon the petrologist the fact that certain characteristic minerals are associated with certain rock types, and other minerals with other rock types. There are inevitable exceptions to these unstated rules, but they are rare. For example, once a field petrologist has identified a metamorphic rock as a pelitic schist (a metamorphosed mudrock), the list of minerals it might contain is limited to little more than a dozen, most of which are quite easily distinguished. A fine-grained, pale, bladed mineral by itself might be either kyanite or tremolite, but the fact that kyanite can and commonly does occur in pelitic schists makes this mineral more likely than tremolite, which never does. Determining differences in mineralogy is important because it permits any metamorphic terrane, regional or contact, to be subdivided into mineralogic zones that can be used to infer both qualitative and quantitative information about metamorphic grade.

SUMMARY

Metamorphic rocks form as a result of the operation of physical and chemical processes of reequilibration when original igneous or sedimentary rocks are subjected to a new set of environmental conditions, especially increased pressure and temperature. Deformation commonly plays a significant role in the development of metamorphic minerals and fabrics. Although metamorphic effects may be subtle, metamorphic rocks usually differ significantly from their protoliths in mineralogy and texture, and sometimes even in chemistry. Metamorphic rocks form within a P-T range from diagenesis to significant melting. Quantitatively, this range of metamorphic conditions is from pressures of several hundred bars to 10 kbar or more, and from about 150°C to as high as 900°C or more, depending on rock type.

Metamorphic rocks are broadly distinguished as regional or contact, depending on whether the heat source

for metamorphism can be readily identified as an igneous intrusion. Although regional metamorphic rocks are more abundant than contact rocks, some magmatic heat input may be required for much regional metamorphism. Shallow contact metamorphic rocks formed at less than about 2 kbar (7 km depth) tend to be fine-grained and contain unusual low-pressure, high-temperature mineral assemblages, but textural and mineralogic distinctions between regional and contact metamorphic rocks blur significantly at greater crustal depths. There are less common types of metamorphism such as burial or impact metamorphism. The first of these may occur on a basinal or regional scale. Impact metamorphism has rarely been recognized and is restricted to the immediate environs of impact craters.

Metamorphic recrystallization or annealing is a process largely driven by temperature increase. Coarsening of grain size is typical in most metamorphic rocks and is driven by a tendency toward reduced surface energy in a rock composed of larger crystals. Development of unusually large porphyroblasts is coupled with overall coarsening in many metamorphic lithologies. The size and spacing of porphyroblasts are controlled by a delicate balance of nucleation and growth rates. The ultimate control of these rates is not well understood, and they vary considerably from rock to rock, even in the same metamorphic lithology under equivalent pressures and temperatures.

Recognition of typical metamorphic rocks is based on the identification of metamorphic minerals and fabrics. Because most regional metamorphic rocks form during orogenic activity, deformational fabrics such as cleavage, foliation, and lineation are commonly present. Metamorphic rock nomenclature is based largely on a combination of mineralogic and textural characteristics, although some rock names are based on bulk composition (for example, marble). Metamorphic rocks are mapped on the basis of protolith type and chemistry and also of texture and mineralogy of metamorphic origin. In regional metamorphism, regional or local changes in environmental variables P, T, and water pressure are reflected by mineralogic changes that can be mapped as mineralogic zones.

STUDY EXERCISES

1. Metamorphism can result from changes in which physical variables in a rock? Is a temperature increase necessary? How are fluids involved in metamorphism?
2. Contrast the recognized types of metamorphism.
3. Describe the process of grain-size increase that is typical of metamorphism.

4. Using mudrocks as an example, describe the textural evolution of metamorphic rocks as a function of metamorphic grade.

5. What are the most important components of a field investigation of metamorphic rocks?

REFERENCES AND ADDITIONAL READINGS

• • • • • • • • • • • • • • •

Barrow, G. 1893. On an intrusion of muscovite-biotite gneiss in the east highlands of Scotland, and its accompanying metamorphism. *Q. J. Geol. Soc. London, 49*, 330–358.

Eskola, P. 1915. On the relations between the chemical and mineralogical composition in the metamorphic rocks of the Orijärvi region. *Bull. Comm. Geol. Finlande, 44*, 1–277.

Goldschmidt, V. M. 1912. Die gesetze der gesteinsmetamorphose, mit beispielen aus der geologic des sudlichen Norwegens. Videnskabelig Skrifter I. *Math. Naturv. Klasse, 22.*

Miyashiro, A. 1973. *Metamorphism and Metamorphic Belts.* London: Allen and Unwin.

Miyashiro, A. 1994. *Metamorphic Petrology.* New York: Oxford University Press.

Spear, F. S. 1993. *Metamorphic Phase Equilibria and Pressure-Temperature-Time Paths.* Washington, DC: Mineralogical Society of America.

Spry, A. 1969. *Metamorphic Textures.* New York: Pergamon.

Williams, H., F. J. Turner, and C. M. Gilbert. 1982. *Petrography: An Introduction to the Study of Rocks in Thin Section*, 2nd ed. San Francisco: W. H. Freeman.

Yardley, B. W. D. 1989. *An Introduction to Metamorphic Petrology.* New York: Wiley.

Chapter 19

ISOGRADS, METAMORPHIC FACIES, AND PRESSURE-TEMPERATURE EVOLUTION

Metamorphism occurs when rocks are subjected to elevated temperature and pressure within the crust. A modern view of pressure-temperature relationships in metamorphism came about through the early twentieth century work of Goldschmidt and Eskola, who developed the concept of metamorphic facies. Simple and elegant, this concept states that two rocks with the same bulk composition, subjected to the same pressure and temperature in two different metamorphic belts, should produce the same metamorphic assemblage. Implicit in this principle is the assumption that rocks rigorously obey the laws of physical chemistry and thermodynamics. Prior to the advent of precise methods for pressure-temperature estimation, petrologists used semiquantitative techniques to subdivide the *P-T* plane into regions characterized by particular metamorphic assemblages. These regions are known as metamorphic facies. Establishing the metamorphic facies provided petrologists with a framework for characterizing the pressures and temperatures of metamorphism on the basis of observed assemblages in rocks.

The traditional view was that metamorphism was caused by simple burial of rocks, which produced pressure increase through weight of overburden and temperature increase along a normal geothermal gradient. The development of quantitative techniques for making estimates of metamorphic temperature and pressure, along with acquisition of a database of such estimates from many metamorphic belts, has altered this simple view. In reality, peak metamorphic pressure and temperature in many, if not most, metamorphic terranes do not actually lie along normal geothermal gradients. Furthermore, plotting the recorded *P-T* points for any terrane that contains a range of metamorphic grades does not accurately reconstruct a normal geothermal gradient. Beginning in the late 1970s, petrologists used heat flow calculations to model the actual paths followed by rocks in the *P-T* plane as they are buried, heated, metamorphosed, up-

lifted, and returned to the surface as a result of erosion. These *actual* paths are not geothermal gradients, but comparison of them with geothermal gradients provides insight into dynamics and heat conduction in the crust, and into the complex interrelationships of metamorphism and tectonism.

METAMORPHIC ISOGRADS

The concept of metamorphic isograds derives directly from the work of Barrow in the eastern Highlands of Scotland in the 1890s, for which he introduced the concept of metamorphic zones. Barrow noted that among rocks of crudely similar composition (mudrocks) near Aberdeen, certain aluminous minerals first appeared in a systematic way that was related to increasing metamorphic grade (Figure 19-1). Later work confirmed his analysis and extended the mapping of metamorphic zones over much of Scotland and beyond. The mineralogic zones described below occur in pelitic rocks that progressively change from slates to phyllites, mica schists, and gneisses as metamorphic grade increases:

Chlorite zone. The typical mineral assemblage is quartz + chlorite + muscovite + albite.
Biotite zone. The typical assemblage is quartz + muscovite + biotite + chlorite + albite.
Garnet zone. Typical assemblage: quartz + muscovite + biotite + garnet + sodic plagioclase.
Staurolite zone. Typical assemblage: quartz + muscovite + biotite + garnet + staurolite + plagioclase.
Kyanite zone. Typical assemblage: quartz + muscovite + biotite + garnet + kyanite + plagioclase ± staurolite.
Sillimanite zone. Typical assemblage: quartz + muscovite + biotite + garnet + sillimanite + plagioclase.

FIGURE 19-1

• • • • • • • • • • • • • • •

Generalized geologic map of
Barrovian metamorphic zones
mapped in metapelites from Glen
Esk and Glen Clova, eastern
Highlands, Scotland. [From
D. W. D. Yardley, 1989, *Introduc-
tion to Metamorphic Petrology*
(New York: Wiley), Fig. 1.2.]

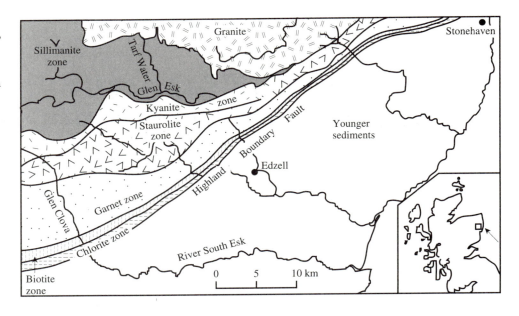

Each mineralogic zone is characterized by the presence of a new mineral that was not present in the previous zone; this new mineral is termed an **index mineral.** After an index mineral has formed, it may persist well beyond the first appearance of the next index mineral. A line drawn on a map to represent the locations on the earth's surface of samples marking the first appearance of an index mineral is called an **isograd.** An isograd is presumed to represent an equivalent metamorphic grade along its entire length because the appearance of an index mineral marks a mineralogic reaction driven by changes in the environmental variables pressure, temperature, and fluid pressure and composition. Note, however, that equivalent metamorphic grade does not imply equal values of the environmental variables. The reactions that lead to the appearance of index minerals are relatively complex functions of mineral chemistry and bulk rock chemistry, as well as of the environmental variables. Some isograds, or zone boundaries, may simply represent shifts (perhaps subtle ones) in the chemistry of the protolith (that is, sedimentary facies changes). It is also possible that pressure, temperature, and water pressure vary independently, perhaps varying significantly and abruptly over short distances. Therefore isograds should not be assumed to provide absolute

values for environmental variables or the relative constancy of values along an isograd.

An isograd, when drawn as a line on a map, represents the intersection of a three-dimensional metamorphic reaction *surface* with the earth's surface (Figure 19-2). Both of these surfaces are likely to be complex curved surfaces and not true planes, so isograds are typically curved lines on a map. Mapping mineralogic

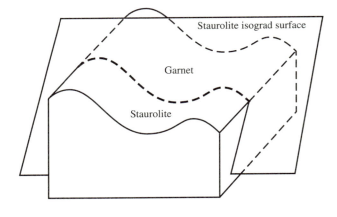

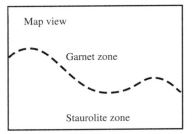

FIGURE 19-2

• • • • • • • • • • • • • • • • •

Three-dimensional schematic illustration showing a dipping,
planar isograd surface intersecting the ground surface to create
a curved isograd as in a metamorphic map.

changes in metamorphic rocks by using isograds is usually accomplished fairly easily because the changes are generally rather abrupt and the isograds tightly constrained. Sometimes there may be practical difficulties such as limited exposures or a paucity of rocks of appropriate bulk composition. Subtle shifts in rock bulk composition can cause the reaction surfaces to be "smeared out" rather than sharp, and the mineralogic zone boundaries can thus expand into narrow zones themselves. Despite the potential difficulties, isograd mapping is a useful and standard field technique widely done by metamorphic geologists because it provides reconnaissance-level information about variation in metamorphic grade that serves as a guide to more detailed study. The concept of the isograd and mineral reaction surfaces is an important one and is discussed in more detail in Chapters 20 and 21.

METAMORPHIC FACIES AND FACIES SERIES
• • • • • • • • • • • • • •

Metamorphic Facies

It is hard to exaggerate the importance of the concept of metamorphic facies on the twentieth-century development of metamorphic petrology. It has provided a "uniformitarian" principle that underlies the development of the systematic characterization of metamorphism. The fundamental basis of facies is the observation that the number of minerals that occur abundantly in metamorphic rocks is strictly limited, both within individual rocks and across the spectrum of rock compositions. Metamorphic rocks are thus equilibrium chemical systems and not random accumulations of minerals that formed at different stages in the development of the rock.

Early in this century, the Norwegian geologist V. M. Goldschmidt examined a group of pelitic, calcareous, and sandy hornfelses in the Oslo region of Norway. These Paleozoic rocks had been intruded and contact metamorphosed by silicic plutons. In spite of their wide range of chemical compositions, the hornfelses of the inner contact zone tended to be mineralogically simple, that is, each of the quartz-bearing hornfelses contained no more than five or six major minerals. In addition, certain mineral pairs were consistently present (for example, orthopyroxene and anorthite), whereas other chemically equivalent pairs such as cordierite and diopside were absent. Goldschmidt noted that mineral assemblages in the pelitic hornfelses typically contained minerals rich in aluminum (plagioclase, cordierite, and an Al_2SiO_5 polymorph, either andalusite or sillimanite),

whereas calcareous hornfelses contained minerals rich in calcium, magnesium, and iron and poor in aluminum (for example, diopside, hornblende, and wollastonite), except for calcic plagioclase, which contained most of the aluminum in the rock. Noting this close correspondence of mineral assemblage with rock composition, Goldschmidt inferred that these rocks had achieved chemical equilibrium under metamorphic conditions.

Goldschmidt's observations near Oslo were confirmed in a study of contact metamorphic rocks from the Orijärvi mining region of southwestern Finland by the Finnish geologist Pentti Eskola (1915). Eskola found a simple and consistent relationship between mineral assemblages and rock compositions in a wide range of metamorphic rock types. Like Goldschmidt, he used this observation as a basis for inferring chemical equilibrium at elevated pressure and temperature. Some of the rocks in the Orijärvi and Oslo districts that were chemically equivalent produced the same assemblages in both places. However, other chemically equivalent rocks produced a different suite of minerals. For example, metapelites from Oslo contained potassic feldspar ($KAlSi_3O_8$) and cordierite [$(Mg,Fe)_2Al_4Si_5O_{18}$], whereas in Orijärvi they contained muscovite [$KAl_3Si_3O_{10}(OH)_2$] and biotite [$K(Mg,Fe)_3AlSi_3O_{10}(OH)_2$]. Eskola concluded that because the chemistries of the rocks were similar, only differences in intensity or character of metamorphism could explain the differences in mineralogy. He correctly predicted that the Orijärvi rocks (which contained abundant hydrated minerals) had formed at lower temperature and higher pressure than the relatively dehydrated rocks at Oslo.

On the basis of these observations, and the additional knowledge that essentially identical metamorphic mineral assemblages occur in similar lithologies worldwide, Eskola developed and elaborated the concept of **metamorphic facies.** This is a scheme for describing and classifying metamorphic rocks on the basis of their mineralogy. As stated by Eskola, "In any rock of a metamorphic formation which has arrived at a chemical equilibrium through metamorphism at constant temperature and pressure conditions, the mineral composition is controlled only by the chemical composition. We are led to a general conception which the writer proposes to call metamorphic facies" (1915, pp. 114–115).

Eskola intended his system as a largely descriptive way of establishing the *relative* pressures and temperatures of different metamorphic assemblages. He had no way of knowing precisely what range of pressure and temperature applied to each of his facies; that information was to come later from both field and laboratory research that used techniques not available to Eskola. He did recognize *relative* pressure-temperature ranges of different facies, however, by making the correct assumption that denser minerals occurred at higher pressures of metamorphism and that progressively less hydrous (and ultimately anhydrous) minerals occurred at higher temperatures. Although metamorphic facies were defined descriptively on the basis of recurring mineral assemblages, these assemblages and the defined facies should now be thought of as indicators of ranges of pressures and temperatures within which different rock compositions yield different but consistent mineral assemblages.

Eskola's five original facies were designated mainly on the basis of names of metamorphosed basaltic rocks because the terrane in which he worked consisted largely of various basaltic volcanic protoliths. His original short list of facies has since been expanded considerably by later workers to include a number of other facies named largely on the basis of significant mineralogy. Figure 19-3 shows the currently accepted major facies and their approximate regions of the *P-T* plane. It should be emphasized that there are no rigid pressure-temperature

boundaries to designated facies—all boundaries tend to be gradational and subjective. Some are rather specifically mineralogically defined, even though the pressure-temperature limits remain fuzzy. For example, the greenschist-amphibolite boundary is commonly defined as the first appearance of hornblende, in addition to actinolite, in mafic schists or gneisses. The amphibolite-granulite facies boundary is a particularly problematic one. Some petrologists define it as the first occurrence of potassic feldspar plus sillimanite (resulting from muscovite breakdown), whereas others use potassic feldspar plus orthopyroxene (from biotite breakdown); there may be a difference of 50°–100°C between these two definitions. Although the granulite facies is commonly thought to be characterized by anhydrous minerals only, the hydrous minerals hornblende and other amphiboles are certainly stable within the lower temperature part of the granulite facies, and perhaps biotite as well, depending on defined boundaries. Table 19-1 lists the typical minerals that occur within the various facies for the four major types

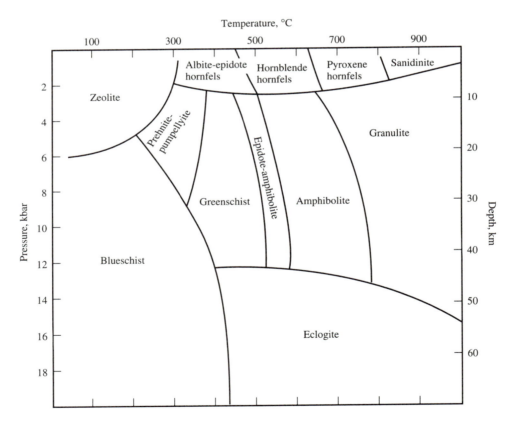

FIGURE 19-3
• • • • • • • • • • • • • • • •

Plot of the *P*(depth)-*T* plane showing current status of the distribution of the metamorphic facies. Note that the low-pressure hornfels facies is subdivided into three subfacies: albite-epidote, hornblende, and pyroxene; the highest contact metamorphic facies is the sanidinite facies. Various subfacies names have been proposed for most of the major metamorphic facies. Note that all boundaries are approximate and gradational.

TABLE 19-1 Generalized guide to characteristic minerals for principal rock compositional types in the various metamorphic facies[a]

• •

Facies	Mafic rocks	Ultramafic rocks	Mudrocks	Calcareous rocks
Zeolite	Analcime, Ca-zeolites, prehnite, zoisite, albite	Serpentine, brucite, chlorite, dolomite, magnesite	Quartz, clays, illite, albite, chlorite	Calcite, dolomite, quartz, talc, clays
Prehnite-pumpellyite	Chlorite, prehnite, albite, pumpellyite, epidote	Serpentine, talc, forsterite, tremolite, chlorite	Quartz, illite, muscovite, albite, chlorite (stilpnomelane)	Calcite, dolomite, quartz, clays, talc, muscovite
Greenschist	Chlorite, actinolite, epidote or zoisite, albite	Serpentine, talc, tremolite, brucite, diopside, chlorite, (magnetite)	Quartz, plagioclase, chlorite, muscovite, biotite, garnet, pyrophyllite, (graphite)	Calcite, dolomite, quartz, muscovite, biotite
Epidote-amphibolite	Hornblende, actinolite, epidote or zoisite, plagioclase, (sphene)	Forsterite, tremolite, talc, serpentine, chlorite, (magnetite)	Quartz, plagioclase, chlorite, muscovite, biotite, (graphite)	Calcite, dolomite, quartz, muscovite, biotite, tremolite
Amphibolite	Hornblende, plagioclase, (sphene), (ilmenite)	Forsterite, tremolite, talc, anthophyllite, chlorite, orthopyroxene, (magnetite)	Quartz, plagioclase, chlorite, muscovite, biotite, garnet, staurolite, kyanite, sillimanite, (graphite), (ilmenite)	Calcite, dolomite, quartz, biotite, tremolite, forsterite, diopside, plagioclase
Granulite	Hornblende, augite, orthopyroxene, plagioclase, (ilmenite)	Forsterite, orthopyroxene, augite, hornblende, garnet, Al-spinel	Quartz, plagioclase, orthoclase, biotite, garnet, cordierite, sillimanite, orthopyroxene	Calcite, quartz, forsterite, diopside, wollastonite, humite-chondrodite, Ca-garnet, plagioclase
Blueschist	Glaucophane, lawsonite, albite, aragonite, chlorite, zoisite	Forsterite, serpentine, diopside	Quartz, plagioclase, muscovite, carpholite, talc, kyanite, chloritoid	Calcite, aragonite, quartz, forsterite, diopside, tremolite
Eclogite	Mg-rich garnet, omphacite, kyanite, (rutile)	Forsterite, orthopyroxene, augite, garnet	Quartz, albite, phengite, talc, kyanite, garnet	Calcite, aragonite, quartz, forsterite, diopside
Albite-epidote hornfels	Albite, epidote or zoisite, actinolite, chlorite	Serpentine, talc, tremolite, chlorite	Quartz, plagioclase, muscovite, chlorite, cordierite	Calcite, dolomite, quartz, tremolite, talc, forsterite
Hornblende hornfels	Hornblende, plagioclase, orthopyroxene, garnet	Forsterite, orthopyroxene, hornblende, chlorite, (Al-spinel), (magnetite)	Quartz, plagioclase, muscovite, biotite, cordierite, andalusite	Calcite, dolomite, quartz, tremolite, diopside, forsterite
Pyroxene hornfels	Orthopyroxene, augite, plagioclase, (garnet)	Forsterite, orthopyroxene, augite, plagioclase, Al-spinel	Quartz, plagioclase, orthoclase, andalusite, sillimanite, cordierite, orthopyroxene	Calcite, quartz, diopside, forsterite, wollastonite
Sanidinite	Orthopyroxene, augite, plagioclase, (garnet)	Forsterite, orthopyroxene, augite, plagioclase	Quartz, plagioclase, sillimanite, cordierite, orthopyroxene, sapphirine, Al-spinel	Calcite, quartz, diopside, forsterite, wollastonite, monticellite, akermanite

[a]Note that these are not necessarily assemblages, but only the minerals likely to occur in each facies. Especially characteristic trace minerals are given in parentheses.

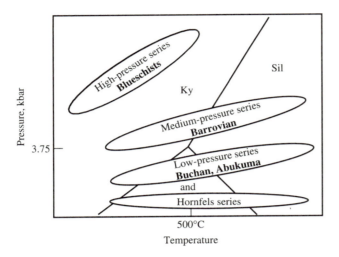

FIGURE 19-4
● ● ● ● ● ● ● ● ● ● ● ● ● ● ● ●

Illustration of approximate metamorphic field gradients for various facies series. The aluminum-silicate triple point is shown for reference, and its temperature (500°C) and pressure (3.75 kbar) are indicated. Note that the low-pressure facies series (Abukuma from Japan, Buchan from Scotland) are characterized by an andalusite (And) → sillimanite (Sil) progression in metapelites, whereas the Barrovian facies series characteristically shows a kyanite (Ky) → sillimanite (Sil) progression. Regional occurrence of cordierite also typifies low-pressure facies series, whereas garnet is typical of higher pressure series.

of rock compositions. Note that this is not a list of specific assemblages; rather it indicates the key minerals that may occur in appropriate rock types within the limits of any facies.

As a research tool, the facies concept began to fade somewhat in the 1980s, particularly for the higher grade metamorphic rocks, because of the development of quantitative techniques for pressure-temperature estimation. However, it remains important and quite useful when relatively precise pressure-temperature determination is not routinely available and serves as a valuable field tool for quick characterization of relative metamorphic grade. The facies or zonal concept still finds considerable application in very low to low grades of metamorphism (see, for example, Turner 1981) where quantitative thermometers and barometers are not generally available. The petrologic literature is replete with references to metamorphic facies, and a thorough understanding of this concept is essential for all students of petrology.

Facies Series

Some metamorphic petrologists found it useful to extend the metamorphic facies concept one step further by a consideration of *sequences* or *series* of facies that occur within particular regions or terranes. Akiho Miyashiro (1961, 1973, 1994) and F. J. Turner (1981) have been particularly active in defining these series. The basis of the facies series idea is that the peak metamorphic conditions within any region without special tectonic complexities tend to lie in a fairly narrow band in the *P-T* plane, with gradually increasing pressures as temperature increases, that is, a shallow positive slope (Figure

19-4). A particular metamorphic terrane may thus be characterized as a "low-pressure," "medium-pressure," or "high-pressure" terrane, depending on how high in pressure this pressure-temperature band lies.

Commonly, facies series have been given geographic names corresponding to the regions where they occur, with one prominent exception. The classic medium-pressure (5–8 kbar) facies series occurring in the eastern Scottish Highlands was originally described by George Barrow and therefore is called the *Barrovian* facies series. One relatively low-pressure facies series was named for a magmatic arc terrane characterized by low-pressure metamorphic mineral assemblages, the *Abukuma* Plateau in Japan. The determination of pressure-temperature trends for many different areas was summarized by Turner (1981) (Figure 19-5). Carefully note that these pressure-temperature gradients are *not* geotherms, nor are they *actual P-T* paths taken during metamorphism, as we show in the next section. They are simply lines connecting the recorded pressures and temperatures of the individual rocks across each terrane. They therefore bear no relation to the variation of temperature *with depth* at any place in that metamorphic terrane during metamorphism. They have been called "metamorphic trajectories," "metamorphic geotherms," or "metamorphic arrays" (Yardley 1989), but it is perhaps best to use another term coined for them, **metamorphic field gradients,** to avoid confusion with actual geotherms. The principal value of both facies series and metamorphic field gradients, as we will see, is to show graphically whether high temperatures were achieved at shallow or great depths in a particular terrane.

(A)

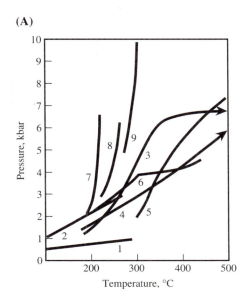

(B)

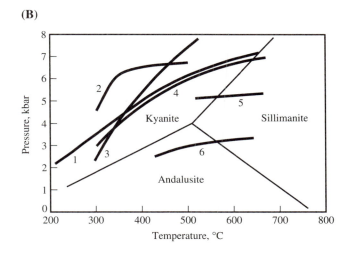

FIGURE 19-5
● ● ● ● ● ● ● ● ● ● ● ● ● ● ● ● ●

Estimated metamorphic field gradients for different metamorphic terranes worldwide. **(A)** Low-temperature, low- to high-pressure terranes. 1, Puerto Rico; 2, Taringatura Hills, New Zealand; 3, blueschist belt, New Caledonia; 4, northern and eastern portions, Haast Schist, New Zealand; 5, Greenschists, Sanbagawa belt, Japan; 6 and 7, northwestern portion, Haast Schist, New Zealand; 8 and 9, sectors of Franciscan blueschist belt, California. **(B)** Medium- to high-temperature terranes. The aluminum silicate triple point is shown for reference. 1, High-temperature portion of the Haast Schist, New Zealand; 2, higher temperature part of the blueschist belt, New Caledonia; 3, high-grade zones of the Sanbagawa belt, Japan; 4, Lake Wakatipu, New Zealand; 5, Lepontine Alps, Italy; 6, northwestern Maine, United States. (Compare with Figure 19-4.) [After F. J. Turner, 1981, *Metamorphic Petrology*, 2nd ed. (New York: McGraw-Hill), Figs. 11-5 and 11-6.]

PRESSURE-TEMPERATURE-TIME MODELS FOR METAMORPHISM
● ● ● ● ● ● ● ● ● ● ● ● ● ● ● ●

Some years ago, it became clear to petrologists that, to understand the interrelationships between metamorphism and tectonism, it was necessary to characterize the actual *P-T* path taken by a particular rock during metamorphism. To understand how both temperature and pressure evolve during the life cycle of a metamorphic event, time, a variable that we have so far ignored, must be considered. Obviously, time is implicit as a variable when the evolution of anything is contemplated, and it is apparent that pressure and temperature are environmental variables that can behave very differently over time. This can be demonstrated quite readily with two simple examples. If an elastic material such as a tennis ball is squeezed, that is, a compressional stress is applied, it will instantly deform into a nonspherical shape. Upon removal of the stress, the ball will immediately return to its originally spherical shape. (If the same thing were done to a plastic material, it would immediately deform, but not return to its original shape after the stress was released.) The response of materials to pressure changes is very rapid, in most cases essentially instantaneous. For example, the elastic pressure (shock) wave from a seismic event travels around the earth in a matter of minutes.

However, if a hand is placed on one side of a sheet of steel 1 cm thick and a blowtorch is applied to the other side, it takes some time, perhaps a few minutes, before the steel becomes too hot to touch. In contrast to stress application, there is a substantial delay for heat (thermal energy) to move through a material. The rate of heat transfer or diffusion through any material is called the *rate of thermal conduction*, and it is widely variable in different materials. Heat conduction is a generally slow process in which thermal energy (in the form of thermal

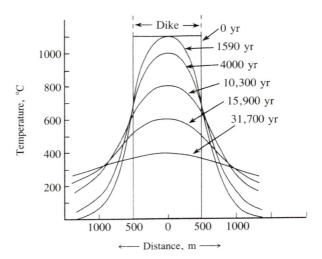

FIGURE 19-6

● ● ● ● ● ● ● ● ● ● ● ● ● ● ● ● ●

Calculated thermal gradients adjacent to a tabular intrusive body, as a function of time. A vertical magma dike 1000 m thick is assumed to have been intruded at 1100°C, the crystallization interval is assumed to be 1100°–800°C, and the country rock was initially at 0°C. Average values of thermal conductivity and heat capacity are assumed. [Based on data from J. C. Jaeger, 1968, in *Basalts*, ed. H. H. Hess and A. Poldervaart, Vol. 2 (New York: Wiley), Fig. 5.]

vibrations) is transferred physically from an atom to adjacent atoms. In general, therefore, it is the density of packing of atoms in a material that controls the thermal conductivity: Most solids conduct heat more rapidly than most liquids, which in turn conduct heat more rapidly than most gases. Metals conduct heat rapidly relative to other solids, whereas rocks are notoriously slow to transfer heat. It is this low thermal conductivity of rocks, coupled with their rapid responses to changes in pressure, that lead to many of the observed characteristics of regional metamorphism.

A Thermal Model for Contact Metamorphism

As an introduction to the quantitative treatment of the interaction of heat and rocks, consider the cooling of an intrusive body and creation of a thermal aureole. This was one of the first petrologic problems to be treated

with thermal modeling that used fairly straightforward techniques of heat flow geophysics. Jaeger (1968) calculated temperature distributions adjacent to a basaltic dike 1000 m thick as a function of time (Figure 19-6), using realistic assumptions about magma temperature and thermal conductivity; all heat transfer is assumed to be by conduction. Note that the temperature rises to no more than 400°C at 1000 m from the contact, and it takes almost 16,000 years for the temperature to increase this much. By 30,000 years, both the dike and the surroundings are beginning to cool down significantly. This thermal event is relatively short in geologic terms but is slow relative to the time it would take to propagate pressure through 1000 m of rock. A larger intrusive body would take longer to cool and would produce a wider and less steep thermal aureole. Note that the maximum temperature at the intrusive contact (700°C) does not approach the initial magma temperature (1100°C), and temperature falls off quite sharply with distance from the contact. This outcome is in good agreement with the distri-

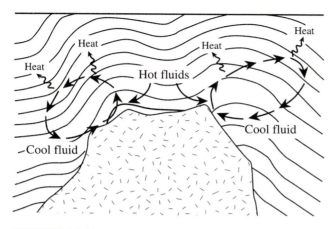

FIGURE 19-7

● ● ● ● ● ● ● ● ● ● ● ● ● ● ● ●

Cartoon of heat and fluid transfer by convection in the proximity of a shallow crustal pluton. Convection cells of circulating meteoric fluids pick up heat from the pluton at or near a contact, then disperse the heat through the aureole during a flow cycle.

bution of contact metamorphic assemblages in observed contact aureoles. A comprehensive treatment of contact metamorphism can be found in the review edited by D. M. Kerrick (1991).

This simple model for heat transfer assumed pure **conduction** of heat, but heat transfer in thermal aureoles and in general may occur through other, more rapid, means as well (*radiation*, which is probably not petrologically important except for cooling of lava flows, will not be considered further). A second mechanism, **convection**, is the cyclical transfer of heat in a liquid or fluid, typically in the form of a closed cell, driven by a local thermal gradient. In the case of a thermal aureole around a pluton, convection cells can be set up if the country rocks surrounding the pluton are permeable enough to allow fluid flow (through either connected porosity or fracture permeability), with the pluton as the heat source that drives the flow (Figure 19-7). Cool fluids flow inward toward the pluton, become heated, then flow outward, carrying heat with them. They gradually lose their heat to cooler rocks as they flow outward, then return to the igneous contact to complete the cycle. This transfer of thermal energy in a moving medium (fluid) is considerably more rapid than conduction, and the channelized flow can cause both localized hot spots in some thermal aureoles as well as metasomatic alteration and skarn formation. A major proportion of the fluid circulating through contact aureoles may be derived by direct

emanations from the pluton, particularly for more hydrous magmas such as granite.

A Thermal Model for Regional Metamorphism

A third mechanism for thermal energy transfer in the earth is **advection**, a process similar to convection. In advection, however, there are no closed cells, and heat is moved by essentially unidirectional flow of a moving medium. In some relatively easily envisioned cases of advective heat transfer, heat flow occurs as hot fluids move up through the crust; these hot fluids are either magmas or metamorphic hydrothermal fluids. The existence of both surface hot springs and volcanism indicates that hot aqueous fluids or magmas can move through cooler surrounding material in a channeled fashion, along pipes or conduits, while losing some but not all of their excess heat to the surroundings. Like convection, this form of energy transfer is more efficient and considerably more rapid than conduction. Advection can be combined with conduction as a mechanism for heating larger volumes of crust, depending on flow rates for the fluids. Country rock between the magma or aqueous fluid conduits can be heated slowly by conduction in a process more akin to contact metamorphism.

Advective heat transfer can also occur when deep, hot rocks are moved upward in the crust more rapidly than they can cool by conduction, a somewhat more difficult process to visualize. This heat transfer process occurs when rates of tectonic uplift are rapid relative to upward conductive heat loss. The process results in the attainment of unusually high temperatures at mid- and shallow-crustal depths, and is the basis for exciting developments in thermal modeling of regional metamorphism.

Beginning in the 1970s and early 1980s, a collaboration between geophysicists and petrologists produced the first realistic quantitative models for how regional metamorphic terranes heat up. Particularly important papers that reinterpreted the thermal evolution of metamorphic belts were those of Oxburgh and Turcotte (1974), England and Richardson (1977), and England and Thompson (1984). Good reviews of the topic can be found in Nisbet and Fowler (1988) and Peacock (1989).

The recorded or measured variations of pressure and temperature across a particular metamorphic terrane tend to form a curvilinear array in the *P-T* plane (see Figures 19-4 and 19-5) called a *metamorphic field gradient (MFG)*. However, the actual path in the *P-T* plane taken by an individual rock from initial burial to final return to the surface through erosion looks quite different from the MFG (Figure 19-8). The paths, such as those

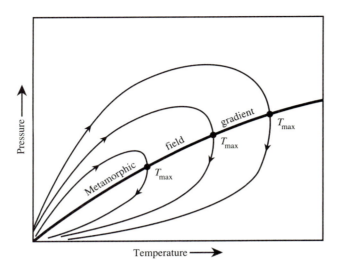

FIGURE 19-8

• • • • • • • • • • • • • • • •

Schematic pressure-temperature diagram showing the actual P-T paths taken by various rocks as they are buried and heated. Burial, heating, and uplift phases of each path are indicated. Also shown is the metamorphic field gradient (see Figure 19-5) that effectively connects the T_{max} points on each P-T path. Many types of petrologic information can be used to constrain these paths: (1) mineral inclusions in porphyroblasts; (2) "peak" P and T from mineral equilibria; (3) reset mineral equilibria, stable isotope exchange, isotopic cooling ages; and (4) range of fluid inclusion entrapment. [Modified from F. S. Spear, J. Selverstone, D. Hickmott, P. Crowley, and K. V. Hodges, 1984, *Geology*, Fig. 1; and E. D. Ghent, M. Z. Stout, and R. R. Parrish, 1988. In E. G. Nisbet and C. M. R. Fowler, *Heat, Metamorphism and Tectonics. Mineralogy Association of Canada Short Course Handbook 14*, Fig. 6.2.]

in Figure 19-8, have been proposed for sedimentary or volcanic rocks progressively buried in deepening depositional basins, but similarly shaped paths can be produced by tectonic thickening through thrust faulting. These paths can be divided into three sections: the burial phase, the heating phase, and the uplift phase. Of course, all paths begin and end at the origin in the P-T diagram, because this point represents conditions at the earth's surface. There is an essentially infinite number of paths for any depositional environment, limited only by the maximum depth of burial of the bottommost beds in the sequence. There is only one point on each path that reflects the pressure and temperature at the time of equilibration of the peak metamorphic mineral assemblage, and it is typically assumed to be the point of maximum temperature (T_{max}). The curved shapes of the paths result from the complex mathematics of the thermal modeling but can be explained in terms of the physical processes involved, by using Figure 19-9.

The steeply sloped *burial phase P-T* paths are due to a suppression of heating during burial. Suppression occurs because cool, relatively fluid-free rocks are buried faster than they can heat up (remember the slowness of conductive heating). Therefore the burial geotherm (increase of temperature as a function of depth, and thus pressure, *at a fixed time*) reflects low temperature, even at deeper levels (see Figure 19-9); in other words, compression (pressure increase) dominates over heating. As the basin matures and both subsidence and deposition rates decrease, conduction of heat from below causes the rocks at any given level to heat up without changing their depth, that is, heating becomes dominant over compression. The rise in temperature along this part of the path is a normal response of the earth, as physical processes attempt to restore the "nor-

mal" or equilibrium geotherm. The geotherms that correspond to this time have thus evolved to a shallower P-T slope, reflecting the input of heat and the higher temperature at any given depth. This phase is the *heating phase* of the metamorphic evolution along a given P-T path.

The third stage in metamorphic evolution—the *uplift phase*—follows the heating phase and results from the isostatic response to crustal thickening. During deposition, the crust loses isostatic balance, and a driving force for isostatic rebound and uplift is built up. As the rocks heat and expand in the heating phase, their density decreases, further contributing to uplift driving forces. As uplift begins, the surface of the sedimentary-volcanic pile ceases to be a topographic low and develops positive relief. It is extremely important to note here that uplift alone cannot cause the pressure decrease illustrated in Figure 19-9, as long as the thickness of the overlying column of rock remains the same. Because pressure is a function of the mass of overlying material, it can decrease only if material is removed at the surface through erosion. The early portion of the uplift phase may actually be isobaric (of constant pressure and thus a horizontal line in Figure 19-9) if mountain building is occurring and before the mountains develop sufficient relief for the erosion rate to balance the uplift rate. Eventually, something approaching a steady-state balance of uplift and erosion rates is established and pressure decreases. Some petrologists have called this process *unroofing*, to distinguish it from uplift, because it involves both physical movement of a rock mass toward the surface and concomitant pressure decrease (decompression) through removal of overburden.

Figure 19-9 also illustrates an interesting phenomenon: Individual rocks continue to heat up *after* they have begun the decompression parts of their paths, for two

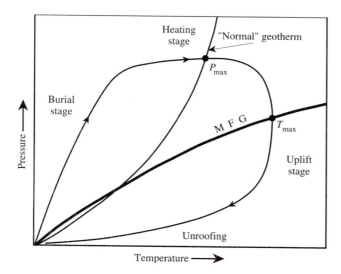

FIGURE 19-9
● ● ● ● ● ● ● ● ● ● ● ● ● ● ● ● ● ●

P-T diagram illustrating actual processes involved in producing a typical "clockwise" metamorphic *P-T* path. Metamorphic field gradient is labeled (MFG).

related reasons. First, the heating phase involves the change of the geotherm toward an equilibrium slope, and this continues after uplift has started. Second, decompression during unroofing proceeds faster than thermal equilibration, which means that the evolving geotherms effectively "overshoot" the equilibrium value and reflect unusually high temperatures for given crustal depths. This is the advective heating process in operation, and it can be thought of as a hot slab of rock being physically moved upward in the crust. Eventually, cooling proceeds and uplift rates subside as isostatic equilibrium is approached. The final portion of the uplift phase is therefore characterized by *P-T* paths in which cooling predominates over decompression. In the absence of this process of advection, only magmatic heating on a regional scale can provide the heat necessary to increase temperatures regionally at midcrustal depths.

FIGURE 19-10
● ● ● ● ● ● ● ● ● ● ● ● ● ● ● ● ●

Counterclockwise *P-T* path produced by regional-scale magmatic heating in typical low-pressure, Buchan-style terranes. This mode of *P-T* evolution has been suggested to explain regional, anhydrous granulite-facies terranes. Metamorphic field gradient is labeled (MFG).

Thermal Evolution of Granulite Facies Terranes

Some petrologists have proposed that certain metamorphic terranes may have undergone a quite different evolution of pressure and temperature, particularly the Precambrian granulite facies terranes such as the Canadian shield and other cratonic shields (Bohlen 1987; Harley 1989). Petrologic evidence in these terranes points to an early rapid increase in temperature at relatively low pressures (shallow depths), succeeded by pressure increase and ultimately by cooling. The shape of this path in the *P-T* plane is illustrated in Figure 19-10, a shape that is often referred to as "counterclockwise" (note that this figure also shows pressure increasing *upward* in the *P-T* plane) to distinguish it from the "clockwise" path shown in Figure 19-9. There is no reasonable thermal scenario for creating a counterclockwise *P-T* path with simple burial. Such paths clearly appear to require magmatic heat input early in their evolution. Crustal extension and thinning have been invoked as precursors for the partial melting of the upper mantle and upward movement of basaltic magmas into the lower crust.

Reasons for the later pressure increase are unclear but probably relate to secondary melting in the crust to form granitoid magmas. The separation of these magmas from their metamorphic residua, and the upward movement and ultimate shallow emplacement of them as batholith-sized plutons, result in a greater thickness of crustal material above the metamorphic rocks, with a concomitant pressure increase. The thermal inversion of the crust also produces the especially slow cooling observed in granulite terranes because of the blanketing

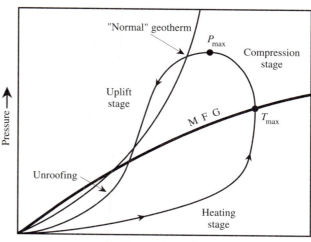

effect of hot magmatic rocks at shallow depths. The special petrologic consequences of ultrahigh-grade metamorphism and melting are examined in Chapter 23.

INFLUENCE OF TECTONICS ON METAMORPHIC *P-T-t* PATHS
• • • • • • • • • • • • • • • •

Variations in tectonic or geodynamic patterns are the ultimate causes for contrasts in *P-T-t* paths between metamorphic terranes. The tectonic patterns reflect the unique combinations of characteristics and processes involved in the plate interactions that result in metamorphism, including lithospheric thicknesses, convergence rates, directions of motion of convergent plates, and detailed subduction or obduction characteristics. Research in metamorphic petrology is largely devoted to calibrating the effects of various plate tectonic factors on *P-T-t* paths in young metamorphic belts such as the Himalaya. Once these influences are better known, reconstruction of the tectonic or geodynamic evolution of older and more complicated (dissected or polydeformed) metamorphic belts

can be accomplished through observation of the detailed metamorphic patterns.

Convergent Margins with Subduction

The majority of active tectonic belts are of this type, with oceanic lithosphere being subducted either under continental lithosphere (for example, Kamchatka, the Andean margin, and the Pacific margins of Oregon, Washington, British Columbia, and southeastern Alaska) or under island arc complexes (for example, New Guinea, the Japanese islands, Indonesia). Miyashiro has used this type of metamorphic environment to define the concept of facies series described above. He also presented the idea that the metamorphic consequence of subduction under an island arc complex or continental margin was the formation of *paired metamorphic belts*, and he cited the Sanbagawa and Ryoke belts of southern Japan as the best example (Figure 19-11). Although Miyashiro proposed additional paired belts, later work has shown that most do not fulfill the strict criteria for paired belts nearly as well as the Sanbagawa and Ryoke belts do.

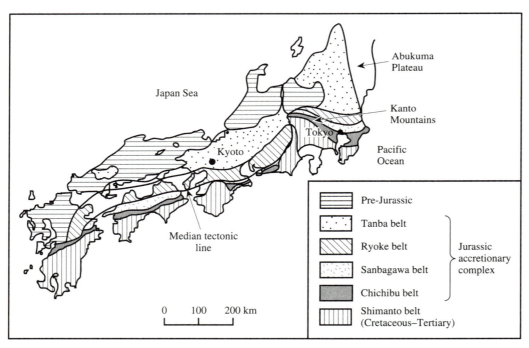

FIGURE 19-11
• • • • • • • • • • • • • • • •

Tectonic sketch map of southwestern Japan. Japan is composed of accretionary complexes of generalized geologic ages as shown. The Jurassic accretionary complex is made up of unmetamorphosed or very weakly metamorphosed Tanba and Chichibu belts, as well as the Ryoke low-pressure metamorphic belt and the Sanbagawa high-pressure belt. [From A. Miyashiro, 1994, *Metamorphic Petrology* (New York: Oxford University Press), Fig. 9.2.]

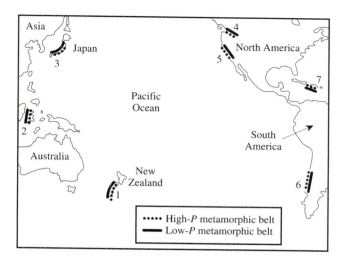

FIGURE 19-12
● ● ● ● ● ● ● ● ● ● ● ● ● ● ● ●

Some of the proposed paired metamorphic belts in the circum-Pacific area, including the Wakatipu (high-*P*) and Tasman (low-*P*) belts in New Zealand (1); unnamed belts in Sulawesi (2); Sanbagawa and Ryoke belts in Japan (3); Shuksan and Skagit belts in Washington State (4); Franciscan and Sierra Nevada belts in California (5); Pichilemu and Curepto belts in Chile (6); and Mt. Hibernia and Westphalia belts in Jamaica (7). [From A. Miyashiro, 1994, *Metamorphic Petrology* (New York: Oxford University Press), Fig. 9.1.]

Paired metamorphic belts are parallel high-pressure and low-pressure belts of the same age and are typically separated by major faults. In Miyashiro's classic model for Japan, the Jurassic-Cretaceous Sanbagawa belt contains a predominance of mafic volcanic lithologies mixed with mudstones, with high-pressure and relatively low-temperature metamorphic assemblages in both mafic and pelitic lithologies. Although it shows many typical characteristics of high-pressure metamorphism, it does not have blueschists, and later work has shown considerably higher pressures in blueschist-eclogite belts in other places. In contrast, the adjacent Ryoke belt of the same age is dominated by metamorphosed clayey sandstones and mudstones, with only minor mafic volcanics. Mineral assemblages are characteristic of the low-pressure, high-temperature metamorphism that is typically driven by magmatic heat input. This belt also has abundant granitic plutons.

The high-pressure belt in Miyashiro's model is typically located in the accretionary wedge between a subduction zone and an island arc or magmatic arc, whereas the low-pressure belt is up to several hundred kilometers farther inland from the plate margin. Notice that the paired belts around the Pacific Ocean (Figure 19-12) are almost always parallel to the plate boundary over extended distances, with a consistent symmetry of low-pressure facies series on the landward side and high-pressure facies series facing the ocean. The distribution of these metamorphic belts must be related to subduction beneath island arc complexes or continental margins. Miyashiro illustrated this with a schematic cross section of a convergent margin (Figure 19-13). Subduc-

tion of old, cold oceanic lithosphere results in a depression of isotherms along and above the subducting slab, with the result that artificially low geothermal gradients are produced in the vicinity of the trench and the upper part of the subduction zone and accretionary wedge (also refer to Figure 9-5). This low geothermal gradient, created and maintained by high subduction rates, allows sedimentary and volcanic material to be subducted to significant depths without the normal heating that such burial would produce. Preservation of high-pressure, low-temperature metamorphic rocks apparently requires rapid exhumation. If they are held at depth until subduction ceases, heating would ultimately destroy the high-pressure assemblages and produce more normal amphibolite-facies rocks. This process is one likely reason that very ancient blueschists are rarer than more modern examples.

The low-pressure belt lies within the island arc and backarc region (in the case of subduction under an island arc) or in the magmatic arc (in the case of continent-edge subduction). Figure 19-13 shows the island arc model, with magmas rising off the deeper levels of the subduction zone. As we discussed in Part I, these basaltic and andesitic magmas are formed by partial melting of both the slab itself and the overlying metasomatized mantle, and they very efficiently transfer large amounts of heat into the crust. Secondary melting in the lower crust generates granitoids that complete the heat transfer process and ultimately cause the shallow, low-pressure metamorphism.

In modern island arc regions above subduction zones, it is usual to find a nonvolcanic, relatively undisturbed

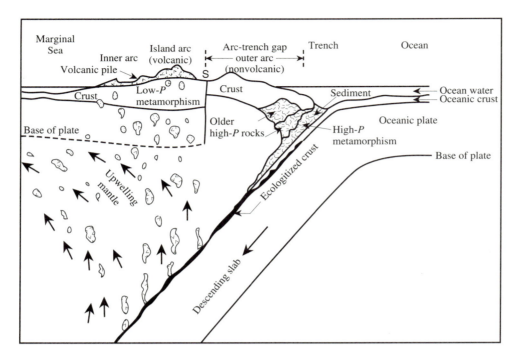

FIGURE 19-13

● ● ● ● ● ● ● ● ● ● ● ● ● ● ● ●

Schematic cross section of a typical western Pacific modern subduction zone with
arc-trench petrologic features. This figure is particularly relevant to the Japanese
arc from the Miocene to the present. [After A. Miyashiro, 1973, *Metamorphism and
Metamorphic Belts* (London: Allen and Unwin), Fig. 3-11.]

region about 100 to 250 km wide between the high-
pressure metamorphic rocks in the accretionary wedge
near the trench and the low-pressure rocks associated
with the volcanic arc and backarc regions. This region
has been called the arc-trench gap (see Figure
19-13). This gap may be present as either uplifted mountains or a trough in which sedimentation occurs. Such a
separation of paired belts is found in the western United
States between the high-pressure Franciscan coastal
rocks (blueschists and eclogites) and the Sierra Nevada
igneous-metamorphic complex farther inland. However,
in some older orogenic belts, the paired metamorphic
belts are immediately adjacent to each other. In such
cases it is common to find a major fault located at a
slight angle to the tectonic trend. These faults typically
have major components of strike-slip movement and
probably represent oblique compressional collapse of
the original paired system with continued convergence.
It was noted earlier that ancient blueschists are rare, and
thermal processes are one possible explanation for their
failure to survive without exceptionally rapid exhumation. It is also likely that older, especially Precambrian,
high-pressure metamorphic belts have undergone at
least one younger episode of metamorphic and deformational reworking in their positions at continental edges.

They may thus have been both thermally overprinted and
modified as well as tectonically dissected.

Continent-Continent Collision

The ultimate end result of closure of oceanic basins and
the termination of subduction is the collision of thicker
continental lithosphere on the two opposed plates. The
geologic record shows numerous such examples of collision zones and major crustal shortening, for example,
the Appalachians, the central Rockies, the Pyrenees
(southwestern France), and the Urals (central Russia).
These zones occur on virtually every continent. The most
striking recent example is the closure of the Tethys sea
to create the Alpine-Himalayan chain that extends from
western Europe to southeastern Asia. The metamorphic
patterns in collision zones are typically far more complex than the relatively simple patterns discussed in the
preceding section. The extent of shortening and the long
history of convergence leading up to collision mean that
metamorphic rocks created in several different environments are jumbled together in the final assembly. A thin
leading edge of continental rocks on the subducting plate
may be deeply underthrust, with resultant very high pres-

sure metamorphism of continental rocks at mantlelike pressures as high as 30 kbar. Some rocks, such as dissected bits of paired belts, probably relate to the early convergence events, whereas others were part of collision itself or related to postcollision magmatism.

THE ALPINE EXAMPLE. The young (Cenozoic) Alpine-Himalayan belt has metamorphic features that are useful examples of the characteristics of many older collision zones. Because of the extensive crustal shortening involved in the collision event, many rocks have been expelled from the axial zone of the orogenic belt in the form of structural nappes which traveled to the north and west. The northern- and westernmost parts of the European Alps (the external or Helvetic zone) (Figure 19-14) consist of folded but unmetamorphosed or only weakly metamorphosed sedimentary rocks tectonically mixed with slices of crystalline basement. Weak or incipient metamorphism (no higher than zeolite to greenschist facies) is characteristic of similar foreland thrust belts in other mountain systems, including the Appalachians and Rockies. South and east of the Helvetic zone is the Penninic zone, which contains the highest grade rocks of Alps, generally of amphibolite facies. Scattered areas of ultrahigh-pressure rocks occur within this zone, some of which are among the highest pressure materials yet found at the earth's surface. Some metagranites in the Sesia-Lanzo zone contain potassic feldspar and quartz together with unusual high-pressure pseudomorphic replacements of plagioclase that contain jadeite and

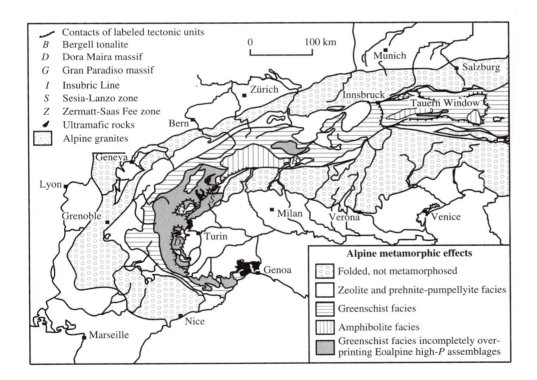

FIGURE 19-14
● ● ● ● ● ● ● ● ● ● ● ● ● ● ● ● ●

Map of the European Alps showing generalized metamorphic zones and some notable localities and tectonic elements. Note the restriction of higher grade metamorphism to the fairly narrow Penninic zone (greenschist- and amphibolite-facies areas, with associated high-pressure massifs). In the areas north and west of Turin, the Penninic zone and high-pressure massifs are incompletely overprinted by a late Alpine greenschist-facies retrograde event. The Helvetic zone (foreland thrust belt) at the northern edge of the Alps is shown as "folded, not metamorphosed," up to prehnite-pumpellyite facies. [From B. W. D. Yardley, 1989, *Introduction to Metamorphic Petrology* (New York: Wiley), Fig. 7.10, after M. Frey et al. (1974), Fig. 1.]

zoisite, an assemblage indicating pressures of at least 10–12 kbar. Pelitic schists contain the high-pressure assemblage talc + phengite, as well as magnesian garnets which have inclusions of coesite, the high-pressure polymorph of silica. This latter coesite occurrence suggests pressures of 30 kbar or more (see Figure 4-1), which corresponds to crustal depths of at least 90–100 km.

The Penninic zone and its ultrahigh-pressure parts (for example, the Sesia-Lanzo zone, the Dora Maira and nearby Monte Rosa massifs, and the Zermatt-Saas Fee zone) include not only metamorphosed platform and ocean basin sedimentary rocks but also oceanic crust and upper mantle (ophiolitic slivers) and slices of old granitic and granulite basement, all mixed tectonically with normal greenschist- and amphibolite-facies rocks. Although very high pressures have been recorded in some rocks, high temperatures have not, which indicates at least transient geothermal gradients with very steep P-T slopes (that is, high $\Delta P/\Delta T$). Late overprinting of many Alpine rocks, including high-pressure ones, by a late, low-grade greenschist facies event is widespread (Figure 19-14) and apparently reflects the termination of collision. Alpine metamorphic events span a significant range of time in western Europe, from some early, apparently subduction-related high-pressure events in Switzerland and Italy as old as 90 Ma, to more typical amphibolite facies metamorphism in Austria as young as 30–35 Ma. Similar long durations of metamorphism have been documented in the Himalayas, with metamorphism (and granitic plutonism) apparently continuing today in some areas.

THE NORTHERN APPALACHIAN EXAMPLE. Both the complexity of metamorphic patterns described earlier and long durations of metamorphism appear to be typical of collision zones. For example, the Paleozoic closing of an ocean basin (the so-called Iapetus Ocean or the ancestral Atlantic) between North America and Africa began with subduction-related metamorphism as old as 525 Ma in the Appalachians. This metamorphism, and associated deformation, is called the Taconic orogeny and occurs along virtually the entire length of the mountain belt from Newfoundland to Alabama. Where subduction-related rock types have been documented, the Taconic subduction zone appears to dip toward the east, although later orogenic events have caused substantial modification of the original geometry and alternate interpretations have been presented. From west to east, there was a trench, an arc-trench gap of unknown width, and an island arc above the east-dipping subduction zone. The involvement of oceanic crust and upper mantle in the collisional event is recorded in ophiolite emplacement (particularly in Newfoundland and Quebec; see Chapter 8). As noted, later tectonic events have substantially obscured the geometry of the Taconic arc-trench system and especially what lay east of

the backarc region. Episodic metamorphism and deformation throughout the entire Appalachian chain continued until the late Paleozoic, including the strong Acadian event at 420–370 Ma (strongest in the northern Appalachians) and culminating in the Alleghanian event throughout the chain at 320–270 Ma. According to the current model, these traditional orogenies actually were not totally separate events but were orogenic pulses within a continuum of convergence that lasted for roughly 200 million years and culminated in continental collision between the North American and either West African or South American cratons. At some places in the Appalachians during these events, convergence appears to have been roughly orthogonal and to have resulted in classic compressional tectonic features such as major thrust faults. In other places, a substantial degree of oblique collision occurred, and major strike-slip and even normal faults dominated. All these events were part of the assembly of the Pangaea supercontinent, which was subsequently rifted apart in the Mesozoic, beginning about 200 Ma, a tectonic event that left behind the current North Atlantic passive margin. Argument continues about which continental plate—Eurasia, Africa, or South America—was involved in each of the various Paleozoic collisions along the length of the Appalachians.

The obvious plate tectonic complexity of the Appalachians is reflected by the metamorphic complexity of the chain, beginning with the Taconic (Ordovician) event. Metamorphic grade ranges from greenschist facies to amphibolite and granulite facies, and relict blueschists have even been found in the Taconic accretionary wedge in Vermont, although they are poorly preserved and have been widely overprinted by later Taconic amphibolite-facies metamorphism. Taconic eclogite-facies rocks occur in western North Carolina, and claims have been made for similar eclogites in the northern Appalachians, although these reports remain controversial. There is little evidence, however, for any low-pressure metamorphic counterpart to the high- and intermediate-pressure facies series. If such low-pressure Taconian rocks existed, they have been either overprinted by Acadian or later amphibolite-facies metamorphism or rifted away in the Mesozoic. Amphibolite-facies metamorphism occurs throughout the Piedmont of the central and southern Appalachians, with scattered greenschist-facies areas (in eastern Pennsylvania and New Jersey) and granulite-facies areas (western North Carolina and southeastern Pennsylvania near Philadelphia). Where P-T paths have been deduced for Taconic rocks, they are virtually always of the "clockwise" or normal type with decompression heating.

Metamorphic patterns of the northern Appalachians are shown in Figure 19-15. Acadian metamorphism in the northern Appalachians shows substantially greater

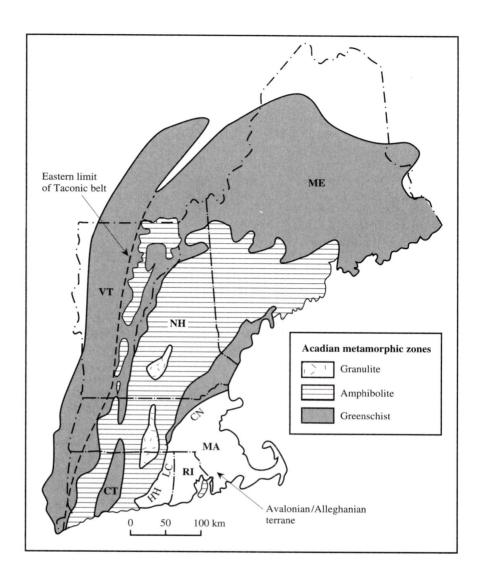

FIGURE 19-15
● ● ● ● ● ● ● ● ● ● ● ● ● ● ● ●

Metamorphic map of New England showing the generalized distribution of peak metamorphic zones during the Acadian (Devonian) orogenic event. Rocks indicated as greenschist include Barrovian chlorite, biotite, and garnet zones; amphibolite includes staurolite through sillimanite zones; granulite indicates sillimanite + potassic feldspar and higher. The western half of the Acadian belt (western Connecticut, western Massachusetts, and Vermont) is of high-pressure Barrovian type, whereas the eastern Acadian (eastern Connecticut, central Massachusetts, New Hampshire, and Maine) is of lower pressure Buchan style. Two areas of granulite-facies rocks in this belt occur from northern Connecticut through southern New Hampshire. At its western edge, Acadian metamorphism is superimposed or overprinted on Taconic (Ordovician) metamorphic rocks of greenschist through upper amphibolite grade. The eastern limit of essentially undisturbed Taconic metamorphism is indicated by the dashed line, but note that Taconic effects originally must have extended farther east before Acadian overprinting. The southeastern limit of Acadian metamorphism is a series of late Paleozoic faults (HH, Honey Hill fault; LC, Lake Char fault; CN, Clint-Newbury fault) that bound the southeastern New England Avalonian (late Precambrian) igneous and metamorphic terrane. Both this terrane and parts of the Acadian terrane (especially the southern edge) are overprinted by greenschist to lower amphibolite facies Alleghanian (Pennsylvanian-Permian) metamorphism. [Adapted from F. S. Spear, 1993, *Metamorphic Phase Equilibria and Pressure-Temperature-Time Paths* (Monograph 1, Mineralogical Society of America), Fig. 21-15, based on an unpublished compilation by W. E. Trczienski.]

diversity of behavior than Taconic metamorphism. In parts of western Connecticut and Massachusetts and southern Vermont, Acadian metamorphism overprints Taconic metamorphic assemblages in older rocks. (This process of metamorphic overprinting seems to be particularly common in continental collision zones.) Subduction zone-related metamorphism (blueschist facies, for example) appears to be absent, but high-pressure metamorphism has been documented along the western margin of the Acadian belt, west of the old, inactive Taconic island arc. This high-pressure metamorphism is, however, both high pressure *and* high temperature (up to 11 kbar and 750°C) and resulted in upper amphibolite, rather than blueschist-facies, rocks. Clockwise *P-T* paths have been documented, and magmatic rocks are rare. In contrast, east of the island arc, Acadian metamorphism is of very different character. In the highest grade rocks, recorded pressures are rarely more than 5–6 kbar at temperatures approaching 800°C. Such metamorphic conditions require magmatic heat input, and indeed the region contains abundant intermediate and granitic plutonic rocks. It has been proposed that at least some of these magmas originated in or above the Taconic subducted lithosphere slab that lies below this terrane. In east-central New England and adjacent Canada, counterclockwise *P-T* paths have been found in granulite-facies rocks, an observation confirming the relationship between magmatism and these very high grade rocks.

The final stage of Appalachian evolution was the late Paleozoic Alleghanian orogeny, of which the most notable feature is the Valley and Ridge Province of the central and southern Appalachians. Although some amphibolite-facies metamorphism is known in Rhode Island (sillimanite zone in metapelites), most Alleghanian metamorphism throughout the Appalachians is low grade, rarely exceeding greenschist or lowermost amphibolite facies. In many areas this pervasive low-grade metamorphism overprints older, higher grade metamorphic rocks of either Taconic or Acadian age. In this regard, the Alleghanian metamorphism is strikingly similar to the late greenschist-facies overprint in the Alps.

SUMMARY

The concept of metamorphic facies grew out of the essentially simultaneous discoveries by Goldschmidt in Norway and Eskola in Finland that rocks rigorously obey chemical laws during metamorphism. Metamorphic assemblages in specific rock compositions thus may serve as predictors of a relatively narrow range of pressure and

temperature of formation. With facies series, Miyashiro extended this concept of metamorphic facies to include the idea that sequences or series of facies within particular regions reflected the relative pressures at which prograde metamorphism occurred. He based his notion on observation of paired metamorphic belts in the Pacific rim, where quite different sequences of assemblages formed during metamorphism in adjacent low-pressure and high-pressure belts.

Refinement of theories of thermal evolution during metamorphism has led to the prediction of actual pressure-temperature-time curves for metamorphism in different tectonic environments. Crustal thickening by repeated thrust faulting or sedimentary basin filling results in clockwise *P-T* paths with initial burial and compression at anomalously low temperatures, followed by passive heating at depth and finally by isostatic uplift, unroofing, and consequent cooling and decompression. This metamorphic cycle ultimately brings an individual metasedimentary or metavolcanic rock back to the surface where it started, but in a mineralogically transformed state. Where metamorphic heating has occurred through magmatic heat input, a reverse or counterclockwise *P-T* path can occur, with development of high temperatures at unusually shallow crustal depths. This latter pattern appears to be typical of many regional-scale granulite-facies terranes. Modern or geologically young convergent plate margins tend to show relatively simple pressure-temperature-time behavior of these types, whereas older belts may be obscured by multiple events. Continental collision zones, which are typical of many of the earth's most classic mountain chains such as the Alps, Himalaya, and Appalachians, show very complex pressure-temperature behavior as the result of long (multi-hundred-million-year) tectonic evolution and consequent telescoping and juxtaposition of diverse original metamorphic environments.

STUDY EXERCISES

1. What are the classic Barrovian metamorphic zones, and what mineralogic effect characterizes each boundary between the zones? What are these boundaries called, and what do they represent?
2. What principle is the basis for the metamorphic facies concept? Do metamorphic facies provide precise information on *P* and *T* of metamorphism, or more relative or general values?
3. How high can contact metamorphic temperatures get, relative to the temperature of the magma?

4. Which are the three most important mechanisms for heat transfer within the earth? Which is slowest and which is fastest?

5. What is a "metamorphic field gradient"? How does it relate to metamorphic facies series? How does MFG for a particular metamorphic terrane depend on the actual *P-T* paths taken by individual rocks?

6. How do the different types of metamorphism (for example, high-*P*, low-*T*; high-*T*, low-*P*) relate to the different plate tectonic regimes in which metamorphic belts can occur?

7. Contrast the general characteristics of Alpine and northern Appalachian metamorphic patterns.

REFERENCES AND ADDITIONAL READINGS
• • • • • • • • • • • • • • •

Barnicoat, A. C., and P. J. Treloar. 1989. Himalayan metamorphism—An introduction. *J. Metamorph. Geol.*, *7*, 3–8.

Bohlen, S. R. 1987. Pressure-temperature-time paths and a tectonic model for the origin of granulites. *J. Geol.*, *95*, 617–632.

England, P. C., and S. W. Richardson. 1977. The influence of erosion upon the mineral facies of rocks from different metamorphic environments. *J. Geol. Soc. London*, *134*, 201–213.

England, P. C., and A. B. Thompson. 1984. Pressure-temperature-time paths of regional metamorphism. Part 1. Heat transfer during the evolution of regions of thickened continental crust. *J. Petrol.*, *25*, 894–928.

Ernst, W. G. 1988. Tectonic history of subduction zones inferred from retrograde blueschist *P-T* paths. *Geology*, *16*, 1081–1084.

Eskola, P. 1915. On the relations between the chemical and mineralogical composition in the metamorphic rocks of the Orijärvi region. *Bull. Comm. Geol. Finlande*, *44*, 109–145.

Frey, M., J. C. Hunziker, W. Frank, J. Boquet, G. V. Dal Piaz, E. Jager, and E. Niggli. 1974. Alpine metamorphism of the Alps. *Schweiz. Mineral. Petrogr. Mitt.*, *54*, 248–290.

Goldschmidt, V. M. 1911. Die kontaktmetamorphose im Kristianiagebeit. *Vidensk. Skrifter. I. Mat-Naturv. Kl.* 1911, No. 11.

Harley, S. L. 1989. The origins of granulites: A metamorphic perspective. *Geol. Mag.*, *126*, 215–331.

Jaeger, J. C. 1968. Cooling and solidification of igneous rocks. In *Basalts*, ed. H. H. Hess and A. Poldervaart, Vol. 2. New York: Wiley, pp. 503–546.

Kerrick, D. M. 1990. The Al_2SiO_5 polymorphs. *Rev. Mineral.*, *22*.

Kerrick, D. M., ed. 1991. Contact metamorphism. *Rev. Mineral.*, *26*.

Miyashiro, A. 1961. Evolution of metamorphic belts. *J. Petrol*, *2*, 277–311.

Miyashiro, A. 1973. *Metamorphism and Metamorphic Belts*. London: Allen and Unwin.

Miyashiro, A. 1994. *Metamorphic Petrology*. New York: Oxford University Press.

Nisbet, E.G., and C. M. R. Fowler. 1988. *Heat, Metamorphism and Tectonics. Mineralogical Association of Canada Short Course Handbook 14*.

Oxburgh, E. R., and D. L. Turcotte. 1974. Thermal gradients and regional metamorphism in overthrust terrains with special reference to the eastern Alps. *Schweiz. Mineral. Petrogr. Mitt.*, *54*, 641–662.

Peacock, S. M. 1989. Thermal modelling of metamorphic pressure-temperature-time paths: A forward approach. In *Metamorphic Pressure-Temperature-Time Paths. Short Course in Geology*, ed. F. S. Spear and S. M. Peacock. Washington, DC: American Geophysical Union, pp. 57–102.

Spear, F. S. 1993. *Metamorphic Phase Equilibria and Pressure-Temperature-Time Paths*. Washington, DC: Mineralogical Society of America.

Turner, F. J. 1981. *Metamorphic Petrology*, 2nd ed. New York: McGraw-Hill.

Yardley, B. W. D. 1989. *An Introduction to Metamorphic Petrology*. New York: Wiley.

Chapter 20

ASSEMBLAGES, REACTIONS, AND EQUILIBRIUM

The basic unit of any rock is, of course, the individual mineral, and the maximum information about metamorphic rocks is gained through characterization of the different ways that individual minerals combine with one another to make rocks. Detailed examination of metamorphic rocks over the decades has demonstrated that these combinations are far from random. In fact, they are the direct result of the operation of metamorphic reactions and equilibria that obey fundamental laws of chemistry and physics, particularly thermodynamics. In Chapter 4 we discussed equilibrium and thermodynamics in relation to igneous rocks. It was noted there, and should be reemphasized here, that thermodynamics is simply a codification of the rules of the interaction of matter and energy. In particular, it describes how the chemical constituents of a rock parcel themselves into different combinations of minerals, magmas, and fluids under different conditions of pressure and temperature. Thermodynamics is the most important basic tool for the petrologist, and in this chapter we discuss some of the simple rules for its application in metamorphism, particularly pointing out how these rules help us to interpret the minerals and textures that we observe in metamorphic rocks.

COMPONENTS, PHASES, AND ASSEMBLAGES
•••••••••••••••

The minerals found in any rock can be described loosely as a "mineral association" or "mineral assemblage." However, in metamorphic petrology the term **mineral assemblage** has a very precise chemical meaning, specifically referring to a thermodynamically stable configuration of minerals. In order to use the principles of chemistry to understand metamorphic processes, it is necessary to review the basic terminology and principles of chemical thermodynamics and their application to rocks. A chemical *system* is a defined mass of constituent atoms that occupies a certain volume, has definable properties, and interacts in certain ways with its environment. Any metamorphic (or, for that matter, igneous or sedimentary) rock is a chemical system, with certain necessary and useful qualifications. Neither the chemical composition, pressure, nor temperature should vary across the system. If a rock is defined as a chemical system, an appropriately sized piece of rock must be selected so that neither pressure nor temperature could vary in it during metamorphism; in virtually all cases, a normal-sized hand sample is acceptable.

From the point of view of composition, however, the appropriate size of a rock system depends on the nature of the rock. A cube of gneissic rock 1 cm on a side that contains multiple 1-mm compositional layers violates the compositional test for homogeneity. A uniform and homogeneous composition is very important because of the connection between the composition of the system and the number and identity of phases (minerals) that should occur. There are two fundamentally different types of chemical systems of relevance to metamorphic petrology (Figure 20-1). **Closed systems** are those which exchange energy but not chemical mass with their surroundings; an example is a typical regional metamorphic rock, which is presumed to be isochemical during metamorphism but clearly experiences changes in temperature and pressure. **Open systems** exchange both energy and chemical constituents with their surroundings; an example would be a contact metamorphic skarn. For petrologic purposes, volatile species such as water and carbon dioxide are not constrained to remain within a system because of their mobility as fluids, and they can move freely in and out of both types of systems.

The fundamental chemical constituents of any system, including rocks, are referred to as **components.**

Depending on needs, they could be defined in any number of arbitrary ways, but the most typical petrologic method is to use simple oxides such as SiO_2 and Al_2O_3, either by themselves or in more complex combinations. These components are parceled out in the system among the discrete physical entities that are called **phases.** Phases are defined as physically separable parts of the system, and the individual minerals in a rock are as good an example of phases as any, although a phase can be in any physical state: solid, liquid, or gas. If a rock is disaggregated, the grains of different minerals can be physically separated by using any of several methods that take advantage of their differing physical properties, for example, by using a magnetic separator, by using heavy liquids, or simply by hand sorting on the basis of color or appearance. The coexisting phases in a system typically have different chemical compositions, but they need not. If an ice cube tray is partially frozen or thawed, there are two phases, ice and liquid water, and they have an identical composition, H_2O.

A further word about components is in order. Defining components in the precise thermodynamic sense can be arbitrary and complex, but several rules apply. The first and most important is that the *minimum* number of components that adequately describe the system must be defined; by this rule, there can *never* be more components than phases. (There are cases where the number of phases exceeds the number of components, but components can never exceed phases.) As a first simple example, consider a pure quartzite. We could define the components as silicon and oxygen, but because there is only one phase, quartz, actually only one component, SiO_2, can correctly define the system. Another example is a dunite consisting only of olivine ($Fo_{90}Fa_{10}$). There are four possible ways to define components: Mg, Fe, Si, and O; MgO, FeO, and SiO_2; Mg_2SiO_4 and Fe_2SiO_4; and $Fe_{0.2}Mg_{1.8}SiO_4$. Only the last possibility is correct, because there is only one phase in our rock system, olivine. The second rule is that no component can be an algebraic combination of other components. This rule sounds complex, but it is actually straightforward. Suppose a rock consists of the three mineral phases periclase (MgO), enstatite ($MgSiO_3$), and quartz (SiO_2). The three mineral compositions could be defined as three components, according to the rule of not having more components than phases, but this is not correct. Examination shows that $MgSiO_3$ is actually an algebraic combination of the other two: $MgO + SiO_2 = MgSiO_3$. Thus we have only a two-component system $MgO–SiO_2$.

Summarizing briefly, any metamorphic rock can be defined as a chemical system as long as it represents a single protolith bulk composition that had no gradients of pressure and temperature spanning it. Most hand sam-

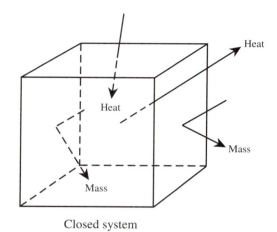

Closed system

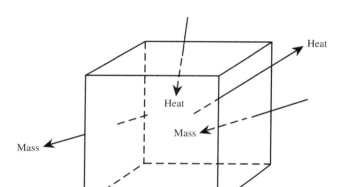

Open system

FIGURE 20-1
● ● ● ● ● ● ● ● ● ● ● ● ● ● ● ● ●

Closed and open chemical systems with relevance to petrology. Closed systems allow thermal energy to pass in and out, but mass is not exchanged with the surroundings. Both energy and mass can be exchanged between an open system and its surroundings.

ples meet this basic criterion. A rock system consists of at least one component, and can have many more: The correct number defined depends on the rock. Typically, however, simple oxides (for example, SiO_2, Al_2O_3, MgO) can be used as components for most rocks. These components are proportioned into phases (minerals), whose identity and number are dictated by thermodynamic relationships discussed below. The term *mineral assemblage* in the thermodynamic sense thus refers to the stable group of mineral phases in a metamorphic rock in a state of chemical equilibrium.

THE CONCEPT
OF EQUILIBRIUM
• • • • • • • • • • • • • • •

As a general concept, the word *equilibrium* connotes stability, and in this sense it is used in fields as diverse as political science, economics, physics, and chemistry. A dictionary definition might suggest that anything in a state of equilibrium, from the exchange rate between currencies to a croquet ball on a flat lawn, has no tendency to change spontaneously. For a change to occur, some net driving force is required. Perhaps more important, for the change to remain permanent, there must be no spontaneous driving force for a return to the initial state or a further shift to a third state.

Consider the example of the croquet ball (Figure 20-2). If a lawn were absolutely flat and horizontal, ball 1 could be pushed in any direction and it would remain where it stopped. Any position on the lawn would be an equally good equilibrium state because the only driving force, other than pushing or hitting the ball, is the force of gravity, and gravity cannot make the ball roll on a horizontal surface. Here the idea of energy can be introduced, which in the physical example is of two kinds: kinetic and potential. Kinetic energy is used to move the ball, but potential energy actually determines equilibrium. All positions on the flat part of the lawn confer equal potential energy to the ball because they are at the same elevation within the earth's gravitational field. The higher the ball is (that is, the farther from the earth's center of mass it is), the greater its potential energy. In the example, the kinetic energy required to move the ball does not contribute to the potential energy but is completely dissipated in moving the ball around.

Suppose, however, that part of the lawn is planar but slopes slightly (see the left side of Figure 20-2). If ball 1 is hit directly uphill, it might well roll back to its initial position when the pushing force stops. This movement pattern occurs because the potential energy of the ball has been increased by moving it upward in the earth's gravitational field. Remember that all spontaneous physical or chemical processes tend to occur in order to *decrease* energy to a minimum possible level, which is defined as the state of equilibrium. If a ball were dropped in the middle of the slope (for example, ball 2), it would roll down the slope until it reached its minimum potential energy on the flat part of the lawn. Ball 2 was in an *unstable* state before it rolled down the slope. Its elevation *h* above the flat part of the lawn contributes to its greater potential energy. Suppose that a third ball (ball 3) is on the sloping part of the lawn, but does not move. Why is the ball there in the first place? It might have been sitting in a small hole or divot, which keeps it from

rolling and allows it to remain indefinitely in what is called a *metastable* state. This condition is not a truly stable equilibrium state but only a temporary state of *apparent* equilibrium. It might only take a small input of energy (called *activation energy*) to activate the process that would lead to an equilibrium state, in this case rolling downhill. Note in Figure 20-2 that ball 3 could roll down to the flat if a small amount of potential energy (activation energy) were added, thereby raising it over the lip of the hole it sits in.

In chemical as well as physical systems, the terms given above are used as expressions of the tendency (or lack of it) for a system to change spontaneously. **Equilibrium** (or stable equilibrium) is the state of a system in which no perceptible spontaneous change occurs. (Of course, in geologic systems one must be cautious in observing perceptible spontaneous change, because "perceptible" in terms of geologic time can be quite different from a human time scale.) It is important to understand that this does not mean that nothing is going on when a system is at equilibrium. Atoms may constantly be leaving one phase and joining another. But equilibrium implies that there is an equal and opposite movement of atoms that *balances* the process and maintains the proportions of the phases without bulk change. A system can easily be changed by perturbing it in some way, but if the conditions are allowed to return to the initial ones,

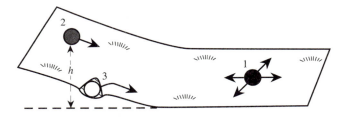

FIGURE 20-2
• • • • • • • • • • • • • • • •

Concepts of equilibrium illustrated by croquet balls on a lawn. Ball 1 on the flat part of the lawn has no tendency to spontaneous movement. It will only move (in any direction) if pushed and will only move as long as the pushing force is maintained. Ball 2 is dropped on the sloping part of the lawn. It will not remain in place but will spontaneously roll downward onto the flat lawn. Its elevation *h* above the flat lawn gives ball 2 excess potential energy, which promotes spontaneous movement. Ball 3 is located in a shallow depression on the slope and remains in place indefinitely until sufficient potential energy is added to it to boost it over the lip of the depression (note arrow), after which it will behave as ball 2 does. Ball 1 represents stable equilibrium; ball 2, an unstable state; and ball 3, a metastable equilibrium.

the system will spontaneously return to its initial equilibrium state. Filling an ice cube tray with liquid water is a simple example of this principle. If thermal energy is removed by placing the tray in a freezer, the water turns to ice. This is the new equilibrium state at a temperature below 0°C. However, if the tray is brought back to room temperature, the ice returns to being liquid water, and no difference from the initial state can be perceived. It is sometimes possible with very pure water to chill it well below 0°C without having it freeze. The liquid water is then in a state of *metastable equilibrium*. Adding a very small amount of energy, perhaps by stirring, almost always causes the water to freeze spontaneously by providing the *activation energy* for the water to proceed to its stable equilibrium state.

APPLICATION OF EQUILIBRIUM CONCEPTS TO METAMORPHIC ROCKS
••••••••••••••••

It is perhaps a long stretch from croquet balls and ice cube trays to metamorphic rocks, but it has become apparent to metamorphic petrologists over the last century that fairly strict rules of chemical equilibrium apply to many, if not most, metamorphic rocks. That is to say, the assemblages of minerals found in metamorphic rocks represent a close approach to a state of equilibrium *under the conditions of metamorphism*. It is clear, of course, that metamorphic rocks exposed at the surface in outcrops are not in equilibrium at the conditions of the surface. We have the opportunity to observe and study them only because the kinetics of their transformation into equilibrium surficial materials are so imperceptibly slow, even in terms of geologic time, until they reach a near-surface weathering horizon. The discoveries of Goldschmidt, Eskola, and their successors that the same rock protoliths have consistently produced similar mineral assemblages when metamorphosed at similar conditions in metamorphic belts around the world have provided strong support for the idea that equilibrium during metamorphism is a normal occurrence. Recent petrologic studies have contributed to the picture by showing just how close is the approach to equilibrium, the processes through which this has occurred, and the kinetics of these processes. Close examination of petrologic processes has consistently shown that with the typically slow kinetics of silicate systems, rapid or abrupt events are more likely to result in disequilibrium phenomena. Lack of equilibrium (that is, disequilibrium) is thus more common in contact metamorphism than in the more gradual process of regional metamorphism.

How do petrologists know that the metamorphic rocks they observe represent equilibrium states? Initial examination of most metamorphic rocks reveals that the textures, structures, and minerals characteristic of sedimentary and igneous rock protoliths have been either substantially or completely obliterated. Profound changes of this sort are not sufficient evidence alone that the rock has completely adjusted to changing conditions, and a deeper look is thus required. Thermodynamic rules specify how many minerals should coexist in any particular rock if the minerals are all in equilibrium with one another. Failure of a rock to achieve chemical equilibrium is commonly reflected by the presence of too many minerals, because of the incomplete transformation of reactant minerals into product minerals in metamorphic reactions. This is a surprisingly rare situation in metamorphic rocks, except in contact metamorphic environments. Another approach to demonstrating equilibrium is based on experimental and theoretical calibration of the stability regions in *P-T* space of different metamorphic minerals and mineral assemblages. This approach requires a database of thermodynamic parameters for calculating mineral stabilities, a computer to do the calculations, and chemical analyses of mineral compositions. The uses for such sophisticated information and techniques will be examined later, but in their absence what can be deduced with only hand samples and a hand lens, or thin sections and a microscope?

The first and most obvious approach is to search in hand sample or thin sections for textural and compositional evidence of equilibrium. One criterion involves examining the thin section to determine whether all the major minerals are in physical contact with one another. Mutual grain boundary contacts of all minerals without obvious evidence of reaction is a first-order indication of equilibrium. Reaction textures such as obvious replacement textures, reaction rims, or scalloped, resorbed rims of porphyroblasts indicate at least some response to changing conditions and may raise the possibility of disequilibrium. Note, however, that a reaction relationship may actually be an indicator that reequilibration has been "frozen" in progress. Reactions clearly occur during metamorphism, and in some cases the cessation of metamorphism and the "freezing in" of assemblages may happen while a reaction is occurring, since it may take hundreds of thousands to millions of years for metamorphic reactions to go to completion.

One texture that can violate equilibrium among all minerals in a rock involves the presence of so-called **armored relics** as mineral inclusions inside poikiloblasts. These apparently occur as a poikiloblast rapidly enlarges and overgrows smaller neighboring grains. In most cases, the inclusions are minerals that also occur in the rock matrix (for example, quartz or iron-titanium oxides) and thus do not present a problem. In a few instances, these armored relics are minerals that no longer exist as stable phases in the rock matrix but apparently did at the time of porphyroblast growth. The general assumption is that they have been completely reacted away in the matrix, where they were not protected by the enclosing mineral. Despite the fact that these minerals occur in the rock, they cannot be shown to be in contact with minerals other than the host poikiloblast, and therefore they cannot be included in the mineral assemblage. They can, however, provide important clues to the rock's earlier metamorphic history that cannot be directly deduced in any other way.

In a strict sense, equilibrium also implies that there should be an absence of compositionally zoned grains. The individual phases, as well as the system, must be chemically homogeneous. By definition, a crystal that is not homogeneous cannot be at equilibrium with itself because of the chemical gradients that exist within it. Such gradients are driving forces for equilibration and prove that a rock containing such zoned minerals has not reached its final equilibrium state. Worse, the central portion of the crystal probably has a composition that is out of equilibrium with the other minerals in the matrix of the rock, analogous to the case of armored relics. Compositional zoning in some solid solution minerals can be seen optically in thin section (for example, in plagioclase) or through differences in color, pleochroism, birefringence, or extinction position. In some cases, however, it is *cryptic*, or hidden, and can be deduced only by using microanalytical instruments such as the electron microprobe.

There are certain metamorphic minerals in which substantial intracrystalline zoning is virtually universal, especially garnet and plagioclase. In all solid solution minerals that become zoned during growth, atoms can move around in the crystals after growth and ultimately erase the zoning. As a result of the natures of the garnet and plagioclase crystal structures, however, atomic movement within them is exceptionally slow, even in geologic time, and zoning typically remains. Some degree of minor compositional variation is virtually universal and indicates simply that equilibrium was closely approached but not perfectly achieved. Where major zoning occurs, it is important to recognize it and account for it in determining the mineral assemblage in the rock. Usually the presence of disequilibrium as mineral zoning does not rule out a thermodynamic interpretation of metamorphic history, as long as it is taken into account. In fact, this extra information can be exploited, as in the case of armored relics, for deduction of certain aspects of the metamorphic evolution of the rock, particularly the P-T path during metamorphism.

It has been assumed that metamorphic mineral assemblages represent equilibrium "at metamorphic conditions." But most metamorphic rocks experience a complete cycle of pressure and temperature from low values at the surface (or near it) to maximum conditions of pressure and temperature, and then back to the surface (see Chapter 19). So *which* set of pressure-temperature conditions is represented by the assemblage in the rock? It is typically assumed that assemblages represent the approximate maximum temperatures achieved during metamorphism, with perhaps minor readjustment in the initial stages of cooling. The common preservation of mineralogic zones and isograds in many metamorphic terranes and the relative regularity of these zones provide evidence for this. Both textural and mineralogic evidence indicate a progressive nature of metamorphism, resulting in what is referred to as **prograde** metamorphism. **Retrograde** metamorphism, or the spontaneous reversion to a lower grade state after the peak of metamorphism, appears to be more unusual. The principles of chemical dynamics suggest that reaction and reequilibration are more likely to occur during heating as the thermal energy in a rock is increasing. Furthermore, the processes of prograde metamorphism during heating typically involve devolatilization reactions as hydrous and carbonate minerals break down. The water and carbon dioxide that are produced are highly mobile and typically leave the rock as a supercritical fluid; thus they are not available for recombination during cooling. In some terranes, patchy or localized retrograde rehydration appears to be associated with entry of meteoric water (essentially deep groundwater) along fractures or fissures.

REACTIONS
AND THE PHASE RULE
• • • • • • • • • • • • • • • •

The assumption that metamorphic mineral assemblages represent at least a close approach to chemical equilibrium allows the use of the phase rule to assess whether a particular assemblage represents a reaction relationship among the minerals. There are several different kinds of metamorphic reactions that have different implications for the interpretation of temperature and pressure in a metamorphic rock. The phase rule was discussed in some detail in Chapter 4 and the reader is referred to this discussion as a review. In this section, we apply the phase rule to metamorphic processes.

On the basis of accumulated data on metamorphic assemblages, V. M. Goldschmidt noted early in this century that certain common metamorphic assemblages can be found worldwide. He reasoned that it would be far too strong a coincidence for all the occurrences of an assemblage to have formed at precisely the same pressure and temperature. So he proposed instead that the observations indicated a small *range* of pressure and temperature for stability of a particular, randomly selected, assemblage. In a phase rule sense, this situation represents **divariant equilibrium,** that is, pressure and temperature can each vary independently without perturbing or changing the mineral assemblage. Petrologists commonly cast this notion in a geometric or analog way. Pressure, temperature, and chemical composition (commonly abbreviated as X) are the three environmental variables of greatest interest to metamorphic petrologists. Boundaries of mineral assemblage stability fields can be shown as a function of their mutual dependence on the variables pressure, temperature, and composition (X) in a three-dimensional P-T-X diagram. Assuming a constant composition, so that X is no longer variable (a reasonable thing to do if only one kind of rock is being considered), the P-T plane becomes a very useful two-dimensional way to portray assemblage relationships (Figure 20-3). In this type of phase diagram, a divariant assemblage occurs over an area in the P-T plane. In contrast, the lower variance assemblages can exist only under more geometrically restricted P-T conditions: **univariant** assemblages along lines and **invariant** ones at points.

To illustrate the way the P-T diagram works, consider a one-component system that is of importance to metamorphic petrology, the system Al_2SiO_5 illustrated in Figure 20-3. Equal molecular amounts of the simple oxides Al_2O_3 and SiO_2 can be combined into this new single component because there are several mineral phases that have the composition of the combined component, specifically, sillimanite, andalusite, and kyanite. Each of these Al_2SiO_5 polymorphs has a region of stability in the P-T plane, and the phase rule specifies how these stability fields intersect. The phase rule is written $F = c - p + 2$, and if divariance ($F = 2$) is specified, it becomes $2 = c - p + 2$. Therefore, in a divariant system, the number of phases should equal the number of components, or $p = c$, which in our case is 1, and this single phase should occur stably over an *area* of the P-T plane. Each different phase of the same composition must occur in a different region of the plane. These *divariant regions* are separated by *univariant lines* (see Figure 20-3) along which pairs of phases can coexist (if there is one component and two phases, the phase rule says the variance is 1). Univariance in the P-T plane means that *both* pressure and temperature cannot be changed inde-

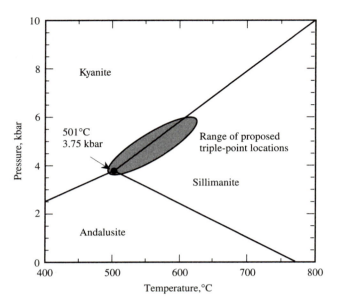

FIGURE 20-3
● ● ● ● ● ● ● ● ● ● ● ● ● ● ● ● ●

P-T diagram for the system Al_2SiO_5. The positions of the kyanite $\rightleftharpoons$ sillimanite, kyanite $\rightleftharpoons$ andalusite, and andalusite $\rightleftharpoons$ sillimanite reactions are taken from Holdaway (1971), the most widely used triple point. Other suggested locations for the triple-point intersection lie within the shaded area.

pendently if the rock system is to remain the same. If temperature is changed, then pressure must be changed simultaneously by the exact value that restores the P-T point to the line, or a mineral phase will be lost by reaction. Similarly, the even greater restriction of invariance means that *neither* pressure *nor* temperature can be changed at all and have the rock remain unchanged. Geometrically this is a P-T point, referred to as an *invariant point*. In a one-component system, where three phases coexist at an invariant point, this point is commonly called a *triple point*.

The P-T diagram for the system Al_2SiO_5 thus consists of three divariant regions, separated by three univariant lines that must intersect in one invariant point (see Figure 20-3). The univariant lines represent the three metamorphic reactions andalusite $\rightleftharpoons$ kyanite, kyanite $\rightleftharpoons$ sillimanite, and andalusite $\rightleftharpoons$ sillimanite. The phase rule itself says nothing about where the triple point is located precisely in the P-T plane or how the univariant lines are oriented (that is, their slopes). The physical and thermodynamic properties of the individual minerals (particularly entropy and molar volume) control the exact locations of mineral stability fields. In the case of Al_2SiO_5, for example, it is known from experiments that kyanite is the high-pressure polymorph because it is the densest of the three; and sillimanite is the high-temperature one because it has the highest entropy. The locations of the univariant lines and triple point shown in Figure 20-3 have been determined experimentally. The simplest possible chemical-rock system has only one component (like that in Figure 20-3), but more complex systems behave in similar ways. Adding extra components and phases simply requires adding more invariant points and more univariant reaction lines. The result is called a **petrogenetic grid** (Figure 20-4); and although it looks complex, it obeys the same basic rules as the one-component system. Realistic and complex petrogenetic

grids for metamorphosed pelites or for calcareous rocks can have as many as a hundred or more univariant lines and tens of invariant points. The areas between the lines and points are the divariant stability fields of typical metamorphic assemblages. Because these areas may be quite limited regions in the *P-T* plane (see Figure 20-4), a well-constrained petrogenetic grid is a very useful tool for estimating the *P-T* conditions for metamorphic rocks.

A typical simplified example of a sequence of metapelite assemblages for a prograde *P-T* path is as follows. For the horizontal *P-T* path shown in Figure 20-4,

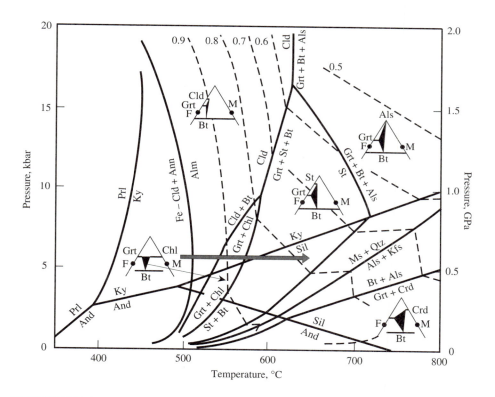

FIGURE 20-4

• • • • • • • • • • • • • • •

Petrogenetic grid for metapelites in the six-component system $K_2O–FeO–MgO–Al_2O_3–SiO_2–H_2O$. All listed reactant and product assemblages may coexist with any of the phases quartz, muscovite, potassic feldspar, or aqueous fluid. Reactants are shown on low-temperature sides of equilibria and products on high-temperature sides. Ferromagnesian minerals shown are considered to be iron-magnesium solid solutions, except for any curve listed as a mineral end member. Dashed lines give labeled ratios of $Fe/(Fe + Mg)$ in garnet that coexists with biotite in any divariant field where garnet and biotite coexist stably. AFM diagrams show assemblages containing garnet + biotite with other AFM minerals, as shown. For explanation of the horizontal *P-T* path (arrow), see text. Mineral abbreviations: Alm, Fe-garnet; Als, aluminum silicate; And, andalusite; Ann, Fe-biotite; Bt, biotite; Chl, chlorite; Cld, chloritoid; Crd, cordierite; Grt, garnet; Ky, kyanite; Kfs, potassic feldspar; Ms, muscovite; Prl, pyrophyllite; Qtz, quartz; Sil, sillimanite; St, staurolite. [From F. S. Spear and J. T. Cheney, 1989, *Contrib. Mineral. Petrol.*, *101*, Fig. 4.]

the initial assemblage (at about 480°C and 6 kbar) might be chloritoid (Cld) + biotite (Bt) (all assemblages given here include quartz and muscovite as well). On crossing the reaction labeled Fe − Cld + Ann = Alm at 500°C, iron-rich garnet (Grt) joins the assemblage, which is now garnet (Grt) + chloritoid (Cld) + biotite (Bt). The dashed lines show how the Fe/(Fe + Mg) of garnet decreases gradually as the temperature increases. Next, the reaction Cld + Bt = Grt + Chl is crossed at about 530°C; this represents the disappearance of chloritoid from biotite-bearing rocks, and indicates a common assemblage of garnet (Grt) + chlorite (Chl) + biotite (Bt). At 570°C, staurolite (St) appears through the reaction Grt + Chl = St + Bt. As all these reactions proceed, the garnet gradually becomes much less iron-rich than the first crystals that formed back at 500°C. At the high-temperature end of the P-T path (about 640°C at 6 kbar), the likely assemblage is garnet (Grt) + staurolite (St) + biotite (Bt) (plus muscovite and quartz); if the composition is aluminous enough for an aluminum silicate mineral to be present as well, it would be sillimanite rather than kyanite. Note that the mineral assemblage changes through reaction as the path passes through the different areas of the grid, and the assemblage in a particular rock could thus be used to predict where on this path that rock formed.

According to the phase rule, no equilibrium assemblage should contain more phases than the number of components plus 2 (an invariant assemblage). Even this kind of assemblage should be very rare, because it is extremely unlikely that a system would exist at exactly that combination of pressure and temperature that matches those at the invariant P-T point. Similarly, univariant lines, which represent reaction assemblages (that is, a reaction has been caught in progress), reflect assemblages that contain one phase in excess of the number of components. A univariant reaction assemblage should also be unusual, although not as rare as an invariant one, because of the special requirement for P-T conditions that fall on a line in the P-T plane. The most general and probable situation is to find divariant assemblages in metamorphic rocks. A geologist who goes into the field to collect a metamorphic rock should therefore expect to find a divariant assemblage with the same number of phases as components, because divariant assemblages cover a far greater area in P-T space than the more restricted lower-variance ones. This is not to say that a univariant or invariant assemblage cannot be found, but it does mean that the probability of finding one is low. Goldschmidt and Eskola recognized this probability argument: The so-called Goldschmidt mineralogic phase rule (as distinct from the true Gibbs phase rule) is written $p \geq c$, which means that we normally expect the number of phases to equal the number of components; but in rare cases, p can exceed c (but by no more than 2).

As rare as they are, invariant and univariant mineral assemblages are of special value to petrologists because they indicate very restricted pressure and temperature for formation of these assemblages. In some regional and contact metamorphic situations, isograds mapped in the field can represent the reactions (univariant lines or, rarely, invariant points) that separate divariant assemblages in adjacent mineralogic zones. With an accurate and detailed petrogenetic grid, these so-called low-variance assemblages (univariant or invariant) can be used to place fairly precise limits on the P-T variation across a mapped metamorphic terrane.

GRAPHICAL REPRESENTATION OF ASSEMBLAGES AND REACTIONS

The compositional relationships between metamorphic minerals are best and most completely described by using algebraic and geometric techniques, and a major emphasis in metamorphic petrology has been the development and application of these techniques, especially those using computers. However, even prior to Eskola, there is a history of using graphical techniques to show mineralogic relationships in metamorphic petrology. The development and use of compositional phase diagrams for metamorphic rocks essentially paralleled the use of liquidus phase diagrams in igneous petrology, and the principles are essentially the same for both.

Fundamental Principles

The basic principle for graphical analysis of phase equilibrium is that one spatial dimension is required for each compositional degree of freedom, and the number of compositional degrees of freedom is one less than the number of components. The number of compositional degrees of freedom simply means how many variables it takes to completely describe the bulk composition of the rock. This concept sounds more complicated than it is. Consider a one-component system, for example, Al_2SiO_5. A one-component system has a fixed composition and no possible compositional variability; thus it has zero compositional degrees of freedom. It is represented geometrically as a point (a zero-dimensional geometrical figure) (Figure 20-5A). If a second component is added—say, SiO_2—there is now *one* compositional degree of freedom; in other words, specifying the ratio of the amounts of the two components describes the rock adequately. Geometrically, two components are represented as a one-dimensional figure, that is, a line of arbitrary

(A)

(B)

(C)

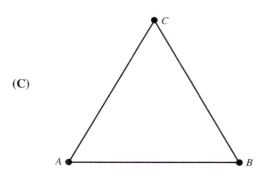

(D)

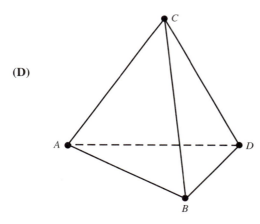

FIGURE 20-5

• • • • • • • • • • • • • • • • • •

(A) Geometric representation of a one-component or zero-dimensional system: a point. (B) Geometric representation of a two-component or one-dimensional system: a line. (C) Geometric representation of a three-component or two-dimensional system: a triangle. (D) Geometric representation of a four-component or three-dimensional system: a tetrahedron.

length with one of the components at each end (see Figure 20-5B). A rock bulk composition can lie anywhere along this line and can be described by using only one variable, X Al_2SiO_5 [the ratio $Al_2SiO_5/(Al_2SiO_5 + SiO_2)$], which varies from 0 to 1.0. This type of system is similar to the binary systems examined in Part I of the book and may or may not have intermediate compounds between

the two end points (there are none in the system $Al_2SiO_5–SiO_2$).

Adding a third component gives *two* compositional degrees of freedom, and therefore two dimensions are required to show the compositional variability. The geometrically simplest two-dimensional figure is a triangle, and an equilateral triangle is typically used to portray a three-component, or ternary, system (see Figure 20-5C). Each component is shown at a corner (or *apex*), and a composition within the ternary system can be specified by using the amounts of any two components (or the ratios of any two pairs). Only two must be specified, because once these two are fixed, the third is not independent but is equal to 100% minus the sum of the other two. This process can continue indefinitely, by adding components and dimensions. For example, a four-component system is shown as a tetrahedron in three dimensions, with the four components at the corners of the tetrahedron (see Figure 20-5D). Unfortunately, it is difficult to show mineralogic relationships graphically in three dimensions (serious artistic abilities are required), and impossible for four or more dimensions.

There are tricks for compressing multicomponent rock systems into two-dimensional triangles. In ternary systems, mineral compositions can involve any combination of the three components, from being pure components (*unary* phases) to being mixtures of two (*binary* phases) or all three components (*ternary* phases). Minerals are typically plotted on the triangular diagram on the basis of the molecular proportions of components in them. For example, in the ternary system $CaO–Al_2O_3–SiO_2$, there are several possible minerals: quartz, corundum, andalusite, kyanite, sillimanite, anorthite, wollastonite, and grossular. This system is shown in Figure 20-6. Quartz and corundum plot on the diagram at the SiO_2 and Al_2O_3 corners, respectively, but there is no mineral with the composition CaO, and thus nothing plots at the CaO corner. It is important to remember that minerals commonly plot at the corners, but a component is not required to be represented by a phase. If, however, carbon dioxide is considered to be present as a *mobile* component (which is not required to be represented graphically as a component), then the mineral calcite plots at the CaO corner (in the presence of carbon dioxide, $CaO + CO_2 = CaCO_3$). Water could be treated similarly, but it has been ignored in this treatment, which can be regarded as simulating possible assemblages in very high temperature anhydrous rocks.

The Al_2SiO_5 minerals andalusite, kyanite, and sillimanite all contain one mole of Al_2O_3 to one mole of SiO_2 in their formulas, so they plot on the edge of the triangle exactly halfway between Al_2O_3 and SiO_2. Note especially that all three aluminum silicate polymorphs plot at this same point, but ordinarily only one of them at a time

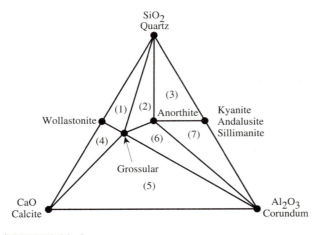

FIGURE 20-6

● ● ● ● ● ● ● ● ● ● ● ● ● ● ● ● ●

Ternary diagram for the anhydrous system $CaO–Al_2O_3–SiO_2$ (with CO_2 in excess). The arbitrary arrangement of tie lines is consistent with assemblages in metamorphic rocks. Numbers refer to assemblages discussed in the text.

is found in a given rock, because the mineralogical phase rule specifies that coexistence of two or three would indicate *special* univariant or invariant conditions. Wollastonite contains one mole each of CaO and SiO_2, so it plots halfway between these components. Anorthite ($CaAl_2Si_2O_8$) is a ternary mineral, that is, it contains some of each of the three components: one mole of CaO, one mole of Al_2O_3, and two moles of SiO_2. Thus anorthite contains 25 mol% CaO (one mole out of four total moles), 25 mol% Al_2O_3, and 50 mol% SiO_2. It therefore plots halfway from the SiO_2 apex toward the opposite side and one-quarter of the way toward each of the other two apexes. Grossular is also ternary but is more complicated than anorthite: $Ca_3Al_2Si_3O_{12}$. There are three moles of CaO, one mole of Al_2O_3, and three moles of SiO_2, or seven total moles of material. The mole fraction of SiO_2 is thus 3/7, that of Al_2O_3 is 1/7, and of CaO 3/7 (all adding up to 7/7, of course). Grossular, therefore, plots near anorthite but farther from the Al_2O_3 apex (see Figure 20-6). Because they have the same ratio of CaO to SiO_2, grossular and wollastonite plot along the same line radial from Al_2O_3. Readers should refer to Figure 3-1 for a reminder of how to plot compositions in ternary diagrams.

Once the positions of the minerals have been plotted, it is necessary to indicate assemblages. Tie lines are drawn between minerals if they coexist stably, that is, if they touch each other in the rock and no reaction occurs along their margins. Indicating assemblages on a compositional diagram makes a fundamental assumption of

equilibrium and thus assumes a fixed pressure and temperature for the assemblage. Compositional phase diagrams are therefore commonly called *isothermal-isobaric* diagrams. Figure 20-6 shows *seven* possible three-phase assemblages in this ternary system, and the one found in a particular rock depends on the subtriangle within which the rock composition falls. (If there are polymorphs that plot at the same point, as in the case of Al_2SiO_5, the one found in an assemblage is the one that is stable under the particular fixed *P-T* conditions of the assemblage.) Assemblages portrayed in such diagrams are limited to restricted ranges of pressure and temperature that fall in the divariant areas of a petrogenetic grid. If pressure and temperature are changed so that a univariant line in the grid is crossed, the assemblages must change and tie lines shift.

One of the basic rules for isothermal-isobaric compositional diagrams is that tie lines must not cross under divariant equilibrium conditions. Crossing tie lines means that four minerals have been described as being in equilibrium in a three-component system. The phase rule specifies that coexistence of four phases in a three-component system indicates univariance or, in other words, indicates that a reaction is occurring. In many cases, there are no choices about where to draw tie lines, but where there are alternative crossing possibilities, as in several cases in Figure 20-6 (for example, grossular + quartz versus anorthite + wollastonite), rock hand samples or thin sections must be carefully examined to decide which of the two alternatives is correct. In Figure 20-6 an anorthite-corundum tie line has been chosen (rather than grossular-Al_2SiO_5) as well as a grossular-quartz line (rather than anorthite-wollastonite). Thus there are seven possible divariant assemblages in the system: (1) quartz-grossular-wollastonite; (2) quartz-grossular-anorthite; (3) quartz-anorthite-Al_2SiO_5; (4) calcite-wollastonite-grossular; (5) calcite-grossular-corundum; (6) grossular-anorthite-corundum; and (7) anorthite-Al_2SiO_5-corundum. The assemblage that would occur in any particular rock depends on where the rock bulk composition falls. In the very unlikely event that a rock composition fell exactly on a line, that rock would have an assemblage consisting only of the two minerals on the tie line.

To summarize, it is convenient to use graphical diagrams to show the possible assemblages that may occur at a given pressure and temperature for different rock compositions within an overall chemical system. For the typical three-component system used in petrology, an assemblage would consist of three minerals. The overall ternary system is thus subdivided into assemblage triangles created by connecting the mineral compositions with tie lines. A compositional diagram of this type applies to a limited region of the *P-T* plane in which all of the divariant assemblages are stable. If pressure and temperature are changed and a univariant line is crossed, some tie lines will shift and a new compositional diagram will apply.

More Complex Systems: ACF, A'KF, and AFM Diagrams

Mineral assemblages in relatively simple rock compositions can be shown graphically, as discussed above, but unfortunately rocks in nature are considerably more complicated. A typical mudrock contains significant proportions of as many as nine or more components (SiO_2, TiO_2, Al_2O_3, FeO, MgO, CaO, Na_2O, K_2O, H_2O) and may contain nine or more minerals. Relationships between phases and components can be dealt with algebraically by using a computer, but most petrologists prefer to find a way to *visualize* these relationships. There are tricks that petrologists have developed to allow compression of the total chemical-mineralogic system into three of four plotting components that allow graphical portrayal in triangles or tetrahedra. Some oversimplification obviously results, but the approach has proved valuable in showing general principles of metamorphic mineral equilibria in progressive metamorphism.

Eskola pioneered the simplified approach with his ACF diagram. Here, both rock and mineral compositions are plotted in terms of three idealized components *A*, *C*, and *F*, which are each combinations of actual components. Eskola's philosophy in creating this diagram was to show only the minerals that appeared or disappeared during metamorphism and thus were useful as indices of metamorphic grade. Coupled with this approach was the use only of components that described the phase relations of these minerals. Minerals that are stable through-out a wide range of metamorphism are thus ignored, and their contributions to the rock composition are ignored. The minerals left out include albite, potassic feldspar, micas, magnetite, sphene, and apatite. Silica can be neglected as a component by assuming silica oversaturation, that is, ubiquitous presence of quartz and enough SiO_2 to create the most silica-rich minerals. The ACF diagram therefore strictly applies only to quartz-bearing rocks, although it is commonly successfully used for rocks in which no quartz occurs, but there are also no silica-undersaturated minerals such as olivine or feldspathoid. The ACF components were defined by Eskola as follows:

$$A = Al_2O_3 + Fe_2O_3 - (Na_2O + K_2O)$$
$$C = CaO - 3.3\,P_2O_5$$
$$F = FeO + MgO + MnO$$

Al_2O_3 and Fe_2O_3 are combined because they are geochemically similar and substitute for each other in solid solution minerals (such as epidote). A similar argument is used for combination of FeO, MgO, and MnO, which are roughly interchangeable in many ferromagnesian minerals. ($Na_2O + K_2O$) is subtracted from the A component to account for the equivalent aluminum that is in albite and orthoclase; if this is not done, a rock composition will plot erroneously high in the A component. The quantity 3.3 P_2O_5 is subtracted from CaO to correct for the number of moles of CaO in the minor mineral apatite. Water and carbon dioxide are neglected for both mineral and rock compositions because they are assumed to be perfectly mobile. Plotting is done by using molecular proportions, which are obtained by dividing the weight percentage of each oxide in the rock chemical analysis by its molecular weight. After the A, C, and F components are calculated, they must be normalized to a

(A)   **(B)**

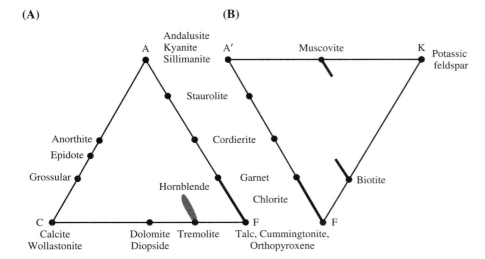

FIGURE 20-7

●●●●●●●●●●●●●●●●●

(A) ACF diagram, with common minerals plotted. **(B)** A'KF diagram, with common minerals plotted. Note that lines indicate solid solution compositional variation in minerals, for example, garnet, muscovite, and biotite.

percentage basis for plotting in a triangle. Thus $A(\%) = 100 \times [A/(A + C + F)]$; $C(\%) = 100 \times [C/(A + C + F)]$; $F(\%) = 100 \times [F/(A + C + F)]$. Mineral compositions are also plotted on the ACF diagram in order to allow interpretation of the rock composition. Figure 20-7A shows and explains the plotting positions of many common metamorphic minerals. Note that minerals can plot either as points, lines, or regions, depending on the degree of solid solution and compositional variability in each mineral: chlorite and hornblende are especially variable, particularly in terms of aluminum content.

A commonly used companion diagram for ACF is the A′KF diagram, developed later by Eskola. It is especially useful for showing phase relationships of potassic minerals (micas, potassic feldspar) in potassium-bearing rocks, particularly in metamorphosed mudrocks (metapelites) and potassic volcanics. A is designated A′ here because it is calculated slightly differently from A in the ACF diagram:

$$A' = Al_2O_3 + Fe_2O_3 - (Na_2O + K_2O + CaO)$$
$$K = K_2O$$
$$F = FeO + MgO + MnO$$

Minerals are plotted in the A′KF diagram in a manner similar to that of the ACF diagram (see Figure 20-7B). The reader is encouraged to do calculations of ACF and A′KF plotting positions for some common minerals to check how these are plotted in Figure 20-7.

A third special compositional diagram was developed by Thompson (1957) and has proved extremely useful in showing mineralogic relationships in metapelites. It is based on the tetrahedron $K_2O-Al_2O_3-FeO-MgO$ and is known as the AFM or AKFM diagram. The advantage of this diagram is that it allows FeO and MgO to be separated and dealt with as separate components. In many metapelites, there are coexisting ferromagnesian minerals with differing values of $Fe/(Fe + Mg)$. Deciphering the phase relations in such rocks depends on portrayal of varying ratios of iron to magnesium.

Representation of a rock or mineral composition in terms of four components (K_2O, Al_2O_3, FeO, and MgO) requires plotting within a three-dimensional tetrahedron with components at the corners of the tetrahedron (Figure 20-8). Because this is difficult to show or to plot accurately on a two-dimensional sheet of paper, Thompson chose to project points within the tetrahedron onto one face (and its extension). Geometric projection is a widely used technique for reducing the dimensionality of a portrayed object. For example, a Mercator-projected map is a projection of a three-dimensional Earth onto a sheet of paper, and a landscape painting uses perspective, a form of projection, to project real objects onto a canvas. Thompson's technique, shown in

Figure 20-8, was to use a point of projection (the muscovite composition) to project all rock and mineral compositions onto the AFM face of the AKFM tetrahedron. This is equivalent to the use of SiO_2 (quartz) as a "projection point" in the ACF diagram and is chemically and graphically correct as long as all rocks and assemblages plotted contain muscovite (that is, are saturated with muscovite). In fact, the AKFM tetrahedron also implies quartz saturation, so the AFM projection should be used only for rocks that contain both muscovite and quartz. Fortunately, this category includes almost all meta-

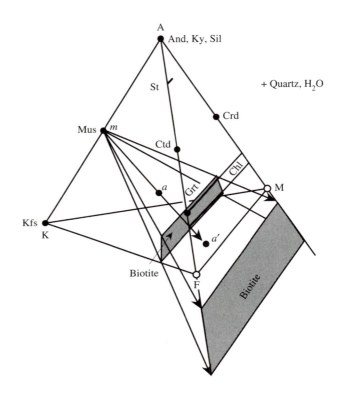

FIGURE 20-8
● ● ● ● ● ● ● ● ● ● ● ● ● ● ● ●

$Al_2O_3-KAlO_2-FeO-MgO$ (AKFM) tetrahedron, with quartz and water in excess. Plotting positions of a number of common metapelite minerals are shown. In order to create the AFM projection, mineral or rock compositions within the tetrahedron are projected from muscovite (Mus) onto the AFM plane, as shown for the arbitrary composition a. Most minerals are already in this AFM plane except for biotite, which projects into a field at negative values of A, as shown. Potassic feldspar projects *away* from the AFM plane, and so is represented in the AFM projection as being at negative infinity. Compare the AFM plane to Figure 20-9A. Mineral abbreviations: And, andalusite; Chl, chlorite; Crd, cordierite; Ctd, chloritoid; Grt, garnet; Kfs, potassic feldspar; Ky, kyanite; Mus, muscovite; Sil, sillimanite; St, staurolite.

(A)   **(B)**

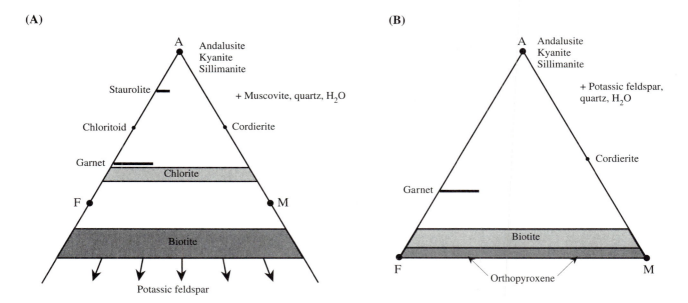

FIGURE 20-9
• • • • • • • • • • • • • • • • •

(A) AFM projection from muscovite in the AKFM tetrahedron. Plotting positions of
common metapelite minerals are shown. Referring to Figure 20-8, note that potassic
feldspar plots at negative infinity when projected from muscovite. **(B)** AFM projection from potassic feldspar in the AKFM tetrahedron. This special projection is used
for high-grade metapelites where muscovite is no longer stable. Common high-
grade (upper amphibolite and granulite-facies) metapelite minerals are shown.

pelites, except at very high grades where muscovite is
unstable and replaced by potassic feldspar. In this case,
potassic feldspar can be used as an alternate projection
point.

The projection technique works as follows (referring
to Figure 20-8). A rock or mineral composition falls
either inside the tetrahedron or only on a face or edge.
Rock compositions typically fall inside, such as that
shown at a. Projection from muscovite, point m, involves
drawing a straight line from m through a until it intersects the AFM plane at a'. Compositions relatively poor
in potassium will project into the positive part of the
AFM plane (the part that makes up one face of the tetrahedron), whereas those richer in potassium (including
the mineral biotite) plot onto the extension of the AFM
face at negative values of A. Many ferromagnesian minerals that have variable ratios of Fe/(Fe + Mg) plot along
horizontal lines at relatively constant values of A. Figure
20-9 shows the AFM projection with a number of common metapelite minerals. Potassic feldspar is the only
metapelite mineral that does not plot at finite values in
the AFM projection. Observe in Figure 20-8 that a line
from muscovite through potassic feldspar plots away
from the AFM plane rather than toward it. Potassic feld-

spar is therefore arbitrarily assigned a position beyond
biotite at negative infinity (see Figure 20-9).

One further point about AFM diagrams should be
made. In ACF and A′KF diagrams, the plotting positions
of minerals are typically fixed by general stoichiometry
of the minerals and do not require a chemical analysis of
the mineral. Prominent exceptions are hornblende, chlorite, biotite, and muscovite (phengite), in which variable
amounts of aluminum can affect the plotting position
(see Figure 20-7A and B). In the AFM diagram, all the ferromagnesian minerals have plotting positions that are
dependent on knowledge of Fe/(Fe + Mg) ratio, as well
as on aluminum content. Because this ratio is typically
not deducible from hand sample or thin section properties of the minerals, a chemical analysis is required for
precise plotting. For schematic representation, an arbitrary value can generally be assumed, one based on
experience of typical relative Fe/(Fe + Mg) values in
metapelites: garnet > staurolite > biotite > chlorite >
cordierite.

Consider how all three types of diagrams can be
applied to representation of the phase equilibria in a rock
such as a metapelite. A relatively common medium-grade
assemblage consists of staurolite, garnet, muscovite,

(A)

(B)

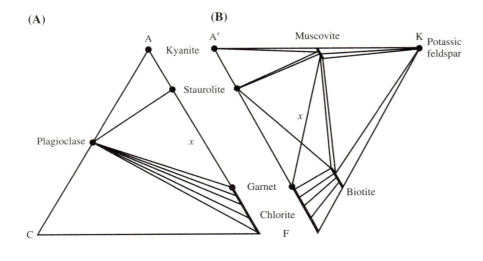

FIGURE 20-10
● ● ● ● ● ● ● ● ● ● ● ● ● ● ● ●

Typical metapelite assemblages at a single middle amphibolite-facies grade, shown on **(A)** ACF, **(B)** A′KF, and **(C)** AFM diagrams. Note that the greatest detail is shown on the AFM diagram, where variable Fe/(Fe + Mg) ratios for ferromagnesian minerals in the various assemblages are explicitly represented.

(C)

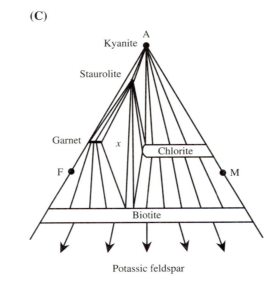

biotite, quartz, and plagioclase (with trace amounts of accessory minerals such as ilmenite, rutile, and apatite). Assuming that all these phases formed in equilibrium at the peak of metamorphism, we can plot the compositional points for garnet, staurolite, and plagioclase on the ACF diagram (Figure 20-10A). Tie lines are used to connect the compositions of the three plotted minerals. The rock composition (arbitrarily shown as x) should fall within the so-called *compatibility triangle* or *three-phase triangle* formed by the minerals; if it does not, then disequilibrium or some other problem must be suspected. Plotted in the AFM diagram (see Figure 20-10C), the rock bulk composition x again falls within a three-phase triangle, this time made up of garnet, staurolite, and biotite (invariably in nature biotite has a lower Fe/(Fe + Mg) than coexisting staurolite, so the triangle has been oriented this way). Remember, however, that

muscovite is also present, and the three-phase AFM compatibility triangle is actually a projection of a four-phase irregular subtetrahedron within the overall AKFM tetrahedron. Finally, these same four minerals (muscovite, biotite, garnet, and staurolite) can be plotted within the A′KF diagram (see Figure 20-10B). Notice that here there is an ambiguity: The rock composition x plots within a four-cornered area made up of the four minerals, and the assemblage does not allow us to choose one unique triangle of three phases; this ambiguity is examined below.

The significance of these results now must be assessed in terms of the earlier discussions of equilibrium and the phase rule. Assuming that a metapelite represents a normal condition of divariant equilibrium, the phase rule ($F = c - p + 2$) reduces to $p = c$. By means of a representation such as the ACF diagram, the assumption is made that the number of components can be legitimately reduced to three—our "compressed" components A, C, and F—such that the diagram behaves according to the graphical rules for a three-component system. If the assumption is correct and $p = c$, then this three-component system should contain assemblages of three coexisting phases under conditions of divariant equilibrium (because $c = p$ is $3 = 3$). The rock bulk composition should fall within the three-phase compatibility triangle whose corners are the composition points of the three coexisting minerals. This is commonly the case; and for specific conditions of pressure and temperature, the ACF triangle can be subdivided into several three-phase triangles (as seen in Figure 20-10A) that represent stable assemblages for different rock compositions under these P-T conditions.

The general validity of the various assumptions is indicated by the fact that it is the normal situation rather than the exception to find no more than three ACF minerals in a rock, rather than four or five. Similar relationships are generally true with the AFM and A′KF diagrams. Mineral compatibilities usually involve three or fewer coexisting phases, but exceptions can be found, such as in Figure 20-10B. In this case, the four problematic minerals are garnet, staurolite, biotite, and muscovite. No problem arises with representing them in the AFM diagram, and the rock composition lies within the garnet-staurolite-biotite three-phase triangle. Because muscovite and quartz are assumed to be present in all AFM assemblages, there is no conflict. But the four minerals garnet, staurolite, muscovite, and biotite are all explicitly plotted in the A′KF diagram (see Figure 20-10B), which places a typical rock composition x within a quadrilateral rather than the usual triangle. This implies either (1) that equilibrium has not been achieved; (2) that the assemblage is actually univariant rather than divariant, and perhaps contains both products and reactants for a reaction such as

garnet + muscovite = staurolite + biotite + H_2O

or (3) that the A′KF system cannot be truly regarded as behaving according to the graphical rules of a three-component system.

The common occurrence of this metamorphic assemblage argues against the first two possibilities: Both disequilibrium and the occurrence of univariant assemblages in metamorphic rocks have been observed to be quite unusual. Close reexamination of the AFM diagram indicates that the third possibility is correct. The most common reason for assemblages in any of these diagrams to disobey the geometric rules is a failure of the assumption that the "compressed" ternary system behaves like a true ternary system. Specifically, the component F does not actually behave like a single component

if its constituent parts FeO, MgO, or MnO have different ratios in coexisting ferromagnesian phases. It can be clearly seen in Figure 20-10C that staurolite and biotite have different values of Fe/(Fe + Mg). Because these two oxides are partitioned unequally between staurolite and biotite, they are not acting in an equivalent manner and cannot be combined into a single F component as is done in A′KF. Similarly, the presence of "extra" components in more than trace amounts in phases in any of these diagrams may allow four or more plotted phases to coexist in apparent divariant equilibrium, and in apparent violation of phase diagram rules. An especially common example of this involves four-phase AFM assemblages in which garnet contains CaO or MnO and biotite contains TiO_2.

SUMMARY

Metamorphic rocks can be considered as chemical systems that rigorously obey the rules of chemical thermodynamics. By properly defining the chemical components that adequately describe any rock and the phases that accommodate these components, we can deduce whether the minerals in a rock exist in an equilibrium state or are undergoing a reaction. Observations in metamorphic rocks have clearly shown that most involve very close approaches to true chemical equilibrium under metamorphic conditions. Preservation of metamorphic assemblages in rocks exposed at the surface is due to very slow kinetics of reequilibration to postmetamorphic conditions. Metamorphic mineral assemblages are chemically stable arrangements of minerals that typically represent divariant equilibria in which minor shifts of pressure and temperature will not cause shifts in mineralogy. The P-T plane is subdivided into regions of stability of divariant assemblages. These regions are separated by univariant lines representing metamorphic reactions that transform the stable assemblages of minerals. Petrogenetic grids are P-T diagrams that are based on experiments or thermodynamic calculations and allow determination of a limited range of pressure and temperature for a particular metamorphic assemblage.

Information on metamorphic assemblages is in the form of algebraic and geometric subdivisions of multidimensional P-T-X space, which can be treated mathematically. Petrologists typically use graphical methods to simplify and visualize this assemblage information. Chemically complex rocks can normally be reduced to no more than three idealized components for which the phase relations can be plotted and displayed in ternary diagrams in two dimensions. The most common of these diagrams are the ACF, A'KF, and AFM diagrams.

STUDY EXERCISES
• • • • • • • • • • • • • • •

1. How does the concept of the *mineral assemblage* differ from a simple listing of minerals found in a sample?
2. Suppose you have found a metamorphic rock that contains the pure minerals anorthite ($CaAl_2Si_2O_8$), grossular garnet ($Ca_3Al_2Si_3O_{12}$), quartz (SiO_2), and wollastonite ($CaSiO_3$). How many components are needed to describe this assemblage? What is the phase rule variance of the assemblage?
3. If you find more minerals in a rock than you should have under equilibrium conditions, what textural or other criteria might you use to try to understand what is going on?

REFERENCES AND ADDITIONAL READINGS
• • • • • • • • • • • • • • •

Brown, T. H., R. G. Berman, and E. H. Perkins. 1988. GEO–CALC: Software package for calculation of and display of pressure-temperature-composition phase diagrams using an IBM or compatible computer. *Comput. Geosci.*, *14*, 279–289.

Bucher, K., and M. Frey. 1994. *Petrogenesis of Metamorphic Rocks* (6th ed. of Winkler's textbook). New York: Springer-Verlag.

Eskola, P. 1915. On the relations between the chemical and mineralogical composition in the metamorphic rocks of the Orijärvi region. *Bull. Comm. Geol. Finlande*, *44*, 109–145.

Fisher, G. W. 1989. Matrix analysis of metamorphic mineral assemblages and reactions. *Contrib. Mineral. Petrol.*, *102*, 69–77.

Goldschmidt, V. M. 1911. *Die Kontaktmetamorphose im Kristianiagebeit*. Oslo: Vidensk. Skr. I, Math-Nat. Kl., No. 11.

Holdaway, M. J. 1971. Stability of andalusite and the aluminum phase diagram. *Am. J. Sci.*, *271*, 97–131.

Hyndman, D. W. 1985. *Petrology of Igneous and Metamorphic Rocks*, 2nd ed. New York: McGraw-Hill.

Kerrick, D. M. 1990. The Al_2SiO_5 Polymorphs. *Rev. Mineral.*, *22*.

Ramberg, H. 1952. *The Origin of Metamorphic and Metasomatic Rocks*. Chicago: University of Chicago Press.

Ricci, J. E. 1966. *The Phase Rule and Heterogeneous Equilibrium*. New York: Dover.

Spear, F. S. 1993. *Metamorphic Phase Equilibria and Pressure-Temperature-Time Paths*. Monograph 1. Washington, DC: Mineralogical Society of America.

Thompson, J. B., Jr. 1957. The graphical analysis of mineral assemblages in pelitic schists. *Am. Mineral.*, *42*, 842–858.

Thompson, J. B., Jr. 1982. Composition space: An algebraic and geometric approach. *Rev. Mineral.*, *10*, 1–31.

Thompson, J. B., Jr. 1982. Reaction space: An algebraic and geometric approach. *Rev. Mineral.*, *10*, 33–51.

Yardley, B. W. D. 1989. *An Introduction to Metamorphic Petrology*. New York: Wiley.

Chapter 21

CONTROLS OF METAMORPHIC REACTIONS

The basic principles that govern the occurrence of metamorphic assemblages have been presented, along with techniques for portraying these relationships graphically. The nature of metamorphic reactions themselves can now be examined in some detail. It is especially important to understand how metamorphic reactions are controlled by environmental variables such as pressure, temperature, and the composition of metamorphic fluids, because reactions, including recrystallization, are the principal mechanisms that control the mineralogy and textures of metamorphic rocks and create the isograds that are mapped in the field.

There are a number of different types of reactions, some of which involve only solids, but most of which involve both fluids and solids, either consuming or producing water or carbon dioxide, or both. These reactions may be univariant, and thus occur along a line in the P-T plane and produce abrupt changes in assemblages in the field, or they may be divariant and occur over a range of temperatures and pressures. The methods of portraying various reaction types differ according to the nature of the reaction. This chapter examines the basic principles of portrayal of reactions both in the P-T plane and in compositional diagrams, and reviews the implications for development of metamorphic textures.

ENVIRONMENTAL CONTROLS

Temperature and Pressure

The most fundamental controls of metamorphic reactions are changes in the two basic environmental variables, temperature and pressure. More specifically, changes in temperature and pressure both control and reflect the energy content of rocks, which in turn determines, through appropriate thermodynamic properties, which minerals appear in reactions, which remain stable over long intervals, and which react and disappear. Composition also plays a significant secondary role in determining which minerals are possible under a given set of metamorphic conditions. A comprehensive examination of assemblages and reactions therefore involves extensive use of both compositional and P-T phase diagrams.

Temperature increase in metamorphic rocks, as in all materials, is a function of the input of thermal energy. The ultimate source of heat for metamorphism could be heat diffusing outward from the earth's deep interior or heat released by cooling plutons. The temperature of a rock or mineral is determined by the interaction of thermal energy with the actual crystal structures. The specific heat, or heat capacity, of any rock or mineral determines just how much the temperature will increase with a given input of energy. All of the energy that enters a rock may not go toward heating it, however. Some reactions are *endothermic*, that is, they consume heat energy in transforming one set of minerals (reactants) into another set (products). This energy of transformation is characterized as enthalpy of reaction (ΔH_r). Melting is a classic endothermic reaction: The energy content of the magma is higher than an equivalent amount of solid minerals at the same temperature. Thus, while an endothermic reaction is occurring, a rock might not increase its temperature despite an input of energy. Other reactions are *exothermic* and actually release energy as part of the mineralogic phase transformation; crystallization of a melt is a classic example. It is important always to keep in mind the complex interplay of reactions, energy, and temperature.

Pressure is exerted in various ways (Figure 21-1). Pressure, in physics, is simply a force applied over an area. An example is atmospheric pressure (nominally 1 bar at sea level): It is the weight of overlying air on 1 cm^2 of material. As rocks become buried beneath accumulating piles of sediment or thrust sheets, they are subject to the weight of the overlying materials. This weight generates a pressure that is hydrostatic in nature, that is, it is undirected and has the same (or very nearly the same) magnitude in all directions. It has variously been referred to as *burial pressure, load pressure,* **lithostatic pressure** (P_{lith}), or *confining pressure*. This is the type of pressure that squeezes mineral grains together in sediments and expels pore water in the process of lithification. Its magnitude is a function of depth and of the density, and thus weight, of the overlying material (overburden). A typical and widely used value of density for crustal materials (2.8 g/cm^3) produces an average lithostatic pressure gradient of about 280 bars per kilometer of depth. A metamorphic rock at a depth of 20 km would therefore be subjected to approximately 5600 bars or 5.6 kbar. In the most recent petrologic literature, there is increasing use of the official SI (*Systeme Internationale*) pressure unit, the *pascal* (abbreviated Pa). One atmosphere (1 bar) is equal to 10^5 Pa, and therefore 1 kbar = 10^8 Pa. Clearly this is cumbersome, so the prefixes *mega-* (M) or *giga-* (G) are used for 10^6 and 10^9 (1 million and 1 billion) pascals, respectively. In this terminology, 1 kbar is either 100 MPa or 0.1 GPa, and 10 kbar would be 1.0 GPa. In current literature, both types of pressure terminology are in use, and it is important to be familiar with both.

(A)

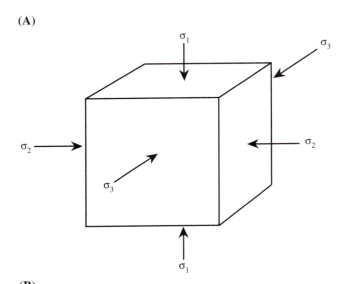

(B)

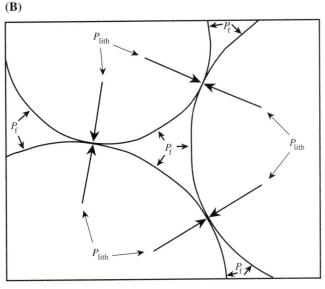

FIGURE 21-1
• • • • • • • • • • • • • • • •

Cartoon illustration of the different types of pressure of interest to petrologists. **(A)** Normal system used by geologists for showing pressures or stresses, with pressure divided into three σ vectors or components at 90° to each other: σ_1 (maximum compressive stress), σ_2 (intermediate stress), and σ_3 (minimum stress). Under conditions of hydrostatic or lithostatic pressure, $\sigma_1 = \sigma_2 = \sigma_3$. Deformation is typically caused when pressures or stresses are nonequal. The orientation of stress directions is arbitrary when they are equal, but when stresses are unequal, their numbers refer to magnitudes as given above. **(B)** In a rock, the solid grains exert lithostatic pressure P_{lith} on each other at grain-grain contacts (shown by the larger arrows). Intergranular fluid in pore spaces exerts fluid pressure, P_{fluid} or P_f, on the surfaces of mineral grains (shown by small arrows). At normal depths and pressures of metamorphism, $P_f = P_{lith}$, but in some shallow sedimentary basin environments, P_f may be 10–20% lower than the P_{lith}. [Modified from B. W. D. Yardley, 1989, *Introduction to Metamorphic Petrology* (New York: Wiley Interscience), Fig. 1.8.]

Lithostatic pressure is not the only type of pressure exerted on the solids. Many rocks undergoing regional metamorphism in orogenic belts are subjected to differential or nonlithostatic stress, which can occur in three forms: compressional, tensional, and shear. The magnitudes of the stress differences are thought to be small in relation to lithostatic pressure (perhaps several hundred bars maximum), but they may be very important in the development of characteristic metamorphic textures such as foliations and lineations.

Compositions of Metamorphic Fluids

In addition to pressures exerted along their contacts by the solid mineral grains, there may also be a hydrostatically pressurized fluid phase present. In a zone extending from the surface to several kilometers deep in sedimentary basins, the hydrostatic pressure on the fluid in a rock is somewhat less than the lithostatic pressure on the mineral grains themselves. Fluid pressurization occurs if the mineral grains all touch one another and are strong enough through these grain-to-grain contacts (a so-called grain-supported framework) to withstand the collapse of void space by lithostatic pressure. Fluids in interconnected pore spaces might then be pressurized only (or mostly) by the weight of overlying sediments, balanced by some leakage mechanism through the rocks bordering the basin. It has typically been observed in this shallow regime (less than 8–10 km) that $P_{fluid} \approx 0.9\ P_{lith}$. At greater depths, however, including the depths of virtually all metamorphism, minerals apparently do not have the strength to support the weight of both rocks and fluids, and everything is generally assumed to be pressurized equally (that is, $P_{fluid} = P_{lith}$). In lithified materials, the top of this deeper zone (which is often called the *geopressured zone*) may coincide approximately with the transition from brittle to ductile behavior in minerals, at depths of roughly 10 km for quartz-rich rocks (around 2.8 kbar or 280 MPa), depending on the geothermal gradient. In some sedimentary basins, however, the top of the geopressured zone may actually come within 1 km of the surface. P_{fluid} generally cannot exceed P_{lith}, except temporarily, because this condition would lead to hydraulic fracturing and the escape of the fluid.

The nature of metamorphic fluids is difficult to assess. They are generally thought to be rich in water, although fluids may be carbon dioxide–rich as the result of metamorphism of calcareous rocks or methane-rich in low- to medium-temperature metamorphism of mudrocks that contain carbonaceous matter or graphite. Minor or trace constituents may include fluorine, chlorine, O_2, H_2, H_2S, and dissolved mineral species. Virtually all metamorphism occurs at temperatures and pressures well above

the critical point of aqueous fluids, and they thus behave as highly compressed supercritical fluids rather than as liquids or gases, although their properties (for example, density) are much closer to those of liquids than to those of gases. There is significant disagreement among petrologists concerning the source of fluids in metamorphic rocks, and even whether fluids are always present, especially at higher metamorphic grades. In contrast to solid minerals, any fluids that might have been present during metamorphism are absent by the time the petrologist collects a hand sample and examines it. It then becomes an article of faith as to whether or not fluid was once there. An indirect but persuasive form of evidence for the presence of fluids is the occurrence of certain metamorphic reactions involving hydrous or carbonate minerals. Temperatures of these types of reactions are very sensitive to the presence and composition of fluids, and considerable research and experience indicate that such reactions occur at temperatures where fluids must have been involved. But evidence from other rocks is more ambiguous.

Several variables are widely used to express the fluid pressure in a rock. P_{fluid} is the most general expression and implies the sum of the partial pressures of all the fluid species. For most purposes, $P_{fluid} = P_{H_2O} + P_{CO_2}$, except for highly carbonaceous rocks, for which P_{CH_4} may be important as well. It is generally assumed that $P_{fluid} = P_{lith}$ for crustal depths where most metamorphism occurs, although this may not necessarily hold for the very lowest grades of incipient metamorphism. In mixed fluids, the most common way of expressing the amounts of different species is by using mole fraction, represented by X. The mole fractions of all species present in the fluid must add up to 1.0, and thus in most cases $X_{H_2O} + X_{CO_2} = 1.0$.

Fluid Inclusions

There is one type of evidence that proves the existence of fluids in at least some metamorphic rocks: **fluid inclusions.** Fluid inclusions are very small (10 micrometers or less) ovoid trapped bubbles of fluid in metamorphic minerals, usually quartz (Figure 21-2). Many fluid inclusions are multiphase, containing a bubble of gas in some liquid; and very small solid inclusions of halide, sulfide, or other minerals are sometimes present as well. Many fluid inclusions in quartz occur along obvious planes, which are assumed to represent healed fractures. Recrystallization of quartz along these fractures apparently trapped small amounts of a fluid that was wetting the surfaces of the fracture. Many of these planar occurrences turn out to be postmetamorphic, forming during the interval of unloading when overlying rocks were

FIGURE 21-2

• • • • • • • • • • • • • • • •

Typical appearance of multiphase fluid inclusions in quartz in medium- to high-grade metamorphic rocks. Horizontal width of field of view is about 0.11 mm (110 μ). The two central fluid inclusions are unusually large; the smaller ones are more typical. Note the vapor bubbles in many of the inclusions. Liquid in these inclusions is water-rich, and the vapor is carbon dioxide–rich. In the large inclusion at right-center, note the secondary solid daughter mineral (halite) with characteristic cubic shape. [Photo by M. Vityk and R. J. Bodnar, Virginia Tech.]

being eroded. Other fluid inclusions, however, occur in random orientations and seem to have formed through trapping of interstitial fluid while the crystal grew during metamorphism. These fluids are important because they probably represent trapped samples of genuine metamorphic fluid.

When examined on a special heating-freezing microscope stage, multiphase fluid inclusions typically homogenize to a single fluid phase at elevated temperature, as both the gas bubble and any daughter crystals are dissolved by the fluid. This temperature, along with the measured freezing temperature, can be used to estimate both the composition of the fluid and a range of entrapment conditions in a P-T diagram. Fluid inclusion studies of metamorphic rocks have identified enough apparently primary (nonplanar) fluid inclusions that formed at or near peak metamorphic conditions to provide persuasive evidence for the presence of fluids during metamorphism. An excellent review by Roedder (1984) is recommended.

METAMORPHIC REACTIONS

• • • • • • • • • • • • • • • •

Textural evidence for prograde mineralogic reactions is sometimes found in metamorphic rocks, even though the crystallization of new, higher grade minerals, along with deformation, tends to obliterate evidence for lower grade precursors. One indication of reaction is pseudomorphic replacement of lower grade minerals by higher grade

ones, for example, the replacement of large andalusite porphyroblasts by fine-grained aggregates of sillimanite (Figure 21-3). Another and even more interesting example is shown in Figure 21-4, where an intergrowth of sillimanite and quartz has partially replaced and embayed a large muscovite crystal, leaving several isolated islands of muscovite in optical continuity. This second example reveals a puzzling problem. Because muscovite is a potassic mineral, $KAl_3Si_3O_{10}(OH)_2$, its breakdown should have produced another potassic mineral as a reaction product, but none is apparent in the vicinity. Petrologists who have observed this phenomenon have interpreted the texture to reflect transport of the K^+ ions in aqueous fluid away from the reaction site, probably to precipitate another potassium-bearing mineral somewhere else. Reactant and product minerals may thus not be conveniently located side by side in a thin section, a situation that makes it difficult to interpret reactions. D. M. Carmichael (1969), who described the above texture, has even suggested that the transported potassium probably reacted with sillimanite, quartz, and water somewhere else in the rock to create a new crystal of muscovite! Carmichael proposed the idea that metamorphic rocks constantly undergo this circularity of reaction and dissolution under conditions of divariant equilibrium. In the overall rock system, equilibrium applies because mineral proportions do not change, but the rock is constantly undergoing complicated recrystallization.

Metamorphic reactions are typically written with the reactant minerals on the left-hand side and the product minerals (and fluid, if necessary) on the right. A

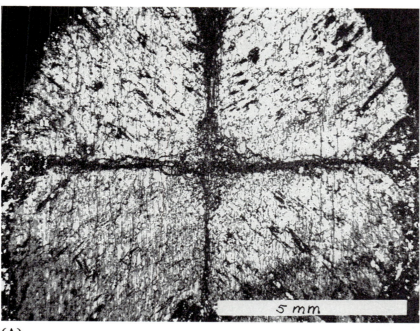

(A)

FIGURE 21-3
● ● ● ● ● ● ● ● ● ● ● ● ● ● ● ●

(A) Large andalusite (chiastolite variety) porphyroblast that has been pseudomorphically replaced by small, diamond-shaped crystals of sillimanite. Note the residual carbonaceous material (now graphite) that constituted the cross in the original chiastolite. The view is down the *c*-axis of both the original orthorhombic andalusite and the sillimanite replacement. Similarity of crystal structures of these two minerals makes this type of replacement possible. Sample from Gap Mountain, southern New Hampshire.
(B) Enlarged view of the sillimanite prisms pictured in part (A). Note the scatter in the orientation of (010) cleavage planes in the different individual sillimanite prisms. [From J. L. Rosenfeld, 1969, *Amer. J. Sci.*, *267*, Plate 3.]

(B)

good example is the reaction originally discussed by Goldschmidt, with calcite and quartz reacting to form wollastonite and CO_2:

$$CaCO_3 + SiO_2 = CaSiO_3 + CO_2$$
$$\text{calcite} \quad \text{quartz} \quad \text{wollastonite}$$

Reactions can be written by using either the mineral chemical compositions, their names, or both (as above). This reaction is relatively simple, because the minerals have a simple chemistry and the balanced reaction has stoichiometric coefficients of one mole of each reactant and product. Most reactions are actually considerably more complicated and, using realistic mineral chemistry, may require hundreds or thousands of moles of reactants and products to balance—a computer is used for the calculation. Although reactions are typically written from left to right in the sense of increasing metamorphic grade, it must be understood that this does not imply simple temperature increase. In some rocks, prograde dehydration reactions can occur under conditions of decreasing pressure at nearly constant temperature.

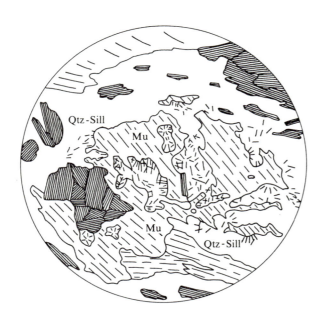

FIGURE 21-4

Muscovite (Mu) embayed by quartz-sillimanite (Qtz-Sill) intergrowth. [After C. F. Tozer, 1955, *Geol. Mag.*, *92*, 310–320; and D. M. Carmichael, 1969, *Contr. Mineral. Petrol.*, *20*, Fig. 4b.]

If a rock has a very simple chemical composition (for example, Al_2SiO_5) and was formed by the reaction of a low-temperature mineral (for example, andalusite) to a high-temperature one (for example, sillimanite), it is clear that there should be no crystals of andalusite left because they would have been completely converted to sillimanite. However, metapelites are more complex mineralogically, and if a typical metapelite had undergone the reaction

9 staurolite + muscovite + 5 quartz

$$= 17 \text{ sillimanite} + 2 \text{ garnet} + \text{biotite} + 9 \, H_2O$$

it is not generally true that all of the reactant minerals should have been eliminated from the rock. (The stoichiometric coefficients of this reaction have been calculated by assuming particular compositions of reactant and product minerals.) For staurolite, muscovite, and quartz to have been exhausted simultaneously, it would have been necessary for them to have been in exactly the molecular proportions of 9:1:5 in the unreacted rock. This is theoretically possible but highly unlikely. Cessation or completion of a reaction such as this is typically caused by the total depletion of only one mineral—the least abundant one. Because all reactants are necessary for reaction to continue, the reaction would cease. It is important to recognize also that any one or two of the product minerals sillimanite, biotite, and garnet could be present in the rock before the reaction begins, because they could have been produced by some other reaction. It is the stable coexistence of all three of the products

that is the key evidence that the reaction has taken place. If the reactant phase eliminated were staurolite, then the divariant assemblage following reaction would be garnet + sillimanite + biotite + muscovite + quartz. Thus assemblages of any five of these minerals could occur on either side of the reaction, depending on which minerals were included. Only the coexistence of all six (probably with textural indications of reaction) indicates that the reaction has been caught in progress.

The preceding discussion can be reviewed by using the graphical techniques from the previous chapter, particularly the AFM diagram. This projection can be used because both muscovite and quartz occur in the rock. Figure 21-5A shows the AFM diagram for assemblages below the temperature of the reaction. Staurolite is stable with muscovite and quartz (these latter two are not shown but are implied by the projection). For rock compositions falling within the triangle made by sillimanite, biotite, and garnet, there are three possible five-phase AFM assemblages (all including muscovite and quartz): garnet-staurolite-biotite (*x*), garnet-staurolite-sillimanite (*y*), and staurolite-sillimanite-biotite (*z*). Note that the assemblage sillimanite-garnet-biotite cannot be present because the reaction has not yet occurred, and this is not a possible three-phase triangle in Figure 21-5A. After the reaction has gone to completion and staurolite has been eliminated (the normal occurrence in metapelites), the AFM diagram in Figure 21-5B applies. Only one five-phase assemblage occurs now: muscovite-quartz-sillimanite-biotite-garnet. All three rock compositions *x*, *y*, and *z* fall within the sillimanite-garnet-biotite triangle. In all three rocks the new stable coexistence of sillimanite, garnet, and biotite demonstrates that the reaction has occurred, although each rock will have different proportions of the three minerals.

In attempting to define an Al_2SiO_5-garnet-biotite isograd based on this reaction in the Whetstone Lake area in southern Ontario, Carmichael (1969) determined a variety of mineralogic associations in metasedimentary and metavolcanic rocks. East of his mapped isograd, he found the lower grade assemblage staurolite-muscovite-quartz with various pairs of garnet, biotite, or sillimanite. Only west of his isograd did he find the higher grade assemblage sillimanite-biotite-garnet. Some samples contained all six phases. The diversity of assemblages Carmichael found reflected the variation of bulk com-

(A)

(B)

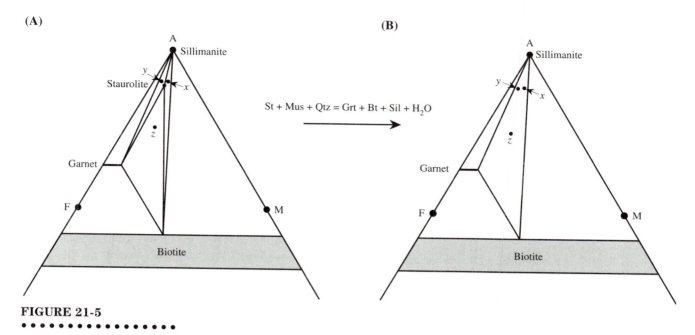

$St + Mus + Qtz = Grt + Bt + Sil + H_2O$

FIGURE 21-5
● ● ● ● ● ● ● ● ● ● ● ● ● ● ● ● ●

Graphical portrayal of staurolite breakdown shown in AFM diagrams. The staurolite point lies within the garnet-sillimanite-biotite triangle, as shown. **(A)** Assemblages before reaction (plus muscovite and quartz): x, garnet + staurolite + biotite; y, garnet + staurolite + sillimanite; z, staurolite + sillimanite + biotite. **(B)** Assemblages after reaction (plus muscovite and quartz). All three rock compositions contain garnet + biotite + sillimanite, although in different proportions. Mineral abbreviations: Bt, biotite; Grt, garnet; Mus, muscovite; Qtz, quartz; Sil, sillimanite; St, staurolite.

positions in the metapelites of the area; most rock units have at least minor variation in bulk chemistry along strike. Some blurring of the isograd from an ideal line is also normal and indicates that the reaction assemblage is not strictly univariant (probably due to slight variations in the compositions of the minerals involved).

REACTION MECHANISMS
● ● ● ● ● ● ● ● ● ● ● ● ● ● ●

Application of experimental or theoretical information on reactions to actual rocks requires characterization of the reactions as completely as possible, including an assessment of the mechanisms or pathways by which the reactant assemblage was transformed into the product one. Returning to a point raised earlier, it must be established whether all the atomic constituents for the reactions are present within the minerals themselves. For example, Figure 21-6 shows isolated islands of optically continuous kyanite and quartz surrounded by muscovite containing needles of sillimanite. Obviously, neither the kyanite nor the quartz contains the K_2O or H_2O neces-

sary for production of muscovite. Consequently, a balanced reaction can be written in terms of both mineralogic and ionic (dissolved) constituents such as

$$3 \text{ kyanite} + 3 \text{ quartz} + 2 \text{ K}^+ + 3 \text{ H}_2\text{O}$$
$$\underset{\text{Al}_2\text{SiO}_5}{} \qquad \underset{\text{SiO}_2}{}$$

$$= 2 \text{ muscovite} + 2 \text{ H}^+$$
$$\underset{\text{KAl}_3\text{Si}_3\text{O}_{10}(\text{OH})_2}{}$$

Where do the necessary ionic constituents come from (or go)? Petrologists have commonly called on an external source, such as an adjacent magma (sometimes exposed as a plutonic rock, sometimes only imagined at depth). In some cases, this is a legitimate approach—granitic magmas, especially as they near solidification, tend to be rich in both alkalis and water and are commonly observed to be the sources of late veins and water-rich pegmatites that intrude the country rocks. Regional scale interaction between magmatic fluids and country rocks has been demonstrated in some places, but granitic plutons are not always conveniently locally exposed in metamorphic terranes.

Carmichael (1969) proposed an alternative scheme, that the source for the missing constituents might lie

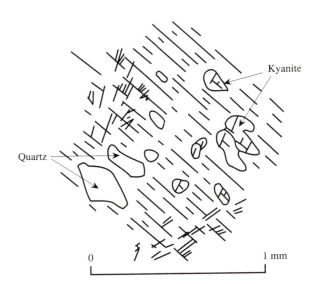

FIGURE 21-6
● ● ● ● ● ● ● ● ● ● ● ● ● ● ● ● ●

Isolated, optically continuous kyanite and quartz islands or blebs
surrounded by a large muscovite flake that contains needles of
sillimanite. [After G. A. Chinner, 1961, *J. Petrol.*, *2*, 312–323; and
D. M. Carmichael, 1969, *Contr. Mineral. Petrol.*, *20*, Fig. 4a.]

in adjacent local domains within the rock that are simultaneously undergoing different reactions, producing a
coupling between domains. The scale of such component transport (or **mass transfer,** as it is called), if it
exists, might be centimeters to meters. Transport is
accomplished by movement of various constituents

down concentration gradients between the reaction
sites. These concentration gradients produce a difference in chemical potential energy (they are thus sometimes called *chemical potential gradients*), which is the
energetic driving force that causes physical movement of
chemical constituents by diffusion. In his analysis of a
variety of such reactions, Carmichael examined the conversion of kyanite to sillimanite. Rarely is there any textural evidence for the direct conversion of any of the
Al_2SiO_5 minerals into one another (see Kerrick 1990),
except for the pseudomorphic replacement of andalusite
by sillimanite (as in Figure 21-3). Instead, when multiple
aluminum silicate minerals coexist, they commonly do
so in apparent textural equilibrium without evidence of
resorption, embayment, or other textural indications of
reaction (Figure 21-7). Carmichael suggested that a polymorphic transition such as kyanite $\rightleftharpoons$ sillimanite is actually accomplished indirectly after a kyanite-bearing
rock has undergone a *P-T* shift into the sillimanite stability field. He suggested that the polymorphic transformation actually involved two simultaneously occurring reactions in adjacent domains of the rock (Figure 21-8):

$$3 \text{ kyanite} + 3 \text{ quartz} + 2 \text{ K}^+ + 3 \text{ H}_2\text{O}$$
$$\quad\; Al_2SiO_5 \qquad\quad SiO_2$$
$$= 2 \text{ muscovite} + 2 \text{ H}^+$$
$$\qquad\qquad KAl_3Si_3O_{10}(OH)_2$$

$$2 \text{ muscovite} + 2 \text{ H}^+$$
$$KAl_3Si_3O_{10}(OH)_2$$
$$= 3 \text{ sillimanite} + 3 \text{ quartz} + 2 \text{ K}^+ + 3 \text{ H}_2\text{O}$$
$$\qquad\quad Al_2SiO_5 \qquad\qquad SiO_2$$

In the rock, the effect of the two reactions occurring
simultaneously (seen by adding the two reactions

FIGURE 21-7
● ● ● ● ● ● ● ● ● ● ● ● ● ●

Photomicrograph of kyanite and
fibrolitic sillimanite in a metamorphosed pelitic rock from Mica Creek,
British Columbia. The straight, unem-
bayed boundary of the kyanite crystal
(K) indicates no resorption by the
adjacent fibrolite (F). Plane-polarized
light. Horizontal field of view is about
1.2 mm. [From D. M. Kerrick, 1990,
Rev. Mineral., *22*, Fig. 8.4. Photo by
D. M. Kerrick and E. D. Ghent.]

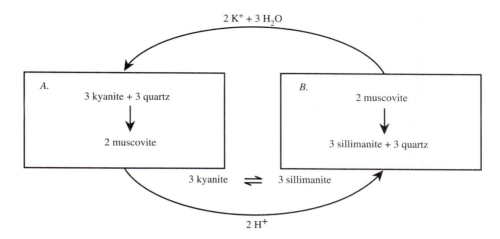

FIGURE 21-8

● ● ● ● ● ● ● ● ● ● ● ● ● ● ● ● ●

A possible mechanism for the reaction kyanite = sillimanite. Separate mineralogic-textural domains exist within the rock at domains A and B. In domain A, kyanite and quartz reacted to form muscovite. In domain B, muscovite has reacted to form sillimanite and quartz. For both reactions to occur in the same rock, mass balance requires that H_2O, K^+, and H^+ must have migrated between the domains. The net result of the two reactions occurring simultaneously (subtracting one from the other) is 3 kyanite = 3 sillimanite. [From D. M. Carmichael, 1969, *Contr. Mineral. Petrol.*, *20*, Fig. 5.]

together algebraically) is shown by crossing out the reactants and products, which are duplicated on opposite sides of the two reactions (such as 2 muscovite or $2 H^+$), leaving the result: 3 kyanite = 3 sillimanite. Therefore, to accomplish this apparently simple inversion of kyanite $\rightleftharpoons$ sillimanite, it is necessary for muscovite, quartz, and ionic constituents in aqueous solution all to have entered the overall reaction to provide a mechanistic pathway for the Al_2SiO_5 component to be "recycled" from kyanite into sillimanite. If the reaction had gone to completion, there would probably be no way to reconstruct this pathway, and all that could be said is that kyanite had reacted to form sillimanite. Carmichael found textural evidence that demonstrated the more complicated reaction mechanism. This, of course, does not prove that kyanite cannot be directly transformed into sillimanite: rather, it shows that in *this* rock the easiest and kinetically most rapid path involved the subsidiary reactions with muscovite, quartz, and fluid. It is still probably correct to specify that the process occurred at *P-T* conditions on or near the kyanite $\rightleftharpoons$ sillimanite univariant line in the *P-T* diagram, but some caution is warranted.

TYPES OF METAMORPHIC REACTIONS

● ● ● ● ● ● ● ● ● ● ● ● ● ● ●

There are a number of different types of metamorphic reactions, including those that involve fluid species and those that do not. Some of these reactions are truly univariant in a phase rule sense, and because they are lines in the *P-T* plane, they mark sharp boundaries between divariant assemblages which can be mapped in the field as sharp isograds. Other, more common, reactions occur *within* stability fields of divariant assemblages and change the proportions but not the identities of the minerals in the assemblage. These reactions are made possible because of solid solution in many minerals (the most common being iron-magnesium solid solution), and mineral compositional change as reactions proceed as a function of gradually changing pressure and temperature. The gradual rather than abrupt nature of these reactions means they form wide bands rather than sharp lines in the *P-T* plane, so the isograds they form are smeared out in the field. Because hydrous or carbonate minerals are virtually ubiquitous in metamorphic rocks, except at very high grades, the reactions that involve the breakdown of these minerals and the production of water or carbon dioxide (*devolatilization reactions*) are the dominant type of metamorphic reaction. The controls of devolatilization are more complex than those for solid-solid reactions, and the techniques for describing and analyzing them are somewhat different.

Discontinuous (Univariant) Reactions

These reactions may or may not be devolatilization reactions but are notable because they are approximately univariant (they occur along lines or very thin bands in the *P-T* plane) and they involve a shift in tie lines in the appropriate compositional diagram. Because of their univariant character, the reaction assemblages have more phases than components ($p = c + 1$) and tend to form very sharp boundaries between mineralogic zones in the field. Truly univariant reactions are actually quite unusual because of the widespread occurrence of compositional variability or solid solution in metamorphic minerals, but some reactions in metapelites are close enough to univariant to make reasonable examples.

The first of these reactions occurs at medium grades of metamorphism and involves the minerals quartz, muscovite, biotite, chlorite, staurolite, and garnet. The use of the AFM projection requires and implies the first two phases, and only the other four are explicitly shown. Figure 21-9 shows a simplified AFM diagram with these minerals portrayed both before and after the reaction. In Figure 21-9A, the four phases constitute two three-phase triangles: garnet-chlorite-biotite and garnet-chlorite-staurolite. The bulk composition of typical metapelites is represented by point x and indicates that under the range of *P-T* conditions for which Figure 21-9A applies, the common stable metapelite assemblage would be quartz-muscovite-garnet-chlorite-biotite. The discontinuous reaction causes a shift in stable tie lines from garnet-chlorite to staurolite-biotite, and the reaction is therefore roughly garnet + chlorite = staurolite + biotite. The reaction must be balanced by using the projected phases muscovite, quartz, and water-rich fluid:

$$\text{garnet + chlorite + muscovite}$$
$$= \text{staurolite + biotite + quartz + } H_2O$$

(Actual stoichiometry is complicated and has been ignored for the moment.) Figure 21-9B shows the new

FIGURE 21-9

● ● ● ● ● ● ● ● ● ● ● ● ● ● ● ● ●

Graphical portrayal of the discontinuous AFM reaction garnet + chlorite + muscovite = staurolite + biotite + quartz + H_2O. A typical metapelite bulk composition, shown at x, would form the assemblage garnet + biotite + chlorite before the reaction (**A**) and the assemblage garnet + staurolite + biotite after the reaction (**B**).

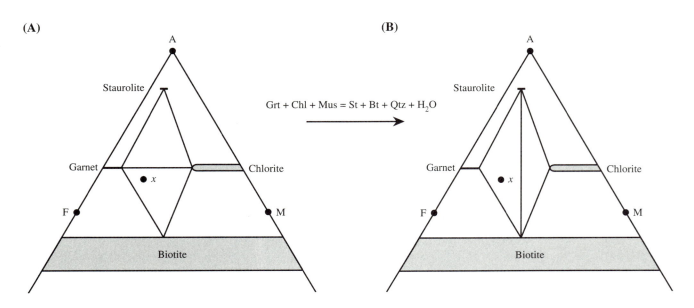

(A)   **(B)**

Grt + Chl + Mus = St + Bt + Qtz + H_2O

arrangement of tie lines and the new three-phase triangles. Note that the typical metapelite composition now lies within the three-phase triangle of garnet-biotite-staurolite. This particular reaction is thought to be responsible for the first appearance of staurolite in many metapelites, which roughly coincides with the disappearance of chlorite and is probably the reaction responsible for the staurolite isograd in many metamorphic terranes.

The second important topological type of discontinuous reaction was discussed in reviewing Carmichael's Ontario study (see Figure 21-5). This reaction is responsible for the ultimate breakdown and disappearance of staurolite from metapelites and is written

staurolite + muscovite + quartz

= garnet + biotite + Al-silicate + H_2O

In contrast to the first type of discontinuous reaction, this type involves not the switching of tie lines, but the disappearance of tie lines as a result of the disappearance of a mineral from the diagram. Proceeding from the tie line arrangement of Figure 21-5A to that of Figure 21-5B, the disappearance of staurolite means, of course, that the three tie lines from staurolite to the other minerals must also disappear. Where there were originally three different three-phase assemblages in Figure 21-5A before the reaction (all involving staurolite), there is only one in Figure 21-5B *after* the reaction. This sort of AFM discontinuous reaction is sometimes called a *terminal disappearance* reaction, because it marks the ultimate limit of stability of a mineral, wherever the rock bulk composition falls within the AFM diagram (that is, as long as quartz and muscovite are still present).

These two types of discontinuous reactions have two things in common: both represent a significant rearrangement of tie lines, and both involve the disappearance of a mineral. Both also represent the "momentary" coexistence of four minerals as the reaction progresses and both reactants and products are present. "Momentary," of course, is relative because of apparently slow kinetics of many reactions. The two types differ, however, in that the *tie line switch* reaction causes the disappearance of a particular mineral only from certain rock compositions (see Figure 21-9), whereas a *terminal* reaction marks its disappearance from all compositions.

Continuous (Divariant) Reactions and Distribution Coefficients

The concept that lies behind this continuous (or divariant) type of reaction is sometimes a little difficult to grasp. When one or more mineral phases portrayed in a compositional diagram have varying compositions within the diagram, and thus varying plotting points, three-

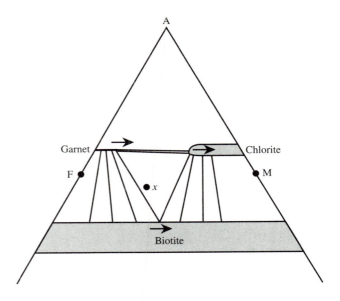

FIGURE 21-10
● ● ● ● ● ● ● ● ● ● ● ● ● ● ● ● ●

Graphical portrayal of the continuous AFM reaction chlorite + muscovite + quartz = Mg-richer chlorite + garnet + biotite + H_2O. The operation of this type of reaction is marked not by a shifting of tie lines but by gradual movement of the three-phase triangle toward the right, the shift reflecting the higher temperature stability of the magnesium-richer minerals in this assemblage. Arrows show the shifts in compositions of garnet, chlorite, and biotite with increasing temperature. For a bulk composition x, the reaction dictates gradual decrease in chlorite abundance until it is eventually totally replaced by garnet and biotite when the garnet-biotite tie line intersects x.

phase triangles can gradually shift positions with changing pressure and temperature, rather than changing abruptly as in discontinuous reactions. Compositional variability of minerals also means that ternary diagrams need not be completely subdivided into three-phase triangles but can have areas in which only two minerals coexist in stable divariant equilibrium. The phase rule allows this topologic situation for solid-solution minerals, and an example is illustrated in Figure 21-10, a partial and simplified AFM diagram for low to medium grades of metamorphism in metapelites (other possible AFM minerals than the ones shown have been ignored for simplicity). All three of the AFM minerals shown—garnet, chlorite, and biotite—can have a wide range of solid solution between iron-rich and magnesium-rich compositions (also aluminum solid solution, in the cases of chlorite and biotite), and they thus plot as horizontal lines or fields in the diagram.

For each particular value of pressure and temperature, there is only one three-phase triangle that connects point compositions of these three (or any three) minerals. The actual compositions are a function of the iron-magnesium *distribution coefficients* among the three minerals (written as K_D). These coefficients are based on thermodynamic relations and reflect the relative Fe/(Fe + Mg) for each mineral as a function of pressure and temperature. For example, the iron-magnesium distribution coefficient between garnet and biotite can be expressed as

$$K_D{}^{\text{Fe-Mg (Grt-Bi)}} = (\text{Fe/Mg})^{\text{Grt}}/(\text{Fe/Mg})^{\text{Bi}}$$

This concept was encountered in Chapter 20 with the relative Fe/(Fe + Mg) values for plotting minerals in the AFM diagram (garnet > staurolite > biotite > chlorite). If iron and magnesium are equally distributed between two AFM minerals, the iron-magnesium distribution coefficient will be 1.0 and the tie lines will be radial from the A apex of the AFM triangle. If iron and magnesium are unequally distributed (that is, iron is preferentially partitioned into one mineral), the tie lines will tilt strongly one way or the other, depending on which mineral is more iron-rich. For example, note that the garnet-chlorite tie lines in Figure 21-10 go across much of the width of the AFM diagram. This geometry reflects the fact that garnet is much more iron-rich than coexisting chlorite (that is, the garnet-chlorite K_D as defined above is much greater than 1.0). In contrast, chlorite-biotite tie lines are more nearly radial from the A corner, meaning that the chlorite-biotite K_D between them is close to 1.0. Rock compositions that do not fall within the single three-phase triangle at a given temperature and pressure will therefore consist of one of the various possible two-phase assemblages (in additon to quartz and muscovite, of course): garnet-chlorite, garnet-biotite, and biotite-chlorite. Each of these *two-phase fields* consists of an infinite number of tie lines connecting the horizontal lines or fields that represent the two minerals. These tie lines obey the same iron-magnesium distribution rules outlined above.

How a continuous reaction occurs can be illustrated by using the relative positions of the one three-phase triangle and the three two-phase fields in Figure 21-10. Consider a rock composition within the three-phase triangle, shown by x in Figure 21-10 and consisting of roughly equal proportions of garnet and biotite, with a smaller amount of chlorite. Experience indicates that with increasing metamorphic grade (that is, increasing temperature), this three-phase triangle will gradually shift toward the right; this movement is indicated in the figure by the right-pointing arrows at each corner of the three-phase triangle. As the three-phase triangle moves toward the right, the position of x within the triangle approaches the garnet-biotite side. Remember that the triangle must shift because its three corner points do, but

x always stays in the same place for a particular rock. Using the triangle lever rule to estimate mineral proportions, we see that the amount of chlorite in the rock will gradually decrease, and the amount of garnet and biotite will increase. At the same time, all of these minerals in the rock will become slightly more magnesium-rich. Clearly a reaction is occurring, but it is proceeding gradually, and no abrupt change in assemblage occurs as the temperature increases. The reaction is written as follows, taking into account the gradual shift in mineral compositions:

chlorite + muscovite + quartz

= Mg-richer chlorite + garnet + biotite + H_2O

because chlorite is gradually decreasing in abundance. If stoichiometric coefficients had been assigned in the above reaction, there would be less chlorite on the right-hand (product) side than on the left-hand (reactant) side; the relative values indicate a decrease in total amount of chlorite in the rock. The amounts of garnet and biotite produced would be added to the amounts of these minerals already present in the rock.

There is a clear limit to how long this reaction can proceed in the rock: When chlorite is completely gone, the reaction must stop (assuming that neither muscovite nor quartz is used up first). Graphically, this means that the three-phase triangle in the AFM diagram has moved to the right until the garnet-biotite edge has coincided with the bulk composition x. The rock now contains a garnet-biotite assemblage (plus muscovite and quartz) until some higher grade reaction comes along. It appears that most of the reaction history of metapelites and other rock types consists of continuous reactions like this, alternating with an occasional discontinuous reaction.

Solid-Solid Reactions

These reactions may be either continuous or discontinuous and are relatively simple to deal with because they do not depend on the presence or composition of fluids. They are most common at high metamorphic grades. There are so many reactive hydrous or carbonate minerals at lower grades that most lower grade reactions are devolatilization reactions. Probably the most important of the solid-solid equilibria that occur over a range of metamorphic conditions are the three univariant equilibria between the Al_2SiO_5 polymorphs. These equilibria may operate directly, as shown in Figure 21-3 with the direct replacement of andalusite by sillimanite, or indirectly, as Carmichael observed in Ontario for the kyanite to sillimanite transition. The important thing, however, is the orientation of the univariant lines in the P_{lith}-T plane. Characteristic of many solid-solid reactions, these are

FIGURE 21-11
• • • • • • • • • • • • • • • • •

The system Al_2SiO_5–SiO_2–H_2O as a function of temperature and pressure. In the absence of H_2O, addition of SiO_2 to the aluminum-silicates does not affect their stability (note the dashed portion of the kyanite-andalusite equilibrium), but under conditions of increased P_{H_2O} at low temperature, pyrophyllite $[Al_2Si_4O_{10}(OH)_2]$ replaces Al_2SiO_5 + 3 SiO_2. [After M. J. Holdaway, 1971, *Am. J. Sci.*, *271*, Fig. 5; and D. M. Kerrick, 1968, *Am. J. Sci.*, *266*, Fig. 4.]

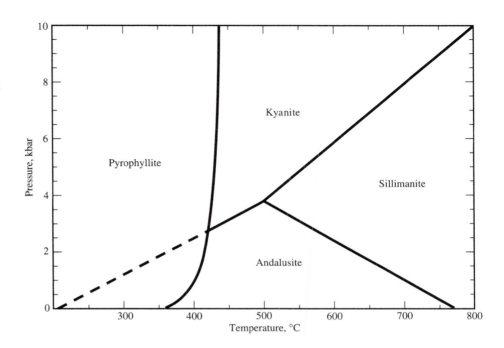

nearly straight lines, and both the position and the slope of the lines are independent of the presence or composition of any metamorphic fluid or other solid phases such as quartz (Figures 21-11 and 21-12). It is also a characteristic of solid-solid equilibria that the slopes of the univariant lines can have virtually any orientation in the P_{lith}-T plane—from vertical to horizontal, with either positive or negative slopes—depending on the minerals involved.

However, if fluid is present, then P_{lith} must be replaced by P_{fluid}, because the fluid and the solids in the rock are equally pressurized. Figure 21-11 illustrates that the presence of aqueous fluid and excess quartz with Al_2SiO_5 *does* affect the P-T diagram by stabilizing the hydrous sheet silicate pyrophyllite, $Al_2Si_4O_{10}(OH)_2$, over a P-T region that replaces parts of the stability fields of kyanite and andalusite and a major part of the kyanite $\rightleftharpoons$ andalusite reaction. The presence of fluid and quartz does not, however, affect either the positions or the slopes of the Al_2SiO_5 phase diagram at temperatures above pyrophyllite stability. If the stability relations of the Al_2SiO_5 minerals had been plotted as a function of temperature and P_{CO_2}, rather than P_{H_2O}, then the diagram would be identical to that with P_{lith}, because no new minerals can form as combinations of Al_2SiO_5 and CO_2.

Dehydration and Decarbonation Reactions

As noted before, the dehydration and decarbonation reactions are by far the most important in the evolution of the more common metamorphic rocks. Hydrous min-

erals and carbonates are virtually ubiquitous constituents of metamorphic rocks of all three major protolith types—mudrocks, calcareous rocks, and mafic volcanic rocks—up to at least 700°C (notably biotite, hornblende, and calcite at the highest metamorphic grades). Prograde metamorphism that involves only anhydrous minerals at $T > 700°C$ (the so-called granulite facies) is not common. It is essential, then, for the student of metamorphic petrology to understand the mechanisms and controls of these reactions and suitable methods for their graphical portrayal.

The presence or absence of fluids during the whole or part of the evolution of a metamorphic rock, and the form this fluid takes, have been the subject of lively debate among metamorphic petrologists for several decades. The very presence of hydrate and carbonate minerals in rocks requires the presence also of at least some molecules of water and carbon dioxide to stabilize these minerals over geologic time, although it is unclear how much water and carbon dioxide would be necessary and whether they could constitute a separate fluid phase. Fluid inclusions prove the presence of fluid species at least part of the time during metamorphism. The release of large numbers of molecules of either water or carbon dioxide (or both) clearly must occur when an active devolatilization reaction is proceeding in the rock, although again the residence time for this fluid is unknown. It is neither possible nor appropriate for this book to critically review the debate over metamorphic fluids. Because many metamorphic petrologists accept the presence of fluids, we assume a pervasive intergranular fluid phase in rocks throughout the whole cycle of

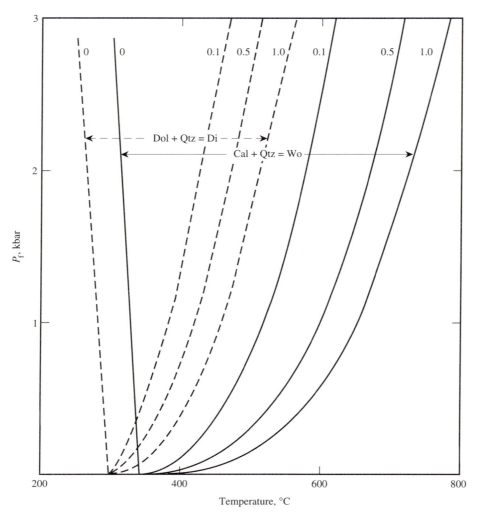

FIGURE 21-12

● ● ● ● ● ● ● ● ● ● ● ● ● ● ● ●

Univariant curves for the decarbonation reactions dolomite + quartz =
diopside + CO_2, and calcite +
quartz = wollastonite + CO_2. The diagram illustrates the effect of dilution
of carbon dioxide by water on decarbonation equilibria. For each reaction,
four curves are shown: X_{CO_2} = 1.0, 0.5,
0.1, and 0. Curves for Dol + Qtz = Di
are dashed, and curves for Cal +
Qtz = Wo are solid. Arrows show the
range in temperature for each reaction curve. Mineral abbreviations: Cal,
calcite; Di, diopside; Dol, dolomite;
Qtz, quartz; Wo, wollastonite. [After
R. J. Tracy and B. R. Frost, 1991,
Rev. Mineral., *26*, Fig. 3.]

metamorphism for our discussion of devolatilization
equilibria. The reader should note that this area of
research is very active and much is yet to be learned.

Devolatilization equilibria may be illustrated in the
P_{fluid}-T plane, starting with the breakdown of the mineral
pyrophyllite (see Figure 21-11). In the presence of excess
quartz and water, pyrophyllite is stable in metapelitic
rocks (and some metabauxites) up to about 400°C, when
it breaks down in the following balanced reaction:

$Al_2Si_4O_{10}(OH)_2$
    $= Al_2SiO_5$ (kyanite or andalusite) + 3 SiO_2 + H_2O

Examination of Figure 21-11 shows that the Al_2SiO_5
product mineral is kyanite above about 3 kbar and andalusite below 3 kbar. As a textural note, it is a common
observation in metapelitic rocks at grades just above the
first appearance of kyanite or andalusite (depending on

pressure) that the aluminum silicate mineral porphyroblasts are filled with small inclusions of quartz, making a
spongy, poikiloblastic texture. The simultaneous production of three moles of quartz for every one mole of
Al_2SiO_5 in the above reaction may mean that some of the
silica cannot diffuse away from the reaction site fast
enough and becomes trapped in a rapidly growing andalusite or kyanite crystal. A similar phenomenon commonly happens with both garnet and staurolite porphyroblasts, both of which are produced in reactions that
also produce abundant quartz.

The shape of the pyrophyllite reaction curve shown in
Figure 21-11 is characteristic for a devolatilization reaction, with a shallow positive slope at low P_{fluid} and a
steeper slope (approaching vertical) at higher pressure,
thus producing a curve that is convex toward higher
temperature. The positive slope is basically due to the Le
Châtelier principle: At higher P_{H_2O}, the thermal stability
of a hydrous mineral is increased. The curvature is

stronger than that found in most solid-solid equilibria and is due to dramatic changes in the properties of the fluid at pressures up to about 3 kbar. The P-T slope of any reaction (P/T) is a function of relative entropy (S) to volume (V) ratios of products and reactants, according to the *Clausius-Clapeyron* relationship, which states that

$$\Delta P/\Delta T = (S/V)_{\text{products}}/(S/V)_{\text{reactants}}$$

Entropy is a thermodynamic variable that expresses the relative randomness of atoms in a material and is always much higher in fluids than in crystalline solids. It tends to increase with increasing temperature. The Clausius-Clapeyron relation says that if the change in volume between reactants and products is large relative to the change in entropy, then the reaction will be mostly a function of pressure and will have a shallow P-T slope ($\Delta P/\Delta T < 1$). Conversely, if entropy change dominates, the reaction will be steep and will be more temperature dependent. At low pressure (several hundred bars), fluids have a very high volume per mole but are very compressible as pressure increases. The volume of the fluid produced in a devolatilization reaction changes rapidly relative to the solids over the first 2–3 kbar of pressure increase, then changes at about the same rate as the solids above 3 kbar. The slope of a devolatilization reaction is thus very pressure dependent below 1 kbar but much more temperature dependent above 3 kbar, as seen in Figure 21-11. This relation holds for decarbonation as well as dehydration. A simple decarbonation reaction in the presence of pure CO_2, for example, calcite + quartz = wollastonite + CO_2 (see below), has exactly the same general shape in a P_{CO_2}-T diagram (see Figure 21-12).

It is possible to have mixed fluids, as well as ones with essentially pure water or carbon dioxide. The composition of the fluid present in a rock exerts a strong influence over the stability of the mineral assemblages and controls the shape and position of devolatilization reactions in the P_{fluid}-T plane. Consider the reaction of calcite and quartz to wollastonite, shown in Figure 21-12:

$$CaCO_3 + SiO_2 = CaSiO_3 + CO_2$$

At atmospheric pressure (1 bar), the reaction occurs at 430°C. At higher P_{CO_2}, the temperature of reaction increases because of the increased stabilization of a carbonate mineral at higher P_{CO_2} (the curve labeled $X_{CO_2} = 1$). If the fluid is mixed and P_{fluid} is due to partial pressure of water as well as carbon dioxide, the position of the reaction curve will change on the P-T diagram. The reaction shifts to lower temperature because the dilution of carbon dioxide in the fluid causes a destabilization in the carbonate mineral; P_{H_2O} contributes nothing to stabilizing a mineral containing carbon dioxide. The amount

of shift depends on how much water enters the fluid (see Figure 21-12). At the limit ($X_{CO_2} = 0$), the reaction behaves in the diagram as if there were no fluid present at all and pressure were simply P_{lith}. Note that at 2 kbar, calcite and quartz would react to form wollastonite (a well-known low-pressure contact metamorphic reaction) at a little over 700°C in the presence of pure carbon dioxide, whereas in the presence of water-rich fluid, this reaction may take place at temperatures as low as 400°C. In the regional metamorphism of carbonate rocks, an ambient carbon dioxide–rich fluid would ordinarily be expected, and the rarity of regional metamorphic wollastonite is thus probably due to its formation near the high end of the possible temperature range. Wollastonite is commonly found in calcareous rocks in the contact aureoles around granitic plutons. These magmas are generally water-rich, and the abundant aqueous fluid they evolve during solidification infiltrates the country rocks and can significantly lower the temperatures of wollastonite formation and other decarbonation reactions.

The same effect occurs with dehydration reactions, except that in this case dilution of water in the fluid by carbon dioxide is what causes reaction temperature to decrease. The shape of the reaction curve changes just like that in Figure 21-12. A handy and simple way to recall this effect is simply to remember that if any component entering the fluid is not included as a reaction product, then it lowers the temperature of a devolatilization reaction.

Finally, there is a class of devolatilization reactions that have become increasingly recognized in importance. This class is the *combined* dehydration-decarbonation reaction, and such reactions are now known to be quite common in the middle grades of metamorphism of calc-silicate rocks. The reaction

$$\text{1 tremolite} + \text{3 calcite} + \text{2 quartz}$$
$$= \text{5 diopside} + \text{1 } H_2O + \text{3 } CO_2$$

which typically causes the first appearance of diopside in marbles and calc-silicate rocks, is a particularly common example. Note that the breakdown of both a hydrous mineral and a carbonate results in production of both water and carbon dioxide, in the ratio of three moles of carbon dioxide to one mole of water, and therefore the fluid produced has $X_{CO_2} = 0.75$. From the principle introduced above, we see that this reaction has its highest temperature when the fluid in the rock has exactly the same composition as the fluid produced in the reaction. For this particular reaction (and a number of similar ones), ambient fluids that are rich in either water or carbon dioxide will reduce the reaction temperature at a given P_{fluid}. Chapter 24 has a detailed discussion of this kind of reaction.

SUMMARY

• • • • • • • • • • • • • •

Metamorphic reactions and recrystallization processes are controlled by pressure, temperature, and fluids. Pressure can be either lithostatic, due to the weight of overlying rocks, or hydrostatic, that is, applied through a fluid medium. P_{lith} and P_{fluid} are thought to be equal at depths greater than 8–10 km in the crust. Temperatures of rocks are controlled by the input of thermal energy. Metamorphic fluids typically consist of mixtures of water and carbon dioxide. Water-rich fluids are typical of rocks containing hydrated minerals, whereas carbon dioxide–rich fluids occur in carbonate or calcareous rocks. Methane may be important in low-grade metamorphism of carbonaceous mudstones or other graphite-rich rocks, and minor species such as fluorine, chlorine, and hydrogen sulfide also occur locally.

There are different types of metamorphic reactions, classified according to whether they involve fluids and how relatively abruptly they occur. Discontinuous reactions are essentially univariant, occur along a line in the *P-T* plane, and involve major flips or rearrangements of tie lines between minerals in compositional phase diagrams. Occurrence of discontinuous reactions produces an abrupt appearance or disappearance of minerals from metamorphic rocks. Continuous reactions involve solid solution minerals of variable compositions. Reactions can occur by shifts in both the compositions and proportions of the minerals, without phases appearing or disappearing. These reactions occur gradually over ranges of pressure and temperature and are thus bands rather than lines in the *P-T* plane. Because of compositional variability, divariant assemblages may consist of only two minerals in a ternary diagram, and the AFM diagram is therefore subdivided into three-phase triangles and two-phase regions. Devolatilization reactions are by far the most common type of metamorphic reactions because of the wide occurrence of hydrous and carbonate minerals; they may be either continuous or discontinuous.

STUDY EXERCISES

• • • • • • • • • • • • • • •

1. The following mineral assemblages are univariant and represent ongoing reactions between the minerals and any possible fluid species. For each assemblage, write the balanced chemical reaction (including H_2O and CO_2 as necessary) that relates the minerals. There is only one possible reaction that can be written for each assemblage. Write your reactions using these pure mineral

compositions: quartz, SiO_2; muscovite, $KAl_3Si_3O_{10}(OH)_2$; orthoclase, $KAlSi_3O_8$; sillimanite, Al_2SiO_5; calcite, $CaCO_3$; dolomite, $CaMg(CO_3)_2$; diopside, $CaMgSi_2O_6$; talc, $Mg_3Si_4O_{10}(OH)_2$; almandine, $Fe_3Al_2Si_3O_{12}$; Fe-biotite, $KFe_3AlSi_3O_{10}(OH)_2$.

a.   muscovite + sillimanite + orthoclase + quartz
b.   dolomite + diopside + quartz
c.   dolomite + quartz + talc + calcite
d.   almandine garnet + muscovite + Fe-biotite + sillimanite + quartz

2. Why do devolatilization reactions (either dehydration or decarbonation) typically have positive slopes in P_f-T diagrams, with convex curvature toward the temperature axis?

3. What kind of metamorphic reaction can occur when you have the same number of minerals as components, but one or more of the minerals have solid solution compositions? In this type of reaction, do the identities or numbers of minerals change, or only the proportions of minerals?

4. Do the temperatures and pressures of solid-solid reactions depend on the presence or partial pressure of H_2O or CO_2? Explain.

5. How many different kinds of pressure can have an influence on metamorphic processes?

REFERENCES AND ADDITIONAL READINGS

• • • • • • • • • • • • • •

Brady, J. B. 1988. The role of volatiles in the thermal history of metamorphic terranes. *J. Petrol.*, *29*, 1187–1213.

Carmichael, D. M. 1969. On the mechanisms of prograde metamorphic reactions in quartz-bearing pelitic rocks. *Contrib. Mineral. Petrol.*, *20*, 244–267.

Eugster, H. P. 1983. Application of the reaction progress variable in metamorphic petrology. *J. Petrol.*, *24*, 343–376.

Frey, M. 1987. Very low-grade metamorphism of clastic sedimentary rocks. In *Low Temperature Metamorphism*, 2nd ed., ed. M. Frey. Glasgow: Blackie, pp. 9–58.

Greenwood, H. J. 1975. Buffering of pore fluids by metamorphic reactions. *Am. J. Sci.*, *275*, 573–594.

Kerrick, D. M. 1974. Review of metamorphic mixed-volatile (H_2O–CO_2) equilibria. *Am. Mineral.*, *59*, 729–762.

Kerrick, D. M. 1990. The Al_2SiO_5 Polymorphs. *Rev. Mineral.*, *22*.

Peacock, S. M. 1987. Thermal effects of metamorphic fluids in subduction zones. *Geology, 15,* 1057–1060.

Roedder, E. 1984. Fluid Inclusions. *Rev. Mineral., 12.*

Spear, F. S. 1993. *Metamorphic Phase Equilibria and Pressure-Temperature-Time Paths. Monograph 1.* Washington, DC: Mineralogical Society of America.

Spry, A. 1969. *Metamorphic Textures.* New York: Pergamon.

Tracy, R. J., and B. R. Frost. 1991. Phase equilibria and thermobarometry of calcareous, ultramafic and mafic rocks, and iron formations. *Rev. Mineral., 26,* 207–290.

Wood, B. J., and J. V. Walther. 1984. Rates of hydrothermal reactions. *Science, 222,* 413–415.

Yardley, B. W. D. 1989. *An Introduction to Metamorphic Petrology.* New York: Wiley.

Chapter 22

METAMORPHISM OF MAFIC AND ULTRAMAFIC IGNEOUS ROCKS

As noted in Chapter 19, metamorphic belts most commonly form in the tectonically active convergent areas at the margins of continents, where lithologic diversity is typical. These are plate boundary environments in which it is usual to have stratigraphic mixing of immature sedimentary rocks such as mudstones and feldspathic sandstones with intermediate to mafic volcanics such as dacites, andesites, and basalts. Tectonic slivers of altered basaltic oceanic crust and the even thicker sections of oceanic lithosphere in ophiolites can become structurally mixed with the locally derived volcanogenic sediments and flows of the island arc, as well as more distal sediments that have moved longitudinally along the trench or been transported as cover on the subducted oceanic crust.

Because mafic volcanics (basalts and andesites) and the mafic and ultramafic rocks of ophiolites commonly constitute a major fraction of lithotypes in convergent metamorphic belts, it is useful to examine systematically the metamorphic assemblages and reactions that occur within these rocks. This chapter discusses the mineralogic transformations that occur between the lowest and highest grades of metamorphism in several lithologic types and emphasizes how this information can be used to decipher the metamorphic history of these rocks. The terms *metabasic* and **metabasite** are commonly used to describe the metamorphosed equivalents of mafic igneous rock compositions.

The assemblages that occur in progressive metamorphism of hydrothermally altered volcanic protoliths also are examined in this chapter. These rocks are sometimes referred to as the "cordierite-anthophyllite" rocks (based on the typical midgrade assemblage in them) and are characterized by unusually high concentrations of magnesium, aluminum, and titanium and low silicon and iron and especially calcium. Geochemical studies have indicated that these bulk compositions have most likely been created by intense hydrothermal alteration of typical ocean floor basalts. Their occurrence in the rock record is indicative of a former mid-ocean ridge or backarc spreading ridge tectonic environment.

SEAFLOOR METAMORPHISM AND HYDROTHERMAL ALTERATION

Oceanic lithosphere consists of a relatively thin (5–10 km) layer of basaltic crust underlain by a variety of ultramafic lithologies of the upper mantle. As seen in the first part of this book, the basaltic rocks of the oceanic crust are of MORB composition and occur as flows and shallow intrusions (dikes and sills) that originated at or near mid-ocean spreading centers. The oceanic crust near spreading centers may also contain shallow magma chambers in which coarser grained gabbroic and even olivine-rich rocks such as harzburgite or peridotite form. These rocks represent initial crystallization of mid-ocean ridge basaltic magma to assemblages of anhydrous, high-temperature minerals such as olivine, pyroxenes, plagioclase, and iron-titanium oxides. Rapid cooling of pillow basalts may even produce some interstitial basaltic glass in pillow crusts. Dredging of samples from the mid-ocean ridges has shown, however, that the pristine anhydrous nature of these rocks is very short-lived in the deep-ocean environment; the formation of hydrated, lower temperature alteration minerals begins almost immediately after crystallization and is virtually ubiquitous.

The anhydrous minerals of unaltered basalt represent an approximate thermodynamic equilibrium at magma

TABLE 22-1 Characteristic mineral associations in metabasites and ultramafic rocks at various grades

Grade	Mafic compositions	High-Mg,K,Al compositions	Ultramafic compositions
Subgreenschist	Chlorite-albite-calcite-prehnite-ilmenite± actinolite±pumpellyite± epidote±Ca-zeolites	Chlorite-albite-ilmenite/rutile± phengite±calcite	Serpentine-chlorite-brucite-magnesite
Lower greenschist	Chlorite-albite-calcite-actinolite-ilmenite± epidote±stilpnomelane± dolomite	Chlorite-albite-ilmenite/rutile± muscovite/phengite	Serpentine-chlorite-talc-magnesite±tremolite
Upper greenschist	Chlorite-albite-calcite-actinolite-ilmenite± epidote±biotite	Chlorite-albite-ilmenite/rutile± muscovite±biotite	Chlorite-olivine-talc-tremolite-magnetite± anthophyllite
Epidote-amphibolite	Actinolite/hornblende-epidote-plagioclase-titanite±calcite±biotite	Chlorite-plagioclase-biotite-ilmenite/rutile± garnet±kyanite±hornblende	Chlorite-anthophyllite-olivine-tremolite-magnetite
Lower amphibolite	Hornblende-plagioclase-titanite±garnet±epidote± cummingtonite±biotite	Anthophyllite-plagioclase-biotite-ilmenite/rutile± garnet±staurolite±kyanite	Chlorite-olivine-anthophyllite-tremolite-magnetite
Middle amphibolite	Hornblende-plagioclase-titanite±garnet± cummingtonite±biotite	Anthophyllite/gedrite-plagioclase-biotite-ilmenite/rutile±garnet± kyanite±staurolite	Olivine-enstatite-tremolite-magnetite±anthophyllite± chlorite
Upper amphibolite	Hornblende-plagioclase-titanite±diopside± garnet±orthopyroxene	Anthophyllite/gedrite-plagioclase-biotite-ilmenite/rutile±sillimanite± cordierite±orthopyroxene± K-feldspar±garnet	Olivine-enstatite-hornblende-diopside± (Mg)chlorite±(Al)spinel
Granulite	Diopside-orthopyroxene-plagioclase-titanite± hornblende±garnet	Orthopyroxene-plagioclase-ilmenite/rutile-garnet± cordierite±gedrite±K-feldspar	Olivine-orthopyroxene-diopside-(Al)spinel-magnetite

solidus temperatures in excess of 1000°C. The bulk compositions of these mixtures of igneous minerals, *with water added*, are represented by more stable assemblages of hydrous and anhydrous minerals at lower temperatures. The rapidity of alteration in the mid-ocean ridge environment indicates that when fresh basalts are exposed to hydrothermal solutions, mineralogic changes are virtually instantaneous. The mechanism of introduction of water-rich solutions into newly formed seafloor rocks is rather straightforward. Observations from deep-sea drilling and deep submersibles, such as *Alvin*, indicate extensive fracturing in and near the ridges, from large-scale faults and fracture zones with spacing of kilo-

meters, to meter-spaced fractures, to centimeter-spaced cooling cracks. Fracture-enhanced permeability is thus substantial, and seawater penetrates pervasively down to several kilometers in the upper part of the oceanic crust.

Passive permeation of seafloor basalt by seawater produces a variety of low-grade metamorphic effects. *Spilites* are pillow basalts containing variable metasomatic alteration. These incipiently metamorphosed basalts contain albite-rich plagioclase and at least some hydrous secondary minerals, most commonly calcic zeolites, epidote, prehnite, chlorite, and carbonates. The replacement of normal basaltic calcic plagioclase by albite reflects two important aspects of the alteration process. First, because seawater is a complex aqueous solution with substantial contents of dissolved sodium, chloride, sulfate, and carbonate ions, interaction between seawater and basalt might be expected to result in addition of sodium to the altered rock. Furthermore, the aqueous solution may be undersaturated in basalt components such as CaO and SiO_2, and thus may dissolve and remove these components as part of the water-rock interaction. Second, the formation of secondary minerals can provide a sink for some elements that occur in the primary igneous phases. For example, the very common secondary minerals epidote and calcic zeolites (such as laumontite and stilbite) incorporate substantial amounts of both CaO and Al_2O_3. These minerals therefore accommodate the components of the anorthite molecule, which becomes unstable and disappears from the plagioclase during alteration. Because epidote also contains ferric iron, there is probably some mild oxidation during alteration. Incipiently metamorphosed basalts commonly display primary igneous textures in both hand specimens and thin section. Close examination reveals that both phenocryst and groundmass minerals may have been passively replaced by secondary minerals (see Figure 22-3). Table 22-1 summarizes characteristic minerals in metabasites for subgreenschist and higher metamorphic grades.

Alteration can also be accomplished through relatively high-temperature aqueous fluids. During the early 1980s, deep-diving submersibles accumulated evidence of hot fluids from the Galápagos Ridge and the East Pacific Rise. Ridge-crest vents release fluids at temperatures that may exceed 400°C while precipitating fine-grained sulfide minerals that accumulate in chimneys. These *black* and *gray smokers* (see Figure 8-5) are essentially seafloor hot springs releasing hydrothermal fluids that have circulated through fractures in hot young basaltic rocks. Fluid-rock interactions metasomatically alter the basalts while at the same time changing the composition of the original seawater. The intensity of rock alteration is probably correlated with proximity to fluid conduits. Approaching the fractures, assemblages progress from chlorite-albite-epidote-actinolite-quartz to chlorite-albite-quartz and finally to chlorite-quartz at sites where the highest fluid to rock ratios existed (Mottl 1983). The more altered basalts with chlorite-albite-quartz assemblages appear to be the protoliths for the unusual cordierite-anthophyllite rocks. These rocks are typically strongly depleted in calcium, relative to MORBs, and mildly to strongly enriched in magnesium, aluminum, and titanium. The predicted local variations in alteration intensity have actually been observed as bulk compositional variations in both greenstones and amphibolites from metamorphosed ancient ridge-crest environments.

BURIAL METAMORPHISM

This low-grade type of metamorphism is caused by burial of piles of sediments and volcanic rocks up to tens of thousands of meters thick at convergent margins or in basins and includes both the zeolite and prehnite-pumpellyite facies. Some petrologists assert that metamorphism at these very low grades is not typical of regional metamorphism. But evidence for metamorphism at these grades can be found where rocks contain reactive materials, either fine-grained or glassy phases. Thus burial metamorphism has typically been documented in volcanic and volcanogenic rocks that contain such reactive phases, and in fine-grained mudstones. As an example, mineralogic and textural evidence of burial metamorphism has been documented in both basaltic flows and tuffs of the Mesozoic rift basins of eastern North America, from the Carolinas to Massachusetts. No orogenic event postdates the formation of these basalts, so the metamorphism must have resulted from burial alone, although there commonly is evidence for the participation of hot (100°–200°C) aqueous fluids or brines in the incipient metamorphic process.

Zeolite facies metamorphism begins at burial depths of 1 to 5 km, depending on geothermal gradient and reactivity; these depths correspond to temperatures of 50° to 150°C. Transition to the prehnite-pumpellyite facies occurs at 3 to 13 km and at temperatures up to about 250°C. Nonmetamorphic or diagenetic rocks commonly contain glass, volcanic minerals and rock fragments, analcite and sedimentary zeolites such as mordenite, clinoptilolite, stilbite, and heulandite. Many of these phases are so fine-grained that they can only be identified by X-ray diffraction techniques. Incipient zeolite-facies metamorphism is indicated by the appearance of the calcium-aluminum zeolites laumontite and wairakite, and perhaps by albite or adularia (potassic feldspar) (see Table 22-1). Mineral identification can be done microscopically in coarse-grained rocks; otherwise X-ray diffraction analysis is necessary. Newly formed metamorphic minerals typically occur in veins, vesicles, or cavity fillings or in the fine matrix. Figure 22-1 illustrates the typical assemblages found in zeolite-facies rocks. Note that the plotting points of rock compositions predict what the assemblage will be. Compositions shown as *a* in the ACF diagram are low in aluminum and thus contain chlorite + calcite (+ quartz) assemblages, whereas compositions *b* contain enough aluminum to have some prehnite as well. The composition points predict not only the assemblage but the proportions of the plotted minerals. Relict grains, incomplete pseudomorphic replacements, and preserved igneous or sedimentary textures

FIGURE 22-1

● ● ● ● ● ● ● ● ● ● ● ● ● ● ● ● ●

Typical mineral assemblages of the zeolite facies, portrayed on ACF and A′KF diagrams. See text for explanation of letters.

(A)

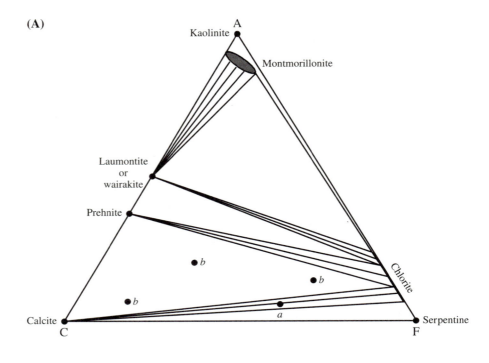

(B)

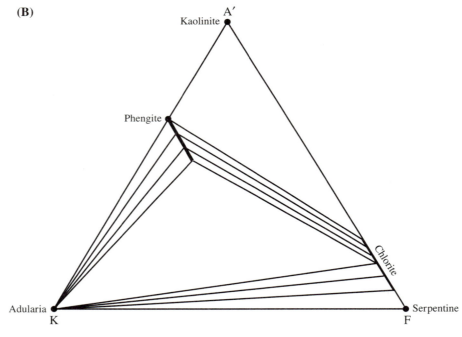

are typically present because very low grade reactions tend to approach equilibrium slowly.

The transition from zeolite to prehnite-pumpellyite facies is marked by the elimination of laumontite in more aluminous compositions and the common occurrence of prehnite, pumpellyite, calcite, and quartz. Epidote with

or without actinolite can appear. Pumpellyite may be present with actinolite if prehnite is not present, particularly in carbonate-free rocks. Typical mineral assemblages in mafic rocks at this facies are illustrated in Figure 22-2. In terranes where these two low-grade facies occur with more abundant higher grade rocks,

(A)

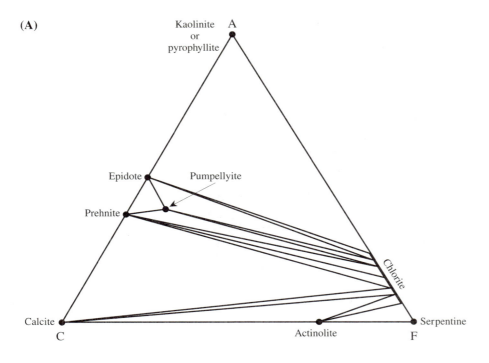

FIGURE 22-2

• • • • • • • • • • • • • • • • •

Typical mineral assemblages of the prehnite-pumpellyite facies in mafic and related compositions, illustrated in ACF and A′KF diagrams.

(B)

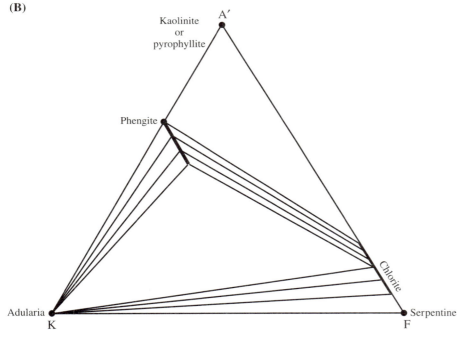

rocks of the low-grade facies are commonly assigned to subgreenschist grade. A typical example of the thin section appearance of a metamorphosed mafic volcanic is shown in Figure 22-3. The upper temperature limit of prehnite-pumpellyite facies cannot be precisely stated because it varies as a function of rock and mineral chem-

istry. Petrologic, fluid inclusion, experimental, and oxygen isotopic studies, however, place the lower limit of greenschist facies at just about 400°C.

Douglas Coombs pioneered the systematic study of these low-grade facies in the 1950s in New Zealand where zeolite- and prehnite-pumpellyite facies rocks are

FIGURE 22-3
• • • • • • • • • • • • • •

Photomicrograph of an incipiently metamorphosed, subgreenschist-facies basalt from the Catoctin Volcanics, Big Meadow, Blue Ridge, Virginia (sample of R. L. Badger). Original igneous textures are preserved: Relict plagioclase laths can be seen, along with larger calcic pyroxene phenocrysts (as at bottom), now replaced by chlorite and epidote. Plane-polarized light. Horizontal field of view is about 1.2 mm.

widespread. There they are most likely associated with the pervasive movement of geothermal fluids through shallow depositional basins. Similar low-grade rocks have now been discovered in other geothermally active regions, notably the Reykjanes thermal field in Iceland and the Salton Sea geothermal area in southern California. In drill cores from the Salton Sea field, active metamorphism has been documented at relatively shallow levels.

LOW- AND MEDIUM-PRESSURE REGIONAL METAMORPHISM
• • • • • • • • • • • • • • •

Greenschist Facies

Rocks of the greenschist facies represent the widely accepted beginning of true regional metamorphism, largely because they begin to show foliation as the effect of applied stress and deformation synchronous with mineral reactions. They are also completely recrystallized, in contrast to lower grade facies that commonly contain relict textures and mineralogy. Greenschist facies rocks are the most common of all the facies and can be found in virtually all metamorphic terranes. The name of the facies actually comes from the rock name greenschist, or **greenstone**, which is applied to metamorphosed and

foliated mafic volcanic rocks that contain the mafic mineral assemblage epidote-chlorite-actinolite.

The greenschist facies comprises the Barrovian chlorite and biotite zones (which actually refer to these minerals in metapelites; see Chapter 23) and can be subdivided into lower and upper greenschist facies. This distinction is more important for metapelites than for metabasites, because little mineralogic change occurs throughout the greenschist facies in metabasites (see Table 22-1). Typical assemblages are shown in both ACF and A'FK diagrams for lower (Figure 22-4) and upper (Figure 22-5) greenschist facies. Note from these diagrams that the upper greenschist facies lacks both the dolomite and stilpnomelane present in the lower greenschist facies; note also that biotite takes the place of stilpnomelane in potassium-bearing metabasites. Ultramafic compositions occur as serpentinites, with their mineralogy dominated by serpentine together with minor amounts of brucite, chlorite, talc, magnesite, and a calcic mineral, typically calcite at lower grades and tremolite at higher ones. Thin section views of greenschist-grade mafic and ultramafic compositions are shown in Figure 22-6.

Formerly, a special facies—the *epidote-amphibolite facies*—that is essentially transitional between greenschist and amphibolite facies was defined. Most petrologists now consider it to be the highest grade subdivision of the greenschist facies and rarely consider it separately. It corresponds loosely to Barrow's lower garnet zone in metapelites and marks the first appearance of almandine-rich garnet in many metabasites (garnet commonly appears in metabasites at a lower grade than in metapelites). At this same grade, actinolite begins to be replaced by blue-green hornblende, whereas plagioclase remains sodium-rich. Mineral assemblages in the epidote amphibolite facies are shown in Figure 22-7.

(A)

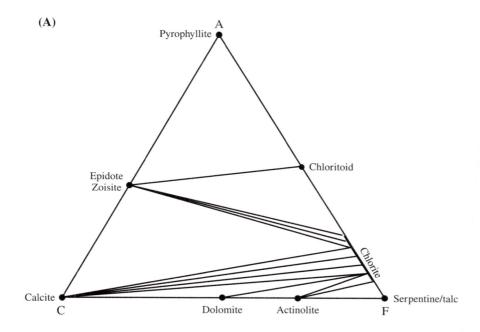

(B)

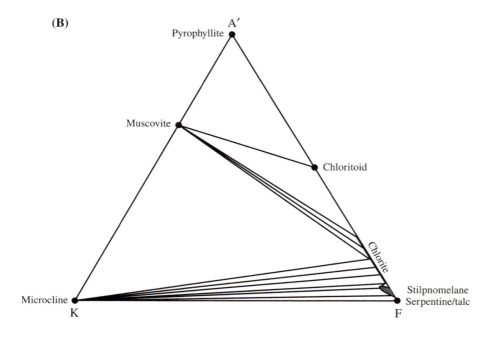

FIGURE 22-4
• • • • • • • • • • • • • • •

Typical mineral assemblages in metamorphosed mafic rocks in the lower part of the greenschist facies, illustrated in ACF and A'KF diagrams.

Amphibolite Facies

The transition from greenschist to amphibolite facies is marked by a significant number of mineralogic changes in metabasites (see Table 22-1). Although these changes do not occur all at precisely the same temperature and pressure, they nonetheless mark a notable mineralogic break. Actinolite, chlorite, and epidote all decrease in abundance, and plagioclase becomes more calcic ($>An_{20}$). Hornblende and garnet increase in abundance.

(A)

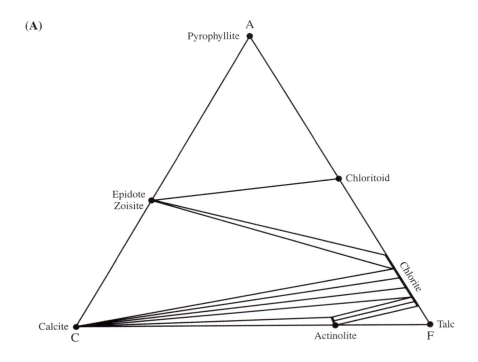

(B)

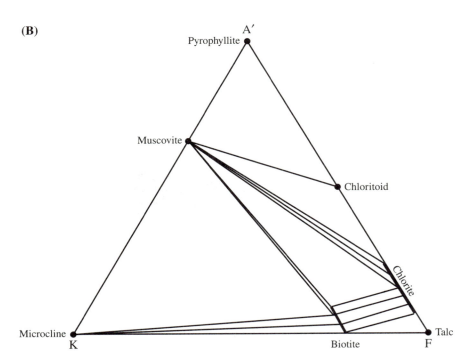

FIGURE 22-5

Typical mineral assemblages in metamorphosed mafic rocks in the upper part of the greenschist facies, illustrated in ACF and A′KF diagrams.

(A)

(B)

FIGURE 22-6

(A) Photomicrograph of a greenschist-facies mafic schist, southern Connecticut. A horizontal band of chlorite with embedded needles of actinolite is enclosed in colorless, low-relief albite and blocky, high-relief epidote. Plane-polarized light. Horizontal field of view is about 0.6 mm. **(B)** Photomicrograph of typical serpentinite, a greenschist-facies ultramafic rock from the Greek island of Naxos. Colorless material is fibrous, magnesium-rich serpentine, and opaque bands are magnetite. The right two-thirds of the photo contains a serpentinized single former olivine crystal. Plane-polarized light. Horizontal field of view is about 1.2 mm.

Actual metamorphic reactions are difficult to deduce or write because there are relatively few phases and multiple continuous reactions operate simultaneously. Progressive metamorphism is thus reflected by gradual shifts in mineral composition rather than by changes in mineral assemblage. Unfortunately, in metabasites there are more changes in mineral assemblage in the lowest part of the amphibolite facies than in most of the rest of it. This asymmetry makes it rather difficult to document changes in metamorphic grade in amphibolite-facies metabasites through petrographic techniques alone.

The amphibolite facies covers a substantial range of both pressure and temperature, but the assemblages in metabasites are not such sensitive indicators of subtle change in these variables as are the assemblages in metapelites. Both epidote and chlorite are eliminated from metabasites in the low-grade part of the amphibolite facies, with concomitant increase in both the anorthite content of plagioclase and the aluminum content of amphibole. In general, most of the amphibolite facies is marked by assemblages of hornblende and plagioclase in metabasites, with minor amounts of garnet in more

FIGURE 22-7

• • • • • • • • • • • • • • • • •

Typical mineral assemblages in metamorphosed mafic rocks in the epidote amphibolite facies, illustrated in ACF and A′KF diagrams.

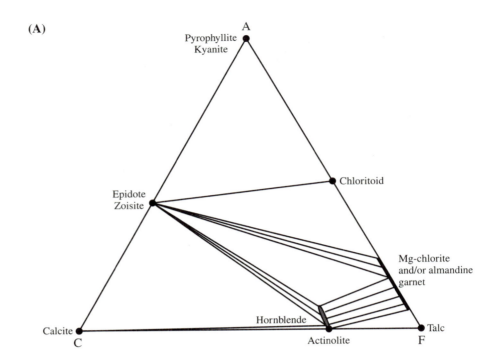

(A)

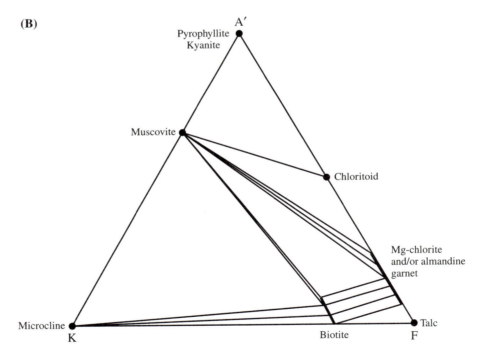

(B)

aluminous lithologies and cummingtonite in subcalcic ones. In the uppermost part of the amphibolite facies, calcic pyroxene (diopside or augite) appears as a result of incipient breakdown of calcic amphibole. These relationships are listed in Table 22-1 and are illustrated in Figure 22-8 for low-, medium-, and high-grade amphibolite-facies rocks. Note that the ACF diagrams in Figure 22-8 are for progressive temperature increase at

medium pressure. Significant shifts in assemblages or mineral compositions may occur under either lower pressure or higher pressure conditions. Figures 22-9, 22-10, and 22-11 illustrate the appearance of low-, medium-, and high-grade amphibolite-facies rocks.

Mineral assemblages in ultramafic compositions change rather dramatically through the amphibolite facies (Figures 22-12 and 22-13; also see Table 22-1). In

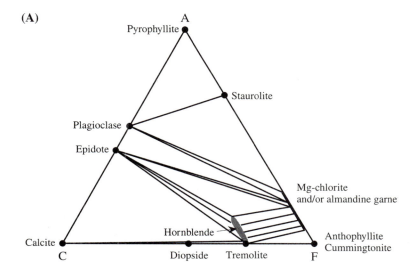

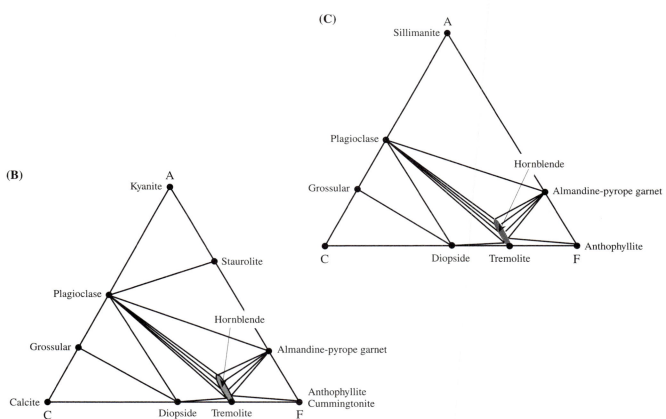

FIGURE 22-8
● ● ● ● ● ● ● ● ● ● ● ● ● ● ● ● ●

Typical mineral assemblages in metamorphosed mafic rocks in the **(A)** lower, **(B)** middle, and **(C)** upper parts of the amphibolite facies, all illustrated in ACF diagrams.

the serpentinites that characterize ultramafic rocks through a good part of the greenschist facies, olivine appears in the upper greenschist facies as a result of the reaction serpentine = olivine + talc + H_2O (going from region D to region E in Figure 22-12). Subordinate minerals include carbonates, magnetite, magnesium-rich chlorite, and aluminum-spinel. Shortly after the first appearance of olivine and talc, these two minerals react to

form anthophyllite: olivine + talc = anthophyllite + H_2O (region E to region F). Toward the upper limit of the amphibolite facies, anthophyllite breaks down to form enstatite: anthophyllite + olivine = enstatite + H_2O (region F to region G). Throughout the middle part of the facies, the calcic mineral in ultramafic compositions is typically tremolite or magnesian hornblende. Under uppermost amphibolite-facies or lower granulite-facies

FIGURE 22-9

• • • • • • • • • • • • • • • • •

Photomicrograph showing a typical lower amphibolite-facies mafic rock. The darker-colored mineral at the top and at the bottom is hornblende, the paler, high-relief mineral in the center is epidote (Ep), and the colorless minerals are plagioclase and quartz. Plane-polarized light. Horizontal field of view is about 2.4 mm.

conditions, calcic amphiboles begin to break down to diopside (region G to region H). At the boundary between amphibolite and granulite facies, therefore, the typical ultramafic assemblage is olivine + enstatite + diopside + spinel ± calcic amphibole (see photomicrograph in Figure 22-14). This mineralogy is surprisingly close to that of an original ultramafic rock before alteration and metamorphism.

In altered volcanic compositions, the lowermost amphibolite facies is marked by the disappearance of chlorite and the appearance of a low-calcium orthoam-phibole, anthophyllite. In contrast to the low-aluminum anthophyllite of ultramafic rocks (see above), orthoam-phiboles anthophyllite or gedrite in the altered volcanic rocks commonly contain substantial aluminum because the bulk compositions are rather rich in aluminum. Through the amphibolite facies, anthophyllite may gradually become richer in sodium, aluminum, and iron and thus become the more compositionally complex mineral gedrite, which has a compositional relationship to anthophyllite that is much like the relationship of hornblende to tremolite-actinolite. The basic assemblage of

FIGURE 22-10

• • • • • • • • • • • • • • • • •

Photomicrograph of typical middle-amphibolite facies mafic rock (amphibolite), Late Precambrian Bassett Formation, southern Virginia. Darker-colored grains are hornblende, and colorless grains are intermediate plagioclase (about An_{50}). Plane-polarized light. Horizontal field of view is about 2.4 mm.

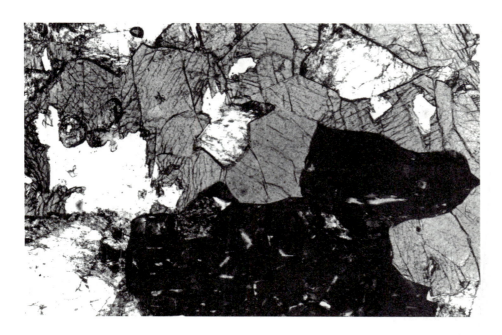

FIGURE 22-11
● ● ● ● ● ● ● ● ● ● ● ● ● ● ● ●

Photomicrograph of upper amphibolite-facies mafic gneiss (amphibolite), central Massachusetts. The dark mineral is hornblende, which is surrounded by augite (pale color) and calcic plagioclase (colorless). Plane-polarized light. Horizontal field of view is about 1.2 mm.

orthoamphibole-plagioclase-biotite persists throughout the amphibolite facies and is joined by various other aluminous minerals, depending on the exact bulk composition. Most commonly, cordierite is part of the assemblage in magnesium-rich compositions (Figure 22-15). More iron-rich compositions can have garnet in addition to, or in place of, cordierite. Quite commonly, other aluminous minerals also occur, including staurolite, kyanite (low to mid-amphibolite facies) and sillimanite (upper amphibolite). A common characteristic of the altered volcanic lithologies is mineralogic diversity. In the upper-

most amphibolite facies and into the lower granulite facies, both biotite and orthoamphibole reach their stability limits and dehydrate to form orthopyroxene and potassic feldspar (in the case of biotite).

Granulite Facies

The granulite facies is found at the extreme upper limits of crustal temperatures (700° to 900°C) and near the limits of crustal pressure (up to about 10 kbar). It is

FIGURE 22-12
● ● ● ● ● ● ● ● ● ● ● ● ● ● ● ● ●

P_{H_2O}-T diagram showing low- to medium-pressure phase relations for metaperidotites as modeled in the chemical system $CaO-MgO-SiO_2-H_2O$. Reaction curves are labeled with the solid reactant and product minerals; H_2O is a product phase on the high-temperature side of each reaction. Mineral abbreviations: Atg, antigorite; Ath, anthophyllite; Brc, brucite; Ctl, chrysotile; Di, diopside; En, enstatite; Fo, forsterite; Tlc, talc; Tr, tremolite. The capital letter in each P-T region corresponds to the chemographic representation of assemblages for that region shown on the $CaO-MgO-SiO_2-H_2O$ diagram in Figure 22-13. [From R. J. Tracy and B. R. Frost, 1991, *Rev. Mineral.*, *26*, Fig. 33.]

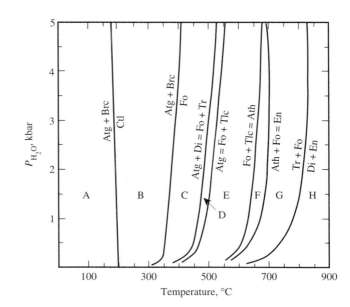

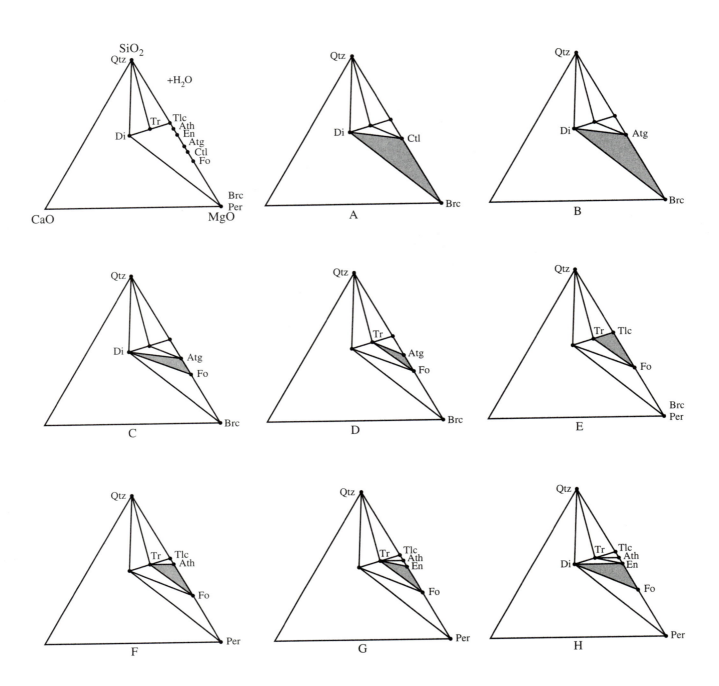

FIGURE 22-13

• • • • • • • • • • • • • • • •

Chemographic relations in the system CaO–MgO–SiO$_2$–H$_2$O
for a variety of ultramafic compositions with increasing metamorphic grade. Letters A–H refer to divariant *P-T* fields in
Figure 22-12. The shaded areas indicate the stable assemblage at
each grade for typical metaperidotite compositions. All diagrams are projected from water (that is, the assemblages all
have water in excess). Except for Per (periclase), abbreviations
are the same as those in Figure 22-12. [From R. J. Tracy and
B. R. Frost, 1991, *Rev. Mineral.*, *26*, Fig. 34.]

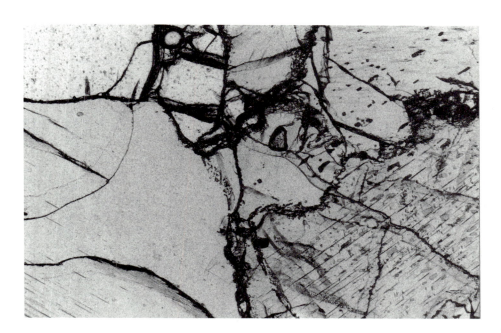

FIGURE 22-14
• • • • • • • • • • • • • • • •

Photomicrograph of a granulite-facies ultramafic rock from an ultramafic xenolith, Tahiti. The left side of the photo consists of colorless olivine grains; a calcic clinopyroxene is in the upper right and an orthopyroxene in the lower right. Both pyroxenes contain very thin, transparent exsolved blebs of aluminous spinel. Plane-polarized light. Horizontal field of view is about 1.2 mm.

presumed that the deep continental crust consists of granulite-facies rocks, and they form only in the hottest and most intensely deformed parts of orogenic belts. Most granulite-facies rocks are relatively coarse-grained through long annealing. The lower temperature portion of the granulite facies is sometimes called the hornblende-granulite facies, because the hydrous minerals hornblende and biotite commonly occur at this grade. Rocks of the upper granulite facies contain only anhydrous minerals. One curious fact about granulites is that they typically occur in significant volumes in discrete terranes, most commonly of Precambrian age.

Many of the earth's continental shield areas are predominantly of granulite facies. It is unclear whether this is the result of great age or of deep erosion (or both). In general, however, shield-type granulite-facies terranes do not represent current lower crust with shallow mantle just below it. Geophysical data typically show relatively normal crustal thickness beneath such granulite terranes. If they indeed originally formed as deep continental crust, new lower continental crust has formed beneath them as they have been unroofed by erosion.

In metabasites, the transition from amphibolite to granulite facies is not sharp but is generally marked by

FIGURE 22-15
• • • • • • • • • • • • • • •

Photomicrograph of typical orthoamphibole-cordierite gneiss, Orijärvi, southwestern Finland. Large, pale-colored blades of anthophyllite are surrounded by colorless cordierite. Dark, very high relief euhedral grains in both anthophyllite and cordierite are rutile, a very common accessory mineral in this rock type. Plane-polarized light. Horizontal field of view is about 1.2 mm.

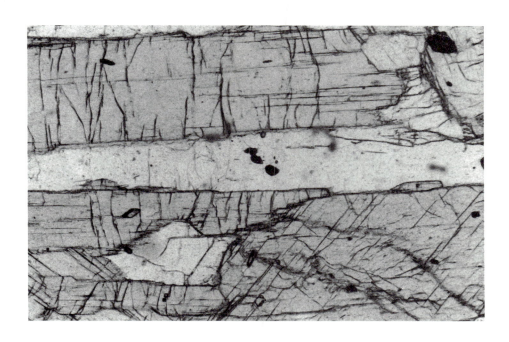

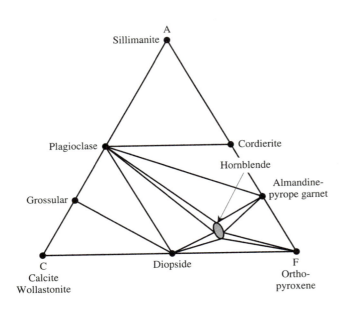

FIGURE 22-16
● ● ● ● ● ● ● ● ● ● ● ● ● ● ● ●

Typical granulite facies assemblages in metamorphosed mafic rocks illustrated in an ACF diagram.

the coexistence of calcic pyroxene and orthopyroxene. Thus, the assemblage hornblende-plagioclase-diopside would be considered as amphibolite facies, whereas hornblende-plagioclase-diopside-hypersthene would be granulite facies. Garnet is a common accessory mineral in granulite-facies metabasites. The typical granulite-facies assemblage is shown graphically in Figure 22-16 and illustrated by the photo in Figure 22-17. Note that the breakdown of hornblende results in coexistence of the four phases diopside + orthopyroxene + plagioclase + hornblende in a reaction relationship. In altered volcanic lithologies, the common granulite-facies assemblage is orthopyroxene + plagioclase + potassic feldspar, again with any of a variety of aluminous minerals including sillimanite, cordierite, garnet, and spinel, depending on bulk composition. In ultramafic rocks, calcic amphibole disappears in the granulite facies, creating a spinel peridotite assemblage of olivine + orthopyroxene + diopside + spinel.

One common problem in interpretation of the granulite facies assemblages in metabasites is deciding whether the rocks are actually metamorphic or igneous. Mantle-derived basaltic or gabbroic magmas that intrude the lower continental crust might be expected to crystallize directly from a melt to a granulite-facies mineralogy.

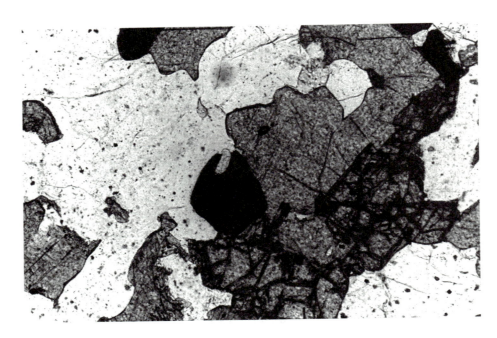

FIGURE 22-17
● ● ● ● ● ● ● ● ● ● ● ● ● ● ●

Photomicrograph of a granulite-facies mafic gneiss from the Adirondacks, New York. The highest relief darker mineral is orthopyroxene, and the slightly lower relief mineral is augite; the colorless, low-relief mineral is calcic plagioclase, and the opaque mineral is magnetite. Plane-polarized light. Horizontal field of view is about 1.2 mm.

HIGH-PRESSURE REGIONAL METAMORPHISM
• • • • • • • • • • • • • • •

Blueschist Facies

The blueschist facies is rare and is formed along convergent margins in environments of high pressure and low temperature. Most blueschist-facies rocks that are actually blue in color are of MORB or oceanic tholeiite composition and contain the unusual mineral glaucophane, a sodium-rich amphibole (Figure 22-18). Glaucophane is blue to blue-black in hand specimen and pale blue in thin section, and its presence gives rise to a notably blue color in rocks. Other minerals typical of this facies are colorless or pale green jadeite (sodic pyroxene), lawsonite, and aragonite, the high-pressure polymorph of $CaCO_3$. Conditions estimated for the formation of blueschists range from 200°C at 6–8 kbar to 550°C at 14 kbar.

Most of the classic blueschist-facies terranes are located in currently or formerly active subduction complexes, many of the more recent ones in the circum-Pacific belts. A high-pressure to ultrahigh-pressure metamorphic belt in eastern China (the Dabie Mountains) that contains both blueschists and eclogites (see below) has long been the source of materials for classic Chinese jade. The Franciscan metamorphic complex of the western Coast Ranges in California, just east of San Francisco, has been widely described as a good example of the relatively higher pressure part of the blueschist facies. The Franciscan complex even contains blocks of eclogite that have been partially retrograded to more normal amphibolites along with adjacent blueschist. The association of blueschist and eclogite also occurs in New Caledonia. The Sanbagawa belt of Japan is a good example of the lower part of the blueschist facies. Older examples of blueschists are relatively rare because of the tendency for such rocks to be overprinted by continuing or later metamorphism under amphibolite-facies conditions. Textural relicts of blueschist mineralogy on a thin section or outcrop scale have been described. Good examples include the occurrences of Ordovician (Taconic) blueschist relicts from central Vermont (Laird and Albee 1981) and of Ordovician high-pressure assemblages in western North Carolina.

FIGURE 22-18
• • • • • • • • • • • • • • •

Photomicrograph of a blueschist-facies mafic schist from Cazadero, northern California. The darker elongated grains are glaucophane; pale, high-relief grains are epidote; and colorless, tabular grains are phengitic muscovite. Plane-polarized light. Horizontal field of view is about 2.4 mm.

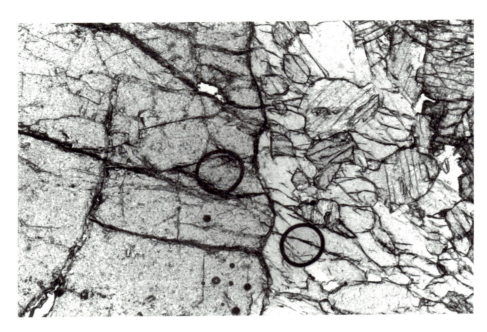

FIGURE 22-19
• • • • • • • • • • • • • • • •

Photomicrograph of eclogite, western Norway. Large magnesium-rich garnets (as on the left) are surrounded by finer-grained omphacitic pyroxene (right). Plane-polarized light. Horizontal field of view is about 1.2 mm.

Eclogite Facies

Eclogite-facies rocks are rare and typically of basaltic composition. Most eclogites are striking in appearance, with coarse reddish brown garnet porphyroblasts set in a fine- to medium-grained matrix of green omphacite (Figure 22-19). The garnet is magnesium (pyrope)-rich, and omphacite is a sodic pyroxene compositionally halfway between jadeite and diopside. The texture is generally massive. Unusually aluminous eclogites may contain kyanite. There are several modes of occurrence of eclogites, which were categorized by Coleman and others (1965) on the basis of typical tectonic associations. Group A samples have been brought to or near the surface as xenoliths in intraplate alkali basalt or kimberlite eruptions and probably have been derived from a layer of eclogite that forms the transition between the continental crust and the upper subcontinental mantle. Group B eclogites are found mixed with high-grade gneissic rocks in orogenic belts, such as the important occurrences along the Norwegian coast. Finally, Group C eclogites are associated with blueschists and other high-pressure, low-temperature rocks at young convergent

margins like those in the Alps and in the Franciscan complex of California. Although some doubt has been cast on Coleman's classification, particularly on the distinctions between Group B and Group C, it has been widely used in eclogite literature.

The disparate occurrences suggest that there is no single explanation for all eclogites and that each occurrence must be evaluated on its own merits. The common association of eclogite bodies with gneissic rocks of the continental crust (like those in Norway) provides an intriguing mechanism for the necessarily rapid uplift of these dense rocks to the surface. Eclogites are significantly denser even than normal metabasites and thus might not be expected to rise through the crust at all. When they are tectonically mixed with much greater volumes of low-density granitic rocks in continental collision zones, however, the density problem is overcome. Eclogites remain an enigmatic rock type and are the subject of much research.

The temperature range for eclogites is not well known, although some that apparently form at temperatures as low as 500°C are interbedded or mixed with blueschists, as seen in the Franciscan complex of

California and in New Caledonia. Others may exceed 700°C. Yardley (1989) notes 12 kbar as an effective lower pressure limit for eclogite formation and points out that pressures as high as 30 kbar have been proposed for some eclogite-facies rocks based on the occurrence in them of coesite, the high-pressure polymorph of SiO_2.

SUMMARY
• • • • • • • • • • • • • •

Metamorphism in many of the mafic and ultramafic rocks found in orogenic belts occurs when these rocks are newly formed volcanics, shallow intrusives, or ophiolitic associations at, or near, a mid-ocean ridge. Deep penetration of seawater into cracks and fissures promotes pervasive hydration of igneous minerals and glassy material, and commonly significant metasomatic alteration. By the time these metabasites and ultramafic rocks become involved in subduction at convergent margins, they contain virtually no relict igneous material and begin metamorphism as fully hydrated, low-grade rocks. Arc volcanics do not undergo this seafloor metamorphism but become hydrated and altered by burial in the arc complex. In contrast, burial of mafic volcanics and volcanogenic sediments in nonconvergent environments results in passive replacement of igneous materials by zeolites and other low-temperature minerals. This burial metamorphism is common in geothermally active areas but rarely is found in orogenic belts.

The first true regional metamorphism is in the greenschist facies. Metabasites contain assemblages of chlorite, epidote, actinolite, and albite; the abundance of green minerals in these rocks gives rise to the name of the facies. Prograde metamorphism through the greenschist and into the amphibolite facies involves progressive dehydration. Chlorite disappears by lower amphibolite facies, as does epidote, and actinolite is gradually replaced by aluminous hornblende. Garnet appears at about the greenschist-amphibolite boundary. Breakdown of other calcic minerals produces anorthite component that is added to albite and makes the plagioclase more calcic. Through most of the amphibolite facies, metabasites are genuinely amphibolites consisting mainly of hornblende and intermediate plagioclase. Pyroxenes then begin to appear, diopside in the upper amphibolite facies and orthopyroxene in the lower granulite facies. Ultimately anhydrous assemblages of pyroxenes and plagioclase (plus or minus olivine) constitute the widespread pyroxene granulites of the granulite-facies Precambrian shield areas.

Unusual high-pressure facies characterize certain convergent margins. Blueschist-facies metabasites are created in subduction zone metamorphism and reflect unusually high pressures (8–14 kbar) at anomalously low temperatures (300°–500°C). Eclogites are commonly associated with blueschists, but some eclogites may actually form in the upper mantle, or as a transitional rock type between mantle peridotites and crustal rocks.

STUDY EXERCISES
• • • • • • • • • • • • • •

1. Most kinds of metamorphism are thought to involve progressive devolatilization of maximally hydrated or carbonated sedimentary rocks. Is this true of typical seafloor metamorphism? What are the protoliths and what are the products of seafloor metamorphism?
2. What mineralogic changes in metamorphosed mafic rocks mark the boundary between greenschist facies and amphibolite facies? Is this an abrupt or gradual transition during heating?
3. In ultramafic rocks, what are typical mineral assemblages for greenschist-, amphibolite-, and granulite-facies conditions, respectively?
4. Eclogites are rare metamorphic rocks that form at unusually high pressure. Does the presence of eclogites in a metamorphic belt convey any special tectonic implication?

REFERENCES AND ADDITIONAL READINGS
• • • • • • • • • • • • • •

Abbott, R. N. 1982. A petrogenetic grid for medium and high grade metabasites. *Am. Mineral.*, *67*, 865–876.

Bucher, K., and M. Frey. 1994. *Petrogenesis of Metamorphic Rocks* (6th ed. of Winkler's textbook). New York: Springer-Verlag.

Coleman, R. G., D. E. Lee, L. B. Beatty, and W. W. Brannock. 1965. Eclogites and eclogites: Their differences and similarities. *Geol. Soc. Am. Bull.*, *76*, 483–508.

Coombs, D. S., A. J. Ellis, W. S. Fyfe, and A. M. Taylor. 1959. The zeolite facies, with comments on the interpretation of hydrothermal synthesis. *Geochim. Cosmochim. Acta*, *17*, 53–107.

Cooper, A. F. 1972. Progressive metamorphism of metabasic rocks from the Haast Schist group of southern New Zealand. *J. Petrol.*, *13*, 457–492.

Ernst, W. G. 1973. Blueschist metamorphism and *P-T* regimes in active subduction zones. *Tectonophysics*, *17*, 255–272.

Ernst, W. G. 1988. Tectonic history of subduction zones inferred from retrograde blueschist *P-T* paths. *Geology, 16,* 1081–1084.

Evans, B. W. 1977. Metamorphism of alpine peridotite and serpentinite. *Ann. Rev. Earth Planet. Sci., 5,* 397–447.

Evans, B. W, and E. H. Brown, eds. 1986. *Blueschists and Eclogites. Memoir 164.* Boulder, CO: Geological Society of America.

Frey, M., C. DeCapatani, and J. G. Liou. 1991. A new petrogenetic grid for low-grade metabasites. *J. Metam. Geol., 9,* 497–509.

Laird, J., and A. L. Albee. 1981. High-pressure metamorphism in mafic schist from northern Vermont. *Amer. J. Sci., 281,* 97–126.

Miyashiro, A. 1973. *Metamorphism and Metamorphic Belts.* New York: Halsted Press.

Miyashiro, A. 1994. *Metamorphic Petrology.* New York: Oxford University Press.

Mottl, M. J. 1983. Metabasalts, axial hot springs and the structure of hydrothermal systems at mid-ocean ridges. *Geol. Soc. Am. Bull., 94,* 161–180.

Ringwood, A. E. 1975. *Composition and Petrology of the Earth's Mantle.* New York: McGraw-Hill.

Ringwood, A. E., and D. H. Green. 1966. An experimental investigation of the gabbro-eclogite transformation and some geophysical implications. *Tectonophysics, 3,* 383–427.

Spear, F. S. 1981. An experimental study of hornblende stability and compositional variability in amphibole. *Am. J. Sci., 281,* 697–734.

Spear, F. S. 1993. *Metamorphic Phase Equilibria and Pressure-Temperature-Time Paths. Monograph 1.* Washington, DC: Mineralogical Society of America.

Starkey, R. J., and B. R. Frost. 1990. Low-grade metamorphism of the Karmutsen Volcanics, Vancouver Island, British Columbia. *J. Petrol., 31,* 167–195.

Tracy, R. J., and B. R. Frost. 1991. Phase equilibria and thermobarometry of calcareous, ultramafic and mafic rocks, and iron formations. *Rev. Mineral., 26,* 207–290.

Yardley, B. W. D. 1989. *An Introduction to Metamorphic Petrology.* New York: Wiley.

Chapter 23

METAMORPHISM OF ALUMINOUS CLASTIC ROCKS

Clastic sedimentary rock protoliths constitute a major proportion of lithic types in regional metamorphic belts. They may represent transported, continentally derived, sedimentary materials that have accumulated either in passive margin sedimentary environments (continental shelf-slope-rise systems) or along active continental margins (magmatic arc-slope-trench systems). Offshore island arcs also contain clastic depositional environments, for example, in forearc basins and the trench, and there may be a substantial contribution of volcaniclastic material derived from intermediate and mafic volcanism. Continental clastic sediments are typically dominated by clays, quartz, feldspars, or lithic fragments, depending on transport distances. Clay-rich mudrocks represent low-energy environments, whereas feldspathic or clayey sandstones (sometimes called graywackes) are more typical of variable energy regimes such as continental slope debris flows or turbidity currents. Consistently higher energy environments can produce sandstones dominated by either feldspars or quartz, although these are less common in the rock record than mudrocks, clayey sandstones, or lithic sandstones.

Regional metamorphism of clay-rich mudrocks produces the family of metamorphic rocks referred to as *pelitic schists*. This name refers to a mudrock or shale protolith as well as to the schistose texture that is typical of these rocks and is due to high mica or chlorite content. At the highest grades, breakdown of micas in pelitic rocks leads to gneissic rather than schistose fabric. Considerable variation in bulk composition occurs (as in most clastic rocks), but pelitic schists are typically high in alumina and alkalis (especially potassium) and low in calcium. Many mineral reactions are possible in pelitic rocks; thus the mineral assemblages can be very sensitive indicators of relative metamorphic grade. Conversely, metamorphosed sandstones tend to be mineralogically simpler and are texturally more massive at all grades because they commonly lack sufficient clays to produce abundant metamorphic micas. Metamorphosed

feldspathic sandstones are sometimes referred to as *granulites* (not to be confused with granulite facies) or *granofelses* and are commonly dominated by quartz and feldspars. Because such feldspathic rocks are relatively close in composition to granite magmas, during ultrametamorphism they may melt almost entirely and probably are the sources of major volumes of the crustally derived granite magmas that form batholiths in orogenic belts.

ALUMINOUS ROCK COMPOSITIONS

The bulk compositions of many chemically analyzed pelitic schists lie relatively close to the accepted standard for mudrocks, the North American Shale Composite (NASC). Table 23-1 gives this composition along with a sampling of aluminous rock compositions. A common feature of most pelitic rocks is moderate to high alumina and potash contents, an artifact of high clay content of mudrock protoliths. This high alumina has several important mineralogic consequences. First, throughout most of the range of metamorphic grades, pelitic schists are rich in the micas muscovite and biotite, along with quartz and plagioclase. Second, excess alumina in the bulk composition necessitates the presence of one or more additional aluminous minerals, many of them ferromagnesian. At low grades, these are typically chlorite and garnet; at medium grades, garnet, staurolite, andalusite, and kyanite; and at high grades, garnet, cordierite, and sillimanite.

The exact limits of occurrence and coexistence of these minerals in pelitic rocks are a complex function of the pressure of metamorphism and subtle aspects of rock composition. For example, high metamorphic pressures favor kyanite plus garnet and low pressures favor andalusite plus cordierite. At moderate metamorphic grades,

TABLE 23-1 Representative chemical compositions of metamorphosed mudrocks and sandstones

Component	1	2	3	4	5	6	7
SiO_2	64.82	60.48	62.58	60.34	56.25	67.3	60.9
TiO_2	0.80	0.91	0.87	0.76	1.05	0.6	0.6
Al_2O_3	17.05	16.58	18.09	17.05	20.18	15.5	16.4
FeO	5.70	8.10	6.85	8.19	10.35	4.2	5.7
MnO	0.25	0.13	0.09	0.09	0.18	0.1	0.1
MgO	2.83	6.35	2.18	2.69	3.23	1.9	3.1
CaO	3.51	2.31	0.16	1.45	1.54	0.6	3.9
Na_2O	1.13	1.80	0.81	1.55	1.80	4.2	4.2
K_2O	3.97	3.17	3.68	3.64	4.02	3.2	0.6
P_2O_5	0.15	0.17	—	0.14	0.19	0.1	0.1
H_2O	—	—	4.18	4.25	3.02	1.8	3.7
CO_2	—	—	0.03	1.05	0.18	0.2	0.1
Total	100.21	100.00	99.52	101.20	101.99	99.7	99.4

1: "North American Shale Composite." Unpublished wet chemical analysis by P. W. Gast, reported in Gromet et al. (1984). (Reported on a volatile-free, reduced-iron basis.) 2: Composite of metamorphosed North American "shales." Unpublished data of P. W. Gast, reported in Gromet et al. (1984). (Reported on a volatile-free, reduced-iron basis.) 3: Average composition of low-grade pelitic rocks from Littleton Formation, New Hampshire (Shaw 1956). 4: Average of roughly 100 published shale and slate analyses (compiled by Ague 1991). 5: Average of roughly 150 published analyses of amphibolite-facies pelitic schists and gneisses (compiled by Ague 1991). 6: Feldspathic sandstone, Franciscan complex, San Francisco Peninsula, California (Bailey, Irwin, and Jones 1964). 7: Lithic sandstone, Franciscan complex, San Francisco Peninsula, California (Bailey, Irwin, and Jones 1964).

more iron-rich compositions are likely to contain garnet and staurolite, whereas andalusite (or kyanite) and cordierite would occur in more magnesium-rich rocks at the same grades. Therefore, both P-T and AFM diagrams must be used to examine the dependence of pelitic mineral assemblages on both pressure and composition. In fact, natural mineral assemblages in various bulk compositions have long been used by petrologists as a natural laboratory to "map out" the tie line arrays on AFM diagrams for a range of metamorphic grades.

The continuum of clastic sedimentary rock chemical compositions creates a wide range of possible metamorphic protoliths, from the highly aluminous pelitic rocks just discussed to the less aluminous feldspathic sandstones and quartz sandstones. Very clay-rich sandstones, especially lithic and feldspathic sandstones, approach the alumina contents of pelitic rocks and thus are sometimes referred to as **semipelites.** Bulk compositions range continuously in alumina content from very high aluminum mudrocks to feldspathic or lithic sandstones. These latter rocks have granoblastic textures after metamorphic recrystallization and are gneissic at essentially all metamorphic grades. Some semipelites are moderately micaceous and schistose and can contain porphyroblasts as rich in alumina as staurolite and garnet,

although aluminum silicates rarely occur in these rocks. Interpretations of sedimentary or tectonic environments that use protolith assumptions for the gneissic rocks must be done with caution because of the potential similarity of bulk compositions of lithic sandstones to intermediate and felsic volcanics.

The organization of the following sections is necessarily complex because of the different sequences of metamorphic reactions that occur in low, medium, and high metamorphic field gradients (MFGs; see Chapter 21), especially in pelites. MFGs are referred to as low or high on the basis of the temperatures reached at a given depth, similar to the nomenclature for geothermal gradients. Remember that a *low MFG* for a terrane is akin to a high-pressure facies series. It represents a family of actual P-T paths along which maximum temperatures are reached at unusually high pressures (similar to blueschist facies), whereas a *high MFG* represents P-T paths with maximum temperatures at relatively low pressures (akin to high-temperature, low-pressure contact metamorphism). Very low grade (sub-greenschist facies) effects are discussed in a single section in this chapter, but separate sections are devoted to metamorphic sequences and mappable isograds for high-pressure and low-pressure MFGs.

VERY LOW GRADE METAMORPHISM

The inception of metamorphism in mudrocks was discussed in Chapter 20. To recapitulate, both textural and mineralogic reactions occur as metamorphism succeeds diagenesis and lithification. These changes occur as a response both to temperature increase and development of directed pressure during burial. Thermal annealing of clay-sized quartz and sodic plagioclase grains produces larger new individual crystals that ultimately become visible with the microscope. Clay mineral grains also anneal while simultaneously undergoing mineralogic reactions that ultimately lead to production of sericitic (fine-grained) white mica, chlorite, and, possibly, stilpnomelane, a ferromagnesian sheet silicate that is a precursor of biotite. The weight of overlying rocks during burial causes planar alignment of these platy minerals into a roughly horizontal fabric (the texture called foliation), although realignment to nonhorizontal orientation commonly occurs during folding in slates. An excellent and detailed review of low-grade metamorphism of mudrocks can be found in Frey (1987).

An essential aspect of the initiation of metamorphism is the onset of wider scale mineralogic reactions between detrital grains and diagenetic minerals, if any, as the rock begins to act as an equilibrium intergranular chemical system. Observation of low-grade rocks indicates that kinetics at such low temperatures are apparently sluggish and reactions incomplete, but it is an important observation that intergranular chemical processes are occurring. Low-grade minerals that can be reacted away before the conditions of the greenschist facies are reached include kaolinite, iron hydroxides, and calcic zeolites. At present, petrologists do not entirely understand the mechanisms by which such reactions occur, but they apparently involve diffusional communication through intergranular fluids. Because diffusion rates are direct functions of temperature, slow or incomplete reactions are typical at low temperatures. Furthermore, the process of low-grade metamorphism is probably not isochemical. The pore fluids that are gradually expelled as pressure increases are aqueous fluids, which can carry away significant amounts of dissolved constituents such as chlorides and ions derived from mineral-fluid reactions, especially silica and alkalis.

Zonal mapping of metamorphic grade within very low grade rocks is generally not possible with standard field or microscopic techniques because of the fine grain size and subtle mineralogic shifts. As noted in Chapter 20, however, X-ray analysis of the degree of crystallinity of the clay mineral illite can be used as a monitor of relative metamorphic grade (Frey 1987). This technique has now been successfully applied in the foreland metamorphic zones of the Alps and other belts in western Europe and North America. With temperature increase, illite is gradually converted to a sericitic white mica rich in the ferromagnesian phengite component, a muscovite molecule containing excess silicon, with iron and magnesium substituting for aluminum. Thin section views of protometamorphic aluminous rocks are shown in Figure 23-1.

The typical cumulative result of very low grade reactions in metapelites is to produce the assemblage quartz-sodic plagioclase-white mica-chlorite by the lowermost part of the greenschist facies. Other commonly occurring minerals can include calcite, ankerite, graphite, stilpnomelane, potassic feldspar, and pyrite. The abundance of carbonates, graphite, and pyrite varies widely among the spectrum of pelitic compositions, depending on the sedimentary geochemistry of the depositional environment. Admixed shelly material in nearshore muds can produce an unusual class of calcic mudrocks, called **calc-pelites,** in which calcite or ankerite are anomalously abundant. These rocks contain metamorphic assemblages that represent part of a spectrum with the so-called calc-silicates (see Chapter 24). Graphite and pyrite appear to be abundant in deeper water basinal mudrocks characterized by low sedimentation rates and chemically reducing conditions (so-called black shales), as demonstrated by R. A. Berner (1981) and others. Graphite results from recrystallization of organic matter during incipient metamorphic heating and typically persists as a metamorphic mineral to the highest metamorphic grades. Metapelites rich in graphite may also contain anomalous abundances of uranium- and thorium-bearing minerals such as zircon and monazite because of the tendency for these heavy radioactive elements to adsorb to sedimentary organic matter.

METAMORPHISM AT MODERATE MFG: THE BARROVIAN SEQUENCE

This variety of metamorphism is typical of tectonic regions where rates of burial are rapid or heat flux is low (or both) and uplift commences rapidly. Sedimentary or volcanic rocks are thus metamorphosed along P-T trajectories on which maximum temperatures occur within the kyanite stability field or near the kyanite-sillimanite boundary. The mineral assemblages in pelites in this type of metamorphism therefore include kyanite or other high-pressure aluminous porphyroblast minerals up to the highest grades, where sillimanite can occur, especially at moderate MFG. The Barrovian facies series of

Scotland falls within this metamorphic type, and the Barrovian metamorphic sequences can be used as a general model (Table 23-2).

The following discussion focuses on the hypothetical reactions by which progressive aluminous minerals are initially stabilized. The general P-T field of stability for any mineral is bounded at low temperature by one or more initial reactions of formation and at high temperature by ultimate breakdown reactions. Minerals commonly appear or disappear in a given rock, however, by reactions that occur within the stability limits and thus are not the same as those that bound the minerals'

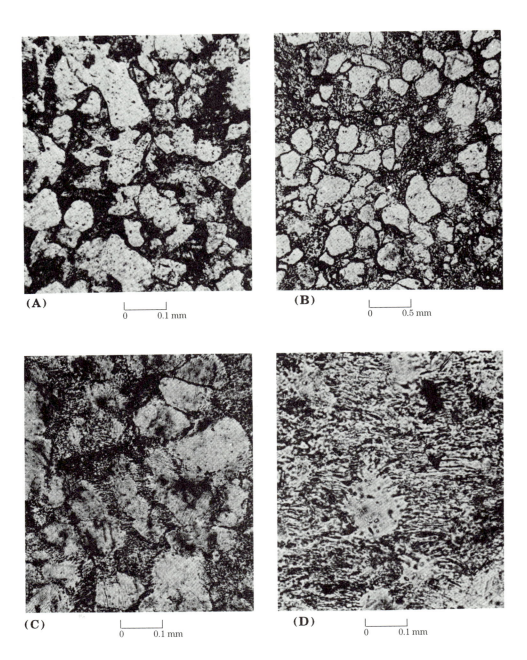

(A)
0   0.1 mm

(B)
0   0.5 mm

(C)
0   0.1 mm

(D)
0   0.1 mm

FIGURE 23-1

• • • • • • • • • • • • • • • •

Textural changes in the transition zone from diagenesis to metamorphism. (A) Zone of unaltered clay cement in marl. (B) Zone of altered argillaceous and authigenic quartz cement in shale. (C) Zone of chlorite-hydromica cement and quartzite-like structures in slate. (D) Zone of spiny structures and chlorite-micaceous cement at the transition from slate to phyllite. [From M. Frey, 1970, *Sedimentology, 15*, Fig. 19.]

TABLE 23-2 Typical silicate minerals in aluminous clastic rocks in the Barrovian zones

Barrovian zone	Typical pelites[a]	High-Al pelites[a]	Low-Al semipelites[a]
Subchlorite	Albite, microcline, clays, zeolites	Albite, microcline, clays, zeolites	Albite, microcline, clays, zeolites
Chlorite	Albite, phengitic muscovite, chlorite	Albite, phengitic muscovite, chlorite, pyrophyllite	Albite, microcline, chlorite, phengitic muscovite
Biotite	Albite, muscovite, chlorite, biotite	Albite, muscovite, chlorite, pyrophyllite	Plagioclase, microcline, chlorite, muscovite, biotite
Garnet	Plagioclase, muscovite, biotite, garnet, chlorite	Plagioclase, muscovite, chlorite, garnet, chloritoid, pyrophyllite, ±kyanite	Plagioclase, microcline, biotite, muscovite, garnet
Staurolite	Plagioclase, muscovite, biotite, garnet, staurolite	Plagioclase, muscovite, garnet, chloritoid, staurolite, kyanite	Plagioclase, microcline, biotite, muscovite, garnet
Kyanite	Plagioclase, muscovite, biotite, garnet, kyanite, ±staurolite	Plagioclase, muscovite, biotite, garnet, staurolite, kyanite	Plagioclase, orthoclase, biotite, muscovite, garnet
Sillimanite	Plagioclase, muscovite, biotite, garnet, sillimanite	Plagioclase, muscovite, biotite, garnet, sillimanite	Plagioclase, orthoclase, biotite, muscovite, garnet
Sillimanite + potassic feldspar	Plagioclase, orthoclase, sillimanite, garnet, biotite, ±cordierite, ±hypersthene	Plagioclase, orthoclase, sillimanite, garnet, cordierite, ±biotite	Plagioclase, orthoclase, biotite, garnet, sillimanite, ±hypersthene

[a] All assemblages include quartz.

absolute stability limits. These reactions generally involve the rearrangement of tie lines within phase diagrams such as the AFM and allow typical bulk compositions to "see" minerals that were inaccessible at lower grades.

Chlorite Zone

The lowest grade Barrovian zone is the chlorite zone, in which pelitic rocks are generally fine-grained slates or phyllites (incipiently metamorphosed rocks are called subchlorite zone). The metamorphic assemblage contains phengitic muscovite and chlorite, along with quartz, albite, and accessory phases such as sulfides, oxides, calcite, or graphite. There is not much point in making a graphical portrayal of this assemblage on the ACF or AFM diagrams because there is only one mineral, chlorite, that plots on each of these diagrams. The coexis-

tence of chlorite and muscovite can be shown in the A′FK diagram, however (Figure 23-2). The extensive range of both chlorite and muscovite compositions is reflected by a broad two-phase field or tie line bundle. Typical pelite bulk compositions fall within this broad field (for example, the composition indicated at *a*) and are thus represented by the stable two-phase assemblage muscovite + chlorite, with each of these minerals having the composition at the end of the tie line that passes through the bulk composition. Particularly aluminous bulk compositions (shown as *b* in Figure 23-2) would contain the three-phase assemblage chlorite + muscovite + pyrophyllite with fixed, aluminum-rich compositions of both chlorite and muscovite. Aluminum-deficient compositions (lithic or clayey sandstones, for example, composition *c*) consist of the three-phase assemblage chlorite + muscovite + potassic feldspar with low-aluminum chlorite and muscovite.

Biotite Zone

The defined boundary between the chlorite and biotite
zones is the first appearance of biotite porphyroblasts
in pelites. The biotite isograd can be recognized and
mapped in the field by using the presence of the fairly
obvious small dark biotite porphyroblasts. Several reactions can account for the initial stabilization of biotite.
One is not seen in muscovite-bearing pelites because it
involves only chlorite and potassic feldspar: chlorite +
potassic feldspar $\rightleftharpoons$ biotite + H_2O. Another possible
biotite-forming reaction in calc-pelites involves ankeritic
carbonate and muscovite: ankerite + muscovite $\rightleftharpoons$ biotite + calcite. The reaction by which biotite first appears
in pelites most commonly involves the destabilization of
phengite-rich muscovite compositions with increasing
grade. The graphical portrayal of this effect is shown in
Figure 23-3A, in which the composition field of muscovite shrinks toward the (Fe,Mg)-free muscovite composition on the A-K join (compare this figure to Figure
23-2). This reaction is of the *continuous* type and means
that the biotite isograd is in fact dependent on bulk composition. As the two-phase muscovite-chlorite field
shrinks upward toward the A corner, bulk composition *a*
emerges into a new three-phase triangle with chlorite,
less-phengitic muscovite, and biotite at the corners. With
further decrease in the phengite content of muscovite,
the amount of biotite gradually increases as the three-
phase triangle expands.

In the AFM diagram, the array of tie lines following
the first appearance of biotite in pelites is shown
in Figure 23-3B. Remember that the projection is from
muscovite, so only chlorite and biotite are shown. A typical pelite bulk composition is shown (*a*) and indicates
that chlorite predominates over biotite at this grade. As

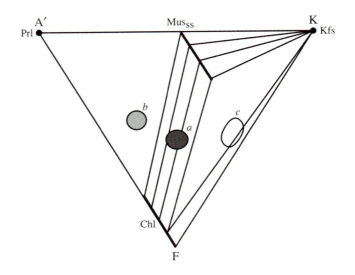

FIGURE 23-2
● ● ● ● ● ● ● ● ● ● ● ● ● ● ● ●

A′KF diagram for the chlorite zone of the Barrovian sequence in
metapelites. The three general compositional types of aluminous
clastic rocks are shown: *a*, typical pelites; *b*, aluminum-rich
pelites; *c*, low-aluminum semipelites (for example, lithic sandstones). Mineral abbreviations: Prl, pyrophyllite; Mus$_{ss}$, phengitic
muscovite solid solution; Kfs, potassic feldspar; Chl, chlorite.

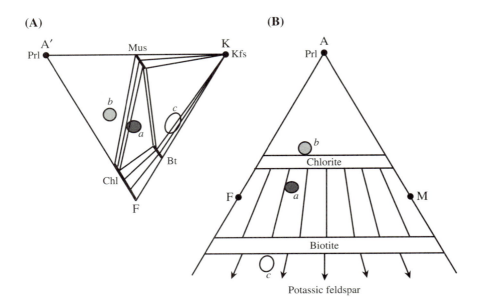

FIGURE 23-3
● ● ● ● ● ● ● ● ● ● ● ● ● ● ● ●

(**A**) A′KF diagram for the biotite
zone. (**B**) AFM diagram for the
biotite zone. The three rock compositional types are shown in both
diagrams, as in Figure 23-2. Typical
assemblages, including quartz and
sodic feldspar, are *a*, muscovite +
chlorite + biotite; *b*, muscovite +
chlorite + pyrophyllite; *c*, muscovite + biotite + potassic feldspar.

shown in Figure 23-3B, the assemblage in unusually aluminous compositions (*b*) remains muscovite-pyrophyllite-chlorite-quartz, whereas that in subaluminous compositions (*c*) reflects the new stability of biotite and becomes muscovite-biotite-potassic feldspar-quartz. This difference illustrates a very important point: The appearance of index minerals at "isograds" does not happen simultaneously in all compositions, except by unusual coincidence.

Garnet Zone

The next Barrovian isograd marks the first appearance of small porphyroblasts of almandine-rich garnet in pelites. It is important to recognize that the sequence first defined by Barrow in Scotland applies to many but not all pelites. Sometimes garnet has been observed forming at a lower metamorphic grade than biotite, particularly in rocks that are rich in manganese or are highly oxidized, because garnet can be stabilized by minor components such as manganese, which forms spessartine garnet, or Ca^{2+} and Fe^{3+}, which form andradite. In the most common type of pelites, however, garnet is almandine-rich and is the next index mineral after biotite.

The initial stabilization of typical iron-rich (almandine) garnet is marked by the reaction between iron-chloritoid (an unusual aluminous mineral) and iron-biotite to produce garnet and muscovite, but this iron end member reaction is rarely observed in nature. More typically in pelites, iron-rich garnet first appears as the result of a continuous reaction in which iron-rich chlorite and biotite break down to produce garnet and muscovite: chlorite + muscovite + quartz $\rightleftharpoons$ garnet + magnesium-richer chlorite + biotite + H_2O. This reaction is illustrated in Figure 23-4 in a series of AFM diagrams at progressively increasing grade (the A'FK diagram is irrelevant for showing reactions that depend on iron-magnesium shifts). In Figure 23-4A, the new three-phase field for garnet-chlorite-biotite has moved slightly toward the right in the AFM diagram, a shift reflecting the increase in magnesium that is allowed in these coexisting minerals with increasing temperature. However, it has not yet intersected the bulk composition for typical pelites (shaded area), which retains the biotite-chlorite two-phase assemblage. Note that unusually iron-rich pelitic bulk compositions thus can have garnet in them even though other pelites do not. Parts B and C of Figure 23-4 show the progressive rightward shift in the three-phase triangle that eventually produces garnet in all compositions. For most pelitic compositions, the appearance of garnet is typically at a temperature of about 450°C.

The three-phase triangle shift explains the disappearance of chlorite as well as the appearance of garnet. Remember that the bulk composition remains fixed. As the

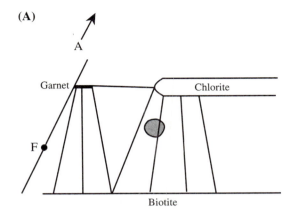

(A)

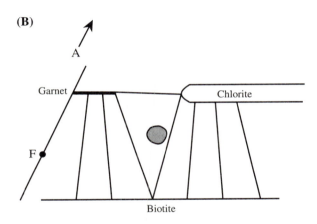

(B)

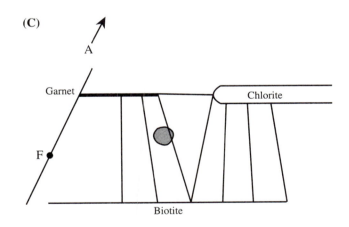

(C)

FIGURE 23-4
• • • • • • • • • • • • • • • • •

Sequence of schematic AFM diagrams showing the movement to the right of the garnet-biotite-chlorite three-phase triangle with increasing metamorphic grade. The typical metapelite bulk composition is shown (shaded area). The movement of the triangle reflects the reaction chlorite + quartz + muscovite $\rightleftharpoons$ garnet + Mg-richer chlorite + biotite + H_2O that is responsible for the garnet isograd (see text for details).

FIGURE 23-5

• • • • • • • • • • • • • • • •

Photomicrograph of a large, inclusion-filled staurolite porphyroblast from the staurolite zone, Glen Esk, Scotland. Note several flakes of biotite around the staurolite. Most of the inclusions within the porphyroblast are quartz. Plane-polarized light. Horizontal field of view is about 4.8 mm.

triangle moves to the right in Figure 23-4, the garnet-biotite tie line will eventually intersect the bulk composition, at which point chlorite disappears from the assemblage. This process means that the first reactant mineral to be exhausted in the reaction is chlorite, leaving garnet, biotite, and muscovite. The fact that a relatively consistent garnet isograd can be mapped in pelitic rocks of the Scottish Highlands reflects the unusually homogeneous chemical compositions of aluminous rocks there, particularly the $Fe/(Fe+Mg)$ ratio. Substantial bulk chemical variation would obviously lead to garnet appearance over a range of metamorphic grade, thus producing a highly diffuse or irregular isograd.

The appearance of garnet in subaluminous gneissic rocks is governed by a different reaction from that observed in pelitic rocks. For low-grade potassic feldspar-bearing rocks, bulk compositions plot below the biotite field on the AFM diagram. The only way that such compositions can "see" garnet is for iron-rich biotite to break down, thus creating the three-phase field garnet-biotite-potassic feldspar in the iron-rich side of the diagram (see Figure 23-4). The reaction is written biotite + muscovite + quartz $\rightleftharpoons$ garnet + potassic feldspar + H_2O. The appearance of garnet in such gneissic rocks can roughly coincide with the first garnet in pelites, but this is only coincidental.

Staurolite Zone

In the classic Barrovian sequence in pelites of the Scottish Highlands, the narrow garnet zone is succeeded upgrade by the staurolite zone. The appearance of stau-

rolite marks the breaking of the prominent "east-west" tie line between garnet and chlorite (see Figure 23-4) and the establishment in its place of a staurolite-biotite tie line at about 500°C. This new, more vertical arrangement of tie lines (see Figure 23-6B) allows common pelite bulk compositions to contain the more aluminous mineral staurolite. When staurolite first appears in metapelites, it commonly displays a sievelike texture with abundant quartz inclusions (Figure 23-5). It must be emphasized that staurolite is stable at metamorphic grades lower than those of its typical first appearance, but it does not ordinarily occur in typical pelites because of the garnet-chlorite tie line. The tie line switch between Figures 23-4 and 23-6 is expressed by the reaction garnet + chlorite + muscovite $\rightleftharpoons$ staurolite + biotite + quartz + H_2O. The garnet zone AFM metamorphic assemblage garnet + chlorite + biotite in typical pelites is thus replaced by either garnet + staurolite + biotite in more iron-rich compositions or by chlorite + staurolite + biotite in less iron-rich compositions. Note that the former assemblage indicates that chlorite was the first reactant phase to be exhausted in the reaction, whereas the latter assemblage reflects exhaustion of garnet.

The mineral staurolite rarely occurs in subaluminous rocks at this or any other grade. The reason for this is apparent in Figure 23-6. For a potassic feldspar-bearing assemblage to "see" staurolite topologically, it would be necessary to break the tie line between garnet and biotite. Considerable observation of various metamorphic terranes worldwide has shown that the stable coexistence of garnet and biotite lasts throughout a very wide range of metamorphic grades, from the first appearance of garnet all the way up to the granulite facies where

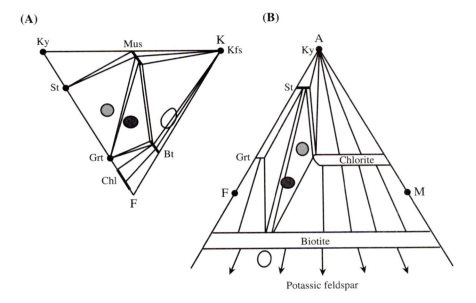

FIGURE 23-6
● ● ● ● ● ● ● ● ● ● ● ● ● ● ● ●

A'KF (**A**) and AFM (**B**) diagrams for the staurolite zone. Note that the breaking of the tie line from garnet to chlorite has now allowed a new set of tie lines from staurolite to biotite, making it possible for staurolite to occur in typical metapelite compositions. The assemblage for this zone is generally quartz + garnet + biotite + staurolite + muscovite + chlorite, with the extra phase stabilized by minor components such as manganese or calcium in garnet, titanium or zinc in staurolite, or titanium in biotite. High-aluminum metapelites have the same assemblages as typical pelites, but with different proportions of minerals, particularly more staurolite. Mineral abbreviations: Bt, biotite; Chl, chlorite; Grt, garnet; Kfs, potassic feldspar; Ky, kyanite; Mus, muscovite; St, staurolite.

biotite ultimately becomes unstable. Another, simpler way of stating this is to note that an aluminum-rich mineral like staurolite typically would not occur in a relatively aluminum-poor rock.

Kyanite Zone

The kyanite zone marks the first occurrence of an Al_2SiO_5 polymorph in medium- to high-pressure pelitic rocks. In most pelites, kyanite appears while staurolite is still stable, and the two minerals coexist through most or all of the kyanite zone (Figure 23-7). Thermometric estimates indicate that kyanite first appears at about 550°C. The reaction that is generally credited for the appearance of kyanite in pelites involves the breaking of the staurolite-chlorite tie line in favor of a kyanite-biotite tie line (Figure 23-8A). Note the similarity of this reaction topology to that which allowed staurolite to appear: A relatively horizontal AFM tie line breaks and a more vertical tie line forms in its place. The reaction is formally written staurolite + chlorite + muscovite + quartz ⇌ kyan-

ite + biotite + H_2O. Examination of Figure 23-8A indicates that common pelite bulk compositions should contain either the assemblage garnet-staurolite-biotite or staurolite-biotite-kyanite (both with muscovite, quartz, and probably plagioclase, of course). Observation of kyanite-grade pelites in a number of metamorphic terranes has indicated the common and apparently stable occurrence over wide areas of the four-phase assemblage garnet-biotite-staurolite-kyanite, however. Although this assemblage appears to violate the rules for divariant assemblages—a maximum of three AFM minerals—there is a valid reason for its occurrence. The topologic rules strictly hold only when all the minerals are pure AFM phases, that is, they do not contain components other than potash, iron, magnesia, alumina, silica, and water. The presence of greater than trace amounts of other components in only one AFM mineral, for example, titanium in biotite, calcium or magnesium in garnet, or zinc in staurolite, can serve to stabilize at least one extra mineral in a nonreaction assemblage. This principle should always be kept in mind when assessing the

FIGURE 23-7

• • • • • • • • • • • • • • • • •

Photomicrograph of a coarse kyanite porphyroblast (center) in a quartz-muscovite-garnet-biotite-staurolite-kyanite schist, Waterbury, Connecticut. Note the subhedral garnet left of center, several large biotite flakes (pale-colored), and swirls of fine-grained muscovite and graphite (left and lower right). Colorless material at the top is a quartz segregation. Plane-polarized light. Horizontal field of view is about 4.8 mm..

variance of assemblages by using compositionally simple projected diagrams like AFM.

An important reaction observed in relatively high-pressure pelites is the disappearance of staurolite by reaction within the upper part of the kyanite zone. The staurolite-out reaction is quite sensitive to pressure and occurs within the kyanite stability field at high pressure and within the sillimanite field at lower pressure. At the pressures of the classic Barrovian MFG (about 6 kbar, as observed in Scotland), staurolite disappearance coincides almost exactly with the transition from kyanite to sillimanite, that is, the sillimanite isograd. The important reaction responsible for staurolite disappearance is staurolite + muscovite + quartz $\rightleftharpoons$ kyanite + garnet + biotite + H_2O. The AFM topology of this reaction is clearly seen in the transition from Figure 23-8A to 23-8B.

(A)   **(B)**

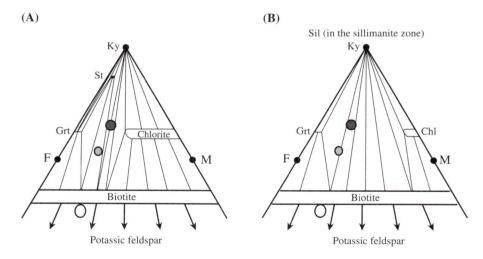

FIGURE 23-8

• • • • • • • • • • • • • • • • •

(A) AFM diagram for the lower part of the Barrovian kyanite zone, in which staurolite is still stable. The terminal disappearance of staurolite within the kyanite zone is reflected by the shift from **(A)** to **(B)**, which shows the AFM assemblages on the high-grade side of the staurolite-out reaction staurolite + muscovite + quartz = garnet + kyanite + biotite + H_2O. Mineral abbreviations: Chl, chlorite; Grt, garnet; Ky, kyanite; Sil, sillimanite; St, staurolite.

(A)

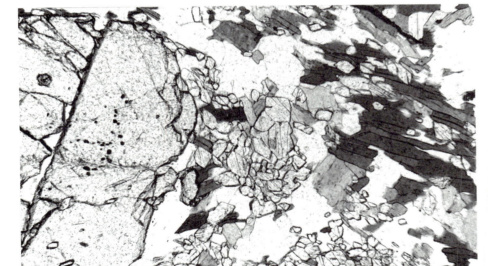

(B)

FIGURE 23-9

● ● ● ● ● ● ● ● ● ● ● ● ● ● ● ●

Plane-polarized light photomicrographs showing the two principal textural types of the mineral sillimanite. **(A)** Fibrolitic sillimanite from a quartz-muscovite-sillimanite-biotite-garnet schist, central Massachusetts. Note the incipient coarsening of some needlelike fibrolite into coarse sillimanite, especially at the lower left. Horizontal field of view is about 1.2 mm. **(B)** Coarse, prismatic sillimanite from a quartz-biotite-garnet-sillimanite-potassic feldspar schist, central Massachusetts. Note the characteristic diamond-shaped cross sections of the sillimanite crystals and the well-developed diagonal cleavage (center). A very large garnet crystal is at the left, and several large red-brown biotite flakes can be seen. Horizontal field of view is about 2.4 mm.

This so-called terminal reaction of staurolite involves no tie line switches, but only the disappearance of tie lines from staurolite to biotite, kyanite, and garnet as staurolite itself disappears from the AFM diagram.

Sillimanite Zone

The highest grade zone typically described for classic Barrovian metamorphism, the sillimanite zone, is marked by the occurrence of sillimanite as the Al_2SiO_5 polymorph in pelites. The AFM diagram for the sillimanite zone is essentially the same as that for the upper part of the kyanite zone following the staurolite-out reaction

(see Figure 23-8B), except that sillimanite rather than kyanite occurs at the A apex. The first appearance of sillimanite is typically as the fibrolitic variety, which occurs as matted bunches or swirls (Figure 23-9A). Higher in the sillimanite zone, coarser prismatic sillimanite is more common (Figure 23-9B). Interestingly, fibrolitic sillimanite is virtually ubiquitous in pelites of the sillimanite zone, whereas kyanite is considerably less common in kyanite-grade pelites. It is unclear whether this puzzling behavior is related to minor bulk compositional differences, nucleation kinetics, or some other factor. Some petrologists have suggested that abundant fibrolite results from Al_2SiO_5-producing reactions involving micas that only occur at sillimanite-grade conditions and above. What is apparent, as noted by Kerrick (1990) in a comprehensive review of the Al_2SiO_5 minerals, is that a relatively simple inversion of kyanite to sillimanite to produce pseudomorphs occurs rarely, if at all. As noted in Chapter 21, Carmichael (1969) outlined a complicated circuitous reaction pathway by which kyanite can invert *indirectly* to sillimanite in a series of reaction steps.

Sillimanite-Orthoclase and Higher Grade Zones

Although Barrow did not describe these high-grade zones from the eastern Scottish Highlands, they have been found in numerous other metamorphic belts at MFGs similar to a Barrovian one. The coexistence of sillimanite and potassic feldspar is made possible by the reaction at about 650°–680°C that limits the stability of muscovite plus quartz: muscovite + quartz $\rightleftharpoons$ sillimanite + potassic feldspar + H_2O. This reaction is portrayed with the Al_2SiO_5 triple point in Figure 23-10. Note that the breakdown of muscovite produces kyanite + potassic feldspar at very high pressures (greater than about 8 kbar). The association of kyanite plus potassic feldspar is rare but has been described in the Alps, the northern Appalachians, and Norway. At very low pressures (below about 3.5 kbar), the association andalusite + potassic feldspar is possible and is discussed in some detail below as a possible contact metamorphic assemblage. The reaction of muscovite leaves biotite as the only stable sheet silicate in pelites and marks the onset of the transition from schistose to gneissic fabric that characterizes highest grade pelites. Many petrologists regard muscovite breakdown as the lower limit of the granulite facies.

The disappearance of muscovite at high grades causes a problem because it invalidates the use of the AFM-muscovite projection for showing assemblage topology. However, it is equally valid to project from potassic feldspar within the AKFM tetrahedron. This is in fact a simpler projection because it is from a tetrahedron apex; thus the negative portion present in the muscovite projection is eliminated. The assemblage in the lower part of the sillimanite-orthoclase zone is shown in this new potassic feldspar projection in Figure 23-11. Typical pelites contain the AFM assemblage garnet-sillimanite-biotite, with quartz, potassic feldspar, plagioclase, and accessory minerals (for example, graphite, ilmenite, pyrrhotite). Note in Figure 23-11A that unusually magnesian pelites can contain cordierite with sillimanite and biotite. Cordierite is typically the most magnesium-rich of all the common aluminous ferromagnesian minerals. It may rarely occur at lower grades in extremely unusual magnesium-rich aluminous rocks, but in general, it is

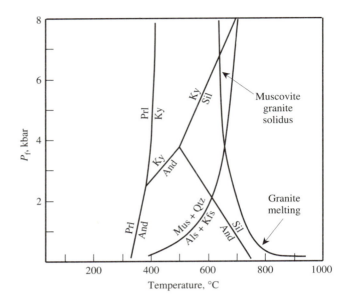

FIGURE 23-10
• • • • • • • • • • • • • • • •

P_{fluid}-T diagram showing the aluminum-silicate triple point, the stability field of pyrophyllite, the muscovite breakdown reaction, and the beginning of melting of pelitic rocks. Mineral abbreviations: Als, aluminum silicate (andalusite, kyanite, or sillimanite); And, andalusite; Kfs, potassic feldspar; Ky, kyanite; Mus, muscovite; Prl, pyrophyllite; Sil, sillimanite. Note that the lower pressure limit for direct crystallization of muscovite from granitic melts is based on the crossover of muscovite breakdown and granite melting reactions.

a classic very high grade mineral that appears in metapelites in the sillimanite-orthoclase zone.

The next reaction occurs at about 700°C and involves the shift of the sillimanite-biotite tie line to one between garnet and cordierite (see Figure 23-11B). The most important implication of this reaction is that it allows cordierite to occur with garnet in common pelite bulk compositions, as has been reported from Scotland, central New England, the Massif Central in France, Antarctica, and Australia. These very high metamorphic grades are unusual; thus garnet-cordierite assemblages are rather rare.

The highest grade pelitic rocks observed are marked by the breakdown of biotite to form orthopyroxene plus potassic feldspar. The very high temperatures that characterize orthopyroxene-garnet-sillimanite cordierite-orthoclase assemblages (800°C or higher) are rare indeed, and the most notable examples of such granulite-facies pelitic rocks have been reported from Enderby Land in Antarctica. Other occurrences (for example, central New England, the Adirondacks, North Africa) are clearly related to magmatic activity, and these rocks are definitely in the temperature range where an overlap with igneous processes is likely. Evidence for partial melting of pelitic rocks in the sillimanite-orthoclase and higher grade zones is common, and this process is unquestionably important in the generation of migmatites and other crustal melts. Because of its importance, the melting of aluminous rocks is discussed in more detail below.

Influence of Minor Components

There are a few chemical constituents that typically occur in minor concentrations but can strongly influence the metamorphic assemblages in pelites. Some of these, for example, sulfur, calcium, and titanium, can even add additional minerals under appropriate conditions. Carbon, occurring typically as graphite, can influence the oxidation state of the rock, the composition of the ambient fluid, and the temperatures at which minerals react.

In the case of calcium, the presence of unusual amounts of calcite in a mudrock can shift it into the compositional category called calc-pelite. The calcite tends to react out early in the metamorphic sequence, but the excess calcium, in the presence of abundant alumina, may form epidote at low grade. Epidote is eliminated by reactions that form hornblende and grossular, the calcic garnet component, both of which persist to high grade. The presence of calcic minerals such as epidote, hornblende, or grossular-rich garnet in a schist that otherwise appears to be a normal pelite indicates the calc-pelite tendency.

The abundance of sulfur in mudrocks is quite variable, from a virtual absence to significant amounts of sedimentary or diagenetic pyrite in black shales. The sulfide can become involved with silicates in metamorphic reactions, particularly around 500°C, at which point a major conversion of pyrite to pyrrhotite occurs. This reaction is written $FeS_2 + (Fe) \rightleftharpoons 2 FeS$. To go to completion, it requires nonsulfide iron, which is removed from silicates and oxides, driving the compositions of remaining ferromagnesian minerals toward magnesium enrichment. The more iron that becomes tied up in nonreactive pyrrhotite, the more magnesium-rich the *effective* bulk composition becomes on the AFM diagram. The bulk composition point thus becomes shifted directly away from the iron point and toward the right-hand side of the AFM diagram. Several petrologic studies have shown that this shift can be a major one in sulfide-rich protoliths. The process is of special interest to metamorphic petrologists because it produces bulk compositions (exclusive of sulfides) that lie in parts of the AFM diagram that are rarely observed, and the natural assemblages can be used to understand tie line arrays involving rarely seen magnesian minerals.

METAMORPHISM AT HIGH MFG– LOW-PRESSURE SEQUENCES

Petrologists have described numerous terranes where metamorphic heating occurred at shallow crustal depths and thus at lower pressures than the classic Barrovian sequence. In most cases, this anomalously shallow heating was accomplished through obvious magmatic heat transport—in effect, a regional-scale contact metamorphic event—because abundant synmetamorphic plutons are distributed throughout the terrane. Classic examples include the northeastern corner of Scotland (the Buchan area), northwestern Maine and eastern Canada, the southwestern part of Finland, and the Abukuma Plateau in Japan. A comprehensive summary of low-pressure phase relations in pelitic rocks can be found in Pattison and Tracy (1991).

(A)

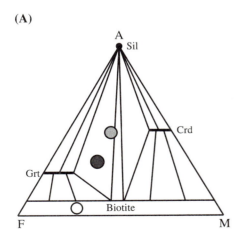

(B)

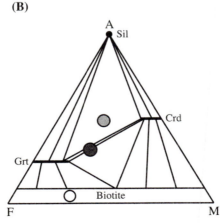

FIGURE 23-11

AFM diagrams projected from potassic feldspar, quartz, and H_2O, illustrating assemblage relationships for the lower **(A)** and upper **(B)** parts of the sillimanite-potassic feldspar zone. Mineral abbreviations: Crd, cordierite; Grt, garnet; Sil, sillimanite.

In general, the sequence of metamorphic isograds and zones in this low-pressure metamorphism differs from the classic Barrovian pattern. Mineralogic and assemblage differences can typically be explained by the arrangement of reaction curves and assemblage boundaries in *P-T* space (Figure 23-12): (1) Kyanite is rarely found, and Al_2SiO_5 occurs as either andalusite or sillimanite, or more rarely as sillimanite pseudomorphs after andalusite (see Figure 21-3). At the lowest pressures—those typical of shallow crustal contact aureoles—muscovite breakdown can even occur in the andalusite stability field, yielding the characteristic contact metamorphic assemblage andalusite + potassic feldspar (see Figure 23-12). (2) At medium pressures, staurolite persists into the sillimanite field and ultimately breaks down to sillimanite + garnet + biotite rather than disappearing within the kyanite zone as shown above for typical Barrovian metamorphism. At sufficiently low pressures, staurolite can even break down to andalusite + biotite + cordierite. (3) Garnet is more typical of higher pressure metamorphism and is replaced at low pressure by cordierite, even at relatively low metamorphic grade.

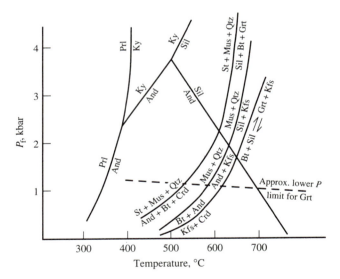

FIGURE 23-12
● ● ● ● ● ● ● ● ● ● ● ● ● ● ● ●

P_{fluid}-*T* diagram showing the sequence of reactions that occurs in metapelites at low pressures. The dashed line indicates the approximate lower limit of stability of iron-magnesium garnet relative to cordierite. Mineral abbreviations: And, andalusite; Bt, biotite; Crd, cordierite; Grt, garnet; Kfs, potassic feldspar; Ky, kyanite; Mus, muscovite; Prl, pyrophyllite; Qtz, quartz; Sil, sillimanite; St, staurolite.

LOW-PRESSURE METAMORPHISM IN NORTHEAST SCOTLAND. The arrangement of metamorphic zones in pelites at low pressures has been described in some detail for the so-called Buchan area of Scotland by Harte and Hudson (1979). The Buchan area is one of the most classic and well studied low-pressure metamorphic terranes in the world (Figure 23-13). Metamorphic zones are particularly well displayed in coastal exposures of the Dalradian Series along the coast of the North Sea west of Aberdeen. The lowest grade rocks form a *biotite zone* similar to that of the Barrovian sequence, with the same assemblage muscovite + chlorite + biotite + quartz + albite. At these low grades, the effect of lower pressure is seen mainly in the chemistry of the individual minerals; for example, muscovite has lower phengite content and chlorite is less aluminous. The first obvious metamorphic reaction going up in grade causes the appearance of spots or lumps of cordierite as a result of the continuous reaction muscovite + chlorite = cordierite + biotite + quartz + H_2O. This *cordierite zone* is unique to low-pressure metamorphism and is characterized by "spotted slates" in which the cordierite lumps are prominent in outcrop. As noted by Yardley (1989), the appearance of cordierite is analogous to that of garnet in the Barrovian garnet zone, except that cordierite forms by a continuous reaction in the magnesium-rich side of the AFM diagram. Cordierite occurrence, like garnet, is thus dependent on rock bulk composition, as a three-phase triangle moves across the AFM diagram (Figure 23-14A and B). In the *andalusite zone*, porphyroblasts of this mineral appear as a result of the reaction chlorite + muscovite + quartz $\rightleftharpoons$ cordierite + andalusite + biotite + H_2O (Figure 23-14C). This is a discontinuous reaction in which chlorite of intermediate iron to magnesium ratio disappears from the AFM diagram in a manner analogous to staurolite disappearance in pelites. Harte and Hudson noted that this reaction marks the first possible coexistence of cordierite and andalusite in typical pelite bulk compositions. In their rocks, however, chlorite was consumed early in the sequence, before this reaction was reached, and andalusite was formed in other reactions at slightly higher grade.

METAMORPHISM AT LOW MFG– HIGH-PRESSURE SEQUENCES
● ● ● ● ● ● ● ● ● ● ● ● ● ● ●

Very high metamorphic pressures at relatively low temperatures can be attained in subduction environments. The effects of such high-pressure metamorphism in

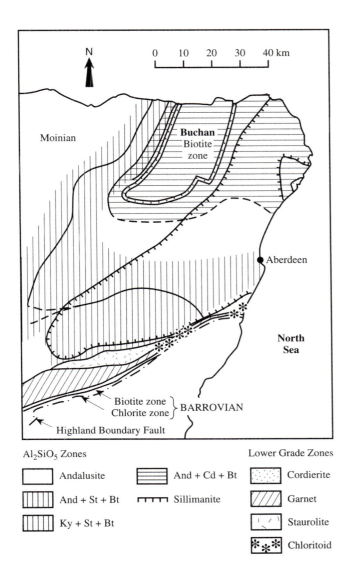

N   0 10 20 30 40 km

Moinian

Buchan
Biotite
zone

Aberdeen

**North
Sea**

Biotite zone ⎫
Chlorite zone ⎬ BARROVIAN

Highland Boundary Fault

Al₂SiO₅ Zones

- Andalusite
- And + St + Bt
- Ky + St + Bt
- And + Cd + Bt
- Sillimanite

Lower Grade Zones

- Cordierite
- Garnet
- Staurolite
- Chloritoid

FIGURE 23-13

• • • • • • • • • • • • • • • •

Distribution of pelitic metamorphic zones in the eastern Scottish Dalradian. Areas on the hatched side of the sillimanite boundary belong to the sillimanite zone within which kyanite and andalusite occur as indicated. The only principal localities of chloritoid are indicated by asterisks. Mineral abbreviations: And, andalusite; Bt, biotite; Cd, cordierite; Ky, kyanite; St, staurolite.

ture to them; but it is also due to the rarity of true pelite compositions in the sedimentary tectonic environments where subduction occurs. In particular, as Yardley pointed out, calcium-poor compositions are unusual in such environments. These compositions are essential, however, to development of aluminum-rich ferromagnesian minerals; in the presence of calcium, calcic-aluminous minerals such as zoisite, calcic plagioclase, and lawsonite form instead. The reader is referred to Yardley (1989), Bucher and Frey (1994), and Spear (1993) for excellent brief discussions of high-pressure metapelites and useful summaries of original sources.

One of the best-known effects of high-pressure metamorphism of pelites is the occurrence of phengitic muscovite instead of biotite. In addition, work in the unusual high-pressure metapelites of the Alps and Afghanistan has uncovered the common coexistence of talc and muscovite. A rare mineral that is characteristic of low-temperature, high-pressure metamorphism of pelites is carpholite, a hydrous mineral similar to chloritoid. Other interesting mineralogic characteristics of these rocks include the presence of very magnesian garnet and absence of cordierite at any grade, and the occurrence of kyanite and even kyanite + potassic feldspar coupled with an absence of sillimanite.

mafic compositions produce blueschists, as discussed in Chapter 22. Mafic rocks are the most common compositions found in subduction regimes, but aluminous rocks have also been found and characterized in such environments. In general, pelite mineral assemblages differ in two important ways. First, a number of minerals occur only in unusually high pressure occurrences, for example, the silica polymorph coesite. Second, unusually high pressure can stabilize the solid solution of certain components in common minerals. Two good examples are phengite component in muscovite and pyrope component in garnet.

As noted by Yardley (1989), there have been few studies of high-pressure metamorphism of pelites. This paucity of research is partly due to the difficulty of studying such rocks and the lack of well-defined zonal struc-

MELTING OF
ALUMINOUS ROCKS

• • • • • • • • • • • • • • • •

Migmatites (from the Greek *migma-*, meaning "mixed") are outcrop-scale mixtures of light- and dark-colored rocks (Figure 23-15). Most commonly, these are layered, with coarse-grained leucocratic layers (*leucosomes*) separated by bands of finer grained dark rock (*melanosomes*). The layering is on a scale of centimeters to more than a meter. More rarely, migmatites are irregular veinlike segregations on a similar scale. Some low-potassium migmatites contain quartz-plagioclase leucosomes separated by amphibolite melanosomes, but the most typical migmatites in high-grade metamorphic

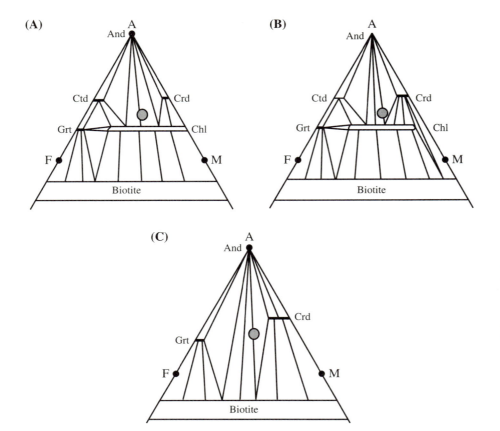

FIGURE 23-14

●●●●●●●●●●●●●●●●

AFM diagrams showing the movements of three-phase triangles that describe the cordierite isograd within the andalusite zone of typical low-pressure metamorphism of pelites such as those of the Buchan area of northeastern Scotland. Metamorphic grade increases from (A) to (B) to (C). Typical bulk compositions of such metapelites are shown by the horizontal lined area. The assemblage in (A) for these compositions is andalusite + chlorite; in (B) it is andalusite + chlorite + cordierite; and in (C) it is andalusite + biotite + cordierite (all assemblages also contain muscovite and quartz). Between (B) and (C), chlorite disappears by the discontinuous metamorphic reaction discussed in the text. Mineral abbreviations: And, andalusite; Chl, chlorite; Crd, cordierite; Ctd, chloritoid; Grt, garnet.

terranes consist of quartz-orthoclase (or microcline)-plagioclase leucosomes separated by biotite-rich dark layers. These melanosomes are commonly also rich in aluminous minerals including garnet, cordierite, and sillimanite.

There has been considerable argument for decades about the origin of migmatites. One school of petrologists has held that they represent injections of granitic magma into high-grade gneissic or schistose country rocks in a *lit-par-lit* (conformable bed-by-bed) fashion. Another school claimed that layering of this sort is the result of metasomatic "self-organization" processes operating in the solid state, aided by fluids. The third and largest group has argued for localized fractional melting of high-grade pelitic rocks, with segregation of the granitic melt into light-colored layers separated by the depleted dark-colored residuum that was left after melt extraction. Calculations have shown that if leucosomes and melanosomes are recombined according to the proportional mass of each, the combination forms a more or less typical pelite composition. This observation is inconsistent with the injection hypothesis but does not discriminate between melting and solid-state segregation. One argument that has been put forward in favor of the

melting hypothesis is that the coarseness of grain size (commonly pegmatitic, up to tens of centimeters) and the nature of grain shapes and grain-to-grain contacts in the leucosomes is more consistent with crystallization of a water-saturated melt than with a solid-state process. This is certainly not an iron-clad argument—very coarse grain sizes can also result from fluid-catalyzed recrystallization—but it supports the melting hypothesis.

One key test is to assess whether melting of a pelitic or semipelitic composition can occur within the temperature range of high-grade metamorphism. Numerous experiments over the last 35 years or so have demonstrated unequivocally that quartzofeldspathic and pelitic rocks can begin to melt at temperatures as low as 650°C, well within the temperature limits discussed above. The petrogenetic grid in Figure 23-16 summarizes some of these experimental results, along with a phase equilibrium analysis that incorporates them into a system of high-temperature reactions for metapelites. The most important element of this grid is the melting curve for granite, which shows a substantial reduction in melting temperature with increasing P_{H_2O}. This melting curve intersects the reaction curve for dehydration of muscovite to sillimanite + potassic feldspar at about 4 kbar.

Below this pressure, melting should not occur until well after muscovite has disappeared from pelites, with serious effects on the amount of melting.

The *simultaneous* breakdown of muscovite and beginning of melting have long been considered crucial to generation of the volumes of leucocratic material observed in most migmatites. The fractional melting process depends on the presence of enough water to saturate the melt, or nearly so, because strongly water-undersaturated granitic magmas require much higher temperatures to form. Muscovite and biotite involved in the melting reaction are important local sources of water. Above 4 kbar, the water liberated by muscovite dehydration can promote more voluminous melting, and indeed the migmatites observed in high-grade muscovite schists of Barrovian sequences are better developed than other migmatites in lower pressure terranes. Classic examples include central New England, the Front Range in Colorado, the Dalradian of the west of Ireland, and the south coast of Brittany in France. The mere presence of muscovite and biotite may not be sufficient water sources, however. Most of the recent work on fractional melting of metapelites and migmatite formation now

indicates that at least some introduction of water from external sources is required to produce the volumes of presumed melt that are seen in most migmatites. Although migmatite formation has been observed in low-pressure potassic feldspar-sillimanite gneisses, particularly in contact aureoles, it always involves less melt production and is rarer than melting in higher pressure muscovite-bearing pelites.

The bulk composition of pelitic and related rocks is also an important determinant of the amount of melting that can occur. Semipelitic or granofelsic rocks (lithic or feldspathic sandstone protoliths) have the greatest potential to generate substantial volumes of melt because these rocks have bulk compositions closest to those of granitic magma. Recall from the discussion of partial melting in Chapter 5 that partial or fractional melting begins with generation of melt at a relatively low-temperature eutectic point. Melting at this temperature can only continue as long as all solid phases remain present in the residuum. If a protolith has a composition close to that of the eutectic, most of it can melt without much increase in temperature. For a rock quite different in composition from the eutectic, only a small amount of

FIGURE 23-15
• • • • • • • • • • • • • • • •

Photograph of an outcrop of folded Precambrian migmatite from the Poudre Canyon, Colorado Front Range, near Fort Collins. The darker layers are enriched in biotite, sillimanite, and garnet, and the coarse, light granite layers consist mainly of quartz, alkali feldspar, and plagioclase. Note the pen for scale; horizontal distance is about 0.5 m. Metamorphic grade at this location is sillimanite + potassic feldspar (about 750°C, 7 kbar), conditions appropriate to localized melting. [Photo by Barbara J. Munn, Stockton State College.]

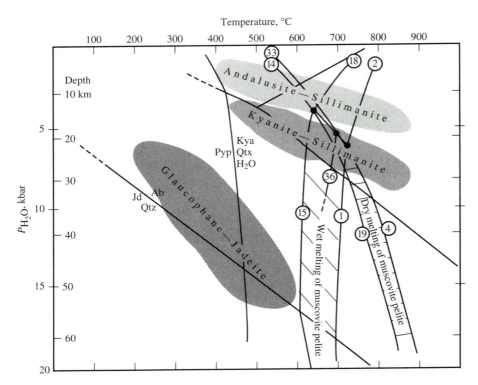

FIGURE 23-16

●●●●●●●●●●●●●●●●

P_{H_2O}-T diagram showing the petrogenetic grid for fluid-present and fluid-absent melting reactions in metapelites. The region labeled "wet melting" indicates metapelite melting with excess water, whereas the "dry-melting" region refers to melting where no external water is available, beyond what occurs in muscovite. Melting reactions (15) and (18) are water-saturated melting of calcium-free granite; reactions (1) and (2) are water-saturated melting of sodium-free granite. Similarly, (19) is water-absent melting of calcium-free granite, and (4) is melting of sodium-free granite. Several types of facies series, as well as the aluminum-silicate triple point of Holdaway (1971) and the jadeite stability curve, are indicated for reference. Note that the maximum overlap between fields of wet and dry melting occurs in the kyanite-sillimanite (Barrovian) type of facies series. [From A. B. Thompson and R. J. Tracy, 1979, *Contr. Mineral. Petrol.*, *70*, Fig. 5.]

melting is likely to exhaust one of the solids and terminate melting. Figure 23-17 shows the water-saturated granite eutectic at 5 kbar in the quartz-plagioclase-orthoclase ternary, along with the plotting positions of typical semipelites and pelites. Note how the semipelite is closer to the position of the eutectic and thus is likely to melt more completely at near-eutectic temperatures.

The presence of migmatites in high-grade metamorphic terranes and the acceptance of a melt origin for these rocks have empasized the potential role of metamorphism in generating large volumes of crustal granitic and related magmas. Some petrologists have regarded migmatitic terranes as source areas for magmas, even referring to them as "baby batholiths." However, no direct genetic connection between migmatite formation and larger plutonic masses has ever been established. An increasing number of petrologists who study migmatites now believe that no connection exists, and that mig-

matite formation is an end point of the metamorphic process rather than the beginning of a magmatic process. Magmas undoubtedly originate in the deep continental crust, but the nature of their source areas has not been established. Unusual nonmigmatitic, alkali-depleted and alumina-enriched lithologies commonly occur over wide areas in granulite-facies terranes, and these may be good candidates for being residua from melting and extraction of the larger volumes of granitic magma required to form batholith-sized plutons.

SUMMARY

●●●●●●●●●●●●●●●●

Aluminous clastic rocks are relatively common in many metamorphic belts. These span a range of composition from sandstones, feldspathic sandstones, and lithic sand-

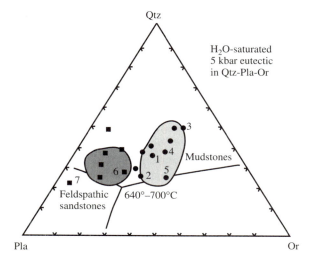

FIGURE 23-17

• • • • • • • • • • • • • • • • •

Water-saturated liquidus relationships in the ternary system plagioclase-orthoclase-quartz at 5 kbar water pressure. The range of eutectic temperature is approximately 640°–700°C, depending on the exact plagioclase composition (An content)—lower for albite-rich plagioclase and higher for anorthite-rich plagioclase. The eutectic also shifts slightly in position with variable plagioclase composition. The eutectic position shown is for plagioclase of An_{25} and is taken from Nekvasil (1988). Approximate bulk compositional ranges for typical feldspathic sandstones (dark shaded area) and mudstones (light shaded area) are based on CIPW norm calculations for published representative bulk compositions. Filled circles indicate published mudstone compositions, and filled squares indicate published feldspathic sandstone compositions. Numbered points represent the compositions given in Table 23-2 and others represent compositions compiled in Blatt, Middleton, and Murray (1980, pp. 375 and 383). The compositional ranges indicate that many feldspathic sandstones have compositions that are closer to the eutectic than the mudstone compositions are and thus can produce significantly greater volumes of low-temperature melt when fractionally melted at eutectic temperatures. Mineral abbreviations: Or, orthoclase; Qtz, quartz; Pla, plagioclase.

stones to siltstones and mudstones. The metamorphosed equivalents of feldspathic sandstones are typically referred to as semipelites, whereas the more aluminous siltstones and mudstones are called pelites. In very low grade metamorphism of aluminous rocks, both textural and mineralogic changes are gradual and result in recrystallization of sedimentary clay minerals into the precursors of white mica and chlorite while quartz and feldspars increase in grain size through annealing. In general, rocks begin to act as equilibrium chemical systems on a scale greater than single grains. By the lower part of the greenschist facies, typical mineral associations include

quartz-albite-white mica-chlorite with possible additional potassic feldspar, calcite, pyrite, and graphite.

At higher metamorphic grades, mineral assemblage sequences differ depending on whether metamorphism occurs at relatively high or low pressures. Higher pressure metamorphism is typified by the Barrovian zonal sequence: chlorite, biotite, garnet, staurolite, kyanite, and sillimanite, with additional higher grade zones in some terranes. The transition from kyanite to sillimanite characterizes Barrovian metamorphism, as does an abundance of garnet and the presence of cordierite only at highest grades. A variety of medium- and lower pressure zonal schemes have been observed, including contact metamorphism; a typical one is the sequence in the Buchan area of northeastern Scotland. Following a biotite zone that is similar to the Barrovian one, higher grade zones include cordierite, andalusite, and sillimanite. Lower pressure metamorphism is characterized by an andalusite-sillimanite transition, the absence of kyanite, and abundant cordierite at all higher grades. Melting is commonly observed at highest grades in both high- and low-pressure metamorphic sequences. Local segregation of melts into coarse-grained pegmatoid layers or lenses generates migmatites, with leucocratic melt layers separated by dark residual melanosomes rich in biotite, garnet, cordierite, and sillimanite.

STUDY EXERCISES

• • • • • • • • • • • • • • •

1. What is the dominant process that occurs during the lowest grades of metamorphism (subgreenschist facies) in metamorphosed aluminous rocks? How do petrologists characterize variations in grade in these very low grade rocks?

2. Does the garnet isograd represent the same temperature and pressure in all metamorphosed aluminous rocks? If not, what controls the first appearance of garnet in metapelites? Approximately what temperature does the garnet isograd typically indicate?

3. Explain why the sequence of mineral zones in progressive metamorphism of metapelites is fundamentally different at low pressures (Buchan-type facies series) from that at high pressures (Barrovian-type facies series). Which two minerals in particular are indicative of low-pressure metamorphism?

4. Sillimanite- and andalusite-bearing peraluminous granites are rare but have been reported. Kyanite-bearing granites are essentially unknown. By referring to Figure 23-10, can you construct an explanation of why kyanite does not occur in hydrous granites?

5. If you are mapping a high-grade metamorphic terrane and encounter migmatites, what kinds of observations might you make in order to decide whether the migmatites were formed by solid-state segregation or by local melting of aluminous rocks?

REFERENCES AND ADDITIONAL READINGS
• • • • • • • • • • • • • •

Ague, J. J. 1991. Evidence for major mass transfer and volume strain during regional metamorphism of pelites. *Geology, 19,* 855–858.

Ashworth, J. R., ed. 1985. *Migmatites.* Glasgow: Blackie.

Bailey, E. H., W. P. Irwin, and D. L. Jones. 1964. *Franciscan and Related Rocks, and Their Significance in the Geology of Western California. Bulletin 183.* California Division of Mines and Geology.

Barrow, G. 1893. On an intrusion of muscovite-biotite gneiss in the SE Highlands of Scotland and its accompanying metamorphism. *Q. J. Geol. Soc. London, 49,* 330–358.

Berner, R. A. 1981. A new geochemical classification of sedimentary environments. *J. Sediment. Petrol., 51,* 359–365.

Blatt, H., G. V. Middleton, and R. C. Murray. 1972. *Origin of Sedimentary Rocks.* Englewood Cliffs, NJ: Prentice Hall.

Bucher, K., and M. Frey. 1994. *Petrogenesis of Metamorphic Rocks* (6th ed. of Winkler's textbook). New York: Springer-Verlag.

Carmichael, D. M. 1969. On the mechanisms of prograde metamorphic reactions in quartz-bearing pelitic rocks. *Contrib. Mineral. Petrol., 20,* 244–267.

Carmichael, D. M. 1978. Metamorphic bathozones and bathograds: A measure of the depth of post-metamorphic uplift and erosion on a regional scale. *Am. J. Sci., 278,* 769–797.

Chopin, C. 1984. Coesite and pure pyrope in high-grade blueschists of the western Alps: A first record and some consequences. *Contrib. Mineral. Petrol., 86,* 107–118.

Ellis, D. J. 1987. Origin and evolution of granulites in normal and thickened crust. *Geology, 15,* 167–170.

Frey, M. 1987. Very low-grade metamorphism of clastic sedimentary rocks. In *Low Temperature Metamorphism,* ed. M. Frey. Glasgow: Blackie, pp. 9–58.

Gromet, L. P., R. F. Dymek, L. A. Haskin, and R. L. Korotev. 1984. The "North American shale composite": Its compilation, major and trace element characteristics. *Geochim. Cosmochim. Acta, 48,* 2469–2482.

Harley, S. L. 1989. The origins of granulites: A metamorphic perspective. *Geol. Mag., 126,* 215–247.

Harte, B., and N. F. C. Hudson. 1979. Pelite facies series and the temperatures and pressures of Dalradian metamorphism in E. Scotland. In *Geological Society of London Special Publication 8,* ed. A. L. Harris et al., pp. 323–337.

Holdaway, M. J., and B. Mukhopadhyay. 1993. Geothermobarometry in pelitic schists: A rapidly evolving field. *Am. Mineral., 78,* 681–693.

Kerrick, D. M. 1990. The Al_2SiO_5 polymorphs. *Rev. Mineral., 22.*

Kisch, H. J. 1992. Development of slaty cleavage and degree of very low grade metamorphism: A review. *J. Metam. Geol., 9,* 735–750.

Nekvasil, H. 1988. Calculated effect of anorthite component on the crystallization paths of H_2O-undersaturated haplogranitic melts. *Am. Mineral., 73,* 966–981.

Pattison, D. R. M., and R. J. Tracy. 1991. Phase equilibria in metapelites. In D. M. Kerrick, ed. Contact Metamorphism. *Rev. Mineral., 24,* 105–206.

Shaw, D. M. 1956. Geochemistry of pelitic rocks. III. Major elements and general geochemistry. *Geol. Soc. Am. Bull., 67,* 919–934.

Spear, F. S. 1993. *Metamorphic Phase Equilibria and Pressure-Temperature-Time Paths. Monograph 1.* Washington, DC: Mineralogical Society of America.

Spear, F. S., and J. T. Cheney. 1989. A petrogenetic grid for pelitic schists in the system SiO_2–Al_2O_3–FeO–MgO–K_2O–H_2O. *Contrib. Mineral. Petrol., 101,* 149–164.

Thompson, A. B. 1982. Dehydration melting of pelitic rocks and the generation of H_2O-undersaturated granitic liquids. *Am. J. Sci., 282,* 1567–1595.

Yardley, B. W. D. 1989. *An Introduction to Metamorphic Petrology.* New York: Wiley.

24

METAMORPHISM
OF CALCAREOUS ROCKS

The calcareous rocks limestone and dolostone, and their impure forms with admixed detrital material (for example, marl), are relatively common in the sedimentary record, particularly where former passive margins have been incorporated into orogenic belts. They are found in these sites because the continental shelf environments on passive margins are prime sedimentary environments for the formation of calcareous sediments. Very pure calcareous rocks maintain their sedimentary mineralogy (calcite, dolomite, or both) to very high metamorphic grades, with little addition of new minerals and with coarsening due only to recrystallization. These metamorphic rocks are the *calcite* or *dolomite marbles*. However, most sedimentary carbonate rocks contain variable amounts of clastic detrital material, particularly quartz, clays, oxides, and organic matter, and of precipitates such as chert and hematite.

Metamorphic reactions between calcite or dolomite and silicate or oxide minerals begin at quite low metamorphic grade and cause the progressive elimination of carbonate minerals, creation of calcium- and magnesium-bearing silicates, and generation of carbon dioxide–rich fluids. Newly formed metamorphic minerals are typically calcic or magnesian, for example, calcic amphibole (tremolite or hornblende), diopside, forsterite, epidote, grossular garnet, and titanite. If abundant, such minerals are commonly incorporated into the metamorphic rock name, as in tremolite-forsterite marble. In some cases, a sufficiently high proportion of detrital noncarbonate minerals results in elimination of carbonates by reaction. Metamorphic lithologies consisting of Ca–Mg–Fe–Al silicates with few or no carbonate minerals are called *calc-silicate* rocks

and are widely assumed to represent original highly impure carbonate rocks. Throughout a wide range of metamorphic grades, these rocks consist mineralogically of hornblende and calcic plagioclase and thus are petrologically similar to the amphibolites that represent metamorphosed mafic volcanics. Caution is therefore required in making protolith assignments when interpretation of tectonic environments is attempted.

The widespread occurrence in calcareous rocks of metamorphic reactions in which both hydrous and carbonate minerals are involved means that these rocks contained fluids with both water and carbon dioxide. Up to now, such mixed fluids have not been a concern, and discussion has assumed the presence of water-rich fluids that had little effect on the temperatures or sequence of reactions. The composition of local fluids in calcareous rocks plays a significant role in determining how reactions occur in these rocks, and the examination of calcareous rock metamorphism can be initiated by discussing the role and behavior of metamorphic fluids.

THE ROLE AND BEHAVIOR
OF METAMORPHIC FLUIDS

Mixed Volatile Equilibria

At the inception of metamorphism, impure calcareous rocks contain a mixture of carbonate minerals and hydrous minerals, typically clays. During low-grade metamorphism, the clay minerals become reorganized into talc, chlorite, micas, or amphiboles, much as happens in the aluminous and mafic rocks. Throughout much of the range of metamorphic grade, mineral reactions will therefore include both CO_2- and OH-bearing minerals, and both of the fluid species carbon dioxide and water will be liberated by progressive devolatilization as temperature increases. Because the breakdown of carbonates and hydrous minerals may be differently affected by changing the proportion of these species in the fluid, fluid composition, along with pressure and temperature, must be considered as an explicit variable that controls metamorphic assemblages and reactions in calcareous rocks.

The importance of fluid composition in calcareous rocks was implicitly recognized by Goldschmidt in his thermodynamic treatment of contact metamorphism in 1911 and explicitly treated in several classic papers by Greenwood (for example, 1962, 1975). Petrologists continue to argue about the chemical and physical nature of metamorphic fluids, and indeed about the very existence of such fluids. Obviously, a metamorphic rock sampled from an outcrop does not typically retain any fluid that might have been present during metamorphism (with the exception of a few tiny trapped bubbles called *fluid inclusions*). The weight of evidence, however, supports the existence of C–O–H fluid (with typically minor amounts of sulfur, fluorine, and chlorine). Experimental geochemical studies have indicated that the fluid is supercritical over much of the range of metamorphic conditions and is dominated by the species water, carbon dioxide, and methane (under reducing conditions). By *supercritical*, we mean that the fluid exists as a single phase and does not physically separate into liquid and gaseous phases.

The volume fraction of fluid present at any given time during metamorphism is probably small because there is very little pore space in metamorphic rocks. Petrographic examination of sedimentary and metamorphic rocks indicates a progressive reduction in pore space with increasing metamorphic grade, from a high of as great as 30% in sedimentary rocks to much less than 1% in medium- to high-grade metamorphic rocks. Metamorphic fluid probably existed as a thin "film" of fluid-species molecules along grain boundaries. The discussion below will refer to *ambient fluid*. This is the very small proportion of free carbon dioxide–water fluid that is presumed to occupy very small pore spaces in the rock or to exist as a grain boundary film. The reader should note that fluid enters or leaves the rock by flow along grain boundaries or through cracks. Therefore, a large volume of fluid can potentially pass through a metamorphic rock during its entire metamorphic history (perhaps several million years), despite the fact that very little fluid is present at any one time. A new and expanding area of metamorphic petrology, which has been called "metamorphic hydrology," seeks to characterize this

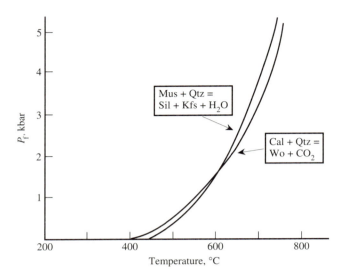

FIGURE 24-1

• • • • • • • • • • • • • • • •

P_{fluid}-T diagram illustrating the devolatilization reactions listed in the text: muscovite + quartz = sillimanite + potassic feldspar + H_2O and calcite + quartz = wollastonite + CO_2. The fluid is assumed to be pure water for the dehydration reaction and pure carbon dioxide for the decarbonation reaction. Note that both types of reactions have positive slopes and that slopes steepen with increased fluid pressure. For each reaction, the effect of diluting the product volatile species in the fluid at a given pressure would be to decrease the temperature of reaction, an effect illustrated by curves (a) and (b) in Figure 24-2.

process of fluid flow through metamorphic rocks in the deep crust.

There are several types of devolatilization reactions that involve loss of carbon dioxide and water from minerals and their incorporation into ambient fluid. The first and simplest type is dehydration or decarbonation. Numerous examples of dehydration were shown in the preceding chapter, and a typical example is muscovite + quartz $\rightleftharpoons$ potassic feldspar + sillimanite + H_2O; a typical analogous decarbonation reaction is calcite + quartz $\rightleftharpoons$ wollastonite + CO_2. On a P_{fluid}-T diagram, these reactions have a positive slope, with a curvature that is convex toward the temperature axis (Figure 24-1). Figure 24-1 illustrates the dependence of these reactions on pressure and temperature in the presence of pure water (in the case of dehydration) and pure carbon dioxide (in the case of decarbonation) but does not show the effect of mixing these volatiles in ambient fluid. One simple principle must be remembered: The highest temperature of reaction (at a given pressure) is for a fluid identical to that produced in the reaction. If ambient fluid differs from this composition, reaction will occur at a lower temperature. This is an example of the Le Châtelier principle: If a carbon dioxide–rich ambient fluid in a calcareous rock is diluted by an influx of water, the perturbation of fluid composition destabilizes the assemblage and can drive a decarbonation reaction. The reaction tends to restore the carbon dioxide–rich initial fluid composition by producing pure carbon dioxide fluid that is added to the ambient fluid.

Greenwood (1962) developed a different type of phase diagram to show the effects of fluid composition. Because only two variables can be shown in two dimensions, Greenwood showed reaction curves as a function of fluid composition and temperature by holding pressure constant. The fluid composition variable is mole fraction of carbon dioxide (X_{CO_2}), or effectively the partial pressure of carbon dioxide divided by the total fluid pressure, assuming a binary CO_2–H_2O fluid (reasonable for metamorphism of most calcareous rocks). This variable thus has a total range of 0 to 1.0. The reactions given above are shown in a T-X_{CO_2} diagram in Figure 24-2. Note that the highest temperature for decarbonation of calcite + quartz is at X_{CO_2} = 1.0, whereas the highest temperature for muscovite + quartz dehydration is at X_{CO_2} = 0 (or X_{H_2O} = 1.0). In the example shown in the figure, note that the wollastonite-forming reaction occurs at about 730°C in the presence of pure carbon dioxide but at only 670°C with X_{CO_2} = 0.5. The curves are asymptotic on their low-temperature sides because it is theoretically impossible to have *no* molecules of carbon dioxide or water in fluids coexisting with calcite or muscovite, respectively.

A number of reactions in calcareous rocks are of the pure decarbonation type discussed above, but many others are *mixed-volatile* reactions, that is, they involve both carbonate and hydrate minerals and thus produce or consume both water and carbon dioxide. One type of mixed volatile reaction is the combined decarbonation-dehydration reaction, in which both carbon dioxide and

water are produced simultaneously (that is, both are on the same side of the reaction). For these reactions, the maximum temperature (at any fixed pressure) is at the fluid composition dictated by the stoichiometry of the reaction. As an example, take the balanced reaction tremolite + 3 calcite + 2 quartz $\rightleftharpoons$ 5 diopside + 1 H_2O + 3 CO_2. The maximum temperature for this reaction will thus be at X_{CO_2} = 0.75, that is, 3 moles of carbon dioxide produced for every 4 moles of total fluid. The reaction temperature is lowered for fluid compositions either richer or poorer in carbon dioxide than this value. The resulting T-X_{CO_2} reaction curve has a parabolic shape as illustrated in Figure 24-2.

The third type of devolatilization equilibrium is one in which water and carbon dioxide appear on opposite sides of the reaction; the two varieties are decarbonation-hydration and dehydration-carbonation (see Figure 24-2). Note that these reactions can have dramatic effects on fluid composition because one fluid species is being removed from the fluid (and incorporated into a mineral) while the other species is being added to the fluid. For example, while both reactants and products are present, a dehydration-carbonation reaction can drive the fluid composition strongly toward water-enrichment with a modest temperature increase. The example shown in Figure 24-2 is 2 zoisite + CO_2 $\rightleftharpoons$ anorthite + 3 calcite + H_2O. A comprehensive review of the graphical analysis of mixed volatile equilibria can be found in Kerrick (1974).

FIGURE 24-2

• • • • • • • • • • • • • • • • •

T-X_{CO_2} diagram, at 2 kbar fluid pressure, illustrating the different types of devolatilization reactions discussed in the text: (a) decarbonation; (b) dehydration; (c) combined decarbonation-dehydration; (d) decarbonation-hydration. Note that dilution of ambient fluid with H_2O lowers the temperature of the decarbonation reaction (a) and, similarly, dilution with CO_2 lowers the temperature of the dehydration reaction (b). Mineral abbreviations: An, anorthite; Cal, calcite; Di, diopside; Kfs, potassic feldspar; Mus, muscovite; Qtz, quartz; Sil, sillimanite; Tr, tremolite; Wo, wollastonite; Zo, zoisite.

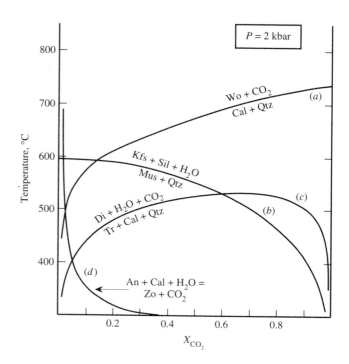

Internal Control of Fluid Compositions

Greenwood (1975) described a process by which the compositions of ambient pore fluids in calcareous rocks could be controlled or buffered by local devolatilization reactions. This is a very important concept in two ways. First, it explains how fluid compositions can evolve throughout prograde metamorphism as a result of iso-chemical internal reactions, without fluid species (or other chemical components dissolved in the fluid) being added or removed metasomatically. Second, it provides a mechanism for *driving* reactions in calcareous rocks through the addition of external fluids. This second process has been called **infiltration metamorphism** and has been extensively described by Ferry (1976) and others. The name derives from the infiltration of water-rich fluids into calcareous rocks, a process particularly observed in contact aureoles around shallow granitoid plutons.

Metamorphic reactions that plot as curves on diagrams such as Figure 24-2 are isobarically univariant. This term means that the coexistence of all reactant and product minerals (plus fluid) dictates that temperature and X_{CO_2} are not independent but must lie along the reaction curve. By the same reasoning, if one or more reactant minerals are missing, the T-X_{CO_2} conditions must lie above the curve; and if one or more product minerals are missing, T-X_{CO_2} must lie below the curve. It is therefore possible to predict qualitatively where conditions lie, relative to a reaction, if there is information on a particular metamorphic assemblage and a grid of possible reactions.

This prediction process is illustrated in the reaction sequence shown in Figure 24-3. Suppose that a calc-silicate rock contains tremolite, calcite, quartz, and fluid at the temperature and X_{CO_2} conditions shown by x. These conditions must lie at a temperature below the reaction tremolite + 3 calcite + 2 quartz $\rightleftharpoons$ 5 diopside + fluid because there is no diopside present in the assemblage. An arbitrary pore fluid composition ($X_{CO_2} = 0.45$) has been chosen and can be assumed to have resulted from all previous reactions or infiltration events. If the reactant assemblage is heated, the rock must take a vertical path on the diagram because there is nothing going on that can change the fluid composition. When the rock path intersects the reaction curve at y, tremolite, calcite, and quartz begin to react to form diopside. As the reaction proceeds and temperature increases, the rock path follows up along the reaction curve to the right, with the internal fluid becoming richer in carbon dioxide. This carbon dioxide enrichment occurs because the fluid being released through reaction and added to the ambient fluid is carbon dioxide–rich relative to this ambient

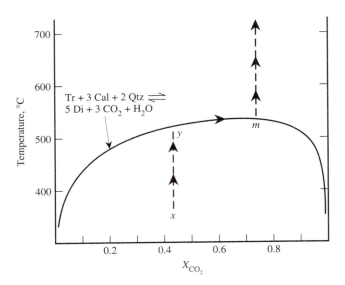

FIGURE 24-3
● ● ● ● ● ● ● ● ● ● ● ● ● ● ● ●

T-X_{CO_2} diagram showing an example of internal buffering or control of fluid composition through reaction. A calcareous rock at x contains the assemblage tremolite + calcite + quartz with an ambient fluid that has an X_{CO_2} of 0.45. Upon heating from < 400°C to about 515°C, the fluid does not change composition. The beginning of reaction at 515°C (at y) produces diopside and fluid with $X_{CO_2} = 0.75$, and this production of carbon dioxide–rich local fluid drives the ambient fluid composition along the curve shown by the arrows until the point m is reached, the maximum T on the reaction curve. As long as reaction continues, the fluid must stay at point m because it can no longer change its composition. As soon as any of the reactant minerals is exhausted, reaction stops and the rock continues to heat up along the dashed line without any further change in fluid composition. Mineral abbreviations: Cal, calcite; Di, diopside; Qtz, quartz; Tr, tremolite.

fluid. As long as the reaction proceeds, the rock path must remain on the curve, heading toward the maximum temperature at $X_{CO_2} = 0.75$.

If all reactants remain when maximum temperature is achieved (point m), the fluid can change no further, and the temperature is fixed at this point until one or more of the reactants is exhausted and reaction stops. (Of course, if this were a simple decarbonation reaction, the equivalent of point m would be at $X_{CO_2} = 1.0$.) There is no change in fluid composition at m for an obvious reason: The ambient fluid and the fluid being produced by reaction have identical compositions, so X_{CO_2} cannot change no matter how much fluid is added by reaction. Eventually, one of the reactants would be used up and reaction would stop at m, and the rock path would leave the reaction curve and move vertically to higher temperature. If one of the reactant minerals should run out between points y and m, the reaction would instantly stop and the rock path would immediately resume a vertical direction with increasing temperature and no fluid composition change. This behavior in which reaction assemblages follow reaction paths on T-X_{CO_2} diagrams has obvious implications for control of fluid compositions in calcareous rocks and has been widely exploited by petrologists as a way of monitoring fluid-rock interactions.

Infiltration and Buffering of Fluid Composition

As previously discussed, internal reaction control of fluid compositions occurs when the only fluid is ambient pore fluid that can be augmented by fluids produced during local reactions. Petrologists have also documented situations in which external fluids penetrate heated calcareous rocks that have a finite permeability. As noted earlier, this process is known as *infiltration*. In most of the studied cases, the external fluids are water-rich and thus are quite different in composition from the local, more carbon dioxide–rich ambient fluids that are controlled by local reactions. Some common sources of such aqueous fluids include dehydration reactions in adjacent pelitic rocks, crystallizing plutons, and veins. Many of the best-known examples are in contact aureoles where thermal gradients can assist in driving fluid flow.

Infiltration of external aqueous fluid can have dramatic effects in calcareous rocks, as illustrated in Figure 24-4. The reaction is the same one used in Figure 24-3: tremolite + 3 calcite + 2 quartz $\rightleftharpoons$ 5 diopside +

fluid, with an unreacted tremolite-calcite-quartz assemblage shown at a. In this case, however, the rock path is horizontal due to a sudden infiltration event of water-rich fluid. When the X_{CO_2} reaches the reaction curve at b, reaction will commence, and diopside and carbon dioxide–rich fluid will be produced. The local production of carbon dioxide–rich fluid will tend to counteract the effect of further infiltration of aqueous fluid and will try to keep the fluid composition from going to the left of the reaction curve. In effect, the reaction **buffers** the X_{CO_2} at a value reflected by b on the reaction curve. Because reaction rates tend to be slow, very high rates of fluid flow can actually overwhelm the buffering capacity of the calcareous rock, but several petrologic studies have indicated that in the absence of massive fluid flow, local buffering can be rather effective at keeping X_{CO_2} from reaching very low values.

There are several important points to note here. First, the horizontal rock path caused by aqueous fluid infiltration leads to reaction at much lower temperatures than would be the case if ambient fluid remained unperturbed (the vertical rock path described above). If a field geologist is mapping the first appearance of diopside as a metamorphic isograd, then the effect of variable fluid composition must be taken into account. Second, the reactive capacity of a calcareous rock may be used up by infiltration-driven reactions. There is typically only a limited supply of one or more reactant minerals in any metamorphic reaction. If any of these is exhausted in a reaction-buffering event, then the typical effect of a reaction sequence producing relatively carbon dioxide–rich fluid in a calcareous metamorphic rock may not happen, and the rock may retain an water-rich ambient fluid. This in turn will affect the temperatures of successively higher grade reactions.

FIGURE 24-4
● ● ● ● ● ● ● ● ● ● ● ● ● ● ● ● ●

T-X_{CO_2} diagram showing the effect of infiltration of externally derived aqueous fluid on the reaction history of a tremolite-calcite-quartz marble. Heating produces no change in fluid up to 460°C (vertical dashed line). At 460°C, the infiltration of aqueous fluid perturbs the fluid composition (a) and isothermally drives fluid toward water enrichment. At b, reaction commences and the production of carbon dioxide–rich fluid keeps ambient fluid composition from moving farther toward the left than b. If fluid flow rate exceeds reaction rate, the buffering capacity of the local assemblage can be overwhelmed and the ambient fluid composition can be driven all the way to essentially pure water. Mineral abbreviations: Cal, calcite; Di, diopside; Qtz, quartz; Tr, tremolite.

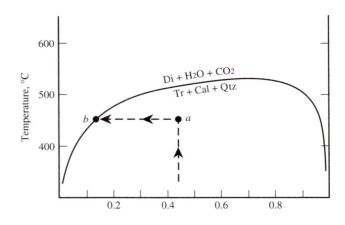

METAMORPHIC REACTIONS AND ASSEMBLAGES IN MARBLES
• • • • • • • • • • • • • • •

Regional Metamorphism

At first glance, it would seem that little could happen mineralogically in calcite marbles with only some admixed quartz. The only reactions between calcite and quartz are the high-grade one that produces wollastonite at about $650°$–$700°C$ and some even higher temperature, low-pressure contact metamorphic reactions that produce the rare minerals spurrite and larnite. Calcite marbles can also contain some graphite that has resulted either from recrystallization of organic matter or inorganic reduction of calcite or carbon dioxide. Coarse marbles of the Grenville Province of the northeastern United States and eastern Canada are well known for their large graphite flakes. Pure calcite marbles are rather unusual, however, and the more common dolomite marbles display a wide range of metamorphic calcium-magnesium silicate minerals. The resulting assemblages and reactions can be useful indicators of relative metamorphic grade (Table 24-1). Observation of assemblages in marbles, and particularly their use as indicators of grade, should always be undertaken with the cautious understanding that the order and temperature of appearance of minerals in these rocks are strongly dependent on both rock bulk composition and especially fluid composition. For a useful detailed case study of progressive metamorphism of relatively pure marbles, readers are referred to a review and discussion of regional metamorphism of dolomitic marbles of the Alps in Yardley (1989, pp. 130–133).

In one of the classic papers in the metamorphic petrology literature, Bowen (1940) characterized the sequence of metamorphic minerals in low- to medium-pressure metamorphism of dolomitic marbles. Because most prograde reactions in the sequence involve decarbonation (alone or in combination with dehydration), Bowen's sequence has become known as **Bowen's Decarbonation Series** in recognition of, and analogy with, his better known Reaction Series in igneous rocks. Bowen even proposed a mnemonic for remembering his series: "Tremble Dire For Peril Walks—Monstrous Acrimony Spurning Mercy's Laws": **tr**emolite-**di**opside-**fo**rsterite-**pe**riclase-**wo**llastonite-**mo**nticellite-**ak**ermanite-**sp**urrite-**me**rwinite-**la**rnite. More recent studies have shown that talc may actually precede tremolite as the first reaction product between detrital quartz and dolomite, but only where the ambient pore fluids are relatively water-rich.

Many relatively pure dolomite marbles have only quartz as an impurity and can thus be represented chemographically on a triangular diagram with CaO, MgO, and SiO_2 on the corners (CMQ; Figure 24-5); FeO is typically quite low in marbles. The CMQ ternary is projected from the other major components, the volatiles carbon dioxide and water. The only other minerals commonly found in such marbles are present because of minor chemical components and generally do not participate in the reactions considered here. Examples include potas-

TABLE 24-1 Typical mineral associations in metamorphosed calcareous rocks

Metamorphic grade	Dolomitic marble	Calc-silicate rock
Very low, protometamorphic	Calcite, dolomite, quartz	Calcite, ankerite, clays, chlorite, quartz, albite, rutile/ilmenite
Low	Calcite, dolomite, quartz, talc	Calcite, ankerite, chlorite, muscovite, albite, quartz, rutile/ilmenite, ±biotite
Middle	Calcite, dolomite, quartz, tremolite, ±diopside	Calcite, biotite, plagioclase, calcic amphibole, titanite, ±quartz, ±epidote, ±dolomite
High	Calcite, diopside, forsterite, ±wollastonite	Calcic amphibole, epidote/zoisite, diopside, calcic plagioclase, titanite, K-feldspar, ±quartz, ±calcite
Very high contact	Calcite, diopside, forsterite, ±monticellite, ±akermanite, ±wollastonite	Calcic garnet, calcic plagioclase, diopside, titanite, ±quartz, ±K-feldspar, ±vesuvianite

(A)

(B)

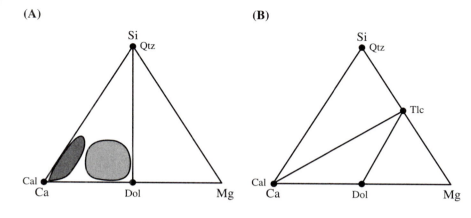

(C)

(D)

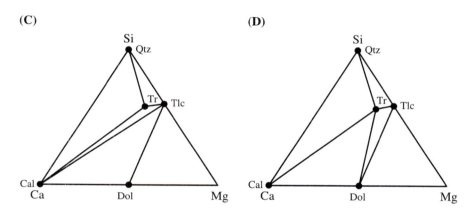

FIGURE 24-5

• • • • • • • • • • • • • • • • •

Graphical representations of low- to medium-grade marble assemblages in the Ca–Mg–Si triangle (CMQ diagram). The approximate ranges of siliceous dolomitic marble and calcite marble bulk compositions are shown by the light and dark shaded areas, respectively, in **(A)**. See text for explanation of reaction sequences going from **(B)** to **(D)**. Mineral abbreviations: Cal, calcite; Dol, dolomite; Qtz, quartz; Tlc, talc; Tr, tremolite.

sic feldspar or phlogopite (present because of potassium and aluminum), plagioclase (present because of sodium and aluminum), titanite (present because of titanium), and pyrite or pyrrhotite (present because of sulfur and minor iron).

The premetamorphic and very low grade metamorphic assemblage in dolomitic marbles is shown in Figure 24-5A; the range of common bulk compositions is indicated. Remember that bulk compositions of typical siliceous dolomitic marbles must lie to the left of the dolomite-quartz tie line because the ratio of magnesium to calcium cannot exceed 1:1, but can be lower. Aside from minor recrystallization and grain coarsening, the first metamorphic effect seen in some marbles (although certainly not all) is the appearance of small flakes of talc growing radially to the margins of larger quartz grains or quartz nodules. The reaction (not balanced) is written dolomite + quartz + $H_2O \rightleftharpoons$ talc + calcite + CO_2 and is represented by a switch from a dolomite-quartz tie line to a talc-calcite tie line (Figure 24-5B). Note that this reaction is of the hydration-decarbonation type. Despite the fact that this reaction seems to be univariant, it is in fact divariant because of its additional dependence on fluid

composition. For this reason, talc-bearing marbles can contain four-phase assemblages comprising calcite, dolomite, quartz, and talc over a range of temperatures. The talc-forming reaction typically occurs under greenschist-facies conditions at about 350° to 400°C.

The next mineral to appear in sequence is tremolite, through the reaction talc + calcite + quartz $\rightleftharpoons$ tremolite + H_2O + CO_2. The reaction is represented by the appearance of tremolite on the projected CMQ diagram in the triangle formed by talc, calcite, and quartz (Figure 24-5C). The assemblage in marbles following this reaction depends on bulk composition, but most commonly is either talc + calcite + tremolite or tremolite + calcite + quartz (Figure 24-6). The final elimination of talc from all marble compositions is due to the reaction talc + calcite $\rightleftharpoons$ tremolite + dolomite + H_2O + CO_2 (Figure 24-5D). Above this reaction, only two assemblages would be found: tremolite + calcite + quartz (in dolomite-poor compositions) and tremolite + dolomite + calcite (in SiO_2-poor bulk compositions). Note that talc remains theoretically stable (for example, in ultramafic compositions) but cannot occur in bulk compositions that contain calcite or dolomite. The appearance of

FIGURE 24-6
● ● ● ● ● ● ● ● ● ● ● ● ● ● ● ●

Photomicrograph of tremolite marble, northwestern Adirondacks, New York. Colorless, elongated, prismatic tremolite crystals (center) lie in a matrix of coarse calcite. Note the rhombohedral cleavage in the large calcite grain at the left. Plane-polarized light. Horizontal field of view is about 2.4 mm.

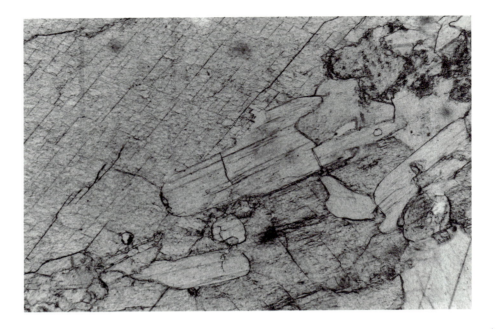

tremolite and the elimination of talc from marbles occur at about 400° to 450°C.

The first appearance of diopside in dolomitic marbles is due to the reaction tremolite + calcite + quartz $\rightleftharpoons$ diopside + H_2O + CO_2. Chemographically, this reaction is represented by the appearance of diopside in the triangle formed by tremolite, calcite, and quartz (Figure 24-7A). Marble assemblages at this grade are calcite + diopside + quartz (in dolomite-poor, quartz-rich compositions), calcite + diopside + tremolite, and calcite + dolomite + tremolite (in quartz-poor, dolomite-rich compositions). At about the same metamorphic grade, forsterite appears in calcite-bearing, SiO_2-poor marbles. The association of calcite plus forsterite is made possible by the reaction tremolite + dolomite $\rightleftharpoons$ forsterite + calcite + H_2O + CO_2 (Figure 24-7B). Note that this reaction eliminates tremolite from dolomitic marbles, producing the assemblages calcite + tremolite + forsterite in SiO_2-richer compositions and calcite + dolomite + forsterite in SiO_2-poor compositions. The final reaction that is commonly observed in regional metamorphism of marbles is tremolite + calcite $\rightleftharpoons$ diopside + forsterite + H_2O + CO_2. This reaction occurs at about 550° to 600°C and finally permits the coexistence of diopside and forsterite in marbles (Figure 24-7C). Diopside + forsterite + calcite assemblages are typical for marbles in upper amphibolite and many granulite-facies regional metamorphic terranes (Figure 24-8). In rare cases, a final regional metamorphic reaction may be observed in some granulite-facies marbles: calcite + quartz $\rightleftharpoons$ wollastonite + CO_2 (Figure 24-7D). Under normal regional metamorphic pressures (>3 kbar), temperatures

approaching 800°C are required to produce wollastonite, in the absence of aqueous fluid infiltration. Most wollastonite marbles have been demonstrated to be the results of shallow contact metamorphism or to involve unusual water-rich fluids.

Contact Metamorphism of Marbles

In general, the assemblages portrayed in Figure 24-7C and D are those observed up to the highest grades of regional metamorphism at elevated pressures. Most of these reactions and assemblages occur in the *lower* grades of contact metamorphism of marbles, however, because of the positive slopes of decarbonation-dehydration reactions in P_{fluid}-T diagrams (see Figure 24-1). Most reactions occur at significantly lower temperatures at pressures below 1.5–2 kbar, and there are a number of reactions and mineral assemblages that occur in marbles only at the low pressures of shallow contact metamorphism. These minerals are typically all those of Bowen's decarbonation series after, and probably including, wollastonite and periclase. Wollastonite of genuine regional metamorphic origin (without any component of magmatic heating) has rarely been unequivocally demonstrated. All the other, higher grade minerals have been rarely observed in nature and only in contact aureoles; the most widely seen include monticellite ($CaMgSiO_4$), periclase (MgO), clinohumite ($Mg_9Si_4O_{16}[OH,F]_2$), and akermanite ($Ca_2MgSi_2O_7$). Chemographic phase relations in contact metamorphosed marbles are summarized in

(A)

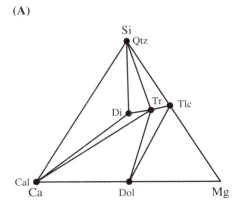

(B)

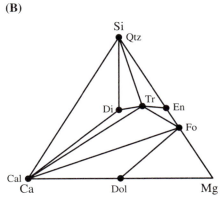

FIGURE 24-7
• • • • • • • • • • • • • • • •

Graphical representation of medium-
to high-grade marble assemblages
in the Ca–Mg–Si triangle (CMQ
diagram). See text for the reaction
sequence going from **(A)** to **(D)**.
Mineral abbreviations: Cal, calcite;
Di, diopside; Dol, dolomite; En,
enstatite; Fo, forsterite; Qtz, quartz;
Tlc, talc; Tr, tremolite.

(C)

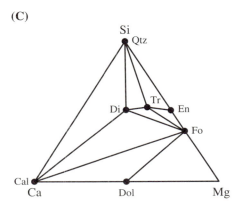

(D)

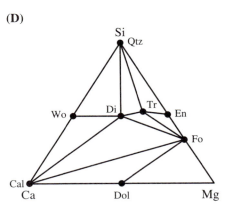

FIGURE 24-8
• • • • • • • • • • • • • • • •

Photomicrograph of coarse diopside-
forsterite marble, from near the
Ausable Gorge, Adirondacks, New
York. Grains of forsteritic olivine (Ol)
and diopside (Cpx) lie in a coarse
matrix of calcite. Note the twinning
of the calcite (left) and incipient
serpentinization of olivine (low-
relief, colorless material surrounding
olivine, center right). Plane-polarized
light. Horizontal field of view is about
4.8 mm.

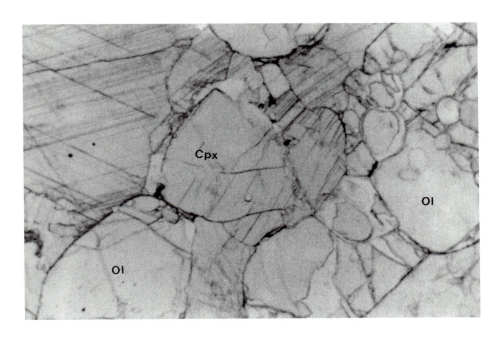

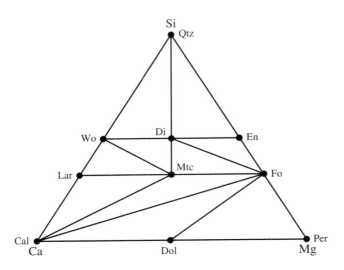

FIGURE 24-9

• • • • • • • • • • • • • • • • • •

Generalized chemographic relationships for the high-temperature CMQ minerals that are found in contact metamorphism of marbles and ultramafic rocks. Note that dolomite may disappear in some very high temperature assemblages, giving rise to a calcite-periclase-forsterite assemblage for SiO_2-poor compositions, and that tie line arrangements may be slightly different as a result of other reactions. Mineral abbreviations: Cal, calcite; Di, diopside; Dol, dolomite; En, enstatite; Fo, forsterite; Lar, larnite; Mtc, monticellite; Per, periclase; Qtz, quartz; Wo, wollastonite.

Figure 24-9, and the interested reader is referred to a comprehensive discussion of the phase equilibria and assemblages of contact metamorphism of siliceous marbles in Tracy and Frost (1991).

As an interesting and relatively typical case study of contact metamorphism of marbles, we will examine the contact aureole surrounding one of several Tertiary stocks about 40 km east of Salt Lake City, Utah (see Moore and Kerrick 1976). The Tertiary Alta stock (granodiorite), Little Cottonwood stock (quartz monzonite), and Clayton Peak stock (diorite) intruded unmetamorphosed Cambrian and Mississippian rocks, with the bulk of the Alta aureole developed in siliceous dolostones. The siliceous dolostone contains two lithologies: nodular dolostone with chert nodules up to 20 cm across in pure dolostone, mixed with massive dolostone that contains disseminated detrital quartz grains. The development of metamorphic textures is quite distinct in the two lithologies: contact metamorphic assemblages developed pervasively and homogeneously in the massive dolostone, whereas concentrically layered zones formed around nodules in nodular dolostone. Moore and Kerrick used an earlier stratigraphic assessment of burial depth of 3 to 4 km as a basis for modeling the metamorphic reactions at 1 kbar pressure.

Figure 24-10 shows a generalized geologic map of the Alta stock and its aureole with isograds mapped by Moore and Kerrick in the massive dolostones. The outermost isograds reflect the first coexistence of calcite + talc (see Figure 24-5B). Because the massive dolostones are generally quartz-poor, exhaustion of quartz in the talc-forming reaction led to this quartz-absent tremolite-forming reaction at the tremolite isograd: talc + calcite $\rightleftharpoons$ tremolite + dolomite + fluid (see Figure 24-5D). In

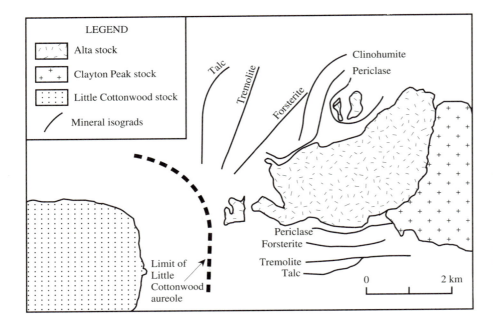

FIGURE 24-10

• • • • • • • • • • • • • • •

Generalized geologic map of the Alta, Utah, area, showing sample locations and isograds in the Alta aureole. The approximate outer limit of the Little Cottonwood aureole is located west of the Alta stock; the Little Cottonwood stock is shown in the lower left. Original mapping sources are listed in the original paper. [From J. N. Moore and D. M. Kerrick, 1976, *Am. J. Sci.*, *276*, Fig. 1.]

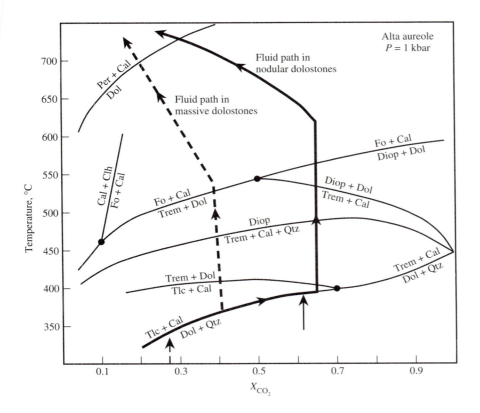

FIGURE 24-11
● ● ● ● ● ● ● ● ● ● ● ● ● ● ● ●

$T\text{-}X_{CO_2}$ diagram at 1 kbar showing postulated $P\text{-}T$ paths for progressive metamorphism of massive and nodular siliceous dolostones in the Alta aureole. Data sources for the reaction curves are listed in the original paper. Note that the shifts toward H_2O enrichment at higher temperature were interpreted by the original authors as being due to infiltration of the aureole by aqueous fluids given off by the crystallizing plutons. These shifts permit the appearance in the inner aureole of periclase and clinohumite, two minerals that require water-rich conditions to form. Mineral abbreviations: Cal, calcite; Clh, clinohumite; Diop, diopside; Dol, dolomite; Fo, forsterite; Per, periclase; Qtz, quartz; Tlc, talc; Trem, tremolite.

more quartz-rich lithologies, textures indicate that tremolite also appeared through the reaction dolomite + quartz + H_2O ⇌ tremolite + calcite + CO_2. The forsterite isograd reflects the reaction tremolite + dolomite ⇌ forsterite + calcite + fluid (see Figure 24-7A and B). Note that the composition of the dolostones precludes observation of the initial diopside-forming reaction that was discussed earlier for regional metamorphism. In the northern part of the aureole, the next isograd marks the appearance of the new mineral clinohumite in assemblages with calcite, dolomite, and forsterite. In the CMQ diagram (see Figure 24-5), clinohumite would plot slightly below forsterite on the QM edge. The probable reaction forming clinohumite is forsterite + dolomite + H_2O ⇌ calcite + clinohumite + CO_2, but the reader should note that virtually all clinohumites contain substantial fluorine. In this case, the source of the fluorine may be fluids emanating from the crystallizing pluton. The final isograd, that closest to the igneous contact, is the periclase isograd that reflects the reaction dolomite ⇌ periclase + calcite + CO_2. This reaction requires very water-rich fluid compositions to occur at reasonable temperatures (Figure 24-11).

Moore and Kerrick estimated temperatures on the basis of the compositions of coexisting calcite and dolomite (see Appendix 2) but were unable to reconstruct the exact metamorphic temperatures for each contact zone because of the variable degrees of reequilibration. Reconstruction of aureole temperatures, using phase equilibrium arguments based on an accurate isobaric $T\text{-}X_{CO_2}$ diagram for 1 kbar pressure (Figure 24-11), suggests incipient talc formation in the outermost aureole, starting at about 300°–350°C and extending over a substantial range of temperatures. In the massive dolostones, Moore and Kerrick argued that the tremolite isograd represents about 400°C, the appearance of diopside about 475°C, the forsterite isograd about 540°C, the clinohumite isograd 575°C, and finally the periclase isograd about 675°C. They further argued that fluid compositions differed substantially between the massive and nodular dolostones, as depicted in Figure 24-11. Their interpretation was that the buffering capacity of the nodular dolostones was greater at low grades, thus causing the fluids to become more carbon dioxide–rich through rock paths that migrated along the talc- and tremolite-forming reaction curves. Other detailed studies of marbles in contact aureoles (reviewed in Tracy and Frost 1991) have revealed a wide variation of fluid compositional behavior from perfectly internally buffered to completely dominated by externally derived fluid.

METAMORPHISM
OF CALC-SILICATE ROCKS
● ● ● ● ● ● ● ● ● ● ● ● ● ● ●

As noted at the beginning of this chapter, relatively pure marbles and siliceous dolostones are not particularly common in metamorphic terranes. The calc-silicate rocks, which are typically defined as rocks rich in Ca–Mg–Fe–Al silicate minerals but poor in carbonates, are much more widespread. These rocks have been variously called calcareous pelites, calc-pelites, argillaceous limestones, and marls; and they show considerably more complicated mineralogy than the marbles do as a direct result of compositional diversity due to a large detrital component admixed with carbonate in the protolith. The detrital component includes quartz and feldspars, as well as clays, ferromagnesian silicates, and oxides. Protolith sedimentary environments probably are near-shore, shallow, warm water banks near major sources of continental or island arc-derived detrital sediments where these are mixed with biogenic or precipitated carbonate material.

In addition to high CaO, important distinguishing aspects of calc-silicate rock compositions include the higher Fe/(Fe+Mg), and the high Al_2O_3, K_2O, and Fe_2O_3 contents (compared with those of most marbles) due to the original detrital minerals. These compositional characteristics give rise to the common occurrence of chlorite, muscovite, biotite, epidote, garnet, and actinolite, metamorphic minerals rarely observed in marbles. Unfortunately, calc-silicate rock compositions can overlap with those of mafic volcanic and volcaniclastic rocks, both compositional types giving rise to epidote amphibolites and amphibolites at middle grades of metamorphism. Thus, for amphibolites, the use of protolith as an indicator of premetamorphic tectonic environment must be carried out with caution, and preferably using sophisticated geochemical indicators such as trace elements or radiogenic isotopes.

Understanding the role of fluids is also crucial in interpreting progressive metamorphism of calc-silicates. In a classic study, Carmichael (1970) first showed a natural example of how shifts in fluid composition in calc-silicates could drive mixed-volatile reactions and dramatically affect the correlation between isograds in metapelites and those in calc-silicates. He examined interbedded pelitic and calc-silicate rocks near Whetstone Lake in Ontario and showed that mapped isograds in the two compositions crossed each other at an acute angle (Figure 24-12). Although the metamorphism was characterized as regional, there are several granitic

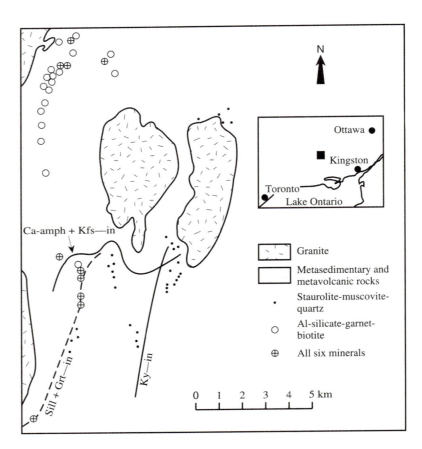

FIGURE 24-12
● ● ● ● ● ● ● ● ● ● ● ● ● ● ● ●

Generalized geologic map of the Whetstone Lake area, Ontario, showing locations of granitic stocks and the intersecting metapelitic and calc-silicate mineral isograds. [After D. M. Carmichael, 1969, *Contrib. Mineral. Petrol.*, *20*, Fig. 8.]

plutons that may have served as local sources of heat or fluid. The crossing isograds indicate that metamorphism could not have been driven by temperature increase alone but must also have been affected by change in another environmental variable, in this case fluid composition in one or both of the rock types. Carmichael noted that the metapelite isograds behave regularly in a typical Barrovian sequence, with a consistent increase in metamorphic grade to the west, culminating in sillimanite-grade rocks adjacent to a larger pluton. In the calc-silicate rocks, the most important mineralogic change is the appearance of calcic amphibole (tremolite-actinolite) and potassic feldspar at the expense of biotite, calcite, and quartz, in a reaction described in more detail below. This amphibole + potassic feldspar isograd cuts across the pelite isograds at an angle, with higher grade rocks to the north of the isograd (Figure 24-12). Carmichael's interpretation was that the release of aqueous fluid from one of the smaller plutons followed by outward infiltration dominated fluid composition in all lithologies. This externally derived fluid had little or no effect on the already water-rich fluid in the metapelites but diluted the carbon dioxide in the ambient fluid of the calc-silicates. As first shown in Figure 24-2, the substantial lowering of X_{CO_2} would likely dramatically lower the temperature of reaction for decarbonation and decarbonation-dehydration equilibria.

Prograde Assemblages in Calc-Silicates

The mineral assemblages that are typical of calc-silicate bulk compositions are listed in Table 24-1. They can best be summarized succinctly by noting (1) the progressive removal of carbonate by reaction, and (2) the dominance of sheet silicates and epidote at low grade, of calcic amphibole at middle grades, and of calcic pyroxene and garnet at high grades, a sequence reflecting progressive dehydration as well as decarbonation. Further, the carbonate minerals shift in composition with prograde metamorphism, from ankerite or dolomite at lowest grades to calcite at middle to high grades. Numerous workers, beginning with Goldschmidt and Eskola, have noted that progressive changes in metamorphic mineralogy of calc-silicates can be correlated with the changes in metapelites and used as a monitor of changing metamorphic grade. For example, Kennedy (1949) observed the general mineralogic shifts referred to above for rocks in the Scottish Highlands and formulated a scheme of calc-silicate zones in which an epidote-biotite-calcite±actinolite zone correlated with the garnet zone in metapelites. At higher grades, his anorthite-hornblende zone correlated with the staurolite and kyanite metapelite zones,

and the anorthite-calcic pyroxene zone was equivalent to the sillimanite zone. Numerous investigators, beginning with Carmichael, have observed how this correlation can be complicated by the introduction of external aqueous fluids into calc-silicates.

Perhaps the most convenient way to illustrate the progressive changes in mineral assemblages in detail in these rocks is to review the comprehensive work of John Ferry (1976, 1983a, 1983b) on the calc-silicate rocks of the Vassalboro Formation of southern Maine. He documented prograde mineral assemblages over a wide area and a substantial range of metamorphic grade in an attempt to use these changes to document fluid infiltration, fluid-rock interactions, and changes in ambient fluid composition. Ferry's map of calc-silicate metamorphic zones and isograds in pelites of the adjacent Waterville Formation is shown in Figure 24-13. The metamorphism apparently occurred at low pressures (2–4 kbar), and the distribution of zones is clearly related to heating caused by the local occurrence of Devonian granitoid plutons, which are also the probable source of aqueous fluids that infiltrated the calc-silicate rocks and drove decarbonation reactions.

ANKERITE ZONE. The lowest grade metamorphism in the Vassalboro Formation (and many other locales) is characterized by the assemblage quartz-muscovite-albite-calcite-ankerite (or dolomite)-chlorite. It is likely that the protometamorphic reaction that produced this assemblage involved clays reacting with ankeritic carbonate to produce muscovite, chlorite, and calcite. At this grade, plagioclase solid solution is invariably almost pure albite because of the instability of the anorthite molecule. Figure 24-14A illustrates this assemblage on ACF and A'FK diagrams. Little study of the lowest grade reactions has been done, mostly because of fine grain size and the difficulty of characterizing the phases mineralogically and chemically.

BIOTITE ZONE. The lowest grade isograd in Figure 24-13 marks the lower limit of the biotite zone and reflects the reaction muscovite + quartz + ankerite $\rightleftharpoons$ biotite + calcite + CO_2. Yardley (1989) suggested that the greater abundance of chlorite in biotite-zone rocks may indicate that some chlorite is also produced by this reaction. There are actually a number of reactions that could be invoked for the appearance of biotite, but this one is typical. In rocks that contain dolomite rather than ankerite, phlogopite is a product of the above reaction, which has been studied experimentally and shown to occur at about 500°C at 6 kbar. A subsidiary reaction that occurs within the biotite zone produces anorthite molecule: muscovite + calcite + quartz $\rightleftharpoons$ potassic feldspar + anorthite + H_2O + CO_2. The progress of this reaction,

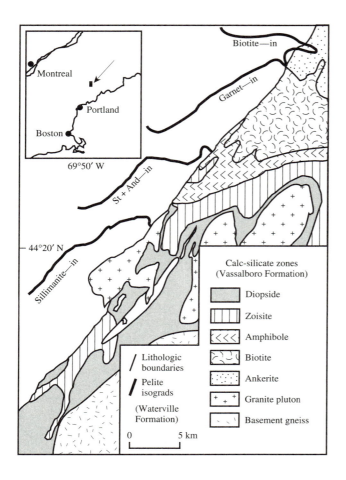

FIGURE 24-13

● ● ● ● ● ● ● ● ● ● ● ● ● ● ● ● ●

Generalized geologic map of the Waterville-Vassalboro area, Maine, showing metamorphic zones in calc-silicate rocks of the Vassalboro Formation, isograds in the adjacent metapelites of the Waterville Formation, and outlines of small granitic stocks that apparently caused much of the metamorphism. Mineral abbreviations: Ca-amph, calcic amphiboles; Grt, garnet; Kfs, potassic feldspar; Ky, kyanite; Sill, sillimanite. [After J. M. Ferry, 1976, *Am. J. Sci.*, *276*, Fig. 3; and J. M. Ferry, 1983, *J. Geol. Soc. London*, *140*, Fig. 2.]

especially in the upper part of the biotite zone, typically shifts the plagioclase toward intermediate compositions (An$_{20}$ to An$_{60}$) by mixing the anorthite component into the preexisting albite.

ACF and A′FK diagrams for the biotite zone are illustrated in Figure 24-14B. Note on the ACF diagram that there is still relatively abundant carbonate for typical calc-silicate bulk compositions.

AMPHIBOLE ZONE. The first appearance of low-aluminum calcic amphibole (tremolite-actinolite) marks the lower boundary of the amphibole zone (Figure 24-15). As grade increases, this amphibole tends to become more aluminous, heading toward becoming hornblende (as also happens in the lowermost part of the amphibolite facies in metabasites). At the same time, continued reaction of calcite with aluminous minerals produces additional anorthite, further moving plagioclase toward anorthite enrichment. One proposed reaction for the amphibole zone is chlorite + calcite + quartz $\rightleftharpoons$ calcic amphibole + anorthite + H$_2$O + CO$_2$. However, the common appearance of microcline in calc-silicates of this zone suggests that the reaction biotite + calcite + quartz $\rightleftharpoons$ calcic amphibole + potassic feldspar + H$_2$O + CO$_2$ is also occurring.

Typically, chlorite disappears from calc-silicates at about this grade. If reactions are being driven by infiltration of aqueous fluid, then much or all of the potassic feldspar produced in the above reaction is dissolved in the aqueous fluid and removed from the rock (especially the potash and silica; most or all of the aluminum is probably insoluble and left behind). Ferry and other petrologists working in calc-silicates have noted metasomatic effects in which the alkali contents of the rocks drop precipitously in the amphibole zone. Under sufficiently water-rich conditions, the residual Al$_2$O$_3$ combines with calcite and silica to produce the common calc-silicate mineral zoisite.

ACF and A′FK diagrams for the amphibole zone are shown in Figure 24-14C. Note that the appearance of calcic amphibole replacing chlorite in the assemblage amphibole + calcic plagioclase + calcite + potassic feldspar dramatically reduces the modal carbonate content of typical calc-silicate bulk compositions.

ZOISITE ZONE. In the Vassalboro Formation, Ferry observed the mineral zoisite growing between plagioclase grains and adjacent calcite. Zoisite is a nearly iron-free epidote-group mineral. This zone is not observed in all calc-silicates, but its presence implies something extremely important about metamorphic fluids. The formation of zoisite above about 400°C in the reaction calcic plagioclase + calcite + H$_2$O $\rightleftharpoons$ zoisite + CO$_2$ can only occur when ambient fluid is virtually pure water ($X_{CO_2} < 0.05$). Because this situation is unlikely in a calc-silicate rock undergoing active decarbonation reactions, Ferry argued that the composition of fluid must be dominated by an external water-rich source, probably fluids given off by nearby crystallizing granitic plutons.

The portrayal of zoisite-zone assemblages on ACF and A′FK diagrams is complicated by the likelihood of variations in fluid composition. Divariant four-phase assemblages can be encountered, in apparent violation

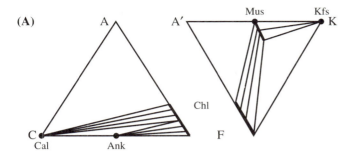

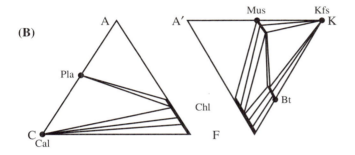

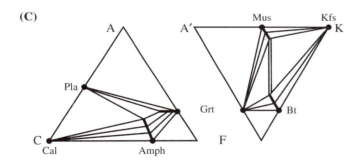

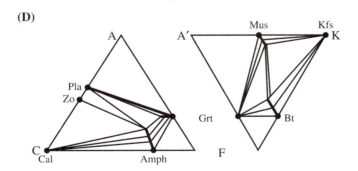

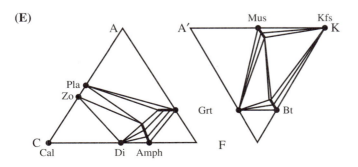

FIGURE 24-14

ACF and A′FK diagrams for the ankerite (**A**) through diopside (**E**) zones of metamorphosed calc-silicate rocks from the Vassalboro Formation, Maine. Mineral abbreviations: Amph, calcic amphibole; Ank, ankerite; Bt, biotite; Cal, calcite; Chl, chlorite; Kfs, potassic feldspar; Grt, Ca–Mg–Fe garnet; Mus, muscovite; Pla, calcic plagioclase; Zo, zoisite. [From data in J. M. Ferry, 1976, *Am. J. Sci.*, *276.*]

of the phase rule, as a result of adding this extra environmental variable. In the ACF diagram of Figure 24-14D, note that the zoisite-zone divariant assemblage is calcic amphibole + calcic plagioclase + calcite + zoisite + potassic feldspar.

DIOPSIDE ZONE. The highest grade calc-silicates observed by Ferry show diopside replacing calcic amphibole as a result of the reaction calcic amphibole + calcite + quartz $\rightleftharpoons$ diopside + H_2O + CO_2, and this is typ-

ically the highest grade mineralogic effect seen in calc-silicates outside obvious, areally restricted, contact aureoles. Because diopside is much less aluminous than hornblende, this reaction indicates that it is probably the tremolite-actinolite component of the calcic amphibole that is breaking down, leaving even more aluminous residual hornblende. This aluminous hornblende can persist in calc-silicates to much higher grades in an extended coexistence with calcic pyroxene. Ferry noted that the growth of zoisite as a result of the reaction of

(A)

FIGURE 24-15

• • • • • • • • • • • • • • • • •

Plane-polarized light photomicrographs of typical calc-silicate rocks. **(A)** Potassic calc-silicate rock, southwestern Connecticut. High-relief, colorless prismatic mineral in the center is tremolite, intergrown with phlogopitic biotite (pale colored mineral). Very high relief mineral at right is sphene, and the colorless material is a mixture of calcite, quartz, and calcic plagioclase. The opaque mineral is graphite. Horizontal field of view is about 1.2 mm. **(B)** Low-potassium aluminous calc-silicate from sillimanite zone, central Massachusetts. Pale colored mineral is magnesian hornblende, which is overgrown by epidote (high relief, colorless). Colorless minerals are calcite (note twinning in the grain at center), quartz, and plagioclase. The rock also contains grossular-rich garnet. Horizontal field of view is about 2.4 mm.

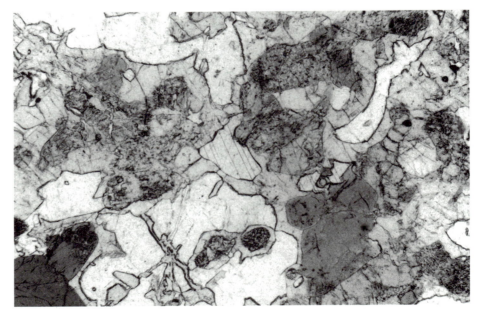

(B)

plagioclase and calcite seems to continue to the highest grades he observed in Maine, an observation suggesting the continued infiltration of aqueous fluid. This proposal makes sense, because his highest grade rocks are in the closest proximity to granitic plutons.

Figure 24-14E shows ACF and A′FK diagrams for the diopside zone. Note that because of extra compositional variables, there can now be as many as five ACF minerals coexisting in an apparent equilibrium divariant assemblage: diopside, calcic amphibole, calcite, zoisite, and calcic plagioclase. Also note that the progressive production of calcium-magnesium silicate minerals closer to the calcium corner (first calcic amphibole, then diopside) reflects the substantial reduction in carbonate content of the rocks. This principle, of course, is also seen in the presence of calcite as a reactant in virtually all the reactions written above. By the time calc-silicate rocks reach the diopside zone, most have little or no calcite left in them, reinforcing the initial definition of calc-silicates.

It is of some interest to know what the overall assemblages are in the total system ACFK, especially the calcite-bearing assemblages. Unfortunately, it is not obvious from the separate ACF and A′FK diagrams what the complete combined assemblages would be. These assemblages are best seen in a series of calcite projections within the ACFK tetrahedron to illustrate the minerals that are in equilibrium with calcite throughout the sequence of calc-silicate zones shown in Figure 24-14A through E. The projected diagrams, labeled A″FK in Figure 24-16, are of assemblages within the ACFK tetrahedron viewed from the C corner. Note that each of the two- or three-phase assemblages shown in these diagrams also contains both calcite and quartz.

For low-potassium, low-aluminum bulk compositions, the normal assemblage within the ankerite zone would be quartz + calcite + ankerite + chlorite + potassic feldspar (Figure 24-16A); within the biotite zone, it would be quartz + calcite + chlorite + biotite (Figure 24-16B); in the amphibole zone, quartz + calcite + calcic amphibole + biotite (Figure 24-16C); in the zoisite zone, quartz + calcite + calcic amphibole + potassic feldspar (Figure 24-16D); in the diopside zone, quartz + calcite + diopside + potassic feldspar + calcic plagioclase or zoisite (Figure 24-16E). Note that progressive assemblages including muscovite, plagioclase, and zoisite generally require higher aluminum contents.

Further Observations on Calc-Silicate Rocks

The zonal sequence described by Ferry for the Vassalboro Formation in Maine has been shown at a map scale of kilometers (see Figure 24-13). Similar zonation has been observed on a much smaller scale of centimeters to meters, where impure carbonate lenses or layers are interbedded with metapelitic rocks or cut by quartz veins or granitic dikes. Even calc-silicate nodules, or boudins, in highly deformed rocks have been described as having centimeter-scale assemblage zonation. In virtually all cases, this zonation has been ascribed to highly localized variations in ambient fluid composition (water to carbon dioxide ratio) produced by diffusion or infiltration of aqueous fluids from sources external to the calc-silicates. Although adjacent metapelites undergoing dehydration reactions have sometimes been identified as a source, zoned calc-silicates are much more typically adjacent either to quartz veins (former conduits for silica-saturated aqueous fluids) or to granitoid plutons. In the case of quartz veins, the development of zonation can be virtually isothermal, with decarbonation or decarbonation-dehydration reactions driven by increased X_{H_2O} alone (see Figure 24-4). In the case of plutons, from meter-sized dikes to considerably larger bodies, metamorphic reactions are likely to have been driven by both heat and fluid flow from the igneous rocks.

One special class of calc-silicate rocks should be mentioned here, the **skarns.** Traditionally of more interest to economic geologists and ore petrologists than to metamorphic petrologists, skarns are the calcareous host rocks to many economically important rare minerals (for example, tungstates, molybdates, rare earth minerals), as well as to common metals such as zinc, lead, and copper. The classic silicate and oxide minerals of skarns include calcic garnet, hydrogarnet, vesuvianite, scapolite, Mg–Al spinel, and periclase, along with more common calc-silicate minerals like calcic plagioclase, diopside, and epidote. Virtually always either adjacent to granitoid plutons or occurring as large xenoliths within such plutons, skarns are the archetypical metasomatic rocks. They represent a continuum between contact metamorphic heating and addition of "juicy" magmatic fluids to aluminous calcareous rocks.

SUMMARY

Metamorphosed calcareous rocks fall into two broad categories. The marbles are those rocks dominated by the carbonate minerals calcite or dolomite at all metamorphic grades. The main detrital contaminant is typically quartz, which leads to metamorphic reactions that produce a variety of magnesium and calcium-magnesium silicates. The calc-silicate rocks represent bulk compositions with a significant noncarbonate detrital content; they consist largely of Ca–Mg–Fe–Al minerals and little

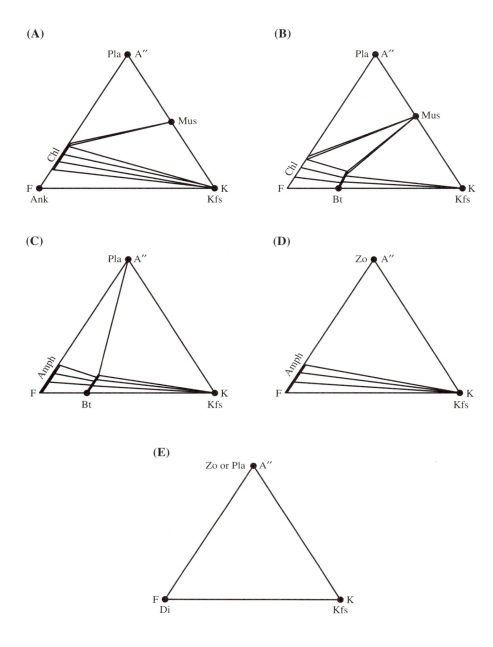

(A)

(B)

(C)

(D)

(E)

FIGURE 24-16

● ● ● ● ● ● ● ● ● ● ● ● ● ● ● ● ●

A″FK diagrams for Vassalboro Formation rocks projected from calcite in the four-component system ACFK. Note that all assemblages portrayed contain calcite and quartz in addition to minerals explicitly shown. **(A)** ankerite zone; **(B)** biotite zone; **(C)** amphibole zone; **(D)** zoisite zone; **(E)** diopside zone. In **(D)** and **(E)**, zoisite only occurs with calcite if ambient fluid is exceptionally water-rich.

or no remaining carbonate at middle to high grades. Many calc-silicate rocks have mineralogic compositions that show considerable overlap with metamorphosed mafic rocks.

Metamorphic reaction histories of both siliceous marbles and calc-silicates are dominated by devolatilization reactions of several types that involve mixing of the volatiles water and carbon dioxide. Decarbonation and dehydration reactions produce pure carbon dioxide and water, respectively. Most calcareous rock equilibria,

however, are of the mixed-volatile type in which both water and carbon dioxide are produced in a ratio dictated by reaction stoichiometry, or in which decarbonation is coupled with hydration or vice versa. Temperatures of reactions can be strongly affected if ambient fluid is substantially different in composition from the fluid produced in the reaction. As an example, a decarbonation reaction will occur at much lower temperature if the fluid is water-rich rather than carbon dioxide–rich. External control and change of fluid composition can thus drive prograde reactions in much the same way that temperature increase does. This concept is well illustrated in contact aureoles around granitoid plutons or adjacent to quartz veins, where an outward flow of aqueous fluid given off by crystallizing magma can cause mineral assemblage zonation. In a relatively few cases, temperature alone has been documented to have caused reaction, and mineral assemblages show internal control of fluid compositions by reactions rather than domination by an external source.

STUDY EXERCISES

1.   There are five important types of devolatilization reactions, two of which are simple dehydration and decarbonation. The other three involve both water and carbon dioxide. What are these other three, and how do they differ from one another? How are they portrayed on metamorphic phase diagrams?

2.   What is the difference between *internal* and *external* metamorphic fluids? How can reactions in a rock counteract the effect of external fluid on local or internal fluid composition? What do petrologists call this process?

3.   The lowest grade metamorphic reaction observed in many siliceous dolostones is the formation of small flakes of talc along grain boundaries between quartz and dolomite. Does this reaction involve addition or removal of water from the rock? What about the next metamorphic step, which involves reaction of talc with calcite to form tremolite?

4.   Why are the highest grade Ca–Mg silicate minerals such as wollastonite, monticellite, and akermanite rarely, if ever, found in regional metamorphism of impure carbonate rocks, but instead are restricted to contact metamorphosed rocks?

5.   How do the calc-silicate rocks differ from marbles? Calc-silicate rocks can be mineralogically similar to metamorphosed mafic rocks, especially when they occur as amphibole-plagioclase-epidote rocks. How might you go about deciding whether such an amphibolite had an igneous or impure carbonate protolith?

REFERENCES AND ADDITIONAL READINGS

Bowen, N. L. 1940. Progressive metamorphism of siliceous limestones and dolomite. *J. Geol.*, *48*, 225–274.

Bucher, K., and M. Frey. 1994. *Petrogenesis of Metamorphic Rocks* (6th ed. of Winkler's textbook). New York: Springer-Verlag.

Carmichael, D. M. 1970. Intersecting isograds in the Whetstone Lake area, Ontario. *J. Petrol.*, *11*, 147–181.

Ferry, J. M. 1976. Metamorphism of calcareous sediments in the Waterville-Vassalboro area, south-central Maine: Mineral reactions and graphical analysis. *Am. J. Sci.*, *276*, 841–882.

Ferry, J. M. 1983a. Regional metamorphism of the Vassalboro Formation, south-central Maine, USA: A case study of the role of fluid in metamorphic petrogenesis. *J. Geol. Soc. London*, *140*, 551–576.

Ferry, J. M. 1983b. Application of the reaction progress variable in metamorphic petrology. *J. Petrol.*, *24*, 343–376.

Greenwood, H. J. 1962. Metamorphic reactions involving two volatile components. *Carnegie Inst. Washington Yearbook*, *61*, 82–85.

Greenwood, H. J. 1975. The buffering of pore fluids by metamorphic reactions. *Am. J. Sci.*, *276*, 573–593.

Kennedy, W. Q. 1949. Zones of progressive regional metamorphism in the Moine schists of the western Highlands of Scotland. *Geol. Mag.*, *86*, 43–56.

Kerrick, D. M. 1974. Review of metamorphic mixed-volatile (H_2O–CO_2) equilibria. *Am. Mineral.*, *59*, 729–762.

Moore, J. N., and D. M. Kerrick. 1976. Equilibria in siliceous dolomites of the Alta aureole, Utah. *Am. J. Sci.*, *276*, 502–524.

Tracy, R. J., and B. R. Frost. 1991. Phase equilibria and thermobarometry of calcareous, ultramafic and mafic rocks, and iron formations. *Rev. Mineral.*, *26*, 207–290.

Yardley, B. W. D. 1989. *An Introduction to Metamorphic Petrology*. New York: Wiley.

CALCULATION
OF THE CIPW NORM

MODE AND NORM—
WHAT AND WHY
· · · · · · · · · · · · · · ·

The composition of an igneous rock can be categorized in several ways, the two most important of which are mineralogical and chemical; and most petrologists extensively use both chemical and mineralogic characterization of igneous rocks. In many cases, subtle mineralogic effects cannot necessarily be deduced from the chemistry alone, nor can chemical effects be deduced from mineralogy. Rock bulk composition can be determined by a variety of techniques (as discussed in Chapter 3) and reported in tabular form in terms of major elements, minor elements, trace elements, and isotopic ratios. Direct mineralogic measurement is typically done in thin section, either by visual estimates or by quantitative point-counting on a mechanical stage. The mineralogic composition, expressed as volume percentages of minerals, is referred to as a mineral *mode*. Point-counting actually measures areal proportions of minerals within a thin section (that is, mineral proportions in two dimensions), but for most rocks this can be directly translated into volume proportions for three dimensions.

A *norm* is a method of expressing mineralogic composition in an idealized way by apportioning chemical components into hypothetical (but hopefully realistic) mineral molecules. A norm is therefore a *synthetic* mineralogy. This is a very useful technique when rocks must be compared on the basis of mineralogic composition. For example, some volcanic rocks are so fine grained that it is virtually impossible to make accurate mineralogic identifications and modal measurements optically. (However, there are some exotic, but not generally available, techniques using scanning electron microscopes or electron microprobes to do this.) Because a mineralogic composition cannot ordinarily be directly measured in these fine-grained materials, rock chemistry must be used to create synthetic mineralogy. A similar situation exists for altered rocks, especially if the alteration consists entirely of secondary hydration or oxidation. Norms provide a means for application of certain mineralogic classification criteria to rocks, for example, silica-saturation index. It is typically not possible to predict silica over- or undersaturation directly from SiO_2 concentration values in chemical analyses. The presence or absence of quartz in a calculated norm for a rock gives an immediate indicator of relative silica saturation.

The first norm to be widely used was that proposed by petrologists Whitman Cross, J. P. Iddings, L. V. Pirsson, and Harry Washington in the early twentieth century; it has come to be known by their last initials as the *CIPW norm*. In its original form, it was calculated by using weight percentages of chemical components and thus recasts these components as weight percentages of hypothetical minerals, that is, as a *weight norm*. Tom Barth and others have proposed later schemes for recasting weight percentages as atom percentages, thus allowing a closer approximation to volume percentages of minerals as presented in a mineral mode. Of course, both weight norms and volume modes can easily be recalculated or normalized to the other form by using mineral densities as a normalizing factor.

The basic philosophy of a norm calculation is to redistribute chemical components from simple oxides of the chemical analysis into mineral molecules representative of actual minerals that occur in rock. Such recombination is not arbitrary but follows rules based on experience of the actual crystallization sequence with typical magmas. These observations are similar to those that inspired Bowen's Reaction Series and follow roughly the same order, in terms of sequence. Normative schemes do not ordinarily create hydrous minerals, so igneous rocks containing such minerals in their modes will yield norms containing anhydrous mineral equivalents. For example, the presence of modal biotite will be reflected by normative orthoclase and pyroxene (and slightly less normative silica): $K(Fe,Mg)_3AlSi_3O_{10}(OH)_2 + 3\ SiO_2 \rightarrow KAlSi_3O_8 + 3\ (Fe,Mg)SiO_3 + H_2O$. Similarly, modal hornblende appears in the norm as diopside, orthopyroxene, olivine, and minor albite. Besides hydrous minerals, several relatively common complex silicates such as garnet and cordierite are not among normative minerals. To interpret norms properly, it is important to remember which possible modal minerals are not among normative minerals.

One peculiarity of the CIPW norm is the typical production of several hypothetical mineral associations that rarely or never occur in nature because of compositional or thermodynamic constraints. One example is the normative association quartz + corundum, which is calculated for all peraluminous granites. Of course, these two minerals are never found together in nature: corundum-normative granites would instead contain (in addition to quartz) one or more of the aluminous minerals muscovite, garnet, or cordierite, none of which is a normative mineral. Another example is the calculated normative occurrence of very iron rich orthopyroxene (ferrosilite) in highly fractionated peralkaline granites. Actual mineralogy would be fayalite + quartz, because ferrosilite is not stable below about 10 kbar. To use the results most effectively, it is important for the user of the CIPW norm to appreciate the few idiosyncrasies of the calculations.

CIPW norms are typically no longer hand calculated using the algorithm presented in the next section; several widely available computer software packages for petrology contain norm calculations. The routine for norm calculation is admirably set up for programming, and interested students are strongly encouraged to practice programming by using the following algorithm guide. For those with access to a norm calculation program, performance of one or two hand calculations is highly desirable to illustrate what the computer program is doing.

CIPW NORM CALCULATION

The following procedure for calculating the CIPW norm is based on the suggestions of the original authors, as restated by Philpotts (1990). The first step is to recalculate *weight* percentage of oxides (or elements) of the original chemical analysis into *mole proportions* of oxides or elements. This conversion is done by dividing the weight percentage of each oxide or element by its molecular weight (the molecular weights of oxides and elements can be found in Table A1-1). (This step can be seen as part of the worked example for an anorthosite in Table A1-3.) Typically, trace or minor amounts of certain elements are combined with major elements to reflect their usual geochemical affinities. For example, mole proportions of manganese and nickel are commonly combined with iron, and barium and strontium are combined with calcium.

The next step is to assign the molar proportions of cations to mineral molecules in a carefully prescribed order. The mineral molecules are the ideal mineral formulas for each normative mineral, for example, $KAlSi_3O_8$ for orthoclase, or $CaMgSi_2O_6$ for diopside. The proportion of orthoclase therefore is based on the molar proportion of K_2O; diopside is based on molar proportions of CaO and MgO. Some cations are only used for one normative mineral (for example, K_2O), whereas others (for example, aluminum and silicon) are used for several. As these ions are progressively used, the remaining amount is appropriately reduced. As Philpotts (1990) suggested, the remaining amounts of

TABLE A1-1   Elemental and oxide molecular weights for rock components

Element or oxide	Molecular weight
Al_2O_3	101.96
BaO	153.33
BeO	25.01
CO_2	44.01
CaO	56.08
Cl	35.45
Cr_2O_3	151.99
F	19.00
FeO	71.85
Fe_2O_3	159.69
H_2O	18.015
K_2O	94.20
MgO	40.30
MnO	70.94
Na_2O	61.98
NiO	74.69
P_2O_5	141.94
S	32.06
SiO_2	60.08
TiO_2	79.88
ZnO	81.38

TABLE A1-2   Normative minerals and formulas

• •

Mineral	Abbreviation	Formula	Mole-to-weight conversion factor[a]
Quartz	qz	SiO_2	$SiO_2 \times 60.08$
Corundum	crn	Al_2O_3	$Al_2O_3 \times 101.96$
Zircon	zrc	$ZrSiO_4$	$ZrO_2 \times 183.30$
Orthoclase	or	$KAlSi_3O_8$	$K_2O \times 556.64$
Albite	ab	$NaAlSi_3O_8$	$Na_2O \times 524.43$
Anorthite	an	$CaAl_2Si_2O_8$	$CaO \times 278.20$
Leucite	lc	$KAlSi_2O_6$	$K_2O \times 436.48$
Nepheline	ne	$NaAlSiO_4$	$Na_2O \times 284.10$
Acmite	ac	$NaFeSi_2O_6$	$Na_2O \times 461.99$
Diopside (di)[b]	wo	$CaSiO_3$	$CaO \times 116.16$
	en	$MgSiO_3$	$MgO \times 100.38$
	fs	$FeSiO_3$	$FeO \times 131.93$
Wollastonite	wo	$CaSiO_3$	$CaO \times 116.16$
Hypersthene (hy)[c]	en	$MgSiO_3$	$MgO \times 100.38$
	fs	$FeSiO_3$	$FeO \times 131.93$
Olivine (ol)[d]	fo	Mg_2SiO_4	$MgO \times 70.34$
	fa	Fe_2SiO_4	$FeO \times 101.89$
Ca-orthosilicate	cs	Ca_2SiO_4	$CaO \times 86.12$
Magnetite	mt	Fe_3O_4	$FeO \times 231.54$
Chromite	chr	$FeCr_2O_4$	$Cr_2O_3 \times 223.84$
Hematite	hem	Fe_2O_3	$Fe_2O_3 \times 159.69$
Ilmenite	ilm	$FeTiO_3$	$TiO_2 \times 151.73$
Sphene (titanite)	tn	$CaTiSiO_5$	$TiO_2 \times 196.04$
Perovskite	pf	$CaTiO_3$	$TiO_2 \times 135.96$
Rutile	rt	TiO	$TiO_2 \times 79.88$
Apatite	ap	$Ca_5(PO_4)_3F$	$P_2O_5 \times 336.21$
Fluorite	fr	CaF_2	$F \times 39.04$
Pyrite	pyr	FeS_2	$S \times 59.98$
Calcite	cc	$CaCO_3$	$CO_2 \times 100.09$

[a]Multiply moles of key oxide shown by the following number.

[b]Mineral contains unit minerals enstatite (en), ferrosilite (fs), and wollastonite (wo).

[c]Mineral contains unit minerals enstatite (en) and ferrosilite (fs).

[d]Mineral contains unit minerals fayalite (fa) and forsterite (fo).

these multiply used cations can be tracked on a balance sheet. Only one oxide, SiO_2, is allowed to go into negative numbers in construction of required normative minerals. If this happens, provisional minerals that have already been made, (for example, enstatite or albite) can be converted into lower-silica alternatives, (for example, forsterite or nepheline), which use the same proportion of the key oxide (in these examples, either MgO or Na_2O) but simultaneously reduce the SiO_2 deficiency. The proportion of each normative mineral is therefore based on the proportion of the key oxide that is in it. Proportions of normative minerals are finally converted back into weight proportions by multiplying mole proportions by mole-to-weight conversion factors that are based on the molecular weights of the ideal mineral formulas (Table A1-2).

The original oxide weight proportions are thus simply redistributed into the form of mineral formulas and so must add up to their original sum, the sum of oxide weight percentages from the original analysis. The final norm is reported as weight percentages of each normative mineral in a tabular form. If a molar norm rather than a weight norm is preferred, then the mole proportions of all the normative minerals can be normalized by adding them, dividing by the sum, and converting to percentages.

The following rules determine the order and method of redistribution. The authors are grateful to A. R. Philpotts for allowing us to adapt his rules with little modification. In the balance sheet method proposed by Philpotts (1990) and shown in Table A1-3, each column reflects the *remaining* proportion of each oxide consumed in making that mineral molecule. Thus a 0.0 entry indicates that none of that particular oxide remains, and no further minerals that require it can be formed. The only exception is SiO_2, which can drop below 0.0 and later be revised to 0.0 by converting higher silica to lower silica minerals, as noted above. The notation below of calculating a *provisional* mineral in any step indicates that this mineral can be recalculated later to account for an SiO_2 deficit. Provisional minerals are indicated with primes (for example, or′).

1. Calcite is formed from CO_2 and an equal proportion of CaO. If there is no CO_2 in the analysis, ignore this step.
2. Apatite is formed from P_2O_5 and CaO; CaO is reduced by $3.33 \times P_2O_5$. If there is no P_2O_5 reported in the analysis, ignore this step.
3. Pyrite is formed from S and FeO; reduce FeO by $0.5 \times S$.
4. Ilmenite is formed from TiO_2 and an equal amount of FeO; reduce FeO by the amount of TiO_2. If TiO_2 exceeds FeO, excess is allocated to provisional titanite and CaO and SiO_2 are both reduced by amount of TiO_2 but is only permanent if some CaO remains after forming anorthite (step 10); otherwise, excess TiO_2 is reported as rutile.
5. Zircon is formed from ZrO_2 and an equal amount of SiO_2; reduce SiO_2 by amount of ZrO_2.
6. Fluorite is formed from F and half as much CaO; reduce CaO by $0.5 \times F$.
7. Cr_2O_3 is allocated to chromite; FeO is reduced by the amount of Cr_2O_3.
8. Orthoclase (or′) is formed from K_2O with equal amounts of Al_2O_3 and six times as much SiO_2; Al_2O_3 is reduced by amount of K_2O, and SiO_2 is reduced by $6.0 \times K_2O$.
9. Albite (ab′) is formed *provisionally* from Na_2O with equal amounts of Al_2O_3 and six times as much SiO_2 to make albite; Al_2O_3 is reduced by Na_2O and SiO_2 is reduced by $6.0 \times Na_2O$.
10. If there is excess Al left over after step 9, it is used to make anorthite by combining the Al_2O_3 with an equal amount of CaO and twice as much SiO_2; reduce CaO and SiO_2 appropriately. If Al_2O_3 exceeds remaining CaO, remaining Al_2O_3 is calculated as corundum.
11. If there is Na_2O left over after step (9) (that is, $K_2O + Na_2O > Al_2O_3$), then the remaining Na_2O is combined with equal Fe_2O_3 and twice as much SiO_2 to make acmite (this, of course, implies that there will be no anorthite in the norm).
12. Any Na_2O that exceeds Fe_2O_3 is calculated as sodium metasilicate; SiO_2 is reduced by amount of Na_2O. (This is very rare and unusual.)
13. Excess Fe_2O_3 is combined with an equal amount of FeO to make magnetite; FeO is reduced by this amount.
14. Pyroxenes and olivines are formed *provisionally* by combining MgO and remaining FeO. This combined proportion will be used in several later calculations.
15. Any CaO remaining from step 10 is combined with an equal amount of (MgO + FeO) and twice as much SiO_2 to make provisional diopside; (MgO + FeO) and SiO_2 are reduced appropriately.
16. If CaO exceeds (MgO + FeO), then wollastonite (wo′) is created by combining CaO with equal SiO_2, which is reduced by this amount.
17. If (MgO + FeO) exceeds CaO, combine the excess with equal SiO_2 to make provisional hypersthene (Fe-Mg orthopyroxene) and reduce SiO_2 appropriately.
18. If there is still positive SiO_2, it is calculated as quartz.
19. If SiO_2 is negative, then provisional minerals have been calculated that are too rich in SiO_2. The order of conversion of higher SiO_2 to lower SiO_2 minerals is hypersthene → olivine; titanite → perovskite; albite → nepheline; orthoclase → leucite; and finally, wollastonite → calcium silicate. In very rare cases, further steps are required; the reader is referred to Philpotts (1990) for details of these steps. In the following steps, let D = silica deficiency; provisional normative minerals are designated by primes.
20. If $D < hy′/2$, ol = D, hy = hy′ − 2D, and the calculation is done. If $D > hy′/2$, all provisional hypersthene is converted to olivine (ol = hy′) and the new silica deficiency is $D_1 = D − hy′/2$.
21. For the unusual case in which a deficiency remains after making all hy into ol, then convert some titanite into perovskite. If $D_1 < tn′$, pf = D_1 and tn = tn′ − D_1. If $D_1 > tn′$, pf = tn′ and the new silica deficiency is $D_2 = D_1 − tn′$. If there was no provisional titanite calculated, $D_2 = D_1$.
22. For the further unusual case in which there is still a deficiency, some or all albite is converted to nepheline. If $D_2 < 4$ ab′, some provisional albite is converted, such that ne = $D_2 > 4$ and ab = ab′ − $D_2 > 4$. If $D_2 > 4$ ab′, all provisional albite is converted to nepheline (ne = ab′) and a new silica deficiency is created: $D_3 = D_2 − 4$ ab′.
23. If $D_3 < 2$ or′, some provisional orthoclase must be converted to leucite: lc = $D_3/2$ and or = or′ − $D_3/2$. If $D_3 > 2$ or′, then all or′ is converted to lc and a new silica deficiency D_4 is created. $D_4 = D_3 − 2$ or′.

Table A1-3 Example of worked CIPW norm calculation with recommended form
Sample is basalt (see Table 3-4, last column)

If mole proportion is affected by the formation of the normative mineral, the new value is entered below.

Oxide	Wt%	Mol. wt. oxide	Mole proportion	ap	ilm	or	ab	an	crn	acm	mt	di^a wo	di^a en	di^a fs	hy en	hy fs	ol fo	ol fa	qz
SiO_2	50.06	60.08	0.8332			0.7642	0.4798	0.2850					0.1496						
TiO_2	1.87	79.88	0.0234		0.0000														
Al_2O_3	15.94	101.96	0.1563			0.1448	0.0974	0.0000											
Fe_2O_3	3.90	159.69	0.0244								0.0000								
FeO	7.50	71.85	0.1044		0.0838						0.0594			0.0421		0.0078		0.0000	
MnO	0.20	70.94	0.0028																
MgO	6.98	40.30	0.1732										0.1228		0.0226		0.0000		
CaO	9.70	56.08	0.1730	0.1651				0.0677				0.0000							
Na_2O	2.94	61.98	0.0474				0.0000												
K_2O	1.08	94.20	0.0115			0.0000													
P_2O_5	0.34	141.94	0.0024	0.0000															
Total	100.51																		
Mole proportion of normative mineral				0.0024	0.0234	0.0115	0.0474	0.0974	—	—	0.0244	0.0677	0.0504	0.0173	0.1002	0.0343	0.0226	0.0078	—
Mole-to-weight conversion factor				336.21	151.73	556.64	524.43	278.20	101.96	461.99	231.54	116.16	100.38	131.93	100.38	131.93	70.34	101.89	60.08
Wt% of normative minerals				0.81	3.55	6.40	24.86	27.10	—	—	5.65	7.86	5.06	2.28	10.06	4.53	1.59	0.80	—
Sum of normative minerals																			

^{a}Calculation of diopside: (en + fs) = wo; en = mol. prop. wo × MgO/(MgO + FeO); fs = mol. prop. wo × FeO/(MgO + FeO).
Source: Philpotts (1990).

24. If $D_4 < wo'/2$ (only in provisional wo, not di), provisional wo' is converted to calcium orthosilicate (cs) such that $cs = D_4$ and $wo = wo' - 2 D_4$. For silica deficiencies beyond this, the reader should refer to Philpotts (1990).

Only the rarest rock types are so deficient in SiO_2 that any steps beyond step 22 must be invoked. A worked example of a CIPW norm calculation is given in Table A1-3. We reiterate that the hand-calculated norm is cumbersome and tedious if many are to be done; we recommend that you use any of various commercially available computer programs. The algorithm for a CIPW norm calculation is easily developed from the rules above, and interested students are encouraged to write their own norm programs or to use spreadsheets for the purpose.

REFERENCES AND ADDITIONAL READINGS

• • • • • • • • • • • • • • •

Cross, C. W., J. P. Iddings, L. V. Pirsson, and H. S. Washington. 1902. A quantitative chemico-mineralogical classification and nomenclature of igneous rocks. *J. Geol.*, *10*, 555–690.

Johannsen, Albert. 1931. *A Descriptive Petrography of the Igneous Rocks*. Vol. 1. *Introduction, Textures, Classifications and Glossary*. Chicago: University of Chicago Press, see esp. pp. 83–99.

Philpotts, A. R. 1990. *Igneous and Metamorphic Petrology*. Englewood Cliffs, NJ: Prentice Hall.

PRESSURE-TEMPERATURE DETERMINATION

In many cases it is useful for the petrologist to make relatively precise quantitative estimates of the pressure and temperature of formation of igneous or metamorphic rocks. Petrologists use the terms *geothermometry* and *geobarometry*, respectively, for the techniques of temperature and pressure estimation in rocks, and *geothermobarometry* for the general process. Geobarometry in plutonic rocks, for example, can help in establishing the depth of intrusion. Geothermometry and geobarometry in metamorphic terranes can constrain actual *P-T* paths during metamorphism and in conjunction with geochronology can be used to characterize *P-T-t* paths, potentially critical information for tectonic reconstructions. In this appendix we summarize some of the techniques that have been developed by petrologists to extract *P-T* information from rocks. Because we do not intend the appendix to be a comprehensive reference, we cite other, more detailed sources for interested readers. Many of the geothermobarometric calculations noted here have been incorporated into computer programs either by original investigators or by others. Readers who wish to pursue geothermobarometric techniques for research should check either the original sources or software notices that are now common in mineralogic and petrologic journals.

GENERAL PRINCIPLES

As we will see later, some techniques do not require any knowledge of the mineralogy or mineral chemistry of igneous or metamorphic rocks, but only geologic information such as depth reconstructed from stratigraphic data. In general, however, quantitative geothermobarometry requires more detailed information such as rock mineralogy and the chemistry of at least some of the minerals. For many mineralogically simple igneous rocks, a relatively imprecise estimate of crystallization temperature can be made by modeling the magma crystallization with published phase diagrams. Certain metamorphic mineral assemblages, particularly in metapelites, are limited to very restricted areas of the *P-T* plane. These small *P-T* ranges can be determined, given the availability of an appropriate petrogenetic grid. Other, more precise techniques depend on a knowledge of mineral chemistry and fall into several different categories.

Exchange Thermometry

A wide array of igneous and metamorphic minerals display solid solution, in particular solid solution between iron and magnesium end members. When two or more ferromagnesian solid solution minerals coexist, the distribution of iron and magnesium between minerals depends on the distribution coefficient, which is a type of chemical equilibrium constant. It is the distribution coefficient that determines the exact slopes of the tie lines in a phase diagram such as AFM (Figure A2-1A). The magnesium-iron distribution coefficient (K_D) between two minerals x and y is defined as K_D (Mg-Fe, x-y) = $(\text{Mg/Fe})^x$ / $(\text{Mg/Fe})^y$. Cation exchange of this sort depends only on the equilibrium coexistence of minerals, not necessarily that they be in an obvious reaction relationship.

(A)   **(B)**

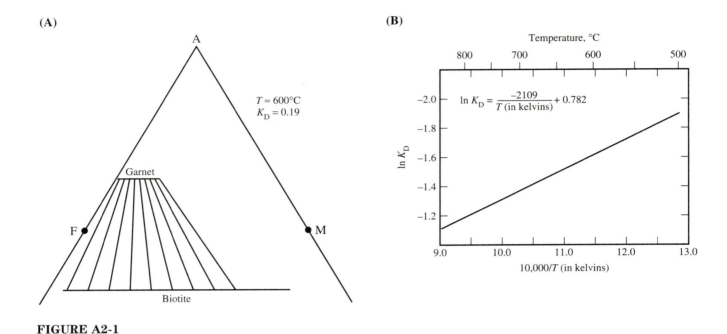

FIGURE A2-1
••••••••••••••••

(A) Distribution of tie lines between garnet and biotite in the AFM (muscovite) diagram at about 600°C, reflecting a K_D(Mg-Fe) of 0.19. Note that as K_D approaches 1.0, the tie lines will approach being radial from the A apex of the diagram. [Adapted from F. S. Spear, 1993, *Metamorphic Phase Equilibria and Pressure-Temperature-Time Paths. Monograph 1.* Mineralogical Society of America, Fig. 15-3.] **(B)** Linear best fit to experimental data on iron-magnesium distribution between garnet and biotite. The slope and intercept of the fit are given in the equation. Experiments were done at 2.07 kbar. [From J. M. Ferry and F. S. Spear, 1978, *Contrib. Mineral. Petrol., 66*, Fig. 2.]

Like all equilibrium constants, K_D is dependent on pressure and temperature and, in fact, is related exponentially to these variables (ln K_D is a linear function of $1/T$ and $1/P$; Figure A2-1B). Iron and magnesium are similar in ionic radius, so the temperature dependence is much greater than the pressure dependence, because pressure has little influence on the preferences of the cations for the two crystal sites of the minerals in exchange equilibrium. This characteristic is typical of most good thermometers, so temperature can be estimated fairly precisely without a concurrent tight constraint on pressure. The dependence of K_D on temperature for a particular mineral pair such as garnet and biotite must be based on thermodynamic considerations and calibrated experimentally to create a useful exchange thermometer, as we will see later.

Despite their widespread use and great utility, exchange thermometers are subject to several problems. The first problem arises from the common tendency of igneous and especially metamorphic minerals to be chemically zoned. To use any of the geothermobarometric techniques that depend on use of mineral composi-

tions for calculation, one particular composition of each mineral must be chosen. Chemical zoning of minerals during growth, or modification of their zoning during cooling, complicates the choice of an actual composition.

The second problem relates to the relative kinetic ease of elemental exchange. If the kinetics are favorable, exchange will continue well past the maximum temperature reached by the rock, especially if cooling rate is slow, as it generally is following regional metamorphism. A calculated exchange temperature must therefore be carefully assessed to decide whether it represents peak temperature (or close to it) or some closure condition during cooling. One of the most important kinetic factors is the diffusivity of exchangeable cations within the crystal structures of individual minerals. Many igneous and metamorphic minerals display relatively rapid internal diffusion, particularly of iron and magnesium. Garnet is a notable exception; and for this reason (as well as its relatively common occurrence), garnet is the basis of the most widely used metamorphic thermobarometers. The tendency of garnet to retain original compositions makes it an attractive mineral on which to base calculations,

FIGURE A2-2

• • • • • • • • • • • • • • • •

Illustration of the way an experimentally calibrated solvus can be used for geologic thermometry. The illustrated solvus is based on experiments in the system calcite-dolomite ($CaCO_3$–$CaMg(CO_3)_2$), recalculated by Anovitz and Essene (1987) using an empirically fit solution model. The miscibility gap at 800°C is shown, as is the composition of magnesian calcite that coexists with dolomite at 800°C, that is, 32 mol% $CaMg(CO_3)_2$ or 16 mol% $MgCO_3$. (Note that the latter value— 16 mol% $MgCO_3$—is the one that should be used in the temperature equation given in the text.) The dolomite in equilibrium with this calcite at 800°C contains 44 mol% $MgCO_3$. [Adapted from L. M. Anovitz and E. J. Essene, 1987, *J. Petrol.*, *28*, Fig. 6.]

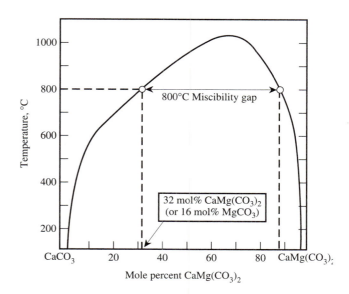

because the slow diffusion characteristics of garnet make it less likely to undergo postgrowth compositional modification. In effect, garnet can act as a "tape recorder" of thermal events in rocks up to temperatures of about 700°C, if a petrologist knows how to interpret the zoning patterns that are present (Tracy 1982; Spear 1993).

The third problem involves error introduced through limits on the precision of analytical data. Any minor or trace cations used for calculations with electron microprobe data are subject to larger analytical errors than are high-concentration cations. Furthermore, if K_D has a limited range close to 1.0 (that is, cation distribution is almost equal), then normal analytical errors can be amplified into large errors in K_D and thus in calculated temperatures. For this reason, magnesium to iron ratios in ferromagnesian mineral pairs should be as different as possible (that is, K_D should be much larger or much smaller than 1.0). Common mineral pairs that are essentially ruled out as thermometers by this criterion include biotite-chlorite, garnet-staurolite, and chlorite-cordierite.

Solvus Thermometry

Readers should recall that some mineral solid solutions are not complete but are interrupted by miscibility gaps that are the results of thermodynamic nonidealities. Examples include feldspars, iron-titanium oxides, pyroxenes, and carbonates. Because the solvus limits are temperature dependent, typically closing (that is, greater solid solution) with increasing temperature, the actual mineral compositions for minerals that coexist across a solvus can be used to estimate temperature based on experimental data on the extent of immiscibility (Figure A2-2). The three most useful igneous thermometers (and practically the *only* igneous thermometers) are all of this type.

Net-Transfer Equilibria

A number of geothermobarometers in metamorphic rocks are based on reactions in which minerals are consumed or produced—that is, net-transfer reactions. These reactions are more complex than the simple exchanges discussed earlier and are typically *continuous* or divariant reactions in which cations are transferred from one mineral to another as mineral proportions shift. Each participating solid solution mineral generally contains other cations that are not involved in transfer. Effectively, solid solution expands the reaction from a line into a band on a *P-T* diagram between two end member reaction lines or curves. The distribution of the exchangeable cation in a particular assemblage, commonly defined in the form of an equilibrium constant, determines a unique single location of the reaction line or curve (called an *isopleth*) between the two end members. Most net-transfer equilibria define *P-T* isopleths with slopes intermediate between vertical and horizontal values. They therefore cannot be used alone for determination of unique *P-T* values but must be combined with a good thermometer if slopes are shallow to moderate or with a good barometer if slopes are moderate to steep (Figure A2-3).

Other Geothermobarometers

Although the most widely applied techniques generally fall into the earlier categories, there are some others that

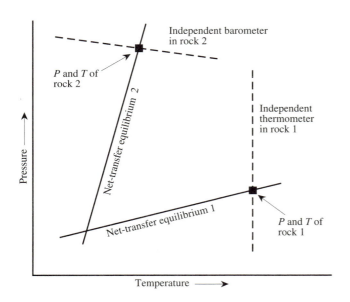

FIGURE A2-3
● ● ● ● ● ● ● ● ● ● ● ● ● ● ● ●

Qualitative pressure-temperature diagram showing two hypothetical isopleths for net-transfer equilibria. Equilibrium 1 has a very shallow *P-T* slope and might serve as an imprecise geobarometer. The intersection of this isopleth with an independent geothermometer (as shown) produces a relatively precise determination of both pressure and temperature of rock 1. Equilibrium 2 has a steep *P-T* slope and could act as a geothermometer. In conjunction with an independent geobarometer, equilibrium 2 generates a *P-T* point for rock 2.

are applicable in special circumstances. The first of these is a chemical technique like those above and also requires a mineral chemical analysis. It differs from earlier approaches in that it depends on the pressure dependence of solubility of some major or minor component in solid solution minerals. Examples include the solubility of aluminum in igneous hornblendes, the solubility of phengite (Fe-Mg) component in muscovites, and the iron solubility in sphalerite (ZnS) that coexists with pyrite and pyrrhotite. Most commonly, pressure-dependent solid solutions like these result from crystal chemical restrictions based on either cation size limits in particular sites or molar volume effects.

Other techniques tend to be more qualitative or semiquantitative than fully quantitative but have been useful in characterizing relative metamorphic grades based on physical characteristics of materials. All show potential for being made more quantitative with careful experimental calibration. A technique referred to as illite crystallinity is mentioned in Chapter 23 and involves the incipient transformation of clay minerals to micas in very low-grade metapelitic rocks. The technique is based on X-ray diffraction analysis of average basal (001) spacings in sheet silicates. Illite crystallinity has been usefully applied in foreland thrust belts of the Alps and Appalachians. Another low-grade technique is called vitrinite reflectance. It depends on the increase in reflectivity (basically microscopic shininess) of organic particles in sedimentary rocks as they are heated to several hundred degrees. It has been used in rocks of the same grade that qualify for illite crystallinity studies. A final example of this type of technique is called conodont color index (CCI). Conodonts are microfossils found in many sedimentary and incipiently metamor-

phosed rocks. They are composed of apatite and tend to display a gradual shift in colors from white or colorless at low grade to very dark or even black at several hundred degrees Celsius. Although not well calibrated quantitatively as a thermometer, CCI has been used qualitatively to constrain degree of thermal maturation in petroleum-bearing sedimentary basins.

IGNEOUS ROCK GEOTHERMOBAROMETRY
● ● ● ● ● ● ● ● ● ● ● ● ● ● ● ●

Temperature Determination

The simplest and most direct way to determine the temperature of a magma is to measure it at the surface or in a drill hole with a thermometer of some sort, but of course this can only be done for near-surface or surface magmas, that is, lavas. Temperatures of various lava flows and lava lakes have been documented this way by volcanologists. For most igneous rocks, however, indirect methods are required to reconstruct magmatic temperatures from solidified, cooled rock exposed in outcrop. As noted earlier, mineralogically simple igneous rocks may be appropriate for estimation of solidus temperature and crystallization interval by using published phase diagrams as models. This technique generally requires a small number of major minerals and the absence of solid solution complexity within these minerals, and thus is not widely applicable.

There are several igneous thermometers based on solvus thermometry. The three most widely used ones are the two-feldspar thermometer (Whitney and Stormer

1977; Brown and Parsons 1985; Fuhrman and Lindsley 1988), the two-pyroxene thermometer (Lindsley and Anderson 1983; Davidson and Lindsley 1985), and the magnetite-ilmenite thermometer (Buddington and Lindsley 1964; Spencer and Lindsley 1981). This last one is actually a hybrid technique, being based on *two* solvi (within both the spinel series and the rhombohedral oxide series) and the gain or loss of oxygen with reequilibration. This last characteristic makes the magnetite-ilmenite calibration also useful as an oxygen barometer.

The two-feldspar thermometer requires the coexistence of a potassic feldspar (orthoclase-albite solid solution, with very minor anorthite) and a plagioclase (albite-anorthite solid solution, with only minor orthoclase). The thermometer is fundamentally based on the distribution of albite component between the coexisting feldspars according to the ternary feldspar solvus (Figure A2-4). However, this technique is typically limited to granitoids (for example, granites, syenites, monzonites, granodiorites), their volcanic equivalents, or high-grade two-feldspar granitic gneisses. A practical difficulty in applying the technique is that many potassic feldspars become perthitic following equilibration with plagioclase at high temperature (see Figure 10-10). An integrated composition of host and lamellae in exsolved perthite must therefore be obtained by analysis or calculated from estimates of proportions. The most recent and widely used calibration of this thermometer can be found in Fuhrman and Lindsley (1988) (a computer program is available from these authors).

The two-pyroxene thermometer is also based on a ternary solvus, in this case between calcic clinopyroxene and orthopyroxene. The technique is based on the temperature dependence of the distribution of wollastonite component between high- and low-calcium pyroxene, and to a lesser extent on iron-magnesium distribution. Like the two-feldspar thermometer, it can be applied to igneous or high-grade metamorphic rocks. It is more widely applicable than the feldspar thermometer because a wider range of igneous rocks contains two pyroxenes. Some igneous clinopyroxenes are exsolved into host and lamellae, which must therefore be integrated for calculation. The most recent calibrations of two-pyroxene thermometry are those of Davidson and Lindsley (1985) and Nickel et al. (1985).

Pressure Determination

There are relatively few techniques available for estimating crystallization pressure for igneous rocks, and those that have been used tend to be less precise than the geobarometers for metamorphic rocks. One of the simplest techniques for shallow plutonic rocks is to make a geologic estimate of the thickness of overburden at the time of intrusion, based on stratigraphic estimates, and this approach has been successfully applied to shallow granitoid plutons in Utah (Moore and Kerrick 1976). It requires (1) that the sedimentary rocks have not been significantly folded, faulted, or tilted prior to intrusion and (2) that the original stratigraphic thickness above the now-exposed pluton can be reconstructed accurately not far away. As an example, if it can be established that the exposed level of a pluton was originally overlain by 10 km of sedimentary and igneous rock, the pressure would be 2.8 kbar, more or less, based on typical density of crustal rocks.

Quantitative pressure estimates for magma crystallization based on mineralogic criteria are limited to the

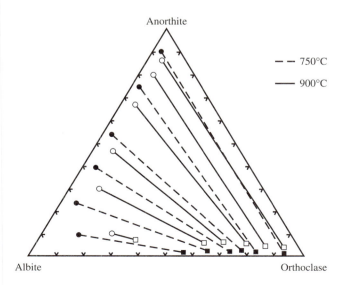

FIGURE A2-4
● ● ● ● ● ● ● ● ● ● ● ● ● ● ● ●

The feldspar ternary showing the calculated distribution of end members between plagioclase (dots) and potassic feldspars (squares) at 750°C, 1 kbar (filled symbols, dashed tie lines) and 900°C, 0.5 kbar (open symbols, solid tie lines), based on thermodynamic model of Fuhrman and Lindsley (1988). This model is the current basis for the two-feldspar thermometer and is applicable both to igneous and high-grade metamorphic rocks. Note the shortening of tie lines with increasing temperature, a change reflecting increased solubility of orthoclase component in plagioclase solid solutions and of anorthite component in potassic feldspar solid solutions. [Modified from M. L. Fuhrman and D. H. Lindsley, 1988, *Am. Mineral.*, *73*, Fig. 2.]

aluminum-in-hornblende technique and the occurrence of magmatic epidote. Both of these can be applied to granitoids, but no dependable barometric techniques are yet available for mafic igneous rocks. The hornblende barometer (Hammarstrom and Zen 1986; Hollister et al. 1987; Johnson and Rutherford 1989) is based on the total aluminum content of hornblende in equilibrium with silicate melt, fluid, biotite, quartz, sanidine, plagioclase ($\approx$ An$_{30}$), titanite, and magnetite or ilmenite, that is, a typical metaluminous granitic magma at a late stage of crystallization. The earlier calibrations of this barometer were empirical and based on the observation that granites crystallized at deeper crustal levels contained more aluminous hornblendes. Johnson and Rutherford performed experiments that confirmed the general observation but significantly improved the precision to $\pm$ 0.5 kbar. The equation for pressure that they proposed is P ($\pm$ 0.5 kbar) = $-3.46 + 4.23$ (Al^{T}), where Al^{T} is the total aluminum content reported as cations per 23-oxygen formula unit.

The presence of magmatic epidote has been used as a semiquantitative barometer for granitic rocks but is more controversial than the hornblende barometer. Most petrologists had not thought of epidote as having a thermal stability extending into the magmatic range until Naney (1983) showed in experiments that epidote appeared above the solidus in hydrous granitic magmas at pressures of 8 kbar or higher. He did not find epidote in lower pressure experiments. Much of the controversy arises from the difficulty of proving a magmatic origin of epidote in granites. Even euhedral epidote crystals have been shown to have a subsolidus, secondary origin through hydration of primary igneous minerals. In any case, many petrologists accept that well-crystallized, euhedral epidote in granites indicates a crystallization pressure of at least 8 kbar (a depth of about 28 km).

METAMORPHIC ROCK GEOTHERMOBAROMETRY

Temperature Determination

A considerably wider range of thermometers and barometers has been developed for metamorphic rocks than for igneous rocks (see the comprehensive review in Spear 1993). In part, this is because the key geothermobarometric mineral garnet is much more common in metamorphic rocks than in igneous ones. In major part, however, it is probably due to the greater interest in documenting temperatures and pressures in metamorphic terranes. Fairly small differences in pressures and temperatures within metamorphic terranes can indicate patterns that have important tectonic implications.

Without doubt, the most widely used thermometer in metamorphic rocks is the garnet-biotite thermometer. For many years, petrologists had been aware that there was a significant temperature dependence to the distribution of iron and magnesium between garnet and biotite, but it was not quantified. The first empirical (Thompson 1976) and experimental (Ferry and Spear 1978) calibrations of the thermometer sparked widespread use. The garnet-biotite thermometer is based on distribution of iron and magnesium between the relatively regular octahedral site in the biotite structure and the larger dodecahedral site in garnet. Because of the size difference in the two sites, the exchange is not ideal and there is a small pressure dependence (Figure A2-5). Stated as a distribution coefficient, K_D (Mg-Fe, gar-bio) = (Mg/Fe)^{gar}/(Mg/Fe)^{bio}. In their experiments, Ferry and Spear found a linear relationship between $\ln K_D$ and $1/T$ [T in kelvins]: $\ln K_D = -(2089 + 0.0096\,P)/T + 0.782$. By calculating the K_D as given above, a researcher

FIGURE A2-5
● ● ● ● ● ● ● ● ● ● ● ● ● ● ● ●

P-T diagram showing lines of constant K_D(Mg-Fe) for the garnet-biotite thermometer of Ferry and Spear (1978). The aluminum silicate triple point of Holdaway (1971) is shown for comparison. K_D is calculated according to the equation given in Figure A2-1B. [From F. S. Spear, 1993, *Metamorphic Phase Equilibria and Pressure-Temperature-Time Path. Monograph 1*. Mineralogical Society of America, Fig. 15-6.]

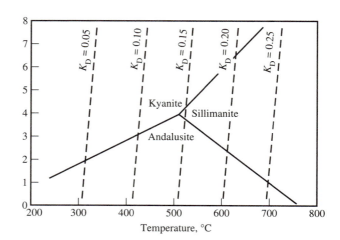

can plug the value into this equation, along with an estimated pressure (in bars), and calculate temperature. Note that calculated T is in kelvins, as is the case for all thermodynamic calculations; T in degrees Celsius is simply T in kelvins -273.15. The experimental calibration of the garnet-biotite thermometer assumes that there are no cations other than iron and magnesium in the exchangeable sites. In real minerals there are always some other cations (for example, manganese or calcium in garnet, especially low-grade garnet; manganese, titanium, or aluminum in biotite). If these other cations occur in sufficiently high concentration (generally above about 10 atom percent), then additional activity terms must be added to the thermometric calculation. Numerous published modifications of the Ferry and Spear calibration have attempted to take into account these effects (see the comprehensive reference list in Spear 1993).

Exchange thermometers based on iron-magnesium exchange between other ferromagnesian mineral pairs have also been developed, and most involve garnet. These include garnet-clinopyroxene, garnet-orthopyroxene, garnet-hornblende, garnet-chlorite, garnet-olivine, and garnet-ilmenite (based on iron-manganese exchange). These alternative thermometers are typically not as well calibrated or tested as garnet-biotite, and many are restricted to unusual compositions or to very high metamorphic grade.

Solvus thermometers that have been used extensively for metamorphic rocks include calcite-dolomite, two-pyroxene, and two-feldspar. The latter two have already been discussed: In metamorphic rocks, they are essentially restricted to anhydrous granulite-facies assemblages. The calcite-dolomite thermometer has been used for many marbles, with mixed results. It is based on a miscibility gap between calcium and calcium-magnesium carbonates (Goldsmith and Newton 1969; Anovitz and Essene 1987) and is the only thermometer that applies to calcareous rocks (Figure A2-2). The dolomite solvus limb is very steep (there is little change in calcium to magnesium ratio in dolomite over a wide range of metamorphic temperatures), so the application of this thermometer typically involves measurement of the calcium to magnesium ratio in calcite and extraction of temperature from the gently sloping calcite limb of the solvus. The equation proposed by Anovitz and Essene, based on an analysis of various experimental studies in the system, is T [in kelvins] $= -2360\ (X^{MgC}_{cc})$ $-0.01345/(X^{MgC}_{cc})^2 + 2620(X^{MgC}_{cc})^2 + 2608(X^{MgC}_{cc})^{0.5} +$ 334, where X^{MgC}_{cc} is the mole fraction of $MgCO_3$ dissolved in calcite.

Common difficulties with this thermometer have generally been ascribed to extensive reequilibration during cooling because of the high reactivity of the minerals. Calculated temperatures as much as several hundred de-

grees below estimated realistic values are common in regional metamorphic marbles and probably represent approximations of closure temperatures during cooling. The thermometer has been more successfully applied in contact aureoles, where rapid cooling generally appears to have precluded reequilibration.

Pressure Determination

Virtually all metamorphic barometers are of the net-transfer reaction type, and most function only with concurrent use of a thermometer like garnet-biotite. This restriction exists because of the normal tendency for net-transfer reactions to have isopleth P-T slopes of intermediate value; consequently, these thermobarometers are very imprecise as thermometers or barometers when used alone. The most ideal approach is to use two or more thermobarometers that have slopes as close as possible to defining a $90°$ intersection. Unfortunately, many reaction-based thermobarometers require the coexistence of a relatively large number of minerals and thus are not applicable to many rocks.

The most widely studied and used net-transfer thermobarometer is referred to as GASP—standing for garnet-aluminum silicate-quartz (silica)-plagioclase. It is based on the distribution of calcium between the grossular component in garnet and the anorthite component in plagioclase. The actual reaction is written $3\ CaAl_2Si_2O_8$ (anorthite) $=\ Ca_3Al_2Si_3O_{12}$ (grossular) $+\ 2\ Al_2SiO_5$ (kyanite) $+\ SiO_2$ (quartz). The calculation is based on an experimental location for the calcium end member reaction (as written above) and an equation for the P-T displacement of the reaction curve because of dilution of anorthite by albite in plagioclase and grossular by almandine, pyrope, and spessartine in garnet. The equilibrium constant for this reaction includes activities and is defined as $K_{eq} = a_{qtz}a_{ky}{}^2a_{grs}/(a_{an})_3$. For most natural assemblages, a_{grs} in garnet and a_{an} in plagioclase are substantially less than 1.0, but the other minerals are essentially pure and activities are thus 1.0. K_{eq} is then combined with the end member location and the thermodynamic variables for the reaction into the full equation for temperature and pressure dependence of K_{eq} for GASP: $0 = -48,357 + 150.66\ T$ [in K] $+ (P - 1$ bar$)$ $(-6.608) + RT \ln K_{eq}$ (Spear 1993). The results of this calculation are shown in Figure A2-6. If T is known independently, for example, from the garnet-biotite thermometer, then P can be calculated from K_{eq}. Activities of grossular and anorthite are calculated from analytically measured mineral chemistry and published activity models.

Other net-transfer equilibria that are applicable to many typical metamorphic rocks include garnet-rutile-aluminum silicate-ilmenite-quartz (GRAIL) (Bohlen et

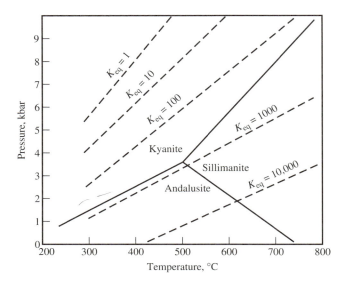

P-T diagram showing the GASP geobarometer, contoured for constant values of the equilibrium constant calculated according to the equation given in the text. Note that the contour labeled "$K_{eq} = 1$" represents the location of the pure-calcium end member reaction. The aluminum silicate phase diagram of Holdaway (1971) is shown for comparison. [Modified from F. S. Spear, 1993, *Metamorphic Phase Equilibria and Pressure-Temperature-Time Paths. Monograph 1.* Mineralogical Society of America, Fig. 15-8.]

al. 1983a), garnet-plagioclase-hornblende-quartz (Kohn and Spear 1990), garnet-plagioclase-muscovite-biotite (Ghent and Stout 1981; Hodges and Crowley 1985; Hoisch 1991), garnet-rutile-ilmenite-plagioclase-quartz (GRIPS) (Bohlen and Liotta 1986), garnet-plagioclase-clinopyroxene-quartz (Moecher et al. 1988), garnet-muscovite-biotite-aluminum silicate-quartz (Hodges and Crowley 1985; Hoisch 1991), and others. A typical research approach is to use as many individual thermometers and barometers as are applicable to a given rock and then to compare and interpret the results.

UNCERTAINTIES
IN *P-T* ESTIMATION
● ● ● ● ● ● ● ● ● ● ● ● ● ● ●

Our discussion of the various techniques for *P-T* estimation has so far avoided the issue of uncertainties or errors. It is most important, however, to understand that all of the techniques we have discussed are subject to at least some error, and that the applicability of a particular technique will depend on the uncertainty of that technique. The magnitude of the potential uncertainty is due to a number of factors, based on the type of thermometer or barometer, and may vary significantly from technique to technique. Errors fall into the typical two categories of random and systematic error, the first more characteristic of measurement resolution problems and the second more likely due to incorrect basic assumptions.

For thermobarometers that depend on a theoretical or experimental calibration and a knowledge of mineral chemical compositions (for example, exchange or net-transfer thermometers), uncertainties generally fall into two categories: errors introduced in the calibration and errors related to measured mineral compositions. Calibration uncertainty can have many sources, but principal among them are uncertainties in thermodynamic data for minerals and in the experimental location of *P-T* positions of reaction curves.

A wider array of problems can contribute to uncertainties in the application of thermobarometers. First, and most fundamental, one or more of the minerals being used for *P-T* estimation either failed to equilibrate compositionally at peak *P* and *T* or underwent some postpeak reequilibration during cooling. This latter case has been shown to be particularly common, especially for high-temperature rocks like granulite-facies ones. This type of error is systematic, and may shift measured temperature or pressure by a significant amount. Mineral zoning can cause problems related to this effect. When zoned crystals produce a range of measured mineral compositions in a rock, it is commonly unclear which individual analysis will yield the closest approximation of accurate peak conditions. In thermometry, such problems generally lead to underestimation of temperature, but not always. Second, random errors in measurement of chemical composition can introduce uncertainties in estimated *P* and *T*, even when the choice of mineral compositions is clear.

Each original source cited in this appendix should be referred to for an estimate of uncertainty associated with that technique. Most thermometers have stated uncertainties of ±25°–50°C; that is, a calculated temperature of 500°C really means a statistically indistinguishable range of about 450°–550°C. Typical mineral barometer uncertainties are ±1–2 kbar; thus a measured pressure of 6 kbar actually means a range of

4–8 kbar. The published uncertainties for each technique are generally quite conservative. Note that these uncertainties are typically those associated with random errors and ordinarily don't have so profound an effect on accuracy as the systematic errors do, particularly postpeak reequilibration. For a thorough discussion of thermobarometric errors, the interested reader is referred to a paper by Hodges and McKenna (1987).

REFERENCES AND ADDITIONAL READINGS
• • • • • • • • • • • • •

Anovitz, L. M., and E. J. Essene. 1987. Phase relations in the system $CaCO_3–MgCO_3–FeCO_3$. *J. Petrol.*, *28*, 389–414.

Bohlen, S. R., and J. J. Liotta. 1986. A barometer for garnet amphibolites and garnet granulites. *J. Petrol.*, *27*, 1025–1034.

Bohlen, S. R., V. J. Wall, and A. L. Boettcher. 1983a. Experimental investigations and geological applications of equilibria in the system $FeO–TiO_2–Al_2O_3–SiO_2–H_2O$. *Am. Mineral.*, *68*, 1049–1058.

Bohlen, S. R., V. J. Wall, and A. L. Boettcher. 1983b. Geobarometry in granulites. In *Kinetics and Equilibrium in Mineral Reactions*, ed. S. K. Saxena. New York: Springer-Verlag, pp. 141–171.

Brown, W. L., and I. Parsons. 1985. Calorimetric and phase-diagram approaches to two-feldspar geothermometry. *Am. Mineral.*, *70*, 356–361.

Buddington, A. F., and D. H. Lindsley. 1964. Iron-titanium oxide minerals and synthetic equivalents. *J. Petrol.*, *5*, 310–357.

Davidson, P. M., and D. H. Lindsley. 1985. Thermodynamic analysis of quadrilateral pyroxenes. II. Model calibration from experiments and applications to geothermometry. *Contrib. Mineral. Petrol.*, *91*, 390–404.

Essene, E. J. 1989. The current status of thermobarometry in metamorphic rocks. In *Evolution of Metamorphic Belts*, ed. J. S. Daly, R. A. Cliff, and B. W. D. Yardley. Oxford: Blackwell, pp. 1–44.

Ferry, J. M., and F. S. Spear. 1978. Experimental calibration of the partitioning of Fe and Mg between biotite and garnet. *Contrib. Mineral. Petrol.*, *66*, 113–117.

Fuhrman, M. L., and D. H. Lindsley. 1988. Ternary feldspar modeling and thermometry. *Am. Mineral.*, *73*, 201–215.

Ghent, E. D., and M. Z. Stout. 1981. Geobarometry and geothermometry of plagioclase-biotite-garnet-muscovite assemblages. *Contrib. Mineral. Petrol.*, *76*, 92–97.

Goldsmith, J. R., and R. C. Newton. 1969. *P-T-X* relations in the system $CaCO_3–MgCO_3$ at high temperatures and pressures. *Am. J. Sci.*, *267-A*, 160–190.

Hammarstrom, J. M., and E-An Zen. 1986. Aluminum in hornblende: An empirical geobarometer. *Am. Mineral.*, *71*, 1297–1313.

Hodges, K. V., and P. C. Crowley. 1985. Error estimation and empirical geothermobarometry for pelitic systems. *Am. Mineral.*, *70*, 702–709.

Hodges, K. V., and L. W. McKenna. 1987. Realistic propagation of uncertainties in geologic thermobarometry. *Am. Mineral.*, *72*, 671–680.

Hoisch, T. D. 1991. Equilibria within the mineral assemblage quartz + muscovite + biotite + garnet + plagioclase, and implications for the mixing properties of octahedrally coordinated cations in muscovite and biotite. *Contrib. Mineral. Petrol.*, *108*, 43–54.

Holdaway, M. J. 1971. Stability of andalusite and the aluminum silicate phase diagram. *Am. J. Sci.*, *271*, 97–131.

Hollister, L. S., G. C. Grissom, E. K. Peters, H. H. Stowell, and V. B. Sisson. 1987. Confirmation of the empirical correlation of aluminum in hornblende with pressure of solidification of calc-alkaline plutons. *Am. Mineral.*, *72*, 231–239.

Johnson, M. C., and M. J. Rutherford. 1989. Experimental calibration of the aluminum-in-hornblende geobarometer with application to Long Valley Caldera (California) volcanic rocks. *Geology*, *17*, 837–841.

Kohn, M. J., and F. S. Spear. 1990. Two new barometers for garnet amphibolites with applications to southeastern Vermont. *Am. Mineral.*, *75*, 89–96.

Lindsley, D. H., and D. J. Anderson. 1983. A two-pyroxene thermometer. *J. Geophys. Res., Proc. 13th Lunar Science Conf.*, Part 2, 88 Suppl. A887–A906.

Moecher, D. P., E. J. Essene, and L. M. Anovitz. 1988. Calculation and application of clinopyroxene-garnet-plagioclase-quartz geobarometers. *Contrib. Mineral. Petrol.*, *100*, 92–106.

Moore, J. N., and D. M. Kerrick. 1976. Equilibria in siliceous dolomites of the Alta aureole, Utah. *Am. J. Sci.*, *276*, 502–524.

Naney, M. T. 1983. Phase equilibria of rock-forming ferromagnesian silicates in granitic systems. *Am J. Sci.*, *283*, 991–1027.

Nickel, K. G., G. P. Brey, and L. Kogarko. 1985. Orthopyroxene-clinopyroxene equilibria in the system CaO–MgO–Al$_2$O$_3$–SiO$_2$ (CMAS): New experimental results and implications for two-pyroxene thermometry. *Contrib. Mineral. Petrol.*, *91*, 44–53.

Spear, F. S. 1993. *Metamorphic Phase Equilibria and Pressure-Temperature-Time Paths. Monograph 1.* Washington, DC: Mineralogical Society of America.

Spencer, K. J., and D. H. Lindsley. 1981. A solution model for coexisting iron-titanium oxides. *Am. Mineral.*, *66*, 1189–1201.

Thompson, A. B. 1976. Mineral reactions in pelitic rocks: II. Calculation of some *P-T-X*(Fe-Mg) phase relations. *Am. J. Sci.*, *276*, 425–454.

Tracy, R. J. 1982. Compositional zoning and inclusions in metamorphic minerals. *Rev. Mineral.*, *10*, 355–397.

Whitney, J. A., and J. C. Stormer. 1977. The distribution of NaAlSi$_3$O$_8$ between coexisting microcline and plagioclase and its effect on geothermometric calculations. *Am. Mineral.*, *62*, 687–691.

GLOSSARY

·······························

aa Lava type characterized by a highly irregular, cindery surface. Reflects high lava viscosity and slow flow.

accessory minerals Those minerals in any rock—igneous, sedimentary, or metamorphic—that occur in relatively small amounts, typically less than 1% of the rock. Synonym for heavy minerals in sedimentary rocks.

activator elements Chemical elements, usually rare earth elements, that produce luminescence in a mineral.

adcumulate texture Texture that results when accumulated crystals (typically plagioclase) are overgrown by material that has a similar composition and fills in interstices between crystals.

adiabatic Thermodynamic term indicating that there is no exchange of heat between a chemical system of interest and its surroundings.

advection Heat transfer by bulk movement of material through crust or mantle. Could include movement of solids, melts, or fluids.

alkali basalt Type of basalt characterized by high total alkalis relative to silica, typically resulting in normative and modal nepheline.

alkali granite Plutonic rock that consists of 20–60% quartz with feldspar that is either all potassic feldspar or a mixture of potassic feldspar and albitic plagioclase (less than An_{10}).

alkali syenite Plutonic rock that consists of either all potassic feldspar or a mixture of potassic feldspar and albitic plagioclase (less than An_{10}) with less than 5% quartz.

aluminocrete A soil composed largely or entirely of aluminum hydroxide.

amorphous Referring to a noncrystalline solid.

amphibolite Metamorphic rock dominated by hornblende and intermediate to calcic plagioclase, with minor sphene and oxides. Likely protoliths include volcanic arc rocks dacite, andesite, and high-aluminum basalt, although some amphibolites are probably derived from calcareous sediments. Also the name of a widely observed metamorphic facies at moderate temperatures and pressures.

amphibolite facies Metamorphic facies characterized by assemblages of hornblende and intermediate to calcic plagioclase in mafic compositions. A common facies in orogenic belts, it represents moderate to high temperature (500°–700°C) and moderate pressures (3–10 kbar).

amygdules Vesicles that occur in lava flows and are filled by low-temperature secondary minerals, commonly calcite, prehnite, and zeolites.

andesite Fine-grained volcanic equivalent of diorite, characterized by silica content greater than 52 wt% and typically by plagioclase more sodic than An_{50}.

anhedral Grain shape that is characterized by little approximation of general crystal form and has no well-developed crystal faces.

annealing Process in high-temperature materials (including rocks) in which exposure to elevated temperatures causes coarsening of grain size and development of more polygonal texture (equant grains). Common in marbles and dunites.

anorthosite   Plutonic rock consisting of greater than 90% plagioclase.

anorthositic gabbro   Plutonic rock transitional between gabbro and anorthosite. Essentially a gabbro that is anomalously enriched in plagioclase.

aplite   Fine- to medium-grained rock, generally of granitic composition, characterized by a lack of dark minerals or mica and having a marked, equigranular "sugary" texture.

arc-trench system   An elongate zone in the marginal parts of an oceanic basin that includes a deep-sea trench and the adjoining group of volcanic islands, for example, the Aleutian Trench and Aleutian Islands.

arkose   A detrital rock that contains an appreciable percentage of feldspar grains, typically at least 20%.

armored relic   Mineral inclusion inside a metamorphic porphyroblast (or, less commonly, an igneous phenocryst) of a mineral species that has been eliminated from the rock matrix by reaction after inclusion of the mineral grain in the porphyroblast.

ash   Loose, unconsolidated volcanic material less than 4 mm in size. Typically glassy, but may consist of crystals or rock fragments as well.

ash flow tuff   Welded compact rock made up of volcanic ash and characterized by small-scale structures indicative of downslope movement.

assimilation   The process of partial or complete melting of country rock or xenoliths and incorporation of these chemical constituents into a magma.

asthenosphere   Layer of the upper mantle about 50–100 km thick that underlies the lithosphere. Also called the low-velocity zone and characterized by anomalously low seismic velocities, the asthenosphere is thought to behave in a ductile fashion because of a small percentage of trapped melt.

A-type granite   Anorogenic granite that does not appear to be associated with regional metamorphism or convergent plate tectonics. Typically of peralkaline type; contains sodic amphibole or pyroxene.

augen gneiss   Blastomylonitic rock that contains large, lenticular or almond-shaped porphyroclasts bounded by microshears. Individual grains (called augen from the German word for "eyes") are most commonly feldspar.

aulacogen   A tectonic trough on a craton, bounded by convergent normal faults oriented normal to the cratonic boundary.

authigenic   Referring to rock constituents that are formed in place and not transported, for example, authigenic clay or quartz cement.

backarc basin   Also called foreland basin or retroarc basin. A basin in a backarc area (that is, the area between an island arc and the continental landmass) that is floored by continental crust and separated from the arc by a system of folds and thrusts.

basalt   Fine-grained volcanic rock characterized by silica content less than 52 wt% and by the presence of abundant calcic plagioclase ($>An_{50}$) and calcic pyroxene, with or without other minerals such as orthopyroxene or olivine.

basalt-andesite-rhyolite association   Characteristic orogenic magmatic association of island arcs and magmatic arcs that reflects calc-alkaline differentiation patterns in which primary basaltic magmas evolve to form derivative magmas.

basanite   Strongly alkaline mafic magma characterized by high Mg:Fe ratio and by significant silica undersaturation that leads to modal nepheline, leucite, and olivine.

batch melting   The process by which melting occurs by equilibrium (mass-balanced) fusion until the proportion of melt reaches an amount sufficiently high to allow the melt to be removed by some physical means.

batholith   Plutonic rock body (typically granite, granodiorite, or diorite) with a surface outcrop area greater than 100 km^2.

bauxite   A rock or soil composed mostly of amorphous or crystalline hydrous aluminum oxides and aluminum hydroxides, typically gibbsite. Aluminous laterite.

beachrock   Limestone formed in the intertidal zone, typically a calcarenite or biosparite.

bentonite   A claystone produced by alteration of glassy igneous material. Composed essentially of the montmorillonite group of clays and colloidal silica. Originally a tuff or volcanic ash.

bimodal   A frequency distribution characterized by two peaks, for example, a sediment composed of sand and clay, but lacking silt.

bimodal magmatism   Where two distinctive magma types that are widely separated along chemical evolutionary trends commingle or are contemporaneously intruded or extruded. Commonly involves basaltic and granitic (rhyolitic) magmas.

black smoker   Name applied to a ridge-crest vent for high-temperature (> 300°C) hydrothermal fluids. The fluids have high concentrations of dissolved components, especially sulfides, which precipitate as the fluids cool by mixing with seawater, thus producing dark gray to black clouds of fine precipitated particles above the vent.

blastomylonite   Mylonite in which strained, shear-bounded larger grains are surrounded by fine-grained sheared matrix. Augen gneiss is a common type.

block   Term for angular volcanic particle greater than 32 mm in diameter.

block lava   Lava flow with a surface consisting of large angular blocks.

blueschist facies   Metamorphic facies characterized by crystallization of sodic blue amphibole (glaucophane), lawsonite, aragonite, and other low-temperature, high-pressure minerals in mafic compositions. Represents low temperatures (200°–500°C) and moderate to high pressures (6 – 15+ kbar).

bomb   Term for rounded, commonly teardrop-shaped volcanic particle greater than 32 mm in diameter.

boninite   Igneous rock type characteristic of forearc areas in volcanic arcs. Essentially a magnesium-rich, primitive andesite containing orthopyroxene, boninite is thought to be primary mantle-derived andesitic magma.

Bowen's Decarbonation Series   Less well known than the Reaction Series, the Decarbonation Series describes the typical sequence of appearance of Ca–Mg silicate minerals and oxides in progressive (especially contact) metamorphism of calcareous rocks.

Bowen's Reaction Series   Scheme devised by N. L. Bowen that divides the sequence of igneous rock crystallization into two simultaneous series, the "discontinuous" series (olivine-augite-hornblende-biotite) and the "continuous" series (calcic to sodic plagioclase). The two series merge with crystallization of biotite and albite and are followed by crystallization of potassic feldspar and quartz.

breccia   Any rock consisting of large angular fragments. Mostly applied to sedimentary and volcanic rocks and to brittlely deformed material in fault zones.

buffering   Generally, the control of some variable by operation of a sedimentary or metamorphic reaction. Common usages refer to internal control of pH in aqueous fluids in sedimentary materials and fluid compositions in calcareous metamorphic rocks through operation of decarbonation-dehydration reactions that counteract the effect of influx of external water-rich fluid.

burial metamorphism   Generally low-grade metamorphism in which mineralogic or textural changes are induced by changes in pressure and temperature along a normal geothermal gradient. Typically anorogenic and occurs in the deeper parts of sedimentary basins.

CIPW norm   Scheme for calculation of synthetic anhydrous mineralogy for igneous rocks based on an analysis of bulk chemistry. Useful for comparisons of fine-grained or glassy rocks in which actual mineralogy is not easily measured.

calc-alkaline trend   Fractionation trend that characterizes magmatic and volcanic arcs. Fractionation produces an increase in alkali content without significant increase in Fe:Mg ratio.

calc-pelite   Compositional term for metamorphic rocks derived from mixtures of carbonate and aluminous sediments.

calcrete   Also called caliche. A limestone precipitated as surface or near-surface crusts and nodules by the evaporation of soil moisture in semiarid climates. A variety of duricrust.

calc-silicate   Category of metamorphic rocks characterized by the predominance of calcium- and magnesium-rich silicate minerals and only minor carbonate, if any. Protoliths for these rocks are probably highly impure carbonate sediments.

caldera   Large volcanic crater centered on a summit vent and typically enlarged by normal faulting as a result of summit collapse following magma withdrawal.

caliche   *See* calcrete

carbonate compensation depth (CCD)   The depth in a body of water at which the rate of dissolution of calcium carbonate equals the rate of supply, that is, the depth below which no calcium carbonate can accumulate.

carbonate rock   A rock composed largely of carbonate minerals, usually calcite or dolomite.

carbonatite   Rare, mantle-derived igneous rock dominated by calcite, dolomite, or sodium carbonate, with minor silicates (typically sodium-, magnesium-, or calcium-rich). Occurs both as intrusions and in lava flows.

catazonal pluton   Deep-crustal intrusion completely surrounded by high-grade metamorphic rocks and, commonly, by migmatites. Metamorphic rocks are of high enough temperature to have been locally melted. Contacts are typically gradational.

cementation   *See* lithification

chadocryst   One of several small, commonly subhedral to euhedral crystals contained within a large crystal (oikocryst) in poikilitic texture.

charnockite   By legend, the rock from which the tombstone of Job Charnock, the founder of Calcutta, was made. Essentially an orthopyroxene-bearing granite with moderately to strongly iron-enriched ferromagnesian minerals. Most known charnockites are Proterozoic and are associated with massif anorthosites, and the name was originally intended only for this type of orthopyroxene granite.

chelate   A metallic cation held by two or more atoms in an organic molecule.

chert   A rock composed of microcrystalline quartz crystals of subequant habit, typically with a diameter of less than 20 μm.

chicken-wire structure   Also called nodular anhydrite. Anhydrite (or gypsum) rock composed of nodules of anhydrite in a matrix of microcrystalline anhydrite.

chilled margin   Zone near the outer contact of an intrusive body that is finer grained than interior portions because of chilling and rapid crystallization adjacent to cold country rocks.

cinder cone   Small volcanic cone (typically not more than several hundred meters high) composed mostly of loose pyroclastic material (cinders, lapilli, or bombs).

clay   Term used to indicate (1) fragmental material less than 4 mm in size and also (2) a group of phyllosilicate minerals with particular compositions. Most clay-sized silicate material is composed of clay minerals.

cleavage   Planar structure in fine-grained (typically lower grade) metamorphic rocks. Distinguished from foliation by a strong tendency for the rock to split cleanly along these planes, as in slate.

closed system   Any rock that exchanges thermal and mechanical energy with its surroundings, but not chemical mass.

compaction   Reduction in bulk volume and decrease in porosity of a sediment as a result of the increasing weight of overburden as deposition continues. Tectonic compaction may occur subsequently.

compatible   Category of chemical elements that favor the highly structured sites in crystal lattices and thus are moderately to strongly fractionated from melts into solids. Examples include magnesium, calcium, chromium, and nickel.

component   Thermodynamic term for a chemical constituent of an igneous or metamorphic rock, which is defined on the basis of the minimum number of constituents required to describe the minerals in the rock. The number of components can never exceed the number of minerals in a rock.

composite volcano   Also called stratiform volcano. Type of volcanic cone that can be several thousand meters high and is composed of mixed or interlayered pyroclastic material and lava flows.

conduction   Form of heat transfer in which thermal energy is transferred essentially from atom to atom in a fixed framework. The slowest form of heat transfer in the earth.

cone sheet   Type of intrusive dike that is intruded in a conical form up to several kilometers across and converges downward. It is probably a result of extension related to subsidence.

conglomerate   Also called rudite. A coarse-grained clastic rock composed of rounded gravel or boulders.

congruent melting   Melting of a crystalline material that produces a melt of the same composition as the solid. The mineral can thus be completely melted at a single temperature.

contact aureole   The roughly concentric zone of contact metamorphosed country rocks surrounding some plutons. Typically shows regular zonation of metamorphic grade, from highest at the igneous contact to lower grades outward.

contact metamorphism   Metamorphism that occurs in the proximity of igneous intrusions (or less commonly beneath lava flows) and has clearly formed as a result of heat transfer outward from the crystallizing igneous rocks. May be superimposed on regional metamorphic rocks in orogenic belts.

contamination   Addition of extraneous material to a magma, typically through either melting or dissolution of country rocks. May include addition of fluid components to a magma.

convection   Form of heat transfer in which thermal energy is transported by moving hot matter (either fluids or ductile solids) in closed cells. Hot material rises in a gravitational field because of density decrease, heats its new, cooler surroundings, then sinks to its place of origin.

coquina   A weakly indurated clastic or detrital limestone composed of mechanically sorted shell fragments that experienced transport and abrasion before reaching the depositional site.

corona texture   Fine-grained multiphase mixture, commonly with a radial fabric, concentrically surrounding the corroded remnants of a crystal or along the boundary of two crystals. Typically produced by solid-state reaction or incongruent melting of a crystal.

cotectic line   Line in a ternary, quaternary, or higher order liquidus phase diagram that represents the multiple saturation of the melt with $(n - 1)$ types of crystals, where n = the order of the system.

craton   Tectonically stable, old continental crust (typically Archean or early Proterozoic) in internal regions of continental lithosphere. Cratons have not undergone significant orogenic activity for a prolonged period. Sometimes called shields; characterized by low topographic relief.

critical point   Point on a phase diagram that marks the highest temperature of a region of immiscibility between two phases (fluid, liquid, or solid).

cross beds   Layers within a stratified unit that are oriented at an angle to the dominant foliation.

cryptic   Subtle, hidden, or not readily visible petrologic effect or process, commonly compositional. Examples include small-scale *cryptic zoning* in individual mineral grains and larger scale *cryptic layering* in mafic layered igneous rocks.

crystal-lithic tuff   Pyroclastic rock consisting of welded particles made up dominantly of single phenocrysts and small, fine-grained rock fragments.

cumulate texture   Texture in igneous rocks that clearly indicates the accumulation of crystals, either through gravitational settling (or flotation) or flow segregation.

cumulus crystal   Core of an individual grain in an igneous rock exhibiting one of the several varieties of cumulate textures. Typically an early-crystallized euhedral liquidus mineral.

dacite   Fine-grained igneous rock type that is the volcanic equivalent of quartz diorite or tonalite. Contains quartz, sodic to intermediate plagioclase, and accessory dark minerals.

decarbonation   Metamorphic reaction in which either calcite or dolomite reacts with oxide or silicate minerals and liberates carbon dioxide as a fluid.

decussate   Metamorphic texture in which annealing of minerals with pronounced nonequant shapes (such as micas or amphiboles) leads to a mixture of polygonal texture and interpenetrating grain relationships.

dehydration   Metamorphic or sedimentary reaction in which water is lost from crystalline sites in hydrous minerals and is liberated as a free aqueous fluid.

detrital   Referring to sediment composed of pieces of preexisting materials derived from outside the depositional basin.

diabase   A rock of basaltic composition that is characterized by ophitic texture, that is, interlocking calcic plagioclase laths with interstitial augite and accessory minerals. Typically occurs in shallow intrusive bodies (dikes, sills) and interiors of very thick lava flows. Synonym for dolerite.

diagenesis   Excluding weathering, all the chemical, physical, and biological changes that a sediment undergoes from the time the grains are deposited until they are metamorphosed or melted.

diapir   A body of rock that has moved upward by plastic flow, piercing the overlying beds, for example, salt diapirs in the Gulf Coast region. May also be used to describe some cylindrical plutons.

diastem   A brief interruption in sediment deposition during which no erosion occurs, for example, a bedding plane.

differentiation   The magmatic process of removal of early-crystallized, higher temperature constituents and enrichment in remaining magma of lower temperature minerals. Typically results in decreasing Mg:Fe ratio and increases in the ratio Na:Ca and in potassium content.

dike   Discordant tabular intrusive form with generally parallel planar walls and a thickness much less than its size in the two perpendicular directions. Generally from less than one to several tens of meters in thickness, although rarely much thicker.

diorite   Medium- to coarse-grained plutonic igneous rock that consists entirely of sodic to intermediate plagioclase ($An_{30}-An_{50}$) and a dark mineral, generally either hornblende or augite.

divariant assemblage   Metamorphic mineral assemblage with the exact number of minerals that one would expect in a randomly sampled rock. Indicates either that no reaction is occurring or that reactions are continuous; that is, that no minerals are abruptly appearing or disappearing but that some minerals are changing their solid solution compositions through reaction.

divariant equilibrium   Equilibrium state in which both temperature and pressure can be independently varied by at least small amounts without changing the number or identity of mineral phases.

dolostone   A sedimentary rock composed largely or entirely of the mineral dolomite.

Dorag dolomitization   Replacement of calcitic limestone by dolomite through the mechanism of mixing freshwater with salt water, presumably along coastlines in subtropical and tropical seas.

dunite   Ultramafic rock that consists of greater than 90% olivine. Chrome-bearing spinel is a common accessory.

duricrust   The hard crust on the surface of a soil, formed by the evaporation of soil moisture. Examples include calcrete and silcrete.

eclogite   High-pressure rock of mafic composition that consists of magnesian (pyrope-rich) garnet and sodium-rich pyroxene (omphacite).

eclogite facies   Metamorphic facies of very high pressure and temperature (>10–12 kbar and 400°–900°C).

Eh   A measure, usually in millivolts, of the relative intensity of oxidation or reduction in a solution. Values range from −1.0 volt (most reducing) to +1.0 volt (most oxidizing).

ejecta   General term for pyroclastic material (ash, lapilli, bombs) that is explosively erupted from a volcanic vent.

endothermic reaction   Any chemical reaction that consumes energy. Therefore, at the same temperature and pressure, the product minerals (plus melt for igneous systems) have a higher free energy content than the reactant minerals. Examples include most prograde metamorphic reactions and melting reactions.

epeiric sea   *See* epicontinental sea

epicontinental sea   Also called epeiric sea. A sea on the continental shelf or within a continent.

epitaxial overgrowth   Phenomenon in both igneous and metamorphic rocks in which a mineral grows episodically and the later overgrowths are in crystallographic continuity with earlier formed material and can be difficult to recognize.

epizonal plutons   Intrusions that are largely discordant with country rocks and typically lack internal flow structure. Represent intrusion at shallow crustal levels.

equilibrium   Term used in petrology in its chemical sense as representing a rock in which no obvious spontaneous reaction is occurring.

equilibrium crystallization   Term for magma crystallization that implies a mass balance; that is, that the overall composition of melt plus crystals remains constant throughout crystallization. Unlikely to occur in its pure form.

equilibrium melting   Indicates that melt remains in contact with residual crystals during melting, and that the overall composition of melt plus residual crystals remains constant and equal to that of the protolith.

euhedral   Crystal that is entirely bounded by perfect crystal faces.

eutectic   Refers to a singular magma composition representing multiple saturation of the magma with a maximum number of solids at the lowest temperature the melt can attain. Crystallization of a melt at a eutectic goes to completion with no decrease in temperature, and melt composition is invariant at a eutectic. Initial melting typically occurs at eutectics during heating, and eutectic melting continues until at least one of the protolith minerals is exhausted.

exothermic reaction   Any chemical reaction in which energy is released; that is, the product minerals (plus melt) have a lower energy content than the reactant minerals (plus melt). Magma crystallization is a classic example.

expandable clay   A clay mineral in which the layers can be readily separated by absorption of water; swelling clay.

fabric   Shape and arrangement of the crystalline (and glassy) parts of a rock, including the geometric relationship of the grains to one another. Fabric determines rock texture, along with degree of crystallinity and grain size. Orientation (or lack of it) in sedimentary grains.

facies   A group of characteristics of a sedimentary rock that distinguish it from other rocks, for example, red bed facies, shale facies, sandy facies.

fenestral fabric   A fabric in a carbonate rock characterized by discontinuities in the rock framework that are larger than grain-supported voids. Fenestrae may be empty or filled. Examples are bird's-eye and stromatactis structures.

fenitization   Special type of metasomatism first associated with intrusion of alkalic rocks near Fen, in Norway. Involves addition of potassium and sodium to the country rocks, along with substantial oxidation.

ferricrete   A soil composed largely or entirely of ferric oxide and hydroxide.

filter pressing   Differentiation process in which fluid pressure in a partially crystallized intrusive magma can promote physical separation of melt and crystals, especially if large crystals form loosely congealed plugs or clots that will not pass through magma conduits.

fissility   The ability of a sedimentary rock to split easily into thin sheets parallel to the bedding of the unit. Caused by parallelism of clay minerals in the unit.

flocculation   A process by which individual flakes of clay minerals are aggregated into clumps (floccules) because of electrostatic charges on clay mineral surfaces. The process is more effective in seawater than in fresh water.

floccule   An aggregate of clay particles produced in salt water.

fluid inclusion   Trapped bubble of aqueous or carbonic fluid in minerals. Typically found in metamorphic rocks and hydrothermal veins, most commonly in quartz, but also in a wide array of other minerals.

foliation   Pervasive planar structure found in most metamorphic rocks. A combination of parallel alignment of platy (mica) grains and fine-scale compositional layering.

forearc basin   A basin located between a deep-sea trench and its adjoining island arc.

foreland basin   *See* backarc basin

fractional crystallization   Magma crystallization process in which all crystals formed are arbitrarily considered to be removed from the melt, thus continuously changing magma composition.

fractional melting   Melting process in which melt is considered to be continuously extracted from the region of melting, thus implying no continuing equilibrium between melt and crystals.

fractionation   Chemical process in magmas and some solid-state metamorphic occurrences in which bulk composition changes through physical removal of a fraction of the rock system. Can occur through crystal settling or flotation in intrusions, or through melting or fluid-rock interaction in metamorphic terranes.

friable   Referring to the state of incomplete cementation in which a rock is coherent but crumbles easily under a small amount of pressure, such as that applied by the fingers.

gabbro   Coarse-grained plutonic equivalent of basalt that consists of calcic plagioclase and augite.

gabbroic anorthosite   Plutonic rock transitional between anorthosite and gabbro. Essentially an anorthosite that is especially rich in ferromagnesian minerals (typically pyroxenes) and falls just below 90% plagioclase.

geopetal structure   Any rock feature that permits distinction between the top and bottom of a bed at the time of deposition, for example, cross-bedding or stromatactis.

geothermal gradient   The rate of increase of temperature with depth in the earth. The geothermal gradient varies in different tectonic regimes, from lowest in continental craton areas to highest at mid-ocean spreading ridges.

geothermobarometry   Use of a variety of techniques for semiquantitative or quantitative estimation of both the temperature and pressure of magma crystallization or equilibration of metamorphic mineral assemblages. Geothermometry is estimation of temperature alone, whereas geobarometry is estimation of pressure alone.

glomeroporphyritic texture   Porphyritic texture (larger crystals in fine groundmass) in which the larger phenocrysts are grouped together in clumps rather than homogeneously distributed.

glowing avalanche   *See* nuée ardente

gneiss   General term for a metamorphic rock that contains prominent layered structure or foliation but does not tend to break along these foliation planes. Typically of high metamorphic grade and consisting mostly of anhydrous minerals.

graded bedding   A layer of sedimentary rock in which the particle sizes change systematically in a direction normal to bedding, usually from coarser at the base to finer at the top.

granite   Plutonic rock consisting of subequal proportions of potassic feldspar, sodic plagioclase, and quartz.

granitic texture   Holocrystalline texture that reflects continuous crystallization of mineral grains that are all of about the same size.

granitoid   Field term for any coarse-grained rock dominated by quartz and feldspars. Can include granites, granodiorites, quartz diorites, and syenites.

granoblastic   Equigranular metamorphic texture that consists of similar-sized grains without any porphyroblasts.

granophyre   Porphyritic igneous rock (typically granitic or syenitic) in which the groundmass minerals are micrographically intergrown.

granulite   Equigranular metamorphic rock typically dominated by quartz and feldspars. Also the name for the highest temperature metamorphic facies in which rocks are virtually anhydrous.

granulite facies   Metamorphic facies characterized by the occurrence of rocks with mostly to totally anhydrous mineral assemblages. Represents temperatures in excess of 700°C at pressures above about 2 kbar.

graphic texture   Intergrown minerals (commonly quartz and potassic feldspar) that maintain rough crystallographic or optical continuity of each component throughout the intergrowth. Common in pegmatites.

gravel   All rounded, clastic particles with diameters greater than 2 mm.

gray smoker   Term for a moderate- to high-temperature (100°–300°C) hydrothermal fluid vent in ridge-crest environments. Named for gray clouds of tiny precipitated mineral grains that are produced as fluid cools by mixing with seawater.

greenschist facies   Metamorphic facies characterized by assemblages including chlorite, epidote, and actinolite in mafic compositions, by muscovite, biotite, chlorite, and garnet in aluminous rocks, and by calcite, dolomite, and tremolite in calcareous rocks.

greenstone   General or field term for low- to medium-grade metamorphosed mafic volcanic rocks. Name refers to typical color of these rocks, which is caused by modal dominance of the green minerals chlorite, epidote, and actinolite.

greenstone belt   Region of metamorphosed mafic volcanic rocks that represents an old volcanic arc. Most are of Precambrian age and occur in concentric belts around cratons.

hardground   A calcium carbonate surface lithified by submarine processes soon after deposition of the sediment.

harzburgite   Ultramafic rock type dominated by olivine and orthopyroxene, with accessory magnetite or spinel.

high-pressure, low-temperature metamorphism   Type of regional metamorphism characteristic of subduction zone environments in which metamorphism occurs at temperatures well below a normal geothermal gradient. Pressures need not be exceptionally high, as long as temperatures are anomalously low at any given depth. Most common type is called blueschist metamorphism, after the presence of the blue amphibole glaucophane.

holocrystalline   Consisting entirely of crystalline minerals with no glass.

holohyaline   Consisting entirely of glass, for example, an obsidian.

hornfels   Fine-grained, dense, compact metamorphic rock that is characteristic of inner parts of shallow contact aureoles.

hornfels facies   Low to very low pressure metamorphic facies representative of contact metamorphic environments. Subdivided into albite-epidote hornfels (low-temperature), hornblende hornfels (moderate-temperature), pyroxene hornfels (high-temperature), and sanidinite (ultrahigh-temperature).

hot spot   Apparently unmoving point heat source in sublithospheric mantle that causes localized melting in the lithosphere passing over it.

hypersolvus   Term for crystallization of alkali feldspars at relatively low water pressure (below about 4 kbar) that results in formation of a single feldspar, which commonly becomes perthitic at subsolidus temperatures.

hypidioblastic   Metamorphic texture in which most mineral grains are anhedral to subhedral.

hypocrystalline   Fine-grained texture with small crystals in a glassy matrix.

hyposometric   Referring to the area of ground surface at any elevation above sea level.

idioblastic   Metamorphic texture in which all minerals show well-formed crystals (subhedral to euhedral).

imbrication   A sedimentary structure consisting of detrital gains (typically gravel) stacked with their flat surfaces at an angle to the main bedding plane; the flat surface dips upstream.

impact (shock) metamorphism   Rare type of metamorphism that is caused by transient ultrahigh pressures (and possibly temperatures) associated with shock waves. Localized melting is a common feature of this type. May be caused either by meteorite or comet impact or by an explosive volcanic eruption.

inclusion   Gas, liquid, or mineral included within a larger crystal. *See also* chadocryst; fluid inclusion; oikocryst; poikiloblast

incompatible   Category of chemical elements that favor the relatively unstructured sites of silicate melts and thus are moderately to strongly fractionated into melt relative to solids. Examples include barium and rare earth elements.

incongruent melting   Melting that is combined with a solid-solid reaction. Results in production of melt and a new crystalline compound that is poor in the components that are enriched in the melt.

index mineral   A newly appearing mineral in a prograde metamorphic sequence that marks the beginning of a new, higher grade metamorphic zone. Examples include garnet (marking the garnet isograd and the beginning of the garnet zone) in aluminous rocks and hornblende (marking the lower limit of the amphibolite facies) in mafic rocks.

induration   *See* lithification

infiltration metamorphism   Metamorphic reactions driven by the pervasive flow of disequilibrium fluid through a rock. Most commonly involves decarbonation-dehydration reactions driven by influx of water-rich fluids into calcareous or calc-silicate rocks.

interarc basin   A basin in a backarc area that is floored by oceanic crust.

intraclast   A carbonate fragment composed of penecontemporaneous limestone or dolostone formed within the basin of deposition or on its fringes, for example, dolostone from a supratidal flat or shallow seafloor torn up by a storm.

intracratonic basin   A tectonic basin within a continent and underlain by continental crust.

invariant assemblage   Metamorphic mineral assemblage containing two more minerals than would normally be expected in a randomly sampled rock. Greatly restricts the possible range of pressures and temperatures.

iron formation   A rock unit consisting of alternating bands of quartz (including chert) and iron-bearing minerals (commonly magnetite and hematite), containing at least 15% iron. Typically of Proterozoic age.

ironstone   A sedimentary rock containing at least 15% iron, typically unbanded and lacking chert. Typically of Phanerozoic age.

island arc   A chain of islands (for example, the Aleutians) rising from the deep-sea floor and located where two oceanic plates converge. The chain is generally curved and convex toward the open ocean.

isograd   Line on a map that indicates roughly equal metamorphic grade.

I-type granite   Most commonly a metaluminous granite (biotite-hornblende granite) that has originated from partial melting of igneous rock protoliths within the crust.

kimberlite   Potassic ultramafic rock, commonly with a brecciated texture. Rich in olivine, with subordinate pyroxenes, oxides, amphiboles, and phlogopite. Well known as an ore of diamonds.

komatiite   Ultramafic lava flows that have been found in Canada, Australia, and southern Africa and are virtually restricted to the Archean. Typically moderately to totally altered to low-temperature minerals.

laccolith   Form of intrusion in which a sill-like magma body domes up its roof to produce a circular or oval outcrop pattern following erosional unroofing.

lamination   Stratification on a scale of less than 10 mm.

lamproite   Ultramafic rock similar to kimberlite and lamprophyre, but uniquely characterized by a high total alkali content that exceeds alumina on a molar basis, thus making lamproites peralkaline. Commonly contains phenocrysts of olivine, pyroxenes, amphiboles, and especially mica, and can contain diamonds.

lamprophyre   Large group of dark gray to black porphyritic dike rocks characterized by large phenocrysts of mafic minerals such as olivine, pyroxenes, amphiboles, or biotite. Many specific names based on detailed mineralogy.

lapilli   Loose volcanic particles between 4 and 32 mm in size.

latite   Volcanic equivalent of monzonite, a rock containing subequal proportions of potassic feldspar and sodic to intermediate plagioclase, with little or no quartz and with accessory mafic minerals.

lava   Magma that has been extruded onto the earth's surface.

lava dome Rounded or bulbous secondary volcanic cone that develops from quiescent lava extrusion within a caldera or blown-out crater.

layered intrusion Medium to large, bowl-shaped or conical intrusive body in which physical processes such as gravity settling have created individual composition-ally distinct layers on a scale of centimeters to many meters thick. Layers are typically subhorizontal and dip inward gently, although steep layering can occur near contacts. Commonly mafic, with ultramafic layers (dunite, harzburgite) at the bottom, feldspathic gabbros in the middle, and granophyres at the top.

leucosome Light-colored, typically coarse-grained por-tion of a migmatite. Commonly of granitic composition and thus represents the melted fraction of aluminous crustal material in high-grade metamorphism.

lherzolite Ultramafic rock type dominated by magne-sian olivine, with minor, subequal proportions of augite and orthopyroxene and accessory spinel or garnet.

limeclast A carbonate rock fragment of clastic or detrital origin contained within a carbonate rock. The fragment may be intrabasinal (intraclastic) or extrabasi-nal (terrigenous).

limestone A sedimentary rock consisting largely or entirely of calcium carbonate (calcite).

liquidus Line or surface on a phase diagram that repre-sents the onset of crystallization of a particular single mineral.

litharenite A sandstone containing an appreciable content of undisaggregated rock (lithic) fragments, typi-cally at least 20%.

lithification Also called cementation or induration. The conversion of unconsolidated sediment into a co-herent aggregate.

lithosphere Outermost, rigid part of the earth that constitutes the tectonic plates. Thickness varies from essentially 0 km at mid-ocean ridge crests to about 70 km in oceanic lithosphere to as much as 150 km in conti-nental lithosphere.

lithostatic pressure Type of pressure at depth in the earth that is equal in all directions and is the result of the weight of overlying rocks and any intergranular fluids.

lopolith Bowl-shaped, large shallow intrusion, com-monly mafic and with layering dipping gently inward.

luminescence petrography The examination of a thin section during bombardment by a broad beam of low-energy electrons. Also called cathodoluminescence.

magma High-temperature physical mixture of silicate melt, crystalline products, and possibly also evolved fluids or gas.

mantle plume Proposed cylindrical upwelling diapir of hot, deep-mantle material, possibly partially molten, that may start its rise from the core-mantle boundary. Invoked to explain hot spots.

mantle wedge Lithospheric mantle that overlies the subducted slab of oceanic lithosphere at convergent margins and underlies island arcs and magmatic arcs. The mantle wedge is assumed to be metasomatized by fluids given off by the descending slab of lithosphere and to be the source region for andesite magmas.

marble Metamorphic rock that consists mostly of carbonate minerals calcite or dolomite, with minor sili-cates, oxides, or sulfides.

marl A friable mixture of subequal amounts of micrite and clay minerals.

mass balance A condition in which the chemical mass of a system (a rock or magma, for example) is divided between several phases but remains constant. During a reaction, the proportions of the phases change, but the bulk composition that represents the weighted or alge-braic sum of individual phase compositions does not change.

mass transfer Transport of chemical mass (atoms, ions, or molecules) through a rock. May occur through diffusion or by flow of intergranular fluids in which chemical constituents are dissolved.

matrix The finer grained material in a sediment with a conspicuous range of grain size, for example, the clay in a texturally immature sandstone.

megacryst Very large phenocryst or xenocryst in an igneous rock of uniformly finer grain size. Lower size limit depends on size of groundmass crystals.

mélange Large-scale breccia deposit found in orogenic belts and assumed to represent tectonically mixed rock above the downgoing slab at a subduction zone. Much of the mélange material is altered or metamorphosed mafic to ultramafic rock derived from the oceanic lithosphere, mixed with sedimentary blocks from the continental shelf and slope, all set in a mudstone matrix.

melanosome Thin layers of very dark rock that sepa-rate leucosome layers in migmatites. Commonly inter-preted to represent the residuum from melting and ex-traction of the granitic leucosome fraction from a pelitic protolith.

mesocumulate Intermediate cumulate texture in which considerable unzoned adcumulate growth has occurred, but the outer portions of cumulate crystals (typically plagioclase) show orthocumulate growth with chemical zoning.

mesoperthite Exsolution texture in potassic feldspar in which potassium-rich host and sodium-rich lamellae are subequal in size and proportions (see Figure 10-10).

mesozonal pluton Midcrustal intrusion surrounded by low- to medium-grade metamorphic rocks that were not hot enough to melt locally. Flow structure is commonly present, and contacts are typically sharp.

meta- Prefix used to indicate a metamorphosed equivalent of a particular igneous or sedimentary rock type, as in metabasalt or metadolostone.

metabasite Term for any metamorphosed mafic igneous or pyroclastic rock.

metaluminous Category of granitoids in which the alumina content (on a molar basis) is greater than sodium plus potassium, but less than sodium plus potassium plus calcium. This chemistry is typically reflected mineralogically by the presence of biotite and either augite or hornblende.

metamorphic facies Concept originated by Eskola, which states that the same protolith will always produce the same rock mineralogically when subjected to the same temperature and pressure. That is, the mineralogic composition is determined by the chemical composition at constant temperature and pressure.

metamorphic facies series Extension of the metamorphic facies concept. States that maximum temperature and pressure for a variety of metamorphic rocks of different grades in any single terrane will tend to lie within a narrow band in the P-T plane that shows little pressure change over a wide range in temperature. Facies series can thus be characterized as low-pressure, medium-pressure, or high-pressure.

metamorphic field gradient Line in the P-T plane that connects the P-T points at which each rock in the terrane developed its peak metamorphic assemblage.

metamorphism Mineralogic or textural change that occurs in a rock in the solid state as a response to changes in environmental variables, especially temperature and pressure.

metapelite Metamorphosed mudrock, shale, or other aluminous clastic rock. Metamorphism of aluminous rocks typically results in production of such aluminous metamorphic minerals as muscovite, Al-silicates, staurolite, garnet, or cordierite.

metasomatism Alteration in the bulk chemistry of a rock during regional or contact metamorphism, typically aided by movement of hot fluids through the rock.

microperthite Exsolution texture in potassic feldspar in which the size of potassium-rich host domains is much larger than that of thin sodic plagioclase lamellae.

migmatite Literally a "mixed" rock with both lighter colored and darker colored portions. May be *stromatic* (layered), *venitic* or *nebulitic* (dikelike light portions), or *chaotic*. Commonly ascribed to local partial melting of metasedimentary rocks but can occur through subsolidus segregation processes.

mineral assemblage Thermodynamically stable configuration of minerals in a metamorphic rock.

mineralogic phase rule Usually written as $p \geq c$. Stated by Goldschmidt, it refers to the great likelihood of finding a divariant assemblage in a randomly sampled rock, and thus substituting F in the phase rule with 2 or, more rarely, 1 or, much more rarely, 0.

mixed-volatile equilibrium In general, a metamorphic reaction that involves both carbonate and hydrous minerals. Mixed volatile equilibria are of several types, which can either consume or release water and carbon dioxide in various combinations.

mode Actual mineralogic composition of a rock, either estimated or measured by various quantitative techniques such as point-counting.

molasse A continental, deltaic, and/or marine sedimentary facies consisting of a thick sequence of crossbedded, fossiliferous detrital rocks with minor coal and carbonate rocks; a postorogenic accumulation.

monocrystalline Referring to a clastic grain consisting of a single crystal.

monzodiorite Plutonic rock type in which there are subequal proportions of potassic feldspar and plagioclase and mafic minerals are augite or hornblende (like monzonite) but in which the plagioclase is typically between An_{30} and An_{50}.

monzonite Plutonic rock type containing subequal proportions of potassic feldspar and sodic plagioclase (An_{10}–An_{30}) and with little or no quartz. Mafic mineral content is typically low.

monzonorite Plutonic rock type with subequal proportions of potassic feldspar and plagioclase (An_{30}–An_{50}) but in which the mafic mineral fraction contains substantial orthopyroxene.

MORB Mid-ocean ridge basalt. Characteristic of the mid-ocean ridges and the seafloor in general. Two main types are N-MORB (normal MORB) and E-MORB (enriched MORB). E-MORB is enriched in certain rare earth and trace elements relative to N-MORB and appears to have resulted from smaller degrees of melting in the mantle.

morphotectonic   A topographic feature produced by a tectonic event.

mud   Sediment in which particles are less than 0.06 mm in diameter.

mudrock   Silicate or carbonate sedimentary rock composed of silt- and clay-sized particles. Commonly texturally massive rather than laminated.

mylonite   Structural term for sheared rocks with significant grain size reduction and a flow texture that occurs at sufficiently elevated temperature that deformation is ductile, that is, crystal-plastic.

nepheline-melilite basalt   Type of alkali olivine basalt characterized by an extreme degree of silica undersaturation that leads to modal olivine and nepheline and to calcic melilite replacing calcic plagioclase.

nepheline syenite   Silica-undersaturated plutonic rock containing potassic feldspar, sodic plagioclase (An_{30}), and nepheline, with minor mafic minerals.

nodular anhydrite   *See* chicken-wire structure.

nonundulatory extinction   Referring to a crystal that has not been plastically deformed and therefore extinguishes as a unit in thin section on rotation of the microscope stage. *See also* undulatory extinction.

nuée ardente   Destructive and dangerous volcanic phenomenon in which a dense magmatic cloud of ash, gases, and fluid is erupted or blown out of a volcanic vent and flows along the surface down the flank of the volcano by following topographic lows such as valleys. Nuées ardentes commonly flow very rapidly (up to several hundred kilometers per hour) and have temperatures of several hundred degrees Celsius.

obsidian   Glassy rock of rhyolitic composition formed from intense compaction and welding at the base of thick piles of glassy shards of pyroclastic debris.

OIB   Ocean island basalt. Typical magmatic suite of the intraplate tectonic environments, including volcanic island chains. A diverse suite containing rock types ranging from highly alkalic rocks (basanite, nephelinite) to tholeiites, trachytes, and phonolites.

oikocryst   Large enclosing crystal in poikilitic texture.

oligomictic   Refers to conglomerates in which the grains consist of only one kind of mineral.

oncolith   Synonym of algal pisolith.

ooid   A spherical or elliptical particle less than 2 mm in diameter, typically composed of calcium carbonate and having a nucleus of mineral matter surrounded by a multilayered rim with concentric or radial fabric.

opaque mineral   A mineral that does not transmit light in thin section, for example, magnetite or pyrite.

open system   Any rock that exchanges chemical mass, as well as thermal and mechanical energy, with its surroundings.

ophiolite   Suite of igneous, metamorphic, and sedimentary rocks thought to represent a cross section of upper oceanic lithosphere formed near mid-ocean ridges. Contains harzburgite at the base, succeeded upward by gabbro, a basalt sheeted-dike complex, pillow lavas, and a capping of deep-sea pelagic sedimentary deposits. Exposed in orogenic belts where oceanic lithosphere is tectonically emplaced at a continental margin.

ophitic   Texture in which subhedral to euhedral plagioclase laths (commonly in radiating masses) are surrounded by anhedral pyroxenes that fill in the interstices. Plagioclase laths form a three-dimensional network. Typical of basalts in dikes, sills, or interiors of thick lava flows.

ortho-   Metamorphic prefix used to denote an igneous protolith, when this is known with some confidence, as in orthogneiss.

oscillatory zoning   Chemical zoning, typically of plagioclase, in which the Na:Ca ratio does not show monotonic increase but fluctuates back and forth from core to rim of the zoned crystal. Gives rise to spectacular concentric banded appearance in cross-polarized thin section view. Common in andesites.

overgrowth   Secondary mineral matter precipitated from solution in crystallographic continuity around a crystal of the same composition; enlargement during diagenesis. *See also* epitaxial overgrowth.

packing   The spatial density of grains in a sedimentary deposit or rock.

pahoehoe   Type of lava characterized by smooth, shiny, ropy appearance, similar to pulled taffy. Reflects low magma viscosity and rapid flow.

para- Metamorphic prefix used to denote a sedimentary protolith, when this is known with confidence, as in paragneiss.

pegmatite Extremely coarse-grained plutonic rock, typically of granitic composition. Originally used for rocks showing simultaneous crystallization of minerals (as in graphic texture). Numerous types are known, from simple quartz-potassic feldspar-mica varieties to some very rich in rare minerals.

pelagic deposit Deep-sea sediment lacking terrigenous material, for example, brown clay and organic oozes.

pelite Term originally used in sedimentary petrology to indicate a clay-sized sediment. However, the term is much more widely used in metamorphic petrology to denote an aluminous schist or gneiss that is probably derived from a mudrock protolith.

peloid A small structureless aggregate of micrite.

peralkaline Category of granitoids in which the sodium plus potassium content (on a molar basis) exceeds alumina content. This composition results in crystallization of a sodium-rich amphibole (for example, riebeckite) or pyroxene (for example, aegerine-augite).

peraluminous Category of granitoids in which the alumina content (on a molar basis) exceeds sodium plus potassium plus calcium content. This composition results in crystallization of some alumina-rich mineral, typically muscovite or garnet.

peridotite General name for ultramafic rocks that contain more than 40% olivine. See Figure 3-4 for particular rock names.

peritectic Point, line, or surface in phase diagrams that reflects incongruent melting behavior of one of the participating minerals.

perthite General name for exsolution texture in feldspars. Describes a host grain (generally potassic feldspar) with variable proportions of exsolved lamellae of sodic plagioclase.

petrogenetic grid Network of metamorphic univariant reactions and invariant points, typically for a restricted compositional system (for example, aluminous, mafic, or calcareous rocks) and shown in the P-T plane. Effectively subdivides P-T space into restricted P-T divariant regions that represent the stability of mineral assemblages.

petrography The microscopic study of rocks in thin section.

pH The negative logarithm of the hydrogen ion activity of a solution; 7 is neutral, less than 7 is acidic, and greater than 7 is basic.

phase Thermodynamic term that refers to a physically separable part of a chemical system. Must have continuous physical and chemical properties and be of a definite composition.

phase rule Relation between compositional constraints and variation in environmental variables for chemical systems that dictates the maximum number of phases that can occur. Stated $F = c - p + 2$, where F is degrees of freedom (independently variable parameters), c is number of components, and p is number of phases. The 2 refers to environmental variables pressure and temperature. If either is held constant, the number in the phase rule is reduced to 1.

phi scale A geometric scale of the sizes of sedimentary particles that uses class boundaries based on negative logarithms to the base 2.0. Extensively used in sedimentology.

phonolite Volcanic rock that is the equivalent of plutonic syenite. Contains potassic feldspar and sodic plagioclase, and, being roughly silica-saturated, can contain minor amounts of either quartz or a feldspathoid, along with accessory mafic minerals.

phosphorite A sedimentary rock composed principally of phosphate minerals, typically carbonate fluorapatite.

photic zone That part of the ocean in which there is sufficient penetration of light to support photosynthesis; in clear ocean water, about 30 m.

phyllite Low-grade metamorphic rock that is similar to slate but is slightly coarser grained and shows silky or shiny cleavage surfaces. Most grains are not visible with a lens, but small porphyroblasts may occur.

phyllonite Mylonite showing recrystallization and films of micaceous minerals.

phytoplankton Microscopic plants that are not free-swimming and are transported solely by prevailing currents.

pillow lava Lava form consisting of large, sacklike or sausagelike masses stacked atop one another. Reflects subaqueous, typically submarine, extrusion.

pisolith In a sedimentary rock, a small, round accretionary body with a diameter greater than 2 mm.

platform facies   A sedimentary facies that contains sediments produced in the neritic environment of shelf seas marginal to a craton.

podsol   A soil characterized by the accumulation of ferric iron in the B horizon.

poikilitic   Texture in which large crystals contain smaller crystals of another, earlier formed mineral. Inclusions typically have no preferred orientation. The name derives from the Greek root *poikilis* ("variegated"), reflecting the spotted appearance that poikilitic rocks commonly have.

poikiloblast   A generally large metamorphic crystal that includes smaller crystals of other minerals.

polycrystalline   Referring to a composite sedimentary particle composed of two or more crystals.

polymictic   Referring to conglomerates in which the particles consist of more than one type of rock; contrasted with oligomictic.

porphyritic   Texture in which an igneous rock has larger phenocrysts set in a finer grained or glassy groundmass.

porphyroblast   Metamorphic mineral that is considerably larger than the rock matrix and is commonly subhedral to euhedral.

porphyroclast   Commonly lenticular, larger strained grain in a blastomylonite.

porphyry   Extrusive or shallow intrusive rock that contains larger phenocrysts in a finer groundmass.

postcumulus crystal   Intergranular, small, discrete crystal in a cumulate-textured rock.

prehnite-pumpellyite facies   Metamorphic facies transitional between zeolite and greenschist facies and thus transitional between burial-type metamorphism and regional metamorphism.

pressure solution   Solution occurring preferentially at the contact surfaces of crystals where the external pressure exceeds the hydrostatic pressure of the interstitial fluid.

prograde metamorphism   Sequence of metamorphic reactions that results in development of the final mineral assemblage. Generally assumed to involve devolatilization or solid-solid reactions during heating, but some prograde reactions can occur during initial stages of cooling and decompression. Typically distinguished from retrograde processes by the localized or nonpervasive character of the latter.

protolith   The rock from which a metamorphic rock was formed.

provenance   The area from which the detrital minerals are derived; the source area.

proximate source   The most recent source of a sedimentary particle (for example, an older sandstone), as contrasted with an ultimate source (for example, a granite).

pseudomorph   Mineralogic term indicating a secondary mineral or multiphase mixture that replaces a former mineral without disrupting the characteristic shape of the earlier mineral.

pumice   Glassy volcanic rock that forms as a froth on lava flows or in silicic pyroclastic material. Volume of former gas bubbles or air pockets may greatly exceed glassy walls, leading to low density and the common pumice behavior of floating on water.

pyroclastic   General term that refers to any volcanic material that is ejected from volcanic vents as loose or fragmental material. Includes many specific terms that refer to shapes or sizes of particles.

pyrolite   Any one of several models for the mineralogic and chemical composition of pristine Earth mantle. Conceived by A. E. Ringwood as a starting material for melting experiments to simulate formation of basaltic magma in the mantle, pyrolite is a hypothetical mixture of residual mantle material (as found in xenoliths) and primitive basaltic magma.

pyroxenite   Ultramafic rock dominated by either augite or orthopyroxene or a mixture of the two. Can contain a minor proportion of olivine or spinel.

quartzarenite   A sandstone in which detrital grains are at least 90% quartz or chert.

recycled sediment   Detrital grains that have been part of an earlier sedimentary rock following ultimate derivation from igneous or metamorphic rock. Synonym of polycyclic sediment.

reflux   A process in which water with a high specific gravity sinks and flows through underlying sediments.

regional metamorphism   Metamorphism that occurs on a regional scale of many square kilometers and shows no obvious heat source.

retroarc basin   *See* backarc basin

retrograde metamorphism   Metamorphic reactions and processes that perturb the apparent peak or final metamorphic assemblage and commonly involve rehydration of a rock. Typically recognized by localized or spotty character and obvious textural evidence of superimposition on the prograde assemblage. May in fact be very difficult to distinguish from late prograde effects. Retrograde metamorphism has generally been ascribed to rehydration during cooling following a prograde metamorphism, but in many cases is apparently due to a later, or even much later, prograde event of lower peak grade that is superimposed on the earlier, higher grade metamorphism.

rhyolite   Volcanic equivalent of granite; consists of potassic feldspar, sodic plagioclase, and quartz; may be partly glassy.

ripple mark   A small ridge of sediment resembling a ripple of water and formed on the bedding surface of a sediment.

rudite   *See* conglomerate

sabkha   A supratidal flat in an arid environment.

sand   A sedimentary particle with a diameter between 0.06 and 2.0 mm.

schistosity   Foliation structure in metamorphic rocks in which moderately coarse platy grains show parallel alignment and the rock breaks irregularly along the planes of schistosity.

scoria   Vesiculated pyroclastic ejecta that commonly forms through fountaining of highly gas-charged magma at vents.

semipelite   Compositional term used for metamorphic rocks derived from feldspathic or lithic sandstones.

serpentinite   Low-grade metamorphic rock of ultramafic composition that consists mostly of serpentine, chlorite, and talc.

shale   A fissile, terrigenous sedimentary rock in which particles are mostly of silt and clay size.

shard   A fragment of glass, usually of volcanic origin, in a pyroclastic or detrital sediment. Characteristically, a shard is bounded by curved fracture surfaces.

shield volcano   Broad, gently sloping volcanic cone composed mostly of thick accumulations of lava flows. Named for resemblance to ancient round military shields. Most of the world's largest volcanoes are shield volcanoes.

silcrete   A siliceous crust of sand and gravel cemented by opal, chert, and/or quartz. Formed by evaporation of water in a semiarid climate. A variety of duricrust, like calcrete.

sill   Concordant tabular intrusive form in which magma intrudes along contacts between sedimentary beds or along foliation planes in metamorphic rocks. Some sills are many hundreds of meters thick, but thinner sills are typical.

silt   A sedimentary particle between 0.004 and 0.062 mm in size.

skarn   Calcium-rich metasomatic rock characteristic of inner parts of contact aureoles. Commonly contains rare and unusual minerals that result from addition of igneous fluids.

slate   Low-grade metamorphic rock of mudrock composition. May show a wide range of colors, from black to gray, red, green, or tan, and have excellent rock cleavage at an angle to the bedding planes. Has dull cleavage surfaces and grains too small to see with a lens.

solidus   Curve or line on a phase diagram that indicates the lowest temperature at which melt or melt plus crystals can exist stably for any particular bulk composition.

solvus   Curve in a temperature-composition diagram that denotes the compositional limits of a region of immiscibility, typically in the subsolidus region. A good example is the alkali feldspar solvus in the orthoclase-albite system (Figure 10-12).

sorting   The selection during transport of particles according to their sizes, specific gravities, and shapes. A well-sorted sediment has only a small amount of variability among the diameters of its particles. *See also* standard deviation

spilite   Term for altered, sodium-enriched basalt. Alteration may be through simple interaction with seawater or by interaction (of lavas) with hydrothermal solutions.

spinifex   Texture that consists of long skeletal blades of calcic pyroxene or olivine in komatiite lava flows. Named for long-bladed wild grasses in the komatiite type locality in South Africa, the texture results from quenching of ultramafic melt.

standard deviation   A departure from the middle of a bell-shaped frequency distribution that includes 34% of the total population of individual values. The relative perfection of sorting is given by the standard deviation of particle sizes.

stock   Small, rounded intrusive body consisting of less than a 100-km^2 outcrop area.

stratiform volcano   *See* composite volcano

stromatolite A laminated and lithified calcareous sedimentary structure formed by sediment-binding blue-green algae that trap silty detritus suspended in the water washing over the algal filaments.

strombolian eruption Semiexplosive eruptive event in which volcanic bombs and blocks are expelled from the vent at great velocities on ballistic trajectories.

stylolite A thin seam or surface of contact between two beds, marked by an irregular mutual interpenetration of the two sides. Relatively insoluble constituents in the rock may be concentrated along the seam. Common in limestones.

S-type granite Peraluminous granite assumed to have originated through melting of aluminous sediments in the crust during high-grade metamorphism.

subduction The process of oceanic lithosphere descending beneath another lithospheric plate, either oceanic or continental.

subhedral Crystal shape in which a tendency to ideal crystal form can be seen, but few, if any, well-formed crystal faces occur.

subophitic Texture in basalts in which well-formed plagioclase laths are surrounded by anhedral clinopyroxenes, but the plagioclase laths do not make the same tight network as in ophitic texture.

subsolvus Crystallization of alkali feldspars at higher water pressures (greater than about 4 kbar) that results in direct formation of discrete potassic feldspar and plagioclase grains. Each type of feldspar may become perthitic upon cooling.

survival potential The likelihood that a detrital sedimentary particle will remain after the rigors of abrasion and dissolution during transport.

sutured contact An irregularly interlocking boundary; intense suturing that affects a large number of grains can produce a stylolite.

syenite Plutonic rock that ranges from just silica saturated to silica undersaturated and contains potassic feldspar, sodic plagioclase, and minor mafic minerals. Some syenites have minor quartz, others have minor feldspathoid, and some have neither.

textural inversion The texture of a sandstone that contains a bimodal mixture of textural maturities.

textural maturity A sequential series of changes in textures that occurs during the transport of sedimentary materials, from the presence of clay (textural immaturity) to loss of the clay (submaturity) to sorting of the nonclay fraction (maturity) to rounding of quartz grains (supermaturity).

texture The size, shape, and geometric arrangement of the crystals or grains in a rock.

tholeiite Type of basalt that consists of phenocrysts of calcic plagioclase and augite in a finer grained groundmass of the same minerals plus iron-titanium oxide.

tholeiite trend Fractionation behavior in basaltic and derivative magmas in which the Fe:Mg ratio of ferromagnesian minerals and fractionated melts increases dramatically before any notable increase in magma alkali content occurs. Typical of layered intrusions and ocean island suites.

tie line Line representing equilibrium coexistence of two or three phases on two-dimensional phase diagrams.

trachyte Volcanic equivalent of syenite. Consists of potassic feldspar and sodic plagioclase, commonly with phenocrysts of augite; typically shows trachytic texture.

trachytic Texture in which a parallel alignment of feldspar laths is due to flow during crystallization. Alignment of laths bends around phenocrysts.

trench A long, narrow depression of the deep-sea floor, oriented parallel to a plate boundary and commonly marking the seafloor expression of the subduction zone.

troctolite Plutonic rock consisting of calcic plagioclase and olivine. Occurs as a rock type in some lopoliths and layered intrusions.

tuff Pyroclastic rock that consists of welded fragmental material, principally including glassy shards, along with individual mineral grains and rock fragments.

turbidity current A current that consists of a suspension of detritus in water and flows down a submarine slope.

ultramafic Term for any rock that contains greater than 90% ferromagnesian minerals, including olivine, pyroxenes, and hornblende.

undulatory extinction Referring to a crystal that has been plastically deformed and therefore does not extinguish as a unit in thin section on rotation of the microscope stage. *See also* nonundulatory extinction

unimodal A frequency distribution that has only a single peak.

univariant assemblage Metamorphic mineral assemblage that contains one more mineral than would normally be expected in a randomly sampled rock. Indicates that a reaction is in progress in a rock and that one or more new minerals is appearing and one or more is disappearing. Somewhat restricts the possible range of pressures and temperatures.

univariant equilibrium   Equilibrium state in which only one of the two environmental variables temperature or pressure can be independently varied without changing the number or identity of mineral phases. When *either* variable is changed, the other is not independent, but must change to a specified value to maintain equilibrium.

vein   Irregular, cross-cutting rock body that generally represents precipitation from aqueous or hydrothermal solutions rather than crystallization of a melt. Many quartz veins are intimately associated with granitoid plutons.

vesicles   Small, round, ellipsoidal or tubular cavities in lavas that represent former gas bubbles trapped by the magma as it crystallized. Elongated shapes reflect continued movement of the lava after formation of the bubble. Vesicles are most common near the tops of thicker flows.

viscosity   Physical parameter that reflects the "stiffness" or resistance to flow of a material. Viscosities of magmas are highly dependent on chemistry, particularly silica and water contents.

vitrophyre   Porphyry in which the phenocrysts are set in a dominantly glassy groundmass.

vug   A large void in a rock, commonly not joined to other voids, for example, the ovoid cavity formed by dissolution of an ooid in a limestone or an open cavity in an igneous rock. Vugs commonly have a lining of euhedral mineral grains.

websterite   Ultramafic rock consisting of subequal proportions of calcic clinopyroxene and orthopyroxene, with minor oxides. Named for the Webster-Addie Complex in North Carolina.

wehrlite   Ultramafic rock consisting of about 60–90% olivine and 10–40% calcic clinopyroxene.

welded tuff   Pyroclastic rock consisting of glassy and crystalline particles that have become welded together by slight melting at particle edges as a result of retained heat and compression by gravity.

whiting   A milky white area of water made white by a dense suspension of calcium carbonate sediment.

xenoblastic   Metamorphic texture in which all mineral grains are anhedral.

xenocryst   Individual crystal from the country rock that has been incorporated into a magma. Called a megacryst if especially large.

xenolith   Fragment of country rock incorporated into a magma.

zeolite facies   Metamorphic facies characterized by the appearance of any of a variety of Ca–Na zeolite minerals in mafic rocks metamorphosed during burial. Represents low pressure (< 5–6 kbar) and temperature (< 300°C). Peak pressure and temperature lie essentially along a normal geothermal gradient.

INDEX

....